Mesostructure and Dynamics in Liquids and Solutions

University of Bristol, UK

18–20th September 2013

FARADAY DISCUSSIONS

Volume 167, 2013

RSC Publishing

The Faraday Division of the Royal Society of Chemistry, previously the Faraday Society, founded in 1903 to promote the study of sciences lying between Chemistry, Physics and Biology.

Faraday Discussions (Print ISSN 1359-6640, Electronic ISSN 1364-5498) is published 8 times a year by the Royal Society of Chemistry, Thomas Graham House, Science Park, Milton Road, Cambridge, UK CB4 0WF. Volume 167 ISBN-13: 978-1-84973-966-5

2013 annual subscription price: print+electronic £765, US $1428; electronic only £727, US $1356. Customers in Canada will be subject to a surcharge to cover GST. Customers in the EU subscribing to the electronic version only will be charged VAT. All orders, with cheques made payable to the Royal Society of Chemistry, should be sent to RSC Order Department, Royal Society of Chemistry, Thomas Graham House, Science Park, Milton Road, Cambridge, CB4 0WF, UK. Tel +44 (0) 1223 432398; E-mail orders@rsc.org

If you take an institutional subscription to any RSC journal you are entitled to free, site-wide web access to that journal. You can arrange access *via* Internet Protocol (IP) address at www.rsc.org/ip. Customers should make payments by cheque in sterling payable on a UK clearing bank or in US dollars payable on a US clearing bank.

PRINTED IN THE UK

Faraday Discussions documents a long-established series of *Faraday Discussion* meetings which provide a unique international forum for the exchange of views and newly acquired results in developing areas of physical chemistry, biophysical chemistry and chemical physics.

Mesostructure and Dynamics in Liquids and Solutions

Faraday Discussions

www.rsc.org/faraday_d

A General Discussion on Mesostructure and Dynamics in Liquids and Solutions was held in Bristol, UK on the 18th, 19th and 20th of September 2013.

RSC Publishing is a not-for-profit publisher and a division of the Royal Society of Chemistry. Any surplus made is used to support charitable activities aimed at advancing the chemical sciences. Full details are available from www.rsc.org

CONTENTS

ISSN 1359-6640; ISBN-13: 978-1-84973-966-5

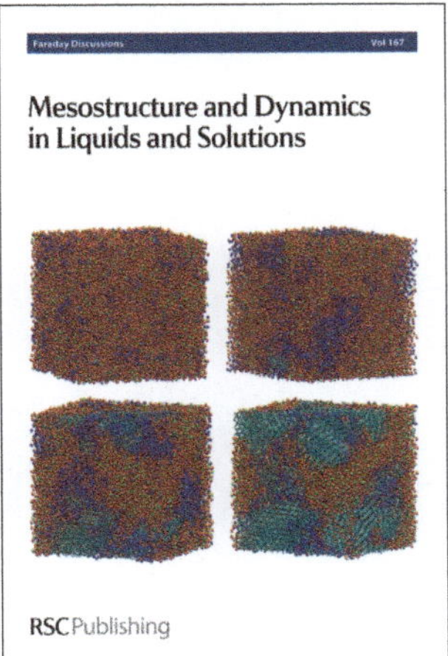

Cover
See Bullock and Molinero, *Faraday Discuss.*, 2013, **167**, 371.

Nucleation and growth of ice in water-salt solutions is preceded by the formation of nanodomains of pure, four-coordinated water molecules within the mixture.

Image reproduced by permission of Valeria Molinero from *Faraday Discuss.*, 2013, **167**, 371.

INTRODUCTORY LECTURE

PAPERS AND DISCUSSIONS

CO-SPONSOR

IOP Institute of Physics

CONCLUDING LECTURE

ADDITIONAL INFORMATION

Faraday Discussions RSC Publishing

PAPER

Importance of many-body orientational correlations in the physical description of liquids

Hajime Tanaka*

Received 23rd October 2013, Accepted 30th October 2013
DOI: 10.1039/c3fd00110e

Liquids are often assumed to be homogeneous and isotropic at any lengthscale and translationally invariant. The standard liquid-state theory is constructed on the basis of this picture and thus basically described in terms of the two-body density correlation. This picture is certainly valid at rather high temperatures, where a liquid is in a highly disordered state. However, it may not necessarily be valid at low temperatures or for a system which has strong directional bonding. Indeed, there remain fundamental unsolved problems in liquid science, which are difficult to explain by such a theory. They include water's thermodynamic and kinetic anomalies, liquid–liquid transitions, liquid–glass transitions, and liquid–solid transitions. We argue that for the physical description of these phenomena it is crucial to take into account many-body (orientational) correlations, which have been overlooked in the conventional liquid-state theory. It is essential to recognise that a liquid can lower its free energy by local or mesoscopic ordering without breaking global symmetry. Since such ordering must involve at least a central particle and its neighbours, which are more than two particles, it is intrinsically a consequence of many-body correlations. Particularly important ordering is associated with local breakdown of rotational symmetry, *i.e.*, bond orientational ordering. We emphasize that translational ordering is global whereas orientational ordering can be local. Because of the strong first-order nature of translational ordering, its growth in a liquid state is modest. Thus any structural ordering in a liquid should be associated primarily with orientational ordering and not with translational ordering. We show that bond orientational ordering indeed plays a significant role in all the above-mentioned phenomena at least for (quasi-)single-component liquids. In this Introductory Lecture, we discuss how these phenomena can be explained by such local or mesoscopic ordering in a unified manner.

1 Introduction

Liquid is one of the most fundamental states of matter, and has unique transport properties, which are absent in the other forms of matter.[1,2] Despite its

Institute of Industrial Science, University of Tokyo, 4-6-1 Komaba, Meguro-ku, Tokyo 106, Japan. E-mail: tanaka@iis.u-tokyo.ac.jp; Fax: +81-3-5452-6126; Tel: +81-3-5452-6125

importance, however, the physical understanding of liquids is very difficult because of its disordered nature (or, the lack of periodicity) and complex many-body interactions due to the high density. As a result, there remain fundamental unsolved problems concerning the basic properties of liquids. They include water's thermodynamic and kinetic anomalies,[3–6] liquid–liquid transitions,[7,8] liquid–glass transitions,[9–14] liquid–solid transitions,[15] liquid structure near interfaces, and structure around solutes and in mixtures,[16] which are the main topics of this Faraday Discussion.

The standard liquid-state theory has been constructed on the assumption that liquids have homogeneous, isotropic, and random structures. The most relevant order parameter for the description of a liquid state is the density field $\rho(\vec{r})$ and its two-body correlation has been believed to be able to characterize the state of liquid and control its static and dynamic properties. This is of course valid in the first approximation and the theory is successful in describing the basic properties of liquids.[1,2] However, to understand the above-mentioned phenomena, we need to go beyond this level of description and explicitly take many-body correlations into account. To do so, we should first recognise that there are two fundamental symmetries to be broken upon ordering: one is translational symmetry and the other is rotational symmetry. The two-body density correlator basically describes how translational order decays spatially. Crystallization accompanies the breakdown of translational symmetry. In two dimensions (2D), there are sequential breakdown of symmetries upon densification: breakdown of rotational symmetry followed by that of translational symmetry.[17] In three dimensions (3D), on the other hand, crystallization has been believed to take place by one step, accompanying the simultaneous breakdown of translational and rotational symmetry. Note that the breakdown of translational symmetry automatically leads to that of rotational symmetry. Here we emphasise that the breakdown of translational symmetry is intrinsically global, but that of rotational symmetry can be local: rotational symmetry can be broken locally before translational symmetry is broken globally. This is a very important point when we consider local or mesoscopic ordering in liquids that preserve global translational invariance.

We argue that any liquids tend to form locally favoured structures, which locally have a lower free energy and a longer lifetime than disordered normal-liquid structures (see Fig. 1). On the basis of this picture, we proposed that the key to solving the long-standing problems in liquid physics is the recognition of the importance of spontaneous breakdown of local or mesoscopic rotational symmetry and thus we need a bond orientational order parameter in addition to density for the physical description of liquids.[18,19] We used this two-order-parameter model to explain water's anomalies,[20–24] liquid–liquid transitions,[18,25] liquid–glass transitions,[19,26–32] and liquid–solid transitions.[33,34] We also recently discussed a possible unified description of all these phenomena.[35] Here we review the current situation of our understanding of liquids and competing views on the phenomena mentioned above.

The organization of this paper is as follows. In section 2, we discuss the origins of mesoscopic ordering in liquids. In section 3, we describe a phenomenological theory, which can capture this feature. In sections 4–7, we discuss water's anomalies, liquid–liquid transitions, liquid–glass transitions, and liquid–crystal transitions, respectively. In section 8, we summarize our paper.

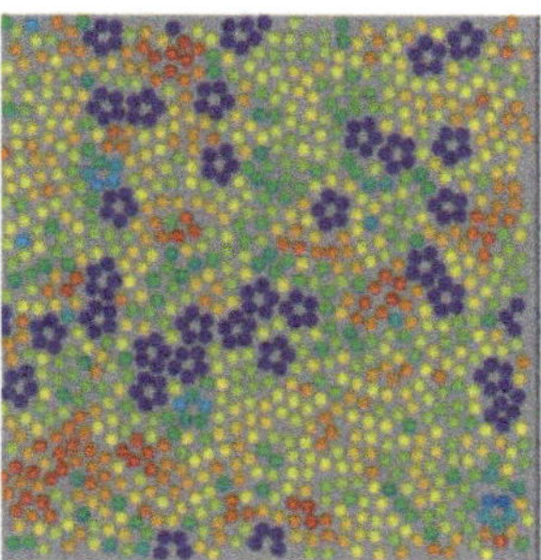

Fig. 1 An example of local bond orientational ordering in a liquid. Here blue pentagons are locally favoured structures spontaneously formed in a sea of normal liquid structures (particles with the other colours). They have finite lifetime and thus are transient. This is obtained by molecular dynamics simulations of spherical particles interacting with a special anisotropic potential, which we call two-dimensional (2D) spin liquid.[36]

2 Local or mesoscopic ordering in liquids

A system always tends to lower its free energy. Local ordering in liquids is also induced to lower the free energy locally. There can be two cases for local ordering: entropy-driven and energy-driven ordering. Here we discuss each case by using typical examples, although in reality there are always both contributions.

2.1 Entropy-driven local ordering

First we consider entropy-driven local ordering. This is particularly important in hard spheres, whose free energy is purely entropic. The entropy of hard spheres is basically composed of the configurational and correlational (or vibrational) part.[37] For example, the crystallization of hard spheres is a consequence of gaining the correlational entropy at the expense of the configurational one, which results in the reduction of the total free energy. Local ordering in hard spheres is a result of dense packing. A system locally tends to form arrangements that can be most densely packed upon compression since such configurations provide large free volumes for vibrational motions in the uncompressed natural state (see Fig. 2). For hard spheres of almost equal sizes, there are three key structures that minimize the local volume: face-centred cubic (fcc), hexagonal close packing (hcp), and icosahedral (ico) structures (see Fig. 3). Thus, these are three relevant locally favoured structures for weakly polydisperse hard spheres. Despite the ability of the structures to be compacted, these structures occupy the same volume as disordered structures in a liquid state. This feature allows the increase of the local correlational (or vibrational) entropy, which is the reason why such structures are favoured. Accordingly, even if locally favoured structures are formed, the density is basically homogeneous in the system and density fluctuations is simply determined by the isothermal compressibility K_T. This is a consequence of the fact that bond orientational ordering in hard spheres is completely decoupled from translational ordering. This is characteristic of entropy-driven local order.

Here it is worth mentioning the important relation between local rotational symmetry and spatial extendability of locally favoured structures. Among fcc, hcp and ico structures, fcc and hcp are spatially extendable and can grow its size,

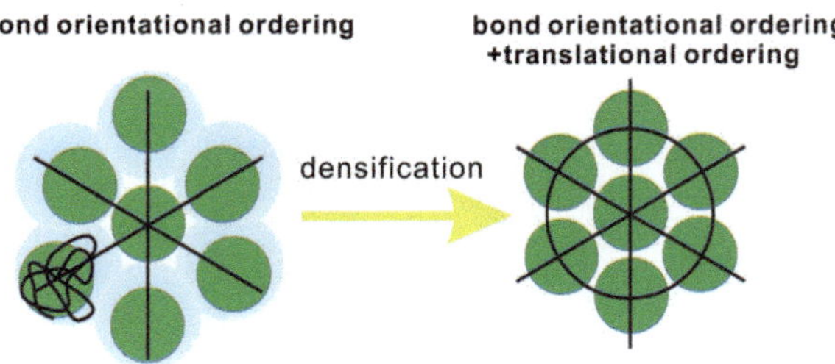

Fig. 2 Schematic figure explaining the relation between bond orientational ordering and translational ordering. A structure having only bond orientational order has a room of a similar amount of thermal fluctuations (or, free volume) for each particle and thus it reduces the correlational entropy at the expense of the orientational configurational entropy. This situation is favoured in an intermediate density region where the density is high enough for a completely random state not to be favoured but low enough for the reduction of the configurational entropy associated with translational ordering not to take place. Without bond orientational order, the free volume for each particle fluctuates too largely, which leads to the loss of correlational entropy at this density regime. Upon its densification, translational order is automatically gained, if the size polydispersity is not so large. However, this happens only for a spatial region where pre-existing bond orientational order has a phase coherency. This mechanism is crucial when we consider crystal nucleation (see section 7).

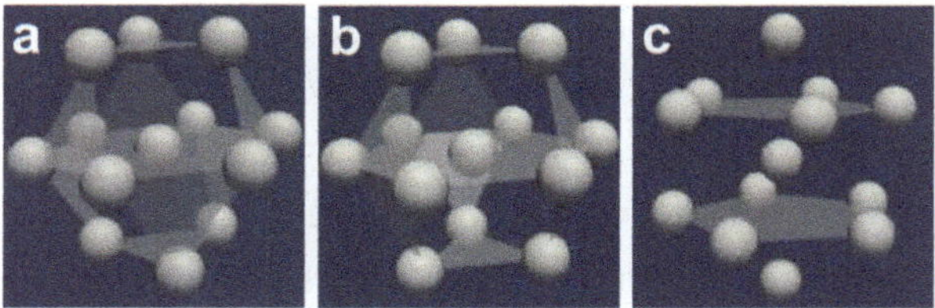

Fig. 3 Schematic figure representing the densely packed structures made of 13 spherical particles, which have fcc, hcp, and icosahedral configurations. (a) fcc, (b) hcp, and (c) icosahedron. This figure is reproduced from Fig. 36 of ref. 35.

whereas ico is not extendable and cannot grow its size. This means that fcc and hcp bond orientational order can be mesoscopic, but ico must be localised, which leads to the essential difference in the nature between them. To what extent fcc or hcp bond orientational order grows is again determined by the competition of the two types of entropy: their growth leads to the loss of the configurational entropy of locally favoured structures and the gain of the correlational entropy coming from the reduction of the free energy associated with the spatial gradient of the order parameter. Frustration is another source disturbing their growth.

2.2 Energy-driven local ordering

In many realistic liquids, locally favoured structures are formed to gain the local energy at the expense of configurational and vibrational entropy. The interaction can be isotropic or anisotropic. A simple isotropic attractive potential leads to locally favoured structures with dense packing, such as fcc, hcp, and ico structures, whereas an anisotropic directional potential can lead to various symmetry depending upon the local symmetry favoured by the potential. In water-type liquids such as water, Si, and Ge, tetrahedral order is locally favoured by hydrogen or covalent bonding. For a 2D spin liquid shown in Fig. 1, pentagons are locally favoured, although antiferromagnetic order is favoured at a longer lengthscale[36] (see Fig. 4).

In this energy-driven "local" ordering, bond orientational ordering usually accompanies the change in the local density, unlike the above entropy-driven ordering. In other words, there is a non-trivial coupling of bond orientational ordering with translational ordering. For example, both hydrogen and covalent bonding favours not only a particular orientation, but also a particular distance between molecules or atoms. Furthermore, the angular constraint between bonds formed by anisotropic interactions can inevitably lead to such a coupling. The formation of a void in the middle of a pentagon formed in 2D spin liquids is such an example (see Fig. 1). In general, thus, locally favoured structures can have specific volumes different from normal-liquid structures. This feature plays a key role in water's anomalies, as will be shown below.

The relationship between local rotational symmetry and spatial extendability of locally favoured structures is the same as in the case of entropy-driven local ordering. For example, in 2D spin liquids, pentagons are non-extendable local structures, whereas antiferromagnetic order is extendable mesoscopic order (see Fig. 4). We note that unlike locally favoured structures mesoscopic ordering does not accompany a density change since translational ordering must be involved to increase the density at such a mesoscopic lengthscale but it never happens in a liquid state.

2.3 How to express local or mesoscopic ordering mathematically

Local or mesoscopic ordering described above is not easy to express by the density field due to its intrinsically many-body nature. Here we use so-called bond orientational order parameter, which can be expressed by the distribution of bonds joining a particle located at $\vec{r}$ to its nearest neighbours.[17] Expanding the density $\rho(\vec{r},\omega)$ of points pierced by these bonds on a small sphere inscribed about $\vec{r}$, we have[17,38,39]

$$\rho(\vec{r},\Omega) = \sum_{l=0}^{\infty} \sum_{m=-l}^{m=l} q_{lm}(\vec{r}) Y_{lm}(\Omega), \tag{1}$$

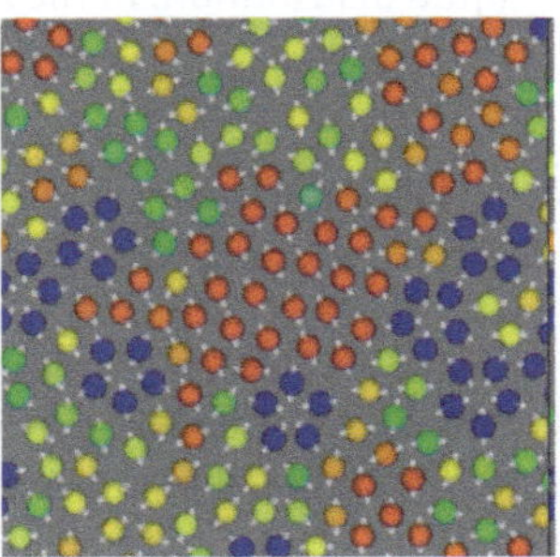

Fig. 4 A snapshot of 2D spin liquid in a supercooled state. Red particles have crystal-like bond orientational ordering (more specifically, antiferromagnetic order), which plays a crucial role in glass transition and crystallization, whereas blue particles are locally favoured structures with pentagonal symmetry, which plays a primary role in water-type anomalies and liquid–liquid transitions. The latter also plays an important role in vitrification if it competes with crystallization, which is linked to the above bond orientational ordering (appeared red). A spin on a particle is also shown by an white arrow in this figure.

where $Y_{lm}(\Omega)$ are spherical harmonics. This order parameter is only sensitive to bond directions and not to the distances. In this sense, it is an ideal order parameter to be complimentary to density order parameter.

We take the normalized average of q_{lm} over a small volume located at $\vec{r}$, which we express by $\bar{q}_{lm}(\vec{r})$. Then, its rotationally invariant combination can be defined as

$$q_l(\vec{r}) = \left[\frac{4\pi}{2l+1}\sum_{m=-l}^{l}|\bar{q}_{lm}(\vec{r})|^2\right]^{1/2}. \tag{2}$$

Note that $l = 6$ for fcc, hcp, and ico, whereas $l = 3$ for tetrahedron.[40] $l = 4$ is also important to pick up a part of the symmetries of fcc and hcp.[41] For tetrahedrality, we can also define a more specific order parameter:[42,43]

$$q_{\text{tetra}} = 1 - \frac{3}{8}\sum_{j=1}^{3}\sum_{k=j+1}^{4}\left(\cos\Psi_{jk} + \frac{1}{3}\right)^2.$$

In the case of water, ψ_{jk} is the angle formed by the lines joining the oxygen atom of a given water molecule and those of its nearest neighbours j and k.

Here we also define other quantities characterizing bond orientational order.

$$w_l = \sum_{m_1+m_2+m_3=0}\begin{pmatrix} \ell & \ell & \ell \\ m_1 & m_2 & m_3 \end{pmatrix} q_{lm_1}q_{lm_2}q_{lm_3}, \tag{3}$$

where the term in brackets in the above third-order rotational invariant is the Wigner 3-j symbol. w_6 is useful to identify ico structures or to distinguish fcc/hcp from body-centred cubic (bcc), whereas w_4 is useful to distinguish fcc from hcp.[44–46]

Following ref. 41 we also use the tensorial bond orientational order parameter coarse-grained over the neighbours:

$$Q_{lm}(i) = \frac{1}{N_i+1}\left(q_{lm}(i) + \sum_{j=0}^{N_i} q_{lm}(j)\right), \tag{4}$$

and define the coarse-grained invariants Q_l and W_l in the same way as the above. Structures with and without spatial extendability are then much easier to tell apart.[41,44] We note that for non-extendable local structures like icosahedra, their Q_l and W_l are buried into the liquid distribution. In the following, we also use Q_l to express q_l unless explicitly stated.

3 Phenomenological two-order-parameter model incorporating local and mesoscopic ordering in a liquid

As discussed above, to express local or mesoscopc ordering in a liquid, we need bond orientational order parameters in addition to the density field. Although we may need more than two bond orientational parameters, we mainly consider the simplest situation that a liquid state is described by two order parameters, density and one bond orientational order parameter.

3.1 Free energy associated with the formation of locally favoured structures

In general, there can be two types of bond orientational orderings, one of which is associated with local structural ordering and the other with medium-range crystal-like bond orientational order (see Fig. 4). We note that each of them may need more than two bond orientational order parameters to specify it (such as Q_l, $Q_{l'}$, W_m, $W_{m'}$,...). Here, focusing only on local structural ordering, we express a liquid state by a simple two-state model with cooperativity of such ordering (see Fig. 5). The first two-state model of liquid–liquid transition (LLT) was developed by Strässler and Kittel.[47] Rapoport[48] used it to explain melting-curve maxima of atomic liquids, such as carbon, at high pressure. Aptekar[49] explained the metal–non-metal transition in germanium and silicon by a two-state model. Some time ago, we generalized this basic idea by introducing the bond order parameter(s) in addition to the density order parameter, and proposed the two- (or multi-) order-parameter model of liquid to explain not only LLT, but also water-like anomalies, liquid–glass transitions and crystallization in a unified manner. Below we present a general framework of our model of liquid to describe these phenomena. We also show how these phenomena, which are apparently independent of each other, can be closely related to each other.[35]

Here we focus on short-range bond orientational ordering, or the formation of locally favoured structures. Our model[19,21–25] relies on a physical picture (see Fig. 1) that (i) there exist distinct locally favoured structures in a liquid and (ii) such structures are formed in a sea of normal liquid structures and its fraction S increases upon cooling since they are energetically (entropically for hard spheres) more favourable by ΔE than normal liquid structures: $\Delta E = E_\rho - E_S$ (see Fig. 5), where E_i is the energy of state i (i = ρ or S). Here normal liquid structures simply mean the background normal liquid structures. The specific volume and the entropy are larger and smaller for the former than for the latter, respectively, by $\Delta v = v_S - v_\rho$ and $\Delta\sigma = k_\text{B}\ln(g_\rho/g_S)$. Here v_i and g_i are, respectively, the specific volume and the degree of the degeneracy of state i (i = ρ or S). Δv can be either positive or negative depending upon a system, whereas $\Delta\sigma$ is positive except for purely repulsive systems such as a hard-sphere liquid, where the gain of correlational entropy is the driving force of local structural ordering. We identify locally favoured structures as a minimum structural unit (symmetry element). It is associated with tetrahedral order for water-type liquids, whereas icosahedron for metallic liquids[28,29,50,51] and hard spheres.[44,52–54] To express such short-range bond ordering in liquids, we introduce the so-called bond orientational order parameter Q_{lm}.

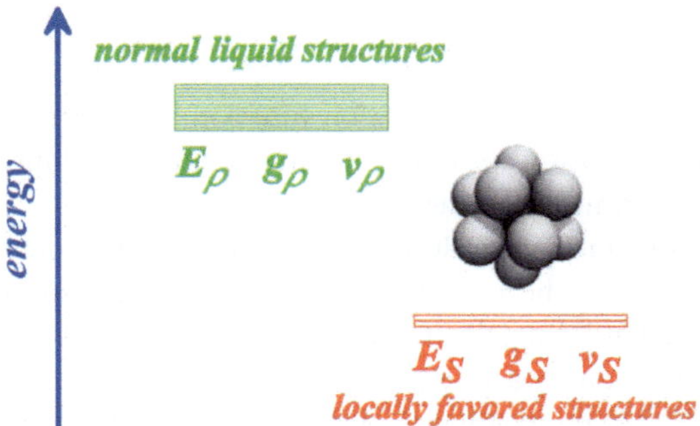

Fig. 5 A two-state model for a liquid: one is normal-liquid structures (energy E_ρ, degeneracy g_ρ, and specific volume v_ρ) and the other is locally favoured structures (energy E_S, degeneracy g_S, and specific volume v_S). For some liquids, there may be more than two distinct energy states.

When locally favoured structures are distinct, there should be a clear threshold value separating the two states. For example, the fraction of blue pentagons is the bond order parameter S in Fig. 1. Then we can use the fraction of atoms (or particles) having $q_l(\vec{r})$ higher than a certain threshold value as the local bond order parameter S (note that S is "not" entropy and instead σ represents entropy throughout this paper). When thermal fluctuation effects make locally favoured structures obscure, we need to decompose the distribution function of the order parameter into two (Gaussian) populations to estimate S.

As can be seen above, both the scalar density field ρ and the tensorial bond orientational order $\mathbf{Q}$ stem from the angle-dependent density field $\rho(\vec{r},\Omega)$. Although $\mathbf{Q}$ is tensorial, the fraction of locally favoured structures, S, is linked to the rotationally invariant scalar order parameter calculated from it. As a function of this scalar order parameter S, the phenomenological liquid-state free energy functional associated with locally favoured structures is given by[19,21–25]

$$F_S = \int d\vec{r}\,[-\Delta GS(\vec{r}) + JS(\vec{r})(1-S(\vec{r})) + k_BT(S(\vec{r})\ln S(\vec{r}) + (1-S(\vec{r}))\ln(1-S(\vec{r})))], \tag{5}$$

where $\Delta G = \Delta E - T\Delta\sigma - \Delta vP$. ΔG is the free energy change associated with the formation of a locally favoured structure. J represents the cooperativity, k_B is the Boltzmann constant, T is the temperature, and P is the pressure. Here J is energetic. We note that the cooperativity of entropic origin has also been considered recently.[55] We stress that this free energy is a function of the scalar order parameter S (see section 7.8 for its implication).

3.2 Free energy associated with density and tensorial bond orientational ordering

3.2.1 Density ordering. The free energy functional, denoted $F\{\rho\}$, is expanded functionally about a density, $\rho = \rho_l$, corresponding to a liquid state lying on the liquidus line of the solid–liquid coexistence phase diagram. The expansion is performed in powers of $\delta\rho = \rho - \rho_l$. Then the Ramakrishnan–Yussouff (RY) free energy density of a single-component system can be written, up to the two-point correlation, as[13,56,57]

$$F_\rho\{\rho\} = k_BT\int d\vec{r}\,\rho(\vec{r})\left[\ln\frac{\rho(\vec{r})}{\rho_l} - 1\right] + \int\int d\vec{r}d\vec{r}'\delta\rho(\vec{r})c(\vec{r}-\vec{r}')\delta\rho(\vec{r}'), \tag{6}$$

where c is the two-point direct correlation function. This free energy functional has widely been used to study not only liquid–crystal transitions, but also liquid dynamics and glass transitions (see, *e.g.*, section 6.11.1).

3.2.2 Bond orientational ordering. In the above, we consider only translational ordering. Partly because translational ordering automatically accompanies orientational ordering, the importance of the latter has been overlooked in theories of solidification for a long time despite the recognition of its importance in the 1980s.[17,38,39,58–63] Interestingly, orientational order has often been used in simulations to detect crystal order in the process of crystal nucleation (see, *e.g.*, ref. 64). The liquid–solid transition accompanies the breakdown of both translational and rotational symmetry. Here we argue that bond orientational order is crucial for our understanding of the liquid state itself as well as its transitions to non-ergodic states such as glass transition and crystal nucleation.

A form of the Landau-type free energy associated with tensorial bond orientational ordering with translational and rotational invariance can be found, *e.g.*, in ref. 63, 65, 66. For simplicity (see also a speculative explanation below), we consider the following free energy form associated with (scalar-like) bond orientational ordering **Q**:

$$F_Q = \int d\vec{r}\left(b\tilde{t}\mathbf{Q}^2 + I_3(\mathbf{Q}) + O(\mathbf{Q}^4) + \frac{1}{2}K(|\nabla\mathbf{Q}|)^2\right) + \cdots$$

where b is a positive constant, $\tilde{t}$ is the reduced temperature $\tilde{t} = 1/\phi - 1/{\phi_0}^b$ (or $\tilde{t} = T - {T_0}^b$), and $\cdots$ represents frustration originating from competing bond orientational orderings (*e.g.*, $\mathbf{Q}_{\rm CRY}$ *vs.* $\mathbf{Q}_{\rm LFS}$), internal frustration (this is the case for icosahedral order[38]), and/or random disorder effects. Here $\mathbf{Q}_{\rm CRY}$ is compatible with the symmetry of the equilibrium crystal (*e.g.*, fcc and hcp order), whereas $\mathbf{Q}_{\rm LFS}$ with incompatible to it (*e.g.*, icosahedral order). Here ${\phi_0}^b$ (or ${T_0}^b$) is the bare transition volume fraction (or temperature). Even though the third order invariant I_3 is suggestive of the first order nature of the transition, the transition might be almost continuous.

Frustration effects originating from competing $\mathbf{Q}_{\rm CRY}$ and $\mathbf{Q}_{\rm LFS}$ orderings and/or random disorder effects, *e.g.*, due to polydispersity may change the nature of the transition from a continuous (characteristic to a tensorial order parameter) to a discrete Ising symmetry (characteristic to a scalar order parameter).[67] We speculate that renormalization of frustration effects changes the symmetry of the transition from the continuous to the discrete Ising symmetry and also shifts the critical point from ${\phi_0}^b$ (or ${T_0}^b$) to ϕ_0 (or T_0), although this should be carefully checked. In relation to this, we note that such transformation of the phase ordering from (Heisenberg-type) continuous to (Ising (Z_2)) discrete symmetry due to frustration and random disorder effects has also been known for spin systems,[68,69] implying the generality of frustration and random disorder effects on the nature of the ordering. We also emphasize that frustration effects may not only change the type of ordering, but also lead to exotic critical phenomena accompanying the growing activation energy towards the hypothetical critical point. We note that the Ising nature of the glass transition has also been recently discussed on the basis of a two-state cluster picture by Langer.[70,71]

3.3 Coupling between density and bond orientational ordering

Now we consider couplings between orderings of ρ and **Q**, whose nature is very important in the following discussion. The lowest order coupling between **Q** and ρ should be given by the rotationally and translationally invariant energy:[38,63]

$$F_{\rm int} = \int dq \sum_{l,m} \alpha_l(q) \int d^2\hat{q}\, Q_{lm} Y^*_{lm}(\hat{\vec{q}})\rho(\hat{\vec{q}})\rho(-\hat{\vec{q}}). \quad (7)$$

Up to the lowest order, ρ is not coupled linearly to **Q**, and $\rho(\vec{q})\rho(-\vec{q})$ is coupled to it. Accordingly, the equilibrium ρ need not have the symmetry of the equilibrium **Q**. This particular type of coupling leads to an asymmetric coupling between the orderings. If the translational ordering temperature T_ρ is higher than the bond orientational ordering temperature T_Q then, because the **Q**–ρ interaction is linear in **Q**, the ordering of ρ at T_ρ will necessarily induce an ordering in **Q**. This seems to justify the theory based on the density field alone, but which may not be the case as we will see later. On the other hand, if $T_Q > T_\rho$ then, because the **Q**–ρ interaction is quadratic in ρ, the ordering of **Q** at T_Q will have the effect of renormalizing the quadratic coupling without necessarily inducing an ordering of ρ. Jaríc proposed that this case of $T_Q > T_\rho$ should correspond to quasicrystal formation.[63]

Here we see the relevance of this form of coupling by looking at the two-dimensional probability distribution of density ρ and bond orientational order Q_6, for a metastable fluid state of hard spheres at pressure $\beta p\sigma^3 = 17$ (before the appearance of the critical nucleus) (see Fig. 6).[72] The probability distribution is related to the Landau free energy, $F(Q_6,\rho) = -k_BT \log P(Q_6,\rho)$. The free energy can be well fitted with a full cubic polynomial, for which the most important term is of the form $Q_6\rho^2$. This term is responsible for the shape of contours lines (black dashed line in Fig. 6): because the interaction is linear in Q_6 and quadratic in ρ, the system can increase its orientational order without an increase of its translational order, but the opposite is not true, and an increase in density inevitably accompanies an increase in the average Q_6. This is fully consistent with the above form of coupling. Note also that a small linear coupling between Q_6 and ρ exists at high Q_6, which can be seen in the small slope of the steepest descent path (white dashed arrow) in Fig. 6.

3.4 Total free energy of a system describing crystallization and vitrification

The total free energy F_{total} may then be given by the sum of translational ordering, local and global orientational ordering, and the couplings between them:

$$F_{\text{total}} = F_\rho + F_S + F_Q + F_{\text{int}}. \tag{8}$$

In the above, however, we need to take special care to avoiding double counting. This may be done with a proper projection procedure.

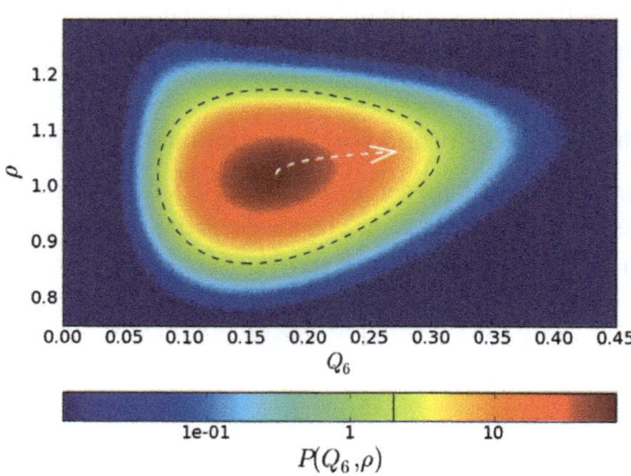

Fig. 6 Probability distribution for a supercooled state of hard spheres in the $\rho - Q_6$ space. The dashed black line is a contour line. The dashed white arrow is a steepest descent path from the maximum to a high Q_6 point of the probability distribution function. This figure is reproduced from Fig. 2 of ref. 72.

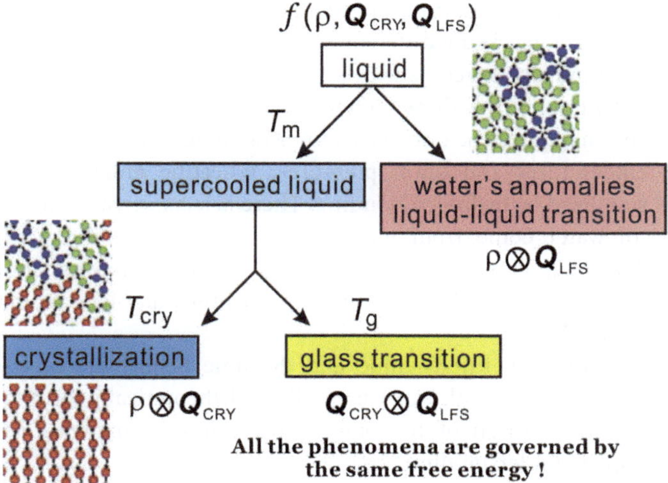

Fig. 7 Schematic figure explaining the relationship between the behaviour of liquid upon cooling and the free energy. A liquid may exhibit water-like anomalies or LLT upon cooling. A liquid also becomes a metastable supercooled state below the melting point T_m and further cooling leads either to crystallization or to glass transition. The former takes place at the crystallization temperature T_{cry}, whereas the latter at the glass transition temperature T_g. The former is a thermodynamic phase transition, but the latter is a kinetic transition. The key fundamental question here is whether the glass transition behaviour is controlled by the same free energy as that for crystallization or a special free energy? We argue that all the phenomena are governed by the same free energy.

3.5 Free energy responsible for water's anomailes, liquid–liquid transitions, liquid–glass transitions, and liquid–solid transitions

In the above we discuss the free energy, mainly focusing on the roles of bond orientational ordering. Here we note that there are new important effects of bond orientational ordering, which have so far been overlooked in the physical description of liquids: (1) thermodynamic effects of short-range bond ordering (a scalar order parameter linked to $\mathbf{Q}_{LFS}$), which can be considered on the basis of the simpler free energy F_S (see eqn (5)), (2) random field effects of $\mathbf{Q}_{LFS}$ on crystallization (long-range translational $\rho(r)$ and bond orientational ordering $\mathbf{Q}_{CRY}$), and (3) long-range crystalline (or quasicrystal) ordering ($\rho(r)$ and $\mathbf{Q}_{CRY}$).

Below, we consider problems of thermodynamic and kinetic anomalies of water-type liquids, liquid–liquid transition, liquid–glass transition, and crystallization, focusing on these three effects (1)–(3). As shown in Fig. 7, we argue that all these phenomena may be described by a common free energy functional (see eqn (8)) in a unified manner.

4 Water's anomalies

4.1 Thermodynamic and kinetic anomalies of water

Liquid water exhibits many anomalous behaviours upon cooling, which include volume expansion (below 4 °C), softening (below 46 °C), and heat capacity increase (below 35 °C).[3,6,9,73–75] In addition to the thermodynamic anomaly, the viscosity η also shows anomalous non-Arrhenius behaviour. Furthermore, at low temperatures η decreases up to 2 kbar and then increases with an increase in pressure,[76] which is markedly different from the behaviour of ordinary liquids, whose η monotonically increases with pressure. All these features are absent in other molecular liquids. It is these unique properties that make water very special and important in nature. The unusual features of liquid water are more enhanced at lower temperatures and lower pressures, reflecting enhanced tetrahedral structures stabilized by hydrogen bonding. There is a consensus that the unique properties of water come from local tetrahedral ordering due to hydrogen bonding. This is supported by the fact that atomic liquids having similar tetrahedral ordering due to covalent bonding also exhibit water-like anomalies and phase behaviours.[24]

These anomalous thermodynamic and dynamic behaviours of water have intensively been studied both experimentally and theoretically for a long time. Nevertheless, the very origin of the water anomalies is still a matter of debate and far from complete understanding.[6,9,35,75]

Many models of water have been proposed to explain the water's anomalies, focusing on the unique features of hydrogen bonding. The models can be classified into three groups:[3,6,9,73–75,77] (a) a stability-limit conjecture,[78] (b) a second-critical-point scenario (see *e.g.*, ref. 3, 79, 80), and (c) a singularity-free scenario.[81,82] Scenario (a) assumes the existence of a retracting spinodal curve and attributes the thermodynamic anomaly to proximity to the spinodal curve. Scenario (b), on the other hand, assumes the existence of a line of first-order transitions between two types of liquid water (low-density and high-density water), terminating at a critical point existing in a metastable state, and attributing the thermodynamic anomaly to critical phenomena associated with the hidden critical point. It is expected that a second critical point exists at a high pressure in the so-called no-man's land.[3] Finally, scenario (c) predicts that the thermodynamic quantities exhibit extrema but no divergence.

In scenario (a), investigation of the thermodynamic properties of water at negative pressure will provide crucial information on its relevance.[77] Scenario (b) is based on (i) experimental evidence of the presence of two amorphous forms of ices and a speculation on their connections to two types of liquids[3,83,84] as well as (ii) support for the presence of LLT in model waters from numerical simulations.[85] However, the connection between two amorphous ices and two liquids is also a matter of debate.[86,87] Whether the transition between the amorphous ices has an equilibrium counterpart, with a first-order phase transition line above the glass transition temperature (T_g) that terminates at a critical point (LLCP), has recently become a matter of much controversy.[88–95] A major source of difficulty lies in the fact that most modern theories of water concentrate on the supercooled region of the phase diagram, which is difficult to access by experiments due to the rapid crystallization of water below its melting line.[6,96] Similar difficulties emerge also in simulations, where the lack of crystallization is sometimes hindered by the limited system sizes and time scales accessible.[95] This is one of the major topics of this Faraday Discussion. We discuss this problem from a different viewpoint in section 7.8, focusing on a fundamental difference in the nature of the relevant order parameter between LLT and crystallization.

Both scenario (a) and (b) predict the anomalies of the thermodynamic and kinetic quantities due to the thermodynamic singularity. In these scenarios, the anomalies of the thermodynamic and dynamic quantities have often been analysed with the form of the power-law divergence, $\varepsilon^{-\gamma}$, where $\varepsilon = (T - T_s)/T_s$ (T_s: mean-field spinodal temperature) and γ is a critical exponent, and found to be well described by such relations. However, it should be noted that the critical exponents are often treated as adjustable parameters and no hyperscaling relations between the exponents have been found so far, unlike the case of the typical critical phenomena. Furthermore, we cannot approach to the mean-field spinodal temperature T_s so closely because homogeneous nucleation of ice crystals takes place far above T_s. Thus, the experimentally accessible range of ε is limited to $\varepsilon > 0.05$ in most cases and accordingly there has been no convincing evidence of the divergence of the thermodynamic quantities at a critical temperature. Note that for ordinary critical phenomena, we may approach a critical point to the order of $\varepsilon \sim 10^{-6}$–10^{-5}. Here it is worth mentioning that the thermodynamics of water has recently been studied in detail on the basis of critical phenomena.[80]

Finally, scenario (c) predicts no divergence of the physical quantities. Focusing on the temperature dependence of hydrogen bonding, many rather complicated functional forms have been proposed to describe the anomalous behaviour of the thermodynamic and kinetic quantities.

Despite these considerable efforts, there has so far been no consensus on which of these three types of scenario is primarily responsible for the above-described anomaly of water or whether we need a new scenario or not.[6] For example, there is still an on-going debate on the presence or absence of enhancement of density fluctuations that is a finger print of the singularity, since the singularity should cause critical-like enhancement of large-scale density fluctuations.[97–101]

Below we consider how the thermodynamic and kinetic anomalies of water-type liquids can be explained in the framework of our two-order-parameter model.

4.2 Prediction of our two-order-parameter model

Here we explain our two-order-parameter model of liquid.[21,22] We first estimate how the average fraction of locally favoured structures, $\bar{S}$, increases with a decrease in T. From the condition $\partial f(S)/\partial S = 0$ (see eqn (5)), $\bar{S}$ can be obtained as

$$\bar{S} = \frac{\frac{g_S}{g_\rho}\exp(\beta(\Delta E - P\Delta v))}{1 + \frac{g_S}{g_\rho}\exp(\beta(\Delta E - P\Delta v))}, \tag{9}$$

where $\beta = 1/k_BT$. In the above, we assume $J = 0$ for simplicity. For $J \neq 0$, the cooperativity plays an important role in inducing a liquid–liquid transition .[18,25] Here $\Delta E = E_\rho - E_S$ and $\Delta v = v_S - v_\rho$. E_i and g_i are the energy level and the number of degenerate states of i-type structure, respectively. $i = \rho$ corresponds to normal liquid structures of water, while $i = S$ to locally favoured structures (see Fig. 5).

The rather unique configurations of locally favoured structures and the existence of many possible configurations for normal liquid structures lead to the relation $g_\rho \gg g_S$. Then, $\bar{S}$ can further be approximated as

$$\bar{S} \sim \frac{g_S}{g_\rho}\exp[\beta(\Delta E - \Delta v P)]. \tag{10}$$

We stress that this relation should hold even for a non-zero J if $\bar{S} \ll 1$.[21,22]

Hereafter we consider thermodynamic anomalies for a case of $\bar{S} \ll 1$. According to the above picture, the unusual decrease in ρ upon cooling below 4 °C in water can simply be explained by an increase in the number density of locally favoured structures, $\bar{S}$, upon cooling. The specific volume v_{sp} and the density ρ are, respectively, given by

$$v_{sp}(T,P) = v_{sp}^{B}(T,P) + \Delta v \bar{S}, \tag{11}$$

$$\rho(T,P) \sim \rho_B(T,P) - \rho_B(T,P)\frac{\Delta v}{v_{sp}}\bar{S}, \tag{12}$$

where $\rho_B(T,P) = M/v_{sp}^{B}(T,P)$ (M: molar mass). Note that the background contributions v_{sp}^{B} and ρ_B almost linearly decrease and increase, respectively, with a decrease in T as for those of ordinary liquids. Then, the isothermal compressibility $K_T = -\frac{1}{v_{sp}}\left(\frac{\partial v_{sp}}{\partial P}\right)_T$ can straightforwardly be calculated from eqn (11) as

$$K_T = -\frac{1}{v_{sp}}\left(\frac{\partial v_{sp}^{B}}{\partial P}\right)_T + \frac{1}{v_{sp}}\left[-\left(\frac{\partial \Delta v}{\partial P}\right)_T + \beta \Delta v^2\right]\bar{S}. \tag{13}$$

The anomalous increase of K_T upon cooling can thus be explained by the following two mechanisms: (a) a decrease in T increases the population of locally favoured structures, which may be softer than normal liquid structures. (b) More importantly, the ability (or the degree of freedom) of the transformation between locally favoured structures and normal liquid structures upon a pressure change provides softness to a system. With an increase in pressure, the anomaly of K_T upon cooling becomes weaker, reflecting the decrease in the population of locally favoured structures, $\bar{S}$.

The anomalous increase in the heat capacity at constant pressure C_P upon cooling can also be explained as follows. The locally favoured structures have rather unique configurations and the associated degrees of freedom are much smaller for them than for the normal liquid structures of water. Thus, the entropy σ decreases upon cooling, reflecting an increase in $\bar{S}$, or short-range tetrahedral ordering:

$$\sigma = \sigma_B(T,P) - \Delta\sigma \bar{S}, \tag{14}$$

where σ_B is the background part of the entropy associated with normal liquid structures. Thus, $C_P = T(\partial\sigma/\partial T)_P$ should increase upon cooling as

$$C_P = T\left(\frac{\partial \sigma_B}{\partial T}\right)_P + \left[-T\left(\frac{\partial \Delta\sigma}{\partial T}\right)_P + \beta \Delta\sigma(\Delta E - P\Delta v)\right]\bar{S}. \tag{15}$$

In this manner, all the thermodynamic anomalies can be expressed simply by the common Boltzmann factor, $\bar{S}$. We confirmed the relevance of these relations by fitting the above functional forms to experimentally measured ρ, K_T, and C_P.[20–22] Here we show only the temperature dependence of the Boltzmann factor, $\bar{S}$, determined by the fitting of our prediction to the experimental data of ρ, K_T, and C_P at various pressures (see Fig. 8).

We also found that the kinetic anomalies of water such as viscosity anomalies can also be described by the same Boltzmann factor $\bar{S}(T,P)$.[22,23] For example, the

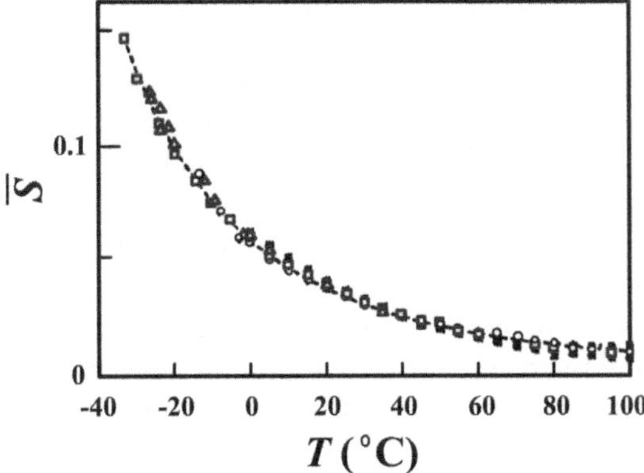

Fig. 8 Temperature dependence of $\bar{S}$ (see the text for its definition) determined by the fitting of our predictions to the experimental data of ρ, K_T, and C_P at various pressures. Open squares, triangles, and circles represent, respectively, data on ρ, K_T, and C_P at ambient pressure. All the other symbols are data at higher pressures. The dashed line is our theoretical prediction for $\bar{S}$. The values of $\bar{S}$ determined from the 23 sets of data of "bulk" liquid water are all collapsed on the master curve, which is described by the single Boltzmann factor. The figure is reproduced from Fig. 1(b) of ref. 21.

viscosity is known to exhibit a minimum as a function of pressure at a low temperature. This is a quite unusual phenomenon, but it can be naturally explained by the competition between the increase of the viscosity of normal liquid structures and the decrease of the fraction of locally favoured structures with an increase in P. We note that the presence of locally favoured structures leads to an extra activation energy for flow to take place and thus the decrease of S with an increase in P lowers this part of the viscosity associated with locally favoured structures.

It is known that the T-dependence of viscosity is well fitted by a power law, which is a prediction of mode-coupling theory (MCT).[102–104] However, we showed that it can be equally well described by the modified Arrhenius law. Furthermore, the pressure dependence can be explained more naturally by our model.[22,23]

4.3 Comparison of our two-state model with previous mixture models

A mixture model of water was first proposed by Röntgen[105] to explain water properties many years ago and then developed by many others (*e.g.*, ref. 73, 106–110). It was recently applied to a water problem by Ponyatovsky *et al.*[79] Their model regards water as a mixture of low-density (LDA) and high-density amorphous ice (HDA) (see also ref. 111). So it may be more appropriate to call this type of model a mixture model rather than a two-state model.

Here we compare our model with such mixture models to clarify what physical factors are key to the description of water's anomalies. The most crucial difference between our model and the mixture models is the value of $\Delta\sigma$. We assume that the difference in entropy, or the degeneracy of states, between the two states is very large, which is a consequence of the disordered nature of normal liquid structures and the more unique nature of locally favoured structures. We note that normal liquid structures are also made of water molecules temporally hydrogen bonded with neighbouring molecules. The important point is that their structural order is still considerably lower than that of locally favoured structures ($g_\rho \gg g_S$). On the other hand, it is assumed (see, *e.g.*, ref. 79) that the difference in the entropy between the two components is small since it is evaluated from the data of solid-state amorphous–amorphous (LDA–HDA) transition. In other words, it is implicitly assumed that both components have unique structures. Considering that a liquid is in a high entropy state and under significant thermal fluctuation effects, our two-state model approach seems to be more reasonable than such mixture model approaches. This subtle, yet crucial difference leads to a drastic difference in the physical picture. In our model, S is very small ($S \ll 1$) at ambient temperature and pressure (see Fig. 8), but in most other models[79,109,110] S (in our terminology) is almost 1/2 or even higher there and in some cases the anomaly was ascribed to critical anomaly associated with the second critical point of LLT (see, *e.g.*, ref. 79). In our case, water's anomalies are explained by a non-critical increase in S with decreasing T: the anomalous parts of physical quantities such as density are proportional to S and can be described by the Boltzmann factor at least in the experimentally accessible region (see eqn (10)).[21,22]

It is worth mentioning another reason why we prefer to use the term "two-state" rather than "mixture". This is because a mixture model gives us an impression that a system is composed of A and B components and thus the order parameter (the fraction of A) is conserved. In reality, however, the order parameter should not be conserved: locally favoured structures are created and annihilated without the constraint from its conservation. This point is crucial when we consider the nature and the dynamics of water-like anomalies and liquid–liquid transition[18,25] (see below).

Finally, we note a possible historical reason why S is estimated to be rather high. It may be related to the fact that the fraction of ice-like structures estimated from spectroscopic measurements such as Raman and infrared spectroscopy is usually rather high (see, *e.g.*, ref. 73, 112, 113). So if we identify this fraction as the fraction of locally favoured structures, S is estimated to be large. However, our study indicates that there may be no direct connection between them and we need to pick up special modes linked to the translational order of the second shell,[114] as will be described below.

4.4 Microscopic support for our two-order-parameter description from numerical simulations

In general, it is difficult to obtain detailed microscopic information on hydrogen bonding in water, and thus we cannot determine the difference in entropy between the two states in a convincing matter. We emphasize that the difference in $\Delta\sigma$ leads to the entirely different scenarios for water's anomalies, as described above. This problem, which is directly related to the microscopic structural identification of normal liquid and locally favoured structures, is the key to our understanding of the physical origin of water's anomalies.

Numerical simulations are obviously very powerful in identifying locally favoured structures. Recent simulation results[115,116] seem to be consistent with our scenario, which predicts that S is rather small in the experimentally accessible region. Anisimov and his coworkers also showed that a two state model with cooperativity of entropic origin describes well the thermodynamic anomalies of mW water.[117] Their results are also basically consistent with ours. However, we should also note that the estimate of the fraction of the LDL-like component by Cuthbertson and Poole is higher.[118] Furthermore, Matsumoto showed that expansion of water upon cooling can be explained without invoking any heterogeneity.[119] Thus, the situation has been quite controversial. The origin of the controversy is due to the lack of a proper structural order parameter for locally favoured structures of water.

Recently we successfully identified a structural order parameter for locally favoured structures of model water (TIP4P/2005) at a microscopic level. It is the degree of translational order in the second shell.[114] The importance of the structure of the second shell was first shown by Soper and Ricci.[120] This new structural order parameter allows us to estimate the T–P dependence of S directly from simulated water structures. We confirmed that the two-state model with the order parameter S that is independently determined in this way can well describe the thermodynamic anomalies. This provides the microscopic basis for our two-order-parameter model. The value of S was found to be low in the experimentally accessible region, consistent with the prediction of our two-order-parameter model (see above). We also showed that we can directly estimate S from the O–O radial distribution function, which provides a method to estimate S experimentally.

4.5 Do anomalies obey the Boltzmann factor or power law?

On the basis of these results, we argue that water's anomalies can basically be explained by our two-order-parameter model, or the Boltzmann factor, and not by power laws. This is supported by the fact that in model waters (TIP4P/2005[114] and mW water[117]), which exhibits all the anomalies characteristic of water, quantities such as isothermal compressibility do not show any divergence at a finite temperature. Furthermore, the effects of cooperativity is very minor in the experimentally accessible region, even if there exists a second critical point. So we conclude that water's anomalies are primarily the consequences of Schottky-type anomaly characteristic to a two-state model, although further studies are necessary to confirm it in an unambiguous manner.

4.6 Water-like anomalies in water-type atomic liquids

Here we briefly consider what makes water so special among 'molecular' liquids. We pointed out[21,22,24] that water is the only molecular liquid for which local bond orientational ordering is basically compatible with a global crystallographic symmetry: the locally favoured tetrahedral structure of water stabilized by hydrogen bonding is basically consistent with the crystallographic symmetry of hexagonal ice I_h and cubic ice I_c, although there is some inconsistency in the symmetry.[114] It is important to recognize that formation of a tetrahedral structure stabilizes hydrogen bonding with a help of local symmetry in a "cooperative" manner.

We argue that all the thermodynamic anomalies of water originate from (i) this dominance of bond orientational ordering below a crossover pressure P_x ($\sim$2 kbar), where the melting point of ice crystals has a minimum, and (ii) an unusually large positive value of Δv. Below P_x, the crystallization is due to bond ordering, while above P_x it is due to density ordering as in ordinary liquids (see Fig. 9). This gives a natural explanation for the unusual pressure dependence of

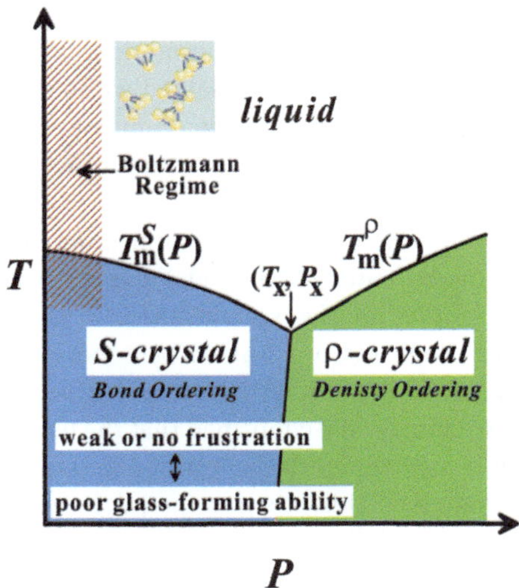

Fig. 9 *T–P* phase diagram of water-type liquids including water itself and water-type atomic liquids (Si, Ge, Bi, Sb, and Ga).

the melting point of ice crystals, including its minimum around 2 kbar. We propose that ice I_h is *S*-crystal, long-range ordering of *S*, while high-pressure ices are ρ-crystals.[21,22,24] The V-shaped *T–P* phase diagram of water-type liquids is just a manifestation of the Clausius–Clapeyron relation.

The above conditions (i) and (ii), which are necessary for having water-like anomalies, should be satisfied for tetrahedral liquids having V-shaped phase diagram. By using this specific shape of the phase diagram as a fingerprint,[24] we classified five elements Si, Ge, Sb, Bi, and Ga into water-type atomic liquids. We showed that our two-order-parameter model of water indeed also explains the thermodynamic and dynamic anomalies of these water-type atomic liquids in a satisfactory manner.[24]

We discuss unique features of these liquids again in relation to the glass transition problem in section 6.13.

5 Liquid–liquid transition

Usually it is considered that atoms or molecules have random disordered structures in the gas and liquid states. This leads to a common sense view that any single-component substance has only one gas and one liquid state. On the other hand, it is widely known that even a single-component liquid can have more than two crystal forms, which is known as "polymorphism". The uniqueness of the state is very natural and correct for gas, where the kinetic energy dominates. However, it is not so obvious for liquid since many-body interactions come into play, reflecting its high density, as we described above.

Recently there has been growing experimental evidence that even a single-component liquid can have more than two liquid states.[3,6,9,74,121–129] The transition between these liquid states is called "liquid–liquid transition" (LLT). There are also experimental indications for the presence of LLT in binary-component liquids such as AsS.[130,131] The existence of liquid–liquid transitions has also been supported by a number of numerical simulations for atomic liquids such as Si[132–135] and molecular liquids such as water.[6,85,136,137] This phenomenon has attracted considerable attention not only because of its counter-intuitive nature but also from the fundamental importance for our understanding of the liquid state of matter. The connection between liquid–liquid transition and polyamorphism is also an interesting issue.

First we describe a simple phenomenological theory, which explains LLT as a transition between a gas state and a liquid state of locally favoured structures. Then we show some experimental pieces of evidence supporting the presence of LLT. However, most of such examples suffer from serious criticisms and thus the situation is quite controversial. Below we also discuss the source of controversies focusing on LLT in molecular liquids.

5.1 Phenomenological two-order-parameter model of liquid–liquid transition

In our model, LLT is described by the free energy $f(S)$ given by eqn (5).[25] The key term there is the coupling term $JS(1-S)$, which represents the cooperativity of formation of locally favoured structures. Thus, we first discuss possible origins of the cooperativity.

One is the microscopic cooperativity of directional bonding, which is related to the change in the electronic state by the formation of locally favoured structures. This effect may be important in liquids have directional bondings such as hydrogen and covalent bonding. Its importance is particularly clear for liquids such as Si and Ge, for which transitions accompany a drastic change in the electronic properties and induce metal–semiconductor transitions. How to incorporate the electronic degrees of freedom into our phenomenological model is an interesting but challenging problem. The second is the modification of the degrees of freedom around locally favoured structures due to the local reduction of configurational and vibrational entropy. The third is a possible role of long range van der Waals forces, which are due to the difference in density between locally favoured structures and normal-liquid structures, $\delta\rho$. The interaction strength may be estimated as

$$U \sim U_{11}(\delta\rho/\rho)^2(b/a), \tag{16}$$

where U_{11} is the interaction between basic units (atoms or molecules) of size a, ρ is the density, and b is the size of locally favoured structures. This interaction might be too weak to cause LLT in the usual situation. Since the origin of cooperativity lies at the heart of our understanding of LLT, further careful studies are highly desirable. First principle calculations may be a promising way to attack this problem. As we see below, the value of J directly determines the location of the critical temperature T_c. Finally, we note that the degeneracies of the two states, or thermal fluctuation effects, produce a sort of renormalization effects on the mean-field interaction parameter J, which may affect the location of a critical point.

The equilibrium value of S is determined by the condition $\partial f(S)/\partial S = 0$, or

$$\beta[-\Delta E + \Delta v P + J(1-2S)] + \ln\frac{g_\rho S}{g_S(1-S)} = 0, \tag{17}$$

where $\Delta E = E_\rho - E_S > 0$, $\Delta v = v_S - v_\rho$, and $\beta = 1/k_BT$. The schematic T-dependence of S is shown in Fig. 10. It is worth noting that the degeneracy of each state, or the entropy difference between the two states, strongly affects the phase behaviour. A critical point is determined by the conditions, $f'_S(S_c) = 0, f''_S(S_c) = 0, f_S^{(3)}(S_c) = 0$, and $f_S^{(4)}(S_c) > 0$, as

$$S_c = 1/2, \tag{18}$$

$$T_c = J/(2k_B), \tag{19}$$

$$P_c = [\Delta E - T_c\Delta\sigma]/\Delta v. \tag{20}$$

A first-order phase-transition temperature T_t is obtained as

$$T_t = (\Delta E - P\Delta v)/\Delta\sigma. \tag{21}$$

Note that a first-order transition occurs only if $T_t < T_c$. For $T_t > T_c$, this T_t is a temperature where $\Delta G = 0$ and thus $\bar{S} = 1/2$. The maximum of K_T is also located near T_t. Δv may be positive for liquids such as water and Si, but it can also be negative for liquids such as triphenyl phosphite (see below and ref. 138, 139). The sign of Δv determines the slope of $T_t(P)$. Liquid I and liquid II are defined as the two possible minima of the liquid-state free energy on the ρ–S plane.

Now we consider the kinetics of LLT. In LLT, the bond order parameter S plays essential roles, and the density order parameter ρ is slaved by S. Using $\delta S = S - \bar{S}$, we introduce the following minimal Landau-type free energy density by expanding $f(S)$ in terms of δS, which governs S fluctuations near a gas–liquid-like critical point or mean-field spinodal lines of bond ordering, where $S = S_{SD}$:

$$\frac{f(\delta S)}{k_BT} = \frac{\kappa}{2}\delta S^2 + \frac{b_3}{3}\delta S^3 + \frac{b_4}{4}\delta S^4 + h\delta S, \tag{22}$$

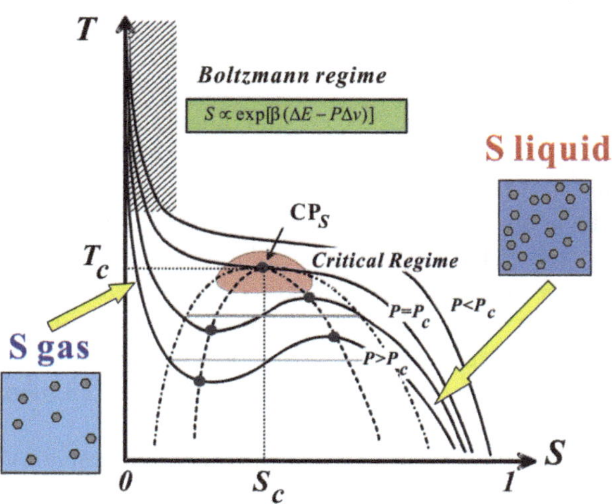

Fig. 10 Schematic phase diagram of liquid–liquid transition in T–S plane.[25] Liquid–liquid transition can be understood as a transition between a S-gas state and S-liquid state.

where $\kappa = \frac{1}{S_{SD}(1-S_{SD})}\Theta$, $b_3 = -\frac{1}{2}\left[\frac{1}{S_{SD}^2} - \frac{1}{(1-S_{SD})^2}\right]$, $b_4 = \frac{1}{3}\left[\frac{1}{S_{SD}^3} + \frac{1}{(1-S_{SD})^3}\right]$, and $h = \left[\ln\frac{g_\rho}{g_S} + \ln\frac{S_{SD}}{1-S_{SD}}\right]\Theta$. In the last relationship, we use $\partial f/\partial S = \partial^2 f/\partial S^2 = 0$ at $T = T^*_{SD}$. In the above, Θ is the scaled temperature: $\Theta = (T - T^*_{SD})/T$, where T^*_{SD} is a critical or spinodal temperature of bond ordering without the coupling to ρ, and b_2 and b_4 are positive constants. By further including the gradient term, we obtain the following Hamiltonian that we believe is relevant to the physical description of a gas–liquid-like transition of locally favoured structures (see Fig. 10):[25]

$$\beta H_S = \int d\vec{r}\left[f(\delta S) + \frac{K_S}{2}|\nabla\delta S|^2\right]. \tag{23}$$

For simplicity, we assume the density ρ is given as a function of S as follows: $\rho(\vec{r}) = \rho_N(1 - S(\vec{r})) + \rho_S S(\vec{r})$, where ρ_N is the density of the normal-liquid structures and ρ_S is the density of the locally favoured structures. For $\Delta v < 0$, which is a case of TPP, an increase in S leads to an increase in ρ.

In our previous papers,[18,25,140] we employed a more complex coupling between ρ and S, which leads to the constraint for the global density. However, experiments are usually performed at constant pressure and there is no constraint for the total density. Indeed, our light scattering experiments on LLT in triphenyl phosphite show that the scattering intensity at the wavenumber $q \to 0$ grows upon LLT, which is characteristic of the correlation function of a non-conserved order parameter. Although we need a more complete description, which takes into account the couplings to the density (mass conservation), velocity fields (momentum conservation), and temperature fields (energy conservation), we here stick to the simplest version of the kinetic theory. Here we note that Takae and Onuki recently investigated the roles of latent heat on LLT. This might also play an important role[141] when local heating induced by LLT takes place due to a weak thermal contact between the sample and the temperature bath.

Since locally favoured structures can be created and annihilated without the constraint of its conservation, the order parameter S should obey the kinetic equation describing the time evolution of the non-conserved scalar order parameter S:[25]

$$\begin{aligned}\frac{\partial\delta S}{\partial t} &= -L_S\left[-K_S\nabla^2\delta S + \frac{\partial f(\delta S)}{\partial\delta S}\right]\\ &= -L_S\left[-K_S\nabla^2\delta S + h + \kappa\delta S + b_3\delta S^2 + b_4\delta S^3\right] + \zeta_S',\end{aligned} \tag{24}$$

where L_S is a kinetic coefficient and ζ'_S represents normalized thermal noises satisfying the fluctuation–dissipation relation.

The type of pattern evolution is grouped into nucleation–growth (NG)-type in a metastable state (above T_{SD}) and spinodal-decomposition (SD)-type in an unstable state (below T_{SD}). Here T_{SD} is the spinodal temperature. NG-type LLT is characterized by nucleation of droplets overcoming the activation barrier by thermal noises and its growth with a constant velocity. SD-type LLT is characterized by spontaneous growth of order parameter fluctuations, whose amplitude grows exponentially in the early stage. In both cases, the final state becomes homogeneous again at a constant pressure condition, which is a consequence of the non-conserved nature of the order parameter S.

However, this mean-field picture suggesting a sharp transition between NG- and SD-type dynamics breaks down under thermal fluctuation effects and the transition becomes broader[142,143] except for systems of long-range interactions.[25]

5.2 Current situations and controversies

Once we accept the formation of locally favoured structures in liquids and its cooperativity, it is natural to assume the presence of LLT. Since it is rather hard to imagine that locally favoured structures are formed completely independently without any cooperativity, we expect that liquid–liquid transitions can exist in many liquids.[25] However, the energy scale of the coupling parameter J is often comparable to the energy scale controlling other cooperative ordering such as crystallization, since both are determined by the common interaction potential. Thus, liquid–liquid transition may often be hidden by crystallization, particularly if the symmetry of locally favoured structures is similar to that of the equilibrium crystal. This may be the source of controversy concerning whether or not liquid–liquid transition exists in many systems including the case of water (see also section 7.8). From a theoretical viewpoint, thus, it is crucial for deeper understanding of LLT to elucidate the *microscopic* mechanism of formation of locally favoured structures and its cooperativity.

Here we note that mixing a target liquid with another liquid that can prevent crystallization may be a good strategy to reveal such a hidden LLT, since LLT may take place even after mixing with other fluids.[144] We actually employed this method in our study of LLTs of aqueous organic solutions[145,146] (see below).

In some model liquids such as the Jagla model,[147,148] a liquid–liquid transition is clearly seen. So the presence of LLT in liquids itself has been accepted at least theoretically. For realistic models, however, the situation is controversial even in numerical simulations, as mentioned above. For example, as we see in this Faraday Discussion, whether LLT exists in model waters or not has recently become a matter of much controversy. Some simulations support the presence,[88–92] whereas the others do not.[93–95,149] This is the case even for the same ST2 water model. A major source of difficulty lies in the fact that LLT exists in a state thermodynamically metastable or unstable against crystallization. Thus the lack of crystallization may be hindered by the limited system sizes and time scales accessible.[95] During the meeting we do not see any consensus on this problem. We discuss this issue from a different viewpoint in section 7.8.

Experimentally, there are also few cases for which there is a consensus on the existence of LLT. This is mainly due to the fact that LLT exists in a region which is difficult to access experimentally: for atomic liquids LLT exists at very high temperature and pressure, whereas for molecular liquids it exists only in a metastable state, where crystallization can take place. Because of such difficulties, the situation still remains very controversial.

First we review the case of atomic liquids. For example, Katayama *et al.* discovered the first order LLT in phosphorus at high pressure and high temperature with synchrotron X-ray scattering.[125,126] They revealed the structure factors for both liquid I and II and confirmed the coexistence of liquid I and II during the transition, which clearly suggests the first-order nature of the transition. The distinct change in the structure factor suggests that LLT in phosphorus is the transformation from tetrahedral to polymeric liquid. This is one of the clearest examples of a transition between two isotropic fluids. Such a structural transition was also observed by the first principle simulation performed by Morishita.[150] However, Monaco *et al.*[127] concluded that the first-order transition in P is between a high-density molecular fluid (not a liquid in the exact sense) and a low-density polymeric liquid. Thus, the transition is now regarded as a 'supercritical fluid'–liquid transition rather than a liquid–liquid transition. This explains an unusually large difference in the density between the two states. The existence of LLT in liquid Si was also suggested by high-pressure experiments[122,151,152] and numerical simulations,[132–134] but the presence of LLT still needs to be checked carefully. LLT was also reported in yttria–alumina.[121,122,153–155] However, there are also still on-going debates on the composition range over which this phenomenon occurs and the experimental conditions required to observe it[156] and even on its existence itself.[157]

For molecular liquids, Mishima *et al.* found an amorphous–amorphous transition in water.[158] The transition has recently been studied in detail.[6] From the presence of the two forms of amorphous states, the presence of LLT has been inferred. Computer simulations also suggested the existence of LLT(s) in water.[3,6,123,136,137,159] On the basis of these findings, the connection of amorphous–amorphous transition and LLT in water was suggested and actively studied.[3,6] Some simulations support this connection,[86] others not.[87] In real water, the LLT is hidden by crystallization, even if it exists. This makes an experimental study on the LLT extremely difficult especially for bulk water. It was also pointed out that the mechanical nature of amorphous–amorphous transition makes its connection to thermodynamic LLT indirect, even if it exists.[21] As mentioned above, even for numerical simulations, difficulties associated with the distinction between LLT and crystallization in a deeply supercooled liquid make the situation very controversial.

This situation has been improved by recent direct observation of LLT at ambient pressure in molecular liquids, triphenyl phosphite (TPP)[160,161] and *n*-butanol.[162] We observed both NG-type and SD-type LLT, which is well explained by our scenario that liquid–liquid transition is a consequence of the cooperative ordering of a non-conserved scalar order parameter, which is the fraction of locally favoured structures, $S(\vec{r})$. However, this phenomenon was also claimed by Hédoux *et al.*[163–169] to be induced by the formation of micro-crystallites rather than LLT. Recently, a similar claim was also made for *n*-butanol.[170–172] In this scenario, what we call liquid II is merely a mixture of liquid I and micro-crystallites.

So strictly speaking, there has been no firm consensus on the existence of LLT for any substance from the experimental side, and it remains a matter of serious debate whether the above-mentioned phenomena are the true evidence of LLT or not. Theoretically, on the other hand, the generality of LLT, or possible existence of LLT in various types of liquids, was recently discussed on the basis of phenomenological[18,25] and analytical models.[173–176]

Below we review our study on LLTs in molecular liquids and then discuss the controversies on the nature of the transition for these examples.

5.3 LLT observed in single-component molecular liquids

Some time ago Kivelson and coworkers reported the following unusual phenomena observed in a supercooled state of TPP.[138,177,178] When TPP is cooled rapidly and deeply enough, it first enters into a supercooled liquid state below the melting point T_m as usual liquids, and then frozen into a glassy state, which we call glass I. This supercooled liquid (liquid I) behaves as a typical fragile glass former. On the other hand, when TPP is quenched to a certain temperature between 213 K and 223 K and then annealed at that temperature, a new apparently amorphous phase (the so-called glacial phase) is formed in a supercooled liquid and thus the system becomes inhomogeneous and optically turbid. Eventually, however, the system completes the transformation into the glacial phase. Surprisingly, the glacial phase is apparently an optically transparent homogeneous amorphous phase, but it is obviously different from ordinary liquid (liquid I) and glass (glass I).

This finding stimulated intensive experimental research on this unusual phenomenon. However, the nature and origin of the glacial phase has been a matter of debate and many different, even controversial, explanations have been proposed for it. The glacial phase was thought to be a new amorphous phase[178–181] or a highly correlated liquid.[182] However, most researchers, including us, have shown that the glacial phase has some crystallinity or anisotropy. Hence the newly formed glacial phase appears to be neither a standard glass nor a liquid. It is this that has led some researchers to conclude that the glacial phase is actually some type of defect-ordered crystals (orientationally disordered or modulated crystal),[138,178,183] liquid crystal,[184] plastic crystal,[181,184] aborted crystallization,[163–169] or nano-clustering.[185]

To access the nature of the transition, we directly observed the process of liquid–liquid transition with optical microscopy for two pure organic liquids, triphenyl phosphite (TPP)[160,161] and *n*-butanol.[162] When we quench and anneal TPP in the metastable region with respect to LLT, droplets of liquid II are randomly nucleated in both space and time in liquid I and the domain size R grows with a constant interface velocity as $R \propto t$ (see Fig. 11(a)).[160–162] In the late stage, droplets of liquid II collide, coalesce, and further grow. This behaviour is characteristic of the NG behaviour. Then, the new phase covers the entire region and eventually the boundary between droplets tends to disappear; and, thus, liquid I almost transforms to homogeneous liquid II (see Fig. 11(a)). This is a consequence of the non-conserved nature of S and the off-symmetric quench. If the LLT were governed by a conserved order parameter, the diameter would grow in proportion to $t^{1/3}$ and the system would never become homogeneous again,[186] as long as we do not cross the binodal lines twice.[187] We also observed SD-type LLT, which occurs when a liquid is quenched into an unstable region below T_{SD}[160,161] (see Fig. 11(b)). The initial stage is reminiscent of the Cahn's linear regime.[186] In the beginning, the amplitude of fluctuations exponentially grows with time and thus the contrast increases. Then, the domain size and the contrast both increase. Later the liquid becomes more homogeneous, which leads to the decrease in the contrast. Finally, the liquid becomes almost homogeneous liquid II. These observations are basically consistent with the prediction of our model. The heat evolution was also measured during LLT. This is also consistent with our model, which assumes that LLT is a consequence of the cooperative formation of locally favoured structures with a lower local free energy. According to our model, provided that the amount of nano-crystallites formed during LLT is negligible, which is the case for LLT at a low temperature below T_{SD}, the heat evolution is proportional to the development of the bond order parameter S, since the heat is

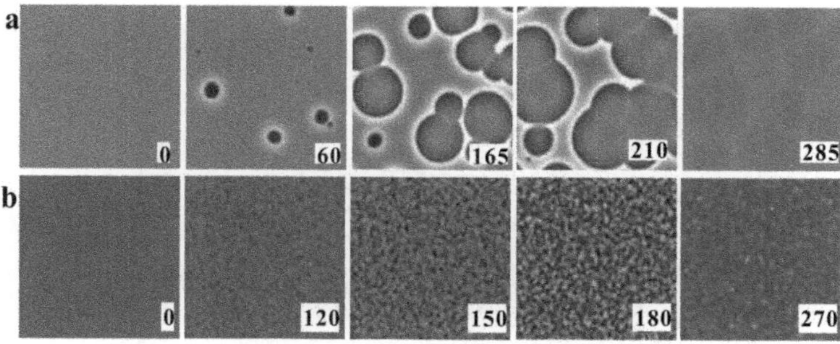

Fig. 11 Pattern evolution observed with phase-contrast microscopy during the annealing of a supercooled liquid at T_a. (a) Experimental results for TPP at $T_a = 219$ K. The intensity is proportional to the refractive index. The size of each image is 120 μm × 120 μm. The contrast between droplets (liquid II) and the matrix (liquid I) decreases with an increase in the distance from the interface (decay length: a few microns). This is due to non-ideality of phase contrast microscopy, and does not mean the change in the refractive-index difference. (b) Experimental results for TPP at $T_a = 214$ K, observed with phase-contrast microscopy. The size of each image is 150 μm × 150 μm. The number in each image indicates the elapsed time in minutes for both (a) and (b).

released in the process of the formation of locally favoured structures. This was supported by the structural study of the process of LLT by X-ray scattering.[188] However, the interpretation is complicated by the presence of micro- or nano-crystallites, which are formed during the transformation. We confirmed that LLT accompanies the formation of micro- or nano-crystallites at rather high temperatures (above 214 K), but at low temperatures (*e.g.*, at 212 K) the amount of crystals becomes very small. On the basis of these experimental results, we concluded that this transformation is actually a transition from a supercooled state of liquid I to a glassy state of liquid II.

We also find that liquid II is stronger than liquid I and the fragility monotonically decreases in the transformation process from liquid I to II. Fig. 12 shows how the upper and lower edge of the glass transition (T_g^H andT_g^L, respectively) and the ideal glass transition temperature T_0 are dependent on the normalized order parameter $\tilde{S}$, which monotonically increases with time during the transformation from 0 to 1. This suggests[189] that the fragility is not a material specific property, but rather controlled by the strength of frustration against crystallization (see section 6). This conclusion was supported by the recent experimental study on the pressure effect on the fragility of liquid II.[139]

Here we also mention the difference in the physical and chemical properties of liquid I and II of TPP. First of all, liquid II has a higher density and a higher refractive index than liquid I. The glass transition point of liquid II is much higher than liquid I, which means the difference in the fluidity between them. As described above, liquid II is less fragile than liquid I. We also found[144] that liquid I is miscible with diethyl ether, but liquid II is not when TPP is mixed with a sufficiently high concentration of diethyl ether. As shown in Fig. 13, whether diethyl ether can mix with TPP depends not only on the concentration of diethyl either Φ, but also the order parameter S. This clearly shows that liquid I and liquid II may have a different miscibility with other liquids. We note that both liquid I and II of TPP are miscible with toluene.[144] We also revealed that liquid I and liquid II have different wettability to a solid substrate.[190] Thus, liquid I and liquid II have

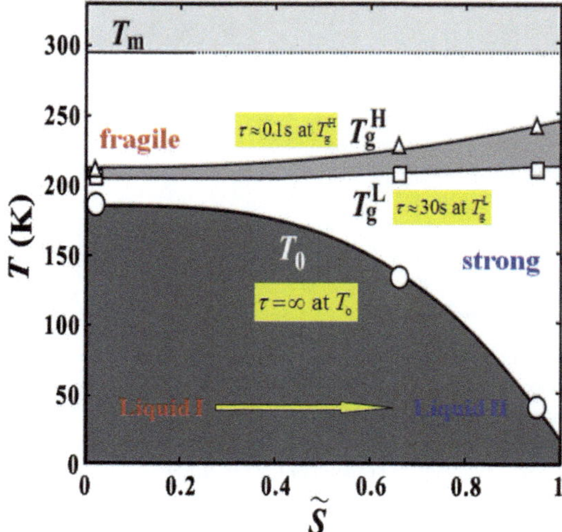

Fig. 12 The dependence of the upper and lower edge of the glass transition (T_g^H and T_g^L, respectively) and the ideal glass transition temperature on the normalized order parameter $\tilde{S}$ for TPP.

differences in density, refractive index, dielectric constant, glass transition point, fragility, fluidity, miscibility, and wettability.

In the case of *n*-butanol, we observed the pattern evolution behaviour almost identical to that in TPP.[162] However, crystallization always occurs even at a low temperature below T_{SD}, the crystallinity is higher than in the case of TPP, and thus the situation is a bit more complicated. For example, Ramos and his coworkers recently claimed that the phenomena observed in *n*-butanol is aborted crystallization and not LLT,[170,171] but we still argue that it is LLT on the basis of the kinetic features of the transformation process (see below).

5.4 LLT in a mixture of water and glycerol

Next we describe our efforts to access a possibly hidden LLT in water experimentally. Unlike the cases of TPP and *n*-butanol, experimental verification of LLT

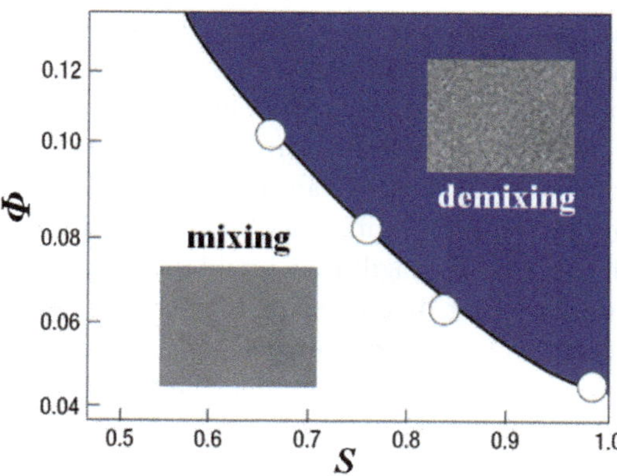

Fig. 13 The phase diagram showing the miscibility of TPP with diethyl ether on the two-order-parameter plane of the concentration of diethyl either Φ, and the order parameter S at $T = 209$ K.

in water is quite difficult due to the interference by instantaneous crystallization. There may still be a few routes to access a hidden LLT in water, if it exists. One strategy is to use water confined in nm-size pores[191,192] or to use protein-hydration water.[193] These works show evidence suggestive of a fragile-to-strong liquid transition and LLT. However, these experiments inevitably suffer from criticisms that water confined into a nm-scale space surrounded by a wall is intrinsically different from bulk water because of the presence of water–wall interactions and the reduced dimensionality.[194–196] It was shown,[194] for example, that (i) without a special care we cannot conclude even whether water inside a nm-size pore is liquid or solid or amorphous or crystalline and (ii) the interaction with the wall makes the confined state inhomogeneous. We also refer the paper presented by Torre in this Faraday Discussion,[197] which indicates significant structuring of water near the hydrophilic surface. Since the presence of the amorphous–amorphous transition in bulk does not prove the presence of LLT either (see above), it may be fair to say that we do not have any firm experimental evidence for LLT in water yet, although there are many implications (see, *e.g.*, ref. 83).

Recently we took a different strategy: mixing water with glycerol to avoid crystallization of water. Note that glycerol is a well-known cryoprotectant non-crystallizable liquid and can cause strong frustration against water crystallization (see, *e.g.*, ref. 199, presented at this Faraday Discussion). In an aqueous glycerol solution we found the direct experimental evidence for genuine (isocompositional) LLT without accompanying demixing.[145] We confirmed that liquid I transforms *via* the two types of kinetics characteristic of the first-order transition of a non-conserved order parameter, NG and SD, towards homogeneous liquid II (see Fig. 14). The processes of pattern evolution are strikingly similar to those observed in TPP, strongly indicating that the nature of the transition should be the same between TPP and water–glycerol mixtures. The state diagram of water–glycerol mixtures is shown in Fig. 15. The liquid–solid phase diagram of water–glycerol mixtures is very similar to the *T*–*P* phase diagram of pure water, which also has a V-shape. We found that liquid I and II differ in the density, the refractive index, the structure, the hydrogen bonding state, the glass transition temperature, and the fragility. We revealed that this transition is mainly driven by

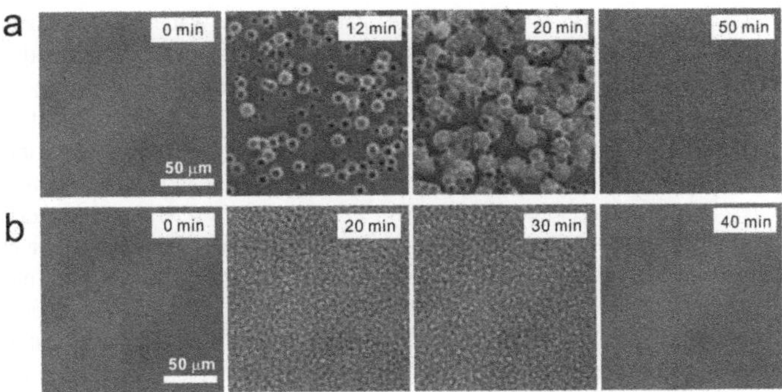

Fig. 14 (a) NG-type pattern evolution during LLT at T = 180 K for c = 0.165. (b) SD-type pattern evolution at T = 173 K for c = 0.165. In both cases, the initial state is liquid I and the final state is liquid II. The patterns were observed with phase contrast microscopy. White bars correspond to 50 μm.

local structuring of water rather than glycerol, suggesting a possible link to LLT in pure water.

In relation to this, it was recently pointed out by Towey and Dougan[200] that glycerol molecules act to "pressurize" water. This further suggests a link between a water–glycerol mixture and pure water. However, we note that there is also a change in the vibrational modes of glycerol molecules upon LLT[145] and thus we cannot deny a possibility that LLT occurs only in solutions and not in pure water. Thus, further study is necessary to clarify whether water has LLT without glycerol or not.

5.5 Controversy on the nature of the transition: LLT or formation of nano-crystallites

5.5.1 The case of TPP. As already mentioned, what we call liquid II for TPP was also interpreted as a mixture of micro- or nano-crystallites and liquid I. Here we explain in detail why we believe that the transition is primarily due to LLT and not due to crystallization, by trying to interpret various behaviours by the two scenarios.

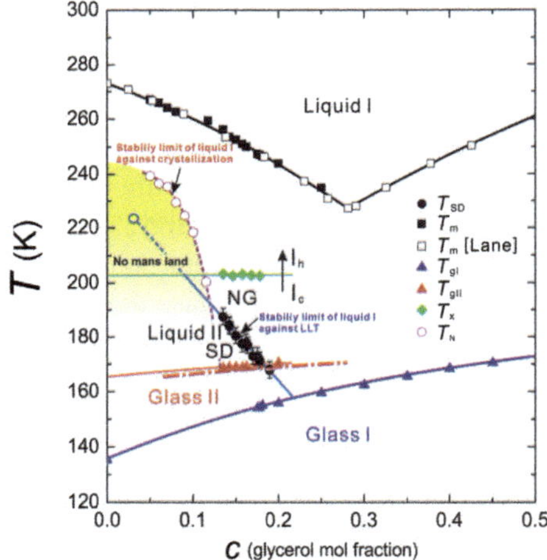

Fig. 15 Glycerol concentration *vs.* temperature (c–T) state diagram of water–glycerol mixtures. T_{SD}: LLT spinodal temperature (black filled circles); T_{gI}: the glass transition temperature of liquid I (blue filled triangles). For pure water ($c = 0$), we use the widely accepted value of 136 K [6] for T_{gI}; T_{gII}: the glass transition temperature of liquid II (red filled triangles). Dot-dashed line indicates T_{gII} of pure liquid II without ice I_c, provided that liquid II contains $\phi_c = 17\%$ of ice I_c, which should result in the increase in the glycerol mole fraction of liquid II by 6.4%; T_H: the homogeneous nucleation temperature (violet open circles) measured for the cooling rate of 100 K min^{-1}; T_x: the transition temperature from ice I_c to I_h, which was determined by microscopy observation (green filled diamonds); T_m: the melting (liquidus) temperature (black filled squares: our data; open squares: the data of Lane[198]). We make a linear extrapolation of T_{SD} to estimate the position of a hypothetical critical point (CP) (light blue open circle), since we cannot access T_{SD} for $c < 0.13$ due to rapid nucleation of ice I_h before reaching the final target temperature in the quench process. For $c > 0.19$, on the other hand, the kinetics of LLT drastically slows down, which also prevents us from accessing LLT during the observation time. Finally we note that the T_{SD} we measured is the stability limit of liquid I, and we could access neither the binodal line nor the stability limit of liquid II because of interference by ice crystallization. This figure is reproduced from Fig. 4 of ref. 145.

First we consider the characteristics of pattern evolution observed by optical microscopy. We observed NG-type and SD-type pattern evolution, which can be naturally explained by our scenario, as already explained above. If we take the crystallization scenario, we may still regard droplets formed in the NG-type process as ordinary spherulites with low crystallinity and growth of fluctuations in the SD-process as homogeneous nucleation. However, it is difficult to explain by this scenario why a system becomes inhomogeneous in the beginning but later becomes homogeneous. We might explain this by replacing locally favoured structures by nano-crystallites in our two-order-parameter model of LLT, or assuming cooperative crystal nucleation. However, this model seems not to be so realistic and we need an intrinsic mechanism preventing the growth of nano-crystallites and stabilizing them at a nm-scale size. It is also not so easy to explain why a non-crystallized remaining liquid becomes a glass even far above the glass transition of the original liquid. One possible explanation is that the dynamics of a liquid confined by nano-crystallites slows down and becomes glassy. However, this interfacial liquid (or, confinement) scenario cannot explain why the glass transition temperature of the glacial phase is not so sensitive to the amount of nano-crystallites, which is largely changed by the annealing temperature. According to this scenario, with a decrease in the amount of nano-crystallites, the glass transition should be more similar to that of pure liquid I, but what is observed is the opposite. In relation to this, we should note that the glass transition of liquid II is very broad, which we interpreted as a signature of the strong nature of liquid II[160,189] and a possible broad distribution of S due to vitrification in the course of LLT (see Fig. 16 and below). In the crystallization scenario, we might be able to explain this broadness by the spatial distribution of the relaxation time, which is controlled by the distance from the surfaces of nano-crystallites. The temporal change in the broadness of the transition during SD-type LLT[189] may be explained by both scenarios at least qualitatively.

In relation to this, it is worth mentioning a mysterious character of liquid II: liquid II is a strong liquid, but exhibits a very broad structural relaxation spectrum. Usually, a strong liquid has a narrow distribution of the relaxation time.[9,201] We explain this as follows: a glassy state of liquid II is formed as a consequence of LLT (see Fig. 16) and the broad distribution of the order parameter S as a consequence of the SD-type ordering[161] is frozen by vitrification before becoming narrow. Within the crystallization scenario, we may interpret this as a consequence of the spatial distribution of the relaxation time (see above).

Next we consider the heat production during LLT. The heat released during LLT is much larger than that expected for the amount of nano-crystallites. Even when we detect little indication of nano-crystallite formation by X-ray scattering (below 215 K), there is a significant release of heat during LLT (see, *e.g.* ref. 160, 202). These are also difficult to explain by the aborted crystallization scenario. We might still be able to argue that nano-crystallites are too small to detect by X-ray scattering, but then it becomes almost impossible to distinguish liquids and crystals any more.

Furthermore, we recently found an indication suggestive of the formation of locally favoured structures whose size is about 3 nm and the increase of the fraction with time during LLT by time-resolved small-angle X-ray scattering measurements.[203] In the process, we observe little change in the wide-angle diffraction pattern. This suggests that this signal is from locally favoured

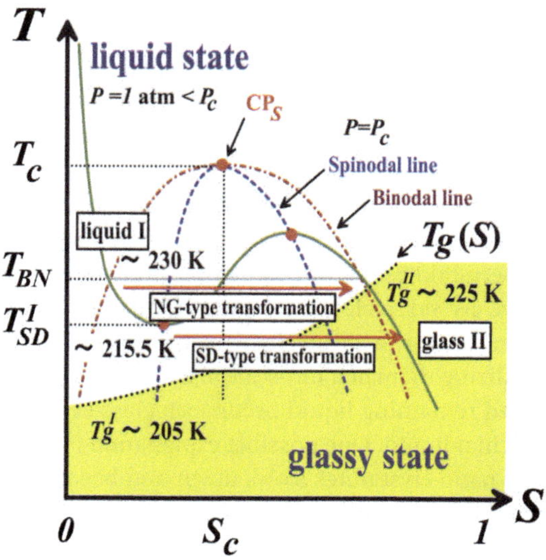

Fig. 16 Schematic *T–S* phase diagram of TPP. The dashed and dot-dashed lines are spinodal and binodal lines, respectively. CP_S is a gas–liquid-like critical point of the bond order parameter *S*, which might exist at a high pressure. The dotted curve is the glass transition line $T_g(S)$ and the yellow region corresponds to a glassy state. T_{BN} and T^I_{SD} represent the binodal temperature and the lower spinodal temperature at atmospheric pressure, respectively. Note that the liquid I → liquid II (glass II) transition inevitably accompanies vitrification, which makes a glass II state non-equilibrium in the sense that both ρ and *S* cannot reach their equilibrium values. This figure is reproduced from Fig. 22 of ref. 35.

structures, and not from nano-crystallites, which supports our scenario rather than the aborted crystallization scenario.

We also found that when we mix TPP with toluene, the final liquid II has a fluidity and has a glass transition temperature T_g different from the initial liquid I.[144] In the NG-type LLT, we also observed droplet formation. On noting that liquid II has a fluidity in the mixture, it is difficult to explain by the crystallization scenario what maintains the interface between the two liquids, or the origin of the interfacial tension. It is not obvious whether the liquid and the same liquid containing nano-crystallites can form the sharp interface or not. We need an exotic mechanism to explain it. A scenario of a colloidal gas–liquid transition might explain the formation of a sharp interface itself, but it cannot explain the final homogeneization in a simple manner. To explain the phenomenon with this scenario, we have to assume that nano-crystallites of TPP, which do not grow in size, are kept formed in the mixture selectively at the interface. We also found that the T_g of liquid II is higher than that of liquid I.[144] We note that toluene is not crystallized in the mixture and the T_g of toluene is lower than that of TPP. In the aborted crystallization scenario, toluene should be enriched in a non-crystallized liquid I region and thus we expect that the T_g of liquid II is lower than that of liquid I, provided that there is a significant amount of micro-crystallites that is large enough to account for the amount of the heat released during LLT. This is inconsistent with our observation: the T_g of liquid II is higher than that of liquid I for TPP–toluene mixtures. Furthermore, we also found that liquid I is miscible with diethyl ether, but liquid II is not.[144] This result is also difficult to explain by the crystallization scenario.

We also observed significant surface wetting effects for liquid II.[190] If this is induced by the ordinary dispersion force, we might be able to explain this by the crystallization scenario. But we found that the wetting effects are induced by specific interactions associated with hydrogen bonding and not by a dispersion force. This strongly indicates that the wetting phenomena are induced by the difference in the microscopic nature between the two liquids, and not by the difference in the macroscopic properties coming from the fraction of micro-crystallites. It also suggests the important role of hydrogen bonding in the transition, supporting our scenario.

Next, we consider why we observed a Maltese cross pattern for droplets of liquid II (see Fig. 17(a)). At first glance, droplets look like ordinary crystalline spherulites, but we note that the strength of birefringence is extremely weak, compared to crystal spherulites formed at a high temperature above the binodal line of LLT (see Fig. 17(b)). To explain the presence of a Maltese cross pattern, we need to explain why crystals have specific orientation along the growth direction in the framework of our scenario. Our explanation is as follows. First we mention that liquid II is completely wettable to crystals (see Fig. 17(b)),[190] indicating the interfacial energy of crystals is significantly lower for liquid II than for liquid I. Another important fact is that liquid II is a glass and not a liquid. In particular, in the NG-type process droplets of high S is directly nucleated and thus it is likely that droplets are nucleated in a glassy state. These considerations lead to a conclusion that a region where crystal nucleation can take place may be exclusively the droplet interface region since only there the two important conditions

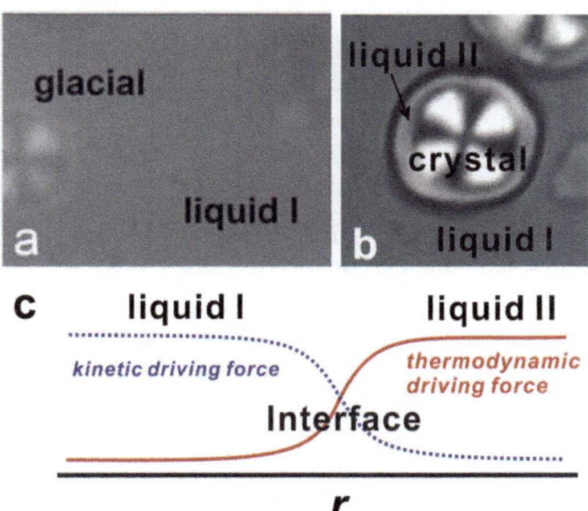

Fig. 17 (a) Droplets of the glacial phase observed with polarizing microscopy under the crossed Nicols condition (60 min at $T_a = 220$ K). (b) Formation of liquid II in the presence of TPP crystal spherulites formed at 237 K, which was observed with polarizing microscopy under the crossed Nicols condition. The layer of liquid II, which has no birefringence, is formed on the surface of the TPP spherulite (60 min after quenching to 220 K). We can clearly see liquid II completely wets the pre-existing crystal spherulite. (c) Schematic figure showing the spatial change in the kinetic driving force for crystal nucleation (mobility) and the thermodynamic driving force determined by the crystal–liquid interfacial tension. The total driving force should be maximum at the interface.

for crystal nucleation, high mobility and low nucleation barrier, are met (see Fig. 17(c)). A low nucleation barrier is a consequence of a low interfacial tension between liquid II and crystals. So we expect crystal nucleation selectively occurs in the interfacial region and then crystals grow towards an outward direction perpendicular to the interface, which is selected by the mobility gradient ∇S. This provides a natural explanation on why nano-crystallites have special orientations in liquid II droplets along the radial direction.

Finally we discuss the difference in the temperature dependence of the growth velocity V and the nucleation frequency Γ between spherulites of the crystal and droplets of the glacial phase (liquid II droplets in our scenario). We found that both V and Γ of the glacial phase obey curves different from those of the crystal (see Fig. 18), indicating that droplets of the glacial phase is distinct from spherulites of the crystal. This might be explained if we assume that the symmetry of the crystal formed in the glacial phase is different from that of the ordinary crystal. However, the X-ray and neutron diffraction peaks of micro-crystallites formed in the glacial phase is identical to that of the ordinary crystal formed at a high temperature.[167,204] Thus, it is difficult to explain the above fact by the crystallization scenario.

On the basis of these considerations, we conclude that the transition observed in TPP is more naturally explained by LLT rather than the crystallization scenario, although further careful study is necessary to completely settle this problem.

5.5.2 The case of water–glycerol mixtures. First of all, we emphasize that the transition behaviour in a water–glycerol mixture is strikingly similar to that in TPP and *n*-butanol. This suggests the commonality of the nature of the transition among these systems.

The transition in a water–glycerol mixture, which we interpreted as LLT, was also interpreted in a different way. Feldman and his coworkers interpreted the

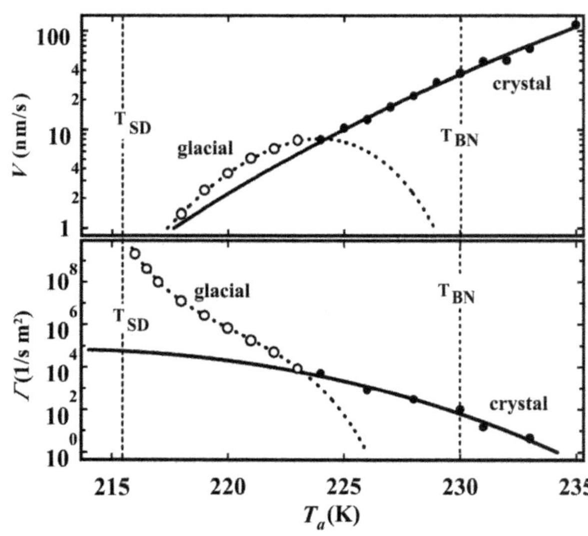

Fig. 18 Dependence of the growth velocity V (a) and the nucleation frequency Γ (b) on the annealing temperature T_a for the crystal (filled circle) and the glacial phase (open circle). This figure is reproduced from Fig. 2 of ref. 160.

final state as a mixture of the solute-rich liquid I phase, ices, and interfacial water around ices, on the basis of dielectric spectroscopy measurements.[205,206] Recently Limmer and Chandler[95] also suggested the possibility that the phenomena we interpret as LLT in a water–glycerol mixture may be a process of the liquid-to-crystal transition coupled with solute concentration fluctuations. This scenario can also explain the fact that the time over which this coarsening occurs is much longer than the relaxation time of the liquid. It also predicts that for NG-type evolution droplets are rich in ices and interfacial water, whereas the matrix is rich in glycerol. This is because ices are more friendly to interfacial water than glycerol, which causes attraction between ice crystals. Since LLTs in the systems we investigated always accompany the formation of cubic ices, these scenarios should be carefully examined. These proposals have similarity to the aborted crystallization scenario proposed for TPP.

We first examine the above-mentioned prediction suggesting possible enrichment of glycerol in the matrix phase. Contrary to it, we recently confirmed by micro-Raman spectroscopy that the glycerol concentration is the same between droplets and the matrix. This means that the glycerol concentration is spatially almost homogeneous in liquid II. This situation is similar to that in a mixture of TPP and toluene.[144] Next we consider the glass transition of the state we call liquid II. For this state we found a single glass transition (see Fig. 19),[145] which is located at a temperature significantly higher than that of liquid I at the same glycerol concentration: T_g of a glycerol solution at $c \sim 0.18$ prepared at $T_a = 164$ K is about the same as T_g of liquid I at $c \sim 0.33$ (see Fig. 15). We know that the mole fraction of cubic ice is 0.14 from X-ray scattering measurements.[145] If we assume that the system is composed of ice crystals, interfacial (pure) water, and liquid I with glycerol ($c \sim 0.33$), we may ascribe the glass transition assigned as T_g of liquid II to that of liquid I with glycerol. Then, the mole fractions of these three components can be estimated as 0.14, 0.31, and 0.55, respectively. On noting that the fraction of interfacial water (~31 mol%) is substantial, we expect two T_g values for interfacial water and liquid I with glycerol, but actually

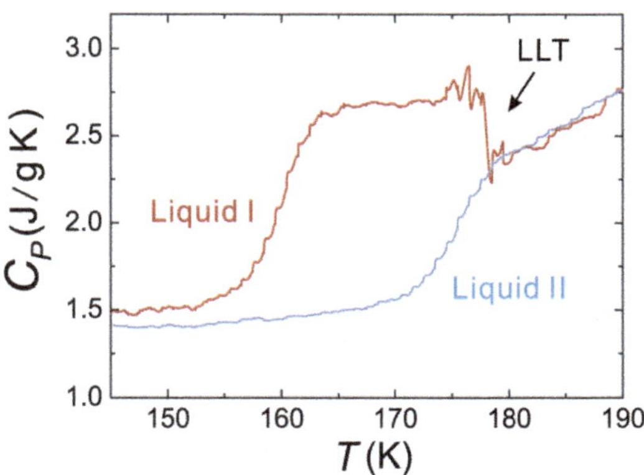

Fig. 19 Isobaric heat capacity C_P of liquid I and liquid II, measured by ac DSC measurements in the heating process (heating rate: 1 K min^{-1}, frequency: 60 s, amplitude: 0.16 K). Stepwise changes around 157 K (liquid I) and 172 K (liquid II) are the onsets of the glass transition. Heating of liquid I induces LLT. The resulting state has the same heat capacity as a liquid state of liquid II, as expected. The figure is reproduced from Fig. 2(d) of ref. 145.

we observe only one T_g. We may still be able to argue that the interfacial water may have a high T_g due to lower mobility near the solid surface (accidentally) near the observed T_g, but then it seems difficult to explain the c-dependence of T_{gII}. As shown in Fig. 19, heating of liquid I induces LLT above the glass transition temperature of liquid I. The resulting state has the same heat capacity as a liquid state of liquid II. If the transition we identify as LLT is crystallization, it seems rather difficult to expect that the state formed by crystallization upon a continuous heating is the same as that formed by crystallization during isothermal annealing.

Another piece of evidence supporting our scenario comes from the following fact: by combining X-ray scattering and ac DSC measurements, we confirmed that the amount of ice I_c in liquid II decreases with decreasing T_a and liquid II become less fragile (stronger) (see the supplementary information in ref. 145 for the details). In other words, the difference in the fragility between liquid I and II becomes more pronounced when liquid II contains less cubic ice. In the above-mentioned crystallization scenarios, however, after transformation the system should increase the fragility towards the fragility of pure liquid I with decreasing T_a, as a consequence of a smaller amount of cubic ice and the resulting decrease of interfacial water. Thus, these scenarios seem difficult to explain why liquid II is stronger when it contains less ice, or is more pure.

Furthermore, we analysed wide-angle X-ray scattering data for liquid I and a mixture of liquid II and cubic ice, as shown in Fig. 20 (see the supplementary information in ref. 145 for the details). This shows that although there is apparently little difference in the line shape between liquid I and liquid II, the line components obtained by the decomposition analysis are significantly different, suggesting a distinct difference in the liquid structure between them, although it may also be possible to claim that the difference comes from the presence of interfacial water.

On the basis of the above considerations, we conclude that the transition we observe should be LLT, although further careful study may be necessary to settle this issue and form a consensus.

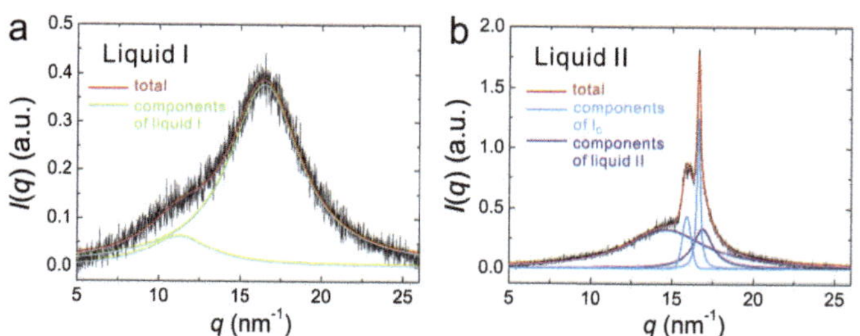

Fig. 20 Examples of the decomposition of the wide-angle X-ray scattering spectra $I(q)$ of liquid I and II. (a) $I(q)$ of liquid I at 167 K ($c = 0.178$). The red curve is the result of the fitting for the total signal. The green curves show the Lorentzian peaks obtained by decomposing the total signal into the individual peaks. (b) $I(q)$ of liquid II at 167 K ($c = 0.178$). The red curve is the result of the fitting for the total signal. The light blue curves represent the Gaussian and Lorentzian peaks from ice I_c (the Bragg peaks at 16.1 and 16.9 nm^{-1}, respectively), whereas the dark blue curves represent those from liquid II. Please refer to the supplementary information of ref. 145 for the details of the decomposition.

6 Liquid–glass transition

6.1 Our basic standpoint

In general, when a liquid is cooled, it is either crystallized or vitrified. Apart from liquids with significant quenched disorder, such as atactic polymers, a single-component liquid can in principle crystallize below the melting point T_m without any accompanying inhomogeneization (phase demixing). Glass transition is thus observed only when crystallization is 'kinetically' avoided. This suggests the presence of a deep link between crystallization and vitrification. Our model of glass transition is based on this physical picture.[19,26,27,30–32,35] However, most previous theories of glass transition did not consider crystallization to be important for the physical description of vitrification itself. In these theories, either a purely kinetic origin for dynamic arrest is sought or the special free energy describing the vitrification branch is newly introduced. In both cases, the crystallization branch has been ignored, or purely kinetic avoidance of crystallization has been assumed. Then the main focus has been put on the origin of slow dynamics in the vitrification branch.[207] This is partly because people who are interested in glass transition are not interested in the crystallization branch but only in the glass transition branch (see Fig. 7). Another reason may come from our intuition linked to a different, but related phenomenon, jamming transition. When we consider slowing down of motion of people in a packed train, we do not care about crystallization. This is also related to the fundamental question concerning the link between glass transition and jamming transition.[34,35,208,209] We believe that the free energy is relevant for the description of glass transition (see Fig. 7), and thus glass transition is essentially different from jamming transition, which is purely of mechanical nature.[35]

The above problem is also related to the most fundamental question of what is the origin of slow dynamics associated with glass transition. There are a few different scenarios: purely dynamical scenarios,[210–213] scenarios based on dynamical correlations due to dense packing,[214] and scenarios based on growing static order ((i) exotic amorphous order,[215–219] (ii) icosahedral order,[17,39,220] and (iii) spatially extendable low free-energy configurations such as crystal-like bond orientational order[19,33–35,67]). We are going to show that at least for weakly frustrated liquids (frustration against crystallization) glassy slow dynamics and dynamical heterogeneity are caused by the development of critical-like fluctuations of static crystal-like bond orientational order.

6.2 Roles of frustration effects on crystallization in glass-transition behaviour

Glass transition takes place if crystallization is avoided upon cooling or increasing density. However, the physical factors controlling the ease of vitrification and the nature of glass transition remain elusive. To answer these fundamental questions, we consider the phase behaviour of a system suffering from frustration effects on crystallization. There are not so many systems in which we can control the degree of frustration on crystallization in a systematic manner. As such examples, we here focus on two-types of systems: polydisperse colloidal systems and 2D spin liquids. The former suffers from random disorder effects (particle size distribution), whereas the latter from energetic frustration in the interaction potential. By comparing these two different systems, we show an intriguing scenario that the strength of frustration controls not only the ease of vitrification but also the fragility of liquid[19] for both colloidal liquids[221,222] and spin liquids.[36,223] Despite the difference in the origin of frustration effects on crystallization (geometrical *vs.* energetic), the behaviour is remarkably similar between them. This implies that there is an intrinsic link between structure and dynamics in glass-forming liquids: slow dynamics may be a consequence of 'glassy structural ordering' towards low local free energy in a liquid.[33–36,67,221] Vitrification may be a process of hidden crystal-like bond orientational (not translational) ordering under frustration[19,26,27,36,67] at least for systems where crystallization is weakly frustrated.

6.2.1 A case of hard-sphere-like systems: geometrical frustration and/or random disorder effects on crystal-like bond orientational ordering. For polydisperse hard-sphere systems, we can control the strength of frustration effects on crystallization in terms of the degree of polydispersity in a systematic manner.[221,222]

A state diagram for 2D polydisperse hard disks is shown in Fig. 21. For a monodisperse case (the polydispersity $\Delta = 0\%$), there are two sequential transitions: bond orientational ordering followed by translational ordering. Above $\Delta \geq 9\%$ (the coloured region in Fig. 21), a system starts to form glass without crystallization even for slow cooling. This shows the increase of glass-forming ability with an increase in Δ. In the glass-forming region, the fragility monotonically decreases with an increase in Δ (see Table 1).

For this 2D system, the only source of frustration against crystallization is polydispersity Δ. This is because hexatic order is the unique bond order parameter for a particle having 6 nearest neighbours on average and this order does not suffer from any competing ordering upon its growth. We stress that bond orientational ordering in hard-sphere-like systems is a direct consequence of dense packing and a manifestation of low configurational entropy.

For a 3D polydisperse hard-sphere system, on the other hand, there are at least two origins of frustration against crystallization: one is local icosahedral ordering

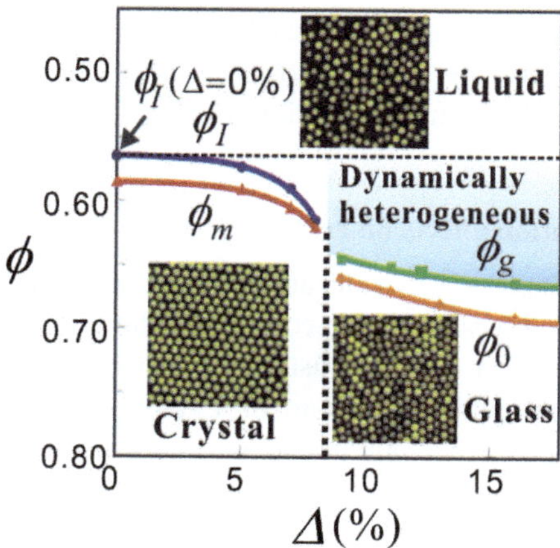

Fig. 21 A state diagram for 2D polydisperse hard-sphere-like systems. Here ϕ is the volume fraction of colloidal particles and Δ is the degree of polydispersity, which can be regarded as the strength of frustration against crystallization. This figure is reproduced from Fig. 1 of ref. 224.

Table 1 Dependence of the ideal glass transition point, ϕ_0(or T_0), and the fragility index, D, on the strength of frustration against crystallization Δ for 2D polydisperse colloids (2DPC) (N = 1024), 3D polydisperse colloids (3DPC) (N = 4096), and 2D spin liquids (2DSL) (N = 1024)

Δ_{2DPC}	ϕ_0	D	Δ_{3DPC}	ϕ_0	D	Δ_{2DSL}	T_0	D
9%	0.78	0.24	6%	0.62	0.70	0.6	0.099	7.4
11%	0.81	0.35	12%	0.65	1.05	0.65	0.090	11
13%	0.82	0.47	16%	0.67	1.29	0.7	0.076	17
16%	0.83	0.66				0.75	0.057	30
						0.8	0.026	84

tendency and the other is random disorder effects originating from the size polydispersity of particles[52] (see Fig. 23 below). Note that for 3D hard spheres a particle having 12 nearest neighbours can have three types of bond orientational order (fcc, hcp, and ico) (see Fig. 3). Among them, local icosahedral ordering is not a major cause of slow dynamics due to its localized nature and the dominant one is crystal-like (fcc-like) bond orientational order. This tells us that spatially extendable structural order is more responsible for slow dynamics than localized order.[44] So the scenario that icosahedral ordering is a major and unique underlying ordering behind vitrification may not be valid at least for a hard sphere system. Nevertheless, local icosahedral structures are formed (see Fig. 23 below)), and their number density increases with an increase in ϕ, which leads to stronger frustration effects on crystal-like bond orientational ordering.[44,45] This situation is similar to that in 2D spin liquids,[36,223] where pentagons prevent crystallization (see below). In this sense, even a monodisperse hard sphere system is not free from frustration effects on crystallization and suffers from self-generated internal frustration controlled by entropy.[35,45,67] This situation might be similar to metallic glass formers,[28,29] although the tendency of icosahedral ordering may be more pronounced for these systems due to the chemical nature of bonding[51] and the matching of atomic sizes.[225]

For 3D polydisperse hard-sphere systems, we also found that the increase in the polydispersity, or the strength of frustration against crystallization, leads to better glass-forming ability and a decrease in the fragility as in the case of the corresponding 2D systems (see Table 1).

6.2.2 A case of 2D spin liquids: competing bond orientational orderings. Next we consider a case of 2D spin liquids (see Fig. 22).[36,223] In this model, we put an anisotropic potential which forces particles having spins to favour the formation of pentagons. The statistical mechanics approaches to both thermodynamics and dynamics have been taken by Procaccia and his coworkers for this model.[226,227] The ground state crystal under the influence of the anisotropic potential has antiferromagnetic order on an uniaxially elongated hexagonal lattice. This crystal has a density higher than a liquid. In a supercooled liquid state of this model system, we found the development of medium-range antiferromagnetic bond orientational order (see Fig. 4), whose characteristic size ξ grows almost as $\xi = \xi_0((T - T_0)/T_0)^{-1}$ when approaching T_0. We confirm that antiferromagnetic bond orientational ordering is almost completely decoupled from density ordering: density change is not accompanied by the crystal-like bond orientational ordering and thus the density of a supercooled liquid is uniform in

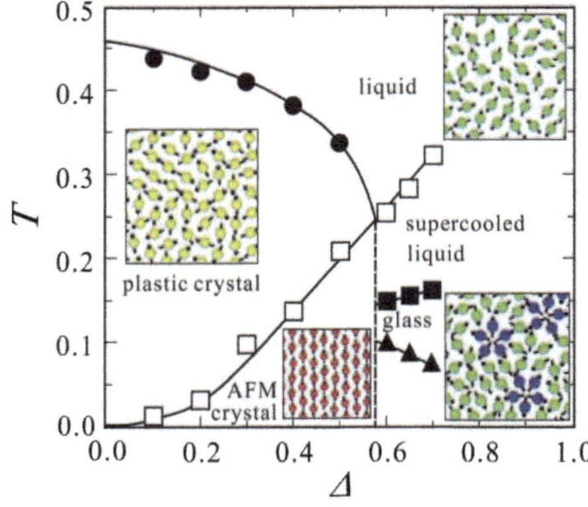

Fig. 22 Phase diagram of 2D spin liquid in the T–Δ plane. Here Δ is a measure of the strength of frustration against crystallization, or the strength of the three-body potential favouring a locally favoured structure of five fold symmetry. Energetic frustration is caused by symmetric mismatch in the interacting potential in this system. The basic structure of the phase diagram is quite similar to that of polydisperse colloids (see Fig. 21). For small Δ, or weak frustration, the glass-forming ability is very low, whereas with an increase in the frustration strength Δ the glass-forming ability is increased and the fragility is decreased, as in the case of polydisperse colloids. This basic trend is also very much consistent with the behaviour of water under pressure and water/salt mixtures.[228] This figure is reproduced from Fig. 1 of ref. 36.

space after a certain level of coarse-graining, irrespective of the degree of antiferromagnetic order. Here it should be mentioned that pentagons have a larger specific volume. There are strong competing orderings between antiferromagnetic and five-fold pentagonal ordering (see Fig. 22).

In this system, the strength of the frustration, which we express by Δ, controls the glass forming ability, fragility, and criticality.[36,223] The state diagram is shown in Fig. 22. For small Δ, a system easily crystallizes into the plastic crystal, where spins can rotate on a hexagonal lattice. For large Δ, where the melting point of the antiferromagnetic crystal becomes higher than that of the plastic crystal, a system can be vitrified rather easily. Thus, the increase in Δ leads to the increase in the glass-forming ability. We also found that the increase in Δ decreases the fragility. This may also be due to the increase in the activation energy E_a dominating the high temperature Arrhenius regime (see section 6.5).

Applying pressure leads to the decrease in pentagons (see eqn (10) and note that $\Delta v > 0$ for a pentagon), which leads to the increase in the fragility.[223] Since pressure does not alter the energy itself, this clearly indicates that the degree of frustration is a controlling factor of the fragility.

It is interesting that the number density of pentagons has a clear correlation with the growth of crystal-like bond orientational order and the fragility, indicating that pentagons disturb the growth of the correlation length of crystal-like bond orientational order.

6.3 Frustration effects on the slow dynamics of glass-forming liquids

Here we show that the disorder (or frustration) effects on crystallization controls the slow dynamics of glass-forming liquids, which is characterized by the ideal glass-transition point T_0 and the fragility index D under the assumption that the structural relaxation time τ_α obeys the following Vogel–Fulcher–Tammann relation: $\tau_\alpha = \tau_0\exp(DT_0/(T - T_0))$ or $\tau_\alpha = \tau_0\exp(D\phi/(\phi_0 - \phi))$ (on the justification of $T = 1/\phi$, see ref. 34). Table 1 summarizes such effects for 2D polydisperse colloids, 3D polydisperse colloids and 2D spin liquids. It tells us that the stronger the disorder effects, the lower the ideal glass-transition temperature T_0, the higher the ideal-glass transition volume fraction ϕ_0, and the stronger the liquid (*i.e.*, the larger D). Despite the difference in the source of frustration and the dimensionality, the behaviour is very similar, suggesting the generality of this scenario at least on a qualitative level.

6.4 Striking similarity between the above two systems

In the above we see that for both polydisperse hard spheres and 2D spin liquids, frustration against crystallization controls the glass-forming ability and the fragility: the stronger the frustration, the higher the glass-forming ability, the stronger the liquid. The phase behaviour is similar between the two systems (compare Fig. 21 and Fig. 22). Furthermore, there is the growth of bond orientational order linked to the underlying crystal structures when approaching the glass transition point. In addition, there are also competing orderings between extendable crystal-like bond orientational order and localized isolated order (icosahedral order or pentagons). The similarity in the structural features of supercooled liquid states between the two types of systems can be clearly seen in Fig. 23. The qualitative behaviour is thus the same between the two systems, despite the large difference in the interaction potential between them. This also suggests the generality of our scenario, at least for weakly frustrated systems.

6.5 Cooperative and non-cooperative origins of slow dynamics

Here we mention that even without any cooperativity or structural ordering the structural relaxation becomes slower either by lowering temperature or increasing density (or pressure). For ordinary liquids, this non-cooperative source of slow dynamics is due to the Arrhenius-type activation process, which leads to non-ergodicity only at $T = 0$ K. For hard-sphere liquids, on the other hand, it has a link to the jamming transition, which leads to the ergodicity breaking at the jamming

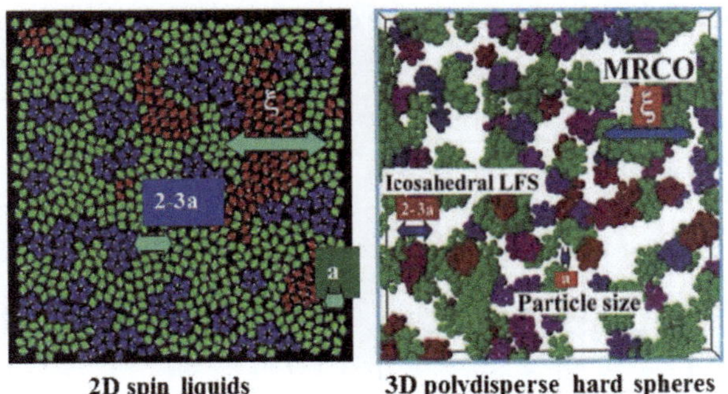

Fig. 23 Typical structures of supercooled liquids for 2D spin liquids and 3D polydisperse colloids. We can see competing orderings between medium-range crystal-like bond orientational ordering and short-range localized ordering in both cases.

point ϕ_J. We emphasize that this jamming transition is a mechanical athermal transition. We argue that even for a thermal system, if there is no structural ordering, the mechanism of slowing down may be similar to the jamming transition. The situations are schematically shown in Fig. 24.

On top of these non-cooperative sources of slow dynamics, the slowness originating from cooperativity contributes, which causes glassy slow dynamics. We proposed a possible empirical functional form describing both contributions.[19,26,27,30–32,35] For example, a stronger liquid has a larger contribution from non-cooperative Arrhenius dynamics. We should always take this contribution into account when analysing slow dynamics. In the case of weakly polydisperse hard spheres, since the contribution of the non-cooperative part is not so significant in the volume fraction ϕ we can access either experimentally or numerically, we may ignore it. For highly polydisperse hard spheres, this contribution may become more important.[34] Furthermore, with an increase in the spatial dimensionality d, the tendency of forming bond orientational order may become weaker since the loss of configurational entropy to gain bond orientational order steeply increases with an increase in d. This may explain why bond orientational ordering becomes less important for a system with higher d.[229]

We note that the presence of structural ordering in a supercooled liquid and the resulting cooperativity can be detected by the Stokes–Einstein violation. The onset of the violation is the reflection of the onset of cooperativity. The degree of the violation is stronger for a more fragile glass-forming liquid. As can be seen in hard-sphere-like liquids, the non-Arrhenius behaviour is not necessarily linked to the fragility in some liquids. Even in such cases, the Stokes–Einstein violation may be used as a finger print of cooperativity (or glassy structural ordering). For example, the Gaussian core model at a high pressure, which exhibits mean-field behaviour,[209] should be regarded as a glass-forming liquid without cooperativity, although it apparently shows the non-Arrhenius behaviour.

6.6 A link between glassy structure ordering, dynamical heterogeneity, and slow glassy dynamics

The above physical picture leads us to the following scenario of slow glassy dynamics. Upon cooling, a liquid enters into a metastable state where long-range orientational and positional ordering is prohibited by frustration effects. However, bond orientational order still develops continuously towards the ideal glass transition point T_0 to lower the free energy of the system. Bond orientational

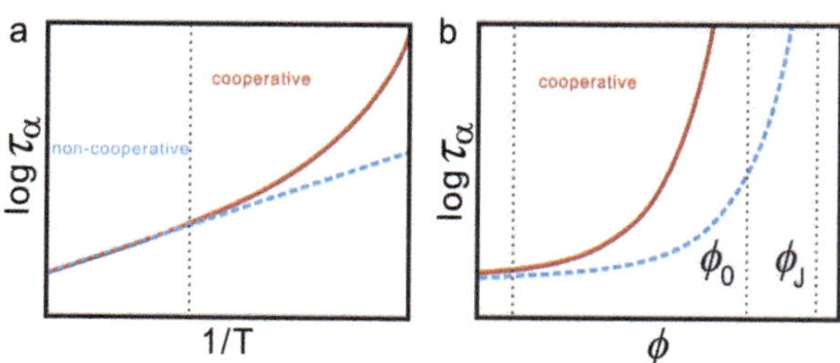

Fig. 24 Schematic figures explaining the non-cooperative and cooperative contributions to slow dynamics for the cases of ordinary liquids (a) and hard-sphere liquids.

ordering is a manifestation of many-body correlations among strongly correlated neighbouring particles around a central particle. For a strongly disordered system, such unique bond orientational order linked to the symmetry of the equilibrium crystal may no longer exist, however, we expect that even for such a system strong correlations may still be represented by some structural signature, which has a link to low free-energy configurations of longer lifetime, namely, lower fluidity.

Hereafter, we consider a case in which bond orientational order is relevant, just for simplicity. In regions of high bond orientational order, particle motion is on average slow since only the coherent motion while keeping bond orientational order is allowed. A distinct correlation between glassy structural order and slowness of particle motion can be clearly seen for a 2D polydisperse colloidal simulation in Fig. 25. We can see a similar structure-dynamics correlation for a 3D polydisperse colloid experiment in Fig. 26. As mentioned above, the length scale of the structural order, or the coherency of particle motion, is a key to the slowness of dynamics. This may also be the origin of dynamic heterogeneity.

Here it may be worth noting a possible difference in the dynamic and static correlation length. In Fig. 25, we can see almost the one-to-one correspondence between static order and mobility. However, this visual comparison is affected by the color codes we employ. There is no proportionality between the static glassy structural order and the local dynamics, which is clear from a strongly nonlinear relation between them. Thus, the bare correlation length can be different between the static and dynamic ones.

Dynamic heterogeneity is an important characteristic feature in supercooled liquids.[230] Now there is a consensus that there is a growing dynamic length in supercooled liquids, although it is not yet clarified whether it is the consequence or the origin of glassy slow dynamics. This length can be estimated, *e.g.*, by bond-breakage correlations[231] or by four-point density correlator.[232] Recently intrinsic differences between the two methods were discussed in detail in connection with underlying vibrational motions by Onuki and his coworkers.[233–235] The detailed

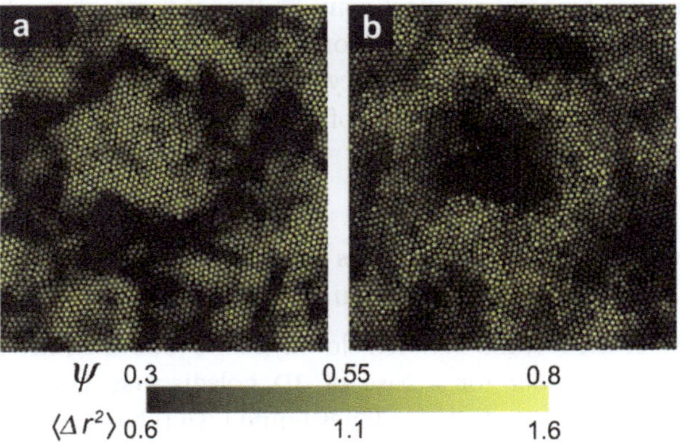

Fig. 25 Relationship between glassy structural order and local mobility in 2DPC ($\phi = 0.740$ and the polydispersity $\Delta = 9\%$). (a) The spatial distribution of the coarse-grained hexatic order parameter ψ. (b) The spatial distribution of the mean-square displacement over $10\tau_\alpha$. We can see the almost one-to-one correspondence between highly ordered regions and regions of low mobility (or low fluidity). This figure is reproduced using a part of Fig. 1 of ref. 67.

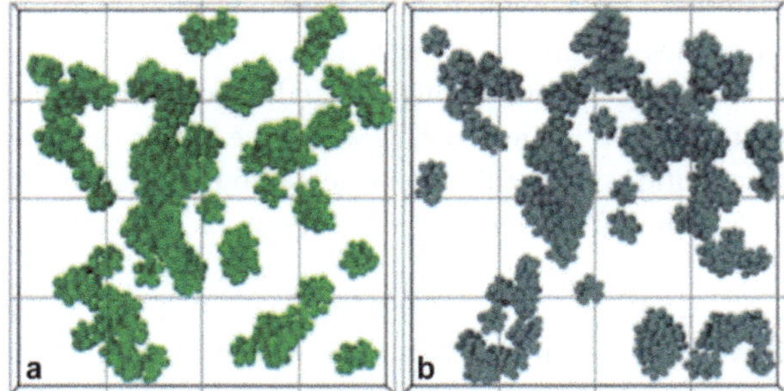

Fig. 26 Computer reconstruction from confocal microscopy coordinates for a polydisperse colloidal suspension ($\phi = 0.575$). Only particles of interest and their neighbours are displayed. The depth of the image is 12 times of diameters. Each particle is plotted with its real radius. (a) Particles having high crystal-like bond orientational order alone (the order parameter was averaged over the order of the structural relaxation time). (b) Slow particles with respect to the coarse-grained displacement. Due to particles going in and out of the field of view, assignment of particles located very near the edges of (a) and (b) were not accurate and have been removed. This figure is reproduced from a part of Fig. 4 of ref. 44.

characterization of dynamic heterogeneity including its lifetime has also been made by Kim *et al.*[236,237] The above observation suggests that the dynamical correlation length ξ_4 is comparable to the bond orientational correlation length ξ_6. Thus, it may be interesting to go a step further and study a link between bond orientational order fluctuations and these new features of dynamic heterogeneity.

The increase in glassy structural order and the resulting decrease in defective structures lead to the decrease in the fluidity and thus to the slowing down of structural relaxation. Such a direct link between glassy structural ordering and slow structural relaxation dynamics is a characteristic nature of glass transition and is absent in ordinary critical phenomena. In the latter, the slowing down of the dynamics is linked to the characteristic size of the order parameter fluctuations alone, but not directly to the transport coefficient such as viscosity. For example, in a critical binary mixture, viscosity exhibits only a very weak logarithmic divergence towards a critical point.[186] This suggests that a link between low free-energy configurations (bond orientational order in the systems we discuss) and local mobility (or, coherency of particle motion) is the key to glassy slow dynamics. In other words, motion of individual particles are constrained by the glassy structural order (see below).

6.7 Correlation between slow dynamics and static glassy structural order: 3D polydisperse Lennard-Jones system as an example

Here we consider the relation between slow dynamics and the correlation length of glassy structural order for a case of 3D polydisperse Lennard-Jones (LJ) system.[67] A 3D polydisperse Lennard-Jones liquid (3DLJ) has been widely used to study glass-transition phenomena (see, *e.g.*, ref. 238–242 and the references therein). We performed microcanonical (NVE) ensemble molecular dynamics (MD) simulations. The diameter of particle i is σ_i and the mass m is the same for all particles. Particles interact with the following Lennard-Jones potential: $U_{ij}^{LJ}(r) = 4\varepsilon\{(\sigma_{ij}/r)^{12} - (\sigma_{ij}/r)^6 + C_{ij}\}$ for $r < r_{ij}^c$ where $\sigma_{ij} = (\sigma_i + \sigma_j)/2$ and $r_{ij}^c = 2.5\sigma_{ij}$. Here C_{ij} is a constant such that $U^{LJ}_{ij}(r^c_{ij}) = 0$. We set the particle number $N = 4096$ and the particle number density $\rho = N/V = 1.2$. All results are shown in reduced units: length unit $= \langle\sigma\rangle$, temperature unit $= \varepsilon/k_B$, and time unit $= \sqrt{m\langle\sigma\rangle^2/\varepsilon}$.

Fig. 27(a) and (b) show, respectively, the T-dependence of the intermediate scattering function $F(q_p,t)$ (q_p: the peak wavenumber of the structure factor $S(q)$) and the structural relaxation time τ_α, which is determined by fitting the stretched exponential function to the slow decay of $F(q_p,t)$. By fitting VFT relation, we obtain $T_0 = 0.73$ and $D = 3.85$. The T-dependence of the stretching exponent β is also plotted in Fig. 27(c). β decreases steeply upon cooling, reflecting growing dynamic heterogeneity. The bond orientational correlation function is shown in Fig. 27(d). To estimate the spatial correlation length of the bond orientational order parameter, ξ_6, we calculated the spatial correlation function $G_6(r)$:

$$G_6(r) = \frac{4\pi}{13}\left\langle \sum_{m=-6}^{6} Q_{6m}(r)Q^*_{6m}(0) \right\rangle \Big/ g(r).$$

The decay is well described by the 3D Ornstein–Zernike function ($\propto r^{-1}\exp(-r/\xi_6)$). Fig. 27(e) shows the T-dependence of the correlation length ξ_6.

We can see that $\xi = \xi_6$ grows as a system approaches the ideal glass-transition point T_0 as $\xi = \xi_0[(T - T_0)/T_0]^{-2/3}$ with $\xi_0 = 3.05$. The relation between τ_α and ξ is shown in Fig. 27(f), which indicates that $\tau_\alpha \propto \exp[D(\xi/\xi_0)^{3/2}]$.

Here we present only the results of a 3D polydisperse LJ system, but the behaviour is essentially the same for 3D polydisperse hard spheres.[67,222] Including 2D polydisperse hard disks,[221,224] 2D driven granular matter,[243] and 2D spin liquids,[36,223] we obtain the following general relations:[67] $\xi = \xi_0[(T - T_0)/T_0]^{-2/d}$ and

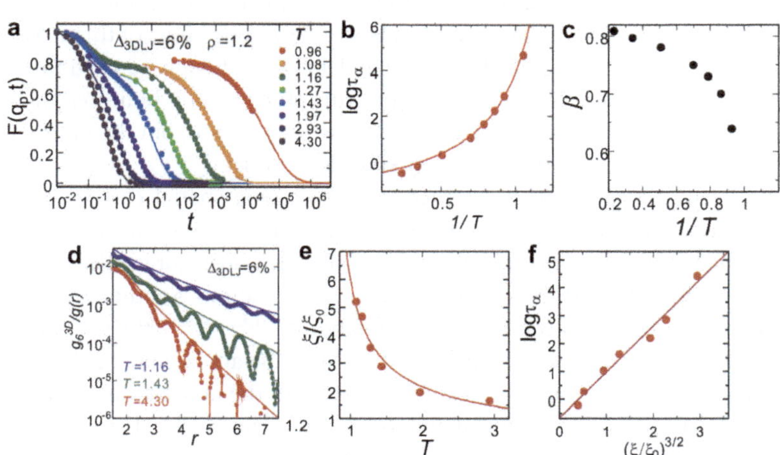

Fig. 27 Dynamical and structural evolution upon cooling in 3D Lennard-Jones system (3DLJ). The density $\rho = 1.2$ and the polydispersity $\Delta_{3DLJ} = 6\%$. (a) T-dependence of the intermediate scattering function $F(q_p,t)$. The solid curve is the stretched exponential function. (b) T-dependence of the structural relaxation time τ_α. It slows down more than five orders of magnitude upon cooling. The solid curve is the VFT relation. (c) T-dependence of the stretching exponent β. (d) The spatial correlation function of the bond orientational order parameter Q_6, $g_6^{3D}(r)$. The solid lines are the fittings by the 3D Ornstein–Zernike function. (e) T-dependence of the correlation length ξ. The solid curve is the power law fitting (see text). (f) The relation between τ_α and ξ.

$\tau_\alpha \propto \exp[D(\xi/\xi_0)^{d/2}]$, where d is the spatial dimensionality. This suggests the Ising-like criticality scenario of glass transition.[35,67,70,71] The Ising-like exponent for the static correlation length was also recently reported numerically[244,245] and theoretically[246–248] on different grounds.

6.8 The key lengthscale for glassy slow dynamics

Here we consider what lengthscale controls the structural relaxation dynamics. There are two candidates: (i) a microscopic cage size and (ii) a mesoscopic length scale. Such a mesoscopic length may be of either static or dynamic origin. Scenario (i) is based on the physical picture that slow dynamics is due to the local caging of particles without any relevant length scale beyond the interparticle distance a. This is the schematic MCT scenario of glassy slow dynamics (or, a mean-field scenario). Other models based on a single particle picture such as hopping dynamics, energy landscape, and free volume concept belong to this category.[249] On the other hand, scenario (ii) is based on the physical picture that slow dynamics is a consequence of cooperative phenomena where single particle dynamics is coherent over the length scale ξ larger than a. For example, Berthier *et al.*[249,250] showed that the length scale marking a crossover from persistent to Fickian diffusion motion has a significant connection to slow dynamics, *i.e.*, dynamic heterogeneity is a central aspect of the dynamics of supercooled liquids in that time and length scales are intimately connected. This is consistent with our scenario.

A recent study by Furukawa and Tanaka[251–253] has clearly shown that the viscous dissipation in a supercooled liquid takes place predominantly in the length scale over ξ_4 (see also ref. 254, 255), which indicates the intrinsic importance of the growing mesoscopic correlation length in the viscous transport. The crossover length ξ_η from macroscopic to microscopic transport was found to be comparable to the dynamical correlation length ξ_4. Implications of this crossover has recently been discussed by Furukawa in detail.[256] Although we need to clarify whether it has a static or a kinetic origin, this study clearly indicates that the mean-field (or microscopic) mechanism may not be relevant, but the mesoscopic spatial correlation is essential to glassy slow dynamics.

The next question is then whether the mesoscopic spatial correlation is of purely kinetic or static origin. One possible scenario is based on the kinetically constrained model (see, *e.g.*, ref. 213, 250, 257), which does not involve any static correlation but still exhibits strong dynamic correlation. Another scenario is based on the presence of static spatial correlation.[67,221,222–244,258–263] The low fluidity of our glassy structural order indicates that slow dynamics may be due to the growing static correlation over ξ, which can explain the above-mentioned crossover from persistent to diffusional motion quite naturally. The point-to-set length[219,229,264–271] based on the random first order transition theory[12,207,216,218,272–274] is another candidate for the static lengthscale. We note that it was proven[275] that if τ_α diverges in a super-Arrhenius manner towards a certain temperature, then $\xi_{\rm PTS}$ must diverge too, faster than the lower bound given by $(T \log\tau_\alpha)^{1/d}$ (d: the space dimension). However, it was reported by many researchers[229,248,265–267,270,276] that the static point-to-set(-like) length $\xi_{\rm PTS}$ is much shorter than ξ_4 and the temperature dependence is different between them, that is, the static mosaic length is decoupled from the dynamical correlation length, which further means its decoupling from the characteristic length of viscous transport ξ_η. Since the viscous transport is the most fundamental nature, we infer that the point-to-set length may not be the relevant length scale for glassy slow dynamics. However, Szamel and his coworkers[277] recently suggested a possible link between the point-to-set and dynamical lengths by combining a relation of the structural relaxation time with the dynamical correlation length ξ_4 found by them and that with the point-to-set length ξ_{PTS} found by Hocky *et al.*[267] This controversial situation might be linked to the delicateness of the estimation of the point-to-set length, or the its dependence on the pinning geometry.[268,278] Interestingly, Biroli *et al.*[279] have recently shown that the point-to-set length ξ_{PTS} estimated by the cavity method coincides with the length ξ_λ where the lowest eigenvalue of the Hessian matrix becomes sensitive to disorder, or plasticity. It was also suggested that $\tau_\alpha = \tau_0\exp(A\xi^\Psi/k_BT)$ $(1 \leq \Psi \leq 2)$, where $\xi = \xi_{PTS} = \xi_\lambda$ and A is a constant. Although the value of Ψ is not yet determined firmly, this is not inconsistent with our scenario together with $\tau_\alpha = \tau_0\exp(A\xi_4^\Psi/k_BT)$ reported by Szamel and his coworkers.[277] Furthermore, Mosayebi *et al.*[244,245] estimated the static correlation length ξ_{nad} from the correlation length of the non-affine displacement field for the corresponding inherent structure of a supercooled state and found its Ising-like divergence towards T_0, similar to our scenario. It is also interesting to note that their correlation length ξ_{nad} may have a connection to ξ_λ since the non-affine nature of deformation is linked to plasticity. We emphasize that the above-mentioned works[244,245,267,277,279] commonly include results of the Kob–Andersen binary Lennards-Jones liquid. Thus, there is a chance to reveal the relationship between all these lengths. This point needs further careful study. We will discuss this problem in detail elsewhere.[280]

We also mention another static length called "patch correlation length",[281,282] which is also expected to be order-agnostic. This length is obtained by computing the entropy of patches of a given size R appearing in a system, which is related to the frequency of occurrence of similar structural motifs. Its usefulness was shown for 2D hard disks frustrated by a curvature of the space.[282] How this works for binary systems is curious, but a possible lack of unique structural motifs for low free-energy configurations may cause a problem.

At this moment, we prefer the simplest physical scenario that there is only one relevant lengthscale, *i.e.*, $\xi_6 = \xi_4 = \xi_\eta$ (see also section 6.11.2). However, we should mention that the relation $\xi_6 = \xi_\eta$ has not been examined yet for polydisperse colloidal systems and thus there might be a possibility that there is no such a link.[256]

6.9 Importance of many-body correlations on slow dynamics

Here we emphasize the importance of many-body correlations in slow dynamics. Our finding suggests that slow dynamics is associated with strong correlation of particle dynamics, or coherency of motion, which originates from the many-body correlations. Thus we need to see many-body correlations to seek the structural origin of slow dynamics in a supercooled liquid.[52] Bond orientational order we focus on in this article is a consequence of many-body correlations since it involves a central particle and its nearest neighbours (6 for 2D and 12 for 3D in hard spheres). Fig. 28 shows the scattering function $S(q)$, which is the two-body density–density correlation function, and the correlation function of relevant

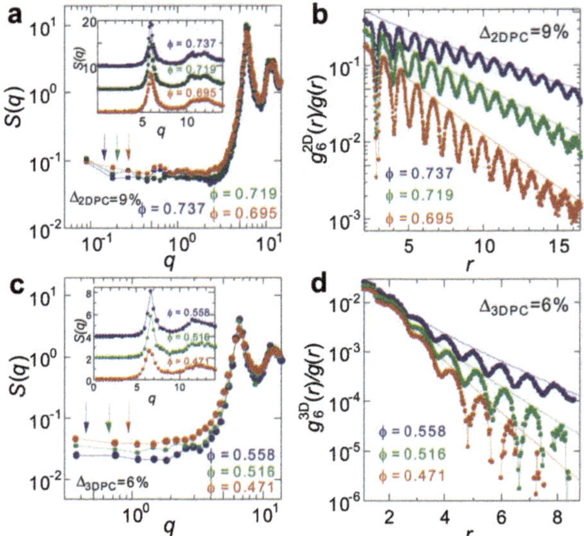

Fig. 28 Importance of many-body correlations. (a) Scattering function $S(q)$ for 2D polydisperse hard disks. The inset shows a wider q-range. (b) The correlation function of ψ_6, $g_6(r)$, for 2D polydisperse hard disks. (c) Scattering function $S(q)$ for 3D polydisperse hard spheres. The inset shows a wider q-range. (b) The correlation function of Q_6, $g_6(r)$, for 2D polydisperse hard disks.

bond orientational order parameters for both 2D and 3D polydisperse systems. We can clearly see that glassy structural order cannot be detected by $S(q)$, but can be detected by $g_6(r)$. This also means that bond orientational ordering is completely decoupled from density field or translational ordering for hard sphere liquids. We confirm that this also holds for 2D spin liquids and 3D polydisperse Lennard-Jones liquids.[67] Thus, this feature has nothing to do with whether the interaction is repulsive or attractive.

6.10 Decoupling between bond orientational order and translational order

In the above, we showed that structural ordering in a supercooled liquid cannot be seen by the two-body density correlation. Below we seek the physical reason behind this by revealing the nature of ordering taking place in a supercooled liquid state.

Here we consider the nature of crystal-like bond orientational ordering in a supercooled liquid and its decoupling with translational order, or density field. As shown above, spatial fluctuations of bond orientational ordering can be seen neither by the two-body density correlator that is a measure of translational order nor by the density field (see also ref. 52). This originates from the asymmetry in the coupling between bond orientational order and translational order (see eqn (7)). Translational ordering must accompany bond orientational ordering, but the latter needs not to accompany the former. A supercooled liquid locally takes a configuration whose rotational symmetry is related to the densest packing configuration. However, particles increase only bond orientational order without enhancement of translational order or densification. This configuration, which has an ability to become much denser yet is not dense, provides extra free volume to particles in that region and lowers local free energy (see Fig. 2). With the expense of configurational entropy associated with orientational ordering, a system gains correlational entropy. This is the case not only for hard spheres but also for systems with attractive interactions, *e.g.*, 2D spin liquids and 3D Lennard-Jones liquids (see above). This absence of translational ordering or densification in a mesoscopic lengthscale in a supercooled state is a direct consequence of the fact that translational ordering or densification beyond a short range take place only when crystallization takes place. We stress that because of the strong first order nature of translational ordering, or crystallization, there is only weak pre-ordering in a liquid state. This also means that the loss of configurational entropy is mainly associated with the loss of bond orientational degrees of freedom. This is the reason why structural ordering cannot be detected by any scattering experiments probing the two-body density correlation function. We argue that any structural ordering in a supercooled liquid should primarily be associated with orientational ordering. This also explains why there is little change in the bulk modulus in a supercooled liquid state.

We note that this decoupling scenario applies to crystal-like bond orientational ordering, but not to locally favoured structures: for example, locally favoured structures (pentagons) formed in 2D spin liquids has a larger specific volume compared to normal liquid structures. See section 2.2 on the origin of coupling.

6.11 Origin of glassy slow dynamics: glassy critical dynamics or RFOT scenario

There are many possible explanations for slowing down of the dynamics near a glass transition point, as discussed above. Here we focus on the scenarios based on the growing static length scale. One of the currently most popular scenarios is the so-called "Random First Order Transition" (RFOT) scenario.[12,207,216,218,219,272–274] Thus, we compare here our scenario with this scenario.

6.11.1 RFOT theory. First we briefly review the microscopic basis of RFOT scenario to elucidate its essence. In this theory, glassiness is explained as a consequence of the emergence of aperiodic crystals using density functionals.[283] For this purpose the Ramakrishnan–Yussouff (RY) free energy (see eqn (6)) was employed. It is well established that crystallization can be described by a periodic $\rho(r)$. For the description of glass transition, this was generalized for aperiodic density. Assuming the harmonic nature of individual cages confining a particle, the following variational density profile was employed:

$$\rho(\vec{r}) = \Sigma_{\mathrm{i}} \left(\frac{\alpha}{\pi}\right)^{3/2} \exp\left(-\alpha\left(\vec{r}-\vec{r}_{\mathrm{i}}\right)^2\right), \tag{25}$$

where $\{\vec{r}_{\mathrm{i}}\}$ represents the atomic coordinates of a particular but generic aperiodic lattice characterized by some average density $n \equiv 1/a^3$, where a is the average lattice spacing. The variational parameter α is a measure of localization within such a many-particle cage. Non-zero α characterizes a localized regime, whereas $\alpha = 0$ represents the completely delocalized regime of the uniform liquid. Singh, Stoessel, and Wolynes[283] showed that for a sufficiently high density, the RY free energy develops a metastable minimum as a function of α at the temperature T_{A} (see also ref. 272, 273). The emergence of this minimum appears as a first order transition, where α plays the role of an order parameter, whose value changes discontinuously during the transition from zero to a non-zero value.

While the transition is first order in α, there are many distinct (both morphologically and spatially) aperiodic states randomly distributed in free energy. Thus this transition is called the "Random First Order Transition" (RFOT).[215,273] Below T_A a system enters into the so-called mosaic state. This T_A has a link to the mode-coupling transition temperature T_{MCT}.[272] It was indicated that this situation is similar to that of the exactly soluble mean-field Potts glass, which also exhibits a dynamic transition at a temperature T_A above its thermodynamic glass transition T_K. It was also shown that for finite range Potts spin glasses and supercooled liquids the transition at T_A would be smeared by droplet-like excitations driven by the configurational entropy.[272,273] These entropic droplet excitations would provide the route to equilibration below the mean field dynamical temperature that corresponds to the mode coupling transition at T_{MCT}. In this way, the difficulty of the mean-field mode-coupling theory that predicts an ergodic-to-nonergodic transition at $T_{MCT} \sim T_A$ can be removed.

As we can see in the above, the RFOT theory is constructed on the basis of the RY free energy functional, indicating that it is basically the two-body level description.

6.11.2 Glassy critical dynamics or RFOT scenario. According to the RFOT scenario,[218,219,274] transport near glass transition is driven by activated processes the driving force for which are 'entropic' in nature. Because the entropy vanishes linearly near the Kauzmann temperature T_K the size of the domains is predicted to grow as $\xi_{PTS} \sim (T - T_K)^{-2/d}$. As shown above, the RFOT theory is at least apparently based on the density functional theory up to two-body correlations (see also ref. 272, 283, 284) and thus does not include many-body correlations, particularly bond orientational correlations, in an explicit manner. Thus, this theory cannot explain the development of bond orientational order in a supercooled liquid. This might explain why the point-to-set length ξ_{PTS} is much shorter than the dynamical correlation length ξ_4,[229,285]which is comparable to the bond order parameter correlation length in our case. In the RFOT scenario, however, the divergences of these two lengths reflect different physical mechanisms: the dynamical correlation length $\xi_4 = \xi_{MCT}$ diverges due to criticality associated with the spinodal at the mode-coupling transition temperature T_{MCT}, whereas the static correlation length ξ_{PTS} diverges at T_K where the two phases, liquid and glass, have the same free energy (see Fig. 29(b)). Thus, in the framework of the RFOT theory the decoupling between the two lengths is natural (see, *e.g.*, ref. 286).

The droplet theory, constructed by balancing the entropic driving force and the opposing cost of creating an interface between glassy states with different configurations, leads to the Vogel–Fulcher–Tammann equation.[218,219,274] Although this form is the same between RFOT and our scenario, the underlying physics is different. Reflecting this, the correlation length also seems to be different between them.

We note that, in the RFOT scenario, droplets are excitation of 'amorphous order', which is apparently not consistent with what we observed in weakly frustrated systems, where crystal-like (not amorphous-like) bond orientational order develops upon cooling. For this type of well-defined order, we cannot expect entropic driving force associated with a large number of amorphous configurations, which is the key to the RFOT scenario. It is not so obvious how many-body

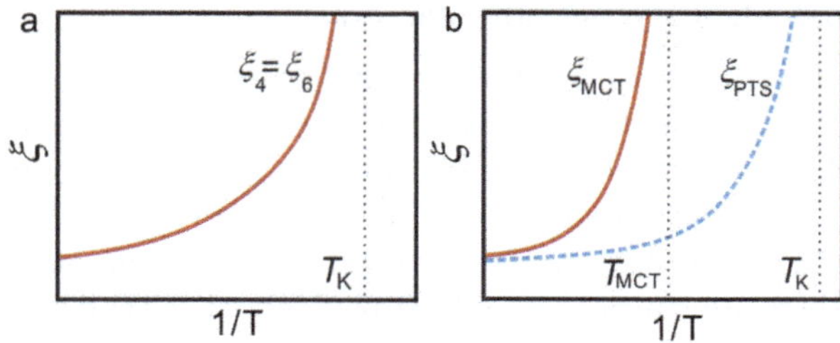

Fig. 29 Comparison of the temperature dependence of the relevant length scale(s) in two scenarios based on the diverging static length scale: (a) our scenario and (b) the RFOT scenario.

(orientational) correlations can be incorporated in the framework of the RFOT scenario. We speculate that the correlation length of RFOT, ξ_{PTS}, is primarily translational rather than orientational. In our scenario, (bond) orientational correlation is a major player. The decoupling between bond orientational order and translational order in a supercooled liquid (see above) may be related to the decoupling between $\xi_6 \cong \xi_4$ and ξ_{PTS}, although it is a matter of controversy whether such a decoupling exists or not (see section 6.8 and ref. 277).

As noted above, we infer that the length scale picked up by the RFOT scenario has translational nature, because of the construction of the theory based on the RY free energy functional that does not include orientational correlations at least apparently. Furthermore, in our simulation and experiments, spatial fluctuations of the crystal-like bond orientational order parameter are not like 'droplets' but rather continuous critical-like fluctuations. Furthermore, the dynamics of the fluctuations is found to be well-described by model A type dynamics (non-conserved order-parameter dynamics).[287] So the behaviour is more second-order-like rather than first-order-like. As discussed above, the point-to-set length ξ_{PTS} that is claimed to be the lengthscale of amorphous order relevant for slow glassy dynamics is always much shorter than the dynamical correlation length ξ_4, which is comparable to the viscous transport crossover length ξ_η and the correlation length of glassy structural order (*e.g.*, crystal-like bond orientational order) ξ. Because of the reason we mentioned above, we infer that the key length scale for glassy slow dynamics is the latter length scales (ξ_4, ξ_η, and/or ξ) (see Fig. 29(a)).

Here we mention a piece of evidence supporting our scenario as well as its weakness. For hard spheres, the mode-coupling theory predicts that $\phi_{MCT} \sim 0.58$. However, our simulation results show that there is no divergence of the dynamical correlation length ξ_4 ($\cong \xi_6$) at $\phi_{MCT} \sim 0.58$. This seems not to be consistent with the MCT–RFOT scenario (see Fig. 29(b)), although it might be ascribed to fluctuation effects. On the other hand, for binary mixtures, we have not so far succeeded in detecting a static length scale comparable to a dynamical length scale.[288–290] This indicates a failure in either detecting a relevant static length scale or our scenario itself. However, it is worth noting that as discussed in section 6.8, Szamel and his coworkers[277] suggests the relation $\xi_4 \propto \xi_{PTS}$ for a few binary systems. This may be a crucial point to elucidate what is the key lengthscale responsible for glassy slow dynamics.

At this moment, we prefer to interpret glassy slow dynamics as a consequence of 'glassy critical dynamics',[35,67,291] although further careful study is necessary to settle this problem.

6.12 Structural origin of cooperative slow dynamics

The origin of slow dynamics lies at the heart of the physics of glass transition. Our current picture on an intuitive level is as follows. Far above the glass transition point, particles move randomly and more or less independently. While approaching the glass transition point, structural order linked to low free-energy configurations develops. Because of the structural nature of this order parameter, the motion of high order regions must be strongly correlated and have a tendency to move together.

Bond orientational order is not resistive to volume deformation. The susceptibility to volume deformation is rather determined by translational order. Bond orientational order is, on the other hand, resistive to transverse shear deformation. In a liquid state, there is little change in translational order, whereas bond orientational order develops when approaching the glass transition point, indicating the importance of the latter in slow dynamics. Furthermore, since shear viscosity, which diverges towards the glass transition point, is linked to the time correlation function of the shear stress tensor, it is natural to expect the important role of bond orientational order in the viscosity increase. The importance of anisotropic correlations in structural relaxation was also pointed out many years ago by Mountain and Thirumalai.[292] On the microscopic level, we can say that types of particle motion that lowers the bond orientational order is not favoured. This constraint that particles in regions of high order whose characteristic length is given by ξ must move coherently may be the origin of slow dynamics, or the enhancement of the viscosity, near glass transition.

6.13 Prediction of our scenario to the glass-forming ability and the fragility of water-type liquids

In our scenario, vitrification is a consequence of frustration against crystallization. This allows us to make a qualitative prediction on the glass-forming ability and the fragility of a liquid on the basis of its phase diagram.[19,26,27,30–32] This is a very unique feature of our model, which stems from the fact that we consider vitrification and crystallization on the same ground[35] (see Fig. 7).

Here we consider an interesting case of competing orderings, which also results in frustration against crystallization. One of the most characteristic cases belonging to this category can be seen in water-type liquids. Water has a very characteristic V-shaped T–P phase diagram (see Fig. 9). We argue that this is closely linked to the anomalies of water[24] and liquids having this V-shaped phase diagram should share characteristics similar to water. By using this specific shape of the phase diagram as a fingerprint, we classified five elements Si, Ge, Sb, Bi, and Ga into water-type atomic liquids.[24] Similarly, some group III–V (*e.g.*, InSb, GaAS, and GaP) and II–VI compounds (*e.g.*, HgTe, CdTe, and CdSe) can also be classified into water-type liquids. As described above, our model provides us with simple analytical predictions for the thermodynamic and dynamic anomalies of these water-type liquids.[24]

In these liquids, at low pressure, a system crystallizes into S crystal, which is favoured by bond orientational (tetrahedral) ordering. It may be worth noting that below P_X there is ‘almost’ no frustration since tetrahedral locally favoured structures having a lower energy than normal liquid structures are basically compatible to S-crystal. In relation to this, it is worth noting that we recently found a weak source of frustration coming from a tendency of the increase of pentagonal rings upon cooling for water.[114] Reflecting the open structure of tetrahedral order, the volume of the system expands upon crystallization of a liquid to S crystal, which leads to the negative slope of the melting point of S crystal in the T–P phase diagram. Under high pressures, a crystal into which a liquid crystallizes generally tends to have a more compact, denser structure. Thus, pressure destabilizes S-crystal and instead stabilizes ρ-crystal. Accordingly, the equilibrium crystal switches from S-crystal to ρ-crystal with increasing pressure at the crossover pressure P_X. In other words, the primary order parameter responsible for crystallization into the equilibrium crystal switches from the bond order parameter S to the density order parameter ρ there, or more precisely, bond order parameters linked to ρ-crystal.

Above P_X, thus, the melting point of ρ crystal becomes higher than that of S crystal. In this situation, we expect that locally favoured structures linked to tetrahedral ordering act as a source of frustration against crystallization to ρ crystal because of the mismatch between the symmetries and thus helps vitrification. We stress that the structure of the first shell is not enough to describe the locally favoured structure and that of the second shell may be necessary to specify it.[114] Thus, water should tend to behave as an ordinary glass-forming liquid at very high pressures, which is consistent with the experimental indication.[76,293] This tendency is difficult to explain in terms of the other existing theories of liquid–glass transition.

At low pressure, local structural ordering simply helps crystallization to S crystal, as explained above. Since an open tetrahedral structure has a specific volume larger than a normal-liquid structure, the increase in the pressure decreases the number density of locally favoured structures, *i.e.*, $\bar{S}$ (see eqn (10)), which should lead to the decrease in the strength of frustration against crystallization to ρ crystal. The situation is thus very similar to the case of 2D spin liquids, where the degree of frustration also decreases with an increase in pressure,[223] as discussed above. Our scenario tells us that the glass-forming ability and the fragility are positively correlated with an increase and a decrease in the degree of frustration against crystallization, respectively. We indeed confirmed this prediction for both polydisperse colloids[221,222] and 2D spin liquids.[36,223] Thus we predict a high glass-forming ability around the minimum of a melting curve, *i.e.*, a triple point, for water-type liquids.[23,24] This enhancement of the glass-forming ability should be more significant in the ρ-crystal side rather than in the S-crystal side. Since the frustration should stronger in the ρ-crystal side. Furthermore, in the glass-forming region, the glass-forming ability should decrease and the fragility should increase with an increase in P.

We confirm this scenario experimentally by using a water–LiCl mixture.[228,294] In this mixture, the addition of the salt leads to the decrease in local tetrahedral order of water and the increase in hydrated structures. We confirmed this by Raman spectroscopy measurements.[228,294] Thus, the salt basically acts as the breaker of locally favoured tetrahedral structures (see Fig. 30(b)), as pressure does.[295] Fig. 30(a) shows the phase diagram of this mixture as a function of the salt concentration ϕ, which has a V-shape similar to the T–P phase diagram of pure water (see Fig. 9). If we replace ϕ by P in the phase diagram in Fig. 30(a), it becomes essentially the same as Fig. 9. We found that the glass-forming ability becomes maximum slightly above ϕ_X, where the melting point has a minimum,

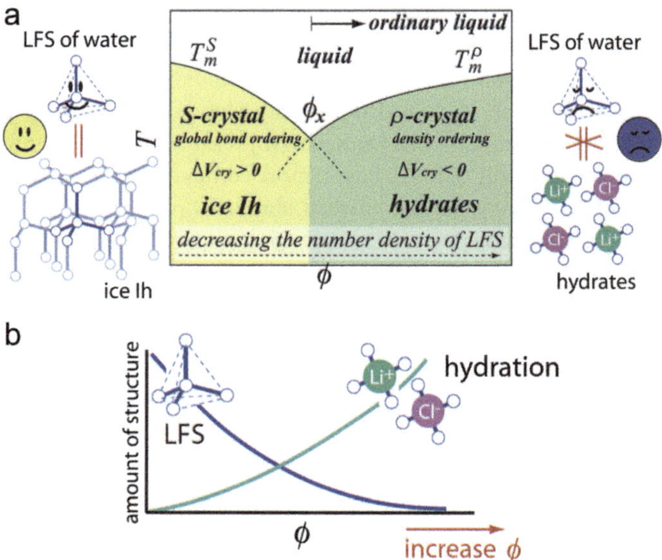

Fig. 30 (a) Schematic phase diagram of water–LiCl mixtures.[228] Locally favoured structures have a symmetry consistent with ice crystals (*S*-crystal), but not with hydrate crystals (*ρ*-crystal). In this figure, if we replace the salt concentration ϕ by pressure, it represents the phase diagram of water or other water-type liquids. Although both addition of salt and application of pressure leads to the decrease in locally favoured tetrahedral structures, their roles should have some difference since the former has local effects whereas the latter has global effects. (b) Schematic representation of the ϕ dependence of the fraction of locally favoured structures and hydration structures. Note that locally favoured structures are the source of frustration against hydrate crystals (*ρ*-crystal). This figure is courtesy of Mika Kobayashi.

and the fragility index D decreases with an increase in ϕ in the glass-forming region. We confirmed, from the ϕ-dependences of the viscosity and the thermodynamic driving force of crystallization which does not include the kinetic factor, that the eutectic-like deep minimum of the melting point and the resulting slow dynamics upon crystallization there alone cannot explain the enhancement of the glass-forming ability, suggesting the importance of a thermodynamic factor (energetic frustration) in glass transition. Furthermore, this conclusion is also supported by the large discrepancy between ϕ_X ($\sim$12 mol%) where the viscosity at T_m has a maximum and ϕ ($\sim$20 mol%) where the glass-forming ability becomes maximum. This discrepancy may be a consequence of the fact that local tetrahedral (S) ordering has random disorder effects only for ρ-crystal and not for S-crystal. This leads to the asymmetry of the glass-forming ability around a triple point (see above).

Consistent with our prediction, Molinero *et al.*[296] succeeded in vitrifying a monoatomic Si-like liquid by weakening the tetrahedrality in the Stillinger–Weber potential in their molecular dynamics simulations: the glass-forming ability increases around the triple point between diamond cubic (dc) crystal, body centred cubic (bcc) crystal, and liquid. We note that also in this case the effect of frustration is more pronounced in the bcc crystal side rather than in the dc crystal side. Furthermore, Bhat *et al.*[297] succeeded in experimentally obtaining a monoatomic 'metallic' glass of Ge at a pressure near the triple point (see also ref. 298,

299). The similar behaviour suggesting a link between crystallization and vitrification was also reported for binary Lennard-Jones mixtures.[300]

Our scenario provides a possibility to predict the glass-forming ability and fragility from the shape of the equilibrium phase diagram. If this is confirmed, it is a clear signature of the importance of the thermodynamics in the physical description of glass transition (see Fig. 7). The key is the relationship between global minimization of the free energy towards crystal and local minimization towards locally favoured structures. Depending upon the consistency of these two symmetries, locally favoured structures can be either a promoter of crystallization or its preventer. A physical factor making water so unusual among 'molecular' liquids is the V-shaped P–T phase diagram: water is only such a molecule! Instead of changing pressure, we can add additives to a liquid to modify the number density of locally favoured structures, which opens up a new possibility to control the glass-forming ability as well as the fragility of a liquid in a systematic way.

6.14 A brief summary of our scenario glass transition

In the above, we proposed that orientational ordering (or, many-body correlations) arising to lower the free energy locally leads to the constraint on particle motion and the resulting motional coherency over the characteristic length ξ, which increases the activation energy for motion and is responsible for slow dynamics associated with glass transition. We showed that such many-body correlations can be expressed by bond orientational order for weakly frustrated quasi-single-component systems. The key to vitrification is frustration against crystallization. This scenario predicts that the strength of frustration is positively correlated with the glass-forming ability and the strong nature of liquids. This prediction is unique to our model, which considers vitrification and glass transition on the same ground. Since the mechanism of glass transition is still elusive, however, further careful study is highly desirable.

7 Liquid–crystal transition

7.1 Crystal nucleation from a supercooled liquid

So far we have focused on the growth of crystal-like bond orientational order in the supercooled state of a hard-sphere system. Previous theories of crystal nucleation are based on the assumption that a supercooled liquid before crystal nucleation is homogeneous (see below). However, our finding of pre-existing structural ordering in a supercooled liquid shows that this assumption is not valid. On this basis, here we consider how a supercooled liquid is destabilized against crystallization and how crystal nuclei are formed.

Crystallization, or more strictly, crystal nucleation in a supercooled liquid, is a process in which a new ordered phase emerges from a disordered state. It is important not only as a fundamental problem of nonequilibrium statistical physics, but also as that of materials science.[13,15,301–303] Crystallization has been basically described by the classical nucleation theory. However, nature provides intriguing ways to help crystallization beyond such a simplified picture. An important point is that the initial and final states are not necessarily the only players. This idea goes back to the step rule of Ostwald,[304] which was formulated more than a century ago. He argued that the crystal phase nucleated from a liquid is not necessarily the thermodynamically most stable one, but the one whose free energy is closest to the liquid phase. Stranski and Totomanow,[305] on the other hand, argued that the phase that will be nucleated should be the one that has the lowest free energy barrier. Later Alexander and McTague[306] argued, on the basis of the Landau theory, that the cubic term of the Landau free energy favours nucleation of a body-centred cubic (bcc) phase in the early stage of a weak first order phase transition of a simple liquid. Since then there have been a lot of simulation studies on this problem, but with controversy (see, *e.g.*, ref. 64, 307 and the references therein). We note that all these approaches are from the crystal side. Here we show a new scenario of crystal nucleation beyond the above classical scenarios, focusing on structural ordering intrinsic to the supercooled state of liquid. Our approach can be regarded as a new approach from the liquid side.

7.2 Density functional theory of crystallization

First we mention a phenomenological approach based on the density functional theory beyond the classical nucleation theory (see ref. 308, 309 for review).

The density functional theory, which is based on the RY free energy functional, treats the solid as an inhomogeneous fluid. The starting point for a calculation of crystal nucleation rates is a Fourier expansion of $\rho(\vec{r})$ in terms of the reciprocal lattice vectors $\vec{G}_i$:[56]

$$\rho(\vec{r}) = \rho_\ell[1 + \mu_s + \Sigma_i \mu_i \exp(i\vec{G}_i \cdot \vec{r})], \tag{26}$$

where ρ_ℓ is the mean-field homogeneous liquid density and μ_s is the average density change on freezing. The parameters μ_i are the amplitudes that describe the periodic structure in the crystal; they are zero in the liquid. The transition is thus characterized by an infinite set of order parameters μ_i instead of the single parameter (the average density) characterizing the gas–liquid transition. The saddle point is found as usual by minimizing the grand canonical potential functional with respect to $\rho(\vec{r})$ with an approximation that the density can be written as as a sum of Gaussians, centred at the lattice sites of the crystal.

Oxtoby and his coworkers[310,311] showed that the classical theory for the free energy of formation of the critical droplet is found to exceed that obtained in the density functional calculation. They introduced an order parameter that continuously distorts a crystal with fcc symmetry into one with bcc symmetry, to allow for the possibility that precritical bcc crystallites form first and then transform to critical fcc droplets. The latter was reported in earlier simulations of a Lennard-Jones system.[64,307] Their calculation of the free energy functional showed a metastable bcc state close to the stable fcc phase. This metastable bcc phase induces a saddle point which serves as the lowest free energy barrier between the liquid and crystal, with the minimum free energy interface passing close to this saddle point. This has significant consequences for nucleation, in that a small critical droplet is largely of bcc structure at the centre and evolves into the stable fcc structure as it grows. We note that a similar framework was also applied to crystallization of hard spheres.[312]

The above approach has some similarity to ours in the sense that both consider the presence of at least two order parameters (density and a structural order parameter which has a link to the crystal structure). At the same time, there is a crucial difference: their order parameter is linked to translational order, whereas ours is linked not to it but to bond orientational order. For example, a supercooled liquid locally has high bond orientational order, but no translational order ($\mu_i = 0$), as we have shown in section 6. In our scenario, the liquid state prior to crystal nucleation is not homogeneous but quite heterogeneous. In the density functional theory, on the other hand, it is treated as completely homogeneous: there the liquid state is simply characterized by a constant density ρ_ℓ (see eqn (26)). Thus, we emphasize that despite the apparent similarity, the physical picture is essentially different between the two scenarios.

7.3 Formation of crystal nuclei in a polydisperse colloidal system

Before going to show the process of crystal nucleation, we discuss a link between crystal nuclei and crystal-like bond orientational order in polydisperse colloids. Fig. 31 shows regions of high crystal-like bond orientational order and crystal nuclei, which are observed in a supercooled polydisperse colloidal suspension by confocal microscopy observation. We can clearly see that crystal nuclei are always embedded in regions of high crystal-like bond orientational order. This also highlights the crucial difference between crystal-like bond orientational order with the average density and crystal order with a high local density. Since the polydispersity of this sample is high enough to avoid crystal nucleation, crystal nuclei are always smaller than the critical nucleus size and thus they have only a finite lifetime and eventually melt and disappear. However, this already suggests an intimate link between crystal-like bond orientational order formed in a supercooled liquid and crystal formation. The consistency of orientational symmetry between regions of high crystal-like bond orientational order and crystal nuclei should significantly reduce the interfacial energy between them, which promotes the formation of crystal nuclei there.

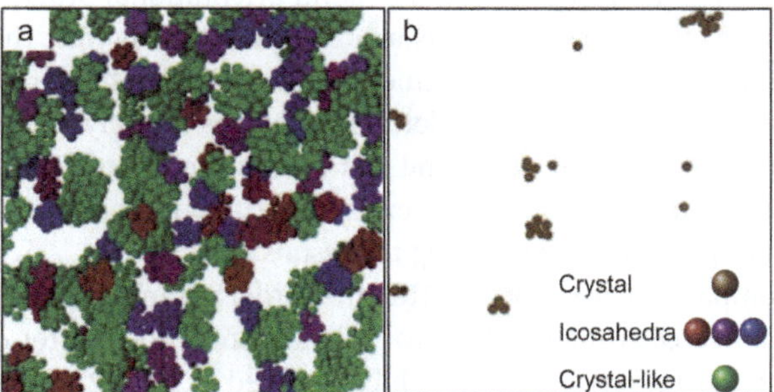

Fig. 31 Computer reconstruction from confocal microscopy coordinates in a deeply supercooled polydisperse colloidal liquid (the polydispersity = 6%, $\phi = 0.575$). Depth is $\sim 12\sigma$. Only particles of interest and their neighbours are displayed. Each particle is plotted with its real radius. (a) A typical configuration of bond ordered particles. Icosahedral particles are shown in the same colour if they belong to the same cluster. If a particle is neighbouring both crystal-like and icosahedral structures, it is displayed as icosahedral. (b) Particles with more than 7 crystalline bonds. These are crystal nuclei, but their size is smaller than the critical nucleus size and thus they never grow and disappear. This figure is reproduced from a part of Fig. 4 of ref. 44.

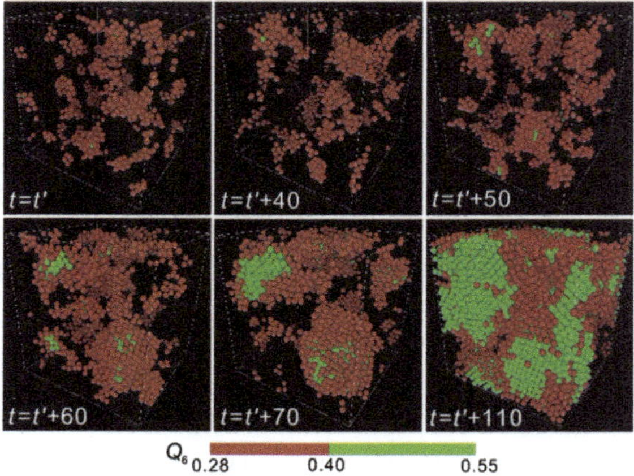

Fig. 32 Birth of crystal nuclei from medium-range structural order in 3D monodisperse hard spheres ($N = 16\ 384$). The process of nucleation of a crystal at $\phi = 0.533$. Particles with intermediate Q_6 ($0.28 < Q_6 < 0.40$) are coloured red, whereas those with high Q_6 ($Q_6 \geq 0.4$) are coloured green. The time unit is the Brownian time of a particle, τ_B. We can see the birth of a crystal and its growth. $t = t_0$ is the time when a supercooled liquid reaches a quasi-equilibrium steady state after the initiation of simulations from a random disordered state. This figure is reproduced from Fig. 2 of ref. 313.

7.4 The process of crystal nucleation in a monodisperse colloidal system

Here we show a typical crystal nucleation process in a monodisperse colloidal liquid in Fig. 32. The process was observed by Brownian dynamics simulations. We can clearly see crystal nuclei are selectively formed in regions of high crystal-like bond orientational order. After a quench, thermal fluctuations of crystal-like bond orientational order are developed and this stage shows a typical behaviour of the supercooled state of a glass-forming liquid.[313] In this stage very small crystal nuclei are created and annihilated in regions of high crystal-like bond orientational order, but they have a finite lifetime and are transient, as shown in Fig. 31. After some time, a nucleus whose size becomes larger than the critical nucleus size is formed and starts to grow continuously. Such a nucleus is always born selectively in a region of high crystal-like bond orientational order,[313] which can be regarded as wetting effects. This is a consequence of the fact that the initial stage of crystallization is the enhancement of the spatial coherence of the phase of crystal-like bond orientational order.[45] This means that crystallization is triggered by bond orientational order and not by positional order for hard-sphere liquids. Once crystals are nucleated, critical-like fluctuations of bond orientational order are pinned due to wetting to crystals.

We also compare the two types of structural order parameters in Fig. 33. One is the coarse-grained $Q_6^{\rm i}$, which is the local orientational order vector $\bar{q}_{6m}^{\rm i}$ averaged over particle i and its surrounding,[41] which we used in Fig. 32. The other is $S_{\rm ij} = \mathbf{q}_6(\mathrm{i})^* \cdot \mathbf{q}_6(\mathrm{j})$ (here * means the complex conjugate) for an equilibrium and a supercooled liquid and bcc, hcp, rhcp, and fcc crystals. This parameter was used by Auer and Frenkel[314] as the order parameter to characterize crystal nuclei and its kinetic pathway of crystallization. We note that both parameters are sensitive to the coherence of the phase of bond orientational order.[52] We obtain almost the same results for the two order parameters, which shows the robustness of our conclusion.

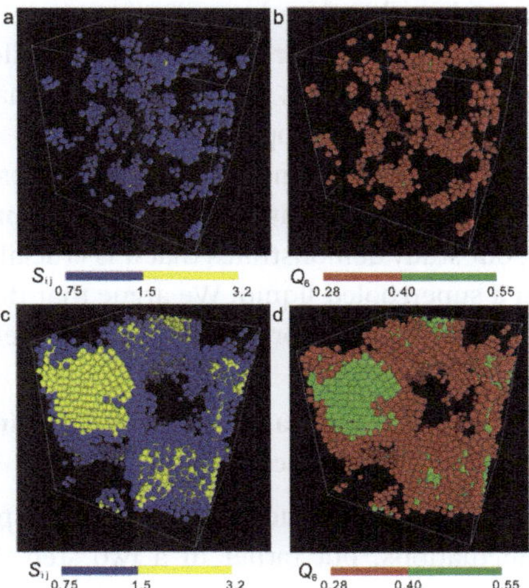

Fig. 33 Spatial distribution of order parameters for a supercooled liquid state and a liquid-crystal coexistence state. Particles coloured by the value of $S_{ij} = \mathbf{q_6}(i)^* \cdot \mathbf{q_6}(j)$ (a) and by the value of Q_6 (b) for a supercooled liquid before nucleation ($\phi = 0.533$ and $N = 16\,384$). Particles coloured by the value of S_{ij} (c) and by the value of Q_6 (d) for a liquid–crystal coexistence state in the same system as (a) and (b). Particles with $S_{ij} < 0.75$, with $0.75 \leq S_{ij} < 1.5$, and with $S_{ij} \geq 1.5$ appear transparent, blue and yellow, respectively, in (a) and (c). Particles with $Q_6 < 0.28$, with $0.28 \leq Q_6 < 0.40$, and with $Q_6 \geq 0.40$ appear transparent, red and green, respectively, in (b) and (d).

7.5 Our scenario of crystal nucleation

Our physical scenario of crystallization can be summarized as follows.[35,45,46,313] After quenching from an equilibrium liquid state to a supercooled state, medium-range bond orientational order whose symmetry has a connection to an equilibrium crystal structure (fcc or hcp in hard-sphere colloids with more weight in fcc[35,44,45]) first develops as spontaneous thermal fluctuations. When high bond orientational regions have high local density as a consequence of thermal density fluctuations, crystal nucleation is initiated with a high probability by accompanying the increase in the coherence of bond orientational order without a discontinuous density jump.[45] Here we note that although regions of high crystal-like bond orientational order do not have high density on average, some regions can have high density as a result of thermal density fluctuations, which allow a system to access even the lower bound density of crystals. The degree of density fluctuations is simply determined by the isothermal compressibility K_T. Thus, crystal nucleation always happens in a region of a supercooled liquid simultaneously having high crystal-like bond orientational order and high density. However, we stress that a factor triggering crystal nucleation is the former and not the latter: the latter is a necessary condition, but not a sufficient condition. The sequence of crystallization from melt induced by a temperature quench is thus summarized as follows: an initial homogeneous equilibrium liquid at a high temperature is transformed into an 'inhomogeneous' supercooled liquid with crystal-like bond orientational order fluctuations after quenching. The phase coherency of crystal-like bond orientational order allows a system to access a high density with the help of spontaneous density fluctuations, which leads to the development of translational order (see Fig. 2 for an intuitive feeling, although the real nucleation process takes place in a larger lengthscale). Finally, a crystalline phase is formed by the development of translational order following the growth of crystal-like bond orientational order. We emphasize that these processes continuously take place at the microscopic level.

Since the Ostwald's seminal argument, intermediate states between the initial liquid and the final crystal state has been searched for from the crystal side.[64,304–307] However, our study demonstrates that it is crucial to consider hidden structural ordering in a supercooled liquid. We argue that the slowness of these structural fluctuations is also crucial for nucleation to efficiently take place.

7.6 Roles of bond orientational and translational ordering in crystallization of hard spheres: continuous or two-step scenario

Recently it was suggested that crystal nuclei are not formed spontaneously in one step from random fluctuations, but rather in a two step through preordered precursors of high density with structural order.[315,316] This two-step crystal nucleation scenario now becomes very popular.[315–321] The importance of locally high density regions as precursors was also pointed out by ref. 316 on the basis of numerical simulation of hard-sphere crystallization.

Thus, we consider which of bond order parameter fluctuations and density fluctuations is crucial for crystal nucleation. We revealed that crystallization starts from crystal-like bond orientational ordering and then density ordering (positional ordering) comes into play later:[45] microscopically, crystallization starts from locally high density regions inside the regions of high bond orientational order, both of which are spontaneously formed by thermal fluctuations.[45] We note that density fluctuations whose amplitude is determined by isothermal compressibility K_T, allow a system to locally access the lower bound of crystal density easily. Contrary to the two-step crystallization scenario, our study[45] shows that a high local density is a necessary condition for crystal nucleation, but not a sufficient condition, as discussed above. On a microscopic scale it is bond order

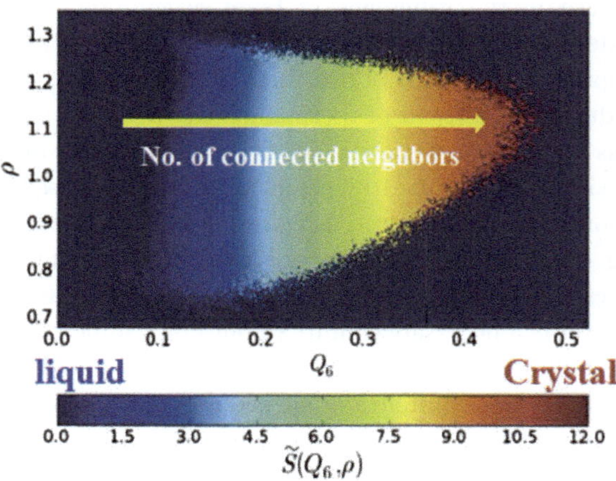

Fig. 34 Probability density for the structural order parameter $\tilde{S}$ in the (Q_6,ρ) plane. The structural order parameter $\tilde{S}$ expresses the number of connected neighbours in a continuous way (for its definition, see ref. 45). The number of connected neighbours grows continuously from 0 to 12 from the fluid to the crystal phase. This figure is reproduced from Fig. 2 of ref. 45.

parameter and neither density nor translational order that triggers crystal nucleation. This can be seen in Fig. 34, where the ordering towards a perfect crystal takes place exclusively along the bond orientational order Q_6 and not along the density ρ. The fact that the colour gradient is almost perpendicular to the ρ axis indicates that density ordering is not a driving force of crystal nucleation, at least in the early stage. As emphasized above, we note that the process of crystal nucleation in hard spheres is 'continuous' rather than made of 'discrete' steps (see Fig. 35). Our finding is markedly different from the conventional view based on macroscopic observation where we can see a discontinuous change in the density upon crystal nucleation. This clearly indicates the crucial role of bond orientational ordering in crystallization.

Crystal nucleation is triggered by the enhancement of the phase coherence of bond orientational order coupled to density fluctuations in a metastable liquid and then translational order follows afterwards.[45] Spontaneous densification due to density fluctuations leads to the enhancement of translational order in a region where bond orientational order has phase coherence (see Fig. 2). This looks quite natural, considering that crystal nucleation starts from a very small size: it is difficult to define translational order for such a small region, since it is characterized by periodicity over a long distance. Translational order can be gradually attained in the growth process of nuclei, but not in the nucleation process. The theory of crystallization may need to be fundamentally modified to incorporate these findings. How universal this scenario is to more complex liquids remains for future investigation, but our preliminary studies on soft sphere and water suggests the generality.[45,46] As we mentioned in section 2, however, for energy-driven locally favoured structures bond orientational ordering and translational

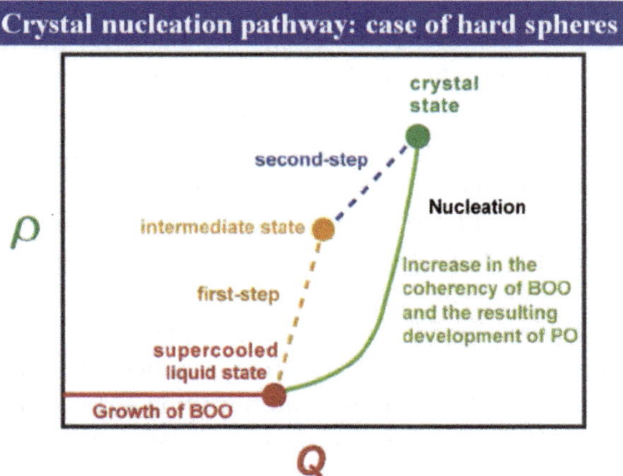

Fig. 35 The microscopic kinetic pathway of crystal nucleation in a two-order-parameter plane. For simplicity, we consider only one type of bond orientational order **Q**. In reality, this process may occur in a multi-dimensional space. The two-step and the continuous scenarios of crystal nucleation are compared. According to the two-step crystallization scenario,[315–321] the formation of precursors accompanies the density change from a liquid state and thus leads to a path along the ρ axis (see orange dashed line). Such behaviour was not observed in our simulations of hard spheres, at least on a mesoscopic scale.[45,67,222,313] On a microscopic scale, on the other hand, there is continuous development of the coherency of crystal-like bond orientational order in high density regions, which accompanies a gradual increase in positional order (PO) and the resulting densification (see text).[45]

ordering can be much more strongly coupled than for entropy-driven one, which may provide a crystal nucleation pathway with some variety.

7.7 Intimate link between glass transition and crystallization

In the above, we showed that (i) there is crystal-like bond orientational ordering in supercooled polydisperse colloidal liquids, which may be the origin of glassy slow dynamics and dynamical heterogeneity, and (ii) crystal nucleation is also triggered by the enhancement of the spatial coherence of this ordering. This finding strongly suggests an intimate link between crystallization and glass transition.[35] Namely, a supercooled liquid is intrinsically heterogeneous and, in this sense, homogeneous nucleation may necessarily be "heterogeneous". The state of a supercooled liquid is prepared, or self-organized, for future crystallization. This feature has been seen in the glass-forming liquids we studied: although crystal nuclei whose size exceeds the critical nucleus size are usually not formed in a good glass former, only small transient nuclei are spontaneously formed selectively in regions of high crystal-like bond orientational order[44] (see Fig. 31). Frustration in crystallization controls the barrier for crystallization, and thus, plays a crucial role in the glass-forming ability. Our study suggests a necessity to develop a theory of glass transition and crystallization based on the free energy including bond orientational order (effects of many-body correlations, particularly, bond angle correlations) as an important factor in addition to translational ordering[19,33–35] (see Fig. 7).

7.8 A possible difference in the type of order parameter between liquid–liquid transition and liquid–crystal (glass) transition

Finally, we discuss a possible difference between the formation of locally favoured structures (local ordering) and crystal-like bond orientational order that is spatially extendable. As stressed in sections 2.2 and 6.10, the former can be coupled to the density field, but the latter cannot. More importantly, the former can be represented by the scalar order parameter S, which is the fraction of locally favoured structures, and have no orientational correlation, whereas the latter has a specific rotational symmetry associated with that of a crystal and cannot be represented by a scalar order parameter. We believe that this causes a crucial difference between the two types of phenomena, liquid–liquid transition *vs.* crystallization.

We argue that the controversy on the presence or absence of LLT in water may be related to this issue. If the relevant order parameter is a scalar one with cooperativity, there should be liquid–liquid transition somewhere in the phase diagram. However, if the relevant order parameter is tensorial bond orientational order, its ordering leads to crystallization. If two types or order parameters coexist, which of the order parameters is more dominant may control which of LLT and crystallization takes place. Since this is speculative, we need to check carefully whether this physical picture is correct or not.

8 Summary

In this article, we have shown that any liquids have a tendency to form low free-energy configurations locally. This low free-energy configurations can often be characterized by bond orientational order, which expresses the degree of the local breakdown of the rotational symmetry.

• **Water's anomaly:** We have shown that water anomalies can be explained solely by the temperature and pressure dependence of the fraction of locally favoured structures, whose cores have tetrahedral structures and further have translational order in the second shell.[114] We have demonstrated that this fraction is proportional to the Boltzmann factor in the experimentally accessible region of the phase diagram.

• **Liquid–liquid transition:** We have explained liquid–liquid transition by cooperative formation of locally favoured structures. We have shown some examples of liquid–liquid transition observed in molecular liquids and discussed the controversy in the interpretation of these phenomena.

• **Glass transition:** We have argued that spatially extendable bond orientational order can develop in a supercooled liquid and growth of its spatio-temporal fluctuations under frustration may be the origin of glassy slow dynamics and dynamic heterogeneity at least for weakly polydisperse hard-sphere and Lennard-Jones systems and 2D spin liquids. We have proposed a scenario of glassy critical phenomena to explain the slow dynamics associated with glass transition.[35,67] When the strength of frustration on crystallization is so strong that crystallization must involve phase separation, glassy order no longer has a direct link to the symmetry of the equilibrium crystal, but may still be associated with low local free-energy configurations.[33–35] The validity of this physical picture is to be checked carefully for various glass-forming systems.

• **Crystallization:** We have shown that crystal-like bond orientational order triggers crystal nucleation with a high probability if regions of high order can reach a density required for crystallization by spontaneous thermal density fluctuations. Frustration on crystallization (polydispersity in colloidal liquids) lowers the probability of crystal nucleation significantly since it reduces the degree of crystal-like bond orientational ordering and increases the degree of icosahedral ordering which has strong frustration effects on crystallization, both of which would significantly increase the crystal–liquid interfacial energy. This strongly suggests an intrinsic link between glass transition and crystallization.

Fig. 36 Schematic figure showing the difference between a classical picture of a liquid (the homogeneous liquid picture) and a picture based on our study (the spatio-temporally inhomogeneous liquid picture). For the latter we used a typical structure of a supercooled colloidal liquid as an example.[44]

Combining these, we argue that any liquid is not a homogeneous state, but has mesoscopic or local spatio-temporal structures (see Fig. 36). We need to change our basic view of liquids and develop a theory based on the recognition that there is intrinsically a spatio-temporal hierarchical structure in liquids as in soft matter. We stress that this feature may not be captured by the standard liquid state theory, which is largely based on the two-body correlation, or translational ordering. Our study indicates the importance of many-body orientational correlations for the physical description of all these phenomena. This also suggests an encouraging possibility that all the phenomena discussed in this article can be understood on the basis of the common free energy functional (see Fig. 7), although we need to incorporate a dynamical set of equations for the description of the dynamical phenomena such as the kinetics of liquid–liquid transition, the slow dynamics associated with glass transition, and crystallization.

Acknowledgements

First the author thanks the organizers of Faraday Discussion 167 for giving him an opportunity to present this work. He is also very grateful to M. Kobayashi for her collaboration on glass transitions in water–salt mixtures, to R. Kurita, H. Mataki, K. Murata, and R. Shimizu for their collaboration on experimental studies on liquid–liquid transitions, to T. Araki, A. Furukawa, T. Kawasaki, M. Leocmach, A. Mulins, C. P. Royall, J. Russo, H. Shintani, and K. Watanabe for their collaboration on glass transition, and to T. Kawasaki, T. Konishi, and J. Russo for their collaboration on crystallization. He also thank S. F. Edwards for his continuous encouragements from the initial stage of this work. This work was partly supported by Grants-in-Aid for Scientific Research (S) and Specially Promoted Research from JSPS and Aihara Project, the FIRST program from JSPS, initiated by CSTP.

References

1 J. P. Hansen and I. R. McDonald, *Theory of simple liquids*, Academic Press, 2006.
2 J. P. Boom and S. Yip, *Molecular Hydrodynamics*, Dover Publications, 1980.
3 O. Mishima and H. E. Stanley, *Nature*, 1998, **396**, 329–335.
4 C. A. Angell, *Annu. Rev. Phys. Chem.*, 2004, **55**, 559–583.
5 P. G. Debenedetti and H. E. Stanley, *Phys. Today*, 2003, 40.
6 P. G. Debenedetti, *J. Phys.: Condens. Matter*, 2003, **15**, R1669–R1726.
7 P. H. Poole, T. Grande, C. A. Angell and P. F. McMillan, *Science*, 1997, **275**, 322.
8 P. F. McMillan, M. Wilson, M. C. Wilding, D. Daisenberger, M. Mezouar and G. N. Greaves, *J. Phys.: Condens. Matter*, 2007, **19**, 415101.
9 C. A. Angell, *Science*, 1995, **267**, 1924–1935.
10 M. D. Ediger, C. A. Angell and S. R. Nagel, *J. Phys. Chem.*, 1996, **100**, 13200.
11 P. G. Debenedetti and F. H. Stillinger, *Nature*, 2001, **410**, 259–267.
12 A. Cavagna, *Phys. Rep.*, 2009, **476**, 51–124.
13 S. P. Das, *Statistical Physics of Liquids at Freezing and Beyond*, Cambridge University Press, 2011.
14 K. Binder and W. Kob, *Glassy Materials and Disordered Solids: An Introduction to Their Statistical Mechanics (Revised Edition)*, World Sci. Pub., 2011.
15 K. F. Kelton and A. L. Greer, *Nucleation in Condensed Matter: Applications in Materials and Biology*, Pergamon, 2010.
16 C. A. Angell, *Chem. Rev.*, 2002, **102**, 2627–2650.
17 D. R. Nelson, *Defects and Geometry in Condensed Matter Physics*, Cambridge University Press, Cambridge, 2002.
18 H. Tanaka, *J. Phys.: Condens. Matter*, 1999, **11**, L159–L168.
19 H. Tanaka, *J. Phys.: Condens. Matter*, 1998, **10**, L207–L214.

20 H. Tanaka, *Phys. Rev. Lett.*, 1998, **80**, 5750–5753.
21 H. Tanaka, *Europhys. Lett.*, 2000, **50**, 340–346.
22 H. Tanaka, *J. Chem. Phys.*, 2000, **112**, 799–809.
23 H. Tanaka, *J. Phys.: Condens. Matter*, 2003, **15**, L703–L711.
24 H. Tanaka, *Phys. Rev. B*, 2002, **66**, 064202.
25 H. Tanaka, *Phys. Rev. E*, 2000, **62**, 6968–6976.
26 H. Tanaka, *J. Chem. Phys.*, 1999, **111**, 3163–3174.
27 H. Tanaka, *J. Chem. Phys.*, 1999, **111**, 3175–3182.
28 H. Tanaka, *J. Phys.: Condens. Matter*, 2003, **15**, L491–L498.
29 H. Tanaka, *J. Non-Cryst. Solids*, 2005, **351**, 678–690.
30 H. Tanaka, *J. Non-Cryst. Solids*, 2005, **351**, 3371–3384.
31 H. Tanaka, *J. Non-Cryst. Solids*, 2005, **351**, 3385–3395.
32 H. Tanaka, *J. Non-Cryst. Solids*, 2005, **351**, 3385–3395.
33 H. Tanaka, *J. Stat. Mech.*, 2010, **2010**, P12001.
34 H. Tanaka, *J. Phys.: Condens. Matter*, 2011, **23**, 284115.
35 H. Tanaka, *Eur. Phys. J. E*, 2012, **35**, 113.
36 H. Shintani and H. Tanaka, *Nat. Phys.*, 2006, **2**, 200–206.
37 M. Baus, *J. Stat. Phys.*, 1987, **48**, 1129–1146.
38 P. J. Steinhardt, D. R. Nelson and M. Ronchetti, *Phys. Rev. B*, 1983, **28**, 784–805.
39 J. F. Sadoc and R. Mosseri, *Geometrical frustration*, Cambridge University Press, 1999.
40 Z. Q. Wang and D. Stroud, *J. Chem. Phys.*, 1990, **94**, 3896–3900.
41 W. Lechner and C. Dellago, *J. Chem. Phys.*, 2008, **129**, 114707.
42 P. L. Chau and A. J. Hardwick, *Mol. Phys.*, 1998, **93**, 511–518.
43 J. R. Errington and P. G. Debenedetti, *Nature*, 2001, **409**, 318–321.
44 M. Leocmach and H. Tanaka, *Nat. Commun.*, 2012, **3**, 974, DOI: 10.1038/ncomms1974.
45 J. Russo and H. Tanaka, *Sci. Rep.*, 2012, **2**, 505, DOI: 10.1038/srep00505.
46 J. Russo and H. Tanaka, *Soft Matter*, 2012, **8**, 4206–4215.
47 S. Strässler and C. Kittel, *Phys. Rev.*, 1965, **139**, A758–A760.
48 E. Rapoport, *J. Chem. Phys.*, 1967, **46**, 2891–2895.
49 L. I. Aptekar, *Soviet Physics Doklady*, 1979, **24**, 993–995.
50 F. C. Frank, *Proc. R. Soc. London, Ser. A*, 1952, **215**, 43–46.
51 T. Tomida and T. Egami, *Phys. Rev. B*, 1995, **52**, 3290–3308.
52 M. Leocmach, J. Russo and H. Tanaka, *J. Chem. Phys.*, 2013, **138**, 12A536.
53 N. C. Karayiannis, R. Malshe, J. J. de Pablo and M. Laso, *Phys. Rev. Lett.*, 2011, **83**, 061505.
54 N. C. Karayiannis, R. Malshe, M. Kröger, J. J. de Pablo and M. Laso, *Soft Matter*, 2011, **8**, 844–858.
55 V. Holten and M. A. Anisimov, *Sci. Rep.*, 2012, **2**, 713.
56 T. V. Ramakrishnan and M. Yussouff, *Phys. Rev. B*, 1979, **19**, 2775.
57 Y. Singh, *Phys. Rep.*, 1991, **207**, 351–444.
58 S. Hess, *Z. Naturforsch. A*, 1980, **35**, 69–74.
59 A. C. Mitus and A. Z. Patashinskii, *Phys. Lett.*, 1982, **87A**, 179–182.
60 A. C. Mitus and A. Z. Patashinskii, *Phys. Lett.*, 1983, **98A**, 31–34.
61 J. Michalski, A. C. Mitus and A. Z. Patashinskii, *Phys. Lett. A*, 1987, **123**, 293–296.
62 A. D. Haymet, *Phys. Rev. B*, 1983, **27**, 1725–1731.
63 M. V. Jarić, *Phys. Rev. Lett.*, 1985, **55**, 607–610.
64 P. R. ten Wolde, M. J. Ruiz-Montero and D. Frenkel, *Phys. Rev. Lett.*, 1995, **75**, 2714–2717.
65 D. R. Nelson and M. Widom, *Nucl. Phys. B*, 1984, **240**, 113–139.
66 S. Sachdev and D. R. Nelson, *Phys. Rev. B*, 1985, **32**, 4592.
67 H. Tanaka, T. Kawasaki, H. Shintani and K. Watanabe, *Nat. Mater.*, 2010, **9**, 324–331.
68 P. Chandra, P. Coleman and A. I. Larkin, *Phys. Rev. Lett.*, 1990, **64**, 88–91.
69 C. Weber, L. Capriotti, G. Misguich, F. Becca, M. Elhajal and F. Mila, *Phys. Rev. Lett.*, 2003, **91**, 177202.
70 J. S. Langer, *Phys. Rev. E*, 2013, **88**, 012122.
71 J. S. Langer, 2013, arXiv:1308.6544.
72 J. Russo and H. Tanaka, *AIP Conf. Proc.*, 2013, 232.
73 D. Eisenberg and W. Kauzmann, *The Structure and Properties of Water*, Oxford University Press, 1969.
74 P. G. Debenedetti, *Metastable Liquids: Concepts and Principles*, Princeton University Press, Princeton, 1996.
75 A. K. Soper, *Mol. Phys.*, 2008, **106**, 2053–2076.
76 K. E. Bett and J. B. Cappi, *Nature*, 1965, **207**, 620–621.
77 F. Caupin, A. Arvengas, K. Davitt, M. E. M. Azouzi, K. I. Shmulovich, C. Ramboz, D. A. Sessoms and A. D. Stroock, *J. Phys.: Condens. Matter*, 2012, **24**, 284110.
78 R. J. Speedy and C. A. Angell, *J. Chem. Phys.*, 1976, **65**, 851–858.

79 E. G. Ponyatovsky, V. V. Sinitsyn and T. A. Pozdnyakova, *J. Chem. Phys.*, 1998, **109**, 2413–2422.
80 V. Holten, C. E. Bertrand, M. A. Anisimov and J. V. Sengers, *J. Chem. Phys.*, 2012, **136**, 094507.
81 H. E. Stanley and J. Texeira, *J. Chem. Phys.*, 1980, **73**, 3404–3422.
82 S. Sastry, P. G. Debenedetti, F. Sciortino and H. E. Stanley, *Phys. Rev. E*, 1996, **53**, 6144–6154.
83 O. Mishima and H. E. Stanley, *Nature*, 1998, **392**, 164.
84 T. Loerting and N. Giovambattista, *J. Phys.: Condens. Matter*, 2006, **18**, R919–R977.
85 P. H. Poole, F. Sciortino, U. Essmann and H. E. Stanley, *Nature*, 1992, **360**, 324–328.
86 N. Giovambattista, T. Loerting, B. R. Lukanov and F. W. Starr, *Sci. Rep.*, 2012, **2**, 390.
87 D. T. Limmer and D. Chandler, 2013, arXiv:1306.4728.
88 Y. Liu, A. Z. Panagiotopoulos and P. G. Debenedetti, *J. Chem. Phys.*, 2009, **131**, 104508.
89 P. Gallo and F. Sciortino, *Phys. Rev. Lett.*, 2012, **109**, 177801.
90 P. H. Poole, R. K. Bowles, I. Saika-Voivod and F. Sciortino, *J. Chem. Phys.*, 2013, **138**, 034505.
91 T. Kesselring, G. Franzese, S. Buldyrev, H. Herrmann and H. E. Stanley, *Sci. Rep.*, 2012, **2**, 474.
92 J. C. Palmer, C. Roberto and P. G. Debenedetti, *Faraday Discuss.*, 2013, **167**, DOI: 10.1039/C3FD00074E.
93 D. T. Limmer and D. Chandler, *J. Chem. Phys.*, 2011, **135**, 134503.
94 D. T. Limmer and D. Chandler, *J. Chem. Phys.*, 2013, **138**, 214504.
95 D. T. Limmer and D. Chandler, *Faraday Discuss.*, 2013, **167**, DOI: 10.1039/C3FD00076A.
96 K. Stokely, M. G. Mazza, H. E. Stanley and G. Franzese, *Proc. Natl. Acad. Sci. U. S. A.*, 2010, **107**, 1301–1306.
97 C. Huang, K. T. Wikfeldt, T. Tokushima, D. Nordlund, Y. Harada, U. Bergmann, M. Niebuhr, T. M. Weiss, Y. Horikawa, M. Leetmaa, *et al.*, *Proc. Natl. Acad. Sci. U. S. A.*, 2009, **106**, 15214.
98 A. K. Soper, J. Teixeira and T. Head-Gordon, *Proc. Natl. Acad. Sci. U. S. A.*, 2010, **107**, E44.
99 G. N. Clark, G. L. Hura, J. Teixeira, A. K. Soper and T. Head-Gordon, *Proc. Natl. Acad. Sci. U. S. A.*, 2010, **107**, 14003–14007.
100 A. Cunsolo, F. Formisano, C. Ferrero, F. Bencivenga and S. Finet, *J. Chem. Phys.*, 2009, **131**, 194502.
101 E. B. Moore and V. Molinero, *J. Chem. Phys.*, 2009, **130**, 244505.
102 F. W. Starr, F. Sciortino and H. E. Stanley, *Phys. Rev. E*, 1999, **60**, 6757–6768.
103 F. W. Starr, M. C. Bellissent-Funel and H. E. Stanley, *Phys. Rev. E*, 1999, **60**, 1084–1087.
104 F. Sciortino, *Chem. Phys.*, 2000, **258**, 307–314.
105 W. K. Röntgen, *Ann. Phys. U. Chim. (Wied)*, 1892, **45**, 91–97.
106 G. Némethy and H. A. Scheraga, *J. Chem. Phys.*, 1962, **36**, 3382–3400.
107 A. Ben-Naim, *J. Chem. Phys.*, 1972, **56**, 2864–2869.
108 A. Ben-Naim, *J. Chem. Phys.*, 1972, **56**, 3605–3612.
109 C. A. Angell, *J. Phys. Chem.*, 1971, **75**, 3698–3705.
110 C. H. Cho, S. Singh and G. W. Robinson, *Phys. Rev. Lett.*, 1996, **76**, 1651–1654.
111 M. C. Bellissent-Funel, *Europhys. Lett.*, 1998, **42**, 161–166.
112 G. E. Walrafen, *J. Chem. Phys.*, 1964, **40**, 3249.
113 G. E. Walrafen, *J. Chem. Phys.*, 1967, **47**, 114–126.
114 J. Russo and H. Tanaka, 2013, arXiv:1308.4231.
115 G. A. Appignanesi, J. A. Rodriguez Fris and F. Sciortino, *Eur. Phys. J. E*, 2009, **29**, 305–310.
116 S. R. Accordino, J. A. R. Fris, F. Sciortino and G. A. Appignanesi, *Eur. Phys. J. E*, 2011, **34**, 48.
117 V. Holten, D. T. Limmer, V. Molinero and M. A. Anisimov, *J. Chem. Phys.*, 2013, **138**, 174501.
118 M. J. Cuthbertson and P. H. Poole, *Phys. Rev. Lett.*, 2011, **106**, 115706.
119 M. Matsumoto, *Phys. Rev. Lett.*, 2010, **103**, 017801.
120 A. K. Soper and M. A. Ricci, *Phys. Rev. Lett.*, 2000, **84**, 2881.
121 S. Aasland and P. F. McMillan, *Nature*, 1994, **369**, 633–639.
122 P. F. McMillan, M. Wilson, M. C. Wilding, D. Daisenberger, M. Mezouar and G. N. Greaves, *J. Phys.: Condens. Matter*, 2007, **19**, 415101.
123 P. H. Poole, T. Grande, C. A. Angell and P. F. McMillan, *Science*, 1997, **275**, 322–324.
124 S. Harrington, R. Zhang, P. H. Poole, F. Sciortino and H. E. Stanley, *Phys. Rev. Lett.*, 1997, **78**, 2409–2412.
125 Y. Katayama, T. Mizutani, W. Utsumi, O. Shimomura and M. Yamakata, *Nature*, 2000, **403**, 170–173.
126 Y. Katayama, Y. Inamura, T. Mizutani, M. Yamakata, W. Utsumi and O. Shimomura, *Science*, 2004, **306**, 848–851.

127 G. Monaco, S. Falconi, W. A. Crichton and M. Mezouar, *Phys. Rev. Lett.*, 2003, **90**, 255701.

128 Y. Katayama and K. Tsuji, *J. Phys.: Condens. Matter*, 2003, **15**, 6085–6103.

129 W. Brazhkin and A. G. Lyapin, *J. Phys.: Condens. Matter*, 2003, **15**, 6059–6084.

130 V. Brazhkin, Y. Katayama, M. Kondrin, T. Hattori, A. G. Lyapin and H. Saitoh, *Phys. Rev. Lett.*, 2008, **100**, 145701.

131 V. Brazhkin, M. Kanzaki, K. Funakoshi and Y. Katayama, *Phys. Rev. Lett.*, 2009, **102**, 115901.

132 S. Sastry and C. A. Angell, *Nat. Mater.*, 2003, **2**, 739–743.

133 P. Ganesh and M. Widom, *Phys. Rev. Lett.*, 2009, **102**, 075701.

134 N. Jakse and A. Pasturel, *Phys. Rev. Lett.*, 2007, **99**, 205702.

135 V. V. Vasisht, S. Saw and S. Sastry, *Nat. Phys.*, 2011, **7**, 549–553.

136 I. Brovchenko and A. Oleinikova, *ChemPhysChem*, 2008, **9**, 2660–2675.

137 H. Tanaka, *Nature*, 1990, **380**, 328–330.

138 B. G. Demirjian, G. Dosseh, A. Chauty, M. L. Ferrer, D. Morineau, C. Lawrence, K. Takeda, D. Kivelson and S. Brown, *J. Phys. Chem. B*, 2001, **105**, 2107–2116.

139 M. Mierzwa, M. Paluch, S. J. Rzoska and J. Ziolo, *J. Phys. Chem. B*, 2008, **112**, 10383–10385.

140 R. Kurita and H. Tanaka, *J. Chem. Phys.*, 2007, **126**, 204505.

141 K. Takae and A. Onuki, *Phys. Rev. E*, 2011, **83**, 041504.

142 K. Binder, *Rep. Prog. Phys.*, 1987, **50**, 783.

143 H. Tanaka, T. Yokokawa, H. Abe, T. Hayashi and T. Nishi, *Phys. Rev. Lett.*, 1990, **65**, 3136.

144 R. Kurita, K. Murata and H. Tanaka, *Nat. Mater.*, 2008, **7**, 647–652.

145 K. Murata and H. Tanaka, *Nat. Mater.*, 2012, **11**, 436–443.

146 K. Murata and H. Tanaka, *Nat. Commun.*, 2013, **4**, 2844.

147 E. A. Jagla, *J. Chem. Phys.*, 1999, **111**, 8980.

148 L. Xu, S. V. Buldyrev, N. Giovambattista, C. A. Angell and H. E. Stanley, *J. Chem. Phys.*, 2009, **130**, 054505.

149 N. J. English, P. G. Kusalik, S. Tse and John, *J. Chem. Phys.*, 2013, **139**, 084508.

150 T. Morishita, *Phys. Rev. Lett.*, 2006, **97**, 165502.

151 P. F. McMillan, M. Wilson, D. Daisenberger and D. Machon, *Nat. Mater.*, 2005, **4**, 680–684.

152 D. Daisenberger, M. Wilson, P. F. McMillan, R. Q. Cabrera, M. C. Wilding and D. Machon, *Phys. Rev. B*, 2007, **75**, 224118.

153 G. N. Greaves, M. C. Wilding, S. Fearn, D. Langstaff, F. Kargl, S. Cox, Q. V. Van, O. Majérus, C. J. Benmore, R. Weber, C. M. Martin and L. Hennet, *Science*, 2008, **322**, 566–570.

154 G. N. Greaves, M. C. Wilding, F. Kargl and L. Hennet, *Adv. Mater. Res.*, 2008, **39–40**, 3–12.

155 G. N. Greaves, M. C. Wilding, S. Fearn, F. Kargl, L. Hennet, W. Bras, O. Majérus and C. M. Martin, *J. Non-Cryst. Solids*, 2009, **355**, 715–721.

156 L. B. Skinner, A. Barnes, P. S. Salmon and W. A. Crichton, *J. Phys.: Condens. Matter*, 2008, **20**, 205103.

157 A. C. Barnes, L. B. Skinner, P. S. Salmon, A. Bytchkov, I. Pozdnyakova, T. O. Farmer and H. E. Fischer, *Phys. Rev. Lett.*, 2009, **103**, 225702.

158 O. Mishima, L. D. Calvert and E. Whalley, *Nature*, 1984, **310**, 393.

159 E. Salcedo, A. B. de Oliveira, N. M. Barraz, C. Chakravarty and M. C. Barbosa, *J. Chem. Phys.*, 2011, **135**, 044517.

160 H. Tanaka, R. Kurita and H. Mataki, *Phys. Rev. Lett.*, 2004, **92**, 025701.

161 R. Kurita and H. Tanaka, *Science*, 2004, **306**, 845–848.

162 R. Kurita and H. Tanaka, *J. Phys.: Condens. Matter*, 2005, **17**, L293–L302.

163 A. Hédoux, Y. Guinet, M. Descamps and A. Benabou, *J. Phys. Chem. B*, 2000, **104**, 11774–11780.

164 A. Hédoux, P. Derollez, Y. Guinet, A. J. Dianoux and M. Descamps, *Phys. Rev. B*, 2001, **63**, 144202.

165 A. Hédoux, Y. Guinet, M. Foulon and M. Descamps, *J. Chem. Phys.*, 2002, **116**, 9374–9382.

166 A. Hédoux, Y. G. amd and M. Descamps, *J. Raman Spectrosc.*, 2001, **32**, 677–688.

167 A. Hédoux, J. Dore, Y. Guinet, M. C. Bellissent-Funel, D. Prevost, M. Descamps and D. Grandjean, *Phys. Chem. Chem. Phys.*, 2002, **4**, 5644–5648.

168 A. Hédoux, Y. Guinet and M. Descamps, *Phys. Rev. B*, 1998, **58**, 31–34.

169 A. Hédoux, Y. Guinet, M. Descamps and A. Bénabou, *J. Phys. Chem.*, 2000, **104**, 11774–11780.

170 I. M. Shmyt'ko, R. J. Jiménez-Riobóo, M. Hassaine and M. A. Ramos, *J. Phys.: Condens. Matter*, 2010, **22**, 195102.

171 M. Hassaine, R. J. Jiménez-Riobóo, I. V. Sharapova, O. A. Korolyuk, A. I. Krivchikov and M. A. Ramos, *J. Chem. Phys.*, 2009, **131**, 174508.
172 A. I. Krivchikov, M. Hassaine, I. V. Sharapova, O. A. Korolyuk, R. J. Jiménez-Riobûo and M. A. Ramos, *J. Non-Cryst. Solids*, 2011, **357**, 524–529.
173 H. K. Lee and R. H. Swendsen, *Phys. Rev. B*, 2001, **64**, 214102.
174 G. Franzese, M. Marqués and H. E. Stanley, *Phys. Rev. E*, 2003, **67**, 011103.
175 G. Franzese, G. Malescio, A. Skibinsky, S. V. Buldyrev and H. E. Stanley, *Phys. Rev. E*, 2002, **66**, 051206.
176 G. Franzese and H. E. Stanley, *J. Phys.: Condens. Matter*, 2007, **19**, 205126–295141.
177 A. Ha, I. Cohen, X. Zhao, M. Lee and D. Kivelson, *J. Phys. Chem.*, 1996, **100**, 1–4.
178 I. Cohen, A. Ha, X. Zhao, M. Lee, T. Fisher, M. J. Strouse and D. Kivelson, *J. Phys. Chem.*, 1996, **100**, 8518–8526.
179 J. Wiedersich, A. Kudlik, J. Gottwald, G. Benini, I. Roggatz and E. Rössler, *J. Phys. Chem. B*, 1997, **101**, 5800–5803.
180 S. Dvinskikh, G. Benini, J. Senker, M. Vogel, J. Wiedersich, A. Kudlik and E. Rössler, *J. Phys. Chem. B*, 1999, **103**, 1727–1737.
181 J. Senker and E. Rössler, *J. Phys. Chem. B*, 2002, **106**, 7592–7595.
182 M. Mizukami, K. Kobashi, M. Hanaya and M. Oguni, *J. Phys. Chem. B*, 1999, **103**, 4078–4088.
183 C. Alba-Simionesco and G. Tarjus, *Europhys. Lett.*, 2000, **52**, 297–303.
184 G. Johari and C. Ferrari, *J. Phys. Chem. B*, 1997, **101**, 10191–10197.
185 B. E. Schwickert, S. R. Kline, H. Zimmermann, K. M. Lantzky and J. L. Yarger, *Phys. Rev. B*, 2001, **64**, 045410.
186 A. Onuki, *Phase Transition Dynamics*, Cambridge University Press, Cambridge, 2002.
187 M. A. Anisimov, Private communication.
188 R. Kurita, Y. Shinohara, Y. Amemiya and H. Tanaka, *J. Phys.: Condens. Matter*, 2007, **19**, 152101–152108.
189 R. Kurita and H. Tanaka, *Phys. Rev. Lett.*, 2005, **95**, 065701.
190 K. Murata and H. Tanaka, *Nat. Commun.*, 2010, **1**, 16, DOI: 10.1038/ncomms1015.
191 L. Liu, S. H. Chen, A. Faraone, C. W. Yen and C. Y. Mou, *Phys. Rev. Lett.*, 2005, **95**, 117802–117805.
192 F. Mallamace, C. Branca, M. Broccio, C. Corsaro, N. Gonzalez-Segredo, J. Spooren, H. E. Stanley and S. H. Chen, *Eur. Phys. J. E*, 2008, **161**, 19–33.
193 W. Doster, S. Busch, A. M. Gaspar, M.-S. Appavou, J. Wuttke and H. Scheer, *Phys. Rev. Lett.*, 2010, **104**, 098101–098104.
194 R. Mancinelli, *J. Phys.: Condens. Matter*, 2010, **22**, 404213.
195 D. Morineau and C. Alba-Simionesco, *J. Phys. Chem. Lett.*, 2010, **1**, 1155–1159.
196 G. H. Findenegg, S. Jähnert, D. Akcakayiran and A. Schreiber, *ChemPhysChem*, 2008, **9**, 2651–2659.
197 A. Taschin, P. Bartolini, A. Marcelli, R. Righinia and R. Torre, *Faraday Discuss.*, 2013, **167**, DOI: 10.1039/C3FD00060E.
198 L. B. Lane, *Ind. Eng. Chem.*, 1925, **17**, 924.
199 J. J. Towey, A. K. Soper and L. Dougan, *Faraday Discuss.*, 2013, **167**, DOI: 10.1039/C3FD00084B.
200 J. J. Towey and L. Dougan, *J. Phys. Chem. B*, 2012, **116**, 1633–1641.
201 K. L. Ngai, *Relaxation and Diffusion in Complex Systems*, Springer Verlag, 2011.
202 R. Kurita and H. Tanaka, *Phys. Rev. B*, 2006, **73**, 104202.
203 K. Murata and H. Tanaka, to be published.
204 A. Hédoux, O. Hernandez, J. Lefebvre, Y. Guinet and M. Descamps, *Phys. Rev. B*, 1999, **60**, 9390.
205 Y. Hayashi, A. Puzenko and Y. Feldman, *J. Phys. Chem. B*, 2005, **109**, 16979–16981.
206 Y. Hayashi, A. Puzenko, I. B. anf, Y. E. Ryabov and Y. Feldman, *J. Phys. Chem. B*, 2005, **109**, 9174–9177.
207 L. Berthier and G. Biroli, *Rev. Mod. Phys.*, 2011, **83**, 587.
208 A. J. Liu and S. R. Nagel, *Annu. Rev. Condens. Matter Phys.*, 2010, **1**, 347–369.
209 A. Ikeda and K. Miyazaki, *Phys. Rev. Lett.*, 2011, **106**, 15701.
210 F. Ritort and P. Sollich, *Adv. Phys.*, 2003, **52**, 219–342.
211 G. H. Fredrickson and H. C. Andersen, *Phys. Rev. Lett.*, 1984, **53**, 1244–1247.
212 M. Merolle, J. P. Garrahan and D. Chandler, *Proc. Natl. Acad. Sci. U. S. A.*, 2005, **102**, 10837.
213 L. O. Hedges, R. L. Jack, J. P. Garrahan and D. Chandler, *Science*, 2009, **323**, 1309.
214 W. Götze, *Complex Dynamics of Glass-Forming Liquids: A Mode-Coupling Theory*, Oxford University Press, Oxford, 2009.
215 T. R. Kirkpatrick, D. Thirumalai and P. G. Wolynes, *Phys. Rev. A*, 1989, **111**, 1045–1054.
216 G. Parisi and F. Zamponi, *Rev. Mod. Phys.*, 2010, **82**, 789–845.

217 X. Xia and P. G. Wolynes, *Proc. Natl. Acad. Sci. U. S. A.*, 2000, **97**, 2990–2994.
218 V. Lubchenko and P. G. Wolynes, *Annu. Rev. Phys. Chem.*, 2007, **58**, 235–266.
219 G. Biroli and J.-P. Bouchaud, *Struct. Glasses Supercooled Liq.*, 2012, 31–113.
220 G. Tarjus, S. A. Kivelson, Z. Nussinov and P. Viod, *J. Phys.: Condens. Matter*, 2005, **17**, R1143–R1182.
221 T. Kawasaki, T. Araki and H. Tanaka, *Phys. Rev. Lett.*, 2007, **99**, 215701.
222 T. Kawasaki and H. Tanaka, *J. Phys.: Condens. Matter*, 2010, **22**, 232102.
223 H. Shintani and H. Tanaka, *Nat. Mater.*, 2008, **7**, 870–877.
224 T. Kawasaki and H. Tanaka, *J. Phys.: Condens. Matter*, 2011, **23**, 194121.
225 M. Shimono and H. Onodera, *Rev. Metall.*, 2012, **109**, 41–46.
226 V. Ilyin, E. Lerner, T. Lo and I. Procaccia, *Phys. Rev. Lett.*, 2007, **99**, 135702.
227 E. Lerner, I. Procaccia and I. Regev, *Phys. Rev. E*, 2009, **79**, 031501.
228 M. Kobayashi and H. Tanaka, *Phys. Rev. Lett.*, 2011, **106**, 125703.
229 B. Charbonneau, P. Charbonneau and G. Tarjus, *J. Chem. Phys.*, 2013, **138**, 12A515.
230 L. Berthier, G. Biroli, J. P. Bouchaud, L. Cipelletti and W. van Saarloos, *Dynamical heterogeneities in glasses, colloids, and granular media*, Oxford University Press, 2011, vol. 150.
231 R. Yamamoto and A. Onuki, *Phys. Rev. E*, 1998, **58**, 3515.
232 N. Lačević, F. W. Starr, T. B. Schroder and S. C. Glotzer, *J. Chem. Phys.*, 2003, **119**, 7372–7387.
233 H. Shiba, T. Kawasaki and A. Onuki, *Phys. Rev. E*, 2012, **86**, 041504.
234 T. Kawasaki and A. Onuki, *J. Chem. Phys.*, 2013, **138**, 12A514.
235 T. Kawasaki and A. Onuki, *Phys. Rev. E*, 2013, **87**, 012312.
236 K. Kim and S. Saito, *J. Chem. Phys.*, 2010, **133**, 044511.
237 K. Kim and S. Saito, *J. Chem. Phys.*, 2013, **138**, 12A506.
238 N. Kiriushcheva and P. H. Poole, *Phys. Rev. E*, 2001, **65**, 011402.
239 R. K. Murarka and B. Bagchi, *Phys. Rev. E*, 2003, **67**, 051504.
240 S. E. Abraham, S. M. Bhattacharrya and B. Bagchi, *Phys. Rev. Lett.*, 2008, **100**, 167801.
241 S. E. Abraham and B. Bagchi, *Phys. Rev. E*, 2008, **78**, 051501.
242 F. Leonforte, R. Boissiére, A. Tanguy, J. P. Wittmer and J. L. Barrat, *Phys. Rev. B*, 2005, **72**, 224206.
243 K. Watanabe and H. Tanaka, *Phys. Rev. Lett.*, 2008, **100**, 158002.
244 M. Mosayebi, E. Del Gado, P. Ilg and H. C. Öttinger, *Phys. Rev. Lett.*, 2010, **104**, 205704.
245 M. Mosayebi, E. Del Gado, P. Ilg and H. C. Öttinger, *J. Chem. Phys.*, 2012, **137**, 024504.
246 S. Yaida, 2012, arXiv:1212.0857.
247 E. Dyer, J. Lee and S. Yaida, 2013, arXiv:1302.2917.
248 E. Dyer, J. Lee and S. Yaida, 2013, arXiv:1309.5085.
249 L. Berthier, *Phys. Rev. E*, 2004, **69**, 020201.
250 L. Berthier, D. Chandler and J. P. Garrahan, *Europhys. Lett.*, 2005, **69**, 320.
251 A. Furukawa and H. Tanaka, *Phys. Rev. Lett.*, 2009, **103**, 135703.
252 A. Furukawa and H. Tanaka, *Phys. Rev. E*, 2011, **84**, 061503.
253 A. Furukawa and H. Tanaka, *Phys. Rev. E*, 2012, **86**, 030501.
254 J. Kim and T. Keyes, *J. Phys. Chem. B*, 2005, **109**, 21445–21448.
255 R. M. Puscasu, B. D. Todd, P. J. Daivis and J. S. Hansen, *J. Chem. Phys.*, 2010, **133**, 144907.
256 A. Furukawa, *Phys. Rev. E*, 2013, **87**, 062321.
257 J. P. Garrahan and D. Chandler, *Proc. Natl. Acad. Sci. U. S. A.*, 2003, **100**, 9710–9714.
258 A. Widmer-Cooper, P. Harrowell and H. Fynewever, *Phys. Rev. Lett.*, 2004, **93**, 135701.
259 A. Widmer-Cooper and P. Harrowell, *J. Phys.: Condens. Matter*, 2005, **17**, S4025–S4034.
260 F. Sausset and G. Tarjus, *Phys. Rev. Lett.*, 2010, **104**, 065701.
261 A. Widmer-Cooper, H. Perry, P. Harrowell and D. R. Reichman, *Nat. Phys.*, 2008, **4**, 711–715.
262 A. Widmer-Cooper, H. Perry, P. Harrowell and D. R. Reichman, *J. Chem. Phys.*, 2009, **131**, 194508.
263 D. Coslovich, *Phys. Rev. E*, 2011, **83**, 051505.
264 J. P. Bouchaud and G. Biroli, *J. Chem. Phys.*, 2004, **121**, 7347.
265 W. Kob, S. Roldán-Vargas and L. Berthier, *Nat. Phys.*, 2011, **8**, 164–167.
266 G. Biroli, J. P. Bouchaud, A. Cavagna, T. S. Grigera and P. Verrochio, *Nat. Phys.*, 2008, **4**, 771–775.
267 G. M. Hocky, T. E. Markland and D. R. Reichman, *Phys. Rev. Lett.*, 2012, **108**, 225506.
268 L. Berthier and W. Kob, *Phys. Rev. E*, 2012, **85**, 011102.
269 C. Cammarota and G. Biroli, *Proc. Natl. Acad. Sci. U. S. A.*, 2012, **109**, 8850–8855.
270 P. Charbonneau and G. Tarjus, *Phys. Rev. E*, 2013, **87**, 042305.
271 S. Karmakar and G. Parisi, *Proc. Natl. Acad. Sci. U. S. A.*, 2013, **110**, 2752–2757.
272 T. R. Kirkpatrick and D. Thirumalai, *J. Phys. A*, 1989, **22**, L149–L155.

273 T. R. Kirkpatrick and D. Thirumalai, *Phys. Rev. B*, 1987, **36**, 5388.
274 T. R. Kirkpatrick and D. Thirumalai, *Struct. Glasses Supercooled Liq.*, 2012, 223.
275 A. Montanari and G. Semerjian, *J. Stat. Phys.*, 2006, **125**, 23–54.
276 A. J. Dunleavy, K. Wiesner and C. P. Royall, *Phys. Rev. E*, 2012, **86**, 041505.
277 E. Flenner, H. Staley and G. Szamel, 2013, arXiv:1310.1029.
278 G. Szamel and E. Flenner, *Europhys. Lett.*, 2013, **101**, 66005.
279 G. Biroli, S. Karmakar and I. Procaccia, *Phys. Rev. Lett.*, 2013, **111**, 165701.
280 J. Russo and H. Tanaka, to be published.
281 J. Kurchan and D. Levine, *J. Phys. A: Math. Theor.*, 2011, **44**, 035001.
282 F. Sausset and D. Levine, *Phys. Rev. Lett.*, 2011, **107**, 045501–045501.
283 Y. Singh, J. P. Stoessel and P. G. Wolynes, *Phys. Rev. Lett.*, 1985, **54**, 1059–1062.
284 P. Chaudhuri, S. Karmakar and C. Dasgupta, *Phys. Rev. Lett.*, 2008, **100**, 125701.
285 B. Charbonneau, P. Charbonneau and G. Tarjus, *Phys. Rev. Lett.*, 2012, **108**, 35701.
286 C. Cammarota, G. Gradenigo and G. Biroli, 2013, arXiv:1305.3538.
287 P. C. Hohenberg and B. I. Halperin, *Rev. Mod. Phys.*, 1976, **49**, 435–479.
288 A. Malins, J. Eggers, C. P. Royall, S. R. Williams and H. Tanaka, 2012, arXiv:1203.1732.
289 A. Malins, J. Eggers, C. P. Royall, S. R. Williams and H. Tanaka, *J. Chem. Phys.*, 2013, **138**, 12A535.
290 A. Malins, J. Eggers, H. Tanaka and C. P. Royall, *Faraday Discuss.*, 2013, **167**, DOI: 10.1039/C3FD00078H.
291 D. S. Fisher, *J. Appl. Phys.*, 1987, **61**, 3672–3677.
292 R. D. Mountain and D. Thirumalai, *J. Chem. Phys.*, 1990, **92**, 6116.
293 E. W. Lang and H. D. Lüdemann, *Angew. Chem., Int. Ed. Engl.*, 1982, **21**, 315–388.
294 M. Kobayashi and H. Tanaka, *J. Phys. Chem. B*, 2011, **115**, 14077–14090.
295 R. Leberman and A. K. Soper, *Nature*, 1995, **378**, 364–366.
296 V. Molinero, S. Sastry and C. A. Angell, *Phys. Rev. Lett.*, 2006, **97**, 075701.
297 M. H. Bhat, V. Molinero, E. Soignard, V. C. Solomon, S. Sastry, J. L. Yarger and C. A. Angell, *Nature*, 2007, **448**, 787–790.
298 F. X. Zhang and W. K. Wang, *Phys. Rev. B*, 1995, **52**, 3113–3116.
299 D. W. He, F. X. Zhang, W. Yu, M. Zhang, Y. P. Liu and W. K. Wang, *J. Appl. Phys.*, 1998, **83**, 5003.
300 A. Banerjee, S. Chakrabarty and S. M. Bhattacharyya, *J. Chem. Phys.*, 2013, **139**, 104501.
301 S. Auer and D. Frenkel, *Adv. Polym. Sci.*, 2005, **173**, 149–207.
302 R. Sear, *J. Phys.: Condens. Matter*, 2007, **19**, 033101.
303 U. Gasser, *J. Phys.: Condens. Matter*, 2009, **21**, 203101.
304 W. Ostwald, *Z. Phys. Chem.*, 1897, **22**, 289–330.
305 N. I. Stranski and D. Totomanow, *Z. Phys. Chem.*, 1933, **163**, 399–408.
306 S. Alexander and J. P. McTague, *Phys. Rev. Lett.*, 1978, **41**, 702–705.
307 P. R. ten Wolde, M. J. Ruiz-Montero and D. Frenkel, *J. Chem. Phys.*, 1996, **104**, 9932–9947.
308 D. W. Oxtoby, *Acc. Chem. Res.*, 1998, **31**, 91.
309 J. D. Gunton, *J. Stat. Phys.*, 1999, **95**, 903–923.
310 C. K. Bagdassarian and D. W. Oxtoby, *J. Chem. Phys.*, 1994, **100**, 2139.
311 Y. C. Shen and D. W. Oxtoby, *J. Chem. Phys.*, 1996, **105**, 6517.
312 L. Gránásy, T. Pusztai, G. Tóth, Z. Jurek, M. Conti and B. Kvamme, *J. Chem. Phys.*, 2003, **119**, 10376.
313 T. Kawasaki and H. Tanaka, *Proc. Natl. Acad. Sci. U. S. A.*, 2010, **107**, 14036–14041.
314 S. Auer and D. Frenkel, *Nature*, 2001, **409**, 1020–1023.
315 J. F. Lutsko and G. Nicolis, *Phys. Rev. Lett.*, 2006, **96**, 46102.
316 T. Schilling, H. J. Schöpe, M. Oettel, G. Opletal and I. Snook, *Phys. Rev. Lett.*, 2010, **105**, 025701.
317 H. J. Schöpe, G. Bryant and W. van Megen, *Phys. Rev. Lett.*, 2006, **96**, 175701.
318 J. R. Savage and A. D. Dinsmore, *Phys. Rev. Lett.*, 2009, **102**, 198302.
319 S. Iacopini, T. Palberg and H. J. Schöpe, *J. Chem. Phys.*, 2009, **130**, 084502.
320 B. OfMalley and I. Snook, *J. Chem. Phys.*, 2005, **123**, 054511.
321 W. Lechner, C. Dellago and P. G. Bolhuis, *Phys. Rev. Lett.*, 2011, **106**, 85701.

Faraday Discussions RSC Publishing

PAPER

The liquid–liquid transition in supercooled ST2 water: a comparison between umbrella sampling and well-tempered metadynamics

Jeremy C. Palmer,[a] Roberto Car[b] and Pablo G. Debenedetti*[a]

Received 4th May 2013, Accepted 17th June 2013
DOI: 10.1039/c3fd00074e

We investigate the metastable phase behaviour of the ST2 water model under deeply supercooled conditions. The phase behaviour is examined using umbrella sampling (US) and well-tempered metadynamics (WT-MetaD) simulations to compute the reversible free energy surface parameterized by density and bond-orientation order. We find that free energy surfaces computed with both techniques clearly show two liquid phases in coexistence, in agreement with our earlier US and grand canonical Monte Carlo calculations [Y. Liu, J. C. Palmer, A. Z. Panagiotopoulos and P. G. Debenedetti, *J Chem Phys*, 2012, **137**, 214505; Y. Liu, A. Z. Panagiotopoulos and P. G. Debenedetti, *J Chem Phys*, 2009, **131**, 104508]. While we demonstrate that US and WT-MetaD produce consistent results, the latter technique is estimated to be more computationally efficient by an order of magnitude. As a result, we show that WT-MetaD can be used to study the finite-size scaling behaviour of the free energy barrier separating the two liquids for systems containing 192, 300 and 400 ST2 molecules. Although our results are consistent with the expected $N^{2/3}$ scaling law, we conclude that larger systems must be examined to provide conclusive evidence of a first-order phase transition and associated second critical point.

1 Introduction

Water is one of the most ubiquitous and important substances on earth, governing the chemical and physical processes essential to life as we know it.[1] As a biological solvent, water influences the evolutionary and physiological processes that determine the very nature of our existence.[2–4] It is critical to energy conversion processes, such as photosynthesis, which are necessary for sustaining life.[5] Due to water's great abundance on earth, it is a major natural driving force that shapes the physical landscape and climate of our planet.[3]

[a]*Department of Chemical and Biological Engineering, Princeton University, Princeton, NJ, 08544, USA. E-mail: pdebene@princeton.edu*
[b]*Department of Chemistry, Princeton University, Princeton, NJ, 08544, USA. E-mail: rcar@princeton.edu*

Water's role in such processes is determined by its distinctive physical properties,[2,3] including its unusually high melting and boiling temperature and heat capacity, and its low compressibility, which set it apart from many other liquids. It also exhibits well-known thermophysical anomalies, such as an increase in the magnitude of its thermal expansion coefficient,[6,7] compressibility[8,9] and heat capacity,[10,11] upon cooling. The anomalous behaviour of water's thermodynamic response functions begins near ambient conditions and becomes significantly more pronounced in the supercooled state, when liquid water becomes metastable with respect to ice.[1,12,13] This trend continues upon cooling and shows no signs of abating as far as can be measured in the supercooled regime.[1]

Several theories have been proposed to explain the anomalies of supercooled water.[1,13] On the basis of computer simulations of the ST2 water model,[14] Poole *et al.*[15] posited the existence of a first-order liquid–liquid phase transition (LLPT) between a high-density and low-density liquid (HDL and LDL, respectively) in deeply supercooled water. The increase in water's response functions upon cooling is explained in the LLPT scenario by the presence of a second critical point terminating the transition between the two liquid phases.[1] In this view, the sharp density change upon interconverting between water's well-known low-density and high-density glassy polymorphs[16,17] is a structurally-arrested manifestation of the LLPT. The thermodynamic implications of the LLPT hypothesis and alternative theories have been discussed thoroughly in recent reviews.[1,13]

Although the LLPT scenario provides a thermodynamically consistent interpretation of the experimental facts regarding water's anomalies,[1] obtaining direct evidence to validate this hypothesis has so far proved to be a significant challenge. Because the region of the phase diagram where this LLPT would occur is below the homogenous nucleation temperature of water, stabilization of the bulk liquid is difficult due to crystallization.[1] Several experimental studies have investigated the behaviour of liquid water stabilized by confinement in hydrophilic nanopores.[18–20] The interpretation of experimental observations, however, is significantly complicated by the fact that the measurements are not performed directly on the bulk liquid.[1] The strongest evidence to date that is consistent with the LLPT hypothesis comes from the measurements of the melting curves of metastable ice polymorphs.[21,22] The thermodynamic interpretation of the measurements, however, leaves open the possibility of singularity-free behaviour that does not involve the existence of a critical point.[23]

The problem of crystallization can often be avoided on the time scales accessible with computer simulation, allowing the properties of model liquids to be examined even under deeply supercooled conditions. Consequently, such methods have played an important role in understanding the thermodynamic behaviour of supercooled water. Following the seminal work of Poole *et al.*,[15] recent computational studies have focused on investigating the LLPT in the ST2 water model.[14] Liu *et al.*[24] were the first to apply state-of-the-art Monte Carlo (MC) techniques specifically designed for studying phase transitions. Using histogram reweighting grand canonical MC,[25] they showed the existence of a LLPT in a variant of ST2 modified for compatibility with the Ewald summation treatment of long-ranged Coulombic interactions.[26] Sciortino *et al.*[27] later showed similar results for ST2 water using successive umbrella sampling grand canonical MC[28] and the reaction field treatment of the long-ranged electrostatics.[26]

Limmer and Chandler[29] further advanced the methodological treatment of the LLPT in computer simulation, applying modern free energy analysis techniques to investigate the phase behaviour of ST2 water modelled with the Ewald summation method. They used a hybrid MC scheme[31] to perform a series of two-dimensional umbrella sampling (US) simulations[32] in the isothermal–isobaric ensemble. Umbrella constraints were applied to the density and a bond-orientational order parameter that distinguishes between liquid- and crystal-like configurations. Time series data from the simulations were analysed with the multi-state Bennett's acceptance ratio method of Shirts and Chodera[33] to obtain an unbiased estimate of the free energy as a function of the two order parameters. The free energy surface revealed the presence of two minima, one corresponding to a high-density liquid and the other to a low-density crystalline polymorph. Similar results were also found for a coarse-grained water model developed by Molinero and Moore.[34] Based on such observations, they concluded that the LLPT in ST2 water reported by Liu *et al.*[24] and Sciortino *et al.*[27] was actually a liquid–crystal transition.[34] Such a scenario would exclude the possibility of a second critical point due to symmetry differences between the liquid and crystal.

In contrast to the Limmer–Chandler[29,30] studies, which found no evidence of a LDL, subsequent investigations using similar free energy analysis methods have provided compelling evidence to support the existence of a LLPT in ST2 water. Using molecular dynamics simulations, Liu *et al.*[35] estimated the equation of state for their Ewald-compatible variant of ST2, including points of L–L coexistence and the limits of stability for each liquid. Umbrella sampling MC calculations were then performed at an estimated point of coexistence, using the same order parameters employed by Limmer and Chandler.[29] In accord with their independent equation of state calculations and previous study,[24] Liu *et al.*[35] observed two distinct liquid basins in the reversible free energy surface. Poole *et al.*[36] used a very similar computational approach to perform an equally-detailed analysis for ST2 water modelled with the reaction field treatment of the long-ranged electrostatics. In agreement with Liu *et al.*,[35] their free energy analysis revealed the presence of two liquid phases at a point of coexistence estimated from independent equation of state calculations. The findings of Liu *et al.*[35] and Poole *et al.*[36] are also consistent with recent molecular dynamics simulations of ST2 water by Kesselring *et al.*,[37] which show so-called "phase flipping" between the HDL and LDL on microsecond time scales. The origin of the differences between our results and those of Limmer and Chandler[29,30] are currently under investigation.

Although a variety of computational techniques have proven useful in understanding the properties of supercooled ST2 water, the free energy analysis approach is particularly appealing because it provides a rigorous framework for systematically exploring complex phase behaviour. In addition to US, which was first applied to study the LLPT in ST2 water by Limmer and Chandler,[29] a number of other computational techniques have been developed for computing free energy in molecular simulation.[38,39] The efficiency and accuracy of such techniques can vary substantially depending on the details of their implementation and the physical properties of the simulated system.[39] Consequently, some techniques may offer significant advantages over others in studying a particular problem.[39]

In this work, we investigate the low-temperature phase behaviour of ST2 water by performing free energy analysis using the well-tempered metadynamics (WT-MetaD) technique of Parrinello and co-workers,[40] which has been successfully applied to investigate phenomena including ice nucleation,[41] solid–solid phase transitions,[42] and protein folding.[43] Several studies have demonstrated that WT-MetaD is more efficient than US, reducing the computational time required to compute free energy while providing comparable accuracy.[44,45] We investigate whether these findings can be generalized to the study of the L–L coexistence in water models by directly comparing WT-MetaD and US. In agreement with Liu *et al.*[35] and Poole *et al.*,[36] we find that free energy calculations from both methods show two distinct liquid phases in coexistence. Although our results show that WT-MetaD is a promising approach, we have identified several challenges in applying the method to study L–L phase transitions. Such challenges are discussed in detail along with recent methodological advances that may improve the performance of WT-MetaD.[46,47]

2 Computational methods

2.1 Model system

The ST2 model of Stillinger and Rahman[14] was one of the first water models capable of qualitatively reproducing many of the anomalous properties observed in real water. Its well-known over-structured tetrahedral order causes it to exhibit anomalies at higher temperatures than many other models (*e.g.*, SPC/E[48]). Although this behaviour is generally considered a limitation, it facilitates computational studies of water's anomalies under conditions where other models suffer from prohibitively slow structural relaxation times. As a result, ST2 has become the principal model used in the computational investigations of the LLPT scenario.[15,24,27,29,35,36]

In the ST2 model,[14] each water molecule is comprised of five interaction sites arranged in a tetrahedral geometry. The oxygen atom at the centre of the tetrahedron is modelled using the Lennard-Jones potential, with a characteristic distance $\sigma = 0.31$ nm and characteristic energy $\varepsilon = 0.31964$ kJ mol^{-1}. The four vertices are occupied by fractional point charges, two positive and two negative, each with a magnitude of $0.2357e$. The negative and positive charges are placed a distance of 0.1 and 0.08 nm from the oxygen site, respectively. Interactions between charged sites on adjacent water molecules are modulated with a function that varies smoothly between 0 at small distances ($r < 0.20160$ nm) and 1 at large distances ($r \geq 0.31287$ nm), where r is the separation between the oxygen centres.

In our implementation of the ST2 model, Lennard-Jones and Coulombic interactions are truncated based on a 0.75 nm cut-off applied to the minimum image distance between oxygen atoms on molecular pairs. Following Liu *et al.*,[24,35] we supplement the original ST2 model described above, treating the long-ranged electrostatic interactions using the Ewald summation method[26] with a spherical wave-vector cut-off and vacuum boundary conditions. Parameters for the reciprocal-space summation are selected based on the real-space cut-off and simulation cell dimensions, using the approach described by Kolafa and Perram[49] to ensure a relative error of less than 10^{-4} in the calculated energy.

2.2 Order parameters and free energy

Phase transitions, such as the hypothesized LLPT, are macroscopic phenomena that occur only in the thermodynamic limit.[50,51] Consequently, computational studies of phase behaviour are inevitably influenced by finite-size effects in the vicinity of such transitions.[51,52] Despite this apparent limitation, however, the influence of finite-size effects on simulated systems can be rigorously described within the framework of statistical mechanics.[50,51] This enables macroscopic phase transitions to be detected and characterized indirectly by probing the behaviour of finite systems using advanced computational techniques.[52,53]

Free energy methods are a powerful class of computational techniques well suited for studying phase behaviour.[39,52,53] Such methods employ a set of phenomenological order parameters to characterize the macroscopic phase behaviour of a system from the microscopic configurations generated during the course of a computer simulation. The order parameters are chosen based on physical intuition and symmetry arguments, such that they are able to distinguish between configurations with features characteristic of different phases. The free energy associated with the order parameters $\phi(\mathbf{x}^N) = (\phi_1(\mathbf{x}^N),\phi_2(\mathbf{x}^N),\ldots,\phi_n(\mathbf{x}^N))$ is given by,

$$F(\phi) = -k_{\mathrm{B}}T \ln[\wp(\phi)], \tag{1}$$

where k_{B} is Boltzmann's constant, T is the temperature, $\wp(\phi)$ is the equilibrium probability of observing a microstate with the specified value ϕ, and $\mathbf{x}^N$ is a vector describing the microscopic coordinates of the N-particle system.

For a finite system near coexistence, eqn (1) exhibits a double-minimum structure similar to that illustrated in Fig. 1. The minima occur near values of the order parameter that are characteristic for each phase (*i.e.*, ϕ^{I} and ϕ^{II}) and have equal depths at coexistence.[50,52] As shown in Fig. 1, the free energy profile is a superposition of a bulk term and a surface term that arises due to the formation of an interface between the phases. Although the bulk term is constant for $\phi^{\mathrm{I}} \le \phi \le \phi^{\mathrm{II}}$, the surface term exhibits a maximum on this interval, giving rise to a free

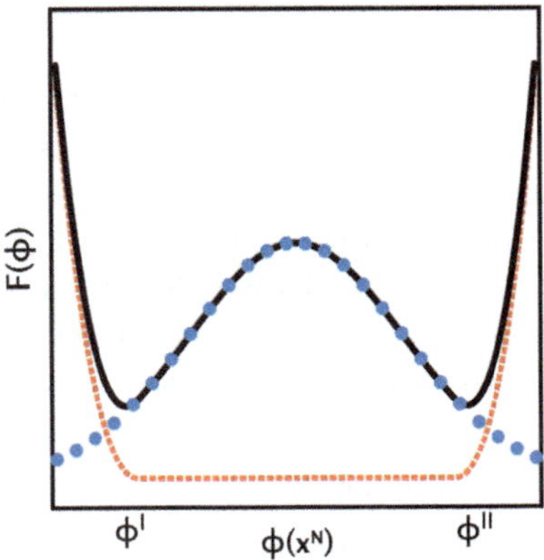

Fig. 1 Schematic showing the double-minimum structure of the free energy (solid line) for two phases at coexistence. The dashed and dotted lines show the bulk and surface contributions to free energy, respectively.

energy barrier.[50,52,54] This barrier separates the minima and must be overcome for the system to transition between the two phases.

Because ϕ is only used to construct a reversible thermodynamic path between the phases, it need not describe the physical mechanism underlying the phase transition. As a result, the free energy difference between the phases will be independent of ϕ, while the barrier height will be sensitive to how it is defined.[55] Although the computed free energy barrier may be sensitive to the order parameter, its finite-size scaling behaviour can be used to characterize the nature of phase transitions.[52,53] By examining the behaviour of the bulk and surface contributions, Lee and Kosterlitz[53] showed that the height of the calculated free energy barrier must increase with system size at a first-order phase transition, scaling as $N^{2/3}$ for sufficiently large systems. Thus, finite-size scaling analysis of the barrier can be used to rigorously detect the presence of macroscopic first-order transitions in computer simulation of finite systems.

Following previous studies of the LLPT in ST2,[29,35,36] we examine the free energy surface parameterized by $\phi = (\rho, Q_6)$ in our investigation, where ρ is density and Q_6 is the bond-orientational order parameter of Steinhardt, Nelson and Ronchetti.[56] The latter parameter is used for quantifying structural order and distinguishing between amorphous and crystalline configurations. To evaluate Q_6, the averaged spherical harmonics of each molecule i with respect to its 4 nearest neighbours are calculated using:

$$q_{l,m}^{i} = \frac{1}{4}\sum_{j \in nn_i}^{4} Y_l^m(\phi_{ij}, \theta_{ij}), \quad -l \leq m \leq l \tag{2}$$

where $Y_l^m(\phi_{ij},\theta_{ij})$ is the l,m spherical harmonic function of the angular coordinates, ϕ_{ij} and θ_{ij}, of the separation vector between molecules i and j. This quantity is summed over each molecule,

$$Q_{l,m} = \sum_{i=1}^{N} q_{l,m}^{i} \tag{3}$$

and subsequently used to construct a rotationally-invariant order parameter,

$$Q_l = \frac{1}{N}\left(\sum_{m=-l}^{l} Q_{l,m} Q_{l,m}^{*}\right)^{1/2} \tag{4}$$

For crystalline configurations, the parameter Q_6 takes on values near 0.5 and exhibits little dependence on system size. For amorphous phases, such as liquids, Q_6 scales as $N^{-1/2}$, taking on small values ($<\sim 0.1$) for the finite systems examined in this study. We note that ρ and Q_6 are global order parameters suitable for distinguishing between the different phases of ST2 water. Additional insight may be gained into the mechanisms governing transitions between such phases using local order parameters to characterize microscopic structure.[57,58]

2.3 Monte Carlo with collective moves

Computational studies of ST2 water have shown structural relaxations are extremely slow in the vicinity of the LLPT.[29,35–37] Under the dynamics of single-particle MC moves, for example, the relaxation time increases by more than an order of magnitude in transitioning from the HDL to LDL phase, reaching time scales as long as 10^6 MC sweeps (1 sweep = N single-particle MC moves). This behaviour can be understood in part by poor sampling of the collective degrees of freedom in the system associated with water's hydrogen-bonded network. To improve sampling of such collective motions, we have implemented a variant of the smart Monte Carlo (SMC) method of Rossky, Doll and Friedman.[59]

In SMC, the system is propagated through configurational phase space by performing simultaneous translations of all N particles in the system. A new configuration n is generated from the current state m by translating each particle i along a vector $\delta\mathbf{r}_i^{nm} = \mathbf{r}_i^n - \mathbf{r}_i^m$. The translation vector for each particle is drawn from a Gaussian distribution shifted by the net force acting on the its centre of mass, $\mathbf{f}_i^m$:

$$P_i(\delta\mathbf{r}_i^{nm}) = \frac{\exp[-(\delta\mathbf{r}_i^{nm} - \beta A^{\mathrm{t}}\mathbf{f}_i^m)^2/4A^{\mathrm{t}}]}{(4A^{\mathrm{t}}\pi)^{3/2}} \tag{5}$$

where $\beta = (k_{\mathrm{B}}T)^{-1}$ and A^{t} is an adjustable parameter that controls maximum attempted displacement. This procedure biases particle translations in the direction of the net force, leading to fewer attempts of energetically unfavourable moves.[59] To ensure that the proper stationary distribution is asymptotically sampled, the bias is removed by accepting the new configuration with the probability:

$$\mathrm{acc}(m \rightarrow n) = \min\left(1, \exp[-\beta\Delta U^{nm}]\frac{\prod_{i=1}^{N}\exp[-(\delta\mathbf{r}_i^{mn} - \beta A^t\mathbf{f}_i^n)^2/4A^t]}{\prod_{i=1}^{N}\exp[-(\delta\mathbf{r}_i^{nm} - \beta A^t\mathbf{f}_i^m)^2/4A^t]}\right) \tag{6}$$

where ΔU^{nm} is the change in potential energy of the system. Collective moves designed to sample molecular orientational degrees of freedom are constructed in an analogous manner, replacing the translation vector and net force in the above equations with a rotation vector and the net torque, respectively. The parameter A^{t} is also substituted with A^{r}, which regulates the magnitude of attempted rotations. As discussed by Karney,[60] torque-biased rotations are most conveniently handled using quaternions to represent molecular orientations. Hybrid translation–rotation moves can also be performed in a collective manner, giving rise to Metropolis-corrected Brownian dynamics.[61] In the implementation used in this work, N-molecule translations and rotations are carried out separately and attempted with equal probability. The parameters A^{t} and A^{r} are adjusted during the simulation to achieve an acceptance of ~25% for each type of collective move. Volume rescaling is also performed every 20 SMC moves on average to maintain isobaric conditions, setting the maximum attempted volume change to achieve ~25% acceptance.

2.4 Free energy techniques

Computing free energy directly from MC or molecular dynamics simulations is possible in principle. However, many systems exhibit large free energy barriers (>$k_{\mathrm{B}}T$) that prohibit the exploration of relevant regions of phase space on the time scales accessible with such techniques. As a result, advanced computational techniques are required in order to obtain statistically reliable estimates of the free energy.

The US technique of Torrie and Valleau[32] relies on stratification to systematically explore order parameter space. In this approach, the order parameter range is divided up into small, overlapping regions or "windows". One or more independent calculations are performed in each window, using a bias potential, $W(\phi)$, to constrain the simulation to its targeted region. This bias can be removed rigorously to estimate the local microstate probability distribution in each window. The free energy may be calculated from eqn (1) using the global probability estimate obtained by combining the local distributions. In practice, the last two steps are often performed simultaneously using well-known statistical techniques, such as the weighted histogram analysis method (WHAM) of Kumar *et al.*[62] or the multi-state Bennett's acceptance ratio (MBAR) technique of Shirts and Chodera.[33] This procedure is illustrated schematically in the left panel of Fig. 2 for the one-dimensional system discussed in section 2.2.

In contrast to stratified US, several techniques have been developed to achieve near-uniform exploration of phase space during the course of a single simulation. Examples include the expanded ensemble method of de Pablo and co-workers,[63] the self-healing umbrella sampling of Marsili *et al.*,[64] and the well-tempered metadynamics (WT-MetaD) technique pioneered by Parrinello and co-workers.[40] Such approaches make use of a repulsive, time-dependent bias potential that keeps track of the states visited by the system, encouraging it to explore new regions of phase space.

In WT-MetaD, the cumulative bias potential, $W(\phi,t)$, is updated at regular intervals with a Gaussian function centred on the point in order parameter space where the simulation resides. For any time t, the cumulative bias potential is computing using:

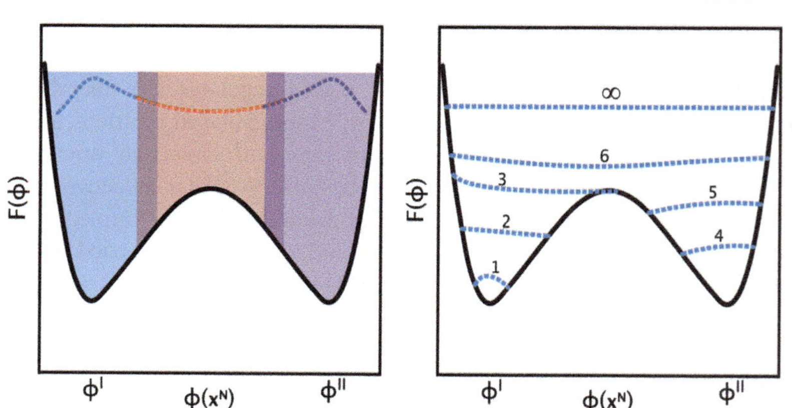

Fig. 2 (*left*) Umbrella sampling applied to a system described by a single order parameter. Independent calculations are performed in each of the shaded regions, using a bias potential to constrain the simulations to their targeted window. Statistical techniques, such as WHAM and MBAR, are used to remove the umbrella bias and combine the local microstate distributions (dashed lines), allowing the free energy (solid line) to be computed using eqn (1). (*right*) Progression of a WT-MetaD simulation on the same system shown in (a). Dashed lines illustrate the transient behaviour of the bias potential at six intermediate times, as well as its asymptotic behaviour as $t \to \infty$. The bias potential fills the first basin at time 3, allowing the system to overcome the free energy barrier and transition to the second basin. Once both basins are filled (time 6), the system becomes ergodic, exhibiting diffusive behaviour in order parameter space. In the long-time limit the bias potential converges and is proportional to the free energy.

$$W(\phi, t) = W_0 \sum_{t' \leq t} \exp\left[-\frac{W(\phi, t')}{k_B \Delta T} \right] \exp\left(-\sum_{i=1}^{n} \frac{[\phi_i - \phi_i(t')]^2}{2\delta\phi_i^2} \right) \quad (7)$$

where n is the number of order parameters, parameters $\delta\phi_i^2$ and W_0 set the width and initial height of the Gaussians, respectively, and parameter ΔT controls the rate of decay of their height.

As illustrated in the right panel of Fig. 2, the bias eventually compensates the underlying free energy surface, smoothing its features and allowing the system to explore the order parameter space in a more uniform fashion. The first exponential term in eqn (7) ensures that the bias becomes constant in the long-time limit (*i.e.*, $\dot{W}(\phi,t\rightarrow\infty) = 0$), allowing it to converge. Because $W(\phi,t)$ keeps track of the visited states, it can be rigorously shown that:[40]

$$W(\phi, t \rightarrow \infty) = -\frac{\Delta T}{\Delta T + T} F(\phi). \quad (8)$$

Eqn (8) provides a simple expression for evaluating the free energy from WT-MetaD simulations. It also reveals that parameter space is effectively sampled at an elevated temperature, allowing the system to explore regions of the free energy surface on the order $k_B(T + \Delta T)$.[40] As a result, ΔT must be chosen carefully to allow the system to overcome energy barriers separating phases of interest.[40]

We performed US and WT-MetaD simulations to investigate the phase behaviour of ST2 water near state conditions previously reported to exhibit metastable liquid coexistence.[35] The SMC scheme described in section 2.3 was used to propagate the trajectories and perform sampling in the isothermal–isobaric ensemble. In the US simulations, constraints were applied to ρ and Q_6, using a harmonic potential:

$$W(\mathbf{x}^N) = \frac{k_\rho}{2}\left[\rho(\mathbf{x}^N) - \rho^*\right]^2 + \frac{k_{Q_6}}{2}\left[Q_6(\mathbf{x}^N) - Q_6^*\right]^2 \quad (9)$$

where k_ρ and k_{Q_6} are spring constants and parameters ρ^* and Q_6^* specify the target window's centre. Simulations were performed in 35 density windows in the range $0.9 \leq \rho^* \leq 1.24$ g cm^{-3}, in steps of 0.01 g cm^{-3}. Two Q_6 windows, with Q_6^* set to 0.05 and 0.1, were used in the low density region (i.e, $\rho^* \leq 1.00$g cm^{-3}), while only the former was used for high densities. Values ranging from 4000 to 12 000 k_BT (cm^3 g^{-1})^2 and 2000 to 4000 k_BT for k_ρ and k_{Q_6}, respectively, proved sufficient to constrain the simulations to the vicinity of their target window. The US simulations were equilibrated for at least 2×10^7 MC sweeps, followed by a production phase of equal or greater duration. Using the method described by Shirts and Chodera,[33] the raw simulation data was re-sampled to remove correlations and produce at least 10^2 statistically independent samples of each biased parameter in each window. Convergence was explicitly checked by comparing histograms generated from three independent generations of simulations in each window inside the low-density region. Finally, the reversible free energy surface in $\rho - Q_6$ space was computed from the uncorrelated data using WHAM.[62]

The WT-MetaD calculations were also carried out using the SMC scheme. Values for the parameters in eqn (7) were systematically determined using the procedure developed by Quigley and Rodger.[44] This resulted in $\Delta T = 3.0T$ and $W_0 = 0.17k_BT$, and values of 0.01 (g cm^{-3})^2 and 0.004 for $\delta\rho^2$ and δQ_6^2, respectively.

Simulations were run for a total of 5×10^7 MC sweeps, updating the bias potential with a new Gaussian every 10^3 MC sweeps. Convergence was assessed using the criterion that the average of $\dot{W}(\phi,t)$ over the last 5×10^6 MC sweeps be less than $\leq 1 \times 10^{-4}$ k_BT per MC sweep. The free energy was estimated from the final bias potential using eqn (8).

3 Results and discussion

3.1 Umbrella sampling

Umbrella sampling was used to calculate the reversible free energy surface parameterized by ρ and Q_6 at 228.6 K and 2.20 kbar for a system containing 192 ST2 water molecules. Results from these calculations are shown in the left panel of Fig. 3. Although Liu *et al.*[35] reported L–L coexistence near this state point, our free energy surface has only one basin at $\rho = 0.9$ g cm^{-3} and $Q_6 = 0.06$, corresponding to the LDL phase. The apparent remnants of a second, high-density basin are also observed at $\rho \approx 1.15$ g cm^{-3} and $Q_6 = 0.06$. Similar behaviour was observed by Liu *et al.*[35] and Poole *et al.*[36] at pressures below the limit of stability for the HDL phase.

To investigate the influence of pressure, we reweighted our free energy surface at 228.6 K using:

$$F(\rho,Q_6;p + \Delta p,T) = F(\rho,Q_6;p,T) + \Delta pN/\rho \tag{10}$$

where Δp is the pressure shift from 2.20 kbar. A second basin corresponding to the HDL phase was observed in the free energy surface upon reweighting to $p >$ 2.20 kbar, confirming that the HDL is below its stability limit at the simulated state point. The centre panel of Fig. 3 shows that bi-stability in the free energy surface is achieved at 2.40 kbar, where the HDL and LDL are in approximate coexistence. Upon further compression, the LDL phase loses stability with respect to the HDL, becoming unstable at 2.72 kbar (*q.v.* Fig. 3. right panel). This sensitivity to pressure is consistent with the increased compressibility expected in the vicinity of the critical point.[35]

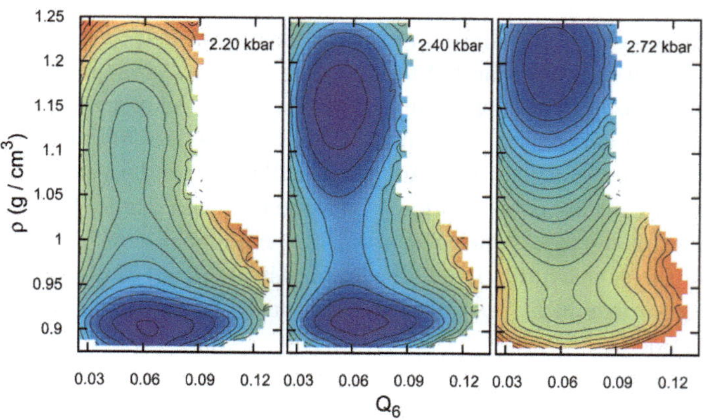

Fig. 3 Reversible free energy surfaces at 228.6 K and 2.20 kbar calculated from US. The results at 2.40 and 2.72 kbar were computed by reweighting using eqn (10). Contours are $1k_BT$ apart.

Although ice is the thermodynamically stable phase under the simulated conditions, the fact that both the HDL and LDL basins are centred around $Q_6 = 0.06$ excludes the possibility that either phase is crystalline. Moreover, because the HDL and LDL phases are not distinguished by Q_6, their relative stability can be easily examined from the free energy projection along ρ:

$$\beta F\left(\rho; Q_6^{\max}\right) = -\ln \int_0^{Q_6^{\max}} \exp[-\beta F(\rho, Q_6)] dQ_6 \tag{11}$$

where $Q_6^{\max}$ is the maximum value of Q_6. Fig. 4 shows this function computed for the free energy surfaces in Fig. 3 using values for $Q_6^{\max}$ of 0.09 and 0.11 for the HDL and LDL regions, respectively.

The profile in Fig. 4 at 2.40 kbar exhibits a double-minimum structure, indicating phase coexistence and confirming our interpretation of Fig. 3. It is trivial to demonstrate using eqn (1) that such behaviour is in accord with the grand canonical MC studies showing bimodal density distributions under similar state conditions.[24,27] The free energy barrier separating the liquids is of the order of ~3–4k_BT, which is in excellent agreement with results obtained by Liu *et al.*[35] and in reasonable agreement with Poole *et al.*'s[36] calculation using the reaction field treatment of long-ranged electrostatics. A barrier of this magnitude is also consistent the with study of Kesselring *et al.*[37] showing phase flipping between the HDL and LDL during μs-long molecular dynamics simulations.

Despite using the same Ewald-compatible variant of ST2 and finding qualitatively consistent behaviour, our estimates of the coexistence pressure and the limits of stability for the HDL and LDL phases are ~ 0.2 kbar higher than those reported by Liu *et al.*[35] This discrepancy is the result of subtle differences between our numerical implementations of the ST2 model. In the present calculations, we have truncated the Lennard-Jones interactions between oxygen atoms at 0.75 nm instead of applying the long-ranged dispersion correction used by Liu *et al.*[35] The pressure contribution from this correction is density-dependent and can be calculated using:[14,26]

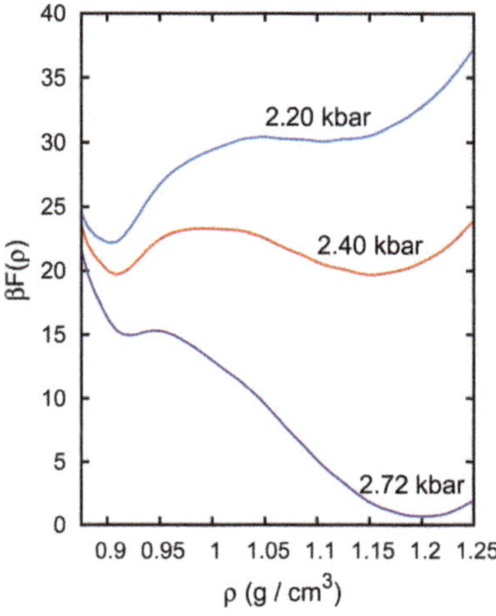

Fig. 4 Reversible free energy profiles calculated from the surfaces in Fig. 3 using eqn (11).

$$\Delta p^{\text{lrc}} = \frac{8}{3}\pi\rho^2\varepsilon\sigma^3\left[\frac{2}{3}\left(\frac{\sigma}{r_c}\right)^9 - \left(\frac{\sigma}{r_c}\right)^3\right] \tag{12}$$

where r_c is the cut-off applied to the Lennard-Jones interactions. Because this contribution is small, we can account for it by reweighting using eqn (10). Fig. 5 shows the free energy surface and profile obtained by applying this correction to simulation data at 2.20 kbar. Comparing the left panels of Fig. 3 and 5, we see that this correction shifts the system from the HDL limit of stability towards coexistence, bringing our results into agreement with the calculations performed by Liu *et al.*[35]

3.2 Well-tempered metadynamics

Our results show that the stratification strategy used in US is effective for computing reversible free energy surfaces in the vicinity of the LLPT in ST2 water, reproducing behaviour that has been observed in previous studies.[24,27,35,36] However, this approach required more than 50 independent SMC simulations, consuming a total of approximately 10 central processing unit years on 3 GHz Xenon processors for a system size of 192 ST2 water molecules. Such calculations may be prohibitively expensive for larger systems, providing a significant challenge to performing the finite-size scaling analysis required to establish the presence of a genuine first-order L–L phase transition. Because WT-MetaD has been reported to show comparable accuracy and greater efficiency in computing free energy for other systems,[44,45] we have explored this method as an alternative to US.

In our initial experimentation phase with WT-MetaD, we identified several challenges in applying it to study the LLPT in ST2. Prior to our US calculations, we performed WT-MetaD simulations at 228.6 K and 2.20 kbar, a state point reported by Liu *et al.*[35] to be near L–L coexistence. Parameter $\Delta T = 3.0T$ was chosen based on the magnitude of the free energy barrier estimated by Liu *et al.*,[35] since such a

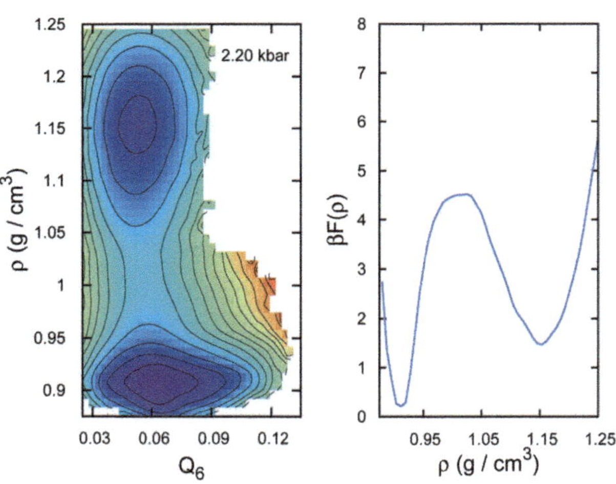

Fig. 5 Reversible free energy surface (*left*) and profile (*right*) at 228.6 K and 2.2 kbar computed by reweighting so as to incorporate the pressure correction from long-ranged Lennard-Jones interactions in eqn (12). Contours on the surface are $1k_BT$ apart.

choice has been shown to optimize the accuracy of WT-MetaD calculations.[40,44] In agreement with our subsequent US simulations, only the LDL basin was observed in the free energy surface at 2.2 kbar. However, we also found that the simulation did not adequately sample the high-density region, frustrating attempts to reweight to coexistence. This behaviour can be understood in hindsight by examining the free energy profile in Fig. 4 at 2.20 kbar obtained from US. Due to the absence of the long-ranged Lennard-Jones correction term, the HDL phase is unstable at 2.20 kbar, requiring the simulations to climb uphill in free energy $\sim 8.0k_BT$ to explore the high-density region. Our choice of ΔT, however, limited the WT-MetaD simulations to explore regions on the order of $\sim 4.0k_BT$ around the LDL basin, leading to under-sampling of the free energy surface at high density.

We subsequently performed an additional round of WT-MetaD simulations at the point of coexistence estimated from our US calculations (*i.e.*, 228.6 K and 2.40 kbar). All of the free energy surfaces from such simulations clearly showed two basins corresponding to the HDL and LDL phases, respectively. However, the relative depths of the basins typically differed by 2–$3k_BT$. Close examination of the data revealed that the system is likely to make only one round trip between the HDL and LDL basins during the course of a typical WT-MetaD trajectory, suggesting that the free energy surface was under-sampled. An example of this behaviour is shown in Fig. 6. Although the simulation was performed at coexistence, the left panel of Fig. 6 shows that the LDL and HDL basins are not of equal depth. The trajectory in the right panel of Fig. 6 shows that the simulation begins and ends in the HDL basin, making only one round trip across the free energy surface along the order parameter ρ.

This behaviour is a well-known limitation of WT-MetaD.[38,46,47] As pointed out by Singh *et al.*,[38,46] transition times between basins may be long in comparison to the length of the simulation, even under the biased dynamics of WT-MetaD. In such cases, convergence of the free energy is not guaranteed.[38,46] Consequently, several strategies have been developed to address this issue. The flux-tempered MetaD of Singh *et al.*[38,46] attempts to maximize the number of round trips made

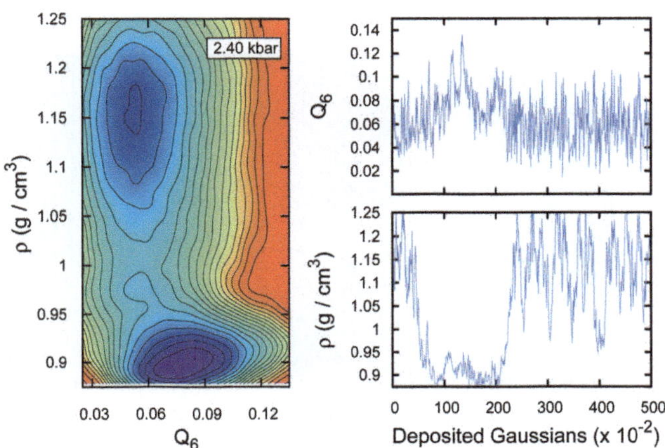

Fig. 6 (*left*) Under-sampled free energy surfaces at 228.6 K and 2.4 kbar from a WT-MetaD simulation. Contours on the surface are $1k_BT$ apart. (*right*) WT-MetaD trajectory in order parameter space (order parameters *vs.* number of deposited Gaussians) used to compute the surface on the left.

across the range of each biased variable by modifying the bias potential to incorporate information regarding the local diffusivity of the system in order parameter space. This method has been shown to dramatically improve the computational efficiency and accuracy of MetaD calculations in bimolecular systems.[65] The dynamically-adapted Gaussian method of Branduardi[47] makes use of a very similar concept, altering the width of the deposited Gaussians based on the system's local dynamics. Alternatively, the multiple walkers scheme of Raiteri *et al.*[66] may be used to efficiently parallelize WT-MetaD. In their method, multiple simulations or "walkers" are used to sample the same free energy surface. During the simulations, each walker's trajectory is used to update the total bias potential, which is shared among the walkers. As a result, each walker is biased to explore a different region of phase space, leading to better sampling and reducing the overall time required to achieve convergence of the free energy surface.

Due to the relative simplicity and ease of implementation of the multiple walkers approach, we applied it in our study of the LLPT in ST2. Simulations were performed at 228.6 K and 2.40 kbar using four walkers. The converged free energy surface was reweighted using eqn (10) to locate the coexistence pressure and the stability limits for the HDL and LDL. Fig. 7 and 8 show the reversible free energy surfaces and the profiles computed using eqn (11), respectively. The results are in overall excellent agreement with our US calculations shown in Fig. 3 and 4, showing two distinct liquid phases in coexistence at ~2.4 kbar, separated by ~3–$4k_BT$ free energy barrier. Estimates of the stability limits are also in very good agreement, indicating that the HDL and LDL phases become unstable at ~2.2 kbar and ~2.7 kbar, respectively. Such results were found to be robust against reasonable perturbations to the parameters used in the WT-MetaD calculations.

Excluding the initial calculations required to understand different aspects of the method, we were able to compute a reversible free energy surface using only one WT-MetaD simulation with four walkers. In contrast, the free energy surface computed with US required more than 50 calculations, each comparable in computational expense to a single walker in the WT-MetaD simulation. Accounting for the fact that four walkers were used, our results indicate that

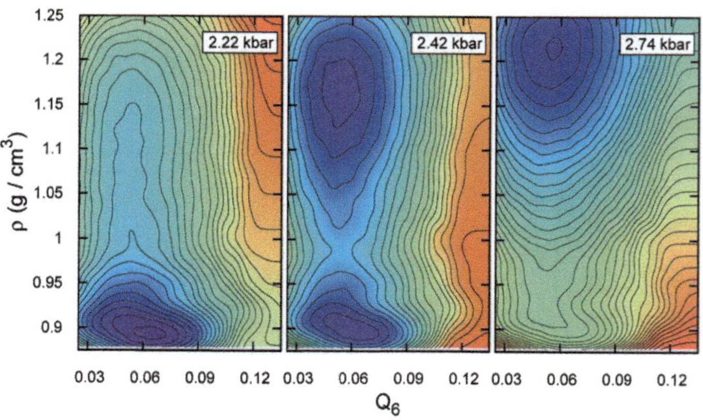

Fig. 7 Reversible free energy surfaces at 228.6 K and 2.42 kbar calculated using WT-MetaD with four walkers. The results at 2.22 and 2.74 kbar were computed by reweighting using eqn (10). Contours are $1k_BT$ apart.

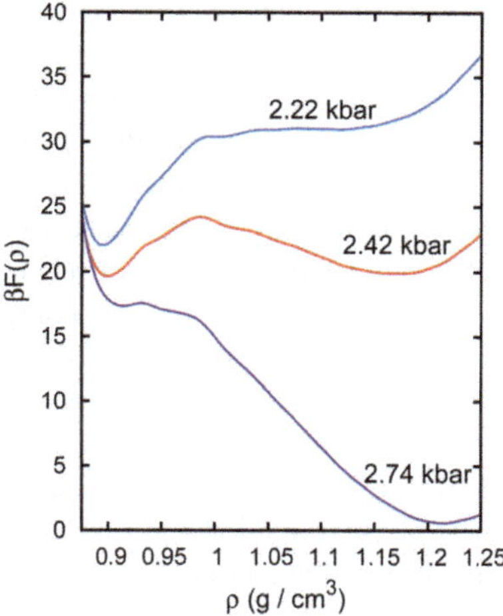

Fig. 8 Reversible free energy profiles calculated from the surfaces in Fig. 7 using eqn (11).

WT-MetaD is roughly an order of magnitude more efficient than US. As a result, we were able to use WT-MetaD to examine the finite-size scaling behaviour of the free energy barrier separating the HDL and LDL. Simulations were performed at coexistence conditions of 228.6 K and 2.4 kbar for systems containing 192, 300 and 400 ST2 molecules. Three independent free energy calculations were performed for each system size. A harmonic wall was imposed at a low density value of $\rho = 0.95$ g cm^{-3} to reduce the parameter space explored by the simulations. This expedited the calculations, while still allowing the barrier region to be sampled.

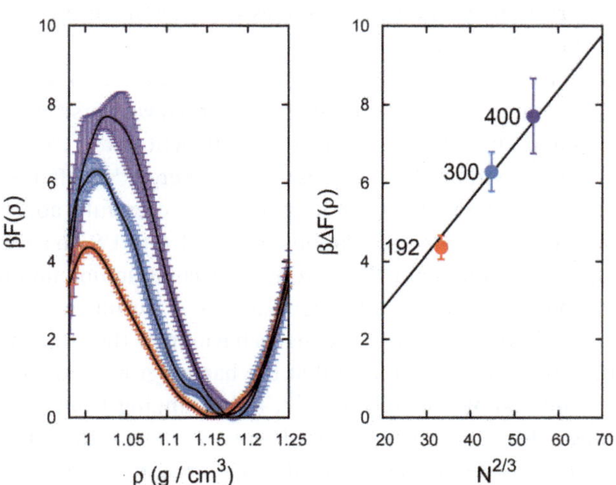

Fig. 9 (*left*) Reversible free energy profiles showing the growth of the barrier height as a function of system size. The bottom, middle and top profiles are for 192, 300 and 400 ST2 molecules, respectively. (*right*) Scaling of the barrier height with $N^{2/3}$. The results are averages computed using data from the three simulations performed for each system size, with uncertainties estimated from the standard deviations.

The left and right panels of Fig. 9 show the free energy profiles computed using eqn (11) and the scaling behaviour of the barrier height, respectively. The results are averages computed using data from the three simulations performed for each system size, along with uncertainties estimated from the standard deviations. We find that the height of the barrier grows monotonically with system size, scaling approximately as $N^{2/3}$. Although this is in accord with the behaviour expected for a first-order phase transition, our results are not entirely conclusive. Due to slow structural relaxations in the LDL phase, it was only computationally feasible to sample the HDL basin and barrier region. As a result, we cannot explicitly demonstrate that the liquids are at coexistence for all three systems. Moreover, we have only been able to examine the scaling behaviour over a modest range of system sizes, spanning less than one order of magnitude. To establish the presence of a genuine first-order phase transition and associated second critical point, significantly larger system sizes will need to be investigated over the full range of densities relevant to the HDL and LDL phases.

4 Conclusions

In this work we have examined the low temperature phase behaviour of a variant of the ST2 water model[14] compatible with the Ewald treatment of long-ranged electrostatics. Using umbrella sampling,[32] we computed free energy at 228.6 K as a function of density and a bond-orientation order parameter that can distinguish between amorphous and crystalline configurations. In agreement with Liu *et al.*[35] and Poole *et al.*,[36] our free energy calculations show very clear evidence of a phase transition between a high-density and low-density liquid. Our findings are also consistent with other investigations of the LLPT in ST2, including those by Poole *et al.*,[15] Liu *et al.*,[24] Sciortino *et al.*[27] and Kesselring *et al.*[37] Although the coexistence pressure for the two liquids was found to be slightly higher than reported by Liu *et al.*,[35] we showed that this discrepancy can be attributed to the absence of the long-ranged corrections to the Lennard-Jones interactions in the present implementation of the ST2 model.

We also performed free energy calculations for ST2 at 228.6 K using the well-tempered metadynamics method of Parrinello and co-workers.[40] Several studies have demonstrated that WT-MetaD is more efficient than US, reducing the computational time required to compute free energy.[44,45] After some initial experimentation with the method, we were able to successfully apply WT-MetaD and reproduce our free energy calculations obtained with US. In agreement with other studies,[44,45] we found that WT-MetaD is more computationally efficient than US by approximately one order of magnitude. As a result, we were able use WT-MetaD to study the finite-size scaling behaviour of the free energy barrier separating the two liquids. We found that the barrier grows monotonically with system size, scaling approximately as $N^{2/3}$, for N ranging from 192 to 400 ST2 water molecules. Although this behaviour is consistent with the expected scaling law, the analysis will need to be extended to larger systems in order to provide conclusive evidence establishing the existence of a first-order phase transition and associated second critical point.

While our results suggest that WT-MetaD is more computationally efficient than US, the analysis did not account for the relative accuracy of both methods. This will require a careful error analysis, which is currently in progress. Our experimentation with WT-MetaD also revealed that it requires significantly more *a priori* knowledge regarding the behaviour of the simulated system than is needed to perform US calculations. We found that the biasing temperature ΔT is a particularly important parameter that must be chosen carefully to facilitate exploration of all relevant regions of phase space. Making this choice is especially challenging when performing calculations under conditions where phase behaviour is very sensitive to thermodynamic state conditions or to numerical aspects of the model.

Acknowledgements

P.G.D. gratefully acknowledges support from the National Science Foundation (grant NSF CHE 1213343). R.C. acknowledges support from the Department of Energy (grant DE-SC0008626). Computations were performed at the Terascale Infrastructure for Ground breaking Research in Engineering and Science (TIGRESS) facility at Princeton University.

References

1 P. G. Debenedetti, *J. Phys.: Condens. Matter*, 2003, **15**, R1669–R1726.
2 P. Ball, *Life's Matrix. A Biography of Water*, Farrar, Strauss and Giroux, New York, 1999.
3 F. Franks, *Water: A Matrix for Life*, 2nd edn, Royal Society of Chemistry, Cambridge, 2000.
4 P. M. Wiggins, *Microbiol. Rev.*, 1990, **54**, 432–449.
5 J. Barber and P. D. Tran, *J. R. Soc. Interface*, 2013, **10**, 1–16.
6 L. Terminassian, P. Pruzan and A. Soulard, *J. Chem. Phys.*, 1981, **75**, 3064–3072.
7 D. E. Hare and C. M. Sorensen, *J. Chem. Phys.*, 1986, **84**, 5085–5089.
8 R. J. Speedy and C. A. Angell, *J. Chem. Phys.*, 1976, **65**, 851–858.
9 O. Mishima, *J. Chem. Phys.*, 2010, **133**, 144503.
10 C. A. Angell, J. Shuppert and J. C. Tucker, *J. Phys. Chem.*, 1973, **77**, 3092–3099.
11 D. G. Archer and R. W. Carter, *J. Phys. Chem. B*, 2000, **104**, 8563–8584.
12 C. A. Angell, *Annu. Rev. Phys. Chem.*, 1983, **34**, 593–630.
13 P. G. Debenedetti and H. E. Stanley, *Phys. Today*, 2003, **56**, 40–46.
14 F. Stillinger and A. Rahman, *J. Chem. Phys.*, 1974, **60**, 1545–1557.
15 P. H. Poole, F. Sciortino, U. Essmann and H. E. Stanley, *Nature*, 1992, **360**, 324–328.
16 O. Mishima, L. D. Calvert and E. Whalley, *Nature*, 1985, **314**, 76–78.
17 O. Mishima, *J. Chem. Phys.*, 1994, **100**, 5910–5912.
18 D. Z. Liu, Y. Zhang, C. C. Chen, C. Y. Mou, P. H. Poole and S. H. Chen, *Proc. Natl. Acad. Sci. U. S. A.*, 2007, **104**, 9570–9574.
19 F. Mallamace, C. Corsaro, M. Broccio, C. Branca, N. Gonzalez-Segredo, J. Spooren, S. H. Chen and H. E. Stanley, *Proc. Natl. Acad. Sci. U. S. A.*, 2008, **105**, 12725–12729.
20 Y. Zhang, A. Faraone, W. A. Kamitakahara, K. H. Liu, C. Y. Mou, J. B. Leao, S. Chang and S. H. Chen, *Proc. Natl. Acad. Sci. U. S. A.*, 2011, **108**, 12206–12211.
21 O. Mishima and H. E. Stanley, *Nature*, 1998, **392**, 164–168.
22 O. Mishima, *Phys. Rev. Lett.*, 2000, **85**, 334–336.
23 P. G. Debenedetti, *Nature*, 1998, **392**, 127–128.
24 Y. Liu, A. Z. Panagiotopoulos and P. G. Debenedetti, *J. Chem. Phys.*, 2009, **131**, 104508.
25 A. Z. Panagiotopoulos, *J. Phys.: Condens. Matter*, 2000, **12**, R25–R52.
26 D. Frenkel and B. Smit, *Understanding molecular simulation: from algorithms to applications*, 2nd edn, Academic Press, San Diego, 2002.
27 F. Sciortino, I. Saika-Voivod and P. H. Poole, *Phys. Chem. Chem. Phys.*, 2011, **13**, 19759–19764.
28 P. Virnau and M. Muller, *J. Chem. Phys.*, 2004, **120**, 10925–10930.
29 D. T. Limmer and D. Chandler, *J. Chem. Phys.*, 2011, **135**, 134503.
30 D. T. Limmer and D. Chandler, *J. Chem. Phys.*, 2013, **138**, 214504.
31 S. Duane, A. D. Kennedy, B. J. Pendleton and D. Roweth, *Phys. Lett. B*, 1987, **195**, 216–222.
32 G. M. Torrie and J. P. Valleau, *J. Comput. Phys.*, 1977, **23**, 187–199.
33 M. R. Shirts and J. D. Chodera, *J. Chem. Phys.*, 2008, **129**, 124105.
34 V. Molinero and E. B. Moore, *J. Phys. Chem. B*, 2009, **113**, 4008–4016.

35 Y. Liu, J. C. Palmer, A. Z. Panagiotopoulos and P. G. Debenedetti, *J. Chem. Phys.*, 2012, **137**, 214505.
36 P. H. Poole, R. K. Bowles, I. Saika-Voivod and F. Sciortino, *J. Chem. Phys.*, 2013, **138**, 034505.
37 T. A. Kesselring, G. Franzese, S. V. Buldyrev, H. J. Herrmann and H. E. Stanley, *Sci. Rep.*, 2012, **2**, 474.
38 S. Singh, M. Chopra and J. J. de Pablo, *Annu. Rev. Chem. Biomol. Eng.*, 2012, **3**, 369–394.
39 C. Chipot and A. Pohorille, *Free energy calculations: theory and applications in chemistry and biology*, Springer, New York, 2007.
40 A. Barducci, G. Bussi and M. Parrinello, *Phys. Rev. Lett.*, 2008, **100**, 020603.
41 D. Quigley and P. M. Rodger, *J. Chem. Phys.*, 2008, **128**, 154518.
42 C. Bealing, R. Martonak and C. Molteni, *J. Chem. Phys.*, 2009, **130**, 124712.
43 S. Piana and A. Laio, *J. Phys. Chem. B*, 2007, **111**, 4553–4559.
44 D. Quigley and P. M. Rodger, *Mol. Simul.*, 2009, **35**, 613–623.
45 F. L. Gervasio, A. Laio and M. Parrinello, *J. Am. Chem. Soc.*, 2005, **127**, 2600–2607.
46 S. Singh, C. C. Chiu and J. J. de Pablo, *J. Stat. Phys.*, 2011, **145**, 932–945.
47 D. Branduardi, G. Bussi and M. Parrinello, *J. Chem. Theory Comput.*, 2012, **8**, 2247–2254.
48 H. J. C. Berendsen, J. R. Grigera and T. P. Straatsma, *J. Phys. Chem.*, 1987, **91**, 6269–6271.
49 J. Kolafa and J. W. Perram, *Mol. Simul.*, 1992, **9**, 351–368.
50 K. Binder, *Rep. Prog. Phys.*, 1987, **50**, 783–859.
51 K. Binder, *Ferroelectrics*, 1987, **73**, 43–67.
52 J. E. Hunter and W. P. Reinhardt, *J. Chem. Phys.*, 1995, **103**, 8627–8637.
53 J. Y. Lee and J. M. Kosterlitz, *Phys. Rev. Lett.*, 1990, **65**, 137–140.
54 R. B. Griffith, *Phys. Rev.*, 1967, **158**, 176–187.
55 D. Frenkel, *Eur. Phys. J. Plus*, 2013, **128**, 10.
56 P. J. Steinhardt, D. R. Nelson and M. Ronchetti, *Phys. Rev. B*, 1983, **28**, 784–805.
57 A. Reinhardt, J. P. K. Doye, E. G. Noya and C. Vega, *J. Chem. Phys.*, 2012, **137**, 194504.
58 D. Donadio, P. Raiteri and M. Parrinello, *J. Phys. Chem. B*, 2005, **109**, 5421–5424.
59 P. J. Rossky, J. D. Doll and H. L. Friedman, *J. Chem. Phys.*, 1978, **69**, 4628–4633.
60 C. F. F. Karney, *J. Mol. Graphics Modell.*, 2007, **25**, 595–604.
61 G. Bussi and M. Parrinello, *Phys. Rev. E*, 2007, **75**.
62 S. Kumar, D. Bouzida, R. H. Swendsen, P. A. Kollman and J. M. Rosenberg, *J. Comput. Chem.*, 1992, **13**, 1011–1021.
63 E. B. Kim, R. Faller, Q. Yan, N. L. Abbott and J. J. de Pablo, *J. Chem. Phys.*, 2002, **117**, 7781–7787.
64 S. Marsili, A. Barducci, R. Chelli, P. Procacci and V. Schettino, *J. Phys. Chem. B*, 2006, **110**, 14011–14013.
65 S. Singh, C. C. Chiu and J. J. de Pablo, *J. Chem. Theory Comput.*, 2012, **8**, 4657–4662.
66 P. Raiteri, A. Laio, F. L. Gervasio, C. Micheletti and M. Parrinello, *J. Phys. Chem. B*, 2006, **110**, 3533–3539.

Faraday Discussions RSC Publishing

PAPER

The thermodynamical response functions and the origin of the anomalous behavior of liquid water

Francesco Mallamace,*[acd] Carmelo Corsaro,[a] Domenico Mallamace,[b] Cirino Vasi[c] and H. Eugene Stanley[d]

Received 3rd May 2013, Accepted 2nd July 2013
DOI: 10.1039/c3fd00073g

The density maximum of water dominates the thermodynamics of the system under ambient conditions, is strongly P-dependent, and disappears at a crossover pressure $P_{cross} \sim 1.8$ kbar. We study this variable across a wide area of the T–P phase diagram. We consider old and new data of both the isothermal compressibility $K_T(T, P)$, the pressure constant specific heat $C_P(T)$ and the coefficient of thermal expansion α_P (T, P). We observe that $K_T(T)$ shows a minimum at $T^* \sim 315 \pm 5$ K for all of the studied pressures, whereas, at the same temperature, $C_P(T)$ has the minimal variation as a function of P in the interval 1 bar–4 kbar. We find the behavior of α_P also to be surprising: all the $\alpha_P(T)$ curves measured at different P cross at T^*. The experimental data show a "singular and universal expansivity point" at $T^* \sim 315$ K and $\alpha_P(T^*) \simeq 0.44$ 10^{-3} K^{-1}. Unlike other water singularities, we find this temperature to be thermodynamically consistent in the relationship connecting the three response functions. By considering also the P–T behavior of the self-diffusion coefficient D_S and of the NMR proton chemical shift δ we have the information that at T^* the water local order points out, with decreasing T, the crossover from a normal fluid to the anomalous and complex liquid characterized by the many anomalies.

1 Introduction

Water is abundant in the universe and on the surface of the earth where it plays very important roles in most natural phenomena and biological systems. Although water is one of the simplest molecules, it has intriguing behaviors that have not been adequately explained, indeed, water remains a complex material with a large number of anomalies that are counterintuitive.[1]

[a]*Dipartimento di Fisica and CNISM, Università di Messina I-98168, Messina, Italy. E-mail: francesco.mallamace@unime.it; Fax: +39 090 395004; Tel: +39 090 6765016*

[b]*Dipartimento di Scienze dell'Ambiente, della Sicurezza, del Territorio, degli Alimenti e della Salute, Università di Messina I-98168, Messina, Italy*

[c]*CNR-IPCF, Istituto per i Processi Chimico-Fisici del CNR, Messina, I-98168, Italy*

[d]*Center for Polymer Studies and Department of Physics, Boston University, Boston, MA 02215, USA*

The best known of water's unusual properties are in the liquid state, at ambient pressure, its density ρ and viscosity: below its density maximum (at 4 °C), water expands and becomes more viscous and compressible. Other important anomalous behaviors of the ambient pressure liquid include those associated with such thermal response functions as isothermal compressibility K_T, isobaric heat capacity C_P, and thermal expansion coefficient α_P. At ambient pressure, when these response functions are extrapolated from their values in the metastable supercooled phase of water (located between the homogeneous nucleation temperature $T_H = 231$ K and the melting temperature $T_M = 273$ K), they appear to diverge at a singular temperature ($T_S \simeq 228$ K).[1,2] Another important anomaly of water is represented by its "polyamorphism" in its disordered amorphous state. Water becomes glassy below $T_g \approx 130$ K and in that region can exist in two distinct amorphous forms (*i.e.*, it is "polymorphous").[3] The low-density-amorphous (LDA) and high-density-amorphous (HDA) phases exist below T_g, and by tuning the pressure the system can be transformed back and forth between the two phases.[3] In addition, immediately above T_g water becomes a highly viscous fluid and at $T_X \approx 150$ K crystallizes. The region between T_X and T_H is a "No-Man's Land" within which water can be studied only if it is confined in small cavities so narrow that the liquid cannot freeze, or if it is located around macromolecules such as the hydration water around proteins.[4]

Water is thus an exciting research topic, and an enormous number of studies have probed the physical reasons for its unusual properties. A convergence of experimental and theoretical results strongly indicates that the key to understanding water's anomalous behavior is the role played by hydrogen bond (HB) interactions between water molecules. All three principal hypotheses proposed to understand water, *i.e.*, the stability-limit,[5] the singularity-free,[6] and the liquid–liquid critical point (LLCP)[7] scenarios agree in this regard.

This latter approach is based on two assumptions: that water "polyamorphism" exists[3] and a clustering process occurs due to the HB in which an open tetrahedrally-coordinated HB network is developed. If we begin with the stable liquid phase and decrease T, the HB lifetime, and the cluster size and stability increase, and this altered local structure continues through the No-Man's Land down to the amorphous phase region (where water is polyamorphic). Hence liquid water has local structure fluctuations, some of which are like those of low density liquid (LDL) and others like those of high density liquid (HDL), with an altered local structure that is a continuation of the LDA and HDA phases.[7] In HDL, which predominates at high T, the local tetrahedrally coordinated HB structure is not fully developed, but in LDL a more open, "ice-like" HB network appears. Water anomalies can reflect the "competition" between these two local forms of liquid. However, it must be stressed that such a model has been the subject of many discussions and controversies from both the experimental[8,9] and theoretical point of view.[10–14]

In addition, the liquid water polymorphic transition is difficult to study as it lies well inside the No-Man's Land, but, as mentioned, the crystallization inside this region can be retarded by confining water within nanoporous structures so narrow that the liquid cannot freeze,[4,15,16] or by using electrolytic solutions.[17,18]

The experiments done on water in nanopores[4,19–21] have shown that, when T is lowered, at a certain point the water HB lifetime increases by approximately six orders of magnitude, suggesting the presence of LDL and HDL inside the supercooled region[22] and the location of the so called Widom line (the locus at which the water response functions are at their maximum values).[19,21] At ambient pressure, the Widom line is crossed at $T_W(P) \simeq 225$ K where: (i) a fragile-to-strong dynamic crossover occurs,[19,23] (ii) the Stokes–Einstein relation is violated,[20,24] and (iii) the LDL local structure predominates over the HDL one.[21,24] These findings on confined water have been confirmed by a number of different experiments[15,25] and MD studies.[23,24] Also this latter dynamical transition has alternative explanations. In fact, in a recent MD simulation study, Moore and Molinero,[12] by using the mW water model (the water molecule is modelled as a single particle for which the HB is represented as a short-range anisotropic interaction), highlight that $T \simeq 225$ K represents the temperature at which the water crystallization rate is maximum.

As yet there has been no definitive proof on the dominance of a specific hypothesis with respect to the others, and at the same time that the physical reality proposed by confined water exists also in bulk water; thus water's anomalous behavior remains an open scientific question, although recent scattering experiments as a function of the wave vector (Q) and of the energy by exploring the power spectrum $S(Q, \omega)$, on bulk and confined water at ambient pressure, propose on the subject an argument of clarification. On decreasing the temperature, the liquid bulk water undergoes a structural transformation with the onset of an extended hydrogen bond network. Such a structure is at the basis of the marked viscoelastic behavior observed as a well defined frequency (ω) and wave vector dependence of the water sound velocity, and thus of the water response functions. All these observed properties appear consistent with the water polymorphism. Under these viscoelasticity conditions (or sound velocity dispersions) the water thermal response functions and their corresponding fluctuations remain finite at ambient pressure.[2]

Despite these numerous experiments and MD simulations focused on investigating water anomalies and polymorphism, there are still many open questions regarding the chemical physics of the stable liquid in the T–P phase diagram.[1] For example: why does bulk water have complex behaviors, and is there or not a region in which water behaves as a normal liquid? The focus of this work is on answering this question by considering experimental data only inside the stable liquid phase far from the metastable supercooled phase.

Here we attempt to clarify such a situation by taking into account the bulk water data of thermodynamical response functions (ρ and K_T, expansivity α_P, pressure constant specific heat C_P), of transport parameters (viscosity η, self-diffusion coefficient D_S) and of the proton NMR chemical shift as a function of both temperature and pressure. In this way we test, across a wide area of the T–P phase diagram, the connection between water anomalies and the local molecular order dominated by HB networking. For the thermodynamical functions we make use of literature data, whereas the NMR data come out from a new experiment.

The starting point of our analysis is the behavior of the density as a function of pressure and temperature. The situation is illustrated in Fig. 1, where it is very interesting to note that one of the most important water anomalies, *i.e.*, the density maximum that dominates system thermodynamics under ambient conditions, is strongly P-dependent. If we increase P, the density maximum moves to a lower T (*e.g.*, at $P = 1$ kbar it is $T \sim 245$ K). Fig. 1 shows the overall changes of the density $\rho(T, P)$ and clearly indicates this behavior; the reported data[26–33] refer essentially to bulk and emulsified water (with water droplets of size 1–10 μm).[33] Note that: i) in addition to being P-dependent, the density maximum disappears

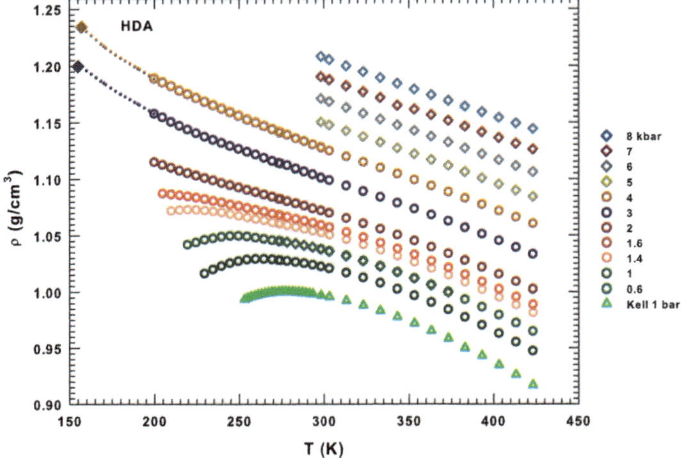

Fig. 1 The bulk water density ρ as a function of T and P, in the intervals $150 < T < 450$ K and 1 bar $< P < 8$ kbar, respectively.[26–34] Two main behaviors can be observed: i) the density maximum temperature is P-dependent and disappears for $P > 2$ kbar; ii) the P increase is accompanied by a complete change in the $\rho(T)$ curvature (from negative to positive) at such pressure. Two density values measured in HDA respectively at 3 kbar and 4 kbar are also reported.

for $1.6 < P < 2$ kbar ($P \simeq 1.8$ kbar); ii) just at this pressure there is a complete change in the $\rho(T)$ curvature from negative to positive. Fig. 1 also shows the two HDA density values at ~155 K measured at 3 kbar and 4 kbar showing a continuity between these values and the bulk water data (dotted lines). Although these HDA densities measured at very high pressures are of the order of 1.2 g cm^{-3} (or even higher), the value of the LDA density measured at 1 bar and 130 K is ~0.94 g cm^{-3}, a value consistent with the confined water ρ values (MCM nano-tubes) inside the No Man's Land, where a density minimum is also seen at $T \sim 200$ K.[34]

From this complex $\rho(T, P)$ behavior we note that, because the water density maximum is strongly P, T dependent and disappears at a certain crossover pressure ($P_{cross} \sim 1.8$ kbar), our understanding of the thermodynamic relevance of the density maximum must be adjusted. Perhaps this crossover pressure and some quantity related to $(\partial\rho/\partial T)_P$ has a physical significance we do not yet understand.

2 Results and discussion

2.1 The isothermal compressibility

On the basis of the complex behavior of the liquid water density (mainly in bulk water) illustrated in Fig. 1, here we consider the isothermal compressibility K_T ($K_T = (\partial \ln\rho/\partial \ln P)_T = -V^{-1}(\partial V/\partial P)_T$) in the same P and T intervals previously reported for $\rho(T, P)$. Fig. 2 shows the literature data of $K_T(T, P)$,[26,28,29,33,35–37] which, as is well-known, is related to volume fluctuations δV as $K_T = \langle\delta V^2\rangle_{P,T}/k_B TV$. Inspecting the data we see (i) two distinct K_T behaviors in the high and low T regimes, (ii) for the pressures in the 1 bar $< P < 4$ kbar range the corresponding K_T (T) curves show a minimum (red dots) that is located at $T^* \sim 315 \pm 5$ K, and (iii) as

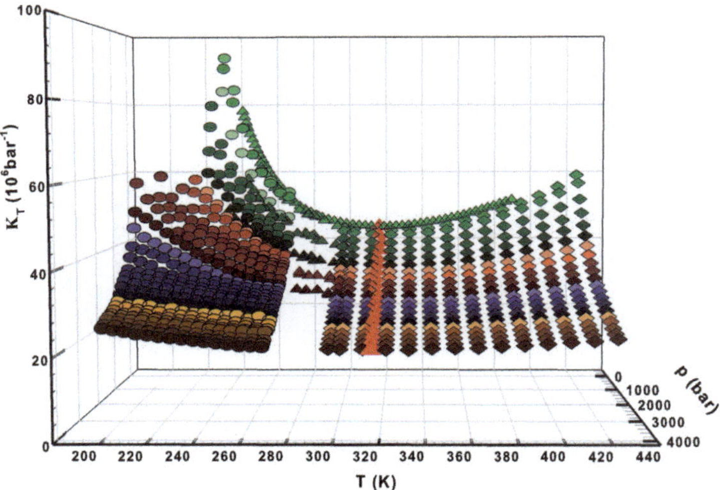

Fig. 2 The bulk water isothermal compressibility K_T (T, P). The colors of the symbols are the same as used in Fig. 1.[26,28,29,33,35–37] A simple data inspection shows: i) two distinct behaviors of the K_T dependence, at the different pressures, in the high and low temperature regimes; ii) at all the reported pressures, the K_T (T) curves present a minimum value located at the same temperature $T^* \sim 315$ K $\pm$ 5 K; iii) also for K_T, like for ρ, it seems that P_{cross} is at the borderline of two regions with different volume fluctuations: one where $\langle \delta V^2 \rangle$ is comparable with that of the liquid in its stable phases and the other ($P < P_{cross}$, and $T < T^*$) with comparatively larger fluctuations.

observed for ρ, for K_T, P_{cross} is the borderline between two regions: one with fluctuations $\langle \delta V^2 \rangle$ comparable to those of liquid in its stable phases and the other with comparatively larger fluctuations in volume ($P < P_{cross}$, and $T < T^*$). Regarding the (i) and the (iii) items, Fig. 2 clearly shows that the P effect on K_T in the low P–T regime (including the supercooled phase) is more and more pronounced than that in the high-T region ($T > T^*$). This is due to the HB network structure (characteristic of the supercooled region and the primary factor behind water's anomalies), which is less dense and more compressible than at high T. This supports the assumption that the LDL water phase is more pronounced in the low T regime, and the HDL in the high T regime.

2.2 The specific heat

Fig. 2 shows data indicating that the onset of the LDL (*i.e.*, the HB network) occurs near T^*. The T-dependence of the specific heat, measured in the same pressure interval (1 bar $< P <$ 4 kbar[38]) of the K_T and reported in Fig. 3, confirms that T^* can have a special thermodynamical role. Fig. 3 also shows, as little blue dots, C_P data measured in the supercooled regime at $P = 1$ bar.[39] As it can be observed just around this latter temperature, C_P has minimal variation as a function of the pressure (ΔC_P) if compared with the corresponding values for $T \lesssim T^*$. In particular, as well evidenced in the figure, the relative pressure variation ΔC_P is larger in the low temperature regime (including the supercooled phase) than that observed in the region $T > T^*$. On considering, hence, that $C_P = (\partial Q/\partial T)_P = T(\partial S/\partial T)_P = \langle \delta S^2 \rangle / k_B$ (where $\langle \delta S^2 \rangle$ are the entropy fluctuations), T^* represents the borderline between two regions: one in which the entropic fluctuations, at the different pressures, increase

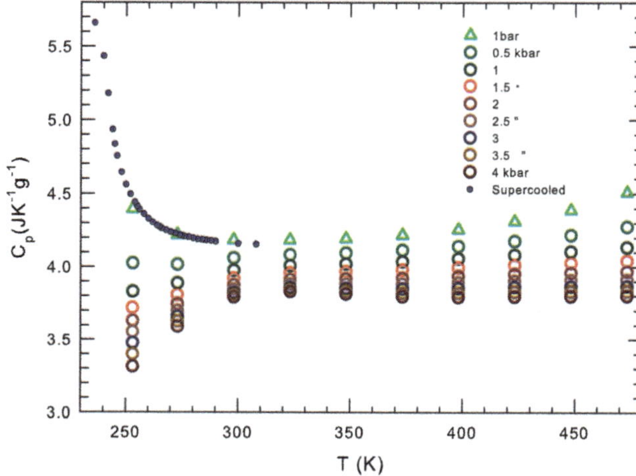

Fig. 3 The pressure constant specific heat $C_P(T)$ in the same T, P intervals of the previous figure. The figure also shows, as little blue dots, the C_P data measured in the supercooled regime at $P = 1$ bar.[39] As it can be observed just around T^*, C_P has minimal variation as a function of the pressure.

by increasing T and another one with opposite behaviors in the low temperature and supercooled regimes. In particular for $T > T^*$ and for $P > P_{\rm cross}$, it can be observed that the C_P behavior is nearly constant; only for $P < P_{\rm cross}$ does the specific heat data show a slow increase by increasing T. Whereas, in the opposite case, $T < T^*$, C_P, and thus $\langle \delta S^2 \rangle$, presents a considerable temperature variation with decreasing T at all the reported pressures characterized by different behaviors below and above $P \sim 0.5$ kbar: whereas in the first case C_P increases by decreasing T, in the second case it decreases.

2.3 The coefficient of thermal expansion

All these three figures (Fig. 1, Fig. 2 and Fig. 3) evidence the role of the thermodynamical response function derivative as a function of T. The latter two (Fig. 2 and Fig. 3) seem to indicate that in the phase region $T > T^*$ and $P > P_{\rm cross}$, liquid water behaves like a normal simple fluid. Hence we consider the coefficient of thermal expansion $\alpha_P = -(\partial \ln\rho/\partial T)_P = -V^{-1}(\partial S/\partial P)_T$, representing the entropy and volume cross-correlations $\langle \delta S \delta V \rangle$ to be $\alpha_P = \langle \delta S \delta V \rangle / k_B TV$. Regarding this function, note that, in simple liquids, δS and δV fluctuations become smaller as T decreases and are positively correlated, whereas in water they become more pronounced and, for $T < 277$ K at ambient P, they are anticorrelated.[1] The local order in water is the microscopic cause of these behaviors. As for compressibility and constant pressure specific heat, the P–T behavior of α_P is surprising and, as shown by Fig. 4, T^* is the border between two different behaviors. In the large P-range explored, all the $\alpha_P(T)$ curves measured at different pressures (in the interval from 1 bar to 8 kbar) cross, within the error bars, at the same temperature T^*. Specifically, the experimental data show a "singular and universal expansivity point" at $T^* \sim 315$ K and $\alpha_P(T^*) \simeq 0.44 \times 10^{-3}$ K^{-1}. From these data we have the confirmation that, for $T > T^*$, the thermodynamic behavior of water is exactly the same as that of a normal fluid for all the available P–T values. A situation

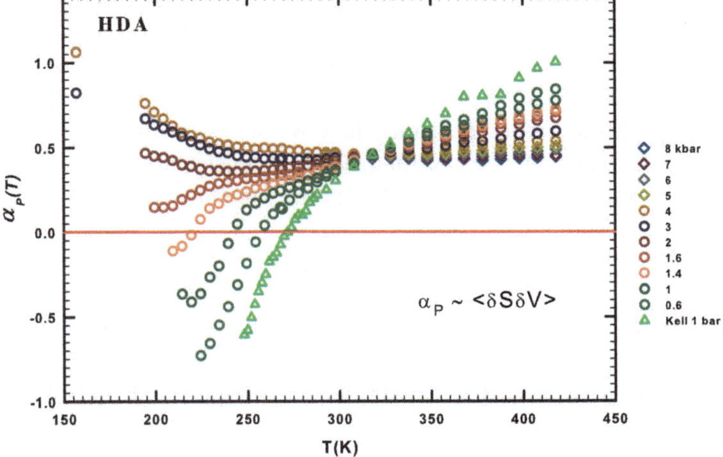

Fig. 4 The bulk water coefficient of thermal expansion $\alpha_P(T)$ in the same T, P intervals of the previous figures. It is clearly observable that all the $\alpha_P(T)$ curves, evaluated at a certain pressure, cross at the same point: $T^* \sim 315$ K with $\alpha_P(T^*) \simeq 0.44(10^{-3}K^{-1})$.

that changes in the remaining regions of the phase diagram where, as a function of P, different behaviors are observed. For $P > P_{\text{cross}}$ the δS and δV fluctuations are positively correlated but for $T < T^*$ they increase as T decreases. For $P = 3$ kbar and $P = 4$ kbar there is an apparent continuity between bulk water and its HDA phase. For $P < P_{\text{cross}}$ the $\alpha_P(T)$ evolution is more complex, *i.e.*, $\alpha_P(T)$ decreases as T decreases and, when $P < 1.6$ kbar, anticorrelation processes appear. According to the data, $\alpha_P(T)$ decreases up to a certain flex point and, after a further decrease in T, goes to a minimum whose value decreases as the pressure increases. The exact values and T-positions of these minima (for $P < P_{\text{cross}}$) are not clearly defined in the bulk water data for $\alpha_P(T)$, but their overall behavior seems fully consistent with a data evolution similar to that observed in confined water. In the case of confined water such a minimum temperature is coincident with that of the fragile-to-strong dynamical crossover and of the Widom line, which at ambient pressure is $T_W(P) \simeq 225$ K, the same temperature in which C_P has a maximum.[2]

The reported data and in particular the expansion coefficient behavior for $T > T^*$ is enough to clarify the water properties from a thermodynamical point of view by considering that these data represent the entropy and volume cross-correlation. As mentioned above, two different behaviors are present in $\langle \delta S \delta V \rangle / k_B TV$ for pressures above and below P_{cross}. Note that anticorrelations are possible only for $P < P_{\text{cross}}$, and that the maximum anticorrelation strength occurs at ambient pressure, decreases with increasing P, and vanishes at P_{cross}. This is clearly linked to the HB networking process that characterizes the local order of water: as T decreases inside the supercooled regime, it affects the growth (with increasing stability) of the molecular water structure and gives rise to a sudden entropy decrease. In contrast, pressure effects cause a progressive decrease in HB clustering. It must also be noticed that, for $T < T^*$, the T-behaviors of $\alpha_P(T)$ and $C_P(T)$, both related with δS, are characterized by analogous curvatures; the only difference is in the opposite sign. The density maximum characterizing water disappears near P_{cross} after which the system behaves as a normal liquid. This is a strong indication that the HB network, *i.e.*, the dynamic water clusters organized in a tetrahedral structure, has a low-density local order. If the presence of this HB network, as far as the behavior proposed by the $\alpha_P(T)$ data (Fig. 3), is or is not consistent with the LLCP approach does not matter with our study; instead, summarizing all the proposed results, here we stress that the water singular temperature T^* has a precise thermodynamical consistence lying in the relationship connecting two of the studied response functions:

$$\left(\frac{\partial \alpha_P}{\partial P}\right)_T = -\left(\frac{\partial K_T}{\partial T}\right)_P \tag{1}$$

Note that T^* represents the liquid bulk water isothermal compressibility minimum temperature and also the crossing point of all the thermal expansion functions in the large phase diagram area, *i.e.*, 200 K < T < 430 K and 1 bar < P < 8 kbar.

2.4 The self-diffusion coefficient and the configurational entropy

To have a further confirmation on the proposed entropy role, now we consider the self-diffusion coefficient $D_S(T, P)$ data. This is a dynamic quantity from which we can determine additional information about T^*. Fig. 5(a) shows D_S measured in bulk water as a function of the pressure (1 bar < P < 10 kbar) at several temperatures also in the supercooled regime 252 K–400 K. The $D_S(T, P)$ data in the interval 252 K < T < 290 K are measured using Nuclear Magnetic Resonance (NMR).[40] The data for T > 300 K assume the validity of the Stokes–Einstein relation and are derived from viscosity data available in the literature.[41] Note that, in the

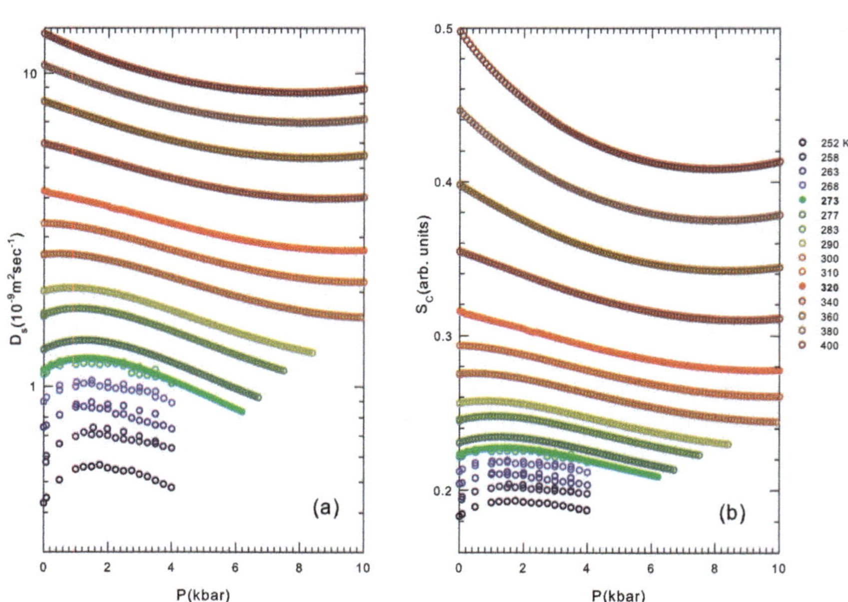

Fig. 5 a) The bulk water self-diffusion coefficient D_S as a function of the pressure in the range 1 bar < P < 10 kbar at different T from the supercooled region 252 K to 400 K. b) The corresponding configurational entropy S_C evaluated according to the Adam–Gibbs approach.[48] Note that, in both cases, $T^* \sim 315$ K and P_{cross} mark the crossover between two different physical realities.

dynamics of the system, $T^* \sim 315$ K marks the crossover between two different physical realities: below T^* the self-diffusion coefficient has a maximum that for $T = 252$ K is located at ≈ 1600 bar and that, as T increases, evolves at the lowest P and disappears near T^*. When $T > T^*$, the $D_S(P)$ behavior is more regular. In a previous study we have considered these data at constant P in an Arrhenius plot ($\ln D_S$ *vs.* $1/T$).[42] We have observed that for all the studied pressure range T^* marks two different regions: for $T > T^*$ the thermal behavior of the self-diffusion coefficient is simply Arrhenius ($D_S = A\exp(E/k_BT)$), but in the temperature range from T^* to the supercooled region (the lowest T is 252 K) the behavior is super-Arrhenius. The Arrhenius activation energy ($T > T^*$) obtained from the data fitting is $E = 15.2 \pm 0.5$ kJ mol^{-1}, *i.e.*, the HB energy value, that fully supports the primary role of HBs in the properties of water. Hence T^* marks a transition from a high-T region characterized by water dynamics with only one energy scale (the Arrhenius energy) to another typical of supercooled glass-forming liquid systems in which the T decreasing causes increasing intermolecular interactions (correlations in the time and length scale, *i.e.*, dynamic clustering). In the water case this is the onset of the HB tetrahedral network.

As in complex liquids, the interaction process originates in the disordered and finite correlation regions (finite polydisperse dynamic clustering) reflected in the transport parameters (relaxation times, viscosity, and self-diffusion) by means of a super-Arrhenius behavior or a multi-relaxation in the time evolution of the density–density correlation functions. Liquid state theory suggests the presence of an onset temperature marking a crossover from normal liquid behavior to supercooled liquid behavior.[43–47] Above that the transport is Arrhenius and below that correlations cause activation barriers to grow with an increasing scale resulting in super-Arrhenius behavior: *i.e.* a change in the explored configuration space.[43,46–48] An analogous behavior can be observed as a supercooled liquid approaches the dynamical arrest, starting from a situation like that of a complex material in which its properties are dominated by local potential minima in its energy landscape.[46,49] A liquid in normal conditions experiences local dynamics in the interaction basins surrounding the minima, and rearranges *via* intra-basin motions and relatively infrequent inter-basin jumps. As the temperature decreases approaching the glass transition, like for the clustering process dominating the complex material dynamics, such jump dynamics become dominant with respect to the intra-basin dynamics and in addition, the molecular interactions impose a levelling in the energy barriers. At this time the dynamics is inverted from super-Arrhenius to pure-Arrhenius, giving rise to a dynamical crossover characterizing the glass-forming liquids.[50]

However, also in the actual case of the transport parameter illustrated in Fig. 5a, the observed behavior can be coherently compared with the case of the thermodynamical functions in terms of the entropic behavior by means of a simple relation originally developed to study glass-forming supercooled liquids: the well known Adam–Gibbs equation.[48] The equation is developed by considering the reduction in the configurational space as the liquid cools by predicting that the configurational entropy, S_C (a sort of measure of the local order sampled by the liquid molecules) is related to the self-diffusion constant D_S, as:

$$\frac{1}{D_S} = \frac{1}{D_{S0}} \exp\left(C/TS_C\right) \tag{2}$$

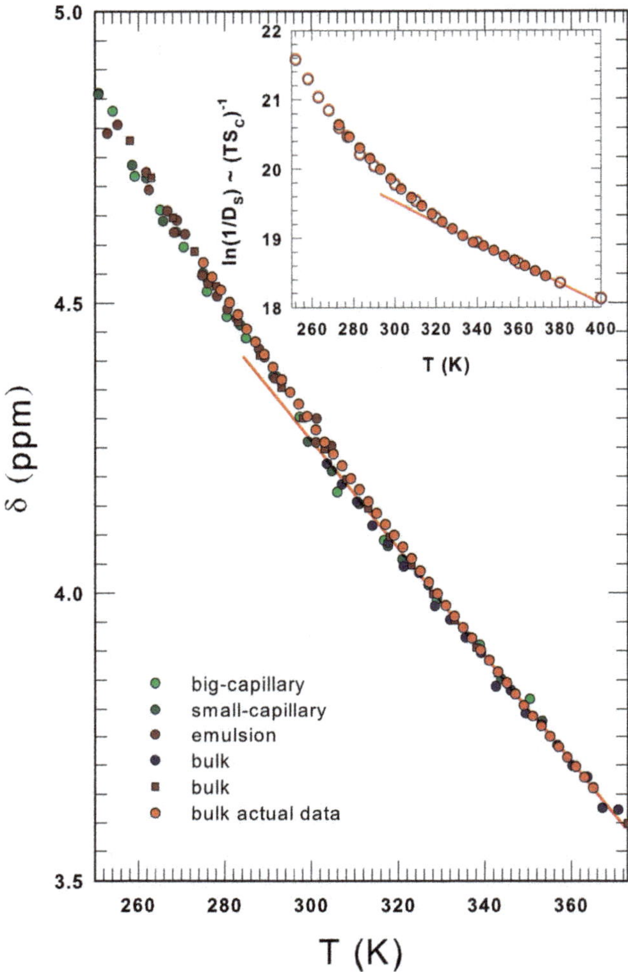

Fig. 6 The bulk water chemical shift $\delta(T)$ measured at ambient pressure in the range 250–370 K in bulk water and in three different samples: big (80–120 μm) and small (10–20 μm) capillaries, and water confined in an emulsion.[59] The inset shows the results of the Adam–Gibbs linearity relation between $\ln(1/D_S)$ and $(TS_C)^{-1}$.[48] As described in the text both $\delta(T)$ and S_C represent the atomic local order.

where $1/D_{S0}$ is a prefactor and C a constant. If we assume that the Adam–Gibbs equation is valid, in the water case also at high temperature we can treat the self-diffusion data of Fig. 5a in terms of the suggestions coming out from the overall behavior characterizing K_T, C_P and α_P for which the HB molecular order governs the water physics. Such an idea is also supported by a MD calculation, in terms of the SPC/E potential, of the configurational entropy at points spanning a large region of the T–ρ plane ($210 < T < 300$ K and $0.9 < \rho < 1.4$ g cm^{-3}), for which the water diffusive dynamics is essentially governed by the S_C.[51] Hence, by using the latter equation we estimate the S_C behavior by using the D_S data of Fig. 5a. The results are plotted in Fig. 5b where an expected behavior can be observed: S_C mimics the P–T changes of D_S. The surprising situation comes out from the comparison with the MD simulation results: the measured $S_C(P, T)$ confirms all the findings of the MD evaluation of $S_C(\rho, T)$. First of all these results ensure the validity also for the high-T regime of the linear regression between $\ln D_S$ and $(T_{S_C})^{-1}$ theoretically proposed and experimentally verified in the supercooled regime in a smaller D_S range.[52] In addition, the $S_C(P, T)$ behavior illustrated in Fig. 5b gives us simple and direct information on the way in which the local order (and hence the HB) changes also showing that T^* represents, even for the configurational entropy, a crossover from two very different regimes. For $T < T^*$ we also see that, in this case, the crossover pressure (P_{cross}) marks two regions of different behavior: by decreasing the pressure from the highest values, the configurational entropy progressively increases with a maximum just around P_{cross} and below which it decreases. Whereas for $T > T^*$, S_C essentially increases by decreasing P. This is a further good representation of the previous results giving a direct illustration on the role of the order (or the disorder) played in all regions of the liquid water phase diagram. In other words the $S_C(P, T)$ behavior illustrated here confirms that ordered HB structures are possible in the water liquid phase only below T^*.

In this section by using the NMR technique (pulsed field gradient spin echo (PGSE) method) by means of a 700 MHz spectrometer we have detailed the self-diffusion coefficient D_S data at ambient pressure in the range $273 < T < 373$ K with steps of 5 K. This is for a direct comparison with the S_C behavior coming out from the measurement, by means of the same technique, of the proton local order discussed in the next section.

2.5 The proton local order

The previous discussion has been based on the consideration that from the self-diffusion data we can obtain information on the local order of a liquid system. There is however another experimental approach that probes more directly the local order observed by a single atom of a given material: the Nuclear Magnetic Resonance (NMR) and more precisely the proton chemical shift δ. It is demonstrated that, if an isolated water molecule in a dilute gas is taken to be the reference for δ, the chemical shift represents the effect of the interaction of water with the surroundings providing, in particular, a rigorous picture of the intermolecular geometry.[53] More precisely, it is widely accepted that δ represents the average number of hydrogen bonds (HB) in which a water molecule is involved at a certain temperature.

It is well known that the chemical shift δ is an assumed linear response of the electronic structure of a system under investigation to an external magnetic field B_0, as $B(j) = (1 - \delta_j)B_0$, where j is an index identifying the chemical environment.[54,55] It is measured in an NMR experiment by the free induction decay (FID), and specifically it is related to the magnetic shielding tensor σ, which in turn relates to the local field experienced by the magnetic moment of the observed nucleus. The magnetic shielding tensor σ, strongly dependent on the local electronic environment, is a useful probe of the local geometry; and in particular for the hydrogen bond structure for water and aqueous systems and solutions.[56] On this basis, δ is experimentally obtained by means of a precise procedure.[53,57] In order to verify the way in which the water local order evolves around T^*, we have performed an NMR experiment in bulk water in the interval $275 < T < 365$ K, at $P =$ 1 bar. This is the first experiment of a series of studies on $\delta(P, T)$ in the entire phase diagram in which the thermodynamical response functions considered here have been studied.

Fig. 6 shows our $\delta(T)$ data (red dots) measured for the present study, after the proper corrections for the density and magnetic susceptibility $\chi(T) = \chi_0\rho(T)$. Fig. 6 together with our data, also shows all the experimentally available $\delta(T)$ data in the temperature range of stable bulk liquid water,[58] as well as the δ values from $T =$ 370 K down to 250 K, of three different samples: big (80–120 μm) and small (10–20 μm) capillaries, and water confined in an emulsion.[59] It can be seen in the figure that there is good agreement, within experimental error, between our data and the previous $\delta(T)$ measurements. From the data behavior as a function of the temperature, a linear behavior can be observed only in the range 320–370 K (the red line is generated by a data fitting in the range 325–370 K), for the lowest temperatures a significant deviation from the linear behavior can be observed that means an increase in the water local order. In the inset of the Fig. 6 we have shown a plot of the $\ln(1/D_S) \sim (TS_C)^{-1}$ *versus* T, evaluated in the range 250–400 K. As seen, the corresponding data from 320 K to 400 K are easily linearly fitted obtaining the red line shown. The correspondence between these results represents, in our opinion, a significant proof from the local atomic point of view, that T^* really represents for water the crossover, by decreasing T, from a normal fluid to the complex liquid characterized by the many anomalies.

3 Conclusions

The reported picture, derived from the thermodynamical functions, transport and local order data, represents the aspect of the chemical-physical reality that characterizes the thermodynamical and structural properties of bulk water by means of the singular temperature and the crossover pressure, respectively T^* and P_{cross}. However the importance of T^* in water can be fully evaluated only by considering in a unitary way all the studied quantities. In particular, on looking to the atomic local order we can clarify that, from the structural point of view, T^* may be the onset temperature of the HB clustering, the point at which liquid water becomes a complex material. In addition, the experimental data, the large P–T phase diagram, and the thermodynamic consistency shown in eqn (1), all indicate that T^* plays a primary role in the physics of water and is the source of its anomalies.

Acknowledgements

The research in Messina is supported by the University of Messina. CC thanks the Fondazione Frisone for its support. The research at Boston University is supported by the NSF Chemistry Division (grants CHE 1213217, CHE 0911389 and CHE 0908218).

References

1 *Water Polymorphism: Advances in Chemical Physics*, ed. H. E. Stanley, Wiley, NY, 2012.
2 F. Mallamace, C. Corsaro and H. E. Stanley, *Proc. Natl. Acad. Sci. U. S. A.*, 2013, **110**, 4899–4904.
3 O. Mishima, L. D. Calvert and E. Whalley, *Nature*, 1985, **314**, 76–78.
4 F. Mallamace, P. Baglioni, C. Corsaro, J. Spooren, H. E. Stanley and S. H. Chen, *La Rivista del Nuovo Cimento*, 2011, **34**, 253–388.
5 R. J. Speedy and C. A. Angell, *J. Chem. Phys.*, 1976, **65**, 851–858.

6 S. Sastry, P. G. Debenedetti, F. Sciortino and H. E. Stanley, *Phys. Rev. E: Stat. Phys., Plasmas, Fluids, Relat. Interdiscip. Top.*, 1996, **53**, 6144–6154; H. E. Stanley, *J. Phys. A*, 1979, **12**, L329–L337.
7 P. H. Poole, F. Sciortino, U. Essmann and H. E. Stanley, *Nature*, 1992, **360**, 324–328.
8 C. Huang, *et al.*, *Proc. Natl. Acad. Sci. U. S. A.*, 2009, **106**, 15214–15218.
9 G. N. I. Clark, G. L. Hura, J. Teixeira, A. K. Soper and T. Head-Gordon, *Proc. Natl. Acad. Sci. U. S. A.*, 2010, **107**, 14003–14007.
10 D. T. Limmer and D. Chandler, *J. Chem. Phys.*, 2011, **135**, 134503.
11 V. Molinero and E. B. Moore, *J. Phys. Chem. B*, 2009, **113**, 4008.
12 E. B. Moore and V. Molinero, *Nature*, 2011, **479**, 506.
13 F. Sciortino, I. Saika-Voivod and P. H. Poole, *Phys. Chem. Chem. Phys.*, 2011, **13**, 19759.
14 Y. Liu, A. Z. Panagiotopoulos and P. G. Debenedetti, *J. Chem. Phys.*, 2009, **131**, 104508; Y. Liu, J. C. Palmer, A. Z. Panagiotopoulos and P. G. Debenedetti, *J. Chem. Phys.*, 2012, **137**, 214505.
15 D. Banerjee, S. N. Bhat, S. V. Bhat and D. Leporini, *Proc. Natl. Acad. Sci. U. S. A.*, 2009, **106**, 11448–11452.
16 S. H. Chen, L. Liu, E. Fratini, P. Baglioni, A. Faraone and E. Mamontov, *Proc. Natl. Acad. Sci. U. S. A.*, 2006, **103**, 9012–9016.
17 D. A. Turton, J. Hunger, G. Hefter, R. Buchner and K. Wynne, *J. Chem. Phys.*, 2008, **128**, 161102–5.
18 D. A. Turton, C. Corsaro, M. Candelaresi, A. Brownlie, K. R. Seddon, F. Mallamace and K. Wynne, *Faraday Discuss.*, 2011, **150**, 493–504.
19 L. Liu, S. H. Chen, A. Faraone, C. W. Yen and C. Y. Mou, *Phys. Rev. Lett.*, 2005, **95**, 117802.
20 S. H. Chen, F. Mallamace, C. Y. Mou, M. Broccio, C. Corsaro, A. Faraone and L. Liu, *Proc. Natl. Acad. Sci. U. S. A.*, 2006, **103**, 12974–12978.
21 F. Mallamace, M. Broccio, C. Corsaro, A. Faraone, D. Majolino, V. Venuti, L. Liu, C. Y. Mou and S. H. Chen, *Proc. Natl. Acad. Sci. U. S. A.*, 2007, **104**, 424–428.
22 F. Mallamace, *Proc. Natl. Acad. Sci. U. S. A.*, 2009, **106**, 15097–15098.
23 P. Kumar, S. V. Buldyrev, S. R. Becker, P. H. Pool, F. W. Starr and H. E. Stanley, *Proc. Natl. Acad. Sci. U. S. A.*, 2007, **104**, 9575–9579.
24 L. Xu, F. Mallamace, Z. Y. Yan, F. W. Starr, S. V. Buldyrev and H. E. Stanley, *Nat. Phys.*, 2009, **5**, 565–569.
25 F. Mallamace, C. Branca, C. Corsaro, N. Leone, J. Spooren, H. E. Stanley and S. H. Chen, *J. Phys. Chem. B*, 2010, **114**, 1870–1878.
26 P. W. Bridgman, *Proc. Am. Acad. Arts Sci.*, 1912, **47**, 441–558.
27 T. Grindley and J. E. Lind, *J. Chem. Phys.*, 1971, **54**, 3983–3989.
28 G. S. Kell, *J. Chem. Eng. Data*, 1975, **20**, 97–105.
29 G. S. Kell and E. Whalley, *J. Chem. Phys.*, 1975, **62**, 3496–3503.
30 C. M. Sorensen, *J. Chem. Phys.*, 1983, **79**, 1455–1461.
31 D. E. Hare and C. M. Sorensen, *J. Chem. Phys.*, 1986, **84**, 5085–5089.
32 D. E. Hare and C. M. Sorensen, *J. Chem. Phys.*, 1987, **87**, 4840–4845.
33 O. Mishima, *J. Chem. Phys.*, 2010, **133**, 144503.
34 F. Mallamace, C. Branca, M. Broccio, C. Corsaro, C. Y. Mou and S. H. Chen, *Proc. Natl. Acad. Sci. U. S. A.*, 2007, **104**, 18387–18391.
35 R. J. Speedy and C. A. Angell, *J. Chem. Phys.*, 1976, **65**, 851–858.
36 H. Kanno and C. A. Angell, *J. Chem. Phys.*, 1979, **70**, 4008–4016.
37 W. D. Wilson, *J. Acoust. Soc. Am.*, 1959, **31**, 1067–1072.
38 C. W. Lin and J. P. M. Trusler, *J. Chem. Phys.*, 2012, **136**, 094511.
39 E. Tombari, C. Ferrari and G. Salvetti, *Chem. Phys. Lett.*, 1999, **300**, 749–751.
40 K. R. Harris and P. J. Newitt, *J. Chem. Eng. Data*, 1997, **42**, 346–348.
41 NIST Chemistry WebBook. http://webbook.nist.gov/chemistry/fluid/2008.
42 F. Mallamace, C. Corsaro and H. E. Stanley, *Sci. Rep.*, 2012, **2**, 993.
43 K. Binder and W. Kob, *Glassy Materials and Disordered Solids*, World Scientific, River Edge, NJ, 2005.
44 J. Jonas, *Science*, 1982, **216**, 1179–1184.
45 H. Tyrrell and K. Harris, *Diffusion in liquids: A Theoretical and Experimental Study*, Butterworth Publishers, London, Boston, 1980.
46 S. Sastry, P. G. Debenedetti and S. H. Stillinger, *Nature*, 1998, **393**, 554–557.
47 V. Lubchenko and P. Wolynes, *Annu. Rev. Phys. Chem.*, 2007, **58**, 235–266.
48 G. Adam and J. Gibbs, *J. Chem. Phys.*, 1965, **43**, 139–146.
49 F. H. Stillinger, *Science*, 1995, **267**, 1935–1939.
50 F. Mallamace, C. Branca, C. Corsaro, N. Leone, J. Spooren, S. H. Chen and H. E. Stanley, *Proc. Natl. Acad. Sci. U. S. A.*, 2010, **107**, 22457–22462.
51 A. Scala, F. W. Starr, E. La Nave, F. Sciortino and H. E. Stanley, *Nature*, 2000, **406**, 166–169.
52 C. A. Angell, E. D. Finch, L. A. Woolf and P. Bach, *J. Chem. Phys.*, 1976, **65**, 3063–3066.

53 M. Matubayasi, C. Wakai and M. Nakahara, *Phys. Rev. Lett.*, 1997, **78**, 2573–2576.
54 E. M. Purcell, H. C. Torrey and R. V. Pound, *Phys. Rev.*, 1946, **69**, 37.
55 F. Bloch, *Phys. Rev.*, 1946, **70**, 460.
56 E. D. Becker, in *Encyclopedia of Nuclear Magnetic Resonance*, ed. D. M. Grant and R. K. Harris, Wiley, Chichester, 1996, p. 2409.
57 F. Mallamace, C. Corsaro, M. Broccio, C. Branca, N. González-Segredo, J. Spooren, S.-H. Chen and H. E. Stanley, *Proc. Natl. Acad. Sci. U. S. A.*, 2008, **105**, 12725–12729.
58 J. C. Hindman, *J. Chem. Phys.*, 1966, **44**, 4582–4592.
59 C. A. Angell, J. Shuppert and J. C. Tucker, *J. Phys. Chem.*, 1973, **77**, 3092–3099.

Faraday Discussions RSC Publishing

DISCUSSIONS

General discussion

DOI: 10.1039/c3fd90036c

Professor Molinero opened the discussion of the paper by Professor Tanaka: You show that a two-state model based on locally favored structures leads to a good representation of the anomalous behavior of TIP4P/2005 water at high temperatures but displays significant deviations in the low-temperature region. Does the the two-state athermal mixture model of Anisimov and coworkers[1,2] lead to a more accurate description of the low temperature behavior of TIP4P/2005 water?

1 V. Holten and M. A. Anisimov, Entropy-driven liquid–liquid separation in supercooled water, *Sci. Rep.*, 2012, **2**, 713.
2 V. Holten, D. T. Limmer, V. Molinero, and M. A. Anisimov, Nature of the anomalies in the supercooled liquid state of the mW model of water, *J. Chem. Phys.*, 2013, **138**, 174501.

Professor Tanaka responded: This is a very interesting issue concerning the physical origin of the cooperativity in the formation of locally favoured structures. In principle, the cooperativity should be a consequence of both energetic and entropic factors. Our model regards it to be mainly energy-driven,[1–3] whereas Anisimov and his coworkers regard it to be mainly entropy-driven.[4,5] We agree that there is a possibility that the latter model might provide a more accurate description for our numerical simulation study of TIP4P/2005.[6] However, since the location or even the presence of the second critical point is not clear for this model,[7–9] we cannot tell at this moment which model is more relevant to the physical description of water. Revealing the origin of cooperativity is key to our physical understanding of liquid–liquid transitions and thus this problem should be addressed properly in the future.

1 H. Tanaka, Thermodynamic anomaly and polyamorphism of water, *Europhys. Lett.*, 2000,**50**, 340–346.
2 H. Tanaka, Simple physical model of liquid water, *J. Chem. Phys.*, 2000, **112**, 799–809.
3 H. Tanaka, Bond orientational order in liquids: Towards a unified description of water-like anomalies, liquid–liquid transition, glass transition, and crystallization, *Eur. Phys. J. E*, 2012, **35**, 113.
4 V. Holten, V. and M. A. Anisimov, Entropy-driven liquid–liquid separation in supercooled water, *Sci. Rep.*, 2012, **2**, 713.
5 V. Holten, D. T. Limmer, V. Molinero, and M. A. Anisimov, Nature of the anomalies in the supercooled liquid state of the mW model of water, *J. Chem. Phys.*, 2013, **138**, 174501.
6 J. Russo, H. and Tanaka, Understanding water's anomalies with locally favored structures, 2013, arXiv:1308.4231.
7 J. L. F. Abascal, and C. Vega, Widom line and the liquid–liquid critical point for the TIP4P/2005 water model, *J. Chem. Phys.*, 2010, **133**, 234502.
8 D. T. Limmer, and D. Chandler, The putative liquid–liquid transition is a liquid–solid transition in atomistic models of water. II, *J. Chem. Phys.*, 2013, **138**, 214504.

9 S. D. Overduin, and G. N. Patey, An analysis of fluctuations in supercooled TIP4P/2005 water, *J. Chem. Phys.*, 2013, **138**, 184502.

Professor Chandler remarked: You have described for us your perspectives on the effects of local structures in super-cooled liquids, especially as they pertain to water and aqueous solutions. I appreciate the clarity of your presentation, which leads me to make a comment and to ask related questions. There is no question that water exhibits local structure that, at low enough temperatures, leads to crystallization. Fluctuations intrinsic to that phase transition (*i.e.*, coarsening of ice) can be arrested through rapid cooling (*i.e.*, hyper quenching). Such arrest produces amorphous ice. These phenomena are all about dynamics of fluctuations. Therefore, I believe that treatments neglecting fluctuations cannot correctly describe how super-cooled water evolves to either glass or crystal states.

The various two-state models presented in your lecture are mean-field models that largely neglect fluctuations. Thus, for example, the critical point and liquid–liquid phase transition exhibited in your phase diagram for the pentamer-forming system seems suspect because a mean-field approximation is employed to derive those features. Is there compelling evidence from numerical simulation that an exact analysis of that particular system would yield those features that you derive by neglecting fluctuations?

Similarly, appropriate order parameters distinguishing liquid from glass must depend upon the time evolution of fluctuations because the state of a glass depends upon the path by which the glass is prepared. Thus, I believe that treatments neglecting the temporal nature of fluctuations cannot correctly describe the formation and stability of glass. Yet the discussion of glass presented in your lecture focuses on ordering associated with local structure without reference to the dynamics during the process of forming the glass. It would seem that such a picture could not be predictive. Is there evidence that average local structure can serve as an order parameter for characterizing glass?

Professor Tanaka replied: We agree that fluctuation effects are certainly important especially if there is criticality. When we describe water's anomalies in terms of our mean-field two state model, critical fluctuation effects are probably not so important since the cooperativity plays few roles. This is because the experimentally accessible region of water is far from the critical point even if it exists. In our 2D spin liquid simulations,[1] where fluctuation effects are naturally included, we found that our mean-field two state model well describes the temperature dependence of the fraction of pentagons.[2] We also found that the mean-field two-state model reasonably describes water's anomalies except at a very low temperature for TIP4P/2005 water by simulations, in which fluctuation effects are naturally incorporated.[3] This may be due to the Gaussian nature of fluctuations in a quasi-equilibrium state. However, when we study structural ordering in model glass-forming systems (*e.g.*, polydisperse hard spheres and 2D spin liquids), we found that the growth of the spatial correlation length of the relevant order parameter obeys the Ising-like criticality rather than the mean-field one.[4] This strongly suggests the importance of fluctuation effects and their renormalization is essential for the description of the phenomena. Furthermore, if we discuss (non-averaged) local quantities, fluctuation effects should be quite significant. When we consider the nonequilibrium process of vitrification, as pointed out by Prof. Chandler, the time evolution of fluctuations becomes an important issue, since the strong non-Gaussian nature of the fluctuations may appear. We found[2,4] that the kinetics in the above-mentioned glass-forming systems obey the model A type dynamical equation in the Hohenberg–Halperin classification.[5] In principle, we may describe the kinetic pathway of vitrification using this dynamical equation. However, the transport coefficient itself should be a strong function of the structural order parameter in this framework, which makes the problem very difficult. Furthermore, how to write down the fluctuation–dissipation theorem for such a situation is far from being obvious. Nevertheless, we recently found in a numerical study of ageing of polydisperse hard spheres, the averaged structural relaxation time can be characterized solely by the correlation length of the structural order parameter at any stage of ageing, at least approximately.[6] So we expect that even for this type of liquid, the state of a supercooled liquid or glass may approximately be characterised by the correlation length of the structural order parameter as long as we concern an averaged quantity. If we consider the distribution of the relaxation time, however, such a simple link may no longer be valid due to non-trivial fluctuation effects. Finally, we should note that whether the structural order parameter is relevant to the description of slow glassy dynamics is a matter of debate and thus further careful study is necessary on such a correlation.

1 H. Shintani, and H. Tanaka, Frustration on the way to crystallization in glass, *Nature Phys.*, 2006, **2**, 200–206.
2 H. Tanaka, Bond orientational order in liquids: Towards a unified description of water-like anomalies, liquid–liquid transition, glass transition, and crystallization, *Eur. Phys. J. E*, 2012, **35**, 1.
3 J. Russo, and H. Tanaka, Understanding water's anomalies with locally favored structures, 2013, arXiv:1308.4231.
4 H. Tanaka, T. Kawasaki, H. Shintani, and K. Watanabe, Critical-like behaviour of glass-forming liquids, *Nature Mater.*, 2010, **9**, 324–331.
5 P. C. Hohenberg, and B. I. Halperin, Theory of Dynamic Critical Phenomena, *Rev. Mod. Phys.*, 1977, **49**, 435–479.
6 T. Kawasaki, and H. Tanaka, to be published.

Dr Limmer asked: Firstly, let me reemphasize a point made by Prof. Chandler. Two state models of the form Prof. Tanaka have presented, as well as those used often in the study of supercooled water, adopt a non-unique partitioning of a continuum of configurations onto a basis of two average local structures. Neglecting fluctuations between these states as well as fluctuations into others has been shown previously to destabilize the liquid[1,2] and qualitatively change the equation of state within the supercooled regime.[3] Therefore, an accurate description of water, or its solutions, must necessarily include such fluctuations.

Secondly, I would like to comment on Prof. Tanaka's experiments on mixed aqueous glycol solutions and specifically his conclusion that his observations support the idea that the mixture undergoes a first-order, reversible liquid–liquid transition with constant composition. The way in which he has characterized these materials assumes that 1) the systems are homogeneous and 2) they are reflective of a metastable equilibrium. However, the presence of ice crystallites that form and coarsen over long times result in a material that is far from homogeneous. Attempts to subtract the effects of these ordered domains from the material must not only include their direct contribution to the material properties but also the effects from the significant amount of interface that is generated under such arrested growth. Prof. Tanaka's analysis neglects this latter contribution. Moreover, the materials that result from the described quenching protocol are out-of-equilibrium and kinetically arrested. This irreversibility means that their properties necessarily depend on the path by which they were prepared. Therefore, conclusions relating to the reversible phase behavior of these materials cannot be addressed.

I submit that an alternative explanation for these observations can be formulated by simply considering two physical processes: 1) frustrated crystal growth due to solute composition fluctuations and 2) super-Arrhenious temperature dependence of the diffusion constant within the supercooled solution. Both of these phenomena are well known and an explanation based on these processes does not invoke a transition that has not yet been directly observed. Calculations based on this hypothesis can be easily performed by adopting a minimal model dynamics that captures this physics and parameterizing it with the microscopic length, energy and timescales appropriate for aqueous glycerol solutions. Specifically, nucleation coupled phase separation is well described by a model first proposed by Rabani *et al*,[4] the super-Arrhenious temperature dependence of the diffusion constant can be quantified using the parabolic law.[5] In this model the liquid is represented as a binary field on a grid, being either locally liquid-like or ice-like, and solutes can only diffuse through liquid-like cells. Parameterizing the model requires only known experimental observables such as the liquid-ice surface tension, melting temperature, *etc.* Details will be described elsewhere.[6]

This model yields a phase diagram in quantitative agreement with the observations of Prof. Tanaka. As an example, Fig. 1 shows snapshots taken from

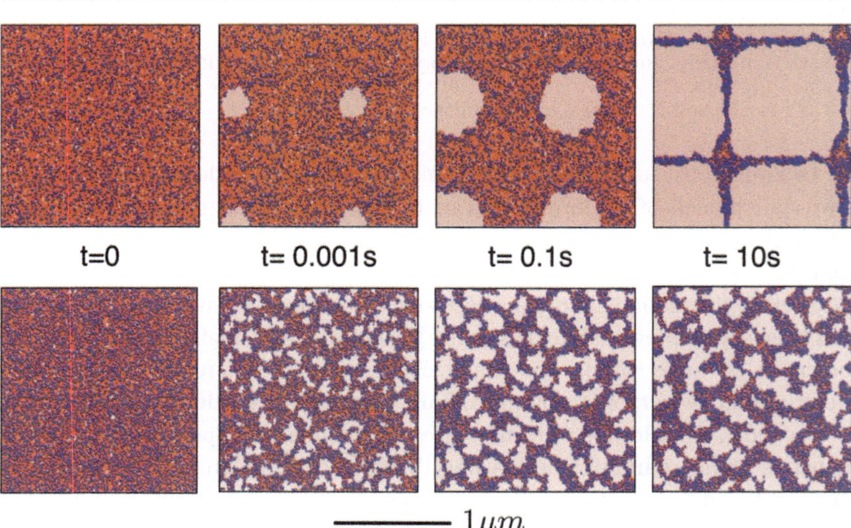

Fig. 1 Nucleation and growth pathways for a two component lattice gas, where blue cells represent solute rich regions and the red cells represent liquid-like regions and the white cells are ice-like regions. The top four panels are a sequence of configurations from a trajectory initialized in the mixed state and quenched instantaneously to below the coexistence temperature at fixed concentration. The bottom four panels are also a sequence of configurations from a trajectory initialized in the mixed state and quenched below the temperature of liquid stability. Periodic boundary conditions are employed, and the images are one and a half periodic boxes.

two different trajectories, initiated in the mixed liquid phase and quenched instantaneously to lower temperatures. In both cases, nucleation of ice crystallites occur. In the top case, where the quench is to moderate supercooling, growth of large domains from initially dilute critical nuclei is facile. Crystallization is dominated by nucleation. In the bottom case, where the quench takes the systems below the regime of stability for the liquid, growth of many critical nuclei is frustrated by slow composition fluctuations. Crystallization is dominated by coarsening, which proceeds by the rearrangement of grain boundaries. Note that in both cases there is a significant amount of interface formation, whose effects are neglected in Prof. Tanaka's analysis. These panels are remarkably similar to those presented by Prof. Tanaka and published in ref. 7 for the same effective conditions. More importantly, the emergent time and length scales are also in good agreement with Prof. Tanaka's observations. This picture offers a simpler explanation of his results than assuming the existence of a liquid–liquid phase transition.

1 D. T. Limmer, and D. Chandler, The putative liquid–liquid transition is a liquid–solid transition in atomistic models of water, *J. Chem. Phys.*, 2011, **135**, 134503.
2 D. T. Limmer, and D. Chandler, The putative liquid–liquid transition is a liquid–solid transition in atomistic models of water. II, *J. Chem. Phys.*, 2013, **138**, 214504.
3 V. Holten *et al.*, Nature of the anomalies in the supercooled liquid state of the mW model of water, *J. Chem. Phys.*, 2013, **138**, 174501.
4 E. Rabani, *et al.*, Drying-mediated self-assembly of nanoparticles, *Nature*, 2003, **426**, 271–274.
5 Y. S. Elmatad, D. Chandler, and J. P. Garrahan, Corresponding states of structural glass formers, *J. Phys. Chem. B*, 2009, **113**, 5563–5567.
6 D. T. Limmer, in preparation.
7 K.-I. Murata, and H. Tanaka, Liquid–liquid transition without macro- scopic phase separation in a water–glycerol mixture, *Nature Mater.*, 2012, **11**, 436–443.

Professor Tanaka remarked: We agree that fluctuation effects are important. However, these effects are automatically included in the free energy functional except for the correction to the cooperativity term. We note that the cooperativity term itself actually turns out not to be so important in the description of water's anomalies, at least in the experimentally accessible region.[1–3] Indeed, we showed in our recent study on TIP2P/2005 water[4] that if we define a proper structural order parameter, which is translational order of the second shell, the distribution function of the order parameter can be decomposed into two Gaussians corresponding to normal-liquid and locally favoured structures. The broadness of the distribution is due to thermal fluctuation effects. From the decomposition, we can estimate the fraction of locally favoured structures, S, study the dependence of S on temperature and pressure, and compare it with the prediction of the two-state model. In this way, we have confirmed that the two-state description can explain water's anomalies in a very natural manner. However, I agree that for a problem in which the cooperativity plays a crucial role, we need careful consideration on the roles of fluctuation effects. On the interpretation of the ordering process we observed in water–glycerol mixtures, we agree that there is a subtlety since this process always involves the formation of nm-size cubic ices.[5] Hereafter we explain our view on this problem, responding to the comments. First we agree that the contribution of the interfaces may be important. Feldman and his coworkers also interpreted what we call liquid II as a mixture of the solute-rich liquid I phase, ices, and interfacial water around ices, on the basis of dielectric spectroscopy measurements.[6] In relation to this, we consider the glass transition behaviour of what we call liquid II. For this state we found a single glass transition (see Fig. 2d in ref. 5), which is located at a temperature significantly higher than that of liquid I having the same glycerol concentration: T_g of a glycerol solution at $c \sim 0.18$ prepared at 164 K is about the same as T_g of liquid I at $c \sim 0.33$. We know the mole fraction of cubic ices is 0.14 from X-ray scattering measurements for this sample.[5] If we assume that the system is composed of ice crystals, interfacial (pure) water, and liquid I with glycerol ($c = 0.33$), we may ascribe the glass transition assigned as T_g of liquid II to that of liquid I with glycerol. Then, the mole fractions of these three components can be estimated as 0.14, 0.31, and 0.55, respectively. On noting that the fraction of interfacial water is substantial, we expect the presence of two T_gs (one for glycerol-enriched water and the other for interfacial water), but actually we observed only one T_g. Furthermore, the thickness of the interface is only a few molecules in ordinary conditions far from the melting point (see, *e.g.*, ref. 7 for the case of water–LiCl mixtures), which is not consistent with the above estimate that suggests a much thicker interface. On the experimental protocol, we emphasize that the final state is a mixture of liquid II and nm-size cubic ices and thus in an ergodic liquid state and not in a glassy state. We confirmed that this state can last for more than hours under certain conditions and be stationary (the fraction of cubic ices is almost constant with time in the late stage). The model proposed by Limmer and Chandler is an interesting one: it resembles our two-order-parameter model of the liquid–liquid transition[8] in the sense that if we replace locally favoured structures by ice crystals the basic theoretical structure is similar. In their model, however, the cooperativity comes from the interaction between crystals and interfacial water, which leads to one crucial difference. During the nucleation–growth-type transformation process, glycerol, unfriendly to ices, is expelled from the ice-rich phase (white domains in the figure of Dr Limmer's comment) and thus accumulated in the matrix phase. This is because ices are more friendly to interfacial water than glycerol, which leads to the cooperative formation of ice crystals. In our model, the concentration of glycerol is almost homogeneous besides its slight enrichment due to ice crystallization. We recently confirmed by micro-Raman spectroscopy that the glycerol concentration is the same between droplets and the matrix. This indicates the spatial distribution of glycerol is rather uniform, which seems to support our scenario. More evidence supporting our scenario comes from the following fact: by combining X-ray scattering and ac DSC measurements, we previously confirmed that the amount of ice Ic in liquid II decreases with decreasing the annealing temperature T_a and liquid II become less fragile (or, stronger) (see Supplementary Information in ref. 5 for the details). In other words, the difference in the fragility between liquid I and II becomes more pronounced when liquid II contains less cubic ices. In the above-mentioned scenarios, however, after transformation the system should increase the fragility towards the fragility of pure liquid I with decreasing T_a, as a consequence of less cubic ice formation and the resulting decrease of interfacial water. Thus, these scenarios seem difficult to explain why liquid II is stronger when it is more pure. On the basis of these considerations, we conclude at this moment that the transition we observe should be LLT. However, since there remains a subtlety due to the lack of clear-cut evidence exclusively supporting our scenario, further careful study may be necessary to settle this issue and form a consensus.

1 H. Tanaka, Thermodynamic anomaly and polyamorphism of water, *Europhys. Lett.*, 2000, **50**, 340–346.
2 H. Tanaka, Simple physical model of liquid water, *J. Chem. Phys.*, 2000, **112**, 799–809.
3 H. Tanaka, Bond orientational order in liquids: Towards a unified description of water-like anomalies, liquid–liquid transition, glass transition, and crystallization, *Eur. Phys. J. E*, 2012, **35**, 113.
4 J. Russo, and H. Tanaka, Understanding water's anomalies with locally favored structures, 2013, arXiv:1308.4231.
5 K. Murata, and H. Tanaka, Liquid–liquid transition without macroscopic phase separation in a water–glycerol mixture, *Nature Mater.*, 2012, **11**, 436–443.
6 Y. Hayashi, A. Puzenko, and Y. Feldman, Ice nanocrystals in glycerol–water mixtures, *J. Phys. Chem. B*, 2005, **109**, 16979–16981.
7 G. Bullock, and V. Molinero, Low-density liquid water is the mother of ice: on the relation between mesostructure, thermodynamics and ice crystallization in solutions, *Faraday Discuss.*, 2013, **167**, DOI: 10.1039/C3FD00085K.
8 H. Tanaka, General view of a liquid–liquid phase transition, *Phys. Rev. E*, 2000, **62**, 6968–6976.

Professor Wynne commented: Based on your two-state model of liquid–liquid phase transitions,[1,2] you have previously predicted[3] that LLPTs could occur in any (molecular) liquid. Why is it that the only molecular liquid with a known LLPT is triphenyl phosphite? Would it be possible to predict in which other liquids LLPTs could be expected?

1 H. Tanaka, *J. Chem. Phys.*, 2000, **112**, 799–809.
2 H. Tanaka, *Faraday Discuss.*, 2013, **167**, DOI: 10.1039/C3FD00110E.
3 H. Tanaka, *Phys. Rev. E*, 2000, **62**, 6968–6976.

Professor Tanaka responded: Our model predicts that if there is cooperativity in the formation of locally favoured structures, there should be a liquid–liquid transition. Since it is rather hard to imagine that locally favoured structures are formed completely independently without any cooperativity, we expect that liquid–liquid transitions can exist in many liquids.[1,2] However, the energy scale of the coupling parameter J is often comparable to the energy scale controlling other cooperative ordering such as crystallization, since both are determined by the interaction potential. Thus, liquid–liquid transitions may often be hidden by crystallization,[1,2] particularly if the symmetry of locally favoured structures is similar to that of the equilibrium crystal. This is also a source of controversy concerning whether there exists liquid–liquid transitions or not in many systems including water.[3] To predict the presence of LLT from molecular or atomic structures, we require deep knowledge on the 'microscopic' mechanism of formation of locally favoured structures and its cooperativity, which is unfortunately not available at this moment. Finally we note that mixing a target liquid with another liquid that can prevent crystallization may be a good strategy to reveal such a hidden LLT, since LLTs may take place even after mixing with other fluids.[4] We employed this method in our study of LLTs of aqueous organic solutions.[5,6]

1 H. Tanaka, General view of a liquid–liquid phase transition, *Phys. Rev. E*, 2000, **62**, 6968–6976.
2 H. Tanaka, Bond orientational order in liquids: Towards a unified description of water-like anomalies, liquid–liquid transition, glass transition, and crystallization, *Eur. Phys. J. E*, 2012, **35**, 1.
3 H. Tanaka, *Faraday Discuss.*, 2013, **167**, DOI: 10.1039/C3FD00110E.
4 R. Kurita, K. Murata, and H. Tanaka, Control of fluidity and miscibility of a binary liquid mixture by the liquid–liquid transition, *Nature Mater.*, 2008, **7**, 647–652.

5 K. Murata, and H. Tanaka, Liquid–liquid transition without macroscopic phase separation in a water–glycerol mixture, *Nature Mater.*, 2012, **11**, 436–443.
6 K. Murata, and H. Tanaka, General nature of liquid–liquid transition in aqueous organic solutions, *Nature Commun.*, 2013, **4**, 2844.

Professor Angell commented: I think it would be appropriate to acknowledge Aptekar's very early prediction of a LL transition in liquid Si and Ge. He was even correct about it occurring at negative pressure.

Professor Tanaka replied: I fully agree. It is a remarkable work. Liquid–liquid transitions in Si and Ge, which sit just on the border between liquid and semiconductor, accompany a drastic change in the electronic properties, as discussed by Aptekar.[1] It may be an interesting and challenging problem how to incorporate the electronic degrees of freedom to a phenomenological two-state model.

1 L. I. Aptekar, Phase transitions in noncrystalline germanium and silicon, *Sov. Phys. Doklady*, 1979, **24**, 993–995.

Dr Soper communicated: In the graph of the diffraction pattern for glycerol–water you showed before and after the liquid–liquid transition, the main peak near 1.7 $Å^{-1}$ does not seem to shift, but simply becomes a bit sharper. Yet we know that for the transition HDA to LDA in heavy water, D_2O, this peak moves a lot, by about 0.4 $Å^{-1}$. How therefore can we be so sure this is indeed a liquid–liquid transition and not simply a freezing in of the disordered structure?

Professor Tanaka communicated in response: This is an important point. We agree that the difference in the X-ray structure factor between liquid I and II is rather subtle[1] and this alone may not be used to claim the presence of a liquid–liquid transition (LLT). However, the line shape analysis tells us that there is a large difference in the spectrum between the two.[1,2] The reasons why the difference in the structure between liquid I and II is more subtle compared to that between LDA and HDA may be explained as follows. Compared to pure water, the X-ray scattering signal is affected by the presence of glycerol molecules, which may make the difference more subtle compared to the difference between LDA and HDA. Furthermore, the difference in the density is also much smaller for LLT in a water–glycerol mixture than for pressure-induced polyamorphic transition in pure water. The X-ray scattering patterns together with dielectric spectroscopy measurements tell us that the newly formed state (what we call liquid II) is mainly composed of a liquid having a disordered structure and a slow, but finite relaxation time, although it contains a small fraction of cubic ices. We also confirmed that this liquid state is stable over several hours at least at a certain condition. Furthermore, Raman spectroscopy suggests that liquid I is HDA-like and liquid II is LDA-like. Then we also found that the glass transition point of liquid II is significantly higher than liquid I.[1] To explain this high glass transition temperature by a simple freezing scenario, we need to suppose a significant enrichment of glycerol in a liquid component. However, the estimated amount of cubic ices is not large enough to explain the amount of the shift of the glass transition point. Furthermore, we found that the less amount of cubic ices leads to the larger difference from the glass transition behaviour of liquid I.[1] From these results, we concluded that the phenomenon we found is a liquid–liquid transition. However, as discussed in my paper,[2] there is a subtlety in the interpretation of the phenomenon (see, *e.g.*, ref. 3 and 4 for scenarios of simple freezing without LLT), and thus further careful studies are certainly necessary to draw a definite conclusion.

1 K. Murata, and H. Tanaka, Liquid–liquid transition without macroscopic phase separation in a water–glycerol mixture, *Nature Mater.*, 2012, **11**, 436–443.
2 H. Tanaka, *Faraday Discuss.*, 2013, **167**, DOI: 10.1039/C3FD00110E.
3 Y. Hayashi, A. Puzenko, and Y. Feldman, Ice nanocrystals in glycerol–water mixtures, *J. Phys. Chem. B*, 2005, **109**, 16979–16981.
4 D. T. Limmer, and D. Chandler, *Faraday Discuss.*, 2013, **167**, DOI: 10.1039/C3FD00076A.

Professor Chandler opened the discussion of the paper by Pablo G. Debenedetti†: I have a comment and then a question concerning the paper that Dr Palmer presents, a paper from Professor Debenedetti and his Princeton colleagues. The paper explains disagreement with the results of Dr Limmer and myself[1] by saying that the Limmer–Chandler calculations were done at a state point outside the relevant regime for examining the possibility of a liquid–liquid transition in the ST2 model. This criticism is not accurate. Dr Limmer and I computed a free energy surface for a temperature below the putative liquid–liquid critical temperature for that model. It is a free energy surface extending over the entire range of relevant densities and crystal order parameters. As such, trivial re-weighting, as explained in our papers ref. 1 and 2, provides the free energy surface for any relevant pressures, including those for which Debenedetti and his co-workers imagine the existence of liquid–liquid coexistence. The re-weighting formula is eqn 10 of Debenedetti's paper, a formula that is also found in various forms in the Limmer–Chandler papers.[1,2] Perhaps Debenedetti and his co-workers should correct their statements regarding this issue. The discrepancies between the Princeton groups results and the Limmer–Chandler results have been explained.[2] Specically, Dr Limmer and I have shown that the results of Debenedetti and co-workers can be reproduced by limiting sampling to irreversible conditions, i.e., by not allowing the crystal order parameter, Q_6, to equilibrate. We have carried out our demonstrations with theoretical analysis, and independently, with simulation. A proper reversible sampling, we also show, removes the prediction of a second liquid phase in cold water. Details are found in our paper.[2]

The physical origin of the problem in the Princeton group's calculations is associated with long time scales of coarsening, time scales that are orders of magnitude longer than those they have considered up to this time. In the early stages of coarsening, density fluctuates widely while the progress of crystal order can seem imperceptible. There are sampling methods to overcome slow progress of crystal ordering, but no such appropriate method has yet been applied by Debenedetti and his co-workers. As a result, even with all the different methods they have applied, calculations by Debenedetti and his Princeton colleagues have been trapped in transient irreversible regimes. Agreement between Limmer–Chandler results and Debenedetti results for reversible free energy surfaces has been found for state conditions where time-scale issues do not appear.

† Prof. Debenedetti's paper was presented by Dr Jeremy Palmer, Department of Chemical and Biological Engineering, Princeton University, Princeton, USA.

Computation on the TIP4P model is one such case. At different conditions, those with corresponding states exhibiting large density fluctuations and coarsening, disagreement between the two groups will persist until the Princeton group carries out proper reversible sampling.

Using appropriate sampling methods, a common story emerges for all reasonable models of super-cooled water: there is one reversible liquid basin; near the boundary to liquid instability (i.e., near the boundary to "no-man's land"), fluctuations in density are large and slow, and ordering of the crystal is even slower. Neglecting the scales of the latter yield the illusion of a second liquid basin, but this impression is transient and irreversible. This common story is found for various models (ST2 water with any of an assortment of boundary conditions; mW water; SW silicon, TIP4P water; etc.), and there is agreement among various different research groups. (Here, I call attention to the work of Overduin and Patey,[3] and also to the interesting and newly published paper by English *et al.* who argue that no interface between two distinct phases of cold water persists when examined over large enough extents in space and time.[4]) So my question to Dr Palmer is this: What theory can you and your coworkers provide for why the Limmer–Chandler calculations are incorrect, and can you apply your theory to reproduce the Limmer–Chandler results and show why those results are found consistently for so many different models?

1 D. T. Limmer, and D. Chandler, *J. Chem. Phys.*, 2011, **135**, 134503.
2 D. T. Limmer, and D. Chandler, Part II, *J. Chem. Phys.*, 2013, **138**, 214504.
3 S. D. Overduin, and G. N. Patey, *J. Chem. Phys.*, 2013, **138**, 184502.
4 N. J. English, P. G. Kusalik and J. S. Tse, *J. Chem. Phys.*, 2013, **139**, 084508.

Dr Palmer offered the following explanation: I present my interpretation of Professor David Chandler's recent published work done in collaboration with Dr David Limmer on the liquid–liquid transition in the ST2 water model.[1,2] Although different versions of the ST2 water model were examined in our respective 2011[1] and 2012[3] studies, I show in Sec. 1 that calculations with our ST2 variant suggest that the state conditions explored in Limmer and Chandler's 2011 study[1] fall outside the predicted region of liquid–liquid coexistence, including both those states that they obtained by direct simulation and those they explored by reweighting. Using thermodynamic arguments based on experimental data for real water, I demonstrate in Sec. 2 that their simulation results suggest coexistence between liquid water and ice Ic at state conditions where it should not be observed. While the origin of the fundamental differences between Limmer and Chandler's calculations[1,2] and our own is not completely clear at present, the fact that their results do not withstand thermodynamic scrutiny, while ours clearly do, indicates that additional work is needed to validate their computational approach. Several scenarios are presented in Sec. 3 that could potentially explain the origin of the discrepancies in Limmer and Chandler's calculations. In light of such discrepancies, I argue that proper due diligence requires performing checks using a variety of different sampling techniques to ensure that consistent results are obtained. The Princeton group has taken this approach, validating our findings with eight different techniques, five of which will be described in our forthcoming publication[4] and were presented both at the Faraday Discussion and at the Liquids Gordon Conference in August. With such methods, we have extended our sampling duration by orders of magnitude with respect to our previously published results[3,5] and verified that our free energy calculations are reversible by performing bi-directional sampling between the low-density liquid and ice phase to explicitly check for irreversible behavior.[4] The results provide definitive evidence confirming that the computations presented in our contribution to Faraday Discussions are devoid of any artifacts arising from restricting sampling to the low-Q_6 region of the free energy surface, or from inadequate equilibration. Crucially, all of our findings are fully consistent with each other and support the existence of a first-order liquid–liquid phase transition in the deeply supercooled regime of the ST2 water model. In Sec. 4, I compare the relative efficiency of several of the Monte Carlo (MC) techniques used in our studies with respect to the method employed by Limmer and Chandler.[1,2] I show that their hybrid MC[6] (HMC) technique is less than one order of magnitude more efficient than the standard MC algorithm employing single-particle moves, disproving Limmer and Chandler's recent claims that it is superior by as much as three orders of magnitude.[2] The efficiency estimates I report are supported by data taken directly from Limmer and Chandler's published work,[2] as well as by a direct comparison I have made by performing simulations with both techniques. Such efficiency estimates are also in accord with other studies in the literature comparing the two methods.[7] In my analysis, I describe how this apparent confusion may have arisen from misinterpreting data[8] presented in our earlier study,[3] and I discuss how efficiency gains at least as large as those attained using HMC can be achieved using several other well-established MC techniques[9] that we have in fact implemented. Although Professor Chandler states that the discrepancies between our respective studies have been explained[2] by the theory alluded to in his comment, I do not find any evidence to support the claim that our results suffer from artifacts associated with transient, irreversible behavior. In Sec. 5, I conclude by showing that the theory posited by Dr Limmer and Professor Chandler[2] is based on assumptions that are not consistent with actual simulation results.

1. State conditions for liquid-liquid coexistence

In a 2011 study[1] examining the liquid–liquid phase transition in the ST2 water model, Dr Limmer and Professor Chandler computed the free energy surface, $F(\rho, Q_6)$, described by density, ρ, and the crystalline order parameter, Q_6. Computations were performed using the hybrid Monte Carlo[6] (HMC) algorithm, which samples phase space by propagating short molecular dynamics (MD) trajectories that are either accepted or rejected based on the Metropolis–Hastings criterion.[9] Simulations for the ST2 model[10] reported in that study were only performed at one state point,[1] 235 K and 2.2 kbar.

Dr Limmer and Professor Chandler also applied a reweighting procedure to examine $F(\rho, Q_6)$ at nearby state conditions.[1] Reweighting was performed in chemical potential at constant temperature,[1] $F(\rho, Q_6;T, \Delta\mu) = F(\rho, Q_6) - \rho\bar{V}\Delta\mu$, where $\bar{V}$ is the mean volume of the system at the simulated condition and $\Delta\mu$ is the shift in chemical potential from the original state point. Figures 6 (b) and (c) in their manuscript[1] show reweighted free energy surfaces at 235 K for $\Delta\mu = 0.55k_BT$ and $\Delta\mu = 0.27k_BT$, respectively. Reweighting using such values for $\Delta\mu$ effectively increases the pressure of the system. Using the reported values of $\Delta\mu$, we have estimated that the state conditions examined for the ST2 water in Limmer and Chandler's 2011 study[1] fall within the pressure range 2.2 to 2.9 kbar at 235 K.

In our 2012 study,[3] calculations were performed to examine the behavior of $F(\rho, Q_6)$ in the vicinity of the reported liquid–liquid phase transition. At 228.6 K and 2.2 kbar, we observed two basins in $F(\rho, Q_6)$ corresponding to a high- and low-density liquid phase, respectively.[3] We also computed the limit of stability for the low-density liquid at 235 K, the same temperature examined in Limmer and Chandler's 2011 study[1] using our variant of the ST2 water model. Our results revealed that the low density liquid is unstable for pressures larger than $\sim$2.0 kbar at 235 K.[3] Consequently, our calculations suggested that Limmer and Chandler[1] explored a range of pressures/chemical potentials where the low density liquid is unstable, providing a plausible explanation for the absence of a second liquid basin in their free energy surfaces.

At the time of submitting our contribution to Faraday Discussions, to our knowledge no evidence had been presented in the peer-reviewed literature contradicting my interpretation of the result presented in Limmer and Chandler's 2011 study.[1] During review of our Faraday manuscript, however, Limmer and Chandler published a second study[2] showing no evidence of a low-density liquid phase at state conditions where we had previously reported liquid–liquid coexistence. Thus, even though the original Limmer–Chandler study explored only state points that according to our estimate lie outside the low-density liquid's domain of stability, the reasons for the discrepancy between our respective results do not originate in this fact. As a result, I have made minor revisions to the introductory section of our manuscript to reflect the current state of the peer-reviewed literature. Professor Chandler's comments about reweighting notwithstanding, the fact remains that our calculations suggest that the reweightings performed by Limmer and Chandler in their original paper were in the wrong direction, namely increasing the pressure and thus taking what we estimate were already unstable conditions for the low-density liquid even further into the unstable regime.

2. Thermodynamic inconsistencies

In Limmer and Chandler's 2013 study,[2] they computed $F(\rho, Q_6)$ at 230 K and 2.2 kbar for the same variant of the ST2 water model used in our previous study,[3] applying identical boundary conditions to the Ewald summation used for treating long-range electrostatic interactions.[2] The free energy surface for this model is shown in Fig. 5 (b) of their paper.[2] Despite the fact that the results presented in Fig. 5 (b) were computed from HMC simulations at 230 K and 2.2 kbar,[2] the free energy surface shows the liquid and crystalline phase (ice Ic) to be near coexistence. From Fig. 5 (b), it would follow that the liquid phase is metastable by $\sim 7.5k_BT$ with respect to cubic ice (Ice Ic) at the simulated conditions, or, equivalently, $\Delta G(\text{ice Ic} - \text{liq}) \approx -67\ \text{J mol}^{-1}$.

A simple thermodynamic integration (TI) calculation for an incompressible fluid, using experimental data for the entropy and enthalpy of fusion of hexagonal ice, as well as the vapor pressure of cubic ice[11] and the enthalpy difference between cubic and hexagonal ice[12–15] yields values of -530 to $-560\ \text{J mol}^{-1}$ for the free energy difference between cubic ice and supercooled water at 228.6 K and 2.2 kbar (the state point used in our calculations). The range of free energy values corresponds to slight differences among various authors between the reported calorimetric data for the enthalpy difference between cubic and hexagonal ice, which in no way affect the present argument. While the TI calculations were performed using experimental data for water, which has a melting temperature of 273.15 K, the established melting temperature for the ST2 water model, which I have verified, is approximately 300 K.[16,17] Using this value for the melting temperature, the TI calculation yields an estimated free energy difference ΔG(ice Ic − liq) between −980 and −1003 J mol^{-1}. As I showed at the Faraday meeting, and will document in a forthcoming publication,[4] we compute a free energy difference of −740 J mol^{-1} between cubic ice and supercooled water at 228.6 K and 2.2 kbar, in very satisfying agreement with the experiment-based TI calculations. Thus, the free energy difference computed by Limmer and Chandler using the same model and at virtually identical conditions (230 *vs* 228.6 K) is one order of magnitude smaller than ours, and in conflict with TI based on experimental data. It follows from straightforward TI that in order for the free energy difference between cubic ice and supercooled water at 2.2 kbar to have the value reported by Limmer and Chandler, the temperature would have to be 262 K. Thus, the calculations reported by Limmer and Chandler in Figure 5(b) of ref. 2 do not withstand thermodynamic scrutiny.

3. Potential technical pitfalls

Based on our thermodynamic analysis, it is clear that one should not expect to find liquid water and cubic Ic near coexistence for the ST2 water model at the state conditions reported by Limmer and Chandler.[2] We note that similar concerns would therefore apply to their previous work on ST2,[1] as well as to free energy surfaces computed for other variants of the ST2 model examined in their 2013 study.[2] To obtain such a free energy surface for the ST2 variant employed in my simulations, I estimate that sampling would have to be performed at 262 K and 2.2 kbar, which is 32 K higher than the purported state point examined by Limmer and Chandler[2] and beyond the range accessible by reweighting techniques. Although the origin of the problem with their calculations is not clear, I can imagine several scenarios that could lead to such behavior.

(i) One possibility, which was posited by Dr David Limmer at the Faraday Discussion meeting, is that there might be some difference in the way the intermolecular potential for ST2 is being treated in our respective calculations. As Dr Limmer suggested, such a difference, for example, could effectively rescale the pressure, altering the melting behavior of the model. Such large deviations from the expected phase behavior, however, would imply that the same water model is not being simulated. It is unclear whether the melting temperature has been computed for Limmer and Chandler's ST2 variant. Such information could potentially validate or falsify Dr Limmer's hypothesis.

(ii) An alternative possibility is that a highly defective ice Ic structure was generated when Limmer and Chandler performed umbrella sampling along the path connecting the liquid and crystal basins. A large number of such defects could result from inadequate equilibration of simulations in the high-Q_6 regime. This would raise the free energy[18] in the region of the ice basin, leading to the incorrect impression that the liquid and crystal phase are near coexistence.

(iii) Finally, I describe a number of potential technical challenges to implementing the HMC algorithm because it is presently unclear to me if these issues have been adequately scrutinized by Dr Limmer and Professor Chandler. First, the chosen MD integrator used in HMC must be both time reversible and volume preserving to ensure that the appropriate equilibrium distribution is sampled.[19]

As a result, there are a limited number of MD integrators available that may be used in conjunction with HMC. Integrator choices are particularly limited in the case of HMC simulations of rigid water models, due to the fact that popular, iterative constraint algorithms used to fix bond lengths and angles in MD simulations, such as SHAKE and RATTLE, are not time reversible.[19] If a poor choice is made for the integration algorithm, it is possible that spurious results may be obtained.[9] We note that Reinhardt *et al.*[7] observed a "catastrophic" divergence from the well-established equation of state for the TIP4P/2005 water model when HMC simulations were performed using the SETTLE constraint[20] algorithm. This is the same constraint algorithm employed in Limmer and Chandler's HMC simulations.[1,2]

Even if a suitable integrator is chosen, there are a number of potential problems that can arise in correctly implementing HMC. In principle, HMC is most efficient when a large integration time step is employed in the short MD simulations used to propagate the system through phase space. However, constraint algorithms such as SETTLE can still fail when using large time steps, leading to significant deviations from the molecules' ideal, rigid structures. If proper checks are not included in the code, such behavior may go unnoticed. In our tests with the SETTLE algorithm, we performed standard MD simulations for several popular water models using GROMACS.[21] We find that SETTLE[20] can fail when used with integration time steps on the order of 5–6 fs. Limmer and Chandler have reported using a 5 fs time step in their HMC simulations.[1,2]

Another potential pitfall lies in the generation of initial velocities for rigid water molecules. Many MD software packages generate initial velocities for each atom in the system by randomly sampling momenta from the Boltzmann distribution.[9] This procedure is incorrect for rigid body simulations because they have a reduced number of degrees of freedom due to the presence of internal constraints.[22] If this were not taken into account, the system would be initialized at essentially the wrong temperature. This effect would likely be negligible for long molecular dynamics simulations in which the system would have time to equilibrate at the appropriate temperature imposed by a thermostat. However, it would have catastrophic consequences for HMC simulations, because the momenta are typically refreshed at the beginning of each MC step.[22] This problem is solved by simply generating angular and center-of-mass velocities for each rigid body instead of momenta for each atom on the molecule. Because the SETTLE algorithm[20] requires the momenta for each atom, such quantities must be computed from the angular and center-of-mass velocities following initialization.

Although I have described only a few potential technical challenges associated with the HMC method, there are many others which space limitations prevent me from addressing in depth, including proper treatment of massless interaction sites on water molecules within the framework of the SETTLE algorithm and sampling (ergodicity) problems that may arise from using very small or large integration time steps. I have also not addressed potential programming errors that may already exist in packaged software such as LAMMPS,[23] which have a long history of such problems, including implementation issues related to water models that have massless interaction sites[24] like TIP4P/2005 and ST2. The likelihood of programming errors affecting a particular calculation is very difficult to assess because such codes are tens of thousands of lines long and would have to be modified in a non-trivial manner to implement the ST2 water model, which has an intermolecular potential that is particularly challenging to implement in many MD software packages. Because of this concern, the computations I have performed[3] have used in-house software that has been specifically developed for simulation of the ST2 model. In addition to using eight different MC techniques, all of which yielded consistent results, we have also verified our results by performing free energy calculations using codes that were written independently by different researchers in the Princeton group.

4. Relative efficiency of HMC

In our 2012 study,[3] free energy calculations were performed using umbrella sampling with MC simulations that employed standard single-particle moves to propagate the system through configuration space. Dr Limmer and Professor Chandler have posited[2] that our calculations, as well as those performed by independent researchers,[25,26] suffer from artifacts resulting from improper equilibration of our simulations, arguing that such inadequate sampling leads to an illusory low-density liquid basin in $F(\rho, Q_6)$. They have suggested the HMC sampling algorithm used in their work avoids such artifacts because of its superiority, claiming that their "... choice of hybrid Monte Carlo moves is computationally more efficient by 1–3 orders of magnitude over single particle moves and by 2 orders of magnitude over molecular dynamics."[2] In this section, I clarify what I believe is a misunderstanding regarding the relative efficiency of the HMC technique.

A useful measure of efficiency to benchmark sampling algorithms is the relative computational cost required to generate statistically independent samples for the observable of interest. The frequency at which independent samples are generated is estimated using the statistical inefficiency, $g \equiv 1 + 2 \times \tau$, where τ is the integrated autocorrelation time for the observable. Note the units of τ depend on how "time" is defined in a simulation. In MC, for example, it is common to define τ in units of "single-particle MC moves" or "MC sweeps", where 1 MC sweep is equal to N single-particle moves and N is the total number of particles in the system. As I will show, misunderstanding how "time" is defined can lead to incorrect estimates of the relative efficiency when comparing two sampling methods. The computational cost required to generate a statistically independent sample is therefore $c = g \times t_{\mathrm{CPU}}$, where t_{CPU} is the actual processor time required to run the calculation for a duration equal to the fundamental unit of time used in reporting τ.

In table I of Limmer and Chandler's 2013 study,[2] the integrated autocorrelation time for the observable Q_6 for their HMC technique is reported to be $\tau_{Q6}^{\mathrm{HMC}} = 10^4$ HMC sweeps, where 1 HMC sweep consists of approximately 10 MD integration steps. In the 2012 study I co-authored,[3] the autocorrelation time for Q_6 using standard single-particle MC moves was estimated to be $\tau_{Q6}^{\mathrm{MC}} = 10^8$ single-particle MC moves. I show that if the units of this value are misinterpreted as "MC sweeps" instead of "single-particle MC moves", as was done in an early version of Limmer and Chandler's 2013 paper posted on arXiv[8] (See Table I in arXiv:1303.3086v1), one could mistakenly arrive at the conclusion that the HMC method is three orders of magnitude more computationally efficient. Using this incorrect interpretation of our results, we find that:

$$\frac{c^{\mathrm{MC}}}{c^{\mathrm{HMC}}}=\frac{g_{\mathrm{Q}_6}^{\mathrm{MC}}}{g_{\mathrm{Q}_6}^{\mathrm{HMC}}}\times\frac{t_{\mathrm{CPU}}^{\mathrm{MC}}}{t_{\mathrm{CPU}}^{\mathrm{HMC}}}\approx\frac{2\times10^{8}\ \mathrm{MC\ sweeps}}{2\times10^{4}\ \mathrm{HMC\ sweeps}}\times\frac{t_{\mathrm{CPU}}^{\mathrm{MC\ sweep}}}{10\times t_{\mathrm{CPU}}^{\mathrm{MD\ Step}}}=10^{3},$$

where I made use of the fact that 1 MD integration step requires the same computational effort as performing 1 MC sweep, noting that approximately the same number of water molecules were used in each study and 1 HMC sweep as is defined as ~10 MD integration steps in Limmer and Chandler's work.[1,2]

On the other hand, a correct estimate of the relative efficiency can be obtained by simply dividing by the number of water molecules (*i.e.*, $N = 192$) to convert our reported autocorrelation from $\tau_{\mathrm{Q6}}^{\mathrm{MC}} = 10^{8}$ single-particle MC moves to $\tau_{\mathrm{Q6}}^{\mathrm{MC}} = 5.2 \times 10^{5}$ MC sweeps,

$$\frac{c^{\mathrm{MC}}}{c^{\mathrm{HMC}}}=\frac{g_{\mathrm{Q}_6}^{\mathrm{MC}}}{g_{\mathrm{Q}_6}^{\mathrm{HMC}}}\times\frac{t_{\mathrm{CPU}}^{\mathrm{MC}}}{t_{\mathrm{CPU}}^{\mathrm{HMC}}}\approx\frac{2\times(5.2\times10^{5})\ \mathrm{MC\ sweeps}}{2\times10^{4}\ \mathrm{HMC\ sweeps}}\times\frac{t_{\mathrm{CPU}}^{\mathrm{MC\ sweep}}}{10\times t_{\mathrm{CPU}}^{\mathrm{MD\ Step}}}=5.2.$$

The correct analysis shows the HMC technique is only computationally more efficient by roughly half an order of magnitude. I have also confirmed this value by directly performing simulations with my own implementation of both methods. From such simulations, I obtain statistically identical values for $\tau_{\mathrm{Q6}}^{\mathrm{MC}}$, $\tau_{\mathrm{Q6}}^{\mathrm{HMC}}$ and the relative efficiency. Reinhardt *et al.*[7] have also obtained a similar efficiency estimate by directly comparing the two methods. They report the HMC technique is approximately 6 times computationally more efficient than standard MC in driving ice nucleation with umbrella sampling.[7]

Although I do not dispute that a factor of 5 is a significant increase in efficiency, there are a number of other MC techniques available, including parallel tempering and Hamiltonian exchange methods, which can offer similar, if not greater, efficiency gains,[9] without involving the potential technical pitfalls associated with HMC. Table I below shows a partial summary of the techniques I have used along with integrated autocorrelation times for density and Q_6 computed in the low-density liquid basin, the sampling duration, and the relative efficiency of each technique in comparison with standard MC employing single-particle moves.

In a forthcoming publication,[4] I shall present free energy surfaces generated using such techniques, where sampling was performed for two orders of magnitude longer than in our current study in Faraday Discussions. The results, which I presented at the Faraday Discussion, are entirely consistent with those included in our Faraday Discussion paper and with the Princeton group's previous work.[3,5] I have also performed bi-directional sampling between the low-

Table 1 Partial summary of sampling methods

Sampling method	τ_ρ, τ_{Q_6}	Sampling duration	Relative efficiency
Monte Carlo[3] (umbrella sampling)	5×10^5, 5×10^5 MCS[a]	$20\tau_{\mathrm{Q}_6}$	1
Monte Carlo (unconstrained)	10^6, 10^6 MCS	$150\tau_{\mathrm{Q}_6}$	0.5
Hybrid Monte Carlo	10^4, 10^4 HMCS[b]	$5 \times 10^3\tau_{\mathrm{Q}_6}$	5
Parallel Tempering (8 replicas)[c]	10^4, 10^4 MCS	$\sim10^4\tau_{\mathrm{Q}_6}$	6

[a] Monte Carlo Sweeps (MCS). [b] Hybrid Monte Carlo Sweep (HMCS) = ~10 MD integration steps [c] Temperature range. 228.6–272 K; bi-directional sampling performed between low-density liquid and ice.

density liquid and ice phase to explicitly check for irreversible behavior.[4] Such calculations, which I presented at the Faraday Discussion, demonstrate that the sampling duration and sampled Q_6 range explored in our current study are adequate to obtain reversible free energy surfaces that do not suffer from the type of artifacts hypothesized by Dr Limmer and Professor Chandler.[2] Finally, as I showed at the Faraday Discussion, when simulations are performed for the ST2 water under the appropriate state conditions using the HMC method with the time reversible, symplectic rigid body molecular dynamics integrator of Miller *et al.*,[19] the computed free energy surface clearly shows evidence of two metastable liquids in coexistence.[4]

5. Theory of artificial polyamorphism

As Professor Chandler mentioned in his comment, he and Dr Limmer have also developed a theory that can produce free energy surfaces exhibiting an artificial basin[2] where others[3,25] have reported observing a low-density liquid phase. This theory predicts that the artificial basin ages as time progresses, becoming less pronounced until it eventually vanishes at sampling times on the order of $10^3\tau_{Q_6}$ (*q.v.*, Fig 2 in ref. 2). Limmer and Chandler[2] argue that the physical origin of this artificial basin is due to large density fluctuations that are decoupled in time from Q_6 relaxation. Support for this argument is provided in Table I of their paper,[2] which shows that the autocorrelation time for density fluctuations (τ_ρ) in their HMC simulations is approximately two orders of magnitude smaller than τ_{Q_6}. Such behavior is used to justify the primary assumption behind their theory: density and Q_6 relaxations are perfectly decoupled in time.[2]

Limmer and Chandler bolster their argument by showing in Table I of their manuscript[2] that values for τ_ρ and τ_{Q_6} obtained from standard MC simulations are also separated by two orders of magnitude, citing our 2012 study[3] as the source for the reported values. Fig. 2 shows the autocorrelation data from our study[3] computed from 32 simulations in the region of the low-density liquid. Notice that the density and Q_6 autocorrelation functions exhibit short-time transient behavior where they are separated by more than one order of magnitude. I have observed that this transient behavior is very sensitive to the sampling technique.

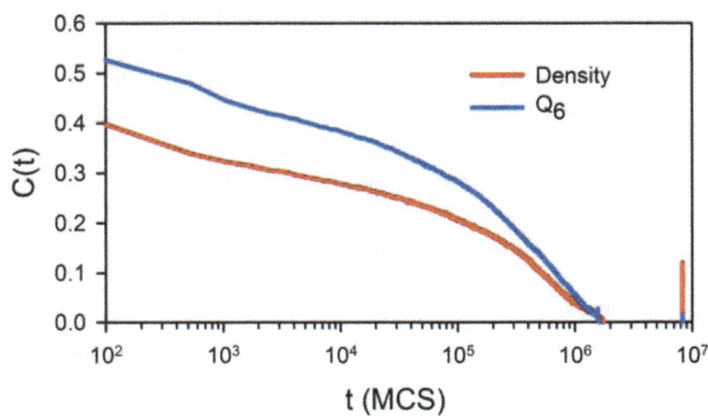

Fig. 2 Density and Q_6 autocorrelation functions computed in the region of the low-density liquid using umbrella sampling MC simulations.[3]

Such behavior is therefore technique-dependent and does not provide a physically meaningful description of the system's fluctuations. However, for every sampling technique I have implemented, including HMC, the computed density and Q_6 autocorrelation functions show similar long-time behavior. It is this long-time behavior that is relevant to sampling the physical properties of the system, especially if the assumption is that density equilibrates instantaneously for any given value of Q_6 that the system slowly samples. While the $1/e$ criterion can in general be used to estimate a typical relaxation time, a more careful analysis is required when comparing correlation functions that are decoupled at shorter times (and in a way that depends on the numerical technique being used!) but invariably decay together at long times, as shown in Fig. 2. Using the inappropriate definition $C(\tau) = e^{-1}$, I obtain relaxation times of 10^4 and 10^6 MCS ($\sim 10^6$ and 10^8 single-particle moves) from the density and Q_6 data presented in Fig. 2, respectively, as reported by Limmer and Chandler[2] for our study.[3] Applying a suitable metric, however, such as the integrated time, τ, which accounts for the slow decay of the functions and captures their long-time behavior, I compute relaxation times on the order of 5×10^5 Monte Carlo sweeps ($\sim 10^8$ single-particle moves) for both density and Q_6. As shown in Table 1, the observation that $\tau_{Q_6} \approx \tau_\rho$ is independent of the sampling technique, if an appropriate metric for the correlation time is used. Professor Peter Poole has also shared results[27] from his 2013 study[25] confirming that $\tau_{Q_6} \approx \tau_\rho$ in his reaction-field implementation of the ST2 water model, suggesting that this finding is quite general. Moreover, in the free energy calculations I have performed with umbrella sampling, the interval between sampling the system's instantaneous values of density and Q_6 is equal to or greater than the statistical inefficiency g, defined as $g \equiv 1 + 2 \times \tau$. As a result, transient short-time behavior, such as that observed in Fig. 2, in not embedded in the statistics used to generate the free energy surface. Finite-time artifacts associated with such transient behavior therefore cannot be used explain the presence of a low-density liquid basin.

The relaxation times computed with different sampling methods show that density and Q_6 fluctuations decay on similar time scales, demonstrating that the underlying assumption behind Limmer and Chandler's theory[2] is not consistent with the behavior actually shown by the ST2 model, as verified using eight different MC techniques. By performing sampling for progressively longer times, I have also verified that no ageing behavior is observed for the low-density liquid basin, as predicted by their theory.[2] In fact the sampling time scales accessible with techniques such as parallel tempering, Hamiltonian exchange and hybrid MC, comfortably exceed $10^3 \tau_{Q_6}$,[4] the time at which the artificial low-density basin should disappear, according to Limmer and Chandler's predictions.[2] I note that the high-Q_6 regime was sampled during these calculations, demonstrating that the low-density liquid basin is not an artifact of restricting sampling to low values of Q_6. Such results were shown at the Faraday Discussion and will be disclosed in a forthcoming publication currently under review.[4]

In conclusion, I offer one additional piece of numerical evidence to demonstrate that there are significant differences between our calculations that cannot be attributed to improper equilibration, as argued by Dr Limmer and Professor Chandler.[1,2] Fig. 3 shows trajectories from five long MC simulations performed using 192 ST2 water molecules at 228.6 K and 2.2 kbar. The simulations were initiated in the region of the low-density liquid and run for $150\tau_{Q_6}$ (1.5×10^8 MC

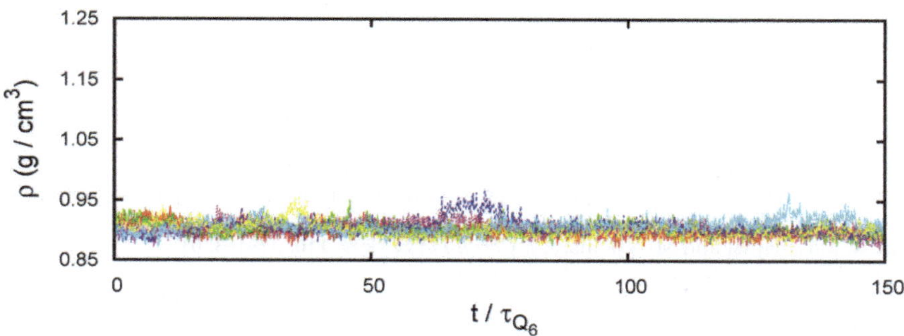

Fig. 3 Trajectories from five long, unconstrained MC simulations performed in the low-density liquid basin at 228.6 K and 2.2 kbar for a duration of $150\tau_{Q_6}$.

sweeps) without applying any constraints.Fig. 3 shows that the unconstrained simulations stay within the low-density liquid basin. The maximum Q_6 value visited by any of the simulations is ~0.13, confirming that the systems are not crystallizing. The trajectories also demonstrate that the low-density liquid is a genuine metastable liquid phase that can be equilibrated and stabilized for many structural relaxation times without crystallizing. In Professor Chandler's comment, he describes how density fluctuates "widely" at such supercooled conditions. Note the absence of this behavior. The density and Q_6 relaxation times computed from these trajectories are statistically indistinguishable and are both on the order of $\sim 10^6$ MC sweeps. Finally, and most importantly, the simulation results shown in Fig. 3 were performed using the exact same ST2 variant reportedly used to compute the free energy surface shown in Fig. 5 (b) of Limmer and Chandler's 2013 study.[2] Examination of that free energy surface reveals that the unconstrained MC simulations shown in Fig. 3 are exploring a region in density space that is predicted by the Limmer–Chandler calculations[2] to be ~40 k_BT uphill in free energy from the high-density liquid phase located at ~1.13 g cm^{-3}. In direct contradiction with the behavior predicted by Limmer and Chandler in their Fig. 5(b),[2] the representative trajectories shown in Fig. 3 exhibit no tendency to explore the supposedly more stable high-density liquid region, which would presumably lie some 40 k_BT below it and along a barrierless path.

1 D. T. Limmer and D. Chandler, *J. Chem. Phys.*, 2011, **135**, 134503.
2 D. T. Limmer and D. Chandler, *J. Chem. Phys.*, 2013, **138**, 214504.
3 Y. Liu, J. C. Palmer, A. Z. Panagiotopoulos and P. G. Debenedetti, *J. Chem. Phys.*, 2012, **137**, 214505.
4 J. C. Palmer, F. Martelli, Y. Liu, R. Car, A. Z. Panagiotopoulos and P. G. Debenedetti, 2013, Submitted.
5 Y. Liu, A. Z. Panagiotopoulos and P. G. Debenedetti, *J. Chem. Phys.*, 2009, **131**, 104508.
6 S. Duane, A. D. Kennedy, B. J. Pendleton and D. Roweth, *Phys. Lett. B*, 1987, **195**, 216–222.
7 A. Reinhardt, J. P. K. Doye, E. G. Noya and C. Vega, *J. Chem. Phys.*, 2012, **137**, 194504.
8 D. T. Limmer and D. Chandler, 2013, arXiv:1303.3086v1.
9 D. Frenkel and B. Smit, Understanding molecular simulation: from algorithms to applications, 2nd edn, Academic Press, San Diego, 2002.
10 F. Stillinger and A. Rahman, *J. Chem. Phys.*, 1974, **60**, 1545–1557.
11 J. E. Shilling, M. A. Tolbert, O. B. Toon, E. J. Jensen, B. J. Murray and A. K. Bertram, *Geophys. Res. Lett.*, 2006, **33**, L17801.
12 E. Mayer and A. Hallbrucker, *Nature*, 1987, **325**, 601–602.
13 O. Yamamuro, M. Oguni, T. Matsuo and H. Suga, *J. Phys. Chem. Solids*, 1987, **48**, 935–942.
14 Y. P. Handa, D. D. Klug and E. Whalley, *J. Chem. Phys.*, 1986, **84**, 7009–7010.

15 J. A. Mcmillan and S. C. Los, *Nature*, 1965, **206**, 806.
16 T. A. Weber and F. H. Stillinger, *J. Chem. Phys.*, 1984, **80**, 438–443.
17 D. T. Limmer and D. Chandler, *Faraday Discuss.*, 2013, **167**, DOI: 10.1039/C3FD00076A.
18 J. S. Vanduijneveldt and D. Frenkel, *J. Chem. Phys.*, 1992, **96**, 4655–4668.
19 T. F. Miller, M. Eleftheriou, P. Pattnaik, A. Ndirango, D. Newns and G. J. Martyna, *J. Chem. Phys.*, 2002, **116**, 8649–8659.
20 S. Miyamoto and P. A. Kollman, *J. Comput. Chem.*, 1992, **13**, 952–962.
21 D. Van der Spoel, E. Lindahl, B. Hess, G. Groenhof, A. E. Mark and H. J. C. Berendsen, *J. Comput. Chem.*, 2005, **26**, 1701–1718.
22 N. Matubayasi and M. Nakahara, *J. Chem. Phys.*, 1999, **110**, 3291–3301.
23 S. Plimpton, *J. Comput. Phys.*, 1995, **117**, 1–19.
24 LAMMPS Molecular Dynamics Simulator, Bug Report: http://lammps.sandia.gov/bug2011.html, Accessed October 18, 2013.
25 P. H. Poole, R. K. Bowles, I. Saika-Voivod and F. Sciortino, *J. Chem. Phys.*, 2013, **138**, 034505.
26 T. A. Kesselring, G. Franzese, S. V. Buldyrev, H. J. Herrmann and H. E. Stanley, *Sci. Rep.*, 2012, **2**, 474.
27 P. H. Poole, Private Communication with P. G. Debenedetti, 2013.

Professor Mallamace noted: When I perform an experiment, I know the limits of the setup and I put error bars on the results. I presume that the MD simulations and the corresponding results can be considered as useful and important suggestions! Because different MD studies give different results on the water system what is the true limit of the simulation "experiment"?

Professor Chandler addressed Professor Mallamace and Dr Palmer: Professor Mallamace expresses concern over the disagreement between the Princeton groups results, presented here by Dr. Palmer, and the results of Dr. Limmer and myself. Professor Mallamace seems to suggest that inherent uncertainties in molecular simulation might be responsible for the discrepancy. In response, let me be absolutely clear: correct independent free energy calculations will give consistent results. Persistent disagreement between presumably competent groups is certainly troubling, but the disagreement can be resolved. Dr. Limmer and I have published a large volume of data to enable resolution.[1,2] We hope the Princeton group will eventually take advantage of that opportunity.

All methods currently known for definitive numerical studies of phase transitions require, either explicitly or implicitly, computation of reversible work functions (*i.e.*, free energies) of order parameters.[3] Our two papers[1,2] detail how it is possible to compute reliable free energy surfaces for models of super-cooled water, even in cases that are generally difficult due to large fluctuations associated with coarsening. We have done so considering several models with differing complexity and boundary conditions, we have estimated statistical uncertainties, and we have demonstrated finite-size scaling in cases of a first-order phase transition. Finally, we have illustrated several examples of errors that are made with incomplete averaging and with inappropriate algorithms that encode undesirable irreversibility.

Before studying one of the more complicated models (*e.g.*, ST2), pitfalls of inaccurate procedures can be detected and then avoided by comparing with our data for one the simpler models (*e.g.*, mW or SW). Dr. Limmer and I have done due diligence in carrying out such quality checks in our own work. Unfortunately, the Princeton group has not done so, at least in cases that would highlight difficulties presented by large slow fluctuations. These cases are where super-cooled water coarsens to form ice and where the Limmer–Chandler results for the ST2 model dffier from those of the Princeton group and their collaborators in Boston and elsewhere.[4]

1 D. T. Limmer, and D. Chandler, *J. Chem. Phys.*, 2011, **135**, 134503.
2 D. T. Limmer, and D. Chandler, *J. Chem. Phys.*, 2013, **138**, 214504.
3 See, for example, D. Frenkel and B. Smit, Understanding Molecular Simulations: From Algorithms to Applications (Academic press, 2001); D. P. Landau and K. Binder, A Guide to Monet Carlo Simulations in Statistical Physics (Cambridge U. Press, 2000).
4 T. A. Kesselring, G. Franzese, S. V. Buldyrev, H. J. Herrmann, and H. E. Stanley, *Sci. Rep.* **2**, 474.

Professor Mallamace responded: First of all I must stress that I consider the Limmer and Chandler work of primary interest for the water community. All of us take advantage from their accurate research. My opinion is the same for the Princeton team and collaborators as far as for the many theoreticians that work in this challenging research area. In my idea the research work and the related discussions are based on an open and fair competition like the one between the Berkeley and Princeton teams. At the same time (I am an experimentalist) I consider only the experimental observations as the true proof of any model. However, I also know (and I have to know) in this frame the absolute importance of the limits of the experimental measurements. My question was proposed on the basis of these considerations. I am indebted with Prof. Chandler for the open and clear reply.

Professor Caupin addressed Dr Palmer: In the debate you are having with David Limmer and David Chandler, the main question is the following: is the low density liquid phase metastable or unstable towards crystallization. Would it be possible to check this issue by inserting a crystal seed in the simulation, and looking if it grows or collapses? This technique has been used by Valeriani *et al.*[1] to study homogeneous crystallization in supercooled water. If there is a finite barrier to nucleation, and if the size of the crystal seed is less than the size of the critical nucleus, then the seed should collapse; otherwise it should grow.

1 E. Sanz, C. Vega, J. R. Espinosa, R. Caballero-Bernal, J. L. F. Abascal, and C. Valeriani, *J. Am. Chem. Soc.*, 2013, **135**, 15008–15017.

Dr Palmer replied: As I discuss in my response to the question posed by Professor Chandler, it is straightforward to demonstrate that the low-density liquid is a genuine metastable liquid phase, separated from the crystal basin by a free energy barrier. In my response to his question, I show that the low-density liquid can be stabilized for more than 100 structural relaxation times without crystallizing or exhibiting density fluctuations leading to the formation of a high-density liquid. Such behavior has also been verified using metadynamics, in which a time-dependent bias is applied to the system along Q_6 to encourage exploration of regions in phase space that have not been previously visited. Even under the action of such a bias, the system does not crystallize. Demonstrating this behavior is essential to confirming that the correct water model is being simulated under state conditions expected to exhibit a liquid–liquid phase transition. In a forthcoming publication,[1] I show that we can reversibly control crystallization by applying a strong umbrella bias, allowing the path connecting the low-density liquid and ice phase to be sampled. The free energy surface obtained from such calculations confirms that there is a barrier to crystallization. As Professor Caupin describes, this behavior could also be validated using unbiased simulations where ice nuclei are seeded in the low-density liquid. I believe that this is an excellent suggestion, one that should be explored in our future work on this topic.

1 J. C. Palmer, F. Martelli, Y. Liu, R. Car, A. Z. Panagiotopoulos and P. G. Debenedetti, 2013, Submitted.

Professor Caupin enquired: As you write in your paper, experiments on bulk supercooled water are limited by homogeneous crystallization of ice, whereas simulations can reach lower temperatures. This helps to find the liquid–liquid transition (LLT), currently at the center of the debate between simulation groups.

However, for the experimentalist, because the LLT lies in a "no man's land", it seems to me that the concept that is more important is that of the Widom line. It is expected that several lines of extrema in response function are related to the Widom line. Therefore my question is: can we define a Widom line, irrespective of the existence or non-existence of a LLT? I mean, even if the LLT is virtual because one liquid is unstable to crystallization as put forward by the work of Limmer and Chandler, won't this virtual transition and its virtual critical point nevertheless induce a Widom line, in the region where the liquid is metastable and not unstable?

As the Widom line is defined as the locus of correlation length maxima, this raises the question of the order parameter. In your study you use the global order parameters ρ and Q_6, but you mention that local order parameters can be defined. Is it possible, by looking at the 2 point correlation function of such local order parameters, to define the correlation length and a Widom line? More specifically, you show that LDL and HDL have the same Q_6, so that they differ only by the density (in your map of the free energy landscape). I believe a local density can be defined. Have you looked at the correlation length for the density? Of course there is an issue with the finite size of the system. But far enough from the LLCP, the correlation length should be small enough for a simulation to capture it. Do you see any other promising local order parameter to look at?

Dr Palmer responded: The Widom line becomes an ambiguous concept if one is not close enough to the critical point. Experimentally one would measure, say, the compressibility, the heat capacity, or the correlation length. The loci of maxima for these response functions only converge asymptotically close to the critical point; away from it there are distinct loci for each response function. Furthermore, even in the absence of a critical point (*i.e.*, the so-called singularity-free scenario) there are response function maxima,[1] so this by itself is not enough to identify a critical point.

Yes, the use of correlation lengths and local order parameters is possible and very interesting. In a forthcoming publication we investigate one such parameter that is able to distinguish HDL, LDL and cubic ice, based on their different local structure.[2] We have not calculated the correlation length associated with this order parameter, but plan to do so in future work. We are especially interested in the growth of this correlation length near the second critical point of ST2, and in comparing it with that of water models that do not exhibit a second critical point, such as mW.[3,4]

1 S. Sastry, P. G. Debenedetti, F. Sciortino and H. E. Stanley, *Phys. Rev. E*, 1996, **53**, 6144–6154.
2 J. C. Palmer, F. Martelli, Y. Liu, R. Car, A. Z. Panagiotopoulos and P. G. Debenedetti, 2013, Submitted.
3 E. B. Moore and V. Molinero, *J. Chem. Phys.*, 2009, **130**, 244505.
4 E. B. Moore and V. Molinero, *Nature*, 2011, **479**, 506-U226.

Professor Evans asked Dr Palmer: In Fig. 9 in your paper, you plot free energy profiles showing the growth of the barrier height with system size. Presumably you can extract (an estimate of) the surface tension following the methods employed by K. Binder and co-workers for fluid–fluid transitions in simple model systems. Have you done this? Is it possible to compare this estimate with that obtained by standard procedures for the tension where one equilibrates the two (fluid) phases and creates an interface? This might provide some additional evidence that the observed transition is indeed one between two (metastable) fluids.

Dr Palmer replied: Professor Evans has made a very useful suggestion, one which I plan to fully address with studies that are currently in progress. As he suggests, the finite size scaling behavior of the free energy barrier between the liquids could be used in conjunction with Binder's approach[1] to extrapolate to the large system limit and obtain an estimate of the surface tension. This value could be compared against surface tension estimates computed by performing direct interfacial simulations, providing evidence to support or falsify the free energy calculations I present indicating liquid–liquid coexistence. While I plan to perform this analysis, it has not been done with the scaling data shown in Fig. 9 of the Faraday Discussions manuscript I co-authored with Professors Debenedetti and Car. As we describe in our manuscript, the scaling calculations were restricted to the high-density liquid and barrier region due to the large computational effort required to sample the low-density liquid basin using larger sample sizes. As a result, we were not able to establish that our finite-size scaling calculations were performed precisely at coexistence conditions, noting that the coexistence pressure is expected to exhibit system size dependence.[2] Our understanding is that this criterion must be satisfied[2] to correctly perform the analysis suggested by Professor Evans.

To correctly perform the suggested analysis, I have recently carried out extensive free energy calculations with umbrella sampling, focusing on both the high-density and low-density liquid regions and ensuring that points of coexistence were accurately determined by reweighting.[3] Such results fully confirm the scaling behavior reported in our Faraday Discussions manuscript and will be detailed in a forthcoming publication currently under review.[4] These results were presented by me at the Faraday Discussion, and by Professor Debenedetti at the recent Liquids Gordon Conference. I have also very recently started to perform direct interfacial simulations of the two liquid phases to compute the surface tension and confirm that a stable interface can be maintained. My results are currently too preliminary for publication. I note, however, that performing such interfacial simulations correctly requires a great deal of care. I believe such care was not exercised in a recent study by English *et al.*,[5] who reported performing computations in which liquid–liquid interfaces were artificially created by joining two simulation cells, purportedly containing a high-density and low-density liquid phase, respectively. Prior to joining the two simulations cells, the authors report relaxing each phase separately for ~5 ns with molecular dynamics at temperatures ranging between 200 and 215 K.[5] At such conditions this amount of time is completely inadequate to equilibrate the high-density liquid, let alone the low-density liquid, which exhibits structural relaxation times on the order of ~100 ns even at much higher temperatures near 240 K.[6] From extensive molecular dynamics simulations at 228 K, I estimate that it takes on the order of ~100 ns for the liquid to even adopt a structure resembling the low-density liquid phase reported in our Faraday Discussion manuscript. Due to inadequate equilibration, English *et al.*[5] noted that the virial pressures in the respective liquid simulation cells differed by 4–5 kbar. As a consequence of this mechanical instability arising from grossly under-equilibrating their liquid samples, English *et al.*[5] observed that their improperly-equilibrated artificial interfaces rapidly disintegrate over the course of only a few nanoseconds. Although the large pressure inequality explains why the interfaces disintegrate, I am admittedly puzzled as to how this process occurs so rapidly in simulations performed at temperatures as low as 200 K. Due to the sluggish dynamics at such conditions, I have found it extremely challenging to perform sampling at 200 K for the ST2 water model even with advanced Monte Carlo methods such as parallel tempering.[7] Learning from the mistakes made by English *et al.*,[5] we have prepared our interfacial simulations with proper care and have not yet observed any signs of such mechanical instabilities. We appreciate Professor Evans' suggestion and we plan to report results from these calculations in due course along with our estimates of the surface tension.

1 K. Binder, *Phys. Rev. A*, 1982, **25**, 1699–1709.
2 J. E. Hunter and W. P. Reinhardt, *J. Chem. Phys.*, 1995, **103**, 8627–8637.
3 Y. Liu, J. C. Palmer, A. Z. Panagiotopoulos and P. G. Debenedetti, *J. Chem. Phys.*, 2012, **137**, 214505.
4 J. C. Palmer, F. Martelli, Y. Liu, R. Car, A. Z. Panagiotopoulos and P. G. Debenedetti, 2013, Submitted.
5 N. J. English, P. G. Kusalik and J. S. Tse, *J. Chem. Phys.*, 2013, **139**, 084508.
6 T. A. Kesselring, G. Franzese, S. V. Buldyrev, H. J. Herrmann and H. E. Stanley, *Sci. Rep.*, 2012, **2**, 474.
7 D. Frenkel and B. Smit, Understanding molecular simulation: from algorithms to applications, 2nd edn, Academic Press, San Diego, 2002.

Professor Evans made a general comment: Dr Palmer, Dr Limmer and Professor Chandler are the main protagonists on this issue.

As a bystander in the debate on a putative liquid–liquid transition in water, can I confirm that everyone agrees that we are not discussing matters of principle, *i.e.* there is no fundamental problem with a one-component system exhibiting a fluid–fluid transition that is metastable w.r.t. the crystal. Perhaps we should recall that many simple pair potential models consisting of a repulsive hard-core plus an attractive short-ranged tail exhibit such behaviour. One of the best known cases is the effective colloid–colloid (depletion) potential in the AO (Asakura–Oosawa) model of colloid–polymer mixtures. By tuning the well-depth one can alter the degree of metastability. A nice overview of the phenomenology is given by D. Frenkel[1]; see Fig.1. The paper by Fortini *et al.*[2] provides some detailed Monte Carlo results for the metastable fluid–fluid coexistence curve and crystallization phenomena for the AO potential and provides references to relevant experiments. Note also that for very short-ranged attraction such models can exhibit (metastable) solid-solid transitions, *e.g.* Fig.1 in ref. 3 for AO, and Fig.14 in ref. 4 for the phase diagram of the depletion potential in an asymmetric binary hard-sphere mixture.

Presumably the issues in water are rather different and relate to whether realistic models that exhibit local tetrahedral order in the liquid and extended tetrahedral order in the stable crystal can exhibit a metastable fluid–fluid transition. This point is made in earlier papers by D. T. Limmer and D. Chandler and probably in papers by P. Debenedetti and co-workers but is probably worth recalling each time the debate arises.

1 D. Frenkel, *Physica A*, 1999, **263**, 26.
2 A. Fortini, E. Sanz and M. Dijkstra, *Phys. Rev. E*, 2008, **78**, 041402.
3 M. Dijkstra, J. M. Brader and R. Evans, *J. Phys.: Condens. Matt.*, 1999, **11**, 10079.
4 M. Dijkstra, R. van Roij and R. Evans, *Phys. Rev. E*, 1999, **59**, 5744.

Professor Anisimov added to the comment of Professor Evans: Some binary fluids demonstrate typical critical-like anomalies without any observable, even metastable, critical point. A well-known example is 3-methyl pyridine–water at about 70 °C. These anomalies are caused by a hidden ("virtual") critical point of demixing, which becomes real and accessible only upon addition of 0.4% NaCl. Therefore, regardless the existence or, alternatively, nonexistence of liquid–liquid separation in metastable supercooled water, a principal question should be addressed: what is the origin of the supercooled-water anomalies? The influence of inaccessible (hidden) liquid–liquid transition (with a critical point) or pre-crystallization fluctuations? Or simply a competition of two alternative structures in water, as suggested by Professor Tanaka in his introductory lecture?

Dr Limmer asked: A question was raised regarding the theoretical possibility of a metastable critical point, given that it is well known that colloidal suspensions that interact over a certain distance can exhibit a metastable critical point between dilute and dense phases.[1] In the case of spherical colloids, fluctuations in two order parameters, density and long range order, are weakly coupled. This means that the while the large density fluctuations associated with the metastable critical point can influence crystallization, instantaneous correlations between density and local order fluctuations are small. In the case of supercooled tetrahedral liquids, these order parameters are strongly coupled and as a consequence correlations between local order and density fluctuations are large.[2] At present it is not known generically what the phase behavior is in the strong coupling case, which is why it is important to clarify what is happening. Water is but one practical example of a system fitting this description. Moreover, the existence of universality in phase transitions would suggest that any model that exhibits correlated fluctuations in these two order parameters would exhibit the same qualitative phase diagram. It is this latter reason that the specific question of whether or not the ST2 water model exhibits a liquid–liquid phase transition is particularly relevant, as understanding it would lend general insight to this class of phase transitions. However, it must be stressed that understanding the phase behavior under such conditions must be done with accurate methods that acknowledge the fluctuation dominance underlying these systems. The strength of these correlations makes it not sufficient to construct mean-field models that ignore this coupling as is done with two state models for example.

1 P. R. tenWolde, and D. Frenkel, Enhancement of protein crystal nucleation by critical density fluctuations, *Science*, 1997, **277**, 1975–1978.
2 D. T. Limmer, and D. Chandler, The putative liquid–liquid transition is a liquid–solid transition in atomistic models of water, *J. Chem. Phys.*, 2011, **135**, 134503.

Dr Limmer continued, addressing Dr Palmer: The paper that Prof. Debenedetti and coauthors have presented suffers from a number of technical shortcomings that I believe have led to their conclusions of a low temperature critical point in water. Most important among these is the use of metadynamics to compute free energy surfaces for supercooled water. Metadynamics is a reliable and effcient method when the slowest variables are treated explicitly. In fact, it is only in this limit that the algorithm can be shown to converge to the equilibrium free energy. For supercooled water, Debenedetti and his co-workers apply metadynamics treating only ρ and Q_6 explicitly. Other slow variables are ignored. Yet, Q_6 changes slowly even in regions where there are no barriers in the free energy $F(\rho, Q_6)$, which implies the existence of at least one additional slow variable. Debenedetti and co-workers have therefore applied metadynamics in a fashion where it is inapplicable. Indeed, they sample only a relatively small range of Q_6, and their neglect of underlying slow variables yield transient non-equilibrium artifacts.

We have shown previously[1] that neglecting the effects of slow variables results in precisely the phenomena presented in the contribution of Prof. Debenedetti, namely the appearance of a transient low density liquid basin. Physically, this transient state manifests the coarsening of ice, in a regime dominated by crystal growth rather than nucleation. A theory for these timescales is presented in Prof. Chandler and my contribution[2] and indeed are found to be long compared those typically accessible to molecular simulation. When appropriate methods are used to access these long timescales, however, these effects disappear in the reversible free energy surfaces. Under such equilibrium conditions, we find free energy surfaces that show basins for a single liquid and a low density crystal.

Along with inaccurate sampling of slow fluctuations, the contribution also contains errors pertaining to its demonstration of finite-size scaling. At coexistence, the free energy barrier separating two phases will scale linearly with $N^{2/3}$. Away from coexistence, however, such interfacial scaling is more complicated. The authors report $N^{2/3}$ scaling in a case where the putative low-density liquid is not demonstrated to be in coexistence with the high-density liquid. Moreover, given the concern over lack of equilibration, it is difficult to interpret this scaling. For sure there is a system-size dependence for even non-equilibrium irreversible work functions of the sort that would emerge from free energy calculations with finite-time artifacts. Such scaling would likely have a non-monotonic dependence on system size, initially increasing due to the $1/\sqrt{N}$ change in the curvature of basins, but ultimately decreasing due to the insignificant statistical weight that such locally correlated configurations have. The possibility of such transient effects have not been explored in Prof. Debenedetti's study.

Finally, while free energy calculations coupled to finite size scaling analysis are the tools that must be used to demonstrate the existence of a phase transition numerically, there are a number of computationally less expensive tests that may be done to show consistency with a hypothesized transition. One example is to calculate the system size dependence of the peak of a fluctuation quantity like the compressibility. As a second moment of the free energy, its convergence is necessarily swifter. When such calculations have been done, as in the case of the ST2 model with reaction field electrostatics for $N = 343$ through $N = 729$ at $T = 237$ K and $p = 2{:}4$ mbar,[3] absolutely no system size dependence has been found, contradicting the existence of a singularity.

1 D. T. Limmer, and D. Chandler, The putative liquid–liquid transition is a liquid–solid transition in atomistic models of water. II, *J. Chem. Phys.*, 2013, **138**, 214504.
2 D. T. Limmer, and D. Chandler, Corresponding states for mesostructure and dynamics of supercooled water, *Faraday Discuss.*, 2013, **167**, DOI: 10.1039/C3FD00076A.
3 E. Lascaris *et al.*, Response functions near the liquid–liquid critical point of ST2 water, *AIP Conference Proceedings*, 2013, **1518**, and personal communication with Erik Lascaris 2013. The compressibility at this state point for both system sizes is 0.012 MPa^{-1}.

Dr Palmer replied: Although techniques such as umbrella sampling guarantee that the appropriate stationary distribution is asymptotically sampled, as correctly argued by Dr. Limmer in his 2013 paper on the liquid–liquid phase transition in supercooled ST2 water,[1] "Free energy methods that do not strictly adhere to this condition, such as metadynamics[2] and Wang–Landau[3] sampling, converge only conditionally in the limit that the biasing degree of freedom is the slowest mode or when the change in the biasing term asymptotes to zero." In the original version of the metadynamics technique,[2] the biasing term does not asymptote to zero in the long-time limit. It has been proved[4] that the original metadynamics technique provides an unbiased estimate of the underlying free energy if the biased coordinate exhibits Langevin-type dynamics. Incidentally, this assumption was invoked in Dr. Limmer's 2013 paper to describe the time evolution of the crystalline order parameter, Q_6, in developing his theory of "artificial polyamorphism".[1] However, it is true that if such slow variables are not biased, erroneous estimates of the free energy may be obtained.

Many of the potential limitations associated with metadynamics have been addressed in the development of more modern versions of the method, including well-tempered metadynamics,[5] which was used to perform the free calculations presented in the Faraday Discussions manuscript that I co-authored with Professors Debenedetti and Car. Well-tempered metadynamics guarantees that the bias asymptotes to zero in the long-time limit,[5] satisfying the latter of the two convergence criteria described in Dr. Limmer's 2013 study.[1] As noted in our manuscript, such behavior was explicitly checked in our calculations. Barducci *et al.*[5] have also explicitly demonstrated that because of this convergence, well-tempered metadynamics provides accurate free energy estimates even in the event that a slow variable is overlooked. Moreover, two slowly-evolving variables were biased in our well-tempered metadynamics simulations, density and the crystalline order parameter, Q_6. We find that the integrated autocorrelation times for both variables are of the same order of magnitude. This result is general and has been confirmed using eight different sampling techniques, including a version of the hybrid Monte Carlo method used in Dr. Limmer's 2013 study.[1] We also show that the free energy surface obtained with well-tempered metadynamics is in excellent agreement with the umbrella sampling calculations presented in our Faraday manuscript, as well as the those from our 2012 study.[6] Finally, the results presented in our Faraday Discussions manuscript have been verified using advanced MC methods that allow us to extend our sampling duration by two orders of magnitude and reversibly sample the high-Q_6 region between the low-density liquid and ice phase. I presented these results at the Faraday Discussion; the calculations will be detailed in a forthcoming publication.[7]

We acknowledge in the original and current version of our Faraday Discussions manuscript that the finite size scaling calculation presented there does not provide definitive proof of a first-order liquid–liquid phase transition in ST2 water. Although we did not establish that each calculation was performed precisely at a point of liquid–liquid coexistence, finite size effects on the coexistence pressure are expected to be relatively small.[8] We have verified this by performing umbrella sampling calculations that demonstrate very clearly that the correct scaling behavior is obtained at conditions were coexistence can be established. I presented these results at the Faraday Discussion, and the calculations will be detailed in a forthcoming publication.[7] Our new results provide strong evidence of a first-order liquid–liquid phase transition and associated second critical point in the ST2 water model.[7] Because I have not been privileged to the data referenced by Dr. Limmer from his private communication with Dr. Lascaris, I cannot offer comments on such results at this time.

1 D. T. Limmer and D. Chandler, *J. Chem. Phys.*, 2013, **138**, 214504.
2 A. Laio and M. Parrinello, *Proc. Natl. Acad. Sci. USA.*, 2002, **99**, 12562–12566.
3 D. P. Landau, S. H. Tsai and M. Exler, *Am. J. Phys.*, 2004, **72**, 1294–1302.
4 G. Bussi, A. Laio and M. Parrinello, *Phys. Rev. Lett.*, 2006, **96**, 090601.
5 A. Barducci, G. Bussi and M. Parrinello, *Phys. Rev. Lett.*, 2008, **100**, 020603.
6 Y. Liu, J. C. Palmer, A. Z. Panagiotopoulos and P. G. Debenedetti, *J. Chem. Phys.*, 2012, **137**, 214505.
7 J. C. Palmer, F. Martelli, Y. Liu, R. Car, A. Z. Panagiotopoulos and P. G. Debenedetti, 2013, Submitted.
8 J. E. Hunter and W. P. Reinhardt, *J. Chem. Phys.*, 1995, **103**, 8627–8637.

Professor Meuwly asked: In his remark on the 2D free energy surface Prof. Chandler pointed out that the existence of the second minimum may be related to the way that the simulations were carried out and analyzed. Can this be related to the findings by P. Bolhuis,[1] that depending on the progression coordinates used, low-dimensional projections of the free energy surfaces can yield very different metastable states and relative energies between them?

1 Peter G. Bolhuis, *Biophys. J.*, 2005, **88**, 50–61.

Professor Chandler answered: You are absolutely correct that Bolhuis' observation is related to the issue under debate concerning Debenedetti's results. I use Fig. 4 to help describe that observation and then show how it relates to Debenedetti's results. The upper part of the figure depicts a free energy function with two basins, one at a small value of Q_6 and another at a high value of Q_6. It is the reversible work function for the two variables x and Q_6. Transitions between the basins require changes in the variable x. If we know nothing of the variable x but nevertheless succeed at sampling its full range of states, we have the contracted free energy $F(Q_6) = -k_BT \int dx \exp[-F(Q_6, x)/k_BT]$. This reversible work function exhibits no barrier between small Q_6 and large Q_6. Considering only Q_6 can therefore give the false impression that dynamics between small and large Q_6 occurs without surmounting a barrier. This point is the essence of Bolhuis' observation.

In the context of Debenedetti's modeling supercooled water, the figure also illustrates another point. Specifically, times between transitions from small Q_6 to large Q_6, and vice versa, are long compared to the time scale with which dynamics

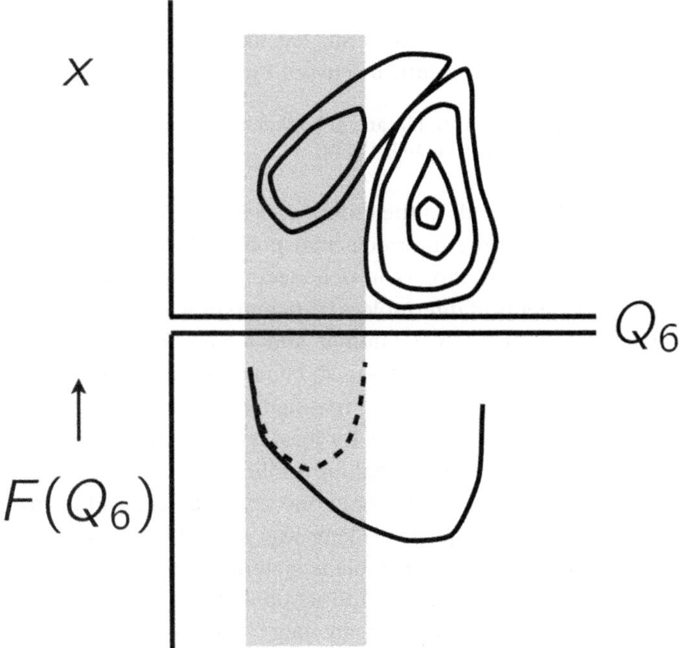

Fig. 4 Schematic depictions of free energy functions. The upper illustrates the topography of a free energy $F(Q_6, x)$ with constant free energy lines. The lower with solid line illustrates the contracted free energy obtained from the equilibrium averaging over the entire range of variable x, *i.e.*, $F(Q_6) = -\ln \int dx \exp[-\beta F(Q_6, x)]$, where $1/\beta$ is temperature times Boltzmann's constant. The lower with dashed line illustrates the non-equilibrium free energy obtained by incomplete averaging over *x*, i.e., averaging over the smaller (shaded) range.

explores states in one or the other basin. If trajectories initiated in the low-Q_6 basin run for times small compared inter-basin transition times, then only the shaded range of *x* will be explored. Averaging under those conditions will give the false impression of a stable basin at low values of Q_6. This artifact of incomplete averaging is the essence of Debenedetti's results.

In particular, the crystal order parameter, Q_6, changes slowly during coarsening, and the transition states are controlled by variables other than just Q6 and the density, ρ. In the early stages of coarsening, the fluctuations in Q_6 are correlated with those of ρ because the birth and death of small crystal-like domains are accompanied by the decrease and increase of density. On the other hand, transitions through which Q_6 increases over long times require alignments of neighboring crystal-like domains, and the collective variables describing such motions are presently unknown. It is like the variable *x* in the figure. Absent of knowledge of *x*, calculations like Debenedetti's will produce the illusion that a transient low-density amorphous state is a metastable state. The way in which it happens is directly analogous to the situation illustrated in Fig. 4. Applications of techniques like meta dynamics, or any other sampling methods, will produce the same illusion if used in ways that remain impervious to long time scales dictated by variables like *x*.

Professor Meuwly addressed Dr Palmer and Professor Chandler: Given that the simulations discussed so far employ relatively simple water models, I wonder how much many-body effects change the findings put forward. In other – but related – fields the explicit inclusion of 3-body terms has led to remarkable improvements in the simulation results as recently discussed by Skinner and coworkers.[1]

1 C. J. Tainter, P. A. Pieniazek, Y.-S. Lin and J. L. Skinner, Robust three-body water simulation model, *J. Chem. Phys.*, 2011, **134**, 184501.

Professor Chandler agreed: In the quest to accurately simulate water, there is plenty of room for improving intermolecular potential models. Concerning the particular issues raised by Professor Debenedetti's paper, however, I do not believe that improvements of intermolecular potentials will play a significant role. My reasoning here is that David Limmer and I have found that "reasonable models" of water and water-like systems all behave similarly when placed at the same corresponding states.[1–3] Here, "reasonable models" stands for models where the liquid state has a preference for local tetrahedral order, and where the crystal globally stabilizes this order, making a solid with ice-like structure. As best we can tell, all such models exhibit only one liquid basin, and at low enough temperatures, this basin disappears giving way to ice-like crystals. Specifically where the liquid basin becomes unstable is system dependent. But when any of these systems are placed close enough to its onset to liquid instability, fluctuations in density and in crystal order are large and slow. Failure in properly sampling these fluctuations is responsible for an assortment of erroneous impressions including the illusion of two distinct liquid phases. Refinements of intermolecular potentials seem to have nothing to do with it.

1 D. T. Limmer and D. Chandler, *J. Chem. Phys.*, 2011, **135**, 134503.
2 D. T. Limmer and D. Chandler, *J. Chem. Phys.*, 2013, **138**, 214504.
3 D. T. Limmer and D. Chandler, *Faraday Discuss.*, 2013, **167**, DOI: 10.1039/C3FD00076A.

Dr Palmer added: Contrary to what others have suggested,[1] I anticipate that various water models may exhibit distinct physical behavior, particularly at extreme state conditions, such as those investigated in our Faraday manuscript. Because such models are typically parameterized to reproduce experimental data near ambient conditions, their ability to qualitatively describe water's behavior in remote regions of the phase diagram is limited. This is evidenced by the exotic, and even unphysical, ice polymorphs observed for some water models.[2] The transferability of each model is determined by the form of its potential function and how this function was parameterized. It has been proved[3] for simple models of water-like fluids that a second critical point may arise simply by adjusting the value of a single parameter in the potential function. I therefore find it very difficult to justify on the basis of statistical mechanical arguments the conjecture[1] that water models described by completely different potential functions should obey a corresponding states principle in the supercooled regime. To devise such a theory, one would need to develop appropriate temperature and pressure scales capable of describing the thermodynamic behaviors of all reasonable water models. At present, this difficult task has not been achieved.

I believe it is therefore difficult, if not impossible, to assess *a priori* the impact that specific contributions, such as many-body interactions, may have on the ability of a water model to exhibit a liquid–liquid phase transition. Consequently, water models with many-body interactions will need to be investigated using the type of free energy analysis methods my colleagues and I have employed for the ST2 water model.[4] I anticipate that such analysis may be prohibitively expensive at present due to the computational cost that is typically associated with calculating many-body interactions. One exception to this is the single-site mW model,[5] which uses a three-body angular term to capture water's tetrahedral nature. At present, results suggest that mW does not exhibit a liquid–liquid transition,[6] such as the one reported in our manuscript for the ST2 water model. However, it has not been established whether the absence of such a transition is related specifically to the functional form of its three-body term or other features of the model.

1 D. T. Limmer and D. Chandler, *Faraday Discuss.*, 2013, **167**, DOI: 10.1039/C1033FD00076A.
2 M. Yamada, S. Mossa, H. E. Stanley and F. Sciortino, *Phys. Rev. Lett.*, 2002, **88**, 195701.
3 T. M. Truskett, P. G. Debenedetti, S. Sastry and S. Torquato, *J. Chem. Phys.*, 1999, **111**, 2647–2656.
4 Y. Liu, J. C. Palmer, A. Z. Panagiotopoulos and P. G. Debenedetti, *J. Chem. Phys.*, 2012, **137**, 214505.
5 V. Molinero and E. B. Moore, *J. Phys. Chem. B*, 2009, **113**, 4008–4016.
6 E. B. Moore and V. Molinero, *Nature*, 2011, **479**, 506-U226.

Professor Barrat asked: This point was already mentioned by Prof. Chandler, but I was puzzled that Fig. 3 in your paper only covers small values of Q_6, and not those that correspond to the solid phase. If the larger values are not sampled while the free energy surface is rather flat, the result for the free energies may be incorrect.

Dr Palmer answered: The free energy surfaces shown in Fig. 3 of our Faraday Discussion paper were computed using umbrella sampling Monte Carlo simulations in which only the low-Q_6 region was sampled. However, at the Faraday Discussions meeting, I also presented unpublished results computed by performing bidirectional sampling between the low-density liquid and ice phase, exploring the high-Q_6 region up to ~0.6. The sampling duration for those calculations was also two orders of magnitude longer than for the results reported in our current manuscript. Such calculations will be reported in a forthcoming publication[1] and show definitive evidence of a liquid–liquid phase transition, in excellent agreement with the results shown in our Faraday Discussion paper, thus confirming that the free energy surfaces shown in Fig. 3 are free of any significant artifacts arising from transient behavior or restricting sampling along Q_6. I also note that no constraints were placed on the maximum value of Q_6 sampled during the metadynamics simulations I performed to compute the free energy surfaces shown in Fig. 7 of our Faraday paper. The free energy surfaces reveal that only the low-Q_6 regime is explored by the system even when a modest bias is applied to drive it away from such regions. I show in my response to the opening question posed earlier by Professor David Chandler that unconstrained Monte Carlo simulations can remain within the low-density liquid basin for a duration ~150 longer than the integrated autocorrelation time for Q_6, τ_{Q_6}. This behavior provides strong evidence that the low-density liquid is a genuine metastable liquid phase separated from the crystal by a free energy barrier that must be overcome before freezing occurs. As I showed in my presentation at the Faraday Discussion meeting, such barriers may be overcome with umbrella sampling, following the path taken by the low-density liquid as it crystallizes into cubic ice. Before investing the computational effort required to compute the free energy surface in the high-Q_6 region, stability tests on the low-density liquid should be performed to ensure that the water model being simulated is at the appropriate state conditions expected to exhibit a liquid–liquid transition. We have provided low-density liquid configurations to Dr. Limmer and Professor Chander so that such tests may be performed using their hybrid Monte Carlo algorithm. While we observed that the low-density liquid is stable for periods of time in excess of $100\tau_{Q_6}$, Limmer and Chandler report[2] results showing that simulations initialized from our configurations exhibit a continuous drift in density, evolving to the high-density liquid region on times as short as $1\text{–}2\tau_{Q_6}$ for pressures in the range of 1.8 to 2.2 kbar at 228 K.[2] Such results demonstrate that the low-density liquid is completely unstable and cannot be sampled under these conditions with their hybrid Monte Carlo algorithm, suggesting that the differences between our respective calculations cannot be simply attributed to the sampling duration or range of Q_6 explored.

1 J. C. Palmer, F. Martelli, Y. Liu, R. Car, A. Z. Panagiotopoulos and P. G. Debenedetti, 2013, Submitted.
2 D. T. Limmer, Private Communication, 2012.

Dr Limmer noted: Dr. Palmer has raised the question of whether our contrasting results are the outcome of a pathology in the choice of integrator we have used in implementing our hybrid Monte Carlo calculations. Specifically, he criticizes the use of the SETTLE algorithm,[1] citing a paper by Juan de Pablo's group (reference unknown). As is clear from the derivation of the hybrid Monte Carlo acceptance criteria,[2] formally all that is required from an integrator is that it is time reversal symmetric and preserves the volume of phase space, both of which are satisfied by the SETTLE algorithm. In practice, however, it is true that the details of an integration algorithm influence the ergodicity of the Markov chain. We have explicitly checked that the SETTLE algorithm as implemented in our calculations samples a uniform distribution of molecular orientations, is translationally diffusive and obeys relevant fluctuation–dissipation relationships for the heat-capacity and compressibility. Proving ergodicity for complex systems is difficult, these results provide ample evidence for the robustness of our methodologies.

1 S. Miyamoto, and P. A. Kollman, SETTLE: an analytical version of the SHAKE and RATTLE algorithm for rigid water models, *J. Comp. Chem.*, 1992, **13**(8), 952–962.
2 D. Frenkel, and B. Smit, Understanding molecular simulation: from algorithms to applications, Elsevier, 2001.

Dr Palmer replied: Dr. Limmer and Professor Chandler dismiss results obtained by several independent researchers[1–6] as artifacts.[7,8] The primary objective of the recent studies by Limmer and Chandler[7,8] is to compute reversible free energy surfaces describing the phase behavior of several water models, including the ST2 model,[9] at conditions reported to exhibit liquid–liquid coexistence.[1] Such free energy surfaces must withstand thermodynamic scrutiny, exhibiting behavior consistent with the reported state conditions. As discussed in my response to Professor Chandler's opening question, the ST2 free energy surfaces presented in Dr. Limmer's studies[7,8] are at odds with reasonable expectations based on thermodynamic arguments. Figure 5(b) in their 2013 study,[8] for example, which is reportedly for the same ST2 variant my co-authors and I have examined,[1] indicates that liquid and ice Ic are near coexistence at 230 K and 2.2 kbar, with the liquid being only slightly metastable with respect to ice by

$\sim$7.5 k_BT. This corresponds to free energy difference ΔG(ice Ic − liq) of $\sim$−67 J mol^{-1}, which is an order of magnitude off the $\sim$−545 J mol^{-1} that my co-authors and I have estimated for real water by performing thermodynamic integration calculations with experimental data. Thermodynamic arguments indicate ΔG(ice Ic − liq) should be even larger for ST2, because it is more deeply supercooled than real water at 230 K and 2.2 kbar due to its higher melting temperature.[10,11] Readers are referred to my response to the opening question posed by Professor Chandler for further details. As reported in my response to the aforementioned question, presented by me at the Faraday Discussion, and documented in a forthcoming publication,[12] my free energy surface is fully in accord with such expectations and provides conclusive evidence of the existence of a liquid–liquid phase transition and associated second critical point in ST2 water.

Because Dr. Limmer's results do not withstand thermodynamic scrutiny and are at odds with the findings presented in numerous studies,[1–6] one would expect that performing due diligence would imply validating calculations with a different sampling technique. I believe that this is particularly important because to my knowledge the only other published computational study employing the SETTLE algorithm to perform hybrid MC simulations noted that spurious behavior described as "catastrophic" was observed in the calculated equation of state for the TIP4P/2005 water model.[13] Such behavior is consistent with the above-described deviations from reasonable thermodynamic expectations observed in Dr. Limmer's free energy calculations.[7,8] I note that the aforementioned study[13] was performed using the GROMACS molecular dynamics package,[14] which arguably has the most well-tested and vetted implementation of the SETTLE algorithm available in the public domain. The LAMMPS software package[15] used by Dr. Limmer to perform free energy calculations already contains a time-reversible and symplectic molecular dynamics integration algorithm developed by Miller *et al.*[16] for use in hybrid Monte Carlo simulations of rigid bodies like the ST2 model. A reasonable check would therefore be to perform free energy calculations with this integrator to validate the result obtained using SETTLE. Other reasonable checks include determining the melting temperature of the ST2 variant used in their simulations and computing the free energy surface using any number of well-established alternative sampling methods comparable in efficiency to the hybrid Monte Carlo technique, such as parallel tempering or Hamiltonian exchange.[17] In performing our due diligence, we have carefully scrutinized our results by carrying out such checks, as well as many others that are significantly more involved and time consuming. As I documented at the Faraday Discussion, the agreement between our calculations performed using eight different techniques is excellent. Our meticulous and exhaustive due diligence exercise will be fully documented in forthcoming publication.[12]

1 Y. Liu, J. C. Palmer, A. Z. Panagiotopoulos and P. G. Debenedetti, *J. Chem. Phys.*, 2012, **137**, 214505.
2 Y. Liu, A. Z. Panagiotopoulos and P. G. Debenedetti, *J. Chem. Phys.*, 2009, **131**, 104508.
3 F. Sciortino, I. Saika-Voivod and P. H. Poole, *Phys. Chem. Chem. Phys.*, 2011, **13**, 19759–19764.
4 P. H. Poole, R. K. Bowles, I. Saika-Voivod and F. Sciortino, *J. Chem. Phys.*, 2013, **138**, 034505.
5 P. H. Poole, F. Sciortino, U. Essmann and H. E. Stanley, *Nature*, 1992, **360**, 324–328.

6 T. A. Kesselring, G. Franzese, S. V. Buldyrev, H. J. Herrmann and H. E. Stanley, *Sci. Rep.*, 2012, **2**, 474.
7 D. T. Limmer and D. Chandler, *J. Chem. Phys.*, 2011, **135**, 134503.
8 D. T. Limmer and D. Chandler, *J. Chem. Phys.*, 2013, **138**, 214504.
9 F. Stillinger and A. Rahman, *J. Chem. Phys.*, 1974, **60**, 1545–1557.
10 T. A. Weber and F. H. Stillinger, *J. Chem. Phys.*, 1984, **80**, 438–443.
11 D. T. Limmer and D. Chandler, Faraday Discuss, 2013, **167**, DOI: 10.1039/C1033FD00076A.
12 J. C. Palmer, F. Martelli, Y. Liu, R. Car, A. Z. Panagiotopoulos and P. G. Debenedetti, 2013, Submitted.
13 A. Reinhardt, J. P. K. Doye, E. G. Noya and C. Vega, *J. Chem. Phys.*, 2012, **137**, 194504.
14 D. Van der Spoel, E. Lindahl, B. Hess, G. Groenhof, A. E. Mark and H. J. C. Berendsen, *J. Comput. Chem.*, 2005, **26**, 1701–1718.
15 S. Plimpton, *J. Comput. Phys.*, 1995, **117**, 1–19.
16 T. F. Miller, M. Eleftheriou, P. Pattnaik, A. Ndirango, D. Newns and G. J. Martyna, *J. Chem. Phys.*, 2002, **116**, 8649–8659.
17 D. Frenkel and B. Smit, Understanding molecular simulation: from algorithms to applications, 2nd edn, Academic Press, San Diego, 2002.

Professor Caupin opened the discussion of the paper by Francesco Mallamace: You emphasize the importance of the line of compressibility minima (LCm) being at 315 $\pm$ 5 K for all pressures studied, which are positive. However, it is possible that the temperature of the LCm decreases when the pressure decreases to negative pressures. A large pressure dependence of the temperature of the LCm is observed in molecular dynamics simulations of ST2.[1] In these simulations, the LCm even crosses the line of density maxima, in a thermodynamically consistent fashion.[2] Do you think that for real water, the LCm remains near 315 K at negative pressure? If it does not, how would this affect your picture?

1 P. H. Poole, I. Saika-Voivod and F. Sciortino, *J. Phys.: Condens. Matter*, 2005, **17** L431–L437.
2 S. Sastry, P. G. Debenedetti, F. Sciortino, and H. E. Stanley, *Phys. Rev. E*, 1996, **53**, 6144–6154.

Professor Mallamace responded: Our observation regards only the behavior of the water compressibility for positive pressures in a very large P range from 1 bar to 8 kbar. The simulation findings you cite certainly propose an interesting scenario but at the same time they should be considered a good suggestion to explore the negative pressure phase diagram. To be honest, after the conflicting opinions just proposed in this Faraday Discussion, I am afraid that MD simulations can be considered as the only and true proof of the chemical-physics research regardless of the experimental results and observations. I presume that only a careful series of experiments can be considered the only datum point in chemical-physics research. In some way, many of us (not only experimentalists) have the feeling that some people working in the MD simulations area discuss research questions with some degrees of self-referentiality.

Professor Ben-Amotz asked: As you pointed out, the compressibility minimum, combined with equation 1 in your paper, implies the fixed point in the thermal expansion coefficient *vs.* pressure. In interpret this to mean that the existence of T^* is a consequence of the compressibility minimum, and thus a consequence of the anomalous thermodynamic properties of water. So, it is not clear to me why you say that the existence of T^* is the "source of the anomalies of water". More specifically, how is it possible to distinguish whether the a parameter such as T^* is a consequence of cause of the anomalies of water?

Professor Mallamace replied: Equation 1 proves the thermodynamical consistency between the compressibility minimum and the fixed point, at T^*, of the thermal expansion coefficient at different pressures. The association of such a temperature with the "source of the water anomalies" is given: (i) by considering both the self diffusion data and the corresponding configurational entropy and (ii) the NMR chemical shift data. The changes, by decreasing temperature, just at about T^* in the configurational entropy and in the "local order" from that of a simple liquid behavior to that of a complex liquid are the signature of the onset of the HB tetrahedral network, the process commonly considered at the basis of the anomalous behavior of water. Such a situation is also reflected in the P–T behavior of the heat capacity as proposed in Figure 3 in our Faraday paper; T^* is just the temperature at which C_P has the minimal variation with P.

Professor Wang enquired: In the conclusion of your paper, you mentioned that "T* may be the onset temperature of the HB clustering." If one tries to plot the population of HB clusters as a function of temperature, do you expect a smooth decrease or do you expect a sharp transition at T^*? The latter seems to hint at the possibility of another phase transition.

Professor Mallamace responded: Certainly a smooth variation. The chemical shift data that we show are considered primarily a measure of the local order (as seen by the hydrogen nucleus) but is also a quantity directly related with the average HB number and thus to the population of the HB clusters. As it can be observed from Figure 6 in our paper, such a quantity displays, on decreasing T, only a slight increase in its curvature.

Professor Wang said: Your observation of P_{cross} and T^* is very interesting. This observation does not seem to conflict with the liquid–liquid phase transition hypothesis. There are many other hypotheses that explain the anomalies of water. Does your observation disapprove some of these hypotheses?

Professor Mallamace replied: Certainly our observations can be coherent with the liquid–liquid phase transition hypothesis but at the same time do not provide any definitive proof on that. In addition, the HB interaction and the HB dynamical network are at the basis of any theoretical model for water.

Mr Wexler communicated: The idea that there is a temperature dependent cross-over in the local proton environment in water is intriguing in relation to biological processes many of which can only occur at temperatures below T^*. Considering the biological toxicity of deuterium have you examined the thermodynamic response functions of D_2O and do you see a consistent shift in the temperature of T^*? Would such a shift indicate that the isotope effect on the local environment interferes with light atom tunnelling due to reduced hydrogen bond zero-point-energy as has been found for supercooled and confined (*i.e.* protein hydration) water?[1]

1 A. Giuliani, M. A. Ricci, and F. Bruni, Water Proton Environment: A New Water Anomaly at Atomic Scale?, in Liquid Polymorphism, 2013, Volume 152 (ed. H. E. Stanley), John Wiley and Sons, Inc., Hoboken, NJ, USA. DOI: 10.1002/9781118540350.ch8.

Faraday Discussions RSC Publishing

PAPER

Fluctuations and micro-heterogeneity in mixtures of complex liquids

Aurélien Perera*[a] and Bernarda Kežić[ab]

Received 3rd May 2013, Accepted 5th June 2013
DOI: 10.1039/c3fd00072a

The role of concentration fluctuation and micro-heterogeneity in aqueous methanol mixtures are examined by two different methods: computer simulations of two systems with sizes of 2048 and 16 384 molecules, and site–site Ornstein–Zernike integral equation theory with two different closures. The focus is on the short and long range behaviour of the correlation functions, their running integrals and the structure factors. It is found that the micro-heterogeneous nature of these mixtures does not allow a proper stabilisation of the long range tail of the correlations as obtained through computer simulations, even over 6 ns runs. Similarly, approximate closures provide poor estimates of the concentration fluctuations and micro-structure of the mixtures. Interestingly, the analysis of the inadequacies of both these studies provide interesting insights about the nature and interplay of the concentration fluctuations and the micro-heterogeneity in these mixtures.

A Introduction

Simple liquids and mixtures are subject to fluctuations of density and concentration, but complex liquids, such as aqueous mixtures, for example, possess in addition a micro-heterogeneous nature which is challenging to distinguish from the intrinsic fluctuations.[1,2] Both manifestations affect the spatial and temporal distribution of the molecules and the species, and the challenge relies on finding observables that would allow one to distinguish between them. For example, the so-called Kirkwood–Buff integrals (KBI), which can be measured from thermodynamic or scattering techniques in mixtures, probe in fact the concentration fluctuations.[3] Yet, it appears that these fluctuations are "enhanced" by the intrinsic micro-heterogeneous nature of some mixtures such as aqueous mixtures,[4–7] for example. This particular observable has been the focus of measurements and discussions over the past two decades, without a clear

[a]Laboratoire de Physique Théorique de la Matière Condensée (UMR CNRS 7600), Université Pierre et Marie Curie, 4 Place Jussieu, F75252, Paris cedex 05, France. E-mail: aup@lptmc.jussieu.fr; Fax: +33 1 44275100; Tel: +33 1 44277291

[b]Department of Physics, Faculty of Sciences, University of Split, Nikole Tesle 12, 21000, Split, Croatia. E-mail: bernarda@pmfst.hr; Tel: +385 21 385286

outcome in the interpretations.[8] Things have become even more complex through the analysis brought by computer simulations. Indeed, the natural micro-segregation of species generates enhanced short range correlations which differ dramatically from one force field model to another.[9] Perhaps the most dramatic feature is the fact that it is often difficult to decide between macroscopic phase separation or microscopic segregation because of the finite size of the simulations,[10] leading to an escalation of computer resources with no clear outcome. We have recently suggested that such mixtures may behave like smaller scale micro-emulsions, which we called molecular emulsions.[11,12] In real micro-emulsions, surfactant molecules form mesoscopic structures such as micelles and lamella.[13,14] In molecular emulsions, the size of the solute molecules does not permit such manifestations,[15] and the micro-heterogeneity is the microscopic equivalent structure.[16] This analogy opens a route to distinguish concentration fluctuation from micro-heterogeneity through the small wave vector behaviour of the structure factor $S(k)$. Fluctuations are defined as usual as the $k = 0$ response of $S(k)$ while micro-heterogeneity is related to small k values in the range of a few inverse Angstroms (instead of a few inverse nanometers for micro-emulsions).

In previous studies of aqueous mixtures such as *t*-butanol–water and acetone–water,[11] we have shown that the water–water correlations tend to show supra-molecular domain oscillations that cover a range around 10 nm, making an incursion in the mesoscopic domain since the diameter of the water molecule is about 3 Å. *t*-Butanol seems to form small micelles, as was predicted from various experimental considerations[13] and was shown in recent simulations.[7] So, the analogy we have put forward is somewhat justified. Does it hold in the case of the even smaller methanol molecule, which is the smallest surfactant molecules available in nature, with only one hydrophobic methyl group?

The importance of micro-segregation in aqueous mixtures gained importance with the publication of the landmark *Nature* paper by the Soper group.[17] This paper focused precisely on the aqueous methanol mixture. What this work showed is a clear micro-phase separation in the equimolar mixture of a system that belongs to the textbook category of solutions and not that of micro-emulsions. Even though they did not explicitly stated it, this work showed that some aqueous mixtures are indeed smaller scale micro-emulsions.

In the present work, we revisit this very same system, with a view to investigate how far the analogy between micro-emulsions and aqueous mixtures can be pushed, mainly in terms of the approach we have put forward in previous papers with larger molecules. The main problem posed by micro-heterogeneity in mixtures is the following. In a disordered liquid, there are various ways of randomly disposing various molecules, but due to preferential interactions, such as the hydrogen bonding for example, molecules will prefer to be disposed in particular positions that optimise the free-energy gain. Since hydrogen bonding between water molecules is so strong,[18] these molecules mostly prefer to gather between themselves, but since they also hydrogen bond to solute molecules, they cannot form macroscopic water domains. The segregation of water molecules from the solute molecules is formed into small domains whose shape and size vary from small clusters of few water molecules to interconnected domains, depending on the concentration of the various species. The way these domains evolve in time depends on the local entropy–energy balance.[5,17,19] The most intriguing and difficult to describe aspect of this micro-segregation is how these domains in equilibrium are maintained without leading to full phase separation. This aspect has strong similarities with what happens in micro-emulsions, for which case there is a consistent literature report that explains the stability of these systems in terms of field theoretic based microscopic description.[16] When it comes to systems such as aqueous methanol mixtures, the very idea of a field based description seems inadequate, since the details of the molecules cannot be neglected any more. No coarse-grained description would be really suitable, and one must resort to either computer simulations or liquid state approaches such as integral equations. In the present work, we examine in detail the feasibility of either approaches. In particular, we focus on how the formation of micro-domains and the existence of concentration fluctuations can be explained in a unified way. We will principally focus on the density–density correlation functions in order to describe the microscopic structure of these mixtures and on the corresponding Kirkwood–Buff integrals that account for the concentration fluctuations in the system.[20]

The remainder of this paper is divided into two principal parts. In the first, we focus on computer simulation, with the question in mind as to if this method is able to account for the stability of these mixtures and at what price. In the second part, we use the site–site Ornstein–Zernike (SSOZ) formalism,[21,22] in order to evaluate correlations and concentration fluctuations in this mixture, which we compare with the simulation results. The last section contains discussions and our conclusions.

B Computer simulation study of micro-heterogeneity

Aqueous methanol mixtures have been previously the focus of many computer simulation studies.[4,17,23–25] One of the issues that is often raised is that of the accuracy of the force fields.[24] In the present work we will use the SPC/E force field for water[26] and the OPLS force field for methanol.[27] We have previously shown that this particular combination is able to account quite accurately the Kirkwood–Buff integrals[25] with system sizes of $N = 2048$ particles. In the present work, we have reinvestigated systems twice the size, in order to sort out the long range part of the molecular correlations. In particular, we would to see if domain oscillations appear in the correlations, just like in mixtures with larger solutes.[7,11] For this reason, we have investigated system sizes of $N = 16\,384$. Instead of using the DLPOLY-2 code,[28] we use now GROMACS-4.6,[29] since it is very fast even on desktop PC type configurations. With this new code, we are now able to investigate these mixtures with statistics in the range of several ns. Just like with our previous DLPOLY runs, we use the isobaric constant NPT Gibbs ensemble molecular dynamics, with Nosé–Hoover thermostat and Parinello–Rahman barostat, with time constants of 0.1 ps and 2 ps, respectively, in order to keep ambient conditions of $T = 300$ K and 1 atm pressure. The time step is kept at 2 fs. We have focused particularly on typical methanol mole fractions of $x = 0.2$, 0.5 and 0.8, which represent water rich, equimolar and methanol rich mixtures. Fig. 1 shows a comparison of the snapshots between the $N = 2048$ and the $N = 16\,384$ systems for each concentrations. These snapshots provide an illustration of the intrinsic micro-heterogeneous nature of these solutions. In particular, one sees that the small system tends to somewhat enhance the clustering, probably due to competition between local segregation and the constraint of keeping periodical conditions of the domains. This is an important issue of the simulation of these systems, and we examine this in some detail below through the behaviour of the correlations.

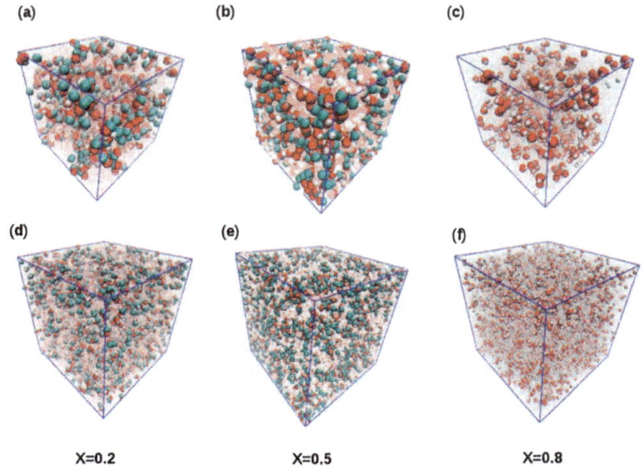

Fig. 1 Snapshots of the aqueous methanol simulation cells for $N = 2048$ (top row (a)–(c)) and $N = 16\,384$ (bottom row (d)–(f)). The left column corresponds to methanol concentration $x = 0.2$, middle column to $x = 0.5$ and right column to $x = 0.8$. For $x = 0.2$ and 0.5 shots, methanol molecules are shown explicitly and water molecules are shown as semi-transparent. For $x = 0.8$ it is the reverse. Carbon is in cyan, oxygen in red and hydrogen sites in white.

With these new system sizes, we are able to calculate correlations up to 50 Å. Fig. 2 shows a comparison between typical correlation functions between the two runs, and it can be seen that the short range differences are virtually indistinguishable. However, long range correlations are very different, as illustrated in Fig. 3(a–c). In these figures we show the following running KB integrals (RKBI):

$$G_{\mathrm{ab}}(r) = 4\pi \int_0^r dss^2 (g_{\mathrm{ab}}(s) - 1) \tag{1}$$

which should converge to the Kirkwood–Buff integrals G_{ab} at large separation. In eqn (1), $g_{\mathrm{ab}}(r)$ is the pair site–site correlation function between site a, belonging to

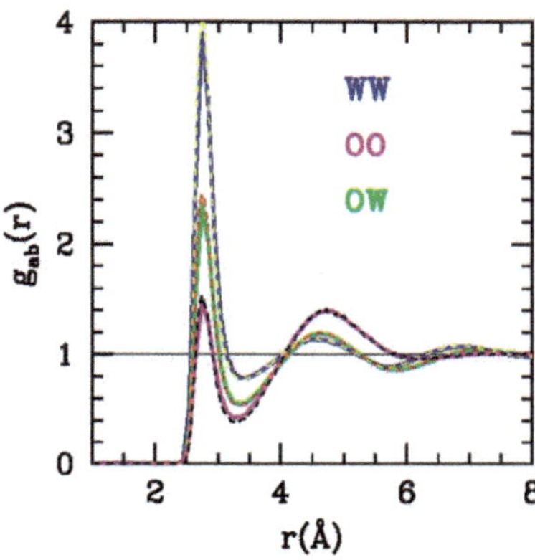

Fig. 2 Oxygen–oxygen site–site pair correlation functions for water–water, methanol–water and methanol–methanol, obtained from the $N = 2048$ system (full lines) and $N = 16\,384$ (dashed lines), at $x = 0.2$.

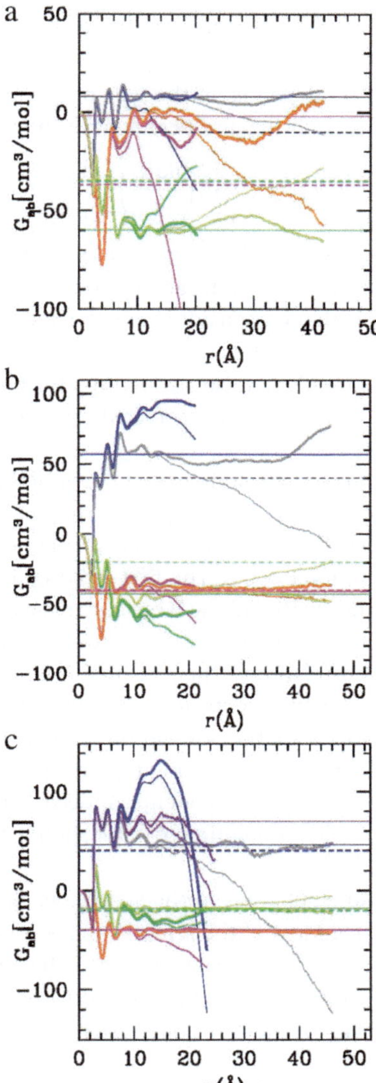

Fig. 3 (a) Running Kirkwood–Buff integrals for methanol concentration $x = 0.2$. Thin lines are uncorrected RKBI and thick lines RKBI corrected for the asymptote. Water–water RKBI in blue (N = 2 K) and grey (N = 16 K). Me–Me RKBI in magenta and orange. Cross RKBI in pale green and bright green. Horizontal dashed lines are expt results[31] with same color codes as N = 2 K systems and full lines are actual predictions from N = 16 K system. (b) RKBI for methanol concentration of $x = 0.5$. Color codes are same as in Fig. 3a. (c) RKBI for methanol concentration of $x = 0.8$. Color codes are same as in Fig. 3a. The purple curves are for the 12 ns run of the N = 2 K system (see text).

molecule of species 1, and b, belonging to molecule of species 2. That way, we show the following functions corresponding to three combinations of the correlations between the oxygen sites of water and methanol molecules $G_{WW}(r)$, $G_{OO}(r)$ and $G_{OW}(r)$. We recall that, in finite size N-constant simulations, the pair correlation functions do not tend asymptotically to 1, but to a slightly different value that depends on the concentration fluctuations[28]

$$\lim_{r\to\infty} g_{ab}(r) = 1 - \frac{1}{N\varepsilon_{ab}} \quad (2)$$

where the exact value of ε_{ab} depends on the type of ensemble. For the constant NVT Canonical ensemble, Lebowitz and Percus have shown[30] that

$$\varepsilon_{ab} = \sqrt{\rho_a \rho_b}\left(\frac{\partial \beta \mu_a}{\partial \rho_b}\right)_{NVT} \quad (3)$$

where ρ_a and μ_a are the density and the chemical potential of species a, respectively, and $\beta = 1/k_B T$ is the Boltzmann factor. However, this formula does not apply in the NPT ensemble that we consider here. We have proposed an empirical modification[25] which consists of shifting the erroneous asymptote to its proper value of 1. Failure to do so leads to incorrect asymptotical behaviour of the RKBI, as illustrated in Fig. 3a–c, where the uncorrected asymptotes are shown in thin lines.

It is obvious that an appropriate shift gives a horizontal RKBI, as witnessed by the thicker lines. However, it is also apparent that in many cases no clear asymptote emerges, especially in the case of the smaller $N = 2048$ systems.

Fig. 3a shows the RKBI for $x = 0.2$. The first striking difference concerns the various trends between the small and the large systems as concerns the minority species, here methanol. These trends are quite similar for water–water correlations, but not the cross-correlations. There seems to be long range oscillations that are obvious in all three components of the large system, but clearly cannot be reproduced with simulations of the smaller $N = 2048$ system. The perfectly matched periodicity of all three component pairs indicate that the domain modulation is the same for both components. Indeed, since both components share the same space, the void modulation of water should match the cluster sizes of methanol.

These domains are quite apparent in the snapshots shown in Fig. 1. Fig. 3b shows the RKBI for $x = 0.5$. Most of the remarks made for the case $x = 0.2$ hold here also. We note the strong discrepancy between the KBI predicted for G_{WW} from both system sizes, while those for G_{OO} are in closer agreement. This alone indicates that the clustering in the $N = 2048$ system is somewhat stronger than that of the $N = 16\,384$ system.

In Fig. 3c, in order to sort out the status of the strong discrepancy between the G_{WW} from the two sizes, we have conducted a longer run of 12 ns on the $N = 2048$ by using the GROMACS code. The corresponding results are reported only for G_{WW} in purple. It is seen that longer statistics on the small system bring it close to that of the larger one, but still with an irreducible difference. It is also apparent that there is lots of noise in the various asymptotes and that runs in the 100 ns range might smooth the results. We also note that the KBI predicted from the larger system are in very good agreement with the experimental ones[31] (shown in horizontal dashes).

These calculations on larger systems indicate that the micro-heterogeneity fluctuates in the range of ns scale, and that it is quite sensitive to system size, and hence to periodical boundaries. Aside from the purely statistical issues and the obvious requirement for intense computational resources, the unsettling of the average in the ns-range indicates that this phenomenon may play a crucial role in system involving aqueous mixtures, such as, for example, biological systems.

C Integral equation studies of micro-heterogeneity

The study of realistic liquid mixtures by liquid state integral equation methods is reliable only for simple liquids that can be represented by Lennard-Jones (12-6) interactions with small dipole moments.[22] These methods lead to inaccuracies both in the correlations and the thermodynamic properties. For example, none of the existing integral equation methods are able to reproduce the properties of liquid water accurately.[32] Since these methods miss entire sets of Mayer diagrams involving higher order correlations,[22] it seems that the presence of strong anisotropies in the pair interactions leads to a local order that requires such higher correlations. In a recent work,[33] we have explored several realistic neat liquids, such as water, acetone and alcohols, with these methods by incorporating the missing correlations by extracting them from computer simulations. However, this methodology cannot be extended to mixtures, since it requires very accurate input of the $g_{ab}(r)$ from the simulations, and we have shown in the previous section that this is not really possible since the asymptote of the correlation function is difficult to stabilize. In the present work, we examine an alternative to this problem, by incorporating the missing correlations from the neat liquids. The theoretical frame that we use is to obtain the various site–site correlations from the SSOZ equation, that is written in matrix form as:

$$H = WCW + WCRH \tag{4}$$

where the H matrix contains the elements $H_{ab} = \tilde{h}_{ab}(k)$, which is the Fourier transform of the pair correlation functions $h_{ab}(r) = g_{ab}(r) - 1$. Similarly, the matrix elements $C_{ab} = \tilde{c}_{ab}(k)$ are the Fourier transform of the site–site direct correlation functions $c_{ab}(r)$. The matrix W contains the Fourier transforms of the site–site rigid links correlations $W_{ab}(r) = \delta(r - r_{ab})$ with r_{ab} being the distance between sites a and b of the same molecule. From this definition, one has $W_{ab} = \tilde{w}_{ab}(k) = \sin(kr_{ab})/(kr_{ab})$ if both sites belong to the same species, and $W_{ab} = 0$ otherwise. Finally, the R matrix is the density matrix, defined as $R_{ab} = \rho_a\delta_{ab}$ where ρ_a is the density of the molecular species containing atom a. Eqn (4) can be rewritten in terms of species–species matrices

$$H_{mn} = W_m C_{mn} W_n + \sum_p \rho_p W_m C_{mp} H_{pn} \tag{5}$$

where (m, n) are species indexes, ρ_p is the density of species p and matrix H_{mn} and C_{mn} contain the site–site correlations functions between sites belonging to species m and n, respectively. The SSOZ equation is solved for both the pair and direct site–site correlation functions, hence another equation is needed. In this work, we have considered the two closure equations that are recurrent in the modern literature of site-site integral equations,[35] namely the hyper-netted closure (HNC) and the KH closure,[34] and in addition a novel closure that we term BHNC. The BHNC closure is defined as:

$$g_{ab}(r) = \exp[-\beta v_{ab}(r) + h_{ab}(r) - c_{ab}(r) + B_{ab}(r)] \tag{6}$$

where $v_{ab}(r)$ is the pair interaction between atomic sites a and b belonging to molecular species 1 and 2, respectively, and $B_{ab}(r)$ is an effective "bridge" function that is taken from that of the neat liquids and used only in conjunction with site–site correlations related to this species component in a mixture. If this function is neglected, then one recovers the usual HNC closure, with $B_{ab}(r) = 0$. For a given concentration of methanol x, the bridge functions are calculated by mixing the pure component bridges:

$$B_{ab}(r) = a_W(1 - x)B_{ab}^{(W)}(r) + a_M x B_{ab}^{(M)}(r) \tag{7}$$

where $B_{ab}^{(W)}(r)$ is the site–site bridge function for pure water, and $B_{ab}^{(M)}(r)$ that for pure methanol. The scaling factors a_W and a_M are such as $0 < a_W, a_M < 1$ and serve to scale the neat liquid bridge functions in order to optimise the numerical solutions, as described below. Needless to say, the cross bridge interactions are totally absent from these considerations. There are empirical ways to include them from the pure components, by mixing them in some way. But we did not attempt any of these for reasons that we explain in the discussion below.

The KH closure[30] is defined as

$$g_{ab}(r) = \begin{cases} \exp[\gamma_{ab}(r)] & \text{if } g_{ab}(r) < 1 \\ 1 + \gamma_{ab}(r) & \text{if } g_{ab}(r) > 1 \end{cases} \tag{8}$$

with $\gamma_{ab}(r) = -\beta v_{ab}(r) + h_{ab}(r) - c_{ab}(r)$. This closure has no diagrammatic proper origins. It is in fact an empirical way to diminish the correlations that tend to grow out of bounds in the exponential form of eqn (8) by linearising this growth. Indeed, the HNC closure often tends to over-enhance correlations in the range of a few molecular diameters, which leads to the growth of long range correlations and the loss of numerical solutions for this equation. This problem has plagued the use of this closure and the "softer" KH version[34,35] is often used in order to avoid such problems. However, this practice comes with a price to pay: even though a numerical solution is obtained, it is almost never in good agreement with the simulation results, often producing poor correlation functions. The BHNC method is used here with the purpose of avoiding linearising part of the correlations, and the expectation of obtaining a better agreement with simulations. This is the issue that we explore here.

We have solved the coupled sets of equations, eqn (4,6) or eqn (4,8) for the aqueous methanol mixture of the same SPC/E water and OPLS methanol models used in Section B. The various functions were discretised on a regular grid of 2048 points, both in r-space and reciprocal k-space. The Coulomb interactions were handled in the standard way described in ref. 33. In the case of the BHNC closure, the neat liquid $B_{ab}(r)$ functions were taken from our previous work on neat liquids[33] where we have extracted these functions from high quality computer simulations, both for SPC/E water and OPLS methanol. All numerical solutions were obtained by standard iteration techniques, from an initial input guess of the $c_{ab}(r)$ functions, starting from either pure methanol or an equimolar system. The initial starting solutions were obtained at very high temperatures, for which all these equations can always be solved, and then the temperature is decreased until room temperature conditions are achieved. Once the solution is obtained at a given methanol mole fraction x, we try to obtain a new solution to another value of x. Typically, x was varied from 1 (pure methanol) to 0 (pure water) by steps of $dx = 0.1$. There are important differences between the closures, which have physical reasons, aside from all technical problems. We were unable to solve the HNC closure below the high temperature regime. The reason for this failure is due to increasing correlations at the rapid growth of the site–site structure factors $S_{ab}(k)$ near $k = 0$.

It is known that the HNC equation has no true spinodal,[36,37] hence $S_{ab}(k = 0)$ cannot diverge, but it attains an arbitrary large value and no numerical solutions can be obtained past this no-solution point.

This situation clearly shows that the solutions of the HNC closure develop large concentration fluctuations and indicate a phase separation scenario between water and methanol. In the absence of a true spinodal, we cannot conclude the certainty of this phase separation, but this is the generic behaviour of this equation in the context where a phase transition is indeed expected. The KH closure is able to give numerical solutions at all concentrations, but these solutions are very poor, as shown below. Finally, when inserting the bridge functions of the neat liquids, we are able to obtain numerical solutions of the BHNC closure.

The aqueous methanol system has been previously studied by other authors,[38] using "extended" SSOZ techniques where the Coulomb interactions are "renormalized" into the correlations, but they did not delve into the problems posed by fluctuations.

We start by very small values of the scaling a_W parameters (typically 0.01) for which a numerical solution is always obtained, while this is impossible for $a_W = 0$ which corresponds to the HNC closure. Then, we increase this parameter until we have problems converging the numerical solutions. Next, we start increasing the a_M parameter until problems are met again. Although empirical, the fact that this method provides a numerical solution attests that the influence of handling correct correlations of the pure liquids. Indeed, as stated in ref. 33, the inclusion of the bridge functions can be seen as changing the pair interaction $v_{ab}(r)$ into an effective one $v_{ab}(r) - B_{ab}(r)$. It was shown that the effective interaction was slightly attractive at contact, as expected from the fact that the hydrogen bonding interaction tends to bring molecules closer than the Lennard-Jones contact range.

What is really surprising is the very good agreement that is generally obtained by including these neat liquid contributions, as witnessed in Fig. 4(a–c) for typical methanol concentrations covering the water rich ($x = 0.2$), equimolar ($x = 0.5$) and methanol rich ($x = 0.8$) regimes. This unexpected agreement shows that most of the short range features are captured from the pure liquids, which is also an indirect proof that there is micro-segregation of the species into domains of about a few molecular diameters. The water–methanol cross corrections are not so good since we miss the corresponding bridge terms, and it turns out that these cross-correlations are relatively the same for both closures. It is also obvious that the KH closure, while always allowing a numerical solution to be obtained, is inaccurate at short range.

However, as we have shown in ref. 33, it is also vital to check in the reciprocal space that the medium to long range correlations are well described.

This is done by calculating the site–site structure factors $S_{ab}(k) = 1 + \sqrt{\rho_a \rho_b} \int d\vec{r} \exp\left(i\vec{r}.\vec{k}\right) h_{ab}(r)$, which are shown in Fig. 5(a–c). The good agreement in the small-k region is a sign of good description of the long range correlations $g_{ab}(r)$ in the direct space. These figures show that the apparent good description of the short range correlations of the BHNC closure is not accompanied by a similar description of the long range correlations.

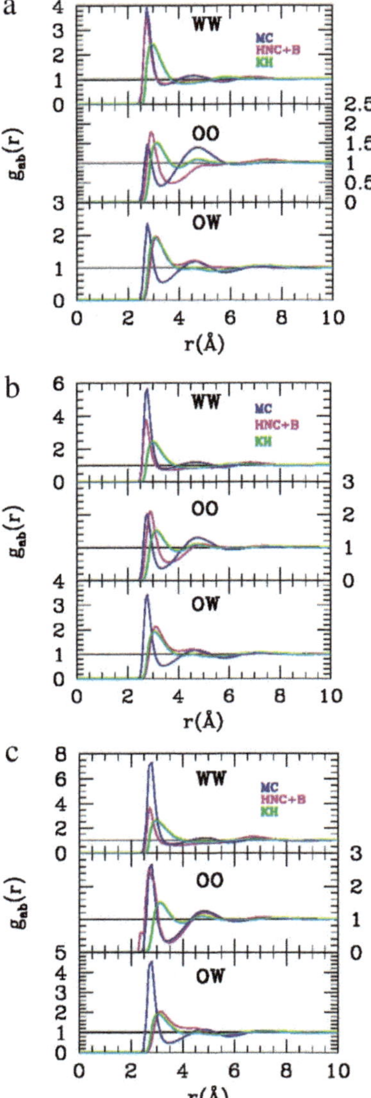

Fig. 4 (a) Oxygen–oxygen site–site correlations between water molecules (top panel), Me molecules (middle panel) and cross correlations (bottom panel), for $x = 0.2$. The simulation results are in blue lines, HNC + B in magenta and KH results in green. (b) Same as in Fig. 4a but for $x = 0.5$. (c) Same as in Fig. 4a but for $x = 0.8$.

What we find is that the predicted concentration fluctuations are too small, in variance with those of the simulations that are quite large. As a consequence, the BHNC closure predicts KBI that are too small. In variance, the KH closure predicts better KBI, even though these are no match to the simulation results. We also note that the BHNC closure shows a better agreement with the simulation near the first peak of the structure factors, which is about 2 Å^{-1} for WW, and which is a direct consequence of good agreement observed in the short-medium range part of the correlations in Fig. 3.

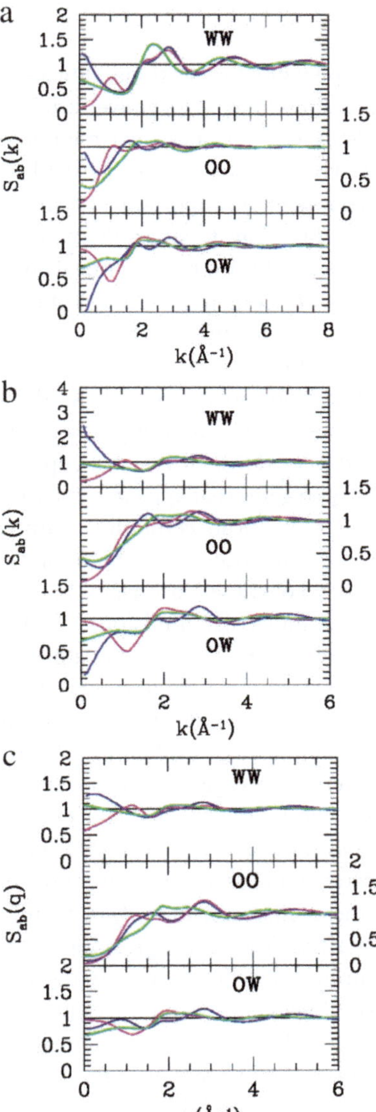

Fig. 5 (a) Site–site structure factor between oxygen sites for $x = 0.2$. The panel and color configurations are the same as in Fig. 4a. (b) Site–site structure factor between oxygen sites for $x = 0.5$. The panel and color configurations are the same as in Fig. 4a. (c) Site–site structure factor between oxygen sites for $x = 0.8$. The panel and color configurations are the same as in Fig. 4a.

The new feature that Fig. 4 shows is the presence of a pre-peak in the BHNC structure factor, mostly for $S_{WW}(k)$, and positioned around 1 $Å^{-1}$. This corresponds to a water–water aggregate structure about 6 Å in real space. This structure is not predicted by the simulations and the KH closure. In view of the relatively good agreement observed in the correlations in Fig. 3, one might wonder why this aggregate structure is predicted by the BHNC closure and not by the simulations. Can we rationalise why the BHNC closure would not be as good at long range as it is at shorter ranges?

D Discussion

The results shown in the previous two sections indicate both the importance of the role played by fluctuations and aggregative structures in simple aqueous mixtures and the problems current statistical methods and theories have in handling them properly. Fluctuations, and in particular concentration fluctuations in the actual context, play an important role in the description of phase transitions.[39] In the present context, since there are no phase transitions, concentration fluctuations should not play a crucial role. Yet, from the lack of stability of the asymptotes of the pair correlation functions as obtained from computer simulations, it is obvious that fluctuations decay at scales larger than the nm and the ns. The formation of self-aggregative structures that are stabilised by hydrogen bonding interactions are responsible for the slow dynamics of these systems. Even though such structures may appear as semi-frozen concentration fluctuations, which "melt" very slowly, it is more appropriate to compare these structures to micelle type aggregates that appear in micro-emulsions. The kinetics of formation of both such structures are certainly very different. In addition, since both formations are not true phase transitions, both can be compared to frozen-in concentration fluctuations.[40] For these reasons, we have previously termed these simple aqueous mixtures as "molecular emulsions".[11,12] From the point of view of the theory of liquids, we have shown that existing approximations are not suited to capture all the features of the micro-heterogeneity. The KH closure appears as a convenient numerical artefact, which allows to obtain solutions where diagrammatically proper methods fail, without explaining the reason why this happens. The BHNC method seems encouragingly accurate for the short range correlations, only to become deceptively inaccurate as concerns the predictions about concentration fluctuations and KBI. However, the features that we have reported here give us a clue to better understand the nature of the fluctuations and the micro-heterogeneity in these mixtures. Indeed, as mentioned in Section C, the numerical solution of the HNC closure fails precisely because this theory predicts a too rapid growth of concentration fluctuations in the small-k region of the site–site structure factors. The inclusion, through the BHNC closure, of the bridge functions from the neat liquids provides a numerical solution by almost suppressing concentration fluctuations because of the very low $k = 0$ values of the site–site structure factors. In addition, this theory predicts an aggregate structure which produces a pre-peak in the water–water structure factor. These two features seem related by this approximate closure: the stabilised aggregate structure diminishes the concentration fluctuations. However, this is in variance with what is predicted by the simulations, that is high concentration fluctuations, with aggregates apparent from the snapshots, but with no apparent pre-peak. Since the theory is missing some sets of cluster diagrams, we deduce that these missing diagrams contains the molecular links that would make the aggregate fluctuate more, in accordance with what is predicted from the simulations. This prediction seems a reasonable way to conciliate both findings. It also shows the complexity of the link between concentration fluctuations and micro-heterogeneity in these mixtures. It is tempting to speculate that these mixtures are not demixed precisely because fluctuations are at the scale of the aggregates and not the molecules. The KH closure predicts some sort of average structure over all ranges, which produces higher concentration fluctuations than the BHNC, but still lower than compared with the raw HNC method. It is striking that the predictions of the correlations of both the KH and HNC closures are indistinguishable when it concerns pure water or pure methanol,[41] but differ so markedly when it comes to the mixtures. This is another indication that the mixture is entirely governed by the cross-correlations, which we do not reproduce correctly by both theories. Finally, we would like to emphasize that there are no simple methods to account for cross correlations from the knowledge of the bridge functions of the pure liquids. Indeed, from inspection of the cross-correlations, it is seen that the self-aggregated domain induces a depletion cross interaction. This effective interaction is not contained in those of the pure liquids. It is possible to introduce an arbitrary repulsive cross interaction, and similarly enhance the self-attraction in order to boost the concentration fluctuations and reach a better agreement with the simulations. We are currently exploring how to introduce such corrections through various thermodynamics self-consistency criteria.

E Conclusions

In biological matter we see spontaneous formation of membranes and other such complex objects, self assembled from elementary molecules. The microscopic mechanism behind self-assembly is the hydrophobic effect, and the aqueous methanol mixture is the most elementary such system at its most rudimentary expression. If we aim at understanding how complex molecular architectures can be formed, we need to explain from first principles the nature of the correlations between all the constituents. Computer simulations are often referred to as the preferred statistical tool to analyse complex systems, since it provides a direct microscopic picture of the system. What we have shown here is that aqueous methanol mixtures reorganise themselves at the scale of the nm and the ns, due to the inherent micro-heterogeneity, which makes statistical analysis rather difficult because true equilibrium needs system sizes larger than 10 K particles. Since these systems are not near a critical point, this is not critical slowing down, rather the existence of aggregative domains that have a slower relaxation time scale. Although this may be considered as a statistical problem – with the obvious remedy to study even larger systems – one can also see this width of the spatial and temporal scales as a characteristic property of aqueous mixtures in general. Consequently, one may ask if this property, hidden behind statistical problems, may not be, in fact, an important one for the biological phenomenon,[18,19] by providing dynamical valleys to continuous re-organisation of the system. The existence of wide relaxation times to reach global equilibrium provides room for the system to reorganise itself locally in several ways, allowing for a certain flexibility or plasticity of the system. In the present work, we have examined how two different statistical methods can account for the complexity of the interplay between concentration fluctuations and self-aggregation, with apparent deceptive results. However, in light of the slow equilibration of these systems, it is also apparent that increased accuracy is required to properly describe these systems with static properties such as correlation functions. Similar problems are encountered by both methods for systems in the vicinity of critical points, in particular in the need for higher order correlations.[22,39] However, while there are accurate methods to handle the problems in this latter context, it is clear that similar methods are missing for handling fluctuations and micro-heterogeneity for systems with strong micro-structure. In this context, ideas developed along the lines of competing attractive and repulsive interactions[16,42] could provide a basis to develop statistical approaches for complex mixtures.

Notes and references

1 A. Perera, in *Fluctuation Theory of Solutions: Applications in Chemistry, Chemical Engineering, and Biophysics*, ed. P. E. Smith, J. P. O'Connell and E. Matteoli, C. R. C. Press (Taylor Francis Group), 2013, pp. 163–190.
2 A. Perera, B. Kežić, F. Sokolić and L. Zoranić in *Molecular Dynamics*, ed. L. Wang, InTech, Rijeka, 2012, vol. 2, pp. 193–220.
3 E. Matteoli and L. Lepori, *J. Chem. Phys.*, 1984, **80**, 2856.
4 J.-H. Guo, Y. Luo, A. Auggustsson, S. Kashtanov, J.-E. Rubensson, D. K. Shuh, H. J. Agren and J. Nordgren, *Phys. Rev. Lett.*, 2003, **91**, 157401-4.
5 S. K. Allison, J. P. Fox, R. Hargreaves and S. P. Bates, *Phys. Rev. B*, 2005, **71**, 024201.
6 R. D. Mountain, *J. Phys. Chem.*, 2010, **114**, 16460.
7 R. Gupta and G. N. Patey, *J. Chem. Phys.*, 2012, **137**, 034509.
8 E. A. Ploetz, N. Bentinis and P. E. Smith, *Fluid Phase Equilib.*, 2010, **290**, 43.
9 M. E. Lee and N. F. A. Van der Vegt, *J. Chem. Phys.*, 2005, **122**, 114509.
10 A. Perera and F. Sokolic, *J. Chem. Phys.*, 2004, **121**, 11272.
11 B. Kežić and A. Perera, *J. Chem. Phys.*, 2012, **137**, 014501; B. Kežić and A. Perera, *J. Chem. Phys.*, 2012, **137**, 134502.
12 B. Kežić, R. Mazighi and A. Perera, *Phys. A*, 2013, **392**, 567.
13 C. Tanford, *Proc. Natl. Acad. Sci. U. S. A.*, 1974, **71**, 1811.
14 D. C. Poland and H. A. Scheraga, *J. Phys. Chem.*, 1965, **69**, 2431.
15 G. D'Arrigo, L. Giordano and J. Teixeira, *Eur. Phys. J.*, 2009, **29**, 37.
16 A. Ciach and W. T. Godz, *Annu. Rep. Prog. Chem., Sect. C*, 2001, **97**, 269.
17 S. Dixit, J. Crain, W. C. K. Poon, J. L. Finney and A. K. Soper, *Nature*, 2002, **416**, 829.
18 P. Ball, *Nature*, 2008, **452**, 291.
19 P. Wiggins, *PLoS One*, 2008, **e1406**, 1.
20 J. G. Kirkwood and F. P. Buff, *J. Chem. Phys.*, 1951, **19**, 774.
21 D. Chandler and H. C. Andersen, *J. Chem. Phys.*, 1972, **57**, 1930.
22 J. P. Hansen and I. R. McDonald, *Theory of simple liquids*, Academic Press, Elsevier, Amsterdam, The Netherlands, 2006.
23 M. Haughney, M. Ferrario and I. R. McDonald, *J. Phys. Chem.*, 1987, **91**, 493.
24 S. Weerasinghe and P. E. Smith, *J. Chem. Phys.*, 2003, **118**, 10663.
25 A. Perera, L. Zoranic, F. Sokolic and R. Mazighi, *J. Mol. Liq.*, 2011, **159**, 52.
26 J. C. Berendsen, J. P. M. Postma, W. F. Von Gusteren and J. Hermans, in *Intermolecular Forces*, ed. B. Pullman, Reidel, Dortrecht, 1981.
27 W. L. Jorgensen, *J. Phys. Chem.*, 1986, **90**, 1276, methanol model.
28 DL_POLY package from CCLRC Daresbury Laboratory, www.ccp5.ac.uk/DL_POLY/.
29 D. van der Spoel., E. Lindahl, B. Hess., G. Groenhof, A. E. Mark and H. J. C. Berendsen, *J. Comput. Chem.*, 2005, **26**, 1701.
30 J. L. Lebowitz and J. K. Percus, *Phys. Rev.*, 1961, **122**, 1675.
31 A. Perera, F. Sokolic, L. Almasy and Y. Koga, *J. Chem. Phys.*, 2006, **124**, 124515.
32 M. Lombardero, C. Martin, S. Jorge, F. Lado and E. Lomba, *J. Chem. Phys.*, 1999, **110**, 1148.
33 B. Kežić and A. Perera, *J. Chem. Phys.*, 2012, **137**, 014501.
34 A. Kovalenko and F. Hirata, *J. Chem. Phys.*, 1999, **110**, 10095.
35 F. Hirata, *Molecular Theory of Solvation Kluwer Academic*, Dordrecht, 2003.
36 M. I. Guerro, G. Saville and J. S. Rowlinson, *Mol. Phys.*, 1975, **29**, 1941.
37 L. Belloni, *J. Chem. Phys.*, 1993, **98**, 8080.
38 H. Tanaka, J. Walsh and K. E. Gubbins, 1992, **76**, p. 1221.
39 P. M. Chaikin and T. C. Lubensky, *Principle of Condensed Matter Physics*, Cambridge University Press, 2000.
40 G. Foffi, C. De Michele, F. Sciortino and P. Tartaglia, *J. Chem. Phys.*, 2005, **122**, 224903.
41 B. Kežić, Doctorate thesis, 2013, Université Pierre et Marie Curie, Paris 6.
42 A. J. Archer and N. B. Wilding, *Phys. Rev. E*, 2007, **76**, 031501.

Faraday Discussions RSC Publishing

PAPER

What happens to the structure of water in cryoprotectant solutions?

James J. Towey,[a] Alan K. Soper[b] and Lorna Dougan*[a]

Received 10th May 2013, Accepted 17th June 2013
DOI: 10.1039/c3fd00084b

Cryoprotectant molecules are widely utilised in basic molecular research through to industrial and biomedical applications. The molecular mechanisms by which cryoprotectants stabilise and protect molecules and cells, along with suppressing the formation of ice, are incompletely understood. To gain greater insight into these mechanisms, we have completed an experimental determination of the structure of aqueous glycerol. Our investigation combines neutron diffraction experiments with isotopic substitution and computational modelling to determine the atomistic level structure of the glycerol–water mixtures, across the complete concentration range at room temperature. We examine the local structure of the system focusing on water structure. By comparing our data with that from other studies of cryoprotectant solutions, we attempt to find general rules for the action of cryoprotectants on water structure. We also discuss how these molecular scale interactions may be related to the macroscopic properties of the system.

1 Introduction

The application of cryopreservation to living cells and tissues has revolutionised areas of biotechnology, plant and animal breeding programmes and modern medicine.[1] The fact that cells can be recovered from temperatures far below the freezing point of water is a remarkable feat of resilience. However, in most of these situations there is an additional ingredient – the cryoprotectants. A cryoprotectant is the term given to any additive which can be provided to cells before freezing and which then yields a higher post-thaw survival rate than can be achieved in its absence.[2] Glycerol was the first cryoprotectant molecule to be discovered, by Polge *et al.* in the 1940s, when it was successfully used to cryopreserve sperm cells.[3] Since this pioneering work, glycerol has become the molecule of choice in many areas of modern science. Cryopreservation is now an effective process and is routinely used to preserve cells or tissues by cooling to sub-zero temperatures. At low temperatures any biological activity, including reactions that would cause cell death is

[a]*School of Physics and Astronomy, University of Leeds, Leeds, United Kingdom, LS9 2JT. E-mail: L.Dougan@leeds.ac.uk*
[b]*ISIS Facility, Rutherford Appleton Laboratory, Chilton, Didcot, Oxon, OX11 0QX, UK*

effectively stopped. The goal of successful cryopreservation is the ability to recover as many cells or tissues as possible, after cooling to sub zero temperatures.

A major challenge is the changes brought about by the transition of water to ice during cooling, resulting in osmotic stress or 'freeze dehydration', as well as physical destruction of membranes by ice and organelle disruption. While glycerol is routinely used as a cryoprotectant, the molecular mechanisms by which the molecule acts to prevent freezing damage still remain unclear. This is limiting given that any future advances in cryopreservation rely on understanding the action of glycerol on water upon cooling and having accurate predictive control over cryopreservation protocols. Advances in biotechnology and in the clinical use of cryopreserved cells or tissues can only be met by moving beyond qualitative descriptions of cryoprotective action. Central to this problem is the action of glycerol on the structure of water. One early and reoccurring hypothesis is that cryoprotectants act by modifying water structure.[4] In the 1960s, Nash attempted to qualitatively explain cryoprotectant action on the ability of cryoprotectant molecules to modulate hydrogen bonding and their interactions with water.[5] Another study postulated that cryoprotectant action was the result of differences in the hydrogen bonding interactions between solute–solute, water–solute and water–water interactions.[6] While conceptually attractive, there has been no direct, experimental structural evidence of this to date. More recently, a number of computer simulations have examined cryoprotectant–water solutions and found a disruption of the tetrahedral arrangement of water molecules near the cryoprotectant due to steric constraints imposed by the solute and its ability to form stable hydrogen bonds with proximal water molecules.[7–9] For example, a molecular dynamics simulation study of hydrogen bonding in glycerol–water suggested that glycerol molecules act to decrease the hydrogen bonding ability of water.[10] Until recently, a comparison between the results of these computational studies and experimental findings has been limited due to lack of complete and direct experimental studies of the microscopic structure of water in cryoprotectant solutions.

Over the last 4 years we have completed a series of studies on glycerol–water solutions using neutron diffraction and isotopic substitution, coupled with computational modelling. These studies have provided atomistic-level structural information on glycerol–water in the liquid state, across the concentration range.[11–14] In addition to this work on glycerol, a growing body of experimental data, using the same experimental and computational modelling approach, has been completed on other cryoprotectant solutions including; dimethyl sulfoxide,[15,16] trimethylamine oxide,[17] methanol,[18–20] tertiary butanol,[21,22] proline,[23] sorbitol[24] and trehalose.[25,26] In this paper we wish to examine and compare the structure of water in each of these cryoprotectant solutions. We aim to determine if there are general rules for the action of cryoprotectants on water structure. In doing so it may be possible in the future to make predictions for the possible action of as yet unstudied cryoprotectant molecules.

2 Cryoprotectant molecules studies using neutron diffraction

2.1 The glycerol molecule

Glycerol is a sugar alcohol with three hydroxyl groups that are thought to be responsible for its high solubility in water. Glycerol has been the subject of

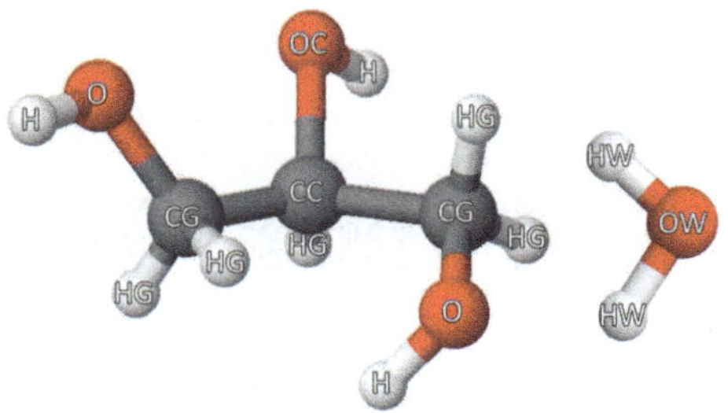

Fig. 1 The glycerol molecule. Single atoms have been labelled according to the symbols used in the simulation and throughout this paper where the carbon atoms are labelled CG and CC, the oxygen atoms O and OC, the hydroxyl hydrogen atoms, H and the methyl hydrogen atoms, HG.

considerable and long-standing scientific interest.[1,3] It is active in maintaining the structure of biological macromolecules and is thought to promote self-assembly through preferential hydration. In nature, organisms use glycerol in high concentrations as a colligate cryoprotectant that is thought to raise the osmolarity of body fluids and reduce the water available to form extracellular ice.[27–29] Glycerol is widely used as a component for storage of biological and pharmaceutical products at low temperature.

On account of molecular flexibility, asymmetry and ability to form hydrogen bonds, glycerol is an intricate substance to study at the molecular level (Fig. 1). In the glycerol molecule all bonds between atoms are single and there is therefore considerable rotational freedom. The traditional classification of the molecular structure of this molecule is based on the concept of backbone conformation, defined by the structure of the heavy atoms (carbon and oxygen) irrespective of the position of the hydrogen atoms.[30] These backbone conformations are defined in terms of the dihedral angles which are formed at the intersection between 2 specified planes within the molecule, where each terminal CH_2OH group can rotate about the other backbone carbon–carbon bond. The conformational structure of glycerol in the crystalline state has been identified by X-ray diffraction which revealed the presence of only $\alpha\alpha$ backbone conformers.[31] For all other phases, the available experimental and theoretical data are contradictory.[32–36] For that reason computational modelling using an acceptable molecular model for glycerol is a highly desirable prospect.

2.2 Other cryoprotectant molecules

As well as our own studies on glycerol, a number of studies have been completed on other cryoprotectant systems (Fig. 2) including dimethyl sulfoxide,[15,16] trimethylamine oxide,[17] methanol,[18–20] proline,[23] tertiary butanol,[21,22] sorbitol[24] and trehalose.[25,26] Dimethyl sulfoxide (DMSO) is an organosulphur compound and is miscible in a wide range of organic solvents as well as water. DMSO is widely used in cryopreservation to reduce ice formation in sperm and embryos that are cold-preserved in liquid nitrogen.[1] Trimethylamine oxide (TMAO) is a naturally occurring osmolyte that is found in saltwater fish, sharks and crustaceans.[28] TMAO counteracts the denaturing effects on proteins of molecules such as urea, as well as high pressure, and low temperatures.[28,37,38] It also thought to increase attractive intermolecular interactions in protein solutions. Methanol is the simplest alcohol and is used as an antifreeze and a cryoprotectant in the preservation of embryos.[39]

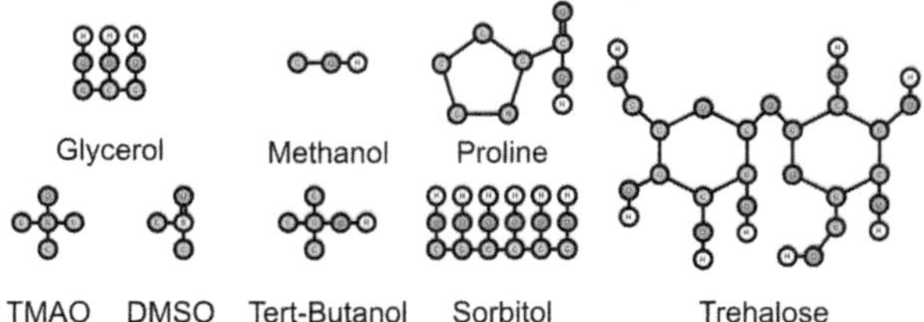

Fig. 2 Schematic diagram of cryoprotectant solutions which have been examined using neutron diffraction and EPSR computational modelling. Carbon atoms are grey spheres labelled C, oxygen atoms are red spheres labelled O, hydroxyl hydrogens are white spheres labelled H, nitrogen are blue spheres labelled N and sulphur are yellow spheres labelled S. The methyl, methylene and amine hydrogen atoms (those bound to the carbon or nitrogen atoms) have been omitted to aid clarity.

The amino acid proline is found to accumulate in the cells of a large variety of water-stressed organisms, where it is thought to have the ability to counter environmental water stress and to protect cellular proteins against denaturation under extreme conditions.[28] Proline is unique among the 20 protein-forming amino acids in that the amino nitrogen binds with the side chain, leading to the formation of a pyrolidine ring. This structure restricts the conformations that proline can adopt within a protein, giving it exceptional conformational rigidity compared to the other amino acids. This may account for the higher prevalence of proline in proteins of thermophilic (heat-loving) organisms.[40] Sorbitol is a sugar alcohol and a common pharmaceutical excipient used to enhance protein stability and drug solubility during freezing and storage.[5,41] Trehalose is a disaccharide sugar and is formed when two glucose molecules are bonded together. It is known to accumulate *in vivo* as a response to temperature stress.[42]

3 Neutron diffraction and computational modelling

3.1 Neutron diffraction experiments

Neutron diffraction coupled with isotopic substitution allows labelling of individual atomic sites in a molecule and the extraction of selected pair radial distribution functions (RDFs), $g(r)$. Neutron diffraction measurements were completed on the SANDALS time-of-flight diffractometer at the ISIS pulsed neutron facility at the Rutherford Appleton Laboratory, UK. SANDALS is a total scattering neutron diffractometer optimised for the study of liquids and amorphous samples containing light elements.[14,22,25,43] The physical quantity measured by the diffractometer is the differential scattering cross section $d\sigma/d\Omega$ as a function of the exchanged wave vector $\boldsymbol{Q}$ (defined as the difference between the incident and the scattered neutron wave vectors). Through the theory of neutron scattering, it is possible to relate this quantity to the static structure factor $S(Q)$, which is the Fourier transform of the atomic pair distribution function $g(r)$. The function $g(r)$ shows how atomic densities vary as a function of radial distance, r, from any particular atom.

Deuteriated samples of water and protiated and deuteriated samples of anhydrous glycerol were obtained from Sigma-Aldrich and used without additional purification. The protiated glycerol sample was molecular biology grade (≥99.5% purity), deuteriated glycerol (98% purity) and partially deuteriated glycerol where the carbonyl atoms are deuterium while the hydroxyl are hydrogen (98%).

Table 1 The different isotopically substituted samples used for each concentration of aqueous glycerol. In each case anhydrous glycerol was obtained from Sigma-Aldrich and used without additional purification. Glycerol–water mixtures ($x_g = 0.05, 0.10, 0.25, 0.50$ and 0.80)

Sample	Description
i	Fully deuteriated glycerol (98 atom% D) with D_2O (99.99 atom% D)
ii	Carbonyl atoms are deuterium and hydroxyl are hydrogen (98 atom% D) with Milli-Q water
iii	1 : 1 mixture of samples i and ii with a 1 : 1 mixture of Milli-Q water and D_2O (99.99 atom% D)
iv	Fully protiated glycerol (99.5%) with Milli-Q water
v	1 : 1 mixture of samples i and iv with a 1 : 1 mixture of Milli-Q water and D_2O (99.99 atom% D)

Experiments were completed across the full concentration range. Specifically, glycerol mole fractions $x_g = 0.05, 0.10, 0.25, 0.50, 0.80$ and 1.00 at 298 K. For each concentration a total of 5 samples were used, making use of relevant isotope substitution of the hydrogen (see Table 1 for full list of samples).

The samples were loaded into the sample container which was made of a titanium zirconium alloy and is a flat plate of internal dimensions 1 mm by 35 mm by 35 mm with a wall thickness of 1.1 mm. This gave negligible coherent scattering signal. The sample container was held in a custom-built loading platform which securely holds the container. Glycerol samples were deposited in the sample container using sterile syringes, after which the sample container was securely sealed using a deformable plastic seal and mounting screws. After the sample container had been sealed they were placed in a vacuum oven. Sample cans were weighed before and after placement in a vacuum oven to ensure the container was sealed and no evaporation had occurred. These cells were mounted on an automated sample changer to cycle through the samples. The differential scattering cross-section for each sample was obtained by normalising to a vanadium standard. Corrections for attenuation and multiple scattering were made using the Gudrun program suite which has been detailed previously.[44] A further correction for inelastic scattering was made and has been described in detail elsewhere.[45] Neutron diffraction on solutions yields the quantity, $F(Q)$, which is the total interference differential scattering cross section, and which is the sum of all partial structure factors $S_{\alpha\beta}(Q)$ present in the sample each weighted by their composition c and scattering intensity b, $F(Q) = \Sigma_{\alpha\beta} c_\alpha c_\beta b_\alpha b_\beta \left(S_{\alpha\beta}(Q) - 1\right)$, where Q is the magnitude of the change in momentum vector by the scattered neutrons ($Q = (4\pi/\lambda) \sin\theta$). Fourier transform of $S_{\alpha\beta}(Q)$ gives the respective atom-atom radial distribution functions (RDFs) $g_{\alpha\beta}(r)$, and integration of $g_{\alpha\beta}(r)$, gives the coordination numbers of atoms α around atoms β between two distances r_1 and r_2.

3.2 Computational modelling

In this paper we refer to 6 distinct atomic components in glycerol molecules (Fig. 1). The carbon atoms are labelled CC and CG and refer to the central and distal carbon atoms respectively. The oxygen atoms are labelled OC and O corresponding to the oxygen atoms attached to the central and distal carbon atoms respectively. The hydrogen atoms in glycerol are labelled H and HG for the hydroxyl and methyl hydrogen atoms respectively. The water oxygen atom is labelled OW and the water hydrogen atoms are labelled HW. A full structural characterisation of the system therefore requires the determination of 36 site–site radial distribution functions, which is well beyond the possibility of any existing diffraction technique by itself.

Thus, to build a model of aqueous glycerol liquid structure, these experimental data are used to constrain a computer simulation. However, unlike conventional simulations, such as molecular dynamics or *ab intio* calculations which use a fixed intermolecular potential, part of the potential used here is obtained directly from the diffraction data. This potential drives the structure of the three-dimensional model toward molecular configurations that are consistent with the measured neutron diffraction data. This process is called Empirical Potential Structure Refinement (EPSR).[46] EPSR aims to produce a model with a simulated differential scattering cross section which fits the experiment results as closely as possible. Given that in this case there are more site–site radial distribution functions than diffraction data sets, extra information is required to define the structure. This is achieved by forcing the molecules in the simulation box to adopt the expected molecular geometries. A reference interaction potential is also used which serves to generate hydrogen bonding between the relevant atom sites and to prevent atomic overlap at unrealistic distance ranges. The combined empirical and reference potentials do not guarantee the final reconstruction of the structure is exactly correct, but they do ensure it is consistent with the diffraction data as well as being physically plausible. We have previously employed this approach to examine dilute glycerol solutions,[11] concentrated glycerol solutions[13,14] and pure glycerol.[12] Here we examine in detail the hydrogen bonding ability of the system across the concentration range.

For the simulations, molecules were contained in a cubic box of the appropriate dimensions to give the measured density at 298 K. The SPC model was employed for water molecules.[47] The intra-molecular structure of glycerol was optimised using the computational chemistry software Ghemical 2.98.[48] A three-dimensional computer model of the solution is constructed and equilibrated using relevant interaction potentials.[14] The charges and Lennard-Jones constants have been described previously.[14] Periodic boundary conditions were imposed and the Coulomb interactions were truncated by means of a derivative of the reaction field method, and other interactions were truncated as described previously, using a radial cut-off of 12 Å in both cases.

Information from the diffraction data is then introduced as a constraint whereby the difference between observed and calculated interference differential scattering cross section enters as a perturbation potential to drive the computer model (*via* Monte Carlo updates of atomic positions) towards agreement with the measured data. The perturbation is refined in successive iterations of the procedure until a satisfactory fit is obtained. In this way an ensemble of three-dimensional molecular configurations of the mixture is generated which exhibit average structural correlations that are consistent with the diffraction data. This approach has previously proved powerful in elucidating structural information on a range of aqueous solutions including methanol,[18–20,49] tertiary butanol,[21,50] DMSO,[15] as well as amino acids[23,43,51,52] and peptides.[53,54]

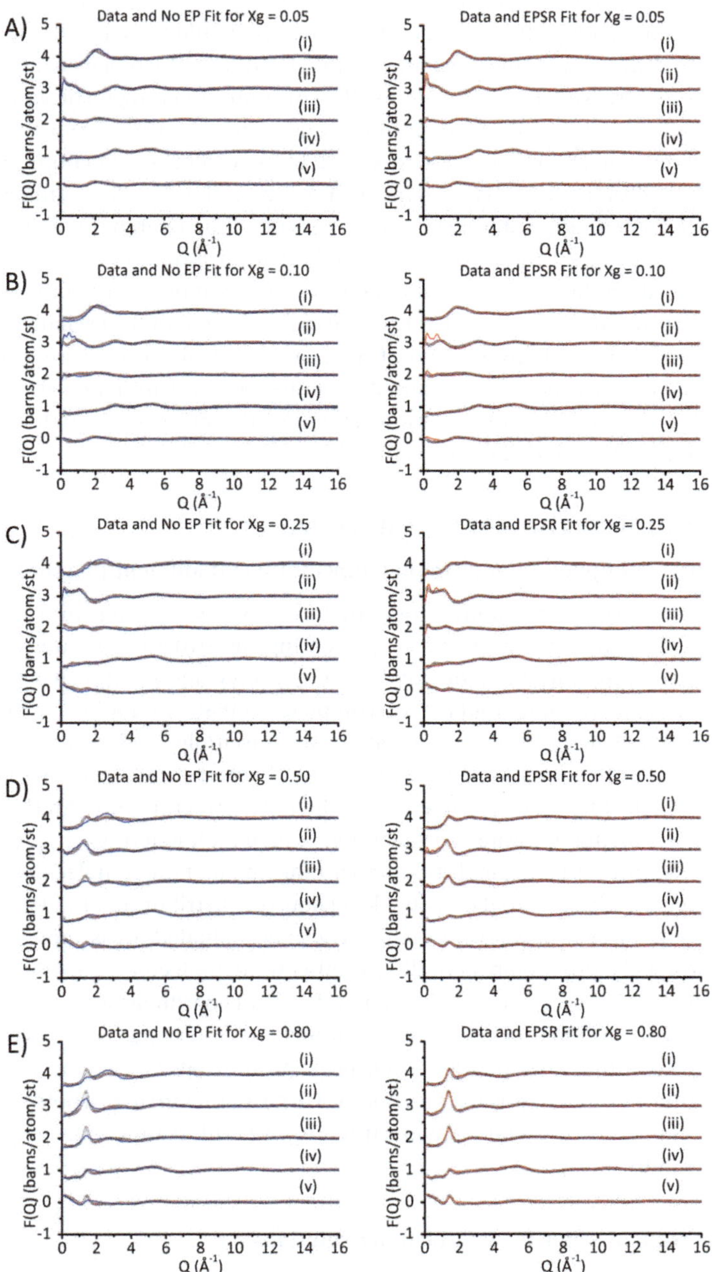

Fig. 3 Diffraction pattern (black circles) and reference potential simulations (left, blue line) EPSR-fitted diffraction pattern (right, red line) for different isotope substitutions of a dilute glycerol water solution (A) mole fraction glycerol $x_g = 0.05$, (B) $x_g = 0.10$, (C) $x_g = 0.25$, (D) $x_g = 0.50$, (E) $x_g = 0.80$ at 298 K. The data and fits are labelled according to Table 1 and have been shifted vertically for improved clarity. The data shown here confirm that this model produces a simulation that is in good agreement with the experimental data.

3.3 EPSR results for glycerol water data

Fig. 3 shows the diffraction data (black circles) and the results of the simulation which employed only the reference potentials (left, blue line) and the EPSR simulation diffraction pattern (right, red line) for different isotope substitutions of a dilute glycerol water solution, across the concentration range. The agreement of the ESPR simulated structure factors is good in all cases. From the EPSR molecular ensembles, we derive site–site radial distribution functions (RDFs), $g_{\alpha\beta}(r)$, which describe the relative density of one type of atom, β, as a function of distance, r, from another atom type α. Comparison of the reference potential only simulation (blue lines) and EPSR simulation (red lines) show how the EPSR model has enhanced the fit to the experimental data. This is more clearly presented in Fig. 4 where the sum of the squared residuals as a function of the exchanged wave vector ($\boldsymbol{Q}$) is shown, where the sum of residuals is calculated by comparing the simulated and experimentally measured structure factors. This demonstrates that EPSR plays a critical role in the improvement of the fit of the model structure factors to the experimental data.

4 Examining the structure of water

4.1 Structure of pure water at room temperature and ambient pressure

Given that we will be examining the structure of water in detail in a range of cryoprotectant solutions, we begin by presenting the structure of pure water at room temperature and ambient pressure. Using previously completed neutron diffraction data on pure water[55] we completed EPSR simulations using the methods described in Section 2 to obtain an ensemble of three-dimensional molecular configurations with average structural correlations which are consistent with the diffraction data. We examine the water structure through the radial distribution function for water oxygen–water oxygen interactions $g_{OwOw}(r)$, shown in Fig. 5. The three dimensional structural arrangement of water molecules can be seen in Fig. 5A, where spatial density plots show the distribution of water oxygen atoms around a central water molecule. The yellow shaded areas represent the regions where the probability of finding another water molecule exceeds a defined threshold. This figure illustrates the first and second coordination shell of water. The radial distribution function of oxygen atoms in water, $g_{OwOw}(r)$, can be seen in Fig. 5b, with a first coordination shell peak position at 2.76 Å and a second coordination shell peak position at 4.56 Å. Given the recurring hypothesis that cryoprotectants act to modify water structure, we will examine the impact of cryoprotectants on $g_{OwOw}(r)$.

4.2 Structure of water in glycerol–water solutions

To investigate the extent to which the water–water interactions are modified by the addition of glycerol, we examined the radial distribution function $g_{OwOw}(r)$ for each of the different glycerol–water solutions; $x_g = 0.05$, 0.10, 0.25, 0.50 and 0.80 at 298 K (Fig. 6). Examination of the RDFs for $g_{OwOw}(r)$ reveals that the first peak position in the glycerol–water solutions shows only small deviations as compared with pure water (Fig. 6 and Table 2). The largest deviation is seen at $x_g = 0.25$ which displays a shift in the radial position of 0.06 Å. The position of the second peak in $g_{OwOw}(r)$ in the glycerol–water solutions shows, however, much larger

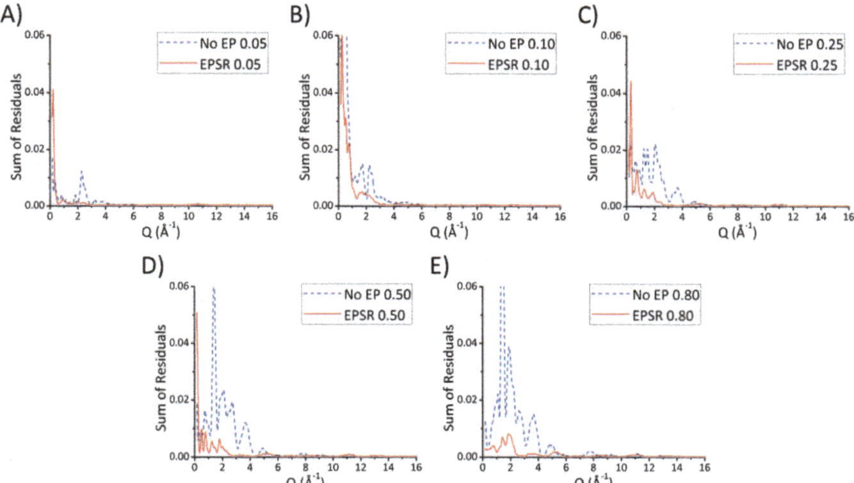

Fig. 4 The sum of the squared residuals as a function of the exchanged wave vector (***Q***), where the sum of residuals is calculated by taking the sum over the different isotopic substitutions of the square of the residual between the simulated and experimentally measured structure factors for each of the EPSR setups as a function of the exchanged wave vector, ***Q***. The simulation using only reference potentials is shown as a dashed blue line and the EPSR data is shown as a solid red line.

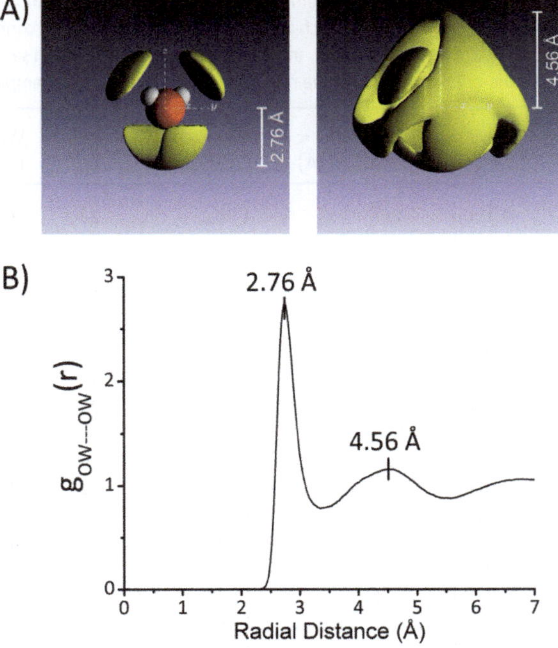

Fig. 5 (A) Spatial density plots showing the distribution of water oxygen atoms around a central water molecule. The yellow shaded areas represent the regions where the probability of finding another water molecule within 5.7 Å of the central oxygen atom exceeds 9% (left) and 20% (right). (B) The radial distribution for water oxygen–water oxygen pair in pure water at 298 K.

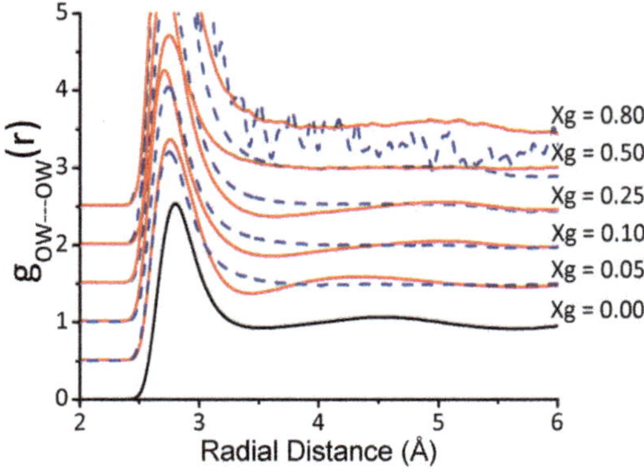

Fig. 6 Water oxygen–water oxygen radial distribution functions in aqueous glycerol at each of the different concentrations studied, mole fraction glycerol $x_g = 0.05$, $x_g = 0.10$, $x_g = 0.25$, $x_g = 0.50$, $x_g = 0.80$ at 298 K. The data taken from the EPSR analysis is shown in solid red line and the data from the reference potential only simulations is shown in dashed blue line. For reference, the radial distribution for water oxygen–water oxygen pair in pure water at 298 K taken from the EPSR analysis is shown in black.

deviations from that of pure water. At the dilute concentration of $x_g = 0.05$ the second peak position shifts inwards from 4.56 Å in pure water to 4.38 Å. While at concentrations of $x_g = 0.10$ and higher the second peak position of $g_{OwOw}(r)$ shifts outwards to larger radial distances and gradually becomes less pronounced.

Table 2 Summary of neutron diffraction experiments and EPSR computational modelling which have been completed on cryoprotectant solutions. Each cryoprotectant is listed with the concentrations which have been studied, in solute mole fraction (m.f.), the peak position of the first and second coordination shell, the reference for the original work and the reference potential which was employed for water

Solute	m.f.	1st Peak Position (Å)	2nd Peak Position (Å)	Ref	Water Potential
Sorbitol	0.1900	2.66	4.06	24	"modified SPC/E"
TMAO	0.0517	2.75	4.26	17	SPC/E
Trehalose	0.0099	2.77	4.24	26	SPC/E
	0.0099	2.76	4.24	25	SPC/E
	0.0385	2.76	4.34	25	SPC/E
Proline	0.0476	2.76	4.63	23	SPC/E
	0.0625	2.78	4.53	23	SPC/E
	0.0909	2.79	4.65	23	SPC/E
DMSO	0.0476	2.79	5.07	16	SPC/E
	0.3333	2.75	5.15	16	SPC/E
Tertiary butanol	0.0600	2.75	4.29	21	SPC/E
	0.1100	2.75	4.51	21	SPC/E
	0.1600	2.76	4.61	21	SPC/E
Methanol	0.0500	2.76	4.20	49	SPC/E
	0.2700	2.76	4.62	18	SPC/E
	0.5400	2.85	4.88	18	SPC/E
Glycerol	0.0500	2.76	4.38	11	SPC
	0.1000	2.73	4.71	14	SPC
	0.2500	2.70	5.04	14	SPC
	0.5000	2.71	5.01	14	SPC
	0.8000	2.73	n/a	13	SPC

4.3 Structure of water in other cryoprotectant–water solutions

Given the wide growing body of experimental data now available on the structure of cryoprotectant–water solutions we examine the RDF $g_{\mathrm{OwOw}}(r)$ for each of these systems (Fig. 7). In Table 2 we show the positions of the first and second peaks in $g_{\mathrm{OwOw}}(r)$ for each cryoprotectant system, for each of the concentrations that have been completed, all at room temperature. These include studies on dimethyl sulfoxide,[15,16] trimethylamine oxide,[17] methanol,[18–20] tertiary butanol,[21,22] proline,[23] sorbitol[24] and trehalose.[25,26] Values for the peak positions were taken from graphs and tables in each of the relevant references. Examination of the RDFs for $g_{\mathrm{OwOw}}(r)$ reveals that the first peak position changes very little for each of the different cryoprotectant–water solutions (Table 2). However, as with glycerol, the second peak position of $g_{\mathrm{OwOw}}(r)$ does undergo significant and differing changes upon addition of the cryoprotectant.

For trehalose–water solutions, at a mole fraction of 0.0099 the second peak shifts inward by 0.32 Å[26] while at a mole fraction of 0.0385 the second peak shifts inward by 0.22 Å.[25] In TMAO–water solutions at a mole fraction of 0.0517 an inward shift in the second peak position of $g_{\mathrm{OwOw}}(r)$ is observed.[17] Sorbitol–water solutions at a mole fraction of 0.19 display a shift inwards by 0.50 Å in the second peak position in $g_{\mathrm{OwOw}}(r)$.[24] For proline–water solutions at mole fractions between 0.0476 and 0.0909, as the proline concentration is increased from 0.0476 to 0.0909 the second peak of the water $g_{\mathrm{OwOw}}(r)$ was shifted to larger radial distances and became less pronounced.[23] In this study, this shift to larger radial distances was attributed to there being an insufficient number of water molecules in the system to form a complete second neighbour shell.[23] Similarly, for DMSO–water solutions at mole fractions of 0.0476 and 0.33 the second peak in $g_{\mathrm{OwOw}}(r)$ shifts outwards to larger radial distances and becomes less pronounced with increasing concentration of DMSO.[15] Finally, for a methanol–water solution at mole fraction 0.54 the second peak in $g_{\mathrm{OwOw}}(r)$ shifts outwards to larger radial distances,[20] while at a

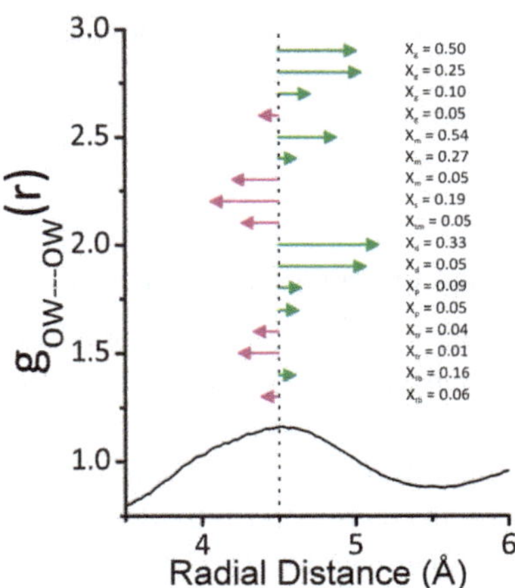

Fig. 7 The second coordination shell of the water oxygen–water oxygen RDF is shown for water at 298 K, black line. The dashed line highlights the peak position of the second coordination shell. The shift in the peak position is shown by arrows for each of the cryoprotectant–water solutions studied at each mole fraction concentration x_g mole fraction glycerol, x_m methanol, x_s sorbitol, x_{TM} TMAO, x_d DMSO, x_p proline, x_{tr} trehalose, x_{tb} tertiary butanol. Green arrows depict a shift to higher radial distances and purple arrows depict a shift to lower radial distance.

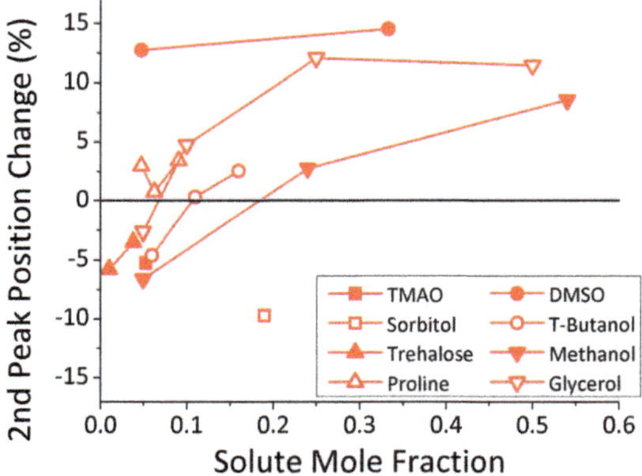

Fig. 8 The movement of the peak position of the second coordination shell of the water oxygen–water oxygen RDF, in relation to pure water. A negative percentage change depicts an inwards shift of the peak position, while a positive percentage change depicts an outwards shift. Data is shown for TMAO (■), sorbitol (□), trehalose (▲), proline (△), DMSO (●), tertiary butanol (○), methanol (▼) and glycerol (▽).

dilute concentration of 0.05 the second peak shifts inwards by 0.36 Å. To help visualise all these changes, with solute and concentration, Fig. 8 shows the movement of the peak position of the second coordination shell of the water oxygen–water oxygen RDF, in relation to pure water. A negative percentage change depicts an inwards shift of the peak position, while a positive percentage change depicts an outwards shift. The addition of small amounts of trehalose, sorbitol, glycerol, methanol, tertiary butanol and TMAO results in an inward shift of the peak position of $g_{\mathrm{OwOw}}(r)$ to smaller radial distances, while small additions of DMSO and proline results in an outward shift of the peak position of $g_{\mathrm{OwOw}}(r)$. The addition of larger quantities of methanol, glycerol and tertiary butanol results in an outward shift of the peak position of $g_{\mathrm{OwOw}}(r)$. Therefore, there does not appear to be a general trend for all cryoprotectants, although glycerol, methanol, tertiary butanol and trehalose all show a concentration dependent inward and then outward shift of the second coordination shell peak position with increasing concentration.

4.4 Water structure at high temperature and pressure

To further examine water structure we consider the effect of increased temperature and pressure on the water oxygen–water oxygen radial distribution function $g_{\mathrm{OwOw}}(r)$. Previous experiments on water have been completed on the SANDALS instrument at ISIS at temperatures between 220 and 673 K and pressures up to 400 MPa.[55] As the pressure was increased from 0.1 to 210 MPa the second peak in $g_{\mathrm{OwOw}}(r)$ shifted to lower radial distances (Fig. 9A). As the temperature increased from 298 K to 423 K the second peak in $g_{\mathrm{OwOw}}(r)$ shifted to larger radial distances (Fig. 9B).

Therefore, the addition of small amounts of trehalose, sorbitol, glycerol, methanol, tertiary butanol and TMAO results in an inward shift of the peak position of $g_{\mathrm{OwOw}}(r)$ to smaller radial distances, similar to the changes observed upon increasing the pressure of pure water. Furthermore, small additions of

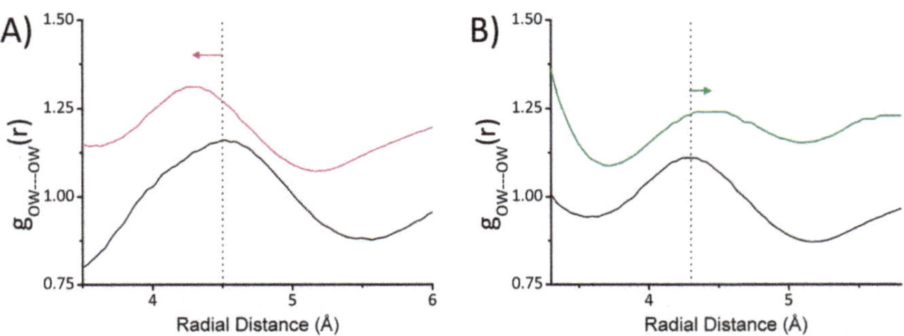

Fig. 9 (A) The effect of increasing pressure. Here, the second coordination shell of the water oxygen–water oxygen RDFs are shown for water at 298 K and two different pressures. The black line shows data taken at 0.1 MPa and the purple line shows data taken at a pressure of 210 MPa. The dashed line highlights the position of the second coordination shell in pure water at 0.1 MPa. (B) The effect of increasing temperature. The datasets here are 298 K at 210 MPa (black) and 423 K at 190 MPa (green).

DMSO and proline and large additions of methanol, glycerol and tertiary butanol result in an outward shift of the peak position of $g_{OwOw}(r)$, similar to the changes observed upon increasing the temperature of pure water.

5 Hydrogen bonded networks

5.1 Examining hydrogen bonded clusters in glycerol–water solutions

We examined the hydrogen bonding ability of glycerol and water by investigating the distribution of water molecules and glycerol molecules in the system. The EPSR generated ensembles were interrogated to extract structural information on glycerol and water clusters in the solution. Previous studies of aqueous glycerol have demonstrated that in these solutions groups of water molecules hydrogen bond together, forming clusters in the solution.[14] Likewise, groups of glycerol molecules exist in glycerol-rich clusters. To quantitatively assess this micro-segregation, we completed a full cluster analysis for both glycerol and water across the concentration range. The cluster analysis is derived from the simulation of the scattering data and provides further structural insight. The cluster size distribution is the probability of finding a cluster of a particular size as a function of cluster size. The glycerol clusters and water clusters are defined by those molecules that participate in a continuous hydrogen bonded network. Using the structural information from the relevant RDFs, two glycerol molecules are hydrogen-bonded if the inter-oxygen contact distance is less than the radial distance of the first minimum of the relevant RDFs; $g_{O\text{-}O}(r)$, $g_{O\text{-}OC}(r)$ and $g_{OC\text{-}OC}(r)$. Two water molecules are hydrogen-bonded if the inter-oxygen contact distance is less than the radial distance of the first minimum of the $g_{Ow\text{-}Ow}(r)$.

Fig. 10 shows the number of clusters containing water molecules (Fig. 10A) and glycerol molecules (Fig. 10B) as a fraction of the total number of clusters $M(i)/M$ [where $M = \Sigma_i M(i)$] against cluster size i. A percolating cluster is defined as having a distribution that crosses the random 3 dimensional percolation threshold given by the power law $N\alpha S^{-2.2}$. Here N is the cluster proportion and S is the size of the cluster. The data show that glycerol and water micro-segregates to form water-rich clusters and glycerol-rich clusters. These regions exist as hydrogen bonded strings, sheets and clusters which differ in size across the concentration range.[14] In the

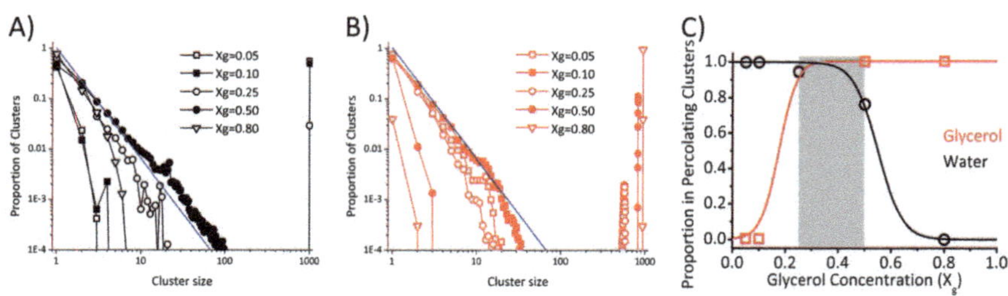

Fig. 10 (A) Cluster size distributions for water at each of the glycerol–water concentrations studied along with the 3D percolation threshold (blue line). (B) Cluster size distributions for glycerol at each of the glycerol–water concentrations studied along with the 3D percolation threshold (blue line). These data show that percolating glycerol clusters are found at all concentrations of $x_g \geq 0.25$ (shown by red data points to the right of the blue line). Similarly, percolating water clusters appear (shown by black data points to the right of the blue line) in all concentrations apart from $x_g = 0.80$. Therefore, there is a bi-percolating mixture at $0.25 < x_g < 0.50$ (both red and black data points to the right of the blue line). (C) Proportion of molecules found in percolating clusters for glycerol (red squares) and water (black circles). The grey region shows the concentration range that the mixture is bi-percolating (both water and glycerol form percolating networks) the solid lines shown are sigmoidal functions to guide the eye.

concentrated glycerol region ($x_g = 0.80$), the majority (56%) of water molecules exist as single monomers. This preference for isolated water molecules results in a well mixed solution with optimal glycerol–water hydrogen bonding. In the dilute glycerol region ($x_g = 0.05$) there is a prevalence of isolated glycerol molecules (40%), again resulting in a well mixed solution with glycerol–water hydrogen bond interactions being favourable. We find that large percolating water clusters exist at glycerol mole fractions of $x_g = 0.50$ and below, while large percolating glycerol clusters exist at glycerol mole fractions of $x_g = 0.25$ and above. Thus at a concentration of $0.25 \leq x_g \leq 0.50$ a solution exists which is composed of a bi-percolating mixture with both glycerol hydrogen bonding clusters and water hydrogen bonded clusters spanning the entire system. We also examine the fraction of molecules in the interface between glycerol clusters and water clusters. The interface is defined by hydrogen bond interactions between glycerol and water and is calculated using the relevant radial distribution functions.[14] Interestingly, we find a maximum in the fraction of molecules in the interface at a mole fraction of $x_g = 0.50$. Therefore, at this concentration the solution is both a bi-percolating liquid mixture and maximizing its hetero-hydrogen bond interactions. The proportion of molecules found in percolating clusters for glycerol and water is shown in Fig. 10C. At all concentrations, glycerol effectively gets in the way of water clusters. This can be seen in the dilute glycerol region, where isolated glycerol molecules exist, maximizing glycerol–water interactions, while in the concentrated glycerol region isolated water molecules exist. The grey region highlights the concentration range over which the mixture is bi-percolating. Within this concentration range the hydrogen bonding ability of glycerol and water allow both species to exist as a percolating network in the system.

5.2 Micro-segregation and clustering in cryoprotectant solutions

We also compared our new glycerol–water results on hydrogen bonded clustering to previously published work on other cryoprotectant solutions. Neutron diffraction experiments and EPSR computational modelling has been used to obtain clustering information for solutions of proline,[23] glycerol,[14] dimethyl sulfoxide[15,16] and methanol.[18–20] Due to the low solubility of proline in water, percolating cluster data only exists at low solute mole fraction.[23] Nevertheless, at a concentration of 0.0909 mole fraction proline a bi-percolating solution exists, composed of percolating proline clusters and percolating water clusters. In the case of DMSO, data exists at dilute and medium concentration and shows that percolating water clusters.[15,16] At a concentration of 0.3333 mole fraction DMSO a bi-percolating solution exists, composed of percolating DMSO clusters and percolating water clusters. Therefore, unfortunately for both DMSO and proline while bi-percolating solutions are evident we cannot determine the full concentration range dependence of clustering and percolation. Like glycerol, methanol–water solutions have been examined across the concentration range. Fig. 11A shows the concentration regions of percolating water clusters (black bar) and percolating cryoprotectant solute clusters (red bar) as a function of solute concentration for glycerol and methanol. For methanol, a bi-percolating mixture exists between 0.27 and 054 mole fraction methanol.[18] As already discussed, for glycerol–water solutions between a concentration of 0.25 and 0.50 a solution exists which is composed of a bi-percolating mixture with both glycerol hydrogen bonding clusters and water hydrogen bonded clusters spanning the entire system. Interestingly, this is the concentration range where many transport properties and thermodynamic excess functions reach extremal values for both of these systems[56,57] (Fig. 11B). The observed concentration dependence of several of these material properties of the solution may therefore have a structural origin, as has been proposed previously for methanol–water solutions.[58] Given glycerol and methanol's role as cryoprotectant it would be interesting to determine if there is a correlation between the hydrogen bonded networks in the system, for example defined through the percolating clusters, and the measured depression in the freezing temperature in the solution. Fig. 11C shows the freezing temperature of aqueous solutions as a function of solute concentration, for glycerol[59] (solid line) and methanol[60] (dotted line). Further studies may shed light on this challenging area.

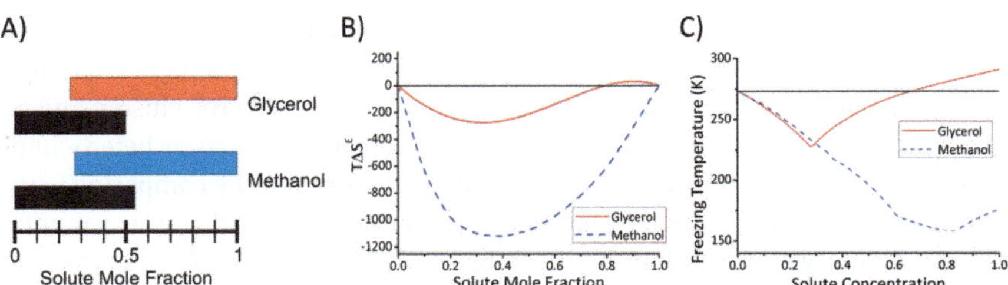

Fig. 11 (A) Schematic depicting the concentration regions of percolating water clusters (black bar) and percolating cryoprotectant solute clusters of glycerol (red bar) and methanol (blue bar) as a function of solute concentration. (B) The excess entropy for aqueous solutions as a function of solute concentration, for glycerol[57] (solid red line) and methanol[56] (dashed blue). (C) The freezing temperature of aqueous solutions as a function of solute concentration, for glycerol[59] (red solid line) and methanol[60] (blue dashed line).

6 Discussion

The structure of glycerol–water solutions at room temperature across the concentration range has been experimentally determined by the combined use of neutron diffraction techniques and an atomistic representation of the system based on a designed reverse Monte Carlo procedure, the EPSR method. This approach, specifically sensitive to the hydrogen atoms in the system, has been key to achieving unambiguous experimental insight into the hydrogen bonding ability in glycerol–water solutions. Fundamental insights have been achieved from the analysis of the present data. We show that the local structure of water in the glycerol–water solutions is sensitive to the concentration of glycerol in the solution (Fig. 6). By comparing a simulation with reference potentials only with that of the EPSR simulation we show that modifications to the water structure, as evidenced by changes to the second coordination shell peak position, is determined by the refinement to the neutron diffraction data. This highlights the importance of obtaining a potential which is informed by experimental information. By comparing our glycerol–water data with neutron diffraction data from the literature on other cryoprotectants (Fig. 7 and 8) and on water at high temperature and pressure (Fig. 9) we attempt to uncover common mechanisms for the action of cryoprotectants on water structure. We find that the addition of small amounts of trehalose, sorbitol, glycerol, methanol, tertiary butanol and TMAO all result in an inward shift of the peak position of $g_{\mathrm{OwOw}}(r)$ to smaller radial distances, similar to the changes observed upon increasing the pressure of pure water. We postulate that the action of glycerol is to drive water toward its structure at high pressure, at which point its freezing temperature reduces, thus protecting it from forming ice. We also find that small additions of DMSO and proline, as well as large additions of methanol, glycerol and tertiary butanol result in an outward shift of the peak position of $g_{\mathrm{OwOw}}(r)$, similar to the changes observed upon increasing the temperature of pure water.

We quantitatively examine the hydrogen bond networks in glycerol–water solutions across the concentration range at room temperature. The data shows that glycerol and water micro-segregates to form hydrogen bonded strings and clusters of water-rich regions and glycerol rich-regions. These clusters differ in size across the concentration range, from small clusters of size two to clusters which account for all the molecules in the system. These percolating clusters exist for glycerol at high glycerol mole fraction ($x_g \geq 0.25$) and for water at low glycerol mole fraction ($x_g \leq 0.50$). Interestingly, we show there is a concentration range $0.25 \leq x_g \leq 0.50$ where glycerol–water solutions exist as a bi-percolating, hydrogen bonded network, composed of a percolating glycerol cluster and a percolating water cluster. This suggests a delicate balance between hetero- and homo-interactions in this system around this concentration. By comparing with other cyroprotectant solutions which have been examined in this way, we note that methanol–water solutions also exhibit a bi-percolating region in which a hydrogen bonded network exists, composed of a percolating methanol cluster and a percolating water cluster (Fig. 11A). Interestingly, the mole fraction range over which simultaneous two-component percolation occurs coincides with the concentration range at which both glycerol and methanol reach a minimum in the excess entropy (Fig. 11B). Furthermore, the concentration region also corresponds with the concentration at which glycerol–water solutions reach their minimum freezing temperature (Fig. 11C). Further work may determine whether the observed concentration dependence of these thermodynamic properties of the solution may have a structural origin.

Based on our data and simulations, we propose a coherent view of the cryoprotectant ability of glycerol–water solutions and other solute systems. At its heart is the fact that cryoprotectants modify water structure, resulting in water local structure which is similar to that of pressurised or heated water. At the same time, at all concentrations, glycerol and methanol segregate the water into clusters. Even where these water clusters are percolating, the combination of pressuring water while imposing water segregation would prevent the formation of ice and the cell damaging volume expansion that would accompany that formation. Indeed, the major damaging event which has been identified during cell freezing is the propagation of intracellular ice crystals. Micro-segregation may act to encapsulate water clusters, thus preventing the propagation of an ice-like water network.[61] We therefore propose that the cryoprotective action of glycerol and methanol is to both pressurize and effectively mix to prevent the formation of a potentially damaging extended ice-like hydrogen bonded network.

Acknowledgements

This work was supported by the Engineering Physical Science Research Council, UK through grant EP/H020616 to Lorna Dougan and through a DTA studentship to James Towey. Experiments at the ISIS Pulsed Neutron and Muon Source were supported by a beam time allocation from the Science and Technology Facilities Council. We are grateful to Dr Daniel Bowron, Dr Silvia Imberti, and Dr Rowan Hargreaves at the ISIS Facility, RAL, U.K. for their support. We thank Prof. Stephen Evans and Prof. Peter Olmsted, Dr Chinmay Das and members of the Dougan group at University of Leeds for useful discussions.

References

1 B. J. Fuller, *Cryoletters*, 2004, **25**, 375–388.
2 J. E. Lovelock, *P Roy Soc Med*, 1954, **47**, 60–62.
3 C. Polge, A. U. Smith and A. S. Parkes, *Nature*, 1949, **164**, 666–666.
4 A. M. Karow, W. R. Webb and J. E. Stapp, *Arch. Surg.*, 1965, **91**, 572.
5 T. Nash, *Cryobiology*, Academic Press, New York, 1966.
6 R. J. Fort and W. R. Moore, *Trans. Faraday Soc.*, 1965, **61**, 2102.
7 G. Bonanno, R. Noto and S. L. Fornili, *J. Chem. Soc., Faraday Trans.*, 1998, **94**, 2755–2762.
8 P. B. Conrad and J. J. de Pablo, *J. Phys. Chem. A*, 1999, **103**, 4049–4055.
9 S. L. Lee, P. G. Debenedetti and J. R. Errington, *J. Chem. Phys.*, 2005, **122**, 204511.
10 C. Chen, W. Z. Li, Y. C. Song and J. Yang, *J. Mol. Liq.*, 2009, **146**, 23–28.
11 J. J. Towey and L. Dougan, *J. Phys. Chem. B*, 2012, **116**, 1633–1641.
12 J. J. Towey, A. K. Soper and L. Dougan, *Phys. Chem. Chem. Phys.*, 2011, **13**, 9397–9406.
13 J. J. Towey, A. K. Soper and L. Dougan, *J. Phys. Chem. B*, 2011, **115**, 7799–7807.
14 J. J. Towey, A. K. Soper and L. Dougan, *J. Phys. Chem. B*, 2012, **116**, 13898–13904.
15 S. E. McLain, A. K. Soper and A. Luzar, *J. Chem. Phys.*, 2006, **124**, 074502.
16 S. E. McLain, A. K. Soper and A. Luzar, *J. Chem. Phys.*, 2007, **127**, 174515.
17 F. Meersman, D. Bowron, A. K. Soper and M. H. J. Koch, *Biophys. J.*, 2009, **97**, 2559–2566.
18 L. Dougan, S. P. Bates, R. Hargreaves, J. P. Fox, J. Crain, J. L. Finney, V. Reat and A. K. Soper, *J. Chem. Phys.*, 2004, **121**, 6456–6462.
19 L. Dougan, J. Crain, J. L. Finney and A. K. Soper, *Phys. Chem. Chem. Phys.*, 2010, **12**, 10221–10229.
20 L. Dougan, R. Hargreaves, S. P. Bates, J. L. Finney, V. Reat, A. K. Soper and J. Crain, *J. Chem. Phys.*, 2005, **122**, 174514.
21 D. T. Bowron, J. L. Finney and A. K. Soper, *J. Phys. Chem. B*, 1998, **102**, 3551–3563.

22 J. L. Finney, D. T. Bowron and A. K. Soper, *J. Phys.: Condens. Matter*, 2000, **12**, A123–A128.
23 S. E. McLain, A. K. Soper, A. E. Terry and A. Watts, *J. Phys. Chem. B*, 2007, **111**, 4568–4580.
24 S. G. Chou, A. K. Soper, S. Khodadadi, J. E. Curtis, S. Krueger, M. T. Cicerone, A. N. Fitch and E. Y. Shalaev, *J. Phys. Chem. B*, 2012, **116**, 4439–4447.
25 S. E. Pagnotta, S. E. McLain, A. K. Soper, F. Bruni and M. A. Ricci, *J. Phys. Chem. B*, 2010, **114**, 4904–4908.
26 S. E. Pagnotta, M. A. Ricci, F. Bruni, S. McLain and S. Magazu, *Chem. Phys.*, 2008, **345**, 159–163.
27 P. H. Yancey, W. R. Blake and J. Conley, *Comp. Biochem. Physiol., Part A: Mol. Integr. Physiol.*, 2002, **133**, 667–676.
28 P. H. Yancey, M. E. Clark, S. C. Hand, R. D. Bowlus and G. N. Somero, *Science*, 1982, **217**, 1214–1222.
29 K. E. Zachariassen and E. Kristiansen, *Cryobiology*, 2000, **41**, 257–279.
30 O. Bastiansen, *Acta Chem. Scand.*, 1949, **3**, 415–421.
31 H. van Koningsveld, *Recl. Trav. Chim. Pays-Bas*, 1968, **87**, 243–249.
32 R. Chelli, P. Procacci, G. Cardini and S. Califano, *Phys. Chem. Chem. Phys.*, 1999, **1**, 879–885.
33 R. Chelli, P. Procacci, G. Cardini, R. G. Della Valle and S. Califano, *Phys. Chem. Chem. Phys.*, 1999, **1**, 871–877.
34 J. L. Dashnau, N. V. Nucci, K. A. Sharp and J. M. Vanderkooi, *J. Phys. Chem. B*, 2006, **110**, 13670–13677.
35 J. Dawidowski, F. J. Bermejo, R. Fayos, R. F. Perea, S. M. Bennington and A. Criado, *Phys. Rev. E: Stat. Phys., Plasmas, Fluids, Relat. Interdiscip. Top.*, 1996, **53**, 5079–5088.
36 M. Garawi, J. C. Dore and D. C. Champeney, *Mol. Phys.*, 1987, **62**, 475–487.
37 T. Y. Lin and S. N. Timasheff, *Biochemistry*, 1994, **33**, 12695–12701.
38 G. B. Strambini and M. Gonnelli, *Biochemistry*, 2008, **47**, 3322–3331.
39 W. F. Rall, M. Czlonkowska, S. C. Barton and C. Polge, *Reproduction*, 1984, **70**, 293.
40 T. Hoffmann, K. M. Tych, D. J. Brockwell and L. Dougan, *J. Phys. Chem. B*, 2013, **117**, 1819–1826.
41 D. M. Piedmonte, C. Summers, A. McAuley, L. Karamujic and G. Ratnaswamy, *Pharm. Res.*, 2007, **24**, 136–146.
42 J. H. Crowe, L. M. Crowe and D. Chapman, *Science*, 1984, **223**, 701–703.
43 N. H. Rhys, A. K. Soper and L. Dougan, *J. Phys. Chem. B*, 2012, **116**, 13308–13319.
44 A. K. Soper, *GudrunN and GudrunX: programs for correcting raw neutron and X-ray diffraction data to differential scattering cross section*, RAL Technical Reports, RAL-TR-2011-013, 2011.
45 A. K. Soper, *Mol. Phys.*, 2009, **107**, 1667–1684.
46 A. K. Soper, *Phys. Rev. B: Condens. Matter Mater. Phys.*, 2005, **72**, 104204.
47 G. W. Robinson, S. B. Zhu, S. Singh and M. W. Evans, *Water in Biology, Chemistry and Physics: Experimental Overviews and Computational Methodologies*, World Scientific, Singaore, 1996.
48 T. Hassinen and M. Perakyla, *J. Comput. Chem.*, 2001, **22**, 1229–1242.
49 S. Dixit, A. K. Soper, J. L. Finney and J. Crain, *Europhys. Lett.*, 2002, **59**, 377–383.
50 D. T. Bowron, J. L. Finney and A. K. Soper, *Mol. Phys.*, 1998, **93**, 531–543.
51 S. E. McLain, A. K. Soper and A. Watts, *J. Phys. Chem. B*, 2006, **110**, 21251–21258.
52 N. H. Rhys and L. Dougan, *Soft Matter*, 2013, **9**, 2359–2364.
53 S. E. McLain, A. K. Soper, I. Daidone, J. C. Smith and A. Watts, *Angew. Chem., Int. Ed.*, 2008, **47**, 9059–9062.
54 S. E. McLain, A. K. Soper and A. Watts, *Eur. Biophys. J.*, 2008, **37**, 647–655.
55 A. K. Soper, *Chem. Phys.*, 2000, **258**, 121–137.
56 R. F. Lama and B. C. Y. Lu, *J. Chem. Eng. Data*, 1965, **10**, 216.
57 Y. Marcus, *Phys. Chem. Chem. Phys.*, 2000, **2**, 4891–4896.
58 A. K. Soper, L. Dougan, J. Crain and J. L. Finney, *J. Phys. Chem. B*, 2006, **110**, 3472–3476.
59 L. B. Lane, *Ind. Eng. Chem.*, 1925, **17**, 924–924.
60 K. Takaizumi and T. Wakabayashi, *J. Solution Chem.*, 1997, **26**, 927–939.
61 S. P. Leibo, J. J. Mcgrath and E. G. Cravalho, *Cryobiology*, 1978, **15**, 257–271.

Faraday Discussions RSC Publishing

PAPER

Distinguishing aggregation from random mixing in aqueous *t*-butyl alcohol solutions

David S. Wilcox, Blake M. Rankin and Dor Ben-Amotz*

Received 13th May 2013, Accepted 24th June 2013
DOI: 10.1039/c3fd00086a

Raman spectroscopic measurements are combined with various multivariate curve resolution (Raman-MCR) strategies, to characterize the aggregation of *t*-butyl alcohol (TBA) in aqueous solutions. The resulting TBA solute-correlated (SC) spectra reveal perturbed water OH features arising from the hydration-shell of TBA as well as shifts in the TBA CH vibrational frequency arising from TBA–TBA interactions. Our results indicate that at low concentrations (below ~0.5 M), there is virtually no TBA aggregation. The first aggregates formed above 0.5 M remain highly hydrated, while those formed above ~2 M are significantly less hydrated. Comparisons with predictions pertaining to a randomly mixed (non-aggregating) solution indicate that below ~1 M there are fewer TBA–TBA contacts than would be present in a random mixture, thus implying that the thermodynamic stability of the first hydration-shell of TBA suppresses the formation of direct contact aggregates at low TBA concentrations. Our results further suggest that microheterogeneous domains containing many water-separated TBA–TBA contacts form near a TBA concentration of ~1 M, while at higher concentrations the TBA-rich domain size distribution may resemble that in a non-aggregating random mixture.

1 Introduction

The present work is motivated in part by experimental[1–3] and theoretical[3–5] evidence suggesting that some aqueous alcohol solutions may have a microscopically heterogeneous structure, composed of mesoscopic alcohol-rich and water-rich domains. One might expect the formation of such structures to be linked to hydrophobic interactions between the oily alcohol groups. However, previous experimental studies have not quantified the degree to which such aqueous alcohol solutions form direct hydrophobic contact aggregates as opposed to water-separated aggregates, or how the structure of the aggregates evolve with increasing alcohol concentration. In an effort to address these questions we have obtained hydration-shell spectra of aqueous *t*-butyl alcohol (TBA, $HOC(CH_3)_3$) solutions, over a concentration range that spans the onset of

Purdue University, Department of Chemistry, West Lafayette, IN 47907, USA. E-mail: bendor@purdue.edu; Tel: +765-494-5256

aggregation and the appearance of two distinct classes of TBA aggregates. We have also compared our results with predictions pertaining to a randomly mixed (non-aggregating) fluid in order to elucidate the influence of hydrophobic hydration on TBA aggregation.

We chose to focus on aqueous TBA solutions in part because some previous studies have suggested that such solutions may have a microscopically heterogeneous structure,[1,2,4,6] and because TBA is one of the largest amphiphilic molecules that remains infinitely miscible in water. Moreover, although numerous previous experimental and theoretical studies have been performed on this system, none have utilized hydration-shell spectroscopy to distinguish monomeric and aggregated TBA species. We have chosen to focus on the concentration range of $0 \leq [\text{TBA}] \leq 4$ M (which corresponds to relatively low TBA mole fractions of $\chi \leq 0.1$) because a concentration of ~2 M is that at which there are just enough water molecules to form a complete hydration shell around each TBA molecule. As a result, above 2 M some degree of direct (or water-separated) contact must occur between TBA molecules, while below 2 M TBA molecules could possibly remain fully hydrated. Thus, the degree of aggregation observed in the above concentration range should provide a sensitive measure of the influence of hydrophobic interactions on TBA aggregation.

Previous studies indicate that TBA aggregation begins to take place below 2 M, while the formation of mesoscale heterogeneities remains a subject of both experimental[1,2,6] and theoretical[4] debate. Aqueous TBA solutions have been investigated using neutron scattering,[7,8] light scattering,[1,2,6] infrared spectroscopy,[9,10] Raman spectroscopy,[11] mass spectrometry,[12] and NMR,[13] as well as various molecular dynamics[4,14–17] and integral equation[18,19] methods. These studies have generally concluded that TBA aggregation begins at a TBA concentration between 1–2 M (with the onset of aggregation often assigned to a concentration of ~1.3 M). Neutron diffraction experiments, combined with empirical potential structure refinement (EPSR) simulations, suggest that a substantial number of direct (as opposed to water-separated) contacts are formed between the methyl groups on two TBA molecules, at a TBA mole fraction of χ ~0.02 (which corresponds to a TBA concentration of ~1 M).[7] However, a more recent very large MD simulation study found that a "significant amount of water remains in the TBA-rich regions," but also noted that the results are "strongly dependent on the particular force field employed, and can be significantly influenced by system size."[4]

Here we report results obtained by combining high signal-to-noise Raman scattering measurements with multivariate curve resolution (Raman-MCR)[20–23] to characterize the aggregation-induced spectral changes in TBA and its hydration shell. More specifically, we have employed various Raman-MCR strategies to quantify concentration dependent changes in both the TBA CH stretch frequency and the OH stretch band arising from the hydration-shell of TBA. In analyzing the results we have also considered how one might distinguish TBA–TBA contacts that are influenced by water-mediated hydrophobic interactions from contacts in a statistically random non-aggregating mixture.

The remainder of this paper begins (in Section 2) by describing a strategy for predicting the number of TBA–TBA contacts in randomly mixed but non-aggregating fluid resembling aqueous TBA solutions. We then describe the experimental and Raman-MCR data collection and analysis strategies that we have employed (in Section 3), and the results we have obtained (in Section 4), followed by our discussion and conclusions (in Section 5), particularly with regard to hydrophobic aggregation and microheterogeneity.

2 Random mixing limit

In order to understand the degree to which hydrophobic interactions contribute to TBA aggregation, it is important to establish what one should expect to observe in a randomly mixed non-aggregating solution. We define such a non-aggregating solution as one in which the concentration of TBA in the first hydration shell of each TBA molecule is equivalent to the bulk concentration. The binomial distribution may be used to characterize TBA–TBA contact statistics in such a random mixture, given the maximum number n of TBA molecules which could occupy the first solvation-shell of each TBA, and the probability p that each of these first solvation-shell sites will be occupied. We may reasonably assume $n \approx 11$, as that is the coordination number in pure liquid TBA.[24] In a randomly mixed solution, the probability that a TBA molecule could occupy a given location in the first solvation-shell may reasonably be taken to be $p \approx [\text{TBA}]\bar{V}$, where $\bar{V} \approx 0.0878$ L mol^{-1} is the partial molar volume of TBA in water.[25] The results and conclusions of relevance to this work are quite insensitive to the precise values of n and $\bar{V}$.

With the above assumptions, the binomial distribution may be used to predict the probability $P(k)$ that the first hydration shell of a given TBA molecule will contain k additional TBA molecules.

$$P(k) = \frac{n!}{k!(n-k)!} p^k (1-p)^{n-k} \tag{1}$$

Thus, $P(0) = (1 - p)^n$ is the probability that a given TBA will not have any nearest neighbour TBA molecules. And so $P(0)$[TBA] is the concentration of non-aggregated TBA molecules and $[1 - P(0)]$[TBA] is the total concentration of TBA molecules that are in contact with at least one other TBA molecule. More generally, $P(k)$[TBA] is the predicted concentration of TBA molecules whose first hydration-shell contain k other TBA molecules (as shown in Fig. 1).

The results shown in Fig. 1 indicate that at a relatively low TBA concentration of 0.5 M (blue histogram) ~60% of the TBA molecules are predicted to have no other TBA molecules in their first hydration shell, and ~40% of the TBA molecules are predicted to be in contact with one or more other TBA molecules. At concentrations of 2 M (red histogram) and 4 M (green histogram) most of the TBA molecules are predicted to be in contact with one or more other TBA molecules. The inset in Fig. 1 shows how the concentrations of TBA molecules that are in contact with k other TBA molecules are predicted to depend on the total TBA concentration. These results clearly indicate that a significant number of TBA–TBA contacts are expected to occur in such random mixtures of non-aggregating molecules, in spite of the fact that there are enough water molecules present in a 2 M solution to form a complete hydration shell around each TBA (and in a 4 M solution each TBA could theoretically retain a complete hydration shell if each water molecule were in contact with two TBA molecules).

Eqn (1) may further be used to predict the coordination number of TBA, as it implies that $\langle k \rangle = \Sigma_k kP(k) = pn$ is the average number of TBA molecules that are in contact with a given TBA. Neutron scattering (and EPSR simulation) studies imply

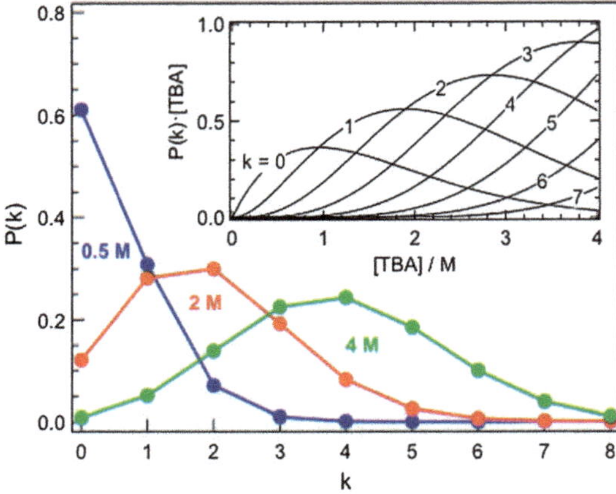

Fig. 1 The three colored curves (and points) represent predicted random mixing histograms of the probabilties $P(k)$ that a given TBA molecule will have $0 \leq k \leq 8$ other TBA molecules in its first hydration shell, when the total TBA concentration is 0.5 M (blue), 2 M (red), or 4 M (green). The inset figure shows the concentrations (in molar units) of TBA molecules whose first hydration shells contain k other TBA molecules.

that the coordination numbers of TBA increases from ~1.5 at 1 M ($\chi = 0.02$) to ~2.9 at 2.6 M ($\chi = 0.06$).[7,8] These coordination numbers are quite close to the corresponding random mixing predictions of $\langle k \rangle \approx 1.0$ at 1 M and $\langle k \rangle \approx 2.5$ at 2.6 M. The fact that the coordination numbers derived from neutron scattering are slightly larger than the random mixing predictions is consistent with the fact that the TBA–TBA radial distribution functions obtained from the neutron scattering and EPSR simulations have a first hydration shell peak that is larger than one, while the above random mixing model essentially implies that the radial distribution function is equal to one at all TBA–TBA separations (beyond the contact separation).

3 Experimental and multivariate signal processing methods

A Raman scattering measurements

All solutions were prepared using ultrapure water (Milli-Q UF plus, Millipore, Inc.), with an electrical resistance of 18.2 MΩ cm. TBA was purchased from Sigma-Aldrich (≥99.5%) and used without further purification. Note that in order to obtain reliable and reproducible MCR results it is critically important that the solutions remain free of any detectable amounts of fluorescent impurities, particularly when using Raman-MCR to distinguish more than two spectroscopically similar components (such as water, hydrated monomeric TBA, and one or more aggregated TBA species).

Raman spectra of aqueous TBA solutions were collected using a home-built Raman system with a 514.5 nm Ar-ion laser as the excitation source. The back-scattered Raman photons were dispersed using either a 300 gr mm^{-1} or a 1200 gr mm^{-1} grating, with resolutions of ~6 cm^{-1} and ~2 cm^{-1}, respectively, in the CH and OH stretching regions. The higher resolution grating was only used for CH frequency shift measurements. Various MCR signal processing strategies, including self-modeling curve resolution (SMCR) and entropy minimization (EM), were used to characterize and quantify the spectroscopically distinct components that are present at various TBA concentrations. More specifically, SMCR and EM were used to decompose experimentally measured spectra in order to obtain both lower and upper bound estimates of the spectral shapes and weights (concentrations) associated with each component. A key feature of this analysis strategy is that it does not require any *a priori* knowledge of the number of spectroscopically distinct components or their spectral shapes or their concentrations. The resulting spectral decompositions and concentration profiles place significant constraints on any model for the hydrophobic hydration and interactions of TBA molecules in water.

Raman spectra of exceptionally high signal-to-noise are required in order to obtain reliable three-component EM results. To achieve the required signal-to-noise ratio (SNR), the back-scattered Raman photons were dispersed using a lower resolution grating (300 gr mm^{-1}) to maximize the number of photons at each wavelength channel. Furthermore, ten consecutive spectra of each solution were collected with an integration time of five minutes per spectrum. To correct for small spectral drifts occurring during these long accumulation times, helium lamp reference peaks were included in each Raman spectrum, and used to correct wavelength drifts as previously described.[23] Raman spectra obtained in this way have a SNR of approximately 11,000 : 1 near the peak of the water OH band (as determined from the ratio of the maximum number of counts in the sum of all ten water spectra and the standard deviation of the difference between the sum of two independently collected sets of five pure water spectra). The most reproducible and reliable three-component MCR results were obtained when all of the input spectra had minimal background (*i.e.* either a vertical offset or baseline) across the spectral region of which the MCR analysis was performed. This was generally true in the CH and OH stretch spectral region, but often not the case at lower Raman shift frequencies, and thus we did not attempt to perform three-component MCR analyses outside the CH and OH region.

B Self-modelling curve resolution

SMCR is an algorithm that can be used to decompose two or more experimental Raman spectra into a linear combination of two spectral components. In dilute solutions, one of the components necessarily represents the spectrum of the solvent while the other non-negative minimum-area solute-correlated (SC) spectrum contains features arising from the solute and any solvent molecules whose spectra are perturbed by the solute. In other words, for dilute aqueous TBA solutions, the solvent spectrum pertains to pure water and the SC spectrum contains the Raman vibrational bands of TBA, such the CH stretch band, as well as features arising from water molecules in the hydrophobic hydration shell of TBA.

Although SMCR is inherently a two-component decomposition algorithm, it may in some cases be used to obtain spectral information from systems with more than two components, as long as only one component concentration is varied when collecting the input Raman spectra.[20] For example one may use SMCR to obtain the SC spectrum of a solute dissolved in an aqueous salt solutions from the spectra of the pure salt solution to which various (small) amounts of the solute are added.[21] Alternatively, one may apply SMCR to higher concentration solutions using a single pair of spectra, one of which is obtained from pure water and the other of which is obtained from an aqueous solution. The SC spectrum that is obtained from each such pair of spectra will necessarily contain features representing the difference between the solution and pure water spectra. More specifically, the latter SC spectrum is equivalent to the difference spectrum obtained when a scaled replicate of the pure water spectrum is subtracted from the mixture spectrum, with a scaling coefficient optimized to obtain a non-negative difference spectrum of the smallest area.

C Entropy minimization

EM may be viewed as a multi-component generalization of SMCR, for the analysis of spectra obtained from systems with more than two independently variable components.[26] Unlike SMCR, EM is a numerical (rather than analytical) method for calculating the component spectra. The spectral components obtained using EM are those that minimize the Shannon entropy of the mixture, under the constraint that both the concentrations (spectral weights) and spectra of each component are positive. In essence, minimizing the Shannon entropy produces component spectra that have the simplest, narrowest, and most separated features. Further details regarding the implementation of EM are the same as those previously described,[27] except that the entropy of the component spectra (rather than derivatives of those spectra) was minimized in the algorithm (although we have found that very similar results could be obtained using either first or second derivative spectra, as long as those input spectra are appropriately smoothed).

Since the Raman spectrum of bulk water is known, the shape of the pure water spectrum is imposed as an additional constraint in our EM analyses. The second component that appears when analyzing dilute TBA solutions necessarily originates from fully hydrated (non-aggregated) TBA monomers. A two-component EM analysis in such a limit will yield a second component that is essentially identical with the corresponding SMCR SC component. A three-component EM analysis in the low concentration limit will again produce two well-resolved components, plus a third component that is essentially indistinguishable from one of the first two components (and has an unphysical concentration profile). Thus, two-component EM of dilute TBA solutions may be used to identify the TBA monomer spectrum, and then the analysis of higher concentration solutions may be performed using this monomer spectrum as an additional constraint. The number of additional components required to describe a given mixture may be determined in a similar manner, by finding the minimum number of spectrally distinct EM components that adequately describe all the input spectra.

In this work, we present results obtained when applying EM with up to three-components. At TBA concentrations up to 0.5 M we find that only two components are present (one of which is bulk water and the other is equivalent to the SMCR SC spectrum of the monomeric TBA species). At concentrations between 0.5 M and 1.2 M we have used EM to obtain the concentration profiles of monomeric TBA and the first aggregated TBA species. Our results indicated that the spectrum of this first aggregated TBA species only differs slightly from that of the TBA monomer (as further discussed in Sections 3A and 3B and shown in Fig. 3 and 4). At higher TBA concentrations, between 1.3 M and 4 M, a quite different aggregated TBA species emerges and grows with increasing TBA concentration (as further described in Section 3B and shown in Fig. 6).

D Rotational ambiguity

Rotational ambiguity is used to describe a situation in which a range of MCR solutions (obtained either using SMCR or EM) may represent the input spectra equally well.[27,28] The range of feasible solutions in a two-component SMCR analysis is bounded by solutions of minimum and maximum area. The minimum area SC spectrum is equivalent to the smallest positive difference spectrum obtained when subtracting as much of the pure water spectrum from the mixture spectrum as possible. A two-component EM analysis produces a solute-correlated spectrum that is equivalent to the SMCR minimum area SC spectrum. The hydration-shell features in the latter SC spectrum have an area that is equivalent to the minimum number of water molecules whose vibrational spectrum is perturbed by the solute. The maximum area SC spectrum may not be as physically relevant, as it corresponds to equating the SC spectrum with the entire spectrum of the highest concentration solution. In other words, the maximum area SC spectrum is equivalent to assuming that the number of water molecules in the solutes' hydration shell is exactly equal to the number of water molecules (per solute molecule) in the most concentrated solution. Unless stated otherwise, all of the SC spectra presented in this work are minimum area spectra.

For multi-component systems, the feasible range of the component spectra (and concentrations) are not uniquely determined by multi-component MCR algorithms, and so the above definition of the minimum and maximum area spectra for each component must be generalized. Various methods for evaluating the range of feasible solutions for multi-component systems, under different constraint conditions, have been described.[29–31] In this work, we have used the results of EM to establish the range of component spectra that are consistent with the measured solution Raman spectra.[31] By analogy with the above discussion of two-component SMCR rotational ambiguity, the sum of the areas of *all* SC spectra (monomer and aggregate) is at a minimum when the contribution of bulk water is maximized in all solutions. In other words, we begin by finding the range of feasible SC solutions when the area of bulk water contribution is maximized in each solution. The resulting SC spectra of the TBA monomer and aggregate all contained prominent CH stretch bands arising from TBA as well as smaller OH stretch features. However, there may still be some residual rotational ambiguity in such TBA monomer and aggregate spectra. One way to resolve this ambiguity is to constrain the results so as to minimize the area of the aggregate spectrum. Since the monomer and aggregate spectra constitute a two-component system (from which a fixed maximum amount of the pure water has already been removed), minimizing the aggregate is equivalent to maximizing the monomer contribution. Thus, this aggregate spectrum represents that portion of the mixture spectrum that cannot be described by a linear combination of the pure water and TBA monomer spectra. This is the procedure that we have used to obtain the results described in Section 3B (and shown in Fig. 4). Alternatively, we may resolve the latter ambiguity by maximizing the area of the aggregated TBA component. This is equivalent to minimizing the monomer contribution to each mixture (as further discussed in Section 3B and shown in Fig. 5 and 6).

3 Results

A Two-component SMCR results

The Raman spectra of aqueous TBA solutions ranging in concentration from 0.5 M to 4 M are shown in Fig. 2 (normalized to unit area). The inset demonstrates that the area of the CH band of TBA is linearly correlated with TBA concentration, and thus is insensitive to concentration dependent changes in the structures of the solution.

Fig. 3 shows results obtained when applying SMCR to spectra such as those shown in Fig. 2. The SC spectra in panel (A) contain CH (~2850–3000 cm^{-1}) and OH (~3100–3600 cm^{-1}) stretch vibrational features. The latter OH features arise from the hydrophobic hydration-shell of TBA, as previous studies have shown that alcohol OH head groups do not contribute significantly to such SC spectra.[22,23] Since all the spectra in Fig. 3A are normalized to the area of the CH band, the OH features would be concentration independent if the hydration-shell structure of TBA was concentration independent. The OH features in panel (A) reveal that there is only a slight decrease in area (and minimal shape change) in the hydration-shell OH band up to TBA concentration of ~1.5 M, but a much more substantial decrease at higher TBA concentrations. Such a decrease in area implies that significantly fewer water molecules are perturbed by each TBA molecule at 4 M than at 2 M.

The symmetric ($\nu_s \sim 2924$ cm^{-1}) and asymmetric ($\nu_a \sim 2981$ cm^{-1}) CH stretch peaks shown in Fig. 3(A) are distinguished by their different Raman

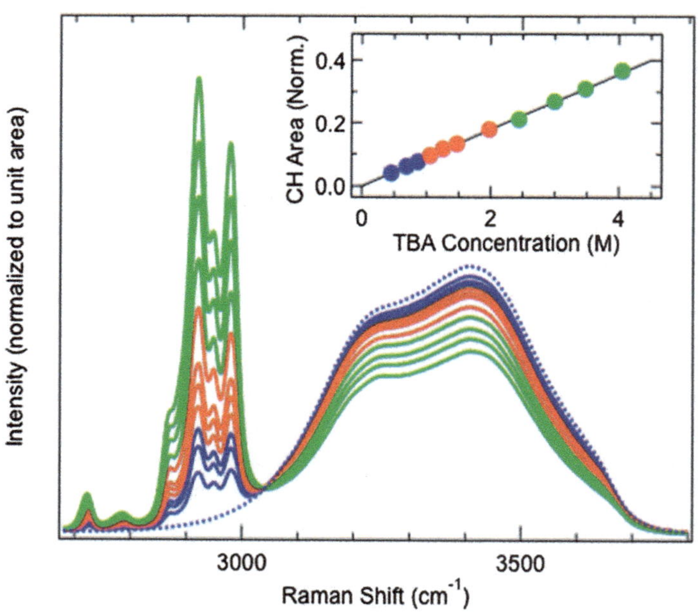

Fig. 2 Raman spectra obtained from aqueous TBA solutions of various concentrations. Each of the spectra are normalized to unit area, but are not baseline subtracted or processed in any other way. The colours of the spectra pertain to different concentration ranges: blue [TBA] ≤ 0.9 M, red 1.1 M ≤ [TBA] ≤ 2 M, and green 2.5 M ≤ [TBA] ≤ 4M. The inset shows the linear correlation between the CH band area and the concentration of TBA (with a slope of 0.0937).

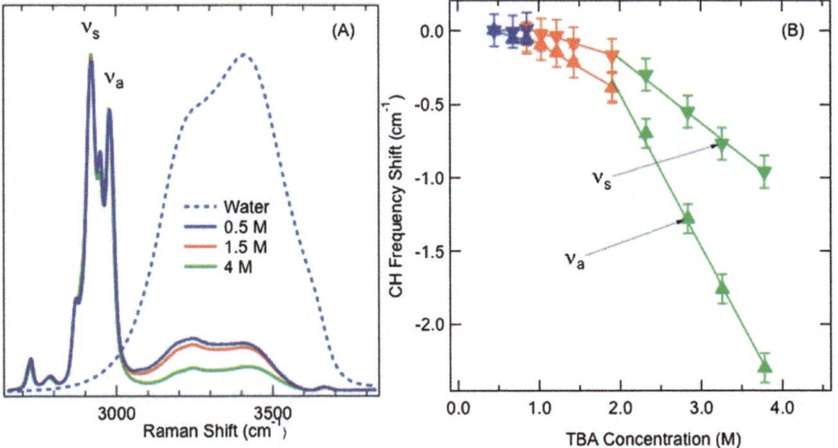

Fig. 3 SC spectra obtained using SMCR from the spectra shown in Fig. 2 (using the same color scheme). Panel (A) shows the resulting pure water (dashed curve) and SC (solid curves) spectra obtained from three TBA solutions of different concentration (normalized to the same CH area). Panel (B) shows the frequency shifts of the symmetric and asymmetric CH stretch peaks, plotted as a function of TBA concentration.

depolarization ratios.[21] Fig. 3(B) shows how the symmetric (ν_s) and asymmetric (ν_a) CH stretch frequencies shift with increasing TBA concentration (relative to their frequencies at a TBA concentration $\leq$0.5 M). Evidently there is essentially no CH shift up to ~1 M, and an increasing red-shift at higher concentrations. Moreover, the slope of the CH shift is smaller between 1 M and 2 M than it is above 2 M. These results are generally consistent with previous IR[32] and Raman[33] measurements performed on ν_a (although the previous Raman measurements had significantly larger error bars). The fact that different frequency shifts slopes are observed in different concentration ranges suggests that these concentration ranges may be associated with a different TBA hydration-shell and/or aggregate structure.

B Multi-component entropy minimization

In order to characterize the structure and concentration of TBA monomers and aggregates that initially form in relatively dilute aqueous TBA solutions, we have applied EM to a series of solutions with 0.5 M $\leq$ [TBA] $\leq$ 1.2 M (as further described in Section 3C). The resulting SC spectra and concentration profiles are shown in Fig. 4. The dashed blue, solid blue, and red spectra in Fig. 4(A) correspond to the pure water, TBA monomer, and the initially formed TBA aggregate species, respectively. As suggested by the SMCR results shown in Fig. 3, the hydration-shell OH features associated with the monomer (blue) and aggregate (red) species are only very slightly different in shape and area, thus suggesting that both species have a similar hydration-shell structure. The corresponding concentration profiles in Fig. 4(B) indicate that the aggregate species begins to emerge above 0.5 M.

In order to better understand the significance of the results shown in Fig. 4 it is instructive to compare them with results obtained using rotational ambiguity

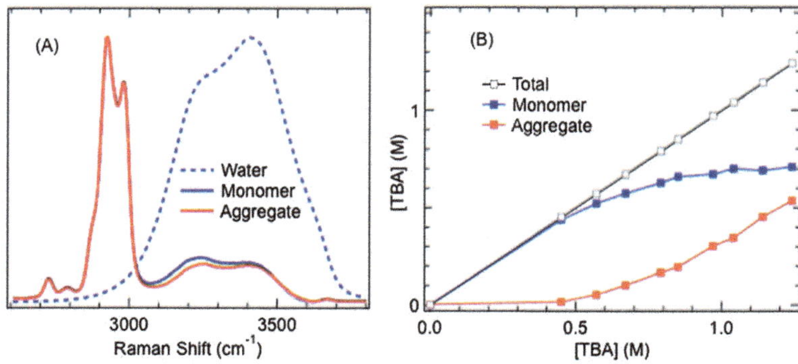

Fig. 4 SC spectra obtained using EM from the spectra shown in Fig. 2. Panel (A) shows the resulting pure water (dashed cuve), monomer (solid blue), and first aggregate (solid red) species (normalized to the same CH area). Panel (B) shows the concentrations of the water, monomer, and first aggregate components as a function of total TBA concentration, obtained from a three-component EM analysis.

analyses, as well as with theoretical predictions obtained using various assumptions regarding the aggregation process, as shown in Fig. 5. The results in Fig. 5A pertain to the TBA monomers and those in Fig. 5B pertain to the corresponding aggregates, and the points are the same as those in Fig. 4. The new hatched regions represent the entire range of Raman-MCR results that may be obtained from the same set of input Raman spectra; the curves with the solid points represent results with the maximum possible number of monomers and minimum possible number of aggregates; the other bounds of the hatched regions represent results with the minimum number of monomer and maximum number of aggregates. The latter mathematical bound is expected to be less physically relevant than the former, as it implies that the highest concentration solution is entirely composed of aggregates (and has no monomers).

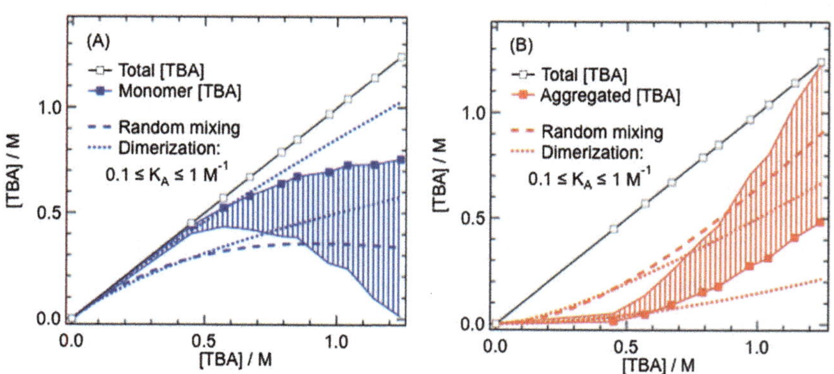

Fig. 5 The spectroscopically feasible range of the TBA monomer (A) and first TBA aggregate (B) concentrations are compared with random mixing (dashed curves) and TBA dimerization model predictions obtained with K_A equal to either 0.1 M^{-1} or 1 M^{-1} (dotted curves). The open points represent the total TBA concentrations and the closed points represent the minimum area monomer (blue) and aggregate (red) concentrations. The hatched area represents the range of feasible monomer (blue) and aggregate (red) concentrations.

Nevertheless, the actual number of monomer and aggregate species is mathematically constrained to lie somewhere within the hatched regions.

The pairs of dotted curves in Fig. 5 represent predictions obtained assuming that the aggregates consist of TBA dimers with an aggregation equilibrium constant of either 0.1 M^{-1} or 1 M^{-1}. Clearly, the shapes of these dimerization equilibrium predictions are significantly different than those obtained from our Raman-MCR results. The larger curvature of our Raman-MCR results implies that the aggregates arise from a higher order aggregation process with significantly more than two TBA molecules per aggregate.

The dashed curves in Fig. 5 indicate the predicted monomer and aggregate concentration profiles pertaining to a non-aggregating random mixture. More specifically, the predicted monomer TBA concentration is $P(0)$[TBA] and the predicted concentration of TBA molecules that are in contact with one or more other TBA molecules is $[1 - P(0)]$[TBA] (see Section 2 for details). Note that the experimental Raman-MCR results (hatched regions in Fig. 5) imply that there are *fewer* aggregates (and *more* monomers) in the actual aqueous TBA solutions than would be expected if the solutions were statistically random mixtures. This implies that in this concentration range, TBA molecules are less likely to aggregate and more likely to retain their hydration shell than in a random mixture.

Fig. 6 shows results obtained when the above analysis is extended to Raman spectra obtained from higher concentration solutions (up to 4 M). Since the monomer and aggregate species formed at low concentrations have quite similar hydration-shell spectra, it is reasonable to group them together into a single low concentration monomer-like species, whose hydration-shell spectrum is indicated by the blue curve in Fig. 6A. More specifically, the latter spectrum was obtained using a two-component EM analysis of spectra obtained when [TBA] $\leq$ 1M. The spectra collected at higher TBA concentrations were analyzed using a three-component EM procedure (with pure water and monomer-like spectral

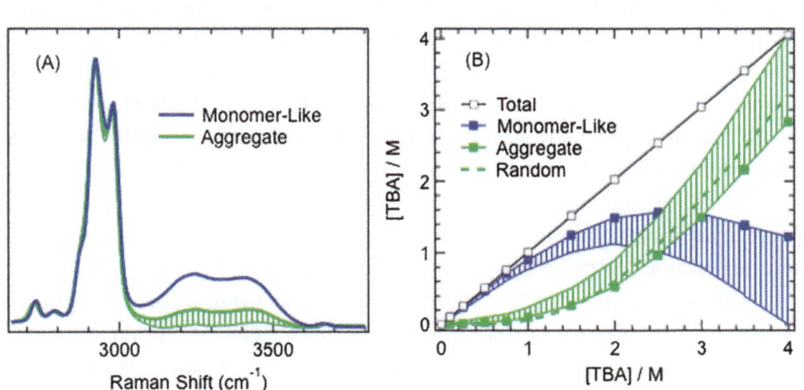

Fig. 6 The analysis of spectra obtained from TBA solutions up to 4 M reveals the presence of TBA species that have far fewer water molecules (green) than the monomer-like species (blue) that predominate at lower concentrations. The SC spectra are shown in (A) and the corresponding concentration profiles are shown in (B). The hatched areas represent the range of feasible monomer-like (blue) and aggregate-like (green) concentrations. The dashed curve represents the concentration of TBA molecules that are predicted to be in contact with three or more other TBA molecules in a non-aggregating random mixture.

shapes obtained from the lower concentration analysis) in order to characterize the spectra and concentration profiles of the TBA aggregates that form at higher concentrations (between 1 M and 4 M), as indicated by the green curves in Fig. 6A. Again, the curves with solid points represent the EM results pertaining to the minimum possible number of aggregates (and maximum number of monomer-like species) and the other bound represents results pertaining to the maximum possible number of aggregates (and minimum number of monomer-like species). The corresponding spectra show that the aggregate species have significantly smaller hydration-shell OH stretch band areas than the monomer-like species, suggesting that they correspond to aggregates in which each TBA molecule is surrounded by far fewer water molecules than a fully hydrated TBA monomer.

4 Discussion and conclusions

We have combined Raman scattering measurements with a variety of MCR analysis strategies to characterize the onset of aggregation in aqueous TBA solutions. In order to quantify the influence of hydrophobic hydration on TBA aggregation we have compared our results with predictions pertaining to the corresponding non-aggregating random mixture. Our results suggest that the standard view of hydrophobic interactions may need to be extended so as to properly account for the thermodynamic stability of hydrated monomeric solute species (as further discussed below).

The CH frequency shifts of TBA (see Fig. 3) imply that there are three distinct TBA hydration or aggregation regimes at low (<1 M), moderate (1–2 M) and high (>2 M) TBA concentration. Previous studies have found evidence of TBA aggregation beginning at $\sim$1.3 M. We have found evidence of the onset of TBA aggregation at a lower concentration near $\sim$0.5 M (although the structure of these first aggregates is apparently quite different than those formed at higher concentrations, as further discussed below). The steeply non-linear concentration profile of the first TBA aggregates (see Fig. 4 and 5) suggests that they result from interactions between more than two TBA molecules. Our results further indicate that these first aggregates remain highly hydrated, as evidenced by the similarity of the hydration-shell OH stretch band of these aggregates and those of TBA monomers observed at lower concentrations (see Fig. 3A and 4A). Our conclusion that the first aggregates are highly hydrated is also consistent with the fact that little or no change in CH frequency is observed at TBA concentrations below $\sim$1 M (see Fig. 2). Thus, the first TBA aggregates are unlikely to be the same as the direct contact hydrophobic dimers inferred from previous neutron scattering and theoretical studies of more concentrated TBA solutions.[7,17,19]

An additional important feature of our low concentration results is that the monomeric TBA species found in dilute aqueous TBA solutions remain the predominant TBA species up to concentrations at which a randomly mixed (non-aggregating) solution would be expected to have a significant number of direct TBA–TBA contacts. More specifically, our Raman-MCR results suggest that TBA remains fully hydrated at [TBA] $\leq$ 0.5 M, while in a randomly mixed (non-aggregating) solution of 0.5 M concentration, about 40% of the TBA molecules are predicted to be in contact with one or more TBA molecules (see Fig. 1). In other words, the observed fraction of TBA molecules that are in direct contact with each other is clearly lower than that expected in a random mixture (see Fig. 5). This implies that the hydrophobic hydration-shell of TBA is sufficiently thermodynamically stable to significantly suppress the formation of direct contact aggregates (relative to the number of TBA–TBA contacts which would have formed in the corresponding random mixture). When combined with the results described in the previous paragraph, our findings suggest that the first aggregates formed in aqueous TBA solutions are primarily water-separated rather than direct-contact aggregates. It is tempting to link this conclusion to the ubiquity of solid clathrate hydrates, whose structure consists of a lattice of hydrophobic solutes in water-separated contact with each other. Note that recent simulations suggest that clathrate nucleation is preceded by the formation of water-separated aggregates in the liquid state.[34] It is also possible that such equilibrium aggregates play a role in clathrate formation that is analogous to the recently proposed non-classical nucleation mechanism for other aqueous crystals.[35]

At somewhat higher TBA concentrations (up to 4 M or $\chi \sim 0.1$) we find evidence of the formation of quite different sorts of TBA aggregates. The hydration-shell spectra of these aggregates have much smaller OH stretch features than the TBA monomers found at lower concentration (see Fig. 6). This observation suggests that at a TBA mole fraction of $\chi \sim 0.1$ much of the water in the first hydration-shell of TBA has been displaced by TBA molecules. This might be taken to imply that the aggregates formed at such concentrations result from hydrophobic interactions between TBA molecules. However, the observed concentration profile of TBA molecules in these aggregates is found to be quite similar to that predicted for TBA molecules that would be in contact with three or more other TBA molecules in a non-aggregating random mixture (see the dashed curve in Fig. 6B). Thus, the number of aggregated TBA molecules that are present when [TBA] $\geq$ 2 M ($\chi \geq 0.04$) appear to be roughly consistent with random mixing predictions, and so may not be significantly influenced by hydrophobic interactions.

Although the present results do not provide quantitative information about the size of the aggregated TBA domains, the strongly non-linear concentration dependence of the first aggregates is not consistent with dimerization predictions (see the dotted curves in Fig. 5B). Thus, the TBA-rich domains formed in low concentration TBA solutions apparently each contain significantly more than two TBA molecules. At higher TBA concentrations, our results suggest that the size distribution of the TBA-rich domains may resemble the domains in a randomly mixed system within which each TBA molecule is in contact with three or more other TBA molecules.

Acknowledgements

Support for this work from the National Science Foundation (CHE-1213338) is gratefully acknowledged, as are useful discussions with professors Gren Patey and Lyudmila Slipchenko.

References

1 T. M. Bender and R. Pecora, *J. Phys. Chem.*, 1986, **90**, 1700–1706.
2 M. Sedlak and D. Rak, *J. Phys. Chem. B*, 2013, **117**, 2495–2504.
3 S. Dixit, J. Crain, W. C. K. Poon, J. L. Finney and A. K. Soper, *Nature*, 2002, **416**, 829–832.
4 R. Gupta and G. N. Patey, *J. Chem. Phys.*, 2012, **137**, 034509.
5 A. B. Roney, B. Space, E. W. Castner, R. L. Napoleon and P. B. Moore, *J. Phys. Chem. B*, 2004, **108**, 7389–7401.

6 D. Subramanian and M. A. Anisimov, *J. Phys. Chem. B*, 2011, **115**, 9179–9183.
7 D. T. Bowron and J. L. Finney, *Phys. Rev. Lett.*, 2002, **89**.
8 D. T. Bowron, A. K. Soper and J. L. Finney, *J. Chem. Phys.*, 2001, **114**, 6203–6219.
9 C. Petersen, A. A. Bakulin, V. G. Pavelyev, M. S. Pshenichnikov and H. J. Bakker, *J. Chem. Phys.*, 2010, **133**, 164514.
10 M. Freda, G. Onori and A. Santucci, *J. Phys. Chem. B*, 2001, **105**, 12714–12718.
11 A. Di Michele, M. Freda, G. Onori, M. Paolantoni, A. Santucci and P. Sassi, *J. Phys. Chem. B*, 2006, **110**, 21077–21085.
12 T. Fukasawa, Y. Tominaga and A. Wakisaka, *J. Phys. Chem. A*, 2004, **108**, 59–63.
13 R. Sinibaldi, C. Casieri, S. Melchionna, G. Onori, A. L. Segre, S. Viel, L. Mannina and F. De Luca, *J. Phys. Chem. B*, 2006, **110**, 8885–8892.
14 M. D. Hands and L. V. Slipchenko, *J. Phys. Chem. B*, 2012, **116**, 2775–2786.
15 A. Fornili, M. Civera, M. Sironi and S. L. Fornili, *Phys. Chem. Chem. Phys.*, 2003, **5**, 4905–4910.
16 S. Paul and G. N. Patey, *J. Phys. Chem. B*, 2006, **110**, 10514–10518.
17 P. G. Kusalik, A. P. Lyubartsev, D. L. Bergman and A. Laaksonen, *J. Phys. Chem. B*, 2000, **104**, 9533–9539.
18 I. Omelyan, A. Kovalenko and F. Hirata, *J. Theor. Comput. Chem.*, 2003, **2**, 193–203.
19 K. Yoshida, T. Yamaguchi, A. Kovalenko and F. Hirata, *J. Phys. Chem. B*, 2002, **106**, 5042–5049.
20 K. R. Fega, D. S. Wilcox and D. Ben-Amotz, *Appl. Spectrosc.*, 2012, **66**, 282–288.
21 B. M. Rankin, M. D. Hands, D. S. Wilcox, K. R. Fega, L. V. Slipchenko and D. Ben-Amotz, *Faraday Discuss.*, 2013, **160**, 255–270.
22 J. G. Davis, K. P. Gierszal, P. Wang and D. Ben-Amotz, *Nature*, 2012, **491**, 582–585.
23 K. P. Gierszal, J. G. Davis, M. D. Hands, D. S. Wilcox, L. V. Slipchenko and D. Ben-Amotz, *J. Phys. Chem. Lett.*, 2011, **2**, 2930–2933.
24 D. T. Bowron, J. L. Finney and A. K. Soper, *Mol. Phys.*, 1998, **93**, 531–543.
25 A. V. Plyasunov, N. V. Plyasunov and E. L. Shock, *The database for the thermodynamic properties of neutral compounds in the state of aqueous solution*, Department of Geological Sciences, Arizona State University, Tempe AZ 85287, 2005.
26 Y. Z. Zeng and M. Garland, *Anal. Chim. Acta*, 1998, **359**, 303–310.
27 K. R. Fega, D. S. Wilcox and D. Ben-Amotz, *Appl. Spectrosc.*, 2012, **66**, 282–288.
28 R. Tauler, A. Smilde and B. Kowalski, *J. Chemom.*, 1995, **9**, 31–58.
29 P. J. Gemperline, *Anal. Chem.*, 1999, **71**, 5398–5404.
30 A. Golshan, H. Abdollahi and M. Maeder, *Anal. Chem.*, 2011, **83**, 836–841.
31 R. Tauler, *J. Chemom.*, 2001, **15**, 627–646.
32 M. Freda, G. Onori and A. Santucci, *J. Phys. Chem. B*, 2001, **105**, 12714–12718.
33 A. Di Michele, M. Freda, G. Onori, M. Paolantoni, A. Santucci and P. Sassi, *J. Phys. Chem. B*, 2006, **110**, 21077–21085.
34 L. C. Jacobson, W. Hujo and V. Molinero, *J. Am. Chem. Soc.*, 2010, **132**, 11806–11811.
35 D. Gebauer and H. Colfen, *Nano Today*, 2011, **6**, 564–584.

Faraday Discussions RSC Publishing

PAPER

Linking molecular/ion structure, solvent mesostructure, the solvophobic effect and the ability of amphiphiles to self-assemble in non-aqueous liquids†

Emmy C. Wijaya,[ab] Tamar L. Greaves[b] and Calum J. Drummond*[b]

Received 7th May 2013, Accepted 8th July 2013
DOI: 10.1039/c3fd00077j

Sixteen non-ionic molecular solvents have been found to exhibit the solvophobic effect and to support the formation of amphiphile self-assembly mesophases. The solvents were low molecular weight polar solvents which contained various combinations of amine, hydroxyl or ether moieties with relatively small proportions of hydrocarbon unit constituents. The studied amphiphiles were hexadecyltrimethylammonium bromide (CTAB), hexadecylpyridinium bromide (C_{16}PyrBr) and tetraethylene glycol monohexadecyl ether ($C_{16}E_4$). Lyotropic liquid crystal mesophases with lamellar, normal hexagonal and normal bicontinuous cubic, with ordered one-, two- and three-dimensional periodic structure respectively, were identified in CTAB and C_{16}PyrBr systems by using cross-polarised optical microscopy (CPOM). Mesophase diversity and thermal stability ranges correlated to the Gordon parameter (*G*) value, a proxy for the solvent cohesive energy density. Infrared spectroscopy confirmed that all the studied molecular solvents were associative liquids. Solvent mesostructure was studied by synchrotron small angle X-ray scattering. The small sub-set of neat solvents which were mesostructured, with polar and non-polar domain segregation, displayed the lowest *G* values, and amongst the lowest mesophase diversity and thermal stability ranges. It has been established that the *G* value is a good indicator of whether or not a molecular solvent is likely to behave as a co-surfactant, residing within the amphiphile–solvent interfacial region of self-assembled objects, thereby influencing specific mesophase structure formation. Structure–property behaviour has been explored and shows that beneficial solvent features for serving as amphiphile-self assembly media, with the potential for rich mesophase diversity, include the presence of hydroxyl > amine > ether moieties, while methyl moieties have an adverse effect larger than that of methylene moieties.

[a]*School of Chemistry, The University of Melbourne, Parkville VIC 3010, Australia*
[b]*CSIRO Materials Science and Engineering, Bag 10, Clayton VIC 3169, Australia. E-mail: calum.drummond@csiro.au*

† Electronic supplementary information (ESI) available. See DOI: 10.1039/c3fd00077j

Introduction

The hydrophobic effect is a very significant factor governing the aggregation of amphiphiles to form lyotropic liquid crystal mesophases in water. The hydrophobic effect is also central to understanding complex biological self-assembly processes such as the protein intramolecular interactions involved in folding. The driving force for the self-assembly process is generally acknowledged as the energy minimisation obtained through segregation of the "oil" and water components.[1]

In Chandler's 2005 review in *Nature*[1] entitled, "Interfaces and the driving force of hydrophobic assembly," he posed the question, "Is water special?" with respect to exhibiting the hydrophobic effect. Chandler answered this question with the statement that the "two physical features responsible for hydrophobicity – the solvent being close to phase coexistence with its vapour, and solute–solvent interactions being significantly less attractive than solvent–solvent interactions are not particularly unusual, at least in the abstract sense."

Indeed, although not often highlighted, it was established by Ray in the period around 1970 that the ability to support amphiphile self-assembly is not unique to water, and they specifically reported twelve solvents which have an analogous solvophobic effect.[2,3] We note there was also a series of relatively unrecognised earlier publications which had shown that the molten salt pyridinium chloride (melting point = 146 °C) supported micellisation of cationic surfactants.[4–8] The ability of ethylammonium nitrate, a protic ionic liquid (PIL), was reported in the 1980s by Evans *et al.*,[9–11] which inspired more recent studies where it has been established as a relatively common property of protic ionic liquids,[12–17] and a less reported property of aprotic ionic liquids.[17–23] Protic ionic liquids result from the stoichiometric combination of a Brønsted acid with a Brønsted base. Amongst molecular solvents it has also been reported that some low molecular weight amides[24–28] and glycols share this ability.[3,29]

The specific solvent properties that appear to be beneficial for supporting amphiphile self-assembly are high cohesive energies, low molecular weights, high polarities and an ability to form hydrogen bonded networks. In years past it was thought that a solvent's ability to form hydrogen bonding networks was essential,[30,31] for example in Evans' 1988 American Chemical Society Langmuir Lecture he stated that "amphiphilic self-assembly is a physical process that occurs in polar, multiple hydrogen-bonding liquids."[31] The requirement for solvent hydrogen bonding has been challenged more recently through reports of amphiphile self-assembly in for example 3-methylsydnone.[26,32,33]

In amphiphile–solvent systems, lyotropic mesophases can include lamellar, and normal and inverse micellar, hexagonal, and discrete and bicontinuous cubic structures. Lamellar, hexagonal and cubic mesophases are ordered with one-, two- and three-dimensional periodicity, respectively. At low amphiphile concentrations the type of mesophase formed is dictated by the effective molecular geometry of the amphiphile. At high amphiphile concentrations the local molecular packing constraints are impacted by global packing constraints linked to the relative volume fraction of amphiphile to solvent. Hence mesophase transitions may occur with increasing amphiphile concentration.

A variety of solvent properties have been considered to classify a solvent's ability to support amphiphile self-assembly. The solvents polarity, as measured by the dielectric constant, does not provide sufficient information to account for changes in solvent conditions.[25,34] Surface tension has been shown to give a better indication[25] but to date a more useful indicative method has been to classify solvents according to their cohesive energy density.[15,17,31,34]

The cohesive energy density is the amount of energy required to completely remove a unit volume of molecules from their neighbours to infinite separation (an ideal gas state), which is equal to the heat of vaporisation divided by the molar volume. For many materials, such as molten salts, ionic liquids and polymers it is not feasible to measure the heat of vaporisation. Instead an approximate relationship between the cohesive energy density, CED, the surface tension at the liquid–vapour interface, γ_{LV}, and the molar volume, V_m, according to eqn (1), where c is a scalar, has been utilised.[35–38] There have been several reported values for c, including 16.1,[36] 14,[38] 1.009.[37] Where it is possible to compare, there has been good consistency between the calculated and experimental CED values, particularly for $c = 16.1$ for polar and strongly hydrogen bonding liquids.[36]

$$\mathrm{CED} = c\gamma_{\mathrm{LV}}/{V_{\mathrm{m}}}^{1/3} \tag{1}$$

The Gordon parameter, G, has been used in the field of amphiphile self-assembly to give a measure of the cohesive energy density, and hence the solvophobic effect.[13,14,31,39–41] G is defined in eqn (2) and is obviously obtained when $c = 1$ in eqn (1).

$$G = \gamma_{\mathrm{LV}}/{V_{\mathrm{m}}}^{1/3} \tag{2}$$

Whether or not G as defined provides the most accurate cohesive energy density values is not important, as the utility of the Gordon parameter is as a relative predictive scale for the solvophobic effect. It is useful to have a widely utilised consistently defined parameter. The standard free energy of micellisation, $\Delta {G_{\mathrm{mic}}}^0$, has been shown to decrease linearly with increasing Gordon parameter, for modifications to an aqueous-amphiphile solution through additions of acetamide,[41] 2-methoxyethanol,[39,40] acetonitrile[39] or formamide.[39] There is also a strong correlation between the Gordon parameter and the CMC,[31] lyotropic liquid crystal mesophase diversity[13,14] and thermal stability ranges[13,14] of these mesophases for amphiphiles in solvents. The Gordon parameter for water is 2.743 J mol$^{1/3}$ m^{-3},[42] with other solvents generally having lower values. Currently, the reported G values for solvents supporting amphiphile self-assembly range between 0.53 to 1.73 J mol$^{1/3}$ m^{-3} for amides,[27] 0.87 to 1.20 J mol$^{1/3}$ m^{-3} for diols,[17] and 0.58 to 1.70 J mol$^{1/3}$ m^{-3} for protic ionic liquids.[43,44] On a cautionary level, it is noteworthy that G values for nominally the same solvent can have large differences in reported values, presumably due to differences in surface tensions which may arise due to differences in impurity (*e.g.*, water) content and/or measurement techniques.

The many PILs which have been reported as amphiphile self-assembly media have enabled structure–property relationships to be developed between their chemical structural components and their ability to support a diverse range of amphiphile self-assembly mesophases, and for them to have a wide thermal stability range.[17] It has been determined that the presence of a hydroxyl moiety on either the cation or anion increases the Gordon parameter and is highly beneficial to obtaining rich mesophase diversity, and in particular for supporting cubic lyotropic liquid crystal mesophase formation.[13,44] The incorporation of ether moieties increases G, but not by as much as hydroxyl moieties.[44] Increasing hydrocarbon unit length, or having a branched hydrocarbon unit tended to lead to lower Gordon parameters, and usually less mesophase diversity, and phases which were more poorly developed.[13,44] PILs containing longer hydrocarbon unit chains may act as co-surfactants and hence modify the lyotropic liquid crystal mesophases.[13,14]

Solvent molecule alignment considerations are not taken into account in eqn (1), which makes it a first-order approximation for many solvents.[45] Using eqn (1) to obtain the cohesive energy density includes the assumption that the solvent molecules are randomly orientated in the bulk and at the liquid–vapour interface. However, it is well known that many molecules will be oriented at the interface such that their polar groups are directed towards the bulk liquid and apolar groups are directed to the vapour phase. Consequently the area per molecule at the interface can be considerably smaller than expected based on the bulk molar volume. Eqn (1) has been modified by Becher to include solvent molecular area and volume terms at the interface to provide eqn (3),[37] where for a solvent molecule length dimension L, the area is given by jL^2 and the volume by kL^3. Therefore, the factor $jk^{-2/3}$ will depend on the shape and orientation of the molecules at the interface. For a highly oriented molecule, such as a primary alcohol which arranges itself to have the OH directed to the bulk and the alkyl chain perpendicular to the interface towards the vapour phase,[46] it would be reasonable to assume that the surface area will be decreased from what the simple model would predict, with the j term in eqn (3) decreasing, and thus leading to a lower calculated CED. It is consistent that as the alkyl chain length increases for a primary alcohol the molecular volume (related to k) will increase, whereas there will be minimal change for the surface area (related to j), and therefore the factor $jk^{-2/3}$ will decrease with increasing alkyl chain length. The surface area of methanol has been used as the approximate surface area for the series of primary alcohols.[47] A similar phenomenon can be expected for other solvent molecules which orientate at the interface, such that their surface area is less than that of randomly orientated molecules, however the specific areas and volumes are mostly unknown.

$$\mathrm{CED} = cjk^{-2/3}\gamma_{\mathrm{LV}}/{V_{\mathrm{m}}}^{1/3} \quad [37] \tag{3}$$

Many neat ionic liquids (ILs) possess a mesostructure consisting of segregated polar and non-polar domains,[48–50] related to those found in primary alcohols[51] and other molecular solvents.[52,53] The use of scattering techniques such as small-angle X-ray scattering (SAXS) and small angle neutron scattering (SANS) have enabled correlation distances for many ILs to be determined.[54,55] A relationship has been observed between increasing mesostructure correlation length and decreasing Gordon parameter of the IL, and hence decreasing ability to behave as a mesophase rich amphiphile self-assembly media.[17]

In this study, we have applied our evolving understanding of amphiphile self-assembly in ILs[17,56] to molecular solvents, with the aim of illustrating that the solvophobic effect is far more widespread than commonly accepted. Further, we seek to extend the structure–property relationships obtained from ILs to include molecular solvents, and to identify differences, if any, in behaviour between these two classes of solvents.

Based on insights obtained from investigating PILs[17,56] we chose molecular solvents with small hydrocarbon unit constituents, and polar functional groups such as amine and hydroxyl moieties. The solvents were selected from widely available low molecular weight polar liquids and are shown in Fig. 1. Of these non-aqueous solvents, amphiphile self-assembly has been previously reported in glycerol, ethylene glycol and 1-amino-2-propanol. These three have been included and studied further since they share many chemical similarities with the others,

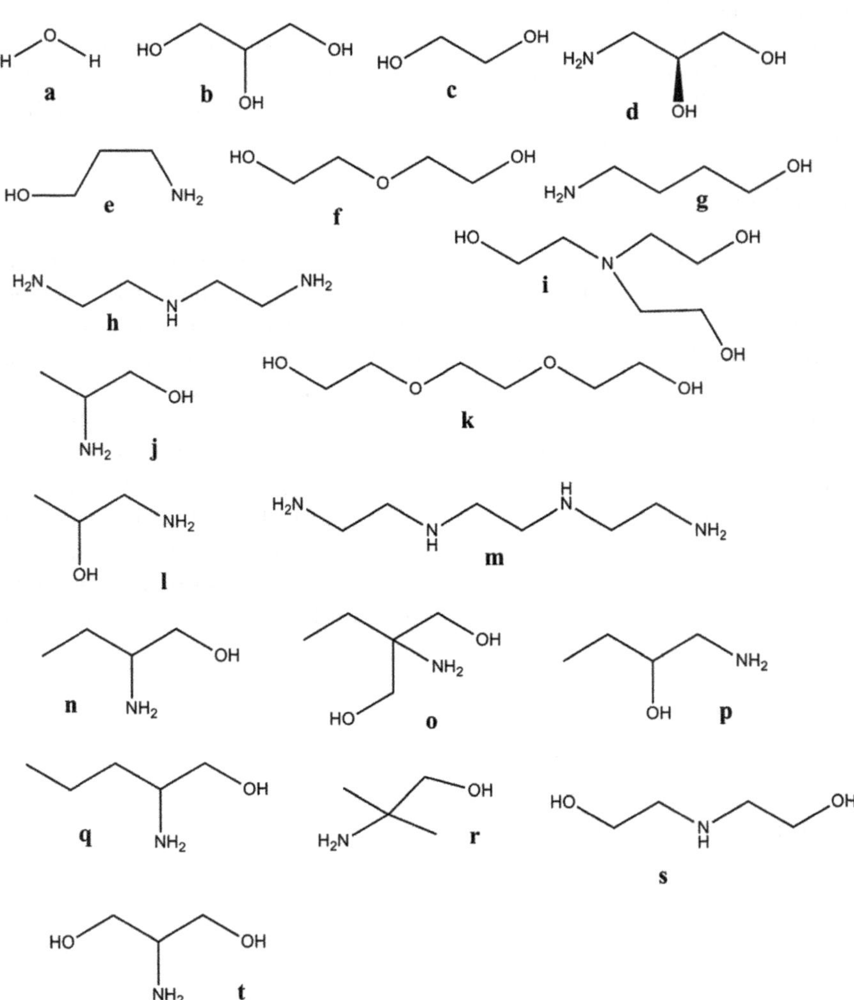

Fig. 1 Chemical structures of the solvents of a) water, b) glycerol, c) ethylene glycol, d) (*S*)-3-amino-1,2-propanediol, e) 3-amino-1-propanol, f) diethylene glycol, g) 4-amino-1-butanol, h) diethylene triamine, i) triethanolamine, j) (*R*)-2-amino-1-propanol, k) triethylene glycol, l) DL-1-amino-2-propanol, m) triethylene tetramine, n) 2-amino-1-butanol, o) 2-amino-2-ethyl-1,3-propanediol, p) 1-amino-2-butanol, q) DL-2-amino-1-pentanol, r) 2-amino-2-methyl-1-propanol, s) diethanolamine, t) 2-amino-1,3-propanediol.

and hence will be useful for developing structure–property relationships. The only report we are aware of for self-assembly in 1-amino-2-propanol was by Ray *et al.* in 1971 where it was reported to support micelle formation of a non-ionic amphiphile *p*,*t*-nonylphenoxy polyethoxy ethanol.[3] While at first assessment ethylamine would have also been logical to include in the molecular solvent series, it has a boiling point below room temperature, and hence could not be readily studied.

The amphiphiles used were hexadecyltrimethylammonium bromide (CTAB), hexadecylpyridinium bromide (C_{16}PyrBr) and tetraethylene glycol monohexadecyl ether ($C_{16}E_4$). These three amphiphiles all have a C_{16} alkyl chain, and vary in head group charge, with two cationic and one non-ionic. The amphiphile chemical structures are provided in Fig. 2. A significant portion of the literature on non-aqueous self-assembly solvents have used CTAB and C_{16}PyrBr which will enable comparisons to this work.

Liquid–vapour surface tension values, calculated G parameter values (eqn (2)), SAXS and infrared (IR) measurements for the neat molecular solvents, and cross-polarised optical microscopy (CPOM) observations for solvent–amphiphile systems informed our study.

Experimental method

The chemicals were used as received, including the solvents of triethylene tetramine (Aldrich, ≥97%), triethylene glycol (Aldrich, 99%), diethylene triamine (Aldrich, 99%), diethanolamine (AnalaR, 99.5%), diethylene glycol (Aldrich, 99%), glycerol (Merck, ≥99%), triethanolamine (Aldrich, 99.5%), 4-amino-1-butanol (Aldrich, 98%), 2-amino-2-methyl-1-propanol (Fluka, ≥99%), DL-1-amino-2-propanol (Aldrich, ≥90%), (±)-2-amino-1-butanol (Fluka, ≥98%), 1-amino-2-butanol (Aldrich), DL-2-amino-1-pentanol (Aldrich, 97%), 2-amino-2-ethyl-1,3-propanediol (Acros organics, 97%), (*R*)-(−)-2-amino-1-propanol (Aldrich, 98%), 2-amino-1,3-propanediol (Aldrich, 98%), (*S*)-3-amino-1,2-propanediol (Aldrich, 97%) and 3-amino-1-propanol (Aldrich, 99+%). The amphiphiles used were hexadecyltrimethylammonium bromide (CTAB) (Kodak), hexadecylpyridinium bromide (C_{16}PyrBr) (Aldrich, ≥97%) and tetraethylene glycol monohexadecyl ether ($C_{16}E_4$) (Nikkol).

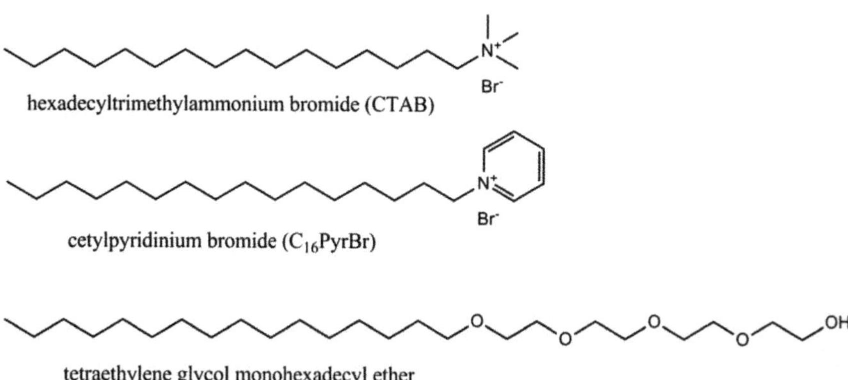

Fig. 2 Chemical structures of the amphiphiles, hexadecyltrimethylammonium bromide (CTAB), hexadecylpyridinium bromide (C_{16}PyrBr) and tetraethylene glycol monohexadecyl ether ($C_{16}E_4$).

The small angle X-ray scattering (SAXS) patterns of the neat solvents which were liquid at room temperature were acquired on the SAXS/WAXS beamline at the Australian Synchrotron.[57] Samples were loaded into 1.5mm glass capillaries and sealed with a silicon rubber (Selleys wet area silicone) to prevent water absorption. Samples were housed in a 60 sample capillary holder mounted on a 3-axis sample stage. The temperature of the sample holder was controlled using a Huber recirculating water bath. Scattering patterns were acquired at 25 °C with an exposure time of 1 s. The SAXS patterns were acquired over the q range of 0.008 to 0.52 $Å^{-1}$.

The cross-polarised optical microscopy (CPOM) images were taken on an Olympus IMT-2 microscope equipped with cross-polarising lenses. The amphiphile (CTAB, C_{16}PyrBr or $C_{16}E_4$) was compressed between a microscope slide and a coverslip. A drop of the solvent was added to the edge of the coverslip and allowed to penetrate. The slides were sealed to prevent water absorption or solvent evaporation and left to equilibrate for approximately one day before use. The samples were heated in a Mettler FP82HT hot stage controlled by a PF90 central processor. Most of the lyotropic liquid crystal mesophases did not require lengthy equilibration times to form, but one day was used with all amphiphile–solvent systems for consistency.

Water contents in the neat solvents were determined using a Metler Toledo DL39 Karl Fischer titrator. The infrared measurements were made on a Nicolet 6700 FT-IR using a diamond ATR attachment. Samples were typically exposed to air for 1–2 s during the measurements. Differential scanning calorimetery (DSC) was performed using a Mettler Toledo 851 system under nitrogen, with a scan rate of 5 °C min^{-1} from −50 to 100 °C. Liquid–vapour surface tension measurements were made on each molecular solvent using a pendant drop method with a KSV CAM 200 instrument.

Results

A series of commercially available, high purity, low molecular weight polar solvents were used in this investigation. These were selected as potential amphiphile self-assembly media based on prior structure–property relationships established from protic ionic liquids (PILs).[13,14,17,44,56] It was previously identified for PILs that short alkyl chains and the presence of hydroxyl groups lead to solvents with an enhanced solvophobic effect, as gauged from the Gordon parameter, and also greater mesophase diversity and greater thermal stability ranges.[13,14,44]

The chemical structures of the solvents used in this investigation are provided in Fig. 1. All of these molecular solvents contain at least two polar moieties of an amine, hydroxyl or ether functional group, and relatively short (or non-existent in the case of the reference solvent water) hydrocarbon unit constituents. The relevant physical properties of these solvents are provided in Table 1, including their water content, liquid–vapour surface tension, γ_{LV}, molecular weight, M_w, Gordon parameter, G, melting point, T_m, and boiling point, T_b. The values for the previously reported amphiphile self-assembly solvents of water, ethylene glycol, glycerol, 1-amino-2-propanol and ethylammonium nitrate (EAN) have been included for comparison. The surface tensions were measured for all the solvents which were liquid at room temperature to provide consistency due to significant

Table 1 Select properties of solvents including water content, surface tension at the liquid–air interface (γ_{LV}), molecular weight (M_w), Gordon parameter (G), melting point (T_m) and boiling point (T_b)

Solvent	Water content (wt%)	γ_{LV} (dyne/cm) at 25 °C	M_w (g mol^{-1})	G (J.mol$^{1/3}$/m^3)	T_m (°C)	T_b (°C)
Water	100	71.99[42]	18.02	2.743[42]	0	100
Glycerol	0.30	63.4[62]	92.09	1.517[62]	17.8[62]	303.85[59]
Ethylene glycol	0.65	47.99[62]	62.07	1.356[62] 1.20[30]	−12.7[62]	197.3[59]
Ethylammonium nitrate	0.22	47.3 (27 °C)[43]	108.10	1.060[14] 1.40[30]	9[43]	242[43,b]
3-Amino-1,2-propanediol	3.23	65.8[a] 56.4[59]	91.11	1.54	[d,e]	264[59]
3-Amino-1-propanol	1.7	45.3[a] 44.70[59]	75.11	1.07	11[59]	187[59]
Diethylene glycol	0.95	44.4[a] 55.10[59]	106.12	0.98	−10.45[59]	244.8[59]
4-Amino-1-butanol	1.88	43.9 43.84[59]	89.14	0.97	8.72[59] 16–18[60]	205[59] 206[60]
Diethylene triamine	0.80	43.4[a] 48.85[59]	103.17	0.91	−39[59] −35[60]	207.1[59] 199–209[60]
Triethanolamine	0.14	46.7[a] 51.48[59]	149.19	0.91	21.2[59] 18–21[60]	335.39[59]
2-Amino-1-propanol	0.15	38.3[a] 47.03[59]	75.11	0.90	13.37[59]	174.50[59] 166–168[60]
Triethylene glycol	0.52	45.1[a] 45.13[58] 64.41[59]	150.18	0.88	−7.36[59] −7[60]	288.35[59]
1-Amino-2-propanol	0.45	36.7[a] 36.5[3] 48.68[59]	75.11	0.86	1.74[59] −2	159.46[59] 160[60]
Triethylene tetramine	1.06	43.4[a] 50.58[59]	146.48	0.82	12[59,60]	266.5[59] 266–267[60]
2-Amino-1-butanol	0.58	32.2[a] 46.65[59]	89.14	0.71	−1[59] −2[60]	178[59] 176–178[60]
2-Amino-2-ethyl-1,3-propanediol[e]	1.87	[e] 33.35[59]	119.17	0.70	[f] 37.5[59] 35–37[60]	176.75[59]
1-amino-2-butanol	1.65	31.1[a] 47.84[59]	89.14	0.67	3[59,60]	168.9[59] 169[60]
2-Amino-1-pentanol	1.04	30.4[a] 44.10[59]	103.68	0.63	21[a] 29.84[59]	194.50[59]
2-Amino-2-methyl 1-propanol[c]	0.29	[c]	89.14	[c]	25.5[59]	165.5[59]
Diethanolamine	1.01	[c]	105.32	[c]	28[59]	268.39[59]
2-Amino-1,3-propanediol[c]	2.05	[c]	91.11	[c]	52–55[60]	277[60]

[a] Values determined during this study. [b] Explosive decomposition. [c] Solvent, gel or solid at room temperature. [d] Not available in the literature. [e] Highly viscous at room temperature. Too viscous to measure surface tension using hanging drop method. [f] No melting point observed using DSC between −50 and 100 °C.

variation in the literature. For example, the largest difference in literature surface tension values for nominally the same solvent was for triethylene glycol which varied between 45.13[58] and 64.41[59] dyne cm^{-1}.

The properties of melting point and boiling point have been included in Table 1 to show the wide useable thermal range for many of these solvents and to provide an indication of the closeness of the solvents to phase coexistence with their vapour. Two of the chosen solvents, 2-amino-1-pentanol and 2-amino-2-ethyl-1,3-propanediol, were viscous liquids at room temperature despite their T_m values in the literature being 28.9 °C[59] and 35–37 °C,[60] respectively. DSC measurements showed a T_m for 2-amino-1-pentanol of 21 °C, where the difference may possibly be attributed to differences in the amount of water present. For 2-amino-2-ethyl-1,3-propanediol no melting point was evident from the DSC which was run over the range −50 to 100 °C. It is known that strongly hydrogen-bonded liquids may form glasses with no distinct solid–liquid melting point.

The Gordon parameters were calculated using eqn (2) utilising the liquid–air surface tension values determined in this investigation and solvent density values sourced from the chemical suppliers.[60] The exception was for 2-amino-2-ethyl-1,3-propanediol which was too viscous at room temperature to enable the surface tension to be measured using the hanging drop method. The literature value was used for this solvent to calculate G. The Gordon parameters, derived in the present work, vary between 0.63 and 1.54 J $mol^{1/3}m^{-3}$, and all are within the range of 0.53 to 2.74 J $mol^{1/3}$ m^{-3} where amphiphile self-assembly has previously been reported in polar solvents.

Previously in our work with surface tension measurements and PILs we have established that PILs preferentially arrange themselves at the liquid–air interface to have relatively long hydrocarbon segments exposed to air and their polar moieties in the bulk liquid.[14] We anticipate similar behaviour for the molecular solvents studied in this current work. Myers reported on a large series of solvents which showed that for primary alcohols there was substantial ordering at the interface, and less in the bulk solvent. In contrast, for straight chained amines and amino containing complexes there was weaker ordering at the interface, but still more than suggested by the association in the bulk solvent.[47] Ethylene glycol was shown to behave similarly to the primary alcohols with significant orientation effect at the interface compared to the bulk,[47] with the OH groups directed towards the bulk solvent and the hydrocarbon segments perpendicular to the interface.[61]

It is anticipated that most of the solvents shown in Fig. 1 will undergo some orientation effects at the liquid–vapour interface. The literature data suggests it will be a greater effect for the hydroxyl containing solvents than those containing amine groups, though we are unaware of information on solvents containing both of these groups. With the exception of water, all of the solvents shown in Fig. 1 have some hydrocarbon component in addition to their polar groups. Consequently it is expected that they will orientate to direct the polar groups towards the bulk solvent and the large hydrocarbon segments towards the vapour phase. The presence of the multiple polar groups and branching will make this much more complex than for linear molecules such as primary alcohols or their amine counterparts. Therefore it is difficult to predict *a priori* whether the molecules will be packed at the interface in such a way as to subvert the Gordon parameter from being a reasonable proxy for the bulk cohesive energy density. The exception is that the presence of methyl groups, or increased numbers of methylene groups in a hydrocarbon unit, will increase the ability of the solvents to arrange themselves at the interface, and increase the packing density of the molecules at the liquid–air interface. Solvents which may behave in this manner include 2-amino-2-methyl-1-propanol, 2-amino-1-propanol, 2-amino-1-butanol and 2-amino-1-pentanol. The solvents of 3-amino-1-propanol and 4-amino-1-butanol can be visualised such that they can arrange themselves to have their terminal hydroxyl and amine moieties directed to the bulk and a hydrocarbon segment on the vapour side of the interface.

Therefore G values listed in Table 1 inherently incorporate varying molecular orientation effects at the liquid–air interface. Nevertheless, because of the simplicity of the G parameter determination, we consider it worthwhile to assess the utility of the G parameter in providing a relative measure of the solvophobic effect for the studied molecular solvents, and to enable a comparison to the solvophobic effect exhibited by other solvents.

Structure–property relationships between their chemical structure and the solvophobic effect as described by the Gordon parameter can be developed for the molecular solvents. The surface tension, and hence G, is highly sensitive to small changes in the alkyl chain length present in the solvents, which is evident for the Gordon parameter of 3-amino-1-propanol being 1.07 compared to 0.97 J $mol^{1/3}$ m^{-3} for 4-amino-1-butanol. Likewise for the series of 2-amino-1-propanol, 2-amino-1-butanol and 2-amino-1-pentanol G is 0.90, 0.71 and 0.63 J $mol^{1/3}$ m^{-3} respectively. Therefore, very small changes in alkyl substituent chain length can potentially have a significant impact on a solvent's ability to serve as an amphiphile self-assembly media.

Diethylene triamine and diethylene glycol had G values of 0.91 and 0.98 J $mol^{1/3}$ m^{-3} respectively, and are analogues with either all amine or all hydroxyl and ether moieties, respectively. Likewise, the pair of triethylene tetramine and triethylene glycol with G values of 0.82 and 0.88 J $mol^{1/3}$ m^{-3}, respectively. This indicates that the surface tension, and hence G, is slightly larger for solvents with hydroxyls compared to amine moieties. In addition, increasing the number of secondary amines or ether groups in the alkyl chain decreased the Gordon parameter, which strongly indicates that the hydroxyls or primary amine moieties have a far greater contribution to G than the secondary amines or ethers, and that increasing the alkyl chain by an ethyl group causes a greater reduction to G than the addition of an ether or secondary amine moiety.

Rearrangements of the structures, such as for the pairs of 2-amino-1-propanol and 1-amino-2-propanol or 2-amino-1-butanol and 1-amino-2-butanol had very little effect on the Gordon parameter. Consequently, it is anticipated that the solid solvent of 2-amino-1,3-propanediol will have a relatively high Gordon parameter of approximately 1.5 J $mol^{1/3}$ m^{-3} based on its analogue of 3-amino-1,2-propanediol. The linear 4-amino-1-butanol had a higher G value compared to its branched counterparts of 2-amino-1-butanol or 1-amino-2-butanol. Likewise, 3-amino-1-propanol had a higher G value than 2-amino-1-propanol or 1-amino-2-propanol.

Neat solvent mesostructure

The mesostructure for the solvents which were liquid at room temperature was investigated by using the SAXS/WAXS beamline at the Australian Synchrotron.

The solvents of 2-amino-1-butanol, 2-amino-1-pentanol and 1-amino-2-butanol displayed features in the SAXS region, as shown in Fig. 3. These features are correlation peaks, corresponding to correlation lengths present in the solvents. The correlation peak for 2-amino-1-pentanol had $q_{max} = 0.385$ Å^{-1}. This can be converted to a correlation distance, d, using Bragg's Law, $d \sim 2\pi/q_{max}$, leading to a correlation distance of 16.3 Å. This is comparable to the correlation distances reported for pentylammonium formate and pentylammonium nitrate of 15.22 and 15.94 Å, respectively.[50] The solvents 2-amino-1-butanol and 1-amino-2-butanol displayed very similar SAXS patterns, with a smaller correlation length (larger q_{max} value) as expected from the shorter alkyl chain.

No mesostructure was detected for the neat solvents when the chain length was decreased, as shown for 1-amino-2-propanol in Fig. 3. No evidence of mesostructure was detected in any of the other neat solvents using SAXS. The lack of observed mesostructure was attributed to the chemical structures containing multiple polar moieties which were located such that the molecules could not arrange themselves into segregated polar and non-polar domains.

Infrared spectroscopy

Infrared spectra were acquired to investigate the hydrogen bonding of the solvents shown in Fig. 1. We anticipated that the bonding present within these solvents will vary considerably due to steric influences, in addition to the number and location of the hydroxyl, ether and amine moieties. To enable comparisons between these solvents we have focussed on the N–H and O–H stretches. The effect of increasing hydrogen bonding strength on either the NH or OH stretches is to cause a redshift to lower wavenumbers.[63] The OH band is affected to a greater extent by hydrogen bonding than the NH stretches, and hence it is a very broad peak where it is difficult to obtain a peak position for many of these solvents. The NH stretches in contrast are usually well defined. The N–H stretches occurred between 3270 and 3360 cm^{-1}, where there are two bands for primary amines, one for secondary amine and none present for tertiary amines. The O–H stretches contributed a very broad band around 3400 cm^{-1}. These band positions are all consistent with the studied molecular solvents possessing considerable hydrogen bonding. Consequently, the IR region of 3000 to 3500 cm^{-1} has been focussed on.

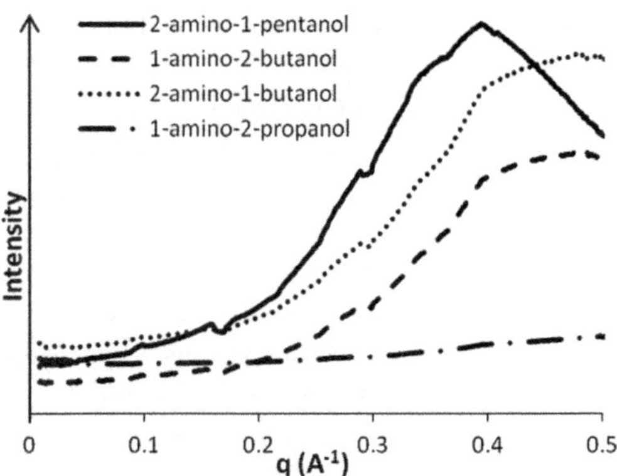

Fig. 3 SAXS 1-d patterns for neat solvents 2-amino-1-pentanol, 1-amino-2-butanol, 2-amino-1-butanol and 1-amino-2-propanol.

Solvents which have hydroxyl and/or amine groups are usually associated solvents in the liquid state. This arises from hydrogen-bonds and other inter- and intra-molecular interactions. There has been considerable work on the intra-molecular bonding in hydrogen bonding associated solvents. This has been conducted using dilute solutions in non-hydrogen bonding media such as carbon tetrachloride or gas phases to remove intermolecular bonds. These solvents,[64] diols of the form $HO(CH_2)_nOH$ with $n = 2$–9,[65] and diethylene glycol monomethyl ether and diethylene glycol monoethyl ether,[66] have been reported to form cylic monomers in dilute solutions or gas phases. Diethylene glycol monomethyl ether and diethylene glycol monoethyl ether have been reported to form a 10-membered dimer ring in concentrated solutions.[66] In comparison, the smaller ammonia[67] and dimethylamine[68] form non-cyclic dimers. A relevant investigation has shown that the amine and hydroxyl containing solvents *N*-methyldiethanolamine and triethanolamine form hydrogen bonds which involve both the hydroxyls and amine groups.[64]

In neat glycerol and glycerol–water solutions it has been shown that there is hydrogen bonding between glycerol CH and water or the hydroxyls in other glycerol molecules.[63] The formation of CH···O hydrogen bonds tends to result in longer C–H bonds, and therefore the CH stretches are red shifted.[69] An exception is if the CH···O bond is from a $F_nH_{3-n}CH$ proton donor then the CH bond is blue shifted.[70] Likewise, in formamide it has been reported that the CH bond length was slightly reduced, even for those involved in hydrogen bonding.[71] These CH···O hydrogen bonds are weaker than conventional hydrogen bonds.[63,69,70] The hydroxyl involved in the hydrogen bonding has a red shift to lower wavenumber of the broad OH stretch around 3400 cm^{-1} with increasing hydrogen bonding.[63,70]

For ammonia in the liquid state there are many ammonium molecules in the solvation shell, and consequently many non-hydrogen bonding interactions, in addition to hydrogen-bonds.[67] Due to this, solvation is dominated by steric packing effects, and not by directional hydrogen-bonding interactions, which is in contrast to water solvation.[67] For formamide in the liquid state there is mainly a N–H···O bonded network with additional C–H···O bonds. While there are dimers present, they are not dominant, nor are isolated molecules.[71]

Through comparison of the IR spectra for the molecular solvents in the current work it was possible to obtain information about the strength of the hydrogen bonding present in the solvents, and the structural features which are influential. The IR spectra are provided for all the solvents in ESI S1,† and for the wavenumber range of 3000–3500 cm^{-1} in ESI S2.†

The IR spectra showing the broad OH stretch present for the glycol series of glycerol, ethylene glycol, diethylene glycol and triethylene glycol are shown in Fig. 4 for 3000 to 3500 cm^{-1}. Since shifts to smaller wavenumber are associated with stronger hydrogen bonding it can be observed that the hydrogen bonding follows glycerol $\sim$ ethylene glycol > diethylene glycol > triethylene glycol. A small red shift for the NH peaks were observed for 2-amino-2-ethyl-1,3-propanediol compared to 2-amino-1-butanol, indicating stronger hydrogen bonding in the former. There was no evident shift in the hydroxyl band. This suggests that the addition of the ethanol group to 2-amino-2-ethyl-1,3-propanediol increases the hydrogen bonding. However, this is in contrast to the series for glycerol, ethylene glycol, diethylene glycol and triethylene glycol. Previously it was reported that there was stronger hydrogen bonding for triethanolamine compared to

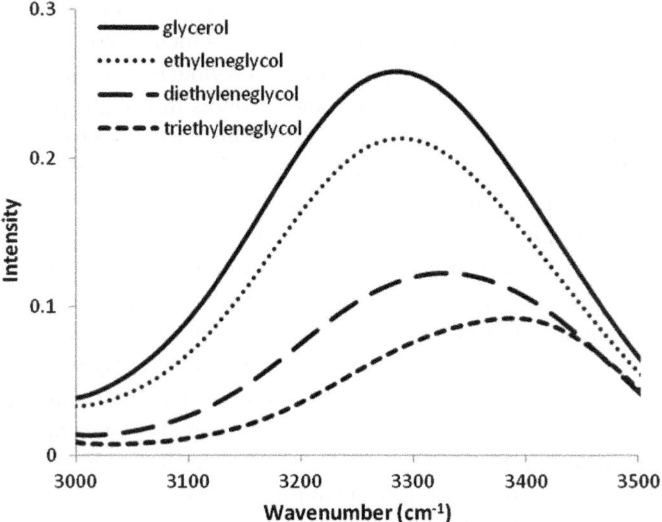

Fig. 4 Infrared spectra for glycerol, ethylene glycol, diethylene glycol and triethylene glycol.

N-methyldiethanolamine based on the larger hydroxyl shift in the IR for triethanolamine in the bulk compared to dilute in tetrachloroethylene.[64]

The structural analogues of 3-amino-1-propanol, 2-amino-1-propanol and 1-amino-2-propanol are shown in Fig. 5. It is evident that the two narrower NH stretches and the broad OH stretch are consistent in their shifts for these solvents, and indicate that the hydrogen bonding follows 2-amino-1-propanol > 3-amino-1-propanol > 1-amino-2-propanol. Similarly, the hydrogen bonding for the related series follows 2-amino-1-butanol > 4-amino-1-butanol > 1-amino-2-butanol.

It is known that the O–H bonds are more polar than the N–H bonds, according to their electronegativities, and therefore there are stronger hydrogen bonds formed with hydroxyls than with amines. It appears that the hydrogen bonding

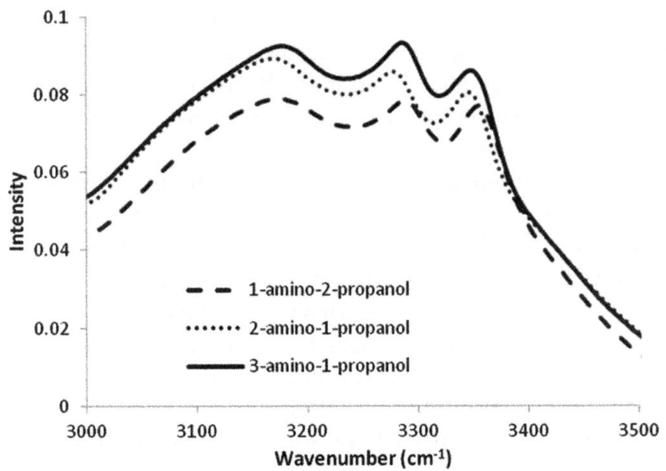

Fig. 5 Infrared spectra for 3-amino-1-propanol, 2-amino-1-propanol and 1-amino-2-propanol.

reflects the steric hindrance of the molecule, with the more constrained the amine group, the greater the hydrogen bonding, which is consistent with the hydroxyls forming stronger hydrogen bonds. Previously, an infrared study of the hydrogen bonding between amines and alcohols in the gas phase reported that the electron donating abilities of the free amines follows $Et_3NH > Et_2NH_2 \sim Me_3N > Me_2NH > EtNH_2 > MeNH_2 > NH_3$.[72] This is consistent with our data, with the greater steric hindrance of the amine group leading to stronger bonds between amine and hydroxyl containing solvents.

The series of 2-amino-1-propanol, 2-amino-1-butanol and 2-amino-1-pentanol had negligible differences in their IR spectra between 3000 to 3500 cm^{-1} for their N–H or O–H bonds, indicating a high similarity in their hydrogen bonding. Similarly, 3-amino-1-propanol and 4-amino-1-butanol were very similar for 3000–3500 cm^{-1}. Overall, this indicates that the addition of a small number of methylene units has little effect on the hydrogen bonding. This is consistent with a previous investigation on solutions of alcohols and amines, where changing the alkyl chain length of the alcohol, or including branching, had negligible difference when combined with the same amine.[72]

Amphiphile self-assembly

The ability of these solvents to support amphiphile self-assembly was trialled with three amphiphiles, whose structures are provided in Fig. 2. These specific amphiphiles were selected due to being commonly used amphiphiles. Two are cationic, CTAB and C_{16}PyrBr, which have a quaternary ammonium and pyridinium head group, respectively, and are paired with the same bromide counter ion. They all have a C_{16} alkyl chain, with the non-ionic $C_{16}E_4$ selected for comparison. Neat CTAB remains in a crystalline form to above 120 °C, and $C_{16}E_4$ melts at 36–38 °C and no liquid crystal phases were observed for either of these in the neat form. The neat C_{16}PyrBr has a melting point of 67–71 °C[60] and forms a lamellar liquid crystal phase from 67–>120 °C.

The lyotropic liquid crystal mesophases were identified according to their characteristic phase texture under cross-polarised optical microscopy (CPOM). A high-throughput flooding technique was used which enables screening across the full amphiphile–solvent composition range, though does not provide detail about the exact composition at phase transitions.[42] The isotropic cubic phase was identified based on the phase progression compared to water and other solvents, and on it being a highly viscous isotropic phase.

A variety of lyotropic liquid crystal mesophases were supported by the solvents in the presence of the amphiphiles of CTAB and C_{16}PyrBr, and a summary is provided in Tables 2 and 3 respectively. The solvents in Tables 2 and 3, like Table 1, are ordered by their Gordon parameter from high to low values down the table. The exceptions are the last three which were solid at room temperature and hence did not have their Gordon parameter determined. Representative CPOM images displaying various characteristic phase textures are provided in Fig. 6 for CTAB and Fig. 7 for C_{16}PyrBr in select solvents.

None of the studied molecular solvents were found to support the formation of lyotropic liquid crystal mesophases with $C_{16}E_4$. The only solvents we are aware of which have been reported to support lyotropic liquid crystal mesophases with

Table 2 Lyotropic liquid crystal mesophases identified from CPOM penetration scans for CTAB in the various solvents

	Lyotropic liquid crystalline mesophase stability regions (°C)		
Solvent	Hexagonal	Cubic	Lamellar
Water	25–190[42]	40–185[42]	50–>200[42]
	25–200[a26,75]	45–180[a26,75]	58–220[a26,75]
Glycerol	55–>150[62]	75–>135[62]	80–>150[62]
	56–>110[a26,75]	76.5–>110[a26,75]	79–>110[a26,75]
Ethylene glycol	55–90[62]	75–85[62]	80–>100[62]
	50.4[26b]	68.7[26b]	76.8[26b]
Formamide	50–125[25]		75–>150[25]
	52–>110[75]	72–>110[75]	82–>110[75]
Ethylammonium nitrate[14]	58–>104	76–>104	76–90
3-Amino-1,2-propanediol	64–>120	68–93	73–>120
3-Amino-1-propanol	60–113	71–101	72–>120
Diethylene glycol	50–79	64–82	68–>120
4-Amino-1-butanol	66–102		71–>120
Diethylene triamine	62–109		68–>120
Triethanolamine	60–93		69–>120
2-Amino-1-propanol	58–110		70–>120
Triethylene glycol			71–>120
1-Amino-2-propanol			80–>120
Triethylene tetramine	63–91		71–>120
2-Amino-1-butanol	65–73		68–>120
2-Amino-2-ethyl-1,3-propanediol	61–78		71–>120
1-Amino-2-butanol	64–85		70–>120
2-Amino-1-pentanol			66–>120
2-Amino-2-methyl-1-propanol			72–105
Diethanolamine	68–94	78–80	78–>120
2-Amino-1,3-propanediol	60–>120	74–>120	79–>120

[a] Upper limit estimated from figure in reference. [b] Upper limit not provided.

$C_{16}E_4$ are water,[42] EAN,[12] formamide[27,73] *N*-methylformamide[27] and dimethylformamide.[74] The phases are provided in Table 4.

CTAB–solvent systems

The lyotropic liquid crystal mesophases for these solvents with CTAB are shown in Table 2, and representative phase textures under CPOM are shown in Fig. 6. These samples were all sealed and left to equilibrate for one day before the CPOM temperature scan was conducted. When the CTAB–solvent systems were characterised without equilibration only two solvents supported the hexaganol, cubic and lamellar mesophases, which were 3-amino-1,2-propanediol and 2-amino-1,3-propanediol. All the other solvents supported only a lamellar or a poorly textured mesophase which could not be assigned. This indicates that for most of the molecular solvents the formation of the mesophases is a slow dynamic process. There was no clear correlation to solvent viscosity, which is particularly evident from the two solvents which supported all three mesophases without equilibration since 2-amino-1,3-propanediol is solid and 3-amino-1,2-propanediol is liquid at room temperature.

Table 3 Lyotropic liquid crystal mesophases identified from CPOM penetration scans for C_{16}PyrBr in the various solvents

Solvent	Lyotropic liquid crystalline mesophase stability regions (°C)		
	Hexagonal	Cubic	Lamellar
Water			34–>64[27]
	35.7[26a]	48.0[26a]	59.5[26a]
Glycerol	36.5[26a]	53.2[26a]	55[26a]
Ethylene glycol	31[26a]	51[26a]	56[26a]
Formamide	34[26a]	53[26a]	56[26a]
Ethylammonium nitrate	22–>120	57–>120	60–>120
3-Amino-1,2-propanediol	49–105	60–105	62–>120
3-Amino-1-propanol	45–105	56–99	56–115
Diethylene glycol	40–91	54–86	54–>120
4-Amino-1-butanol	52–81	58–77	56–>120
Diethylene triamine	53–100	60–94	54–120
Triethanolamine	48–96	57–87	57–92
2-Amino-1-propanol	45–92	53–92	57–>120
Triethylene glycol	47–>110		68–>120
1-Amino-2-propanol	46–100	50–76	54–>120
Triethylene tetramine	58–108		58–111
2-Amino-1-butanol	52–60		54–>120
2-Amino-2-ethyl-1,3-propanediol	48–111	56–111	61–>120
1-Amino-2-butanol	53–85		56–120
2-Amino-1-pentanol			61–>120
2-Amino-2-methyl-1-propanol			55–80
Diethanolamine	49–120	58–111	60–120
2-Amino-1,3-propanediol	49–>120	50–>120	50–>120

[a] Upper limit not provided in the reference.

The inclusion of a wide range of molecular solvents with various numbers of methylene, amine, hydroxyl, and ether groups enables structure–property relationships to be explored. It is apparent from Table 2 that decreasing G value (down the table) generally leads to less phase diversity, and narrower thermal stability ranges for the hexagonal and cubic mesophases, but little change to the lamellar. The last three solvents in the table were solid at room temperature and their G value is not known. However, it is expected that diethanolamine will have a G value higher than triethanolamine, and that 2-amino-1,3-propanediol will have a comparable G value to 3-amino-1,2-propanediol. Consequently, these solid solvents support mesophases consistent with the determined G values listed in Table 1.

Only five of the solvents in this investigation were observed to support a viscous isotropic phase between a hexagonal and lamellar mesophase, which was attributed to a cubic phase. These all correlated to the solvents with the highest G values, or solids which are expected to have high G values. Therefore, it appears that solvents with G values of 0.98 and below do not support cubic phases with CTAB. As shown in Table 2, formamide supports all three phases, though the cubic band has a narrow concentration range of stability according to its phase diagram.[75]

Increasing the non-functionalised hydrocarbon unit chain length in the solvent by one methylene group from 3-amino-1-propanol to 4-amino-1-propanol

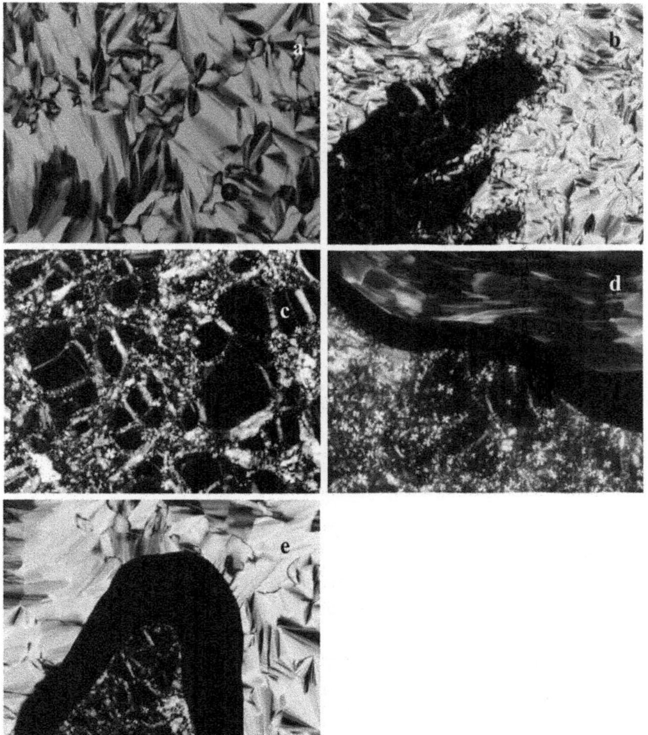

Fig. 6 Cross polarised optical micrographs of CTAB–solvent penetration scans for a) hexagonal mesophase in diethylene glycol at 60 °C, b) lamellar and hexagonal mesophases in diethylene triamine at 75 °C with highest CTAB concentration in bottom left, c) lamellar mesophase in diethylene glycol at 118 °C, d) anisotropic hexagonal, isotropic cubic, and anisotropic lamellar mesophase in 3-amino-1,2-propanediol at 93 °C, with highest CTAB concentration in bottom left and e) anisotropic hexagonal, isotropic cubic, and anisotropic lamellar mesophase in 3-amino-1-propanol at 94 °C with highest CTAB concentration at bottom middle. Magnification ×100.

impacted the solvophobic effect of the solvent and moved the system from supporting hexagonal, cubic and lamellar mesophases to just supporting hexagonal and lamellar, with a narrower thermal stability range for the hexagonal mesophase. Similarly, increasing the hydrocarbon unit chain length for the series of 2-amino-1-propanol, 2-amino-1-butanol and 2-amino-1-pentanol decreased the mesophase diversity and the thermal stability ranges, with the latter solvent only supporting the lamellar mesophase. The solvent may serve as a co-surfactant and modify the curvature at the amphiphile-solvent interface in self-assembled objects.

Generally structural analogues such as 2-amino-1-butanol and 1-amino-2-butanol or 2-amino-1,3-propanediol and 3-amino-1,2-propanediol had comparable mesophases supported and thermal stability ranges, which was consistent with their comparable G values. The exception was 1-amino-2-propanol only supporting the lamellar mesophase, whereas 2-amino-1-propanol also supported the hexagonal mesophase.

The presence of hydroxyls compared to amine moieties in the solvent molecular structure promotes higher Gordon parameters, and also provides richer mesophase diversity and larger stability windows. For example, diethanolamine

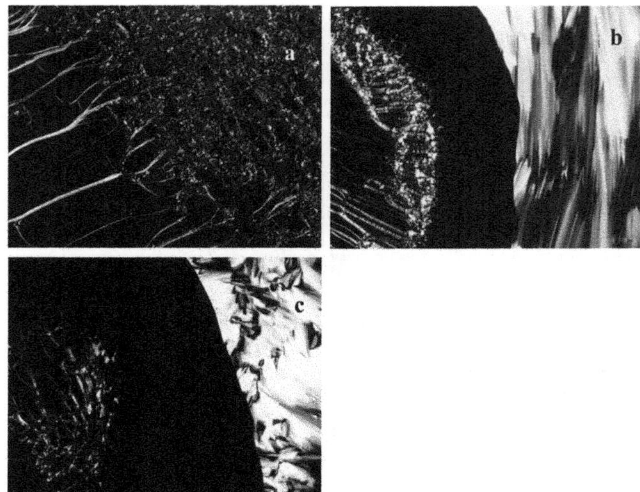

Fig. 7 Cross-polarised optical micrograph of C_{16}PyrBr–solvent penetration scans for a) lamellar mesophase in 2-amino-1-butanol at 67 °C, b) anisotropic hexagonal, isotropic cubic, and anisotropic lamellar mesophase in 3-amino-1,2-propanediol at 96 °C (high C_{16}PyrBr concentration to left) and c) anisotropic hexagonal, isotropic cubic, and anisotropic lamellar phase in diethanolamine at 78 °C (high C_{16}PyrBr concentration to left). Magnification ×100.

Table 4 Lyotropic liquid crystal mesophases from the literature for $C_{16}E_4$ in various solvents

	Lyotropic liquid crystalline mesophase stability regions (°C)		
Solvent	Lamellar	Cubic	Hexagonal (H_{II})
Water	Up to 76[42]		
EAN	Up to >60[12]		
Formamide	28–53[73]		
	28–46[27]		
N-Methylformamide	31–46[27]	44–46[27]	44–46[27]

and diethylene glycol supported hexagonal, cubic and lamellar mesophases, whereas diethylene triamine did not support the cubic mesophase. Increasing the number of ethanol substituents from ethyl-, diethyl- to triethylene glycol made little difference to the mesophase presence or thermal stability range for ethylene glycol and diethylene glycol with all three mesophases supported, whereas triethylene glycol only supported the lamellar mesophase.

The addition of methyl groups causes a dramatic reduction in mesophase diversity, which is highly evident for 2-amino-1-methyl-1-propanol compared to 2-amino-1-propanol. 2-Amino-1-methyl-1-propanol only supported the lamellar mesophase, and within a relatively narrow thermal stability range of 72–105 °C.

C_{16}PyrBr–solvent systems

The lyotropic liquid crystal mesophases for the molecular solvents with C_{16}PyrBr are provided in Table 3, and representative phase textures from CPOM in Fig. 7. It is evident that richer mesophase diversity is supported by more solvents with C_{16}PyrBr compared to CTAB systems. Eleven of the new solvents supported the hexagonal, cubic and lamellar mesophases with C_{16}PyrBr compared to just five with CTAB, and only two solvents solely supported a lamellar mesophase. The greater mesophase diversity with C_{16}PyrBr is attributed to the bulkier amphiphile head group, which leads to a greater effective amphiphile head group area and hence a stronger preference towards mesophases with higher curvature amphiphile–solvent interfaces such as those associated with the cubic and hexagonal mesophases.[76]

Allowing the C_{16}PyrBr–solvent systems to equilibrate for one day led to greater mesophase diversity being observed, and lower onset temperatures for mesophase formation. Without equilibration only the solvents of 2-amino-1,3-propanediol, 3-amino-1,2-propanediol, diethanolamine, 3-amino-1-propanol and diethylene glycol supported the hexagonal, cubic and lamellar mesophases, consistent with these solvents having the higher G values. The other solvents only supported lamellar mesophases. The thermal ranges of the lamellar mesophase had no observable trends with changes to the solvent structure.

The structure–property behaviour observed for the C_{16}PyrBr–solvent systems is consistent with that observed in the CTAB–solvent systems. Increasing the hydrocarbon chain length in the series of 2-amino-1-propanol, 2-amino-1-butanol, and 2-amino-1-pentanol results in a significant decrease in the supported mesophases from hexagonal, cubic and lamellar, to hexagonal and lamellar and finally to just lamellar, respectively, for the three solvents. One possible explanation is that the molecular solvents with larger hydrocarbon unit lengths serve as co-surfactants residing in the amphiphile–solvent interfacial region of amphiphile self-assembly objects.

The hydroxyl moieties have a greater contribution to mesophase diversity than amine moieties, including lower mesophase onset temperatures for the hexagonal and cubic mesophases according to the series of diethylene glycol < diethanolamine < diethylene triamine. Similarly, replacing the amine moieties with hydroxyls lowered the mesophase onset temperatures for triethylene glycol compared to triethylene tetramine and for diethylene glycol compared to diethylene triamine.

Increasing the number of ethanol groups through the sequence ethyl-, diethyl- and triethylene glycol causes an increase in the number of methylene groups and hydroxyl moieties with the overall result being a decrease in the Gordon parameter. As expected, the mesophase diversity decreased with increasing ethanol groups, with triethylene glycol supporting hexagonal and lamellar mesophases, whereas the other two solvents also supported the cubic mesophase. The thermal stability ranges decreased with increasing numbers of ethanol groups. The addition of a methyl group to 2-amino-1-propanol to form 2-amino-1-methyl-1-propanol led to a significant reduction in the number of supported mesophases with only the lamellar mesophase present, and a narrow thermal stability range.

Molecular solvents which were structural analogues generally supported similar mesophases and had similar thermal stability ranges. This is evident through comparing 3-amino-1,2-propanediol with 2-amino-1,3-propanediol, 2-amino-1-butanol with 1-amino-2-butanol, and 1-amino-2-propanol with 2-amino-1-propanol.

Discussion

The Gordon parameter, G, values, serving as a proxy for the solvent cohesive energy density, were calculated by using eqn (2) and found to be in the range of 0.63 and 1.54 J $mol^{1/3}$ m^{-3} for the experimentally studied molecular solvents. The results of the present study suggest that determining the surface tension and density of a molecular solvent can enable a good prediction of a molecular solvent's ability to exhibit the solvophobic effect, and therefore to serve as amphiphile self-assembly media, and can also guide the assessment of the likelihood of mesophase diversity. Specifically, in the present work, the G values correlate with the diversity of lyotropic liquid crystal mesophase formation; higher G values generally corresponding to greater supported mesophase diversity, and larger thermal stability ranges for the mesophases. Relatively low G values, as a consequence of low liquid–air surface tensions, generally reflect a molecular solvent with more non-polar substituents, and higher surface activity, which makes them more likely to behave as a co-surfactant influencing the average effective amphiphile head group area at the amphiphile–solvent interface in the self-assembled state.

The molecular solvents used in this study have enabled structure–property behaviour to be explored. The solvophobic effect is enhanced by the inclusion of hydroxyl moieties in comparison to amine moieties. Any increase in methylene or methyl groups decreases the solvophobic effect. Structural analogues have very similar behaviour. The addition of ethanol groups leads to reductions in the solvophobic effect, indicating that the increase due to the hydroxyl group was less than the reduction from two methylene groups. The structure–property trends are consistent with what we have previously reported for protic ionic liquids as amphiphile self-assembly media.[13,44] Increasing the number of methylene or methyl groups present in the solvents is anticipated to increase the solubility of the amphiphile alkyl chain, thus decreasing the solvophobic effect. Likewise, increasing the number of polar groups, such as amines, hydroxyls or ethers, will decrease the alkyl chain solubility and increase the solvent–amphiphile head group interaction. Previously, solvents with two methyl groups present were reported to only support lamellar phases with either CTAB or C_{16}PyrBr, whereas those with one methyl group supported hexagonal, cubic and lamellar for C_{16}PyrBr, but only lamellar with CTAB.[26]

The hydrogen bonding present within these solvents was investigated using infrared spectroscopy. The N–H stretches between 3270 and 3360 cm^{-1} were prominent as two bands for the solvents containing primary amines, and these were observed to red shift with increasing hydrogen bonding. Similarly, the broad hydroxyl stretch at around 3400 cm^{-1} red shifted with increasing hydrogen bonding. For all the solvents the NH and OH bands were consistent with highly associated liquids. The hydroxyl forms a stronger hydrogen bond than the amine moiety, and hence steric factors which hindered the amine moiety led to overall stronger hydrogen bonds for the solvents containing both functional groups. Consequently, this explains the differences between the cohesive energy density, as gauged from the Gordon parameter, for structural analogues such as 2-amino-1-propanol and 1-amino-2-propanol. In 2-amino-1-propanol the amine moiety is more hindered, and therefore there is less competition between the hydroxyls and the amine moiety. Therefore, the proportion of hydroxyls compared to amine moieties participating in the hydrogen bonding is expected to be larger for 2-amino-1-propanol compared to 1-amino-2-propanol. Since the hydroxyls form stronger hydrogen bonds than amine moieties this leads to an overall higher cohesive energy density for 2-amino-1-propanol compared to 1-amino-2-propanol, and an anticipated enhanced solvophobic effect.

Solvent segregation into polar and non-polar domains leads to a much greater hydrocarbon solubility and hence weaker solvophobic effect.[17] Increasing correlation distances for solvents is a good indication of an increasing non-polar domain size, which facilitates a higher hydrocarbon solubility, and hence weaker solvophobic effect.[17,50] Only three of the studied neat molecular solvents were identified as possessing a mesostructure, which most likely consists of polar and non-polar domains. These were 2-amino-1-pentanol, 1-amino-2-butanol and 2-amino-1-butanol. These three solvents had the lowest Gordon parameters of these solvents of 0.63, 0.71 and 0.67 J $mol^{1/3}$ m^{-3}, respectively (along with 2-amino-2-ethyl-1,3-propanediol). 2-Amino-1-pentanol had the largest correlation distance, which corresponded to it having the weakest solvophobic effect. The differences in cohesive energy density between the analogues such as 4-amino-1-butanol and 2-amino-1-butanol can be predominantly attributed to the mesostructure present in 2-amino-1-butanol, which is absent in 4-amino-1-butanol. There is a greater difference in Gordon parameter between the non-structured 4-amino-1-butanol and its mesostructured analogues of 2-amino-1-butanol and 1-amino-2-butanol than there is for the propanol series where none of them are structured.

The local molecular constraints, which influence the geometry of amphiphile self-assembly objects, can be described by the critical packing parameter, CPP, where CPP $= v/a_o l_c$, where v is the average volume of the amphiphile molecule, a_o is the effective amphiphile head group area and l_c is the effective length of the amphiphile chain.[77] For the lamellar phase CPP = 1, and there is zero curvature. For the normal hexagonal and cubic phases CPP < 1. The effective head group area, and the effective length are dependent on the amphiphile–solvent interactions. It has previously been reported by Auvray *et al.* that the lattice parameters for all the lyotropic liquid crystal mesophases of CTAB and C_{16}PyrBr in the non-aqueous solvents of formamide, glycerol, ethylene glycol, *N*-methylsydnone, *N*-methylformamide and *N*,*N*-dimethylformamide were lower than in water.[26] This suggests that the effective lengths, l_c, of the amphiphiles are reduced in the non-aqueous solvents compared to water, and hence the higher curvature mesophases are more favoured in water than these solvents. Solvents with an enhanced solvophobic effect are likely to have a stronger interaction with the amphiphile polar head group, and a weaker interaction with the alkyl chain. Therefore, they will lead to greater solvation of the head group, and hence an increase in the effective head group area, which in turns decreases the CPP and favours higher curvature mesophases. Consequently, as the Gordon parameter decreases, the solvophobic effect becomes less favourable and fewer high curvature mesophases supported. It has previously been suggested that bulky solvents are likely to prevent the formation of curved interfaces due to geometric constraints.[26]

The amphiphiles CTAB and C_{16}PyrBr differ in the head group between the trimethylammonium present in CTAB and the more bulky pyrdinium group for C_{16}PyrBr. It is evident from Tables 2 and 3 that there are more high curvature phases (hexagonal and cubic) supported by these solvents with C_{16}PyrBr than with CTAB. This behaviour was previously observed for a smaller group of non-aqueous solvents with the same two amphiphiles.[26] We attribute this in part to the larger effective head group area for C_{16}PyrBr. We also expect that the charge delocalisation over the ring of C_{16}PyrBr will aid in it forming a larger solvation area compared to the localised charge on the quaternary ammonium nitrogen for CTAB.

Interestingly, none of these solvents supported any lyotropic liquid crystal mesophases with the non-ionic surfactant $C_{16}E_4$. In contrast, as provided in Table 4, water,[42] EAN[12] and formamide[27,73] supported lamellar phases, *N*-methylformamide supported lamellar, inverse cubic and inverse hexagonal phases,[27] and micelles have been reported in dimethylformamide.[74] The solvents of 3-amino-1,2-propanediol, glycerol and ethylene glycol had Gordon parameters between those of water and EAN, and hence it was anticipated that at least these three would support mesophases. It was reported for various amides that increasing the alkyl chain length of the solvent removed the solvents' ability to support amphiphile self-assembly of $C_{16}E_4$, with *N*-methyformamide able to, but *N*-ethylformamide not able to.[27] Previously it was reported for C_xE_y nonionic surfactants in EAN that similar liquid crystalline mesophases were observed to those seen with water, but that longer alkyl chains were needed to drive the formation.[12] An investigation into micellar aggregates of non-ionic surfactants showed a critical micelle concentration nearly 300 times higher in DMF than in water. This was attributed to the lower cohesive energy density of DMF and that DMF is a better hydrocarbon solvent than water.[74] In a comparison of non-ionic surfactants between the amide containing tetra(ethylene oxide)n-dodecyl amide (TEDAd) and C_xE_y surfactants it was reported that the amide group is more hydrophilic than the ethylene oxide groups.[78] This leads to a better defined boundary between the tail and head group regions.[78]

The four ethylene oxide groups which comprise the $C_{16}E_4$ head group are much less hydrophilic in comparison to the cationic surfactants of CTAB and C_{16}PyrBr. Therefore there may well be weaker molecular solvent interactions for the $C_{16}E_4$ head group. The C_{16} alkyl chain length is consistent for all three surfactants so it is assumed there is negligible difference between the solvent–chain interactions. Therefore one possible explanation is that there is insufficient solvation of the head groups, and the amphiphiles are becoming insoluble rather than aggregating. We suggest that increasing the number of ethylene oxide groups, or replacing an ethylene oxide group with an amide could be trialled to obtain systems where these molecular solvents support non-ionic amphiphile lyotropic liquid crystal mesophases.

Previously we have reported that a large proportion of protic ionic liquids (PILs) are capable of supporting amphiphile self-assembly.[13,14,44,79] The ability of PILs to support the formation of CTAB lyotropic liquid crystal phases has been reported,[14,44] which enables broad comparisons between PILs and molecular solvents. PILs and molecular solvents with G parameters above 0.97 J mol$^{1/3}$ m^{-3} supported the formation of lamellar, cubic and hexagonal phases, with one exception of the PIL 2-(2-hydroxy-ethyoxy)-ethylammonium formate which had a G value of 1.03 J mol$^{1/3}$ m^{-3} but only supported a lamellar phase.[14,44] For PILs with lower G values only lamellar phases were supported (with two exceptions where more phases were observed),[14,44] whereas it is evident from Table 1 that molecular solvents with G values between 0.67 and 0.97 J mol$^{1/3}$ m^{-3} supported both lamellar and hexagonal phases, whereas the solvent with $G = 0.63$ J mol$^{1/3}$ m^{-3} only supported lamellar. In general, there are lower onset temperatures for phase formation in the molecular solvents compared to the PILs. These comparisons suggest that there is a great deal of similarity in the solvophobic effect between these two broad solvents classes, however, the molecular solvents appear to have a greater ability to support higher curvature phases and greater phase diversity at lower G values than the PILs, whereas for high G values there is less difference.

Conclusions

We report amphiphile self-assembly in sixteen molecular solvents for the first time. These solvents promote the formation of lyotropic liquid crystal mesophases with the cationic amphiphiles CTAB and C_{16}PyrBr, while none of them support mesophase formation with the non-ionic amphiphile $C_{16}E_4$.

We have established that the ability to serve as an amphiphile self-assembly medium is widespread within the class of low molecular weight polar associative solvents. We have previously shown that many protic ionic liquids are also able to support amphiphile self-assembly.[13,14] Consequently, the solvophobic effect should not be considered as an 'unusual' effect, instead there is a great variety of solvents, and solvent combinations, which can promote amphiphile self-assembly. Gordon parameter (G) values, a proxy for the cohesive energy density, of the studied molecular solvents were between 0.63 and 1.54 J mol$^{1/3}$ m^{-3}, which are in the range for solvents previously reported to be amphiphile self-assembly media. Indications are that calculating the Gordon parameter is a simple method which can be used to predict whether a solvent will exhibit the solvophobic effect. On this basis, we anticipate that many more polar solvents will be identified in the future as sharing the ability to be amphiphile self-assembly media.

A link between lower Gordon parameter values and molecular solvents behaving as co-surfactants, and residing at the amphiphile–solvent interface of self-assembled objects, has been observed. Consistent with this observation, there is a reasonable correlation between the Gordon parameter and the diversity of lyotropic liquid crystal mesophases supported in CTAB–solvent and C_{16}PyrBr–solvent systems, with solvents of lower G value displaying less mesophase diversity.

Non-aqueous solvents that exhibit the solvophobic effect and support amphiphile self-assembly may find application in a number of processes, including surfactant templated mesoporous inorganic materials synthesis[80] and biocatalysis.

Acknowledgements

Emmy Wijaya acknowledges the receipt of a CSIRO Materials Science and Engineering PhD studentship. Part of this research was undertaken on the SAXS beamline at the Australian Synchrotron, Victoria, Australia.

References

1 D. Chandler, *Nature*, 2005, **437**, 640.
2 A. Ray, *J. Am. Chem. Soc.*, 1969, **91**, 6511.

3 A. Ray, *Nature*, 1971, **231**, 313.
4 H. Bloom and V. C. Reinsborough, *Aust. J. Chem.*, 1967, **20**, 2583.
5 H. Bloom and V. C. Reinsborough, *Aust. J. Chem.*, 1968, **21**, 1525.
6 H. Bloom and V. C. Reinsborough, *Aust. J. Chem.*, 1969, **22**, 519.
7 V. C. Reinsborough, *Aust. J. Chem.*, 1970, **23**, 1473.
8 V. C. Reinsborough and J. P. Valleau, *Aust. J. Chem.*, 1968, **21**, 2905.
9 D. F. Evans, E. W. Kaler and W. J. Benton, *J. Phys. Chem.*, 1983, **87**, 533.
10 D. F. Evans, A. Yamauchi, R. Roman and E. Z. Casassa, *J. Colloid Interface Sci.*, 1982, **88**, 89.
11 D. F. Evans, A. Yamauchi, G. J. Wei and V. A. Bloomfield, *J. Phys. Chem.*, 1983, **87**, 3537.
12 M. U. Araos and G. G. Warr, *J. Phys. Chem. B*, 2005, **109**, 14275.
13 T. L. Greaves, A. Weerawardena, C. Fong and C. J. Drummond, *J. Phys. Chem. B*, 2007, **111**, 4082.
14 T. L. Greaves, A. Weerawardena, C. Fong and C. J. Drummond, *Langmuir*, 2007, **23**, 402.
15 T. L. Greaves and C. J. Drummond, *Chem. Soc. Rev.*, 2008, **37**, 1709.
16 S. Thomaier and W. Kunz, *J. Mol. Liq.*, 2007, **130**, 104.
17 T. L. Greaves and C. J. Drummond, *Chem. Soc. Rev.*, 2013, **42**, 1096.
18 T. Inoue, *J. Colloid Interface Sci.*, 2009, **337**, 240.
19 C. D. Tran and S. F. Yu, *J. Colloid Interface Sci.*, 2005, **283**, 613.
20 J. L. Anderson, V. Pino, E. C. Hagberg, V. V. Sheares and D. W. Armstrong, *Chem. Commun.*, 2003, 2444.
21 J. Tang, D. Li, C. M. Sun, L. Zheng and J. Li, *Colloids Surf., A*, 2006, **273**, 24.
22 C. Patrascu, F. Gauffre, F. Nallet, R. Bordes, J. Oberdisse, N. de Lauth-Viguerie and C. Mingotaud, *ChemPhysChem*, 2006, **7**, 99.
23 J. C. Hao and T. Zemb, *Curr. Opin. Colloid Interface Sci.*, 2007, **12**, 129.
24 H. N. Singh, S. M. Saleem, R. P. Singh and K. S. Birdi, *J. Phys. Chem.*, 1980, **84**, 2191.
25 T. Warnheim and A. Jonsson, *J. Colloid Interface Sci.*, 1988, **125**, 627.
26 X. Auvray, T. Perche, C. Petipas and R. Anthore, *Langmuir*, 1992, **8**, 2671.
27 T. L. Greaves, A. Weerawardena and C. J. Drummond, *Phys. Chem. Chem. Phys.*, 2011, **13**, 9180.
28 A. Lattes, E. Perez and I. Rico-Lattes, *C. R. Chim.*, 2009, **12**, 45.
29 X. Auvray, T. Perche, R. Anthore, C. Petipas, I. Rico and A. Lattes, *Langmuir*, 1991, **7**, 2385.
30 A. H. Beesley, D. F. Evans and R. G. Laughlin, *J. Phys. Chem.*, 1988, **92**, 791.
31 D. F. Evans, *Langmuir*, 1988, **4**, 3.
32 X. Auvray, C. Petipas, T. Perche, R. Anthore, M. J. Marti, I. Rico and A. Lattes, *J. Phys. Chem.*, 1990, **94**, 8604.
33 S. Cassel, I. Rico-Lattes and A. Lattes, *Sci. China: Chem.*, 2010, **53**, 2063.
34 M. L. Moya, A. Rodriguez, M. D. Graciani and G. Fernandez, *J. Colloid Interface Sci.*, 2007, **316**, 787.
35 J. L. Gardon, *J. Phys. Chem.*, 1963, **67**, 1935.
36 I. Vavruch, *J. Colloid Interface Sci.*, 1978, **63**, 600.
37 P. Becher, *J. Colloid Interface Sci.*, 1972, **38**, 291.
38 A. Beerbower, *J. Colloid Interface Sci.*, 1971, **35**, 46.
39 Kabir-ud-Din and P. A. Koya, *Langmuir*, 2010, **26**, 7905.
40 P. A. Koya, D. Kabir ud and K. Ismail, *J. Solution Chem.*, 2012, **41**, 1271.
41 D. Das and K. Ismail, *J. Colloid Interface Sci.*, 2008, **327**, 198.
42 D. J. Mitchell, G. J. T. Tiddy, L. Waring, T. Bostock and M. McDonald, *J. Chem. Soc., Faraday Trans. 1*, 1983, **79**, 975.
43 T. L. Greaves, A. Weerawardena, C. Fong, I. Krodkiewska and C. J. Drummond, *J. Phys. Chem. B*, 2006, **110**, 22479.
44 T. L. Greaves, A. Weerawardena, I. Krodkiewska and C. J. Drummond, *J. Phys. Chem. B*, 2008, **112**, 896.
45 J. L. Gardon, *J. Colloid Interface Sci.*, 1977, **59**, 582.
46 B. Chen, J. I. Siepmann and M. L. Klein, *J. Am. Chem. Soc.*, 2002, **124**, 12232.
47 R. T. Myers, *J. Colloid Interface Sci.*, 2004, **274**, 229.
48 J. N. Canongia Lopes and A. A. H. Padua, *J. Phys. Chem. B*, 2006, **110**, 3330.
49 A. Triolo, O. Russina, H. J. Bleif and E. Di Cola, *J. Phys. Chem. B*, 2007, **111**, 4641.
50 T. L. Greaves, D. F. Kennedy, S. T. Mudie and C. J. Drummond, *J. Phys. Chem. B*, 2010, **114**, 10022.
51 M. Tomsic, A. Jamnik, G. Fritz-Popovski, O. Glatter and L. Vlcek, *J. Phys. Chem. B*, 2007, **111**, 1738.
52 L. Zoranic, R. Mazighi, F. Sokolic and A. Perera, *J. Phys. Chem. C*, 2007, **111**, 15586.
53 L. Zoranic, R. Mazighi, F. Sokolic and A. Perera, *J. Chem. Phys.*, 2009, **130**, 124315.
54 O. Russina, A. Triolo, L. Gontrani and R. Caminiti, *J. Phys. Chem. Lett.*, 2012, **3**, 27.
55 O. Russina and A. Triolo, *Faraday Discuss.*, 2012, **154**, 97.

56 T. L. Greaves and C. J. Drummond, *Chem. Rev.*, 2008, **108**, 206.
57 N. Kirby, J. W. Boldeman, I. Gentle, D. Cookson, in *Synchrotron Radiation Instrumentation, Pts 1 and 2*, ed. J. Y. Choi, S. Rah, Amer. Inst. Physics, Melville, 2007, vol. 879, p. 887.
58 www.dow.com/ethyleneglycol/resources/index.htm, Triethylene Glycol, 2007.
59 C. L. Yaws, *Thermophysical Properties of Chemicals and Hydrocarbons*, William Andrew Inc., Norwich, N.Y., 2008.
60 www.sigma.aldrich.com.
61 R. S. Taylor and B. C. Garrett, *J. Phys. Chem. B*, 1999, **103**, 844.
62 T. Wolff and G. von Bunau, *Bunsenges. Phys. Chem.*, 1984, **88**, 1098.
63 J. L. Dashnau, N. V. Nucci, K. A. Sharp and J. M. Vanderkooi, *J. Phys. Chem. B*, 2006, **110**, 13670.
64 S. T. Mulla and I. J. Chalakkal, *J. Chem. Soc., Faraday Trans. 1*, 1986, **82**, 681.
65 W. K. Busfield, M. P. Ennis and I. J. McEwen, *Spectrochim. Acta, Part A*, 1973, **29**, 1259.
66 L. S. Prabhumirashi and C. I. Jose, *J. Chem. Soc., Faraday Trans. 2*, 1976, **72**, 1721.
67 A. D. Boese, A. Chandra, J. M. L. Martin and D. Marx, *J. Chem. Phys.*, 2003, **119**, 5965.
68 L. Du and H. G. Kjaergaard, *J. Phys. Chem. A*, 2011, **115**, 12097.
69 S. Scheiner, T. Kar and J. Pattanayak, *J. Am. Chem. Soc.*, 2002, **124**, 13257.
70 Y. L. Gu, T. Kar and S. Scheiner, *J. Am. Chem. Soc.*, 1999, **121**, 9411.
71 E. Tsuchida, *J. Chem. Phys.*, 2004, **121**, 4740.
72 M. A. Hussein and D. J. Millen, *J. Chem. Soc., Faraday Trans. 2*, 1974, **70**, 685.
73 M. Jonstromer, M. Sjoberg and T. Warnheim, *J. Phys. Chem.*, 1990, **94**, 7549.
74 G. Atun, M. Tuncay and G. Hisarli, *Colloids Surf., A*, 1996, **113**, 61.
75 X. Auvray, C. Petipas, R. Anthore, I. Rico and A. Lattes, *J. Phys. Chem.*, 1989, **93**, 7458.
76 E. S. Blackmore and G. J. T. Tiddy, *J. Chem. Soc., Faraday Trans. 2*, 1988, **84**, 1115.
77 J. N. Israelachvili, D. J. Mitchell and B. W. Ninham, *J. Chem. Soc., Faraday Trans. 2*, 1976, **72**, 1525.
78 U. R. M. Kjellin, P. M. Claesson and P. Linse, *Langmuir*, 2002, 6745.
79 J. Y. Wang, T. L. Greaves, D. F. Kennedy, A. Weerawardena, G. H. Song and C. J. Drummond, *Aust. J. Chem.*, 2011, **64**, 180.
80 Z. Chen, T. L. Greaves, R. A. Caruso and C. J. Drummond, *J. Mater. Chem.*, 2012, **22**, 10069.

Faraday Discussions RSC Publishing

PAPER

Mesoscale inhomogeneities in aqueous solutions of small amphiphilic molecules

Deepa Subramanian,[ab] Christopher T. Boughter,[a] Jeffery B. Klauda,[a] Boualem Hammouda[c] and Mikhail A. Anisimov[ab]

Received 30th April 2013, Accepted 17th June 2013
DOI: 10.1039/c3fd00070b

Small amphiphilic molecules, also known as hydrotropes, are too small to form micelles in aqueous solutions. However, aqueous solutions of nonionic hydrotropes show the presence of a dynamic, loose, non-covalent clustering in the water-rich region. This clustering can be viewed as "micelle-like structural fluctuations". Although these fluctuations are short ranged ($\sim$1 nm) and short lived (10 ps–50 ps), they may lead to thermodynamic anomalies. In addition, many experiments on aqueous solutions of hydrotropes show the occasional presence of mesoscale ($\sim$100 nm) inhomogeneities. We have combined results obtained from molecular dynamics simulations, small-angle neutron scattering, and dynamic light-scattering experiments carried out on tertiary butyl alcohol (hydrotrope)–water solutions and on tertiary butyl alcohol–water–cyclohexane (hydrophobe) solutions to elucidate the nature and structure of these inhomogeneities. We have shown that stable mesoscale inhomogeneities occur in aqueous solutions of nonionic hydrotropes only when the solution contains a third, more hydrophobic, component. Moreover, these inhomogeneities exist in ternary systems only in the concentration range where structural fluctuations and thermodynamic anomalies are observed in the binary water–hydrotrope solutions. Addition of a hydrophobe seems to stabilize the water–hydrotrope structural fluctuations, and leads to the formation of larger (mesoscopic) droplets. The structure of these mesoscopic droplets is such that they have a hydrophobe-rich core, surrounded by a hydrogen-bonded shell of water and hydrotrope molecules. These droplets can be extremely long-lived, being stable for over a year. We refer to the phenomenon of formation of mesoscopic droplets in aqueous solutions of nonionic hydrotropes containing hydrophobes, as mesoscale solubilization. This phenomenon may represent a ubiquitous feature of nonionic hydrotropes that exhibit clustering in water, and may have important practical applications in areas, such as drug delivery, where the replacement of traditional surfactants may be necessary.

[a]*Department of Chemical and Biomolecular Engineering, University of Maryland, College Park, MD 20742, USA*
[b]*Institute for Physical Science and Technology, University of Maryland, College Park, MD 20742, USA*
[c]*NIST Center for Neutron Research, National Institute of Standards and Technology, Gaithersburg, MD 20899, USA*

1 Introduction

Properties of aqueous solutions of various small amphiphilic molecules such as C1–C4 monohydric alcohols, amines, and ethers are of great chemical and biological importance. Liquid phases of these solutions are generally considered homogeneous on the macroscopic scale.[1,2] However, recent experimental and computational studies show that they are inhomogeneous over smaller length scales.[3–39] Molecular-scale "clusters" (order of 1 nm in size)[3–21] and, occasionally, mesoscale structures (order of 100 nm in size)[22–39] have been reported in such aqueous solutions.

Simple amphiphiles such as ethanol, *n*-propanol, isopropanol, tertiary butyl alcohol, 3-methylpyridine, 1,4-dioxane, and tetrahydrofuran can be referred to as hydrotropes. Hydrotropes are small amphiphilic molecules that are known to increase the "solubility" of sparingly soluble compounds in water.[40] Hydrotropes are different from traditional surfactants as the hydrotrope molecules have a smaller non-polar tail and do not form distinct micelles in aqueous solutions.[41–43] However, nonionic hydrotropes may show the presence of a dynamic, loose non-covalent "clustering" akin to "micelle-like structural fluctuations".[3–22] Such fluctuations can be detected from molecular dynamics simulations[8,9,12–21] and neutron spin echo experiments.[7] The molecular clusters have a size of the order of 1 nm and a lifetime of tens of picoseconds.[44]

In addition to these molecular-scale inhomogeneities, many experimental works also report the presence of mesoscale inhomogeneities with a size of the order of 100 nm.[23–38] The observed mesoscale inhomogeneities have been a subject of a variety of different explanations: from artifacts and impurities[24] to structural "phase transitions",[23] from microphase separation[36] to clathrate-hydrate precursors,[26] from loose supramolecular clusters[32–34] to kinetically arrested gaseous nanobubbles.[35] The lifetime of these inhomogeneities varies greatly, from hours to years. In our recent works, we have shown that long-lived (practically permanent) mesoscale inhomogeneities in aqueous solutions of hydrotropes originate only when the solution contains at least a trace amount of a third, more hydrophobic, component.[37–39]

In this work, we focus on the mesoscopic properties of one specific nonionic hydrotrope, namely aqueous solutions of tertiary butyl alcohol (TBA). TBA is a "perfect" amphiphile. It is the highest-molecular-weight alcohol isomer to be completely soluble in water at ambient conditions.[1,45] When placed at a water–oil interface, a TBA molecule aligns itself equally between the two phases, such that the hydroxyl group of the TBA molecule is in the water phase and the methyl groups are in the oil phase.[46] The story of aqueous solutions of TBA is full of controversies. The earliest light-scattering experiments in aqueous solutions of TBA were carried out by Vuks and Shurupova in 1972.[23] They observed strong scattering in a solution containing about 3 mol% (11 mass%) TBA. The authors attributed this effect to a possible "phase transition" between a clathrate-like structure and a less ordered structure in the macroscopically homogeneous liquid phase. Beer and Jolly repeated these experiments in 1974 and found that the effect depended on the source and prehistory of the sample.[24] Thus, they regarded the observed strong scattering as an "artifact". In 1977, Iwasaki and Fujiyama carried out light-scattering experiments in a range of aqueous TBA solutions and observed strong scattering in a solution close to 10 mol% (31 mass%) TBA.[25,26] They attributed this effect to the possible existence of clathrate hydrate structures in solution. The idea of clathrate hydrate structures in aqueous TBA solutions was also supported by the X-ray diffraction studies of Tanaka *et al.* and Nishikawa and Ijima.[47,48] The earliest dynamic light-scattering experiments in aqueous TBA solutions were carried out by Euliss and Sorensen in 1984.[27] While studying short-ranged concentration fluctuations in aqueous TBA solutions, they observed the presence of larger (mesoscale) inhomogeneities, of the order of 100 nm in size, in a solution of about 7 mol% (24 mass%) TBA. The authors speculated that this effect might be associated with the formation of clathrates, stabilized by trace amounts of another component. In 1986, Bender and Pecora repeated the experiments of Euliss and Sorensen, and did not observe any mesoscale inhomogeneities.[28] However, when studying aqueous 2-butoxyethanol solutions, Bender and Pecora did observe the presence of mesoscale inhomogeneities (~100 nm in size).[29] They concluded that the mesoscopic aggregates could be a result of contaminants in the solution.

Debates on the existence of mesoscale inhomogeneities are not unique to aqueous solutions of TBA or 2-butoxyethanol. In 2000, Georgalis *et al.* reported observation of a mesoscale structure (50–500 nm in size) in aqueous electrolyte solutions, namely sodium chloride, ammonium sulfate, and sodium citrate.[30] They attributed the formation of such a structure to hydrogen-bond interactions between water molecules, which act as a bridge for interactions between like-charged ions. In 2004, Yang *et al.* observed mesoscale inhomogeneities (200–600 nm in size) in aqueous solutions of nonionic hydrotropes such as tetrahydrofuran and 1,4-dioxane.[31] They also observed that these inhomogeneities could be removed by filtration, and thus concluded that the effect could be a result of incomplete mixing of water and the organic solute at the molecular scale. In 2006, Sedlák carried out an extensive static and dynamic light-scattering study of around 100 different solute–solvent pairs and observed the presence of large supramolecular aggregates in aqueous solutions of many ionic and non-ionic species.[32–34] Sedlák explained this phenomenon as the formation of a supramolecular structure due to hydrogen-bond interactions between solute and solvent molecules. Soon thereafter, Jin *et al.* in 2007, carried out a static and dynamic light-scattering study in aqueous solutions of various nonionic hydrotropes such as tetrahydrofuran, ethanol, urea, and α-cyclodextrin.[35] The authors observed mesoscopic aggregates of around 100 nm in these systems. They attributed this to the formation of gaseous nanobubbles, which are kinetically stabilized in solution by the adsorption of small amphiphilic molecules on the surface of the nanobubble. Long-lived, mesoscale inhomogeneities were also observed in aqueous solutions of 3-methylpyridine.[36] It was found that these inhomogeneities depended on the source of the sample and thus might be the result of impurities in the solution. In addition to mesoscale inhomogeneities, aqueous solutions of certain hydrotropes, such as aqueous isobutyric acid, show a subtle "soap-like" third phase at the liquid–vapor interface.[49] The authors attributed the formation of this soap-like phase to the aggregation of solute molecules on the interface, driven by hydrophobic effects.

Motivated by this spectrum of explanations for the observed mesoscale inhomogeneities in aqueous solutions of hydrotropes, we set out to clarify this long-standing issue and carried out a systematic study of aqueous solutions of TBA. We have combined results obtained from molecular dynamics (MD) simulations, small-angle neutron scattering (SANS) and dynamic light scattering (DLS) experiments to elucidate the nature and structure of these mesoscale inhomogeneities. We have studied aqueous solutions of TBA and aqueous solutions of TBA with cyclohexane (a typical hydrophobe with a molecular size similar to that of TBA). Since many of the mesoscale inhomogeneities reported in the literature did not appear to be at equilibrium (disappearing within several hours or days after sample preparation),[33,36] we paid special attention to long-term monitoring and equilibration of our samples. From MD simulations in TBA–water solutions, we see that TBA and water form short-lived (<1 nm in size), short-ranged (lifetime of tens of picoseconds) micelle-like structural fluctuations.[44] These structural fluctuations seem to be responsible for the anomalies in the thermodynamic properties of aqueous TBA solutions.[44] On the addition of a hydrophobic compound, larger aggregates of about 100 nm in size are formed as seen from DLS. These aggregates are not gaseous nanobubbles or solid particles but rather Brownian diffusive droplets. The aggregates consist of a hydrophobe-rich core surrounded by a hydrogen-bonded shell of hydrotrope and water molecules. They can be unusually stable over long periods of time (a year or longer).[39]

2 Materials and methods

2.1 Sample preparation

TBA was procured from two different sources: one (with a labeled purity greater than 99.7%) was purchased from Sigma Aldrich, while another (with a labeled purity greater than 99.8%) was purchased from Alfa-Aesar. Most of the samples described in this work were prepared by using the TBA procured from Alfa-Aesar, unless otherwise specified. Cyclohexane (CHX), with a labeled purity greater than 99%, was purchased from Merck. Heavy water (used for the SANS experiments), with a labeled purity greater than 99.9 atom%, was purchased from Sigma Aldrich. Deionized water was obtained from a Millipore setup.

For the DLS experiments, the binary TBA–water solutions were filtered with 200 nm Nylon filters to remove dust particles. In order to eliminate mesoscale inhomogeneities, additional cold filtrations with 20 nm Anopore filters (at a temperature of $\sim$ 5 °C), were carried out. CHX was used after filtering it through 200 nm Nylon filters. Light-scattering and neutron-scattering measurements were performed after equilibrating the samples for at least 24 h.

2.2 Molecular dynamics simulations

MD simulations on models for pure TBA, on TBA–water mixtures and TBA–water–CHX mixtures were performed. The concentration and number of molecules for each system are presented in Table 1. The TIP4P-Ew water model[50] was used and parameters for TBA and CHX were taken from the CHARMM General Force Field.[51] The systems were built by using the Packmol package,[52] which randomly packs all molecules in a simulation box.

The NAMD simulation program[53] was used to perform all MD simulations with 2 fs time steps for a total of 10 ns–1000 ns (see Table 1). Most simulations were run for 50 ns, while one binary TBA–water simulations was run longer to determine if any clathrate-like structures would form. The van der Waals interactions

Table 1 Concentrations of the samples studied by MD simulations

Sample #	TBA (mol%)	Water (mol%)	CHX (mol%)	T/K	Number of molecules	Time (ns)
Pure - P1	100	0	0	283	192	10
Binary						
B1	1	99	0	285	2125	50
B2	3.8	96.2	0	285	2188	1000
B3	7	93	0	285	2266	50
B4	18	82	0	285	2556	50
B5	40	60	0	285	3503	50
Ternary						
T1	11.84	85.31	2.84	298	8440	50
T2	17.70	79.65	2.65	298	9040	50
T3	27.78	69.44	2.78	298	8640	50
T4	40.05	57.21	2.74	298	8740	50

were smoothly switched off between 8 and 10 Å by a potential-based switching function. Long-range electrostatic interactions were calculated by using the particle-mesh Ewald (PME) method.[54] An interpolation order of 4 and a direct space tolerance of 10^{-6} were used for the PME method. Langevin dynamics were used to maintain constant temperatures for each system, while the Nosé-Hoover Langevin-piston algorithm[55,56] was used to maintain constant pressure at 1 bar. The Visual Molecular Dynamics (VMD) program[57] was used to create snapshots and to calculate the radial distribution functions (RDF).

2.3 Small-angle neutron scattering

SANS experiments were performed by using the NG3 SANS instrument at the NIST Center for Neutron Research. The SANS experiments were carried out on TBA–heavy water solutions and on TBA–heavy water–CHX solutions. The concentrations of the samples studied by SANS are presented in Table 2.

The essential measurement length scale in SANS is the wavenumber q. The wavenumber is related to the length scale l, as $l = 2\pi/q$, with $q = \left(\frac{4\pi}{\lambda}\right)\sin\left(\frac{\theta}{2}\right)$, where $\lambda = 6$ Å is the neutron wavelength, and θ is the scattering angle. In our experiments, q was varied from 0.005 Å^{-1} to 0.5 Å^{-1}, corresponding to the length scales from ~1000 Å to ~10 Å.

Table 2 Concentrations of the samples studied by SANS and results from fits to eqn (4) and (5)

Sample #	TBA (mol %)	Heavy water (mol %)	CHX (mol %)	T/K	ξ_{OZ} (nm)	
Binary						
SB1	3.5	96.5	0	298	0.06	
SB2	5	95	0	298	0.24	
SB3	7.4	92.6	0	283	0.40	
				298	0.50	
				313	0.63	
Ternary						$R_{g(fixed)}$ (nm)
ST1	7.40	92.57	0.03	298	0.5	100

2.4 Dynamic light scattering

DLS experiments were performed with a PhotoCor Instruments setup, as described in ref. 37. Temperature was controlled with an accuracy of ±0.1 °C. For two exponentially decaying relaxation processes, the intensity auto-correlation function $g_2(t)$ (obtained in the homodyning mode) is given by:[58,59]

$$g_2(t) - 1 = \left[A_1 \exp\left(\frac{t}{\tau_1}\right) + A_2 \exp\left(\frac{t}{\tau_2}\right)\right]^2 \tag{1}$$

where A_1 and A_2 are the amplitudes of the two relaxation processes, t is the "lag" (or "delay") time of the photon correlations and τ_1 and τ_2 are the characteristic relaxation times. For a diffusive relaxation process, the decay rate (Γ) is related to the diffusion coefficient D, as:[58,59]

$$\Gamma = \frac{1}{\tau} = Dq^2 \tag{2}$$

where q is the difference in the wavenumber between incident and scattered light, $q = \left(\frac{4\pi n}{\lambda}\right)\sin\left(\frac{\theta}{2}\right)$, n is the refractive index of the solvent, λ is the wavelength of the incident light in vacuum ($\lambda = 633$ nm for our set-up), and θ is the scattering angle. For monodisperse, non-interacting, spherical Brownian droplets the hydrodynamic radius R can be calculated by using the Stokes–Einstein relation:[58,59]

$$R = \frac{k_B T}{6\pi\eta D} \tag{3}$$

where k_B is Boltzmann's constant, T is the temperature, and η is the shear viscosity of the medium.

3 Molecular-scale inhomogeneities

3.1 Phase behavior and thermodynamic anomalies of aqueous solutions of TBA

Fig. 1 shows the solid–liquid phase diagram of TBA–water solutions at ambient pressure.[60] While there exist different solid phases, the liquid phase is homogeneous on the macroscopic scale, with TBA and water completely miscible with each other. Fig. 1 also approximately shows the temperature and concentration domain (shaded in grey) where aqueous TBA solutions exhibit micelle-like structural fluctuations and thermodynamic anomalies. Excess and partial molar properties, heat capacity, and isothermal compressibility all exhibit extrema in the solvent-rich region of TBA–water solutions.[61–72] These anomalies occur in the concentration range of 3 mol% to 8 mol% (11 mass% to 26 mass%) TBA and become enhanced below room temperature.

The thermodynamic anomalies provide insight into solute–solvent interactions and on the structural changes that occur at the molecular scale. For example, as shown in Fig. 2, the enthalpy of mixing is negative in the solvent-rich region with a minimum at ~6 mol% (21 mass%) TBA, and becomes positive as the TBA concentration is increased.[66–68] The excess chemical potential of water shows a similar trend.[69,70] These anomalies indicate that at low TBA concentrations, solute–solvent interactions are favorable, with water and TBA molecules

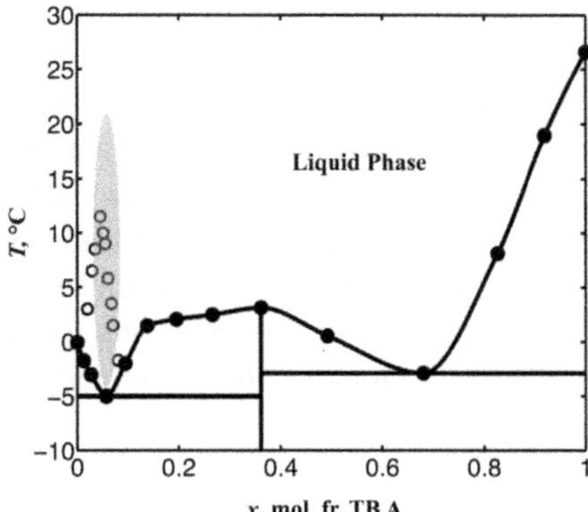

Fig. 1 Solid–liquid phase diagram of TBA–water solutions at ambient pressure (the solidus data points (connected by a guide to the eye) are reproduced with permission from ref. 59). The solid phase shows the presence of two eutectics (at ~6 mol% (21 mass%) and 70 mol% (90 mass%) TBA), while the liquid phase is macroscopically homogeneous. Open circles correspond to maxima of the heat capacity.[71] The grey area schematically represents the region where molecular scale clustering and thermodynamic anomalies are reported. Mesoscale solubilization of hydrophobic compounds is also observed in this region. The horizontal and vertical black lines separate the different phases of TBA–water hydrates.

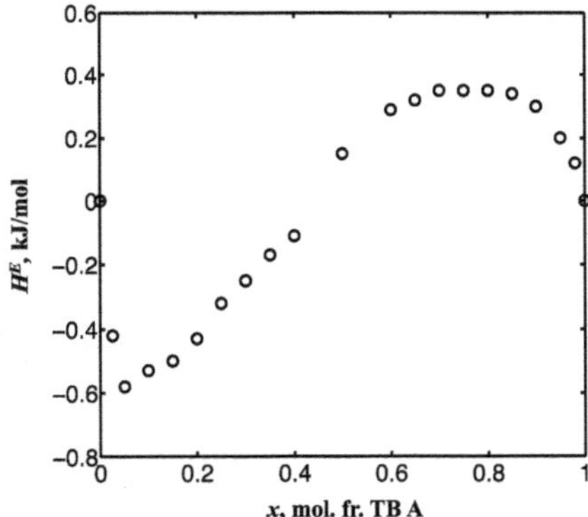

Fig. 2 Excess enthalpy (heat of mixing) of aqueous TBA solutions at $T = 25$ °C (reproduced with permission from ref. 67). The negative values of the excess enthalpy in the water rich region indicate favorable solute–solvent interactions.

preferring to couple with each other. As the TBA concentration is increased, solute–solute and solvent–solvent interactions are favored over solute–solvent interactions, indicating that TBA and water molecules prefer to demix.

The heat capacity of aqueous TBA solutions also exhibits anomalies, with maxima observed in the solute-rich region.[64,65] The maxima become sharper as the temperature is lowered.[72] This could be an indication of a structural change in this region. Remarkably, as most recent experiments have demonstrated, this heat capacity anomaly is rather insensitive to the presence or absence of mesoscale inhomogeneities (the heat capacity anomaly persists even after the mesoscale inhomogeneities have been eliminated by filtration).[73] This indicates that the anomaly is inherent to the molecular structure of binary TBA–water solutions, and is not significantly affected by the presence (or absence) of mesoscale inhomogeneities.

Thermodynamic anomalies have also been observed in aqueous solutions of many other hydrotropes. Aqueous solutions of other alcohols such as methanol, ethanol, *n*-propanol, isopropanol, and 2-butoxyethanol all show similar anomalies in their thermodynamic properties within the water-rich region.[1] In TBA–water solutions, these anomalies are most pronounced.[1] We attribute these thermodynamic anomalies, to structural fluctuations (clustering) occurring on the molecular scale.[44] MD simulations of TBA–water solutions, presented in the following section, support this view.

3.2 Molecular-scale clustering in aqueous solutions of TBA

MD simulations were performed on models of pure TBA and aqueous solutions of TBA. In the aqueous solutions, the concentration of TBA was varied from 1 mol% to 40 mol% (4 mass% to 73 mass%). Fig. 3a shows the radial distribution functions (RDFs) between the central carbons of TBA in pure TBA and in aqueous solutions of TBA. In pure TBA, there exists a peak at a distance of 4.7 Å and a second peak at a distance of 6 Å. The first peak in the RDF corresponds to strong

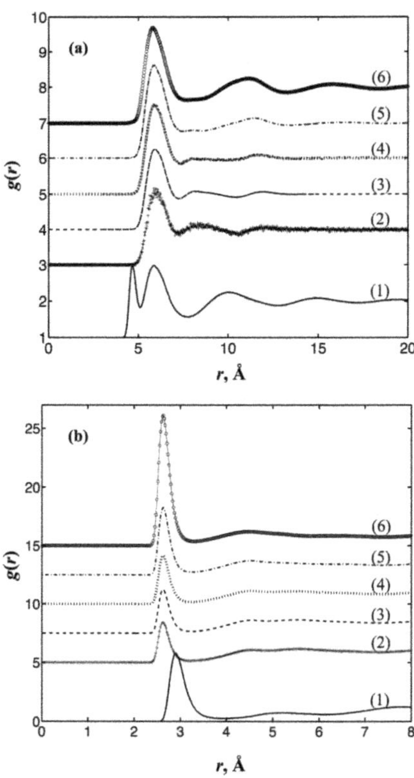

Fig. 3 Radial distribution functions between: (a) Central carbons of TBA and (b) oxygen of TBA and oxygen of water (for pure TBA RDF is between oxygen atoms of TBA). The curves are offset vertically for clarity. (1) Pure TBA. (2) 1 mol% (4 mass%) TBA. (3) 4 mol% (15 mass%) TBA. (4) 7 mol% (24 mass%) TBA. (5) 18 mol% (47 mass%) TBA. (6) 40 mol% (73 mass%) TBA.

van der Waals interactions, while the second peak corresponds to weaker van der Waals interactions, between the methyl groups of TBA.

In the aqueous solutions, the first peak (at 4.7 Å) between the central C atoms of TBA disappears, but the peak at 6 Å becomes enhanced. This indicates that in aqueous solutions, TBA molecules prefer to interact with water rather than with other TBA molecules, in agreement with thermodynamic anomalies. It is also observed that the magnitude of this peak increases as the TBA concentration is increased. This is an indication that at higher TBA concentrations more TBA molecules "cluster" together at this distance. Secondary and tertiary peaks are also seen in aqueous TBA solutions, which differs from simulations of pure TBA. As the concentration of TBA increases, the secondary peak occurring at 8 Å tends to disappear. This could be an indication that at higher concentrations aqueous TBA solutions do not form "isolated clusters" with water or that the clustering with water has changed in a TBA-rich solution.

Fig. 3b shows the RDFs between the oxygen atom of TBA and the oxygen atom of water in aqueous solutions of TBA and between oxygen atoms of TBA in the pure TBA system. This figure shows a large initial peak at 3 Å in pure TBA, which disappears in aqueous solutions. An initial peak between water and TBA is seen in aqueous solutions at 2.6 Å. This again indicates that in aqueous solutions, TBA tends to form hydrogen bonds with water rather than with itself. Moreover, the distance is shorter indicating stronger hydrogen bond with water than between TBA–TBA. As the concentration of TBA increases, the magnitude of the first peak increases. This indicates that there is a stronger preference for water to form a hydrogen bond with the hydroxyl of TBA at these higher concentrations. As the TBA concentration is raised, the secondary peak, which occurs at 4.4 Å, increases while the tertiary and higher peaks almost disappear. This indicates that at higher TBA concentrations the water molecules surrounding the TBA molecules in tertiary shells and beyond are not very well defined and the resultant "cluster" loses its structural integrity.

Snapshots from MD simulations can help in interpreting the RDFs as shown in Fig. 4. At the lower concentrations, TBA forms clusters due to van der Waals interactions between its methyl groups. These TBA clusters are surrounded by a hydrogen-bonded polygonal (either pentagonal (Fig. 4b and c) or hexagonal (not shown)) network formed between TBA–water and water–water molecules. The main RDF peak at 6 Å, as seen in Fig. 3a, corresponds to the distance between central carbon atoms of TBA, which may be dimers, trimers or tetramers of TBA. The secondary peak at 8 Å, as seen in Fig. 3a, corresponds to a distance between the central carbon atoms in oligomerized TBA with its nearest neighbor of unstructured TBA. The water molecules are organized in a specific hydrogen-bonded structure around TBA molecules, with the hydroxyl group of the TBA molecules forming one of the vertices of a hydrogen-bonded polygon.

The structural significance of the peaks calculated in the O(TBA)–O(Water) RDFs of Fig. 3b can be visualized from the snapshots in Fig. 4. The primary peak in Fig. 3b is from 3 waters coordinating the hydroxyl group of TBA. The secondary RDF peak at 4.4 Å and the tertiary peak at 5.6 Å in Fig. 3b correspond to larger distances between the vertices in the pentagon and/or hexagon ring of water. The secondary and tertiary peaks correspond to an additional 17 to 21 surrounding water molecules. This leads to an effective structure, which we refer to as a "micelle-like cluster". The cluster has an inner radius of $\sim$4 Å that constitutes 4 to

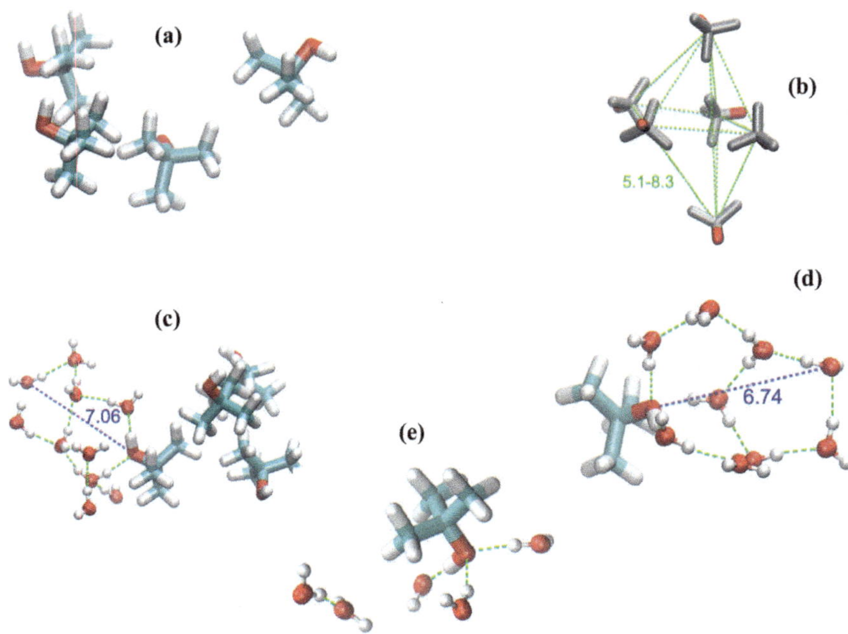

Fig. 4 Snapshots from MD simulations in aqueous solutions of TBA with increasing concentration of TBA. TBA molecules are represented by the licorice model, while water molecules are represented by the ball and stick model. The structure between TBA and water is fairly similar from 1 mol% to 7 mol% TBA. The only difference is the amount and size of TBA clusters with increasing TBA concentration. However, beyond ~7 mol% TBA, the clusters do not seem to be well-defined and lose their structural integrity. (a) 1 mol% (4 mass%) TBA. (b) 4 mol% (15 mass%) TBA. (c) 7 mol% (24 mass%) TBA. (d) 18 mol% (47 mass%) TBA. (e) 40 mol% (73 mass%) TBA.

5 TBAs and an additional distance, including water shells, of ~6.5 Å. Thus, this "micelle-like cluster" constituting TBA and water molecules is ~10.5 Å in radius. Although the water structures surrounding TBA form polygons commonly found in hydrates,[74] the micelle-like clusters are short-lived and appear to be transient, with an estimated lifetime of the order of 10 ps–50 ps.

The effect of TBA concentration seen from the RDFs of aqueous TBA solutions can also be explained from the snapshots in Fig. 4. At low concentrations, 1 mol% to 2 mol% (4 mass% to 8 mass%) TBA, the average cluster sizes of TBA molecules range from 1 to 4 and the number of clusters tends to be small. TBA forms a micelle-like cluster with water. At higher TBA concentrations, 3 mol% to 7 mol% (11 mass% to 24 mass% TBA), the clusters tend to be larger with 4–8 TBA molecules. Bipyramidal structures of TBA clusters (Fig. 4b) transiently exist (20 ps to 50 ps) in the 3 mol% to 7 mol% concentration range. As the TBA concentration is further raised, the clusters of TBA–water do not behave as a micelle, but rather TBA becomes dominant in terms of volume component in the solution (Fig. 5d and e), reflecting a growing tendency to be apart from water molecules. These solutions no longer have the short-lived TBA–water clusters and instead look more like a randomly mixed solution of TBA and water (usual non-ideal solution). At 40 mol% (73 mass%) TBA, the water concentration is too low to form any significant water polygon structure indicative of the reduction and loss of secondary and higher order peaks in the RDFs (Fig. 3).

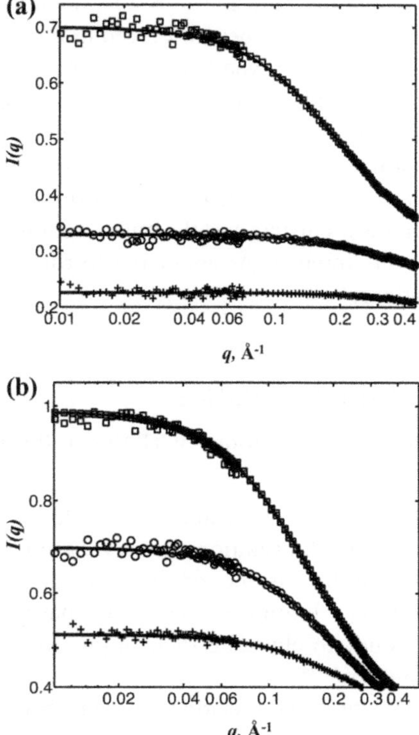

Fig. 5 (a) SANS data from TBA–heavy water solutions at $T = 25$ °C. Crosses: 3.5 mol% (12 mass%) TBA (sample # SB1). Circles: 5 mol% (16 mass%) TBA (sample # SB2). Squares: 7.4 mol% (23 mass%) TBA (sample # SB3). Standard statistical error bars are comparable to the size of the symbols. The solid lines are fits to the data in accordance with eqn (4). The results of the fits are summarized in Table 2. (b) SANS data from a 7.4 mol% (23 mass%) TBA–heavy water solution (sample # SB3). Crosses: 10 °C. Circles: 25 °C. Squares: 40 °C. Standard statistical error bars are comparable to the size of the symbols. The black lines are fits to the data in accordance with eqn (4). The results of the fits are summarized in Table 2.

3.3 SANS in aqueous solutions of TBA

In order to characterize the molecular-scale inhomogeneities in aqueous solutions of TBA, SANS experiments were carried out on TBA–heavy water solutions. The following equation best fits the SANS intensity data:[71]

$$I(q) = \frac{A_3}{1 + (\xi q)^2} + B \tag{4}$$

where A_3 is the amplitude, B is a background parameter, and ξ is the correlation length, which characterizes the length scale of concentration fluctuations; it increases as the temperature is raised and as the TBA concentration is increased. This behavior of the correlation length is expected as the system approaches a "virtual" critical point (hidden here by the vapor–liquid transition, but may become real on the addition of a salt, such as KCl[27]) located at an inaccessible higher temperature and higher TBA concentration. Fig. 5a shows the SANS data for various concentrations of TBA at 25 °C, while Fig. 5b shows the SANS data for 7.4 mol% (23 mass%) TBA solution at different temperatures. The results from the SANS fits are presented in Table 2. The results obtained for the correlation length are consistent with what has been observed in the literature.[27]

We also wanted to investigate whether the micelle-like clusters (structural fluctuations) observed from MD simulations could be seen from SANS data. However, we were unable to unambiguously detect such clusters, although some spectra at low temperatures suggest the existence of a marginally detectable peak in the structure factor at $q \sim 0.1$ Å. The reasons for this could be two-fold: poor SANS contrast between these clusters and the bulk solution or small contribution of these clusters into the intensity as compared to the contribution from the concentration fluctuations.

4 Mesoscale inhomogeneities

4.1 DLS and SANS in aqueous solutions of TBA upon the addition of a hydrophobe

The discussion so far has focused on molecular-scale (order of 1 nm in size) inhomogeneities in aqueous TBA solutions. However as mentioned earlier, there have been many reports on the existence of mesoscale inhomogeneities (order of 100 nm in size) in aqueous solutions of TBA.[22–38] We have carried out a comprehensive experimental study to elucidate their origin and to characterize these mesoscale inhomogeneities.[37–39]

Fig. 6 shows the intensity auto-correlation function observed from an aqueous solution of TBA (purchased from Sigma Aldrich). The correlation function shows the presence of two relaxation processes – a fast process with a relaxation time of 65 μs and a slow process with a relaxation time of 22 ms. The fast process corresponds to molecular diffusion, with a diffusion coefficient of 1.5×10^{-6} cm^2 s^{-1}. In accordance with eqn (3), this corresponds to a hydrodynamic radius of

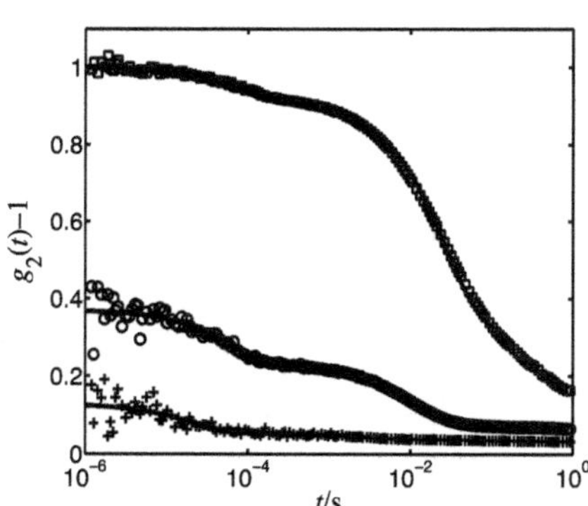

Fig. 6 Time-dependent intensity autocorrelation functions obtained in aqueous solutions of TBA from dynamic light scattering ($\theta = 45°$). The solid lines are fits to the data in accordance with eqn (1). Circles represent the correlation function obtained in ~8 mol% (26 mass%) TBA solution (TBA procured from Sigma Aldrich) at $T = 24$ °C. This correlation function shows the presence of two relaxation modes – the fast mode with a relaxation time of ~65 μs and a slow mode with a relaxation time of ~22 ms. After filtering this solution multiple times with a 20 nm Anopore filter at ~10 °C, the slow mode is almost eliminated (correlation function represented by crosses). Adding trace amounts of a hydrophobic component (0.03 mol% cyclohexane) regenerates the slow mode (correlation function represented by squares). Statistical error bars are comparable to the size of the symbols.

about 0.6 nm. This is consistent with the correlation length (ξ) of the concentration fluctuations obtained from SANS data as discussed in the previous section.

The slower process, with a relaxation time of 22 ms, corresponds to mesoscale inhomogeneities.[37] As the temperature is increased, this slow mode disappears and reappears as the temperature is lowered.[37] In fact, at low temperatures the contribution from the slow mode is enhanced so significantly, that it becomes quite difficult to detect molecular diffusion.[37] The mesoscale inhomogeneities were only observed between the concentrations 3 mol% to 8 mol% (11 mass% to 26 mass%) TBA. Above this concentration range, the mesoscale inhomogeneities disappeared.

In order to understand the origin of the slow mode, this aqueous TBA solution was filtered multiple times by using a 20 nm Anopore filter, at a low temperature (~5 °C), to eliminate the slow mode. The intensity auto-correlation function obtained after filtering the aqueous TBA solutions at cold conditions is also shown in Fig. 6. The resultant correlation function shows no mesoscale inhomogeneities, but only the contribution from molecular diffusion. A controlled "impurity", namely a trace amount (0.03 mol%) of a third, more hydrophobic, component (cyclohexane) was added to an aqueous TBA solution that initially did not show any mesoscale inhomogeneities. Upon the addition of cyclohexane, mesoscale inhomogeneities emerged, and the slow mode was observed. Various hydrophobic additives such as propylene oxide, isobutyl alcohol, and methyl *tert*-butyl ether were also studied. All these experiments showed that the slow mode appears only when the aqueous TBA solution contains a more hydrophobic component.[38] The wavenumber dependence of the relaxation rate of these inhomogeneities (in accordance with eqn (2)) revealed that they are diffusive Brownian droplets.[38] Confocal microscopy images are also consistent with what is observed from dynamic light scattering.[37]

The presence of mesoscopic droplets can also be verified from SANS experiments. Fig. 7 shows the SANS intensity $I(q)$ from a TBA–heavy water solution containing trace amounts of cyclohexane (CHX) as the hydrophobe. The SANS

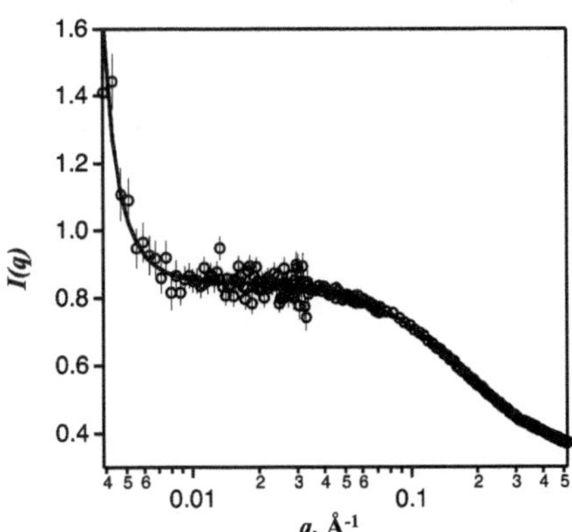

Fig. 7 SANS data from a TBA–heavy water–CHX solution at $T = 25$ °C. 7.4 mol% TBA (23 mass%), 0.03 mol% (0.1 mass%) CHX (sample # ST1 from Table 2). The black line is a fit to the data in accordance with eqn (5). The results of the fit are summarized in Table 2. Statistical error bars represent one standard deviation.

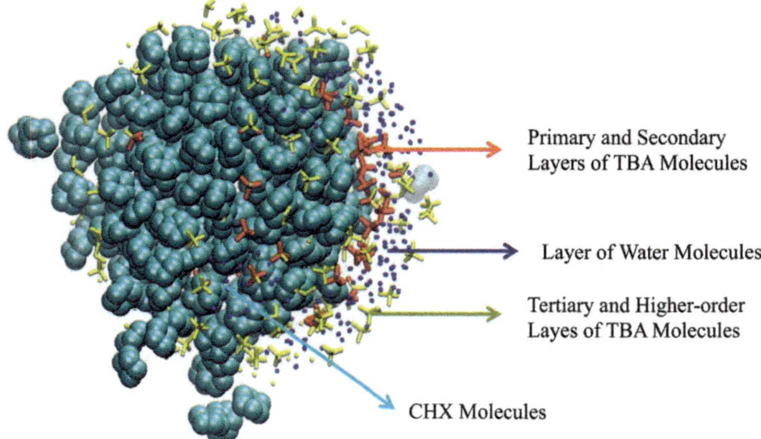

Fig. 8 Snapshots (at ~50 ns) from MD simulations of a TBA–water–CHX solution (sample # T1 from Table 1). TBA molecules are represented by the licorice model, while van der Waals spheres represent CHX molecules. This snapshot demonstrates the formation of a "droplet" with aggregated CHX molecules in the core, surrounded by primary and secondary layers of TBA molecules. These layers are further solvated by a water layer and a tertiary layer of TBA molecules.

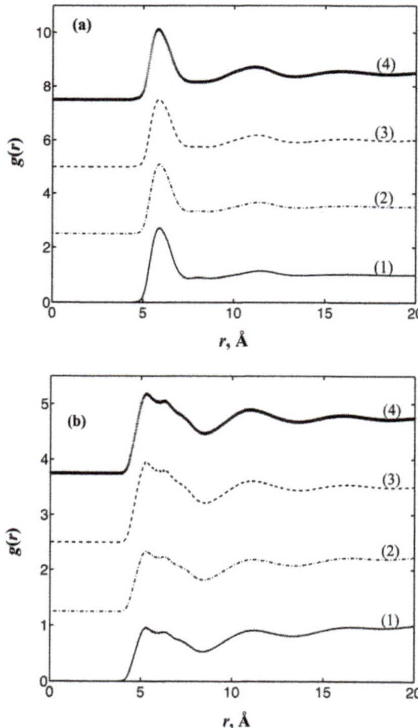

Fig. 9 Radial distribution functions in TBA–water–CHX solutions with increasing concentrations of TBA (samples # T1 to T4 from Table 1). (a) RDFs between central C atoms of TBA (b) RDFs between central C of TBA and C1 on CHX.

data in this TBA–heavy water–CHX system were best fit to an equation of the following form:[75]

$$I(q) = \frac{A_4}{(R_g q)^4} + \frac{A_5}{1 + (\xi q)^2} + B_1 \tag{5}$$

where A_4 and A_5 are the amplitudes, R_g is the radius of gyration of the mesoscopic droplets, ξ is the correlation length of the concentration fluctuations, and B_1 is a background parameter. The above equation includes contributions from the concentration fluctuations and contributions from the much larger, mesoscopic droplets. Since the SANS data do not reach a Guinier region (plateau at low q), the R_g corresponding to mesoscopic droplets was fixed at 100 nm (as was observed from DLS).

4.2 Investigating the structure of mesoscopic droplets by MD simulations

In order to further investigate the nature of these mesoscopic droplets, MD simulations of TBA–water–CHX solutions were carried out. Fig. 8 shows a snapshot from MD simulations where aggregated CHX molecules are surrounded by TBA molecules. The concentration of TBA in the layer surrounding the CHX aggregate is higher than in the bulk solution. The structure of this layer is similar to a "microemulsion" structure, which seems to occur at higher concentrations of TBA.[20] Water molecules form hydrogen bonds with the hydroxyl groups of the TBA molecules. There may be secondary or higher order layers of TBA and water molecules that surround the CHX aggregate (Fig. 8), leading to the formation of a "droplet". The observed droplets remained stable for the length of the simulation.

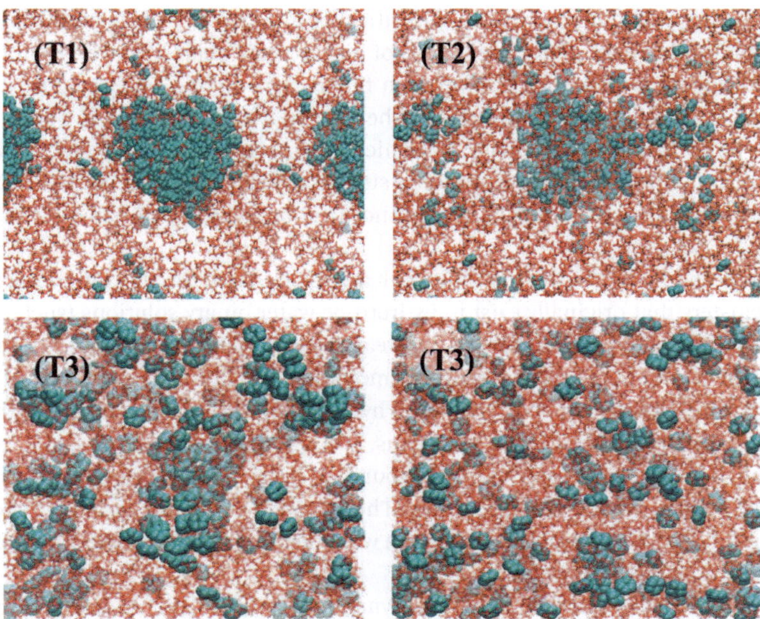

Fig. 10 Snapshots from MD simulations in TBA–water–CHX system with increasing concentrations of TBA (samples # T1 to T4 from Table 1). These snapshots indicate that as the TBA concentration increases, the tendency of CHX to form droplets decreases. At high TBA concentrations, TBA and CHX preferred to remain mixed, rather than form the droplets.

Simulations of TBA–water–CHX solutions with different concentrations of TBA, but almost the same concentration of CHX, were also carried out. Fig. 9a shows the RDFs between central carbon atoms of TBA molecules, while Fig. 9b shows the RDFs between central C of TBA molecule and a carbon on the CHX (C1) molecule. Fig. 10a to d show snapshots from MD simulations. The RDFs and the snapshots indicate that as the TBA concentration increases, the tendency to form "droplets" (as described above) disappears. Fig. 10c and d show that at high TBA concentrations, TBA and CHX prefer to remain mixed with each other rather than form droplets.

5 Mesoscale solubilization – a state between molecular solubility and macrophase separation

We refer to the phenomenon of formation of mesoscopic droplets in aqueous solutions of hydrotropes containing hydrophobes, as the mesoscale solubilization. Mesoscale solubilization is a distinct intermediate state between molecular solubility and macrophase separation.[76] Molecular solubility of nonpolar solutes (hydrophobes) in water can be explained by the phenomenon of hydrophobic hydration,[77–79] where water molecules form a hydrogen-bonded shell around the solute molecule. This shell is similar to a clathrate shell and the water molecules in the shell do not interact strongly with the nonpolar solute in the core.[79] However, molecular solubility in aqueous solutions of nonionic hydrotropes is quite different, where the water molecules strongly interact with the solute molecules through strong hydrogen bonds. This interaction leads to the formation of loose micelle-like clusters in water–hydrotrope solutions. Such clusters may have the same length scale as the correlation length of the concentration fluctuations (when far away from the critical point), but very different dynamics. Clusters, which have a life-time of tens of picoseconds, relax by the reorientation of hydrogen bonds, while concentration fluctuations have a time-scale of about tens of microseconds at $q \sim 10^7\ \mathrm{m}^{-1}$ (when far away from the critical point) and relax by diffusion. The cluster dynamics can be experimentally detected by dynamic neutron scattering techniques such as disc chopper spectrometer (DCS) or neutron spin echo (NSE), whereas concentration fluctuations can be detected by both SANS and DLS.

On the addition of a hydrophobe (such as cyclohexane), the short-lived micelle-like clusters that originally exist in hydrotrope-water binary solutions (such as in TBA–water) seem to be stabilized and rearranged. Over a certain concentration range of hydrophobe, the hydrophobe molecules start to aggregate. Part of the hydrotrope–water clusters surround the hydrophobe aggregates, protecting them from the water-rich environment. Thus, we view the mesoscopic droplets as having a hydrophobe-rich core, surrounded by a hydrogen-bonded "micro-emulsion-like" water–hydrotrope shell. The schematic of such a droplet is shown in Fig. 11. This picture is consistent with what has been observed and interpreted in other aqueous systems.[22,80]

In order to understand the thermodynamic stability of the mesoscopic droplets, we have studied the macroscopic behavior of the ternary system TBA–water–CHX.[76] The ternary phase diagram at ambient conditions is shown in Fig. 12.[76] This figure shows that TBA is completely miscible with water and CHX, while water and CHX are almost completely immiscible with each other. The region

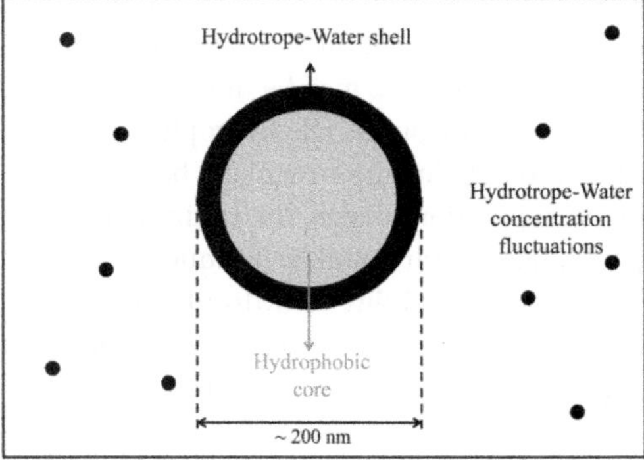

Fig. 11 Schematic representation of mesoscale solubilization in aqueous solutions of hydrotropes containing a hydrophobe. The mesoscopic droplets have a hydrophobic core surrounded by a hydrogen-bonded microemulsion-like hydrotrope–water shell.

where mesoscopic droplets are observed is shown in the inset of Fig. 12. Remarkably, this region corresponds to the concentration range where structural fluctuations and thermodynamic anomalies are observed in binary TBA–water solutions. Mesoscopic droplets are not observed on the addition of CHX to pure water or to TBA–water solutions where the TBA concentration is greater than 25 mol%. In the region around 7 mol% TBA the mesoscale droplets are extremely long-lived, being stable for over a year.[39] Only in the presence of a macroscopic hydrophobe-rich phase (samples studied in the two-phase region of the ternary

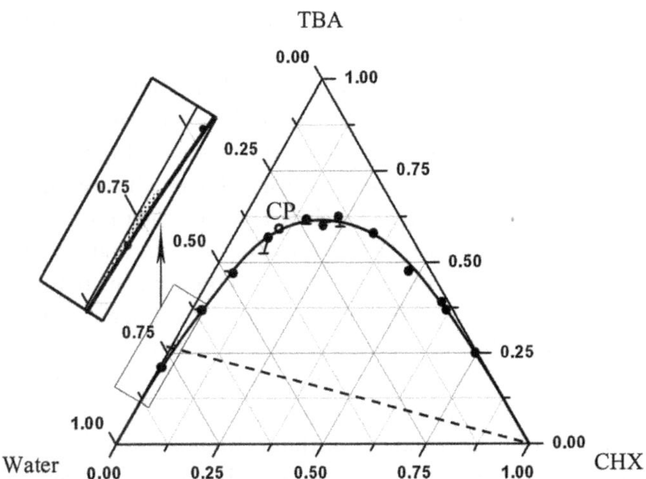

Fig. 12 Phase diagram of TBA–water–CHX system at $T = 21$ °C (reproduced from ref. 39). The smooth line across the points is a guide to the eye. All concentrations are shown as mass fractions. The open circle (CP) represents the approximate location of the critical point. The dashed line from the vertex of the hydrophobe (CHX) represents concentrations with a constant TBA–water ratio (25 : 75 mass basis/7 : 93 mole basis) where thermodynamic anomalies in the binary TBA–water solution are observed. The region inside the solid curve is the two-phase region, while the region outside this curve is the macroscopic one-phase region. The dotted area in the phase diagram shows the region where mesoscopic droplets are observed.

system), these droplets tend to slowly (over a period of months) condense on the water–CHX interface. The instability of the mesoscopic droplets in the presence of the macroscopic water–CHX interface may be due to the destruction of the TBA–water protective shell, which surrounds the hydrophobic CHX core.

Mesoscale solubilization, being intermediate between molecular solubility and macroscopic phase separation, makes the traditional definition of solubility ambiguous. Thus, popular experimental techniques, such as chromatography methods, used for measuring solubility of hydrophobic species in water may be misinterpreted. Moreover, the bulk equilibration may require an unrealistically long time, making the definition of thermodynamic equilibrium also ambiguous.

6 Summary and conclusions

We have investigated molecular-scale and mesoscale inhomogeneities in TBA–water solutions and in TBA–water–CHX solutions. We have combined results obtained from MD simulations with those obtained from SANS and DLS experiments. MD simulations in TBA–water solutions show the presence of short-ranged (~1 nm), short-lived (10 ps–50 ps) clusters, interpreted as micelle-like structural fluctuations. These clusters may have a length-scale similar to the concentration fluctuations, but have very different (much faster) dynamics. Concentration fluctuations relax by diffusion, while clusters relax by the reorientation of hydrogen bonds. Clustering is observed in the low concentration region of TBA (1 mol% to 8 mol% or 4 mass% to 26 mass%) and tends to disappear at higher TBA concentrations. These clusters are likely responsible for the thermodynamic anomalies observed in aqueous TBA solutions. We were unable to unambiguously detect these clusters from SANS experiments, but the presence of concentration fluctuations was clearly detected by both SANS and DLS techniques.

Mesoscale inhomogeneities, which are Brownian diffusive droplets, order of 100 nm in size, are observed in aqueous solutions of TBA containing a hydrophobic component. The hydrophobe tends to stabilize and rearrange the short-ranged, short-lived structural fluctuations initially present in aqueous TBA solutions and leads to the formation of larger (mesoscopic) droplets. The structure of these droplets is such that they contain a hydrophobe-rich core surrounded by a microemulsion-like hydrogen-bonded shell of TBA and water molecules. The shell can be regarded as a "protective layer" consisting of TBA and water molecules, which separate the oily core of the aggregates from the aqueous-rich bulk phase of the solution. We call the formation of mesoscopic droplets in aqueous solutions of hydrotropes, containing hydrophobe, the mesoscale solubilization.

Two peculiar features characterize mesoscale solubilization. The mesoscopic droplets are order of 100 nm in size. This size does not seem to significantly depend on the type of the hydrotrope or the hydrophobe. Lowering the temperature enhances the number of these droplets, but their size remains almost unchanged. Moreover, these droplets are extremely long-lived, being stable for over a year. Only in the presence of a macroscopic hydrophobe-rich phase do these droplets tend to slowly (over a period of months) condense on the water–oil interface. The phenomenon observed in aqueous solutions of TBA may represent a ubiquitous feature of aqueous solutions of nonionic hydrotropes, and may have important practical implications in areas such as drug delivery, where traditional surfactants may need to be replaced by hydrotropes.

Acknowledgements

We acknowledge fruitful discussions with J. Leys and V. Molinero. We also thank J. V. Sengers for useful comments on the manuscript. This research is supported by the Division of Chemistry of the National Science Foundation under Grant No. CHE-1012052 (MAA). Part of this research (MD simulations) was supported by the National Science Foundation through XSEDE resources provided by National Institute for Computational Sciences under grant number TG-MCB100139 (JBK). Additional computational resources were used on the High Performance Computing Cluster at the University of Maryland. SANS work is based upon activities supported by the National Science Foundation under Agreement No. DMR-0944772. The identification of commercial products does not imply endorsement by the National Institute of Standards and Technology nor does it imply that these are the best for the purpose.

References

1 F. Franks and D. G. Ives, The structural properties of alcohol-water mixtures, *Q. Rev. Chem. Soc.*, 1966, **20**, 1–44.
2 F. Franks, *Water: A Comprehensive Treatise*, Volumes 2 and 7, Plenum Press, New York, 1982.
3 G. D'Arrigo and J. Teixeira, Small-angle neutron scattering study of D_2O-alcohol solutions, *J. Chem. Soc., Faraday Trans.*, 1990, **86**, 1503–1509.
4 G. D'Arrigo, R. Giordano and J. Teixeira, Temperature and concentration dependence of SANS spectra of aqueous solutions of short-chain amphiphiles, *Eur. Phys. J. E*, 2009, **29**, 37–43.
5 M. Misawa, T. Sato, A. Onozuka, K. Maruyama, K. Mori, S. Suzuki and T. Otomo, A visualized analysis of small-angle neutron scattering intensity: Concentration fluctuation in alcohol-water mixtures, *J. Appl. Crystallogr.*, 2007, **40**, s93–s96.
6 R. D. Koehler, K.-V. Schubert, R. Strey and E. W. Kaler, The Lifshitz line in binary systems: Structures in water/C_4E1 mixtures, *J. Chem. Phys.*, 1994, **101**, 10843–10849.
7 K. Yoshida, T. Yamaguchi, T. Otomo, M. Nagao, H. Seto and T. Takeda, Concentration fluctuations and cluster dynamics of 2-butoxyethanol-water mixtures by small-angle neutron scattering and neutron spin echo techniques, *J. Mol. Liq.*, 2005, **119**, 125–131.
8 J. L. Finney, D. T. Bowron and A. K. Soper, The structure of aqueous solutions of tertiary butanol, *J. Phys.: Condens. Matter*, 2000, **12**, A123–A128.
9 D. T. Bowron, J. L. Finney and A. K. Soper, Structural investigation of solute-solute interactions in aqueous solutions of tertiary butanol, *J. Phys. Chem. B*, 1998, **102**, 3551–3563.
10 J. L. Finney, D. T. Bowron, R. M. Daniel, P. A. Timmins and M. A. Roberts, Molecular and mesoscale structures in hydrophobically driven aqueous solutions, *Biophys. Chem.*, 2003, **105**, 391–409.
11 D. T. Bowron, J. L. Finney and A. K. Soper, Structural characteristics of a 0.23 mole fraction aqueous solution of tetrahydrofuran at 25 °C, *J. Phys. Chem. B*, 2006, **110**, 20235–20245.
12 K. R. Harris and P. J. Newitt, Diffusion and structure in water-alcohol mixtures: Water + *tert*-butyl alcohol (2-methyl-2-propanol), *J. Phys. Chem. A*, 1999, **103**, 6508–6513.
13 N. Nishi, S. Takahashi, M. Matsumoto, A. Tanaka, K. Muraya, T. Takamuku and T. Yamaguchi, Hydrogen bonding cluster formation and hydrophobic solute association in aqueous solution of ethanol, *J. Phys. Chem.*, 1995, **99**, 462–468.
14 P. G. Kusalik, A. P. Lyubartsev, D. L. Bergman and A. Laaksonen, Computer simulation study of *tert*-butyl alcohol. 2. Structure in aqueous solution, *J. Phys. Chem. B*, 2000, **104**, 9533–9539.
15 J.-H. Guo, Y. Luo, A. Augustsson, S. Kashtanov, J.-E. Rubensson, D. K. Shuh, H. Ågren and J. Nordgren, Molecular structure of alcohol-water mixtures, *Phys. Rev. Lett.*, 2003, **91**, 157401.
16 A. Fornili, M. Civera, M. Sironi and S. L. Fornili, Molecular dynamics simulation of aqueous solutions of trimethylamine-N-oxide and *tert*-butyl alcohol, *Phys. Chem. Chem. Phys.*, 2003, **5**, 4905–4910.

17 M. Kiselev and I. Ivlev, The study of hydrophobicity in water-methanol and water-*tert*-butanol mixtures, *J. Mol. Liq.*, 2004, **110**, 193–199.
18 A. B. Roney, B. Space, E. W. Castner, R. L. Napoleon and P. B. Moore, A molecular dynamics study of the aggregation phenomena in aqueous *n*-propanol, *J. Phys. Chem. B*, 2004, **108**, 7389–7401.
19 S. K. Allison, J. P. Fox, R. Hargreaves and S. P. Bates, Clustering and microimmiscibilityin alcohol-water mixtures: Evidence from molecular-dynamics simulations, *Phys. Rev. B*, 2005, **71**, 024201.
20 B. Kežić and A. Perera, Aqueous *tert*-butanol mixtures: A model for molecular emulsions, *J. Chem. Phys.*, 2012, **137**, 014501.
21 R. Gupta and G. N. Patey, Aggregation in dilute aqueous *tert*-butyl alcohol solutions: Insights from large-scale simulations, *J. Chem. Phys.*, 2012, **137**, 034509.
22 Z. Li, H. Cheng, J. Li, L. Hao, L. Zhang, B. Hammouda and C. C. Han, Large-scale structures in tetrahydrofuran-water mixture with a trace amount of antioxidant butylhydroxytoluene (BHT), *J. Phys. Chem. B*, 2011, **115**, 7887–7895.
23 M. F. Vuks and L. V. Shurupova, The scattering of light and phase transition in solutions of tertiary butyl alcohol in water, *Opt. Commun.*, 1972, **5**, 277–278.
24 C. W. Beer, Jr. and D. J. Jolly, Comments on "the scattering of light and phase transition in solutions of tertiary butyl alcohol in water", *Opt. Commun.*, 1974, **11**, 150–151.
25 K. Iwasaki and T. Fujiyama, Light scattering study of clathrate hydrate formation in binary mixtures of *tert*-butyl alcohol and water, *J. Phys. Chem.*, 1977, **81**, 1908–1912.
26 K. Iwasaki and T. Fujiyama, Light-scattering study of clathrate hydrate formation in binary mixtures of *tert*-butyl alcohol and water: 2. Temperature effect, *J. Phys. Chem.*, 1979, **83**, 463–468.
27 G. W. Euliss and C. M. Sorensen, Dynamic light scattering studies of concentration fluctuations in aqueous *t*-butyl alcohol solutions, *J. Chem. Phys.*, 1984, **80**, 4767–4773.
28 T. M. Bender and R. Pecora, A dynamic light scattering study of the *tert*-butyl alcohol-water system, *J. Phys. Chem.*, 1986, **90**, 1700–1706.
29 T. M. Bender and R. Pecora, Dynamic light scattering measurements of mutual diffusion coefficients of water-rich 2-butoxyethanol/water systems, *J. Phys. Chem.*, 1988, **92**, 1675–1677.
30 Y. Georgalis, A. M. Kierzek and W. Saenger, Cluster formation in aqueous electrolyte solutions observed by dynamic light scattering, *J. Phys. Chem. B*, 2000, **104**, 3405–3406.
31 C. Yang, W. Lei and C. Wu, Laser light-scattering study of solution dynamics of water/cycloether mixtures, *J. Phys. Chem. B*, 2004, **108**, 11866–11870.
32 M. Sedlák, Large-scale supramolecular structure in solutions of low molar mass compounds and mixtures of liquids: I. Light scattering characterization, *J. Phys. Chem. B*, 2006, **110**, 4329–4338.
33 M. Sedlák, Large-scale supramolecular structure in solutions of low molar mass compounds and mixtures of liquids: II. Kinetics of the formation and long-time stability, *J. Phys. Chem. B*, 2006, **110**, 4339–4345.
34 M. Sedlák, Large-scale supramolecular structure in solutions of low molar mass compounds and mixtures of liquids: III. Correlation with molecular properties and interactions, *J. Phys. Chem. B*, 2006, **110**, 13976–13984.
35 F. Jin, X. Ye and C. Wu, Observation of kinetic and structural scalings during slow coalescence of nanobubbles in an aqueous solution, *J. Phys. Chem. B*, 2007, **111**, 13143–13146.
36 A. F. Kostko, M. A. Anisimov and J. V. Sengers, Criticality in aqueous solutions of 3-methyl pyridine and sodium bromide, *Phys. Rev. E*, 2004, **70**, 026118.
37 D. Subramanian, D. A. Ivanov, I. K. Yudin, M. A. Anisimov and J. V. Sengers, Mesoscale inhomogeneities in aqueous solutions of 3-methylpyridne and tertiary butyl alcohol, *J. Chem. Eng. Data*, 2011, **56**, 1238–1248.
38 D. Subramanian and M. A. Anisimov, Resolving the mystery of aqueous solutions of tertiary butyl alcohol, *J. Phys. Chem. B*, 2011, **115**, 9179–9183.
39 D. Subramanian, *Ph.D Dissertation*, University of Maryland, 2012.
40 C. Neuberg, Hydrotropische erscheinungen, *Biochem. Z.*, 1916, **76**, 107–176.
41 V. Srinivas and D. Balasubramanian, When does the switch from hydrotropy to micellar behavior occur?, *Langmuir*, 1998, **14**, 6658–6661.
42 T. K. Hogdgon and E. W. Kaler, Hydrotropic solutions, *Curr. Opin. Colloid Interface Sci.*, 2007, **12**, 121–128.
43 J. Eastoe, M. H. Hatzopoulas and P. J. Dowding, Action of hydrotropes and alkyl-hydrotropes, *Soft Matter*, 2011, **7**, 5917–5925.
44 D. Subramanian, J. B. Klauda, J. Leys and M. A. Anisimov, Thermodynamic anomalies and structural fluctuations in aqueous solutions of tertiary butyl alcohol, *Вестник СПбГу (Herald of St. Petersburg University)*, 2013, **4**, 140–153. Available at: http://arxiv.org/abs/1308.3676.

45 J. B. Ott, J. R. Goates and B. A. Waite, (Solid+liquid) phase equilibria and solid-hydrate formation in water + methyl, + ethyl, + isopropyl, and + tertiary butyl alcohols, *J. Chem. Thermodyn.*, 1979, **11**, 739–746.
46 A. Fiore, V. Venkateshwaran and S. Garde, Trimethylamine N-oxide (TMAO) and tert-butyl alcohol (TBA) at hydrophobic interfaces: Insights from molecular dynamics simulations, *Langmuir*, 2013, **29**, 8017–8024.
47 K. Nishikawa and T. Iijima, Structural study of *tert*-butyl alcohol and water mixtures by X-ray diffraction, *J. Phys. Chem.*, 1990, **94**, 6227–6231.
48 H. Tanaka, K. Nakanishi and K. Nishikawa, Clathrate-like structure of water around some nonelectrolytes in dilute solution as revealed by computer simulation and X-ray diffraction studies, *J. Inclusion Phenom.*, 1984, **2**, 119–126.
49 J. Jacob, M. A. Anisimov, J. V. Sengers, A. Oleinikova, H. Weingärtner and A. Kumar, Novel phase-transition behavior near liquid-liquid critical points of aqueous solutions: Formation of a third phase at the interface, *Phys. Chem. Chem. Phys.*, 2001, **3**, 829–831.
50 H. W. Horn, W. C. Swope, J. W. Pitera, J. D. Madura, T. J. Dick, G. L. Hura and T. Head-Gordon, Development of an improved four-site water model for biomolecular simulations: TIP4P-Ew, *J. Chem. Phys.*, 2004, **120**, 9665–9678.
51 K. Vanommeslaeghe, E. Hatcher, C. Acharya, S. Kundu, S. Zhong, J. Shim, E. Darian, O. Guvench, P. Lopes, I. Vorobyov and J. A. D. Mackerell, CHARMM general force field: A force field for drug-like molecules compatible with the CHARMM all-atom additive biological force fields, *J. Comput. Chem.*, 2010, **31**, 671–690.
52 L. Martínez, R. Andrade, E. G. Birgin and J. M. Martínez, PACKMOL: A package for building initial configurations for molecular dynamics simulations, *J. Comput. Chem.*, 2009, **30**, 2157–2164.
53 J. C. Phillips, R. Braun, W. Wang, J. Gumbart, E. Tajkhorshid, E. Villa, C. Chipot, R. D. Skeel, L. Kale and K. Schulten, Scalable molecular dynamics with NAMD, *J. Comput. Chem.*, 2005, **26**, 1781–1802.
54 T. Darden, D. York and L. Pedersen, Particle Mesh Ewald - an NLog(N) Method for Ewald Sums in Large Systems, *J. Chem. Phys.*, 1993, **98**, 10089–10092.
55 S. E. Feller, Y. Zhang, R. W. Pastor and B. R. Brooks, Constant pressure molecular dynamics simulation: The Langevin piston method, *J. Chem. Phys.*, 1995, **103**, 4613–4621.
56 G. J. Martyna, D. J. Tobias and M. L. Klein, Constant pressure molecular dynamics algorithms, *J. Chem. Phys.*, 1994, **101**, 4177–4189.
57 W. Humphrey, A. Dalke and K. Schulten, VMD: Visual molecular dynamics, *J. Mol. Graphics*, 1996, **14**, 33–38.
58 B. J. Berne and R. Pecora, *Dynamic Light Scattering: With Applications to Chemistry, Biology, and Physics*, Wiley, New York, 1976; Dover Publications, Mineola, New York, 2000.
59 B. Chu, *Laser Light Scattering: Basic Principles and Practice*, Academic Press, Boston1991.
60 K. Kasraian and P. P. DeLuca, Thermal analysis of the tertiary butyl alcohol- water system and its implications on freeze-drying, *Pharm. Res.*, 1995, **12**, 484–490.
61 P. K. Kipkemboi and A. J. Easteal, Densities and viscosities of binary aqueous mixtures of nonelectrolytes: *tert*-butyl alcohol and *tert*-butylamine, *Can. J. Chem.*, 1994, **72**, 1937–1945.
62 K. Nakanishi, Partial molal volumes of butyl alcohols and of related compounds in aqueous solution, *Bull. Chem. Soc. Jpn.*, 1960, **33**, 793–797.
63 G. Roux, D. Roberts, G. Perron and J. E. Desnoyers, Microheterogeneity in aqueous organic solutions: Heat capacities, volumes and expansibilities of some alcohols, aminoalcohol and tertiary amines in water, *J. Solution Chem.*, 1980, **9**, 629–647.
64 M. A. Anisimov, V. S. Esipov, V. M. Zaprudskii, N. S. Zaugol'nikova, G. I. Ovodov, T. M. Ovodova and A. L. Seifer, Anomaly in the heat capacity and structural phase transformation of the ordering type in an aqueous solution of *t*-butanol, *J. Struct. Chem.*, 1977, **18**, 663–670.
65 C. De Visser, G. Perron and J. E. Desnoyers, The heat capacities, volumes and expansibilities of *tert*-butyl alcohol–water mixtures from 6 to 65 °C, *Can. J. Chem.*, 1977, **55**, 856–862.
66 Y. Koga, Excess partial molar enthalpies of water in water-*tert*-butanol mixtures, *Can. J. Chem.*, 1988, **66**, 3171–3175.
67 Y. Koga, Differential heats of dilution of *tert*-butanol in water-*tert*-butanol mixtures at 26.90 °C, *Can. J. Chem.*, 1986, **64**, 206–207.
68 Y. Koga, Excess partial molar enthalpies of *tert*-butanol in water-*tert*-butanol mixtures, *Can. J. Chem.*, 1988, **66**, 1187–1193.
69 Y. Koga, W. Y. U. Siu and T. Y. H. Wong, Excess partial molar free energies and entropies in aqueous *tert*-butyl alcohol solutions, *J. Phys. Chem.*, 1990, **94**, 7700–7706.
70 W. S. Knight, *Ph.D Dissertation*, Princeton University, 1962.

71 M. A. Anisimov, *Critical Phenomena in Liquids and Liquid Crystals*, Gordon & Breach Science Publishers, New York, 1991.
72 K. Tamura, A. Osaki and Y. Koga, Compressibilities of aqueous *tert*-butanol in water-rich region at 25 °C: Partial molar fluctuations and mixing schemes, *Phys. Chem. Chem. Phys.*, 1999, **1**, 121–126.
73 P. Losada-Perez and J. Thoen, KU Leuven, Personal Communication, 2012.
74 E. D. Sloan and C. A. Koh, *Clathrate Hydrates of Natural Gases*, Taylor and Francis, Boca Raton, FL, 2008.
75 G. Gompper and M. Schick, *Self-Assembling Amphiphilic Systems, Volume 16 Phase Transitions and Critical Phenomena*, Academic Press, London, 1994.
76 D. Subramanian and M. A. Anisimov, Mesoscale solubilization in aqueous solutions of hydrotropes, special issue of Fluid Phase Equilibria: *Proceedings of the Conference on Properties and Phase Equilibria for Process and Product Design*, *Fluid Phase Equilib.*, 2013, (in press). Available at: http://arxiv.org/abs/1309.7255.
77 F. H. Stillinger, Structure in aqueous solutions of nonpolar solutes from the standpoint of scales-particle theory, *J. Solution Chem.*, 1973, **2**, 141–158.
78 L. R. Pratt and D. Chandler, Theory of hydrophobic effect, *J. Chem. Phys.*, 1977, **67**, 3683–3704.
79 K. Lum, D. Chandler and J. D. Weeks, Hydrophobicity at small and large length scales, *J. Phys. Chem. B*, 1999, **103**, 4570–4577.
80 A. Häbich, W. Ducker, D. E. Dunstan and X. Zhang, Do stable nanobubbles exist in mixtures of organic solvent and water?, *J. Phys. Chem. B*, 2010, **114**, 6962–6967.

Faraday Discussions RSC Publishing

DISCUSSIONS

General discussion

DOI: 10.1039/C3FD90037A

Professor Anisimov opened the discussion of the paper by Dr Perera: Although two types of fluctuations (structure and concentration) in solutions are fundamentally different, they may occur on the same length-scale. Relaxation of concentration fluctuation is relatively slow diffusion (wavenumber dependent), while a fast rearrangement of H-bonds (tens of ps) may occur at the same length-scale but its dynamics will not depend on the wavenumber. In other words, these two phenomena belong to two different dynamic universality classes.

Dr Vila Verde said: It would be very useful if you could precisely define "correlation" and "concentration fluctuation" in the context of this paper.

Dr Perera replied: In our paper the term "correlation" refers to what the correlation functions – a quantity well defined from statistical physics – measure in a system of particles. The term "concentration fluctuation" (CF) refers again to the statistical physics definition, where N_i is the number of molecules of species i and refers to an ensemble average. From the theory of Kirkwood–Buff, this quantity is also related to the spatial integral over the correlation function $g_{ij}(r)$ between molecules of species i and j.

The $g_{ij}(r)$ contains more information on the short range correlation between molecules than the related CF, in particular about the cluster formation and micro-segregation. Hence, one should not confuse CF for this latter information, which is better spotted in the small-q form of structure factor $S_{ij}(q)$ which are the Fourier transforms of the $g_{ij}(r)$ (see the exact definition in the paper). The asymptote of the RKBI relates directly to $S_{ij}(q=0)$, which is the CF, while it is the small-q part of the $S_{ij}(q)$ that shows the existence of clusters through a non-Ornstein–Zernike behaviour explained in the text.

Dr Vila Verde commented: You show in Fig. 3 in your paper that, in certain cases, the running Kirkwood–Buff integrals calculated from the simulations do not converge to the experimentally determined values. This lack of convergence would suggest that the site–site structure factors calculated directly from the simulation, shown in Fig. 5, have large associated statistical uncertainty, particularly at small q. It would be interesting if you could estimate this uncertainty, using *e.g.* block-averaging,[1,2] and then briefly discuss how this uncertainty limits what we can learn from the observed differences and similarities between the site–site structure factors obtained from integral equation theory (also shown in Fig. 5) and those calculated from simulation.

1 R. Chitra and S. Yashonath, Estimation of Error in the Diffusion Coefficient from Molecular Dynamics Simulations, *J. Phys. Chem. B*, 1997, **101**, 5437–5445.
2 *Physics of simple liquids*, ed. W. W. Wood, H. N. V. Temperley, J. S. Rowlinson and G. S. Rushbrooke, North-Holland Publishing Company, Amsterdam, 1968.

Dr Perera answered: There are two issues here. One is the convergence to the experimental value, which is very much model dependant, and which we do not consider to be the real problem here. The second is the convergence issue, which is one of the main points of the paper, and which witnesses the fact that one is doing two types of averages. The first average is over spatio-temporal scales corresponding to the molecular motions. This average is what standard molecular dynamic simulations deal with, which is not problematic. The second is over the scales corresponding to domain motions. This is the problematic issue. Block averaging is not going to help that, since the problem is about domain statistics, hence the number of domains one can accommodate in the cell. Clearly, the larger the cell, the larger the number of domains, the better the corresponding statistics, but also the heavier the calculations. Integral equations are build in the thermodynamic limit by construction and do not suffer these problems. However, they are inapt to describe the coupling between concentration fluctuations and domain segregation. Our paper shows clearly that we cannot rely on simulations to help improve the shape of the correlations at large distances, since this is precisely where statistical problems are the most severe. The theory must be improved following other paths, such as for example the molecular emulsion concept, which we are currently developing.

Professor Mallamace commented: Methanol has hydrophilic/hydrophobic parts. Discriminated hydrophobicity effects are observed by changing concentration. Also your systems are characterized by hydrophobic/hydrophilic parts. How are these or how can these be related to your static correlation function?

Dr Perera responded: Hydrophobicity, or more generally solvophobicty, is the term coined to this observed fact that some solvent tends to segregate solutes in clusters.

Many books and papers haven been written on this subject, but your question raises an important point unadressed at present, even in the most recent theoretical approaches of solvophobicty: how this property reflects upon correlation functions. What we have observed in many contexts, is that the solvent–solvent correlations tend to increase strongly in a finite range upon solute insertion.

For water this is about 10–12 Å, after which the correlations decay exponentially. This produces a very large volume integral of the function, hence a large Kirkwood–Buff integral. This large integral should not be confused with that appearing near the spinodal when fluid mixtures tend to demix. I hesitate to relate this effect to the correlation to a signature of the hydrophobic effect, principally because the hydrophobic effect is unambiguously attributed for large surfaces plunged in water, and we have not studied such cases.

For the case of methanol–water, we do not observe the large raise of water–water correlations, which would tend to suggest that it is a hydrophilic molecule. This is in contrast with the obvious micro-segregation seen in the snapshots of the 2 K and 16 K systems. This means that, although segregated, both molecules do not have a correlational coherence that translates into the correlation function. We start to observe the correlation-raise effect for *t*-butanol. I am convinced that a formulation of solvophobicity/solvophilicity in terms of correlation function is the ultimate microscopic definition of these effects, which may not always confirm prior classifications.

Professor Mallamace commented: Water–methanol mixtures were studied by us using NMR at several molar fraction and concentrations.[1] The main observation is that at low T the system is dominated by hydrophilicity, whereas in the high T regime it is determined by hydrophobicity. Given your study focused on fluctuations and micro-heterogeneities also in the same systems, have you obtained some results connected with these experimental observations?

1 C. Corsaro, *J. Chem Phys. B*, 2008, **112**, 10449–10454.

Dr Perera responded: We did not study the temperature dependence of the methanol–water mixtures, so I cannot answer your question directly. I would be very careful of invoking concepts of hydrophobicity/hydrophilicity in the case of methanol, which is a small molecule. As I said my previous answer, we do not observe strong raise of short-range correlations, and I presume that thermalisation will only lower them. So interpreting these purely thermal effects in terms of solvent collective coherent effects should be considered with care.

Professor Barrat asked: The distinction between micro heterogeneities and concentration fluctuations as measured by structure factors or pair correlations is quite difficult to understand. Is it mainly a semantic distinction, or are the micro heterogeneities fluctuations particularly strong or long lived? And if not semantics, would the appropriate tool to investigate them not be 4 point correlations of the type used to characterize dynamical heterogeneities in supercooled liquids?

Dr Perera answered: The distinction is hard to understand because you have formulated the problem in terms of time-dependence, and the analogy you suggest with dynamical heterogeneity in super cooled liquids enhances this difficulty. In fact, come to think of if, one may even say that it is difficult to speak of static pair correlations because of the existence of the van Hove function. Yet, and everyone that has well learned the theory of liquids knows, the static correlation function is a well defined concept, outside the realm of its dynamical counter part, while being related to it at zero frequency. This part is just to point out that the difficulty is not inherent to the difference outlined in my talk, but to the way one may want to look at it.

Now, that being said, the distinction between CF and MH is seen in the following simple way. The CF is by definition related to the $k = 0$ part of the partial structure factors (this is the essence of the Kirkwood–Buff theory), just like density fluctuation is related to the structure factor of a single component liquid. The pre-peak in the structure factor, at a wave vector intermediate between $k = 0$ and k_{max}, witnesses the existence of static domains, much just like the main peak at k_{max} witnesses the presence of the particle itself. This concept is independent of any associated kinetics or lifetime of the cluster or domain. As to the pair correlation functions, the domain spatial modulation will introduce a corresponding oscillation in the long range part of the correlation functions. This is the part that is difficult to properly sample in finite size simulations, which was the topic of the first part of my contribution.

Professor Idrissi commented: Although the intermolecular potentials of liquids A and B could be individually "good" (they are parameterized to reproduce the structure and dynamics of the solute A and the solvent B, respectively), the combination of these two intermolecular potentials does not necessarily reproduce the structure and dynamics of the mixture of A and B. The free energy of mixing then provides one of a rigorous test of the force field accuracy, to the extent that all the system degrees of freedom are adequately sampled. Indeed, the free energy of mixing two components is a key physical–chemical quantity, because its sign determines whether the components mix or demix at a given molar ratio. In cases when the mixing of the two components in computer simulation is accompanied by a slight increase of the free energy, demixing might not visibly occur on the simulation time and length scales.[1] Detection of such demixing is also hindered by the use of periodic boundary conditions. Further, besides the qualitative description of the mixing of the components, good reproduction of the experimental free energy of mixing in computer simulation is a pre-requisite of the accurate description of the microscopic structure (*e.g.*, in terms of self-association) of the mixture. On the other hand, the calculation of the free energy of mixing is computationally rather costly, because the Helmholtz free energy is related to the entire phase space on the canonical ensemble, and hence it cannot simply be calculated by sampling its lowest energy domains. The technical issues of such calculations were discussed, and their application is presented for several systems in which the free energy of mixing, resulting from a subtle interplay of the energetic and entropic terms, deviates only slightly from zero.[1–6]

1 P. Jedlovszky, A. Idrissi and G. Jancsó, Can existing models qualitatively describe the mixing behavior of acetone-water mixtures?, *J. Chem. Phys.*, 2009, **130**, 124516.
2 M. Darvas, G. Jancsó and P. Jedlovszky, Free Energy of Mixing of Pyridine and Its Methyl-Substituted Derivatives with Water, As Seen from Computer Simulations, *J. Phys. Chem. B*, 2009, **113**, 7615–7620.
3 M. Mezei and D. L. Beveridge, Free Energy Simulations, *Ann. N. Y. Acad. Sci.*, 1986, **482**, 1–23.
4 A. Pinke and P. Jedlovszky, Modeling of Acetone-Water Mixtures – How Can Their Full Miscibility Reproduced in Computer Simulations?, *J. Phys. Chem. B*, 2012, **116**, 5977–5984.
5 A. Idrissi, K. Polok, M. Barj, B. Marekha, M. Kiselev and P. Jedlovszky, Free Energy of Mixing of Acetone and Methanol – a Computer Simulation Investigation, *J. Phys. Chem. B*, submitted.
6 A. Idrissi, I. Vyalov, M. Kiselev, P. Jedlovszky, Assessment of the potential models of acetone/CO2 and ethanol/CO2 mixtures by computer simulation and thermodynamic integration in liquid and supercritical states, *Phys. Chem. Chem. Phys.*, 2011, **13**, 16272–16281.

Professor Chandler remarked: Following up on Professor Barrat's question to Dr. Perera, I note that integral equation theories are valid only in cases of homogeneous fluids. Spatial density fluctuations in those cases obey Gaussian statistics to a high degree of accuracy,[1,2] and that form of statistics generally underlies integral equations for pair correlation functions.[3]Because Gaussian fields do not support interfaces, it is hard to imagine how integral equation theories can correctly describe mesoscopic assembly. Indeed, the simplest description that might describe mesoscopic assembly would seem to require four-point functions. With Gaussian fields, four-point functions factor into products of two-point functions. One way to assess the presence of mesoscopic clustering is to examine the deviation from the Gaussian factorization of four-point functions.

1 G. Hummer, S. Garde, A. E. Garcia, A. Pohorille and L. R. Pratt, *Proc. Natl. Acad. Sci. U. S. A.*, 1996, **93**, 8951.
2 G. Crooks and D. Chandler, *Phys. Rev. E*, 1997, **56**, 4217–4221.
3 D. Chandler, *Phys. Rev. E*, 1993, **48**, 2898–2905.

Dr Perera replied: It is worth recalling that integral equation theories consist of a set of two exact equations, the first being the Ornstein–Zenike equation, and the second the closure relation $g(1,2 = \exp[-v(1,2)/kT + h(1,2) - c(1,2) + b(1,2)]$ involving the bridge function $b(1,2)$ (this second equation is exact under pairwise additive interactions). The bridge function can be approximated by several well documented methods, thus allowing to study a large variety of liquids including liquid–gas and liquid–solid interfaces (see for example the book by Hansen and McDonald[1]), and contrary to what is stated in the remark by Professor Chandler.

His comment on the inapplicability of integral equation is based on his previous work,[2] which concerns the worst possible closure, namely the MSA closure which has been know decades prior to this work to be of mean-field nature, attesting the Gaussian nature of fluctuations. Recent attempts to extend field theoretic methods beyond MSA did not lead to convincing results beyond previously known formal results.[3,4] Since my paper uses the HNC closure with bridge functions extracted from simulations of pure liquids, its reach is beyond any Gaussian approximation.

Next, the micro-heterogeneity pointed out in the paper is still microscopic in nature for one thing, and not mesoscopic. Moreover applicability to mesoscopic systems such as microemulsion has been demonstrated by Kezic–Perera[5] by showing that the Teubner–Strey approach to these system is embedded in integral equation theories theories with proper closure relations.

1. J. P. Hansen and I. R. McDonald, *Theory of Simple Liquids*, Academic Press, London, 1986.
2 D. Chandler, *Phys. Rev. E*, 1993, **48**, 2898–2905.
3 J.-M. Caillol, *Mol. Phys.*, 2003, **101**, 1617.
4 J.-M. Caillol, *Mol. Phys.*, 2006, **104**, 1931.
5 B. Kežić and A. Perera, *J. Chem. Phys.*, 2012, **137**, 014501.

Professor Molinero opened the discussion of the paper by Dr Dougan:†Your paper suggests that percolation of water in mixtures is related to the onset of ice crystallization at low temperatures. Our study of the crystallization of ice in solutions of water and ions indicates that is not so much the percolation of the water network but the characteristic dimensions of the water domains that determine whether ice can nucleate from the solution.[1] Fig. 10 of your paper shows that the largest water clusters in water–glycerol solutions with water content below the percolation threshold contain about 50 molecules, which is about three times smaller than the size of the critical ice nucleus at the temperature of homogeneous nucleation in pure water.[2,3] Have you analyzed the shape of

† Dr Dougan's paper was presented by Dr Soper, Rutherford Appleton Laboratory, Didcot, United Kingdom

the water domains at concentration above percolation to determine whether they are chain-like or pool-like?

1 G. Bullock and V. Molinero, Low-density liquid water is the mother of ice: on the relation between mesostructure, thermodynamics and ice crystallization in solutions, *Faraday Discuss.*, 2013, DOI: 10.1039/C3FD00085K.
2 E. B.Moore and V. Molinero, Structural transformation in supercooled water controls the crystallization rate of ice, *Nature*, 2011, **479**, 506–508.
3 J. Liu, C. E. Nicholson and S. J. Cooper, Direct measurement of critical nucleus size in confined volumes, *Langmuir*, 2007, **23**, 7286–7292.

Dr Soper and **Mr Towey** responded: This is a topic that we are interested in. We have measured the proportion of molecules that are found at the surface of these clusters to investigate their structure. Here, we define a molecule as being on the surface of a cluster based on the separation of the relevant oxygen and hydrogen atoms. This is similar to the criteria that have been used to investigate methanol–water systems.[1]

From this analysis we find that over 90% of the water molecules are found at the surface of clusters at concentrations of Xg > 0.50. This suggests clusters with either chain or sheet like structures. Similarly we find that the glycerol clusters are chain/sheet-like at glycerol concentrations of Xg <0.25. There is, however, little evidence for "pools" of water or glycerol: such pools would give rise to small Q scattering which we do not observe. These results are unsurprising given that we have previously shown that there is a preference for isolated water molecules at high glycerol concentration[2]and small glycerol concentrations at high water concentration.[3]

1 A. K. Soper, L. Dougan, J. Crain, and J. L. Finney, Excess entropy in alcohol–water solutions: a simple clustering explanation, *J. Phys. Chem. B*, 2006, **110**(8), 3472.
2 J. J. Towey, A. K. Soper and L. Dougan, Preference for isolated water molecules in a concentrated glycerol-water mixture, *J. Phys. Chem. B*, 2011, **115**, 7799.
3 J. J. Towey and L. Dougan, Structural examination of the impact of glycerol on water structure, *J. Phys. Chem. B*, 2012, **116**, 1633.

Dr Henchman asked: Regarding the mechanism of cryoprotect action, is the greater mixing than ideal mixing partly responsible for this behaviour? And do you see a temperature dependence in the diffraction and percolation data?

Dr Soper answered: Our results do not indicate greater mixing compared to ideal mixing. If anything the degree of mixing is less than what we would expect, which is the reason we believe these solutions are segregated at the microscopic level.

We have data at lower temperatures which are currently being prepared for publication.

Professor Mallamace remarked: Percolation processes, characteristic of clustering phenomena at the sol–gel transition, mean that you have a spanning cluster. As you affirm the cryoprotectant solutions you studied, are just governed by this process due to the H-bond? How do you reach such a conclusion?

Dr Soper answered: We use computer simulations (EPSR) to reconstruct the total scattering data (see Fig. 3 in our paper). These simulations indicate some segregation of the mixture in water clusters (defined by Ow–Ow distances) and glycerol clusters (defined also by O–O distances between neighbouring glycerol molecules). Depending on concentration either the water clusters or the glycerol clusters, or both, can percolate. There is however no indication of a sol–gel transition at any concentration studied so far. Such a transition would manifest as pronounced small Q scattering, but there is no such enhanced scattering in any of the scattering data or simulations.

Professor Caupin commented: What would be the mechanism for cryopreservation? The paper by Towey *et al.* proposes that "micro-segregation may act to encapsulate water clusters, thus preventing the propagation of an ice-like water network". Do you mean that one needs to reach the regime where the proportion of water molecules found in percolating clusters fall down? According to your Fig. 10C, this would require quite a large mole fraction of glycerol, around 0.6. In experiments on cryopreservation, what is the typical fraction of glycerol required to get a good efficiency?

Professor Tanaka replied: In our experiments of water–glycerol mixtures, we found that above 13 mol% glycerol concentration, we can practically avoid ice crystallization for a cooling rate of about 100 K min^{-1} (see Fig. 4a of the paper by Murata and Tanaka[1]). But this glass-forming region is crucially dependent on the cooling rate employed. For water–LiCl mixtures, we recently measured a critical cooling rate required for vitrification as a function of the LiCl concentration.[2,3] We think that such quantitative analysis may be important to discuss the glass-forming ability, or cryopreservation ability. Concerning the mechanism for cryopreservation, we proposed that the glass-forming ability is generally controlled by the strength of frustration effects on crystallization.[4–6]

1 K. Murata, and H. Tanaka, Liquid–liquid transition without macroscopic phase separation in a water-glycerol mixture, *Nature Mater.*, 2012, **11**, 436–443.
2 M. Kobayashi, and H. Tanaka, Possible link of the V-shaped phase diagram to the glass-forming ability and fragility in a water-salt mixture, *Phys. Rev. Lett.*, 2011, **106**, 125703 (2011).
3 M. Kobayashi and H. Tanaka, Relationship between the phase diagram, the glass-forming ability, and the fragility of a water/salt mixture, *J. Phys. Chem. B*, 2011, **115**, 14077–14090.
4 H.Tanaka, A simple physical model of liquid-glass transition: Intrinsic fluctuating interactions and random fields hidden in glass-forming liquids, *J. Phys.: Condens. Matter*, 1998, **10**, L207–L214.
5 H. Shintani and H. Tanaka, Frustration on the way to crystallization in glass, *Nature Phys.*, 2006, **2**, 200–206.
6 H. Tanaka, Bond orientational order in liquids: Towards a unified description of water-like anomalies, liquid–liquid transition, glass transition, and crystallization, *Eur. Phys. J.*, 2012, **E35**, 1.

Mr Towey replied: The research that Prof. Tanaka highlights here is consistent with the results regarding cryopreservation that are presented in the manuscript. Namely, it is not the breakdown of the water network but rather the formation of a percolating glycerol cluster that is important. It is conceivable that this glycerol network provides the frustration of crystallisation that Prof Tanaka refers to. This means that the glycerol concentration that is required is much lower than the value (Xg = 0.60) that is referred to in Prof. Caupin's question.

Professor Caupin remarked: In the paper by Warkentin *et al.*,[1] the critical cooling rate required to make a glass was measured for different aqueous solutions. For glycerol, at 5 mol L^{-1}, that is a mole fraction around 0.1, the rate dropped to around 50 K s^{-1}. Is this relevant to cryoprotection and can you make a link between your results (such as the proportion of molecules found in percolating clusters) and this number?

1 M. Warkentin, J. P. Sethna, and R. E. Thorne, *Phys. Rev. Lett.*, 2013, **110**, 015703.

Dr Soper answered: Vitrification is very relevant to cryoprotection. Many of the complications that arise due to storing bio-molecules at low temperatures are due to the formation of ice.[1] This is why vitrification has long been employed to prevent water crystallisation.[2] Interestingly, a glycerol concentration of around Xg = 0.10 is often used in cryoprotection protocols.[1] Fig. 10 in the manuscript shows that the glycerol clusters are close to percolation at Xg = 0.10.[3] It is likely that the extended glycerol networks act to frustrate crystallisation and make vitrification easier to attain.

1 B. J. Fuller, Cryoprotectants: The essential antifreezes to protect life in the frozen state, *Cryoletters*, 2004, **25**(6), 375–388.
2 T. Nash, *Cryobiology*, Academic Press New York, USA, 1966, pp. 179–211.
3 J. J. Towey, A. K. Soper and L. Dougan, What happens to the structure of water in cryoprotectant solutions?, Faraday Discuss., 2013, DOI: 10.1039/C3FD00084B.

Dr Perkin opened the discussion of the paper by Professor Ben-Amotz: My thanks for the clear presentation in your paper of the random mixing statistics. You go on to show that, at low TBA concentrations, the TBA molecules are more dispersed (hydrated rather than aggregated) compared to what might be expected from random mixing. Could you provide a physical interpretation as to why this is the case?

Professor Ben-Amotz responded: Our results suggest that the first hydration shell around hydrophobic groups dissolved in water is more stable than one might have expected. The level of theory required to accurately describe this stability remains to be determined, as it may require including multi-body (classical and/or quantum mechanical) contributions to properly predict the cooperative stabilization of coupled H-bond networks in hydrophobic hydration-shells. The key point is that our experiments point to the enhanced stability of water-separated hydrophobic contacts. This conclusion is also apparently consistent with Professor Molinero's finding that the formation water-separated hydrophobic contact aggregates in liquid water precedes the nucleation of solid clathrate hydrates.[1] In fact, the existence of clathrate hydrates in itself points to the stability of water separated hydrophobic contact structures.

1 L. C. Jacobson, W. Hujo and V. Molinero, J. Am. Chem. Soc., 2010, 132, 11806–11811.

Dr Henchman asked: Do you see a temperature dependence in your mixing data?

Professor Ben-Amotz answered: At high temperature we do see the onset of direct hydrophobic contacts at (slightly) lower TBA concentrations – the difference is relatively subtle. We expect to include our temperature dependent results on aqueous TBA solutions in a forthcoming publication.

Professor Chandler asked: Some of the discussion concerning the papers by Dougan and Ben-Amotz seem to overlook length-scale dependence of hydrophobic effects.[1,2] Specifically, water reorganizes to accommodate solutes in ways that depend upon the size and shape of the solutes. For small hydrophobic solutes, it does so with arrangements that do not break hydrogen bonds; for large hydrophobic solutes, it does so by breaking hydrogen bonds and forming soft water-vapor-like interfaces adjacent to the solute. The crossover length is about 1 nm. Soft interfaces present in the large-length regime lead to hydrophobic forces of assembly. In the absence of such interfaces, the small-length regime, there is no significant solvent induced attraction between oily species. Indeed, it is in the nature of hydrogen bonding geometries that some pairs of small hydrophobic species favor congurations with liquid water between the so-called solvent separated hydrophobic bond.[3] It is a structure about which Dr. Perkin questioned Professor Ben-Amotz. Hydrophobic interactions are not expected to stabilize contact pairs of TBA molecules because the TBA hydrophobic groups are too small to nucleate a soft water interface. On the other hand, hydrophobic interactions do collapse proteins in the initial steps of folding because several hydrophobic side chains can arrange in a large enough cluster to nucleate a soft interface, a cluster that is larger than 1 nm across.[4]

1 K. Lum, D. Chandler and J. D. Weeks, *J. Phys. Chem. B*, 1999, **103**, 4570–4577.
2 D. Chandler, *Nature*, 2005, **437**, 640–647.
3 L.R. Pratt and D. Chandler, *J. Chem. Phys.*, 1977, **67**, 3683–3704.
4 P. R TenWolde and D. Chandler, *Proc. Natl Acad. Sci. U. S. A.*, 2002, **99**, 6539–6543.

Professor Ben-Amotz answered: As you pointed out, our present hydrophobic interaction results pertain to the small size regime, rather than to the larger size regime in which your LCW theory[1,2] has predicted a crossover in both the structure of water and the interactions between hydrophobic groups. Although we have recently obtained experimental evidence of such a crossover in water structure around hydrocarbon chains longer than about 1 nm,[3] extending our present hydrophobic interaction studies to larger hydrophobic groups is challenging because of the low solubility of such amphiphiles. However, our finding that the initially formed TBA aggregates contain more than two TBA molecules, each of which remain highly hydrated, are remarkably consistent with simulations of methane aggregation in water which point to the enhanced stability of water-separated aggregates containing several methane molecules.[4]

Your seminal theoretical work on hydrophobicity with Lawrence Pratt predicted that the mean force potential between two small hydrophobic particles (resembling methane) should contain both a direct and water-separated minima of comparable depth.[5,6] Those predictions have been born out by subsequent classical[7] and quantum[8] investigations of this same mean force potential. These theoretical studies have generally concluded that the potential of mean force minimum for the contact configuration is slightly lower (more negative) than that for the water-separated minimum (except at higher pressures, when the water-separated minimum is predicted to become more stable).[7] However, as you have pointed out,[9] it is also important to keep in mind that relating the potential of mean force to the relative populations of the direct and water-separated species requires integrating the hydrophobic particle pair distribution function over the volumes of configuration space associated with each species, and so favors the larger volume water-separated configurations. Thus, our experimental finding that there are fewer direct hydrophobic contact aggregates in aqueous TBA solutions than in the corresponding random mixture is qualitatively consistent with your predictions pertaining to water-separated and direct contact populations for hard-spheres (the size of methane) dissolved either in water or a in a hard-sphere fluid (of the same number density and core diameter as water).[9]

1 K. Lum, D. Chandler and J. D. Weeks, *J. Phys. Chem. B*, 1999, **103**, 4570–4577.
2 D. Chandler, *Nature*, 2005, **437**, 640–647.
3 J. G. Davis, K. P. Gierszal, P. Wang and D. Ben-Amotz, *Nature,* 2012, **491**, 582–585.
4 L. C. Jacobson, W. Hujo and V. Molinero, *J. Am. Chem. Soc.*, 2010, **132**, 11806–11811; M. Matsumoto, *J. Phys. Chem. Lett.*, 2010, **1**, 1552–1556.
5 L. R. Pratt and D. Chandler, *J. Chem. Phys.*, 1977, **6**, 3683–3704.
6 L. R. Pratt and D. Chandler, *J. Chem. Phys.*, 1980, **73**, 3434–3441.
7 G. Hummer, S. Garde, A. E. Garcia, M. E. Paulaitis and L. R. Pratt, *J. Phys. Chem. B*, 1998, **102**, 10469–10482.
8 J. L. Li, R. Car, C. Tang and N. S. Wingreen, *Proc. Natl. Acad. Sci. U. S. A.*, 2007, **104**, 2626–2630.
9 L. R. Pratt and D. Chandler, *J. Solut. Chem.*, 1980, **9**, 1–17.

Professor Meuwly continued the discussion of the paper by Dr Dougan: In your contribution you "postulate that the action of glycerol is to drive water toward its structure at high pressure...". This seems to be based on comparing how $g(r)$ changes a) if small amounts of trehalose, sorbitol, *etc.* are added or b) upon increasing pressure in pure water. Is the similarity of $g(r)$ really sufficient for such a postulate and can you detail why you would believe that this is the case?

Professor Wang continued: This is a follow up to Professor Meuwly's question. Since you mentioned in your paper that the addition of molecules, such as glycerol, have effects similar to increasing pressure. Have you tried to obtain the RDFs of glycerol water at different glycerol concentrations and of pure water at different pressures to see if there is a quantitative correlation?

Dr Soper answered: What is well known is that when you pressurise water the first neighbour shell remains largely intact, both in terms of near-neighbour distances and spatial arrangement of water around a central molecule.[1] On the other hand the second shell, as signified by the second peak in the oxygen–oxygen radial distribution function, either moves inwards under pressure, or is replaced by a second shell which is closer in and which clearly overlaps with the first shell.[1,2] Given that this movement of the second shell implies at least bond bending between first and second shell molecules, if not actual bond breaking, it is natural to think of this effect of pressure as breaking, or tending to break, the local water network, even when the local, near-tetrahedral, environment is preserved.

However, if one can imagine putting water under tension, the corollary of this movement, *i.e.* the second peak moving to larger radii, will not happen because the second peak is already at its tetrahedral position: if it were to expand further, the tetrahedral network would be destroyed completely. All that would happen is that the second peak would become sharper and better defined as the system readies itself to transform to hexagonal ice.

When some ions are dissolved in water something very analogous appears to happen, but the size of the effect depends on the charge and sizes of the cation and anions involved. Thus Na_2SO_4 has a big effect on water structure while NH_4Cl has only a marginal effect.[3] In this case the effect is caused by the ions drawing in or pushing away the water molecules, bending the H-bonds, and so having an effect similar to pressure. In other work[4] this second peak movement has been used to assign an equivalent pressure to the ionic solution (although of course the actual pressure does not change). The concept may be useful as it gives a macroscopic feel for the effect of ions on water structure at the molecular level.

The effect of non-ionic solutes is much less clear-cut and variable. In the present case (and also in the case of other alcohols in water as referred to in the paper) the initial observation is that at low alcohol concentrations the second peak in the Ow–Ow distribution moves inwards as occurs when pure water is under pressure – so one can think of this as a "pressurising" effect on water structure, and the amount of "pressure" can be gauged from the degree of movement of the peak compared to that observed in the pure liquid. At higher concentrations however the peak moves out again, eventually going beyond the ~4.5 Å expected for tetrahedral order. Hence there is no equivalent pressure in this case, and the concept has little meaning. To try to read more into these results than that stated in the paper seems to us to be taking the idea of an equivalent pressure beyond the point where it is useful. Presumably at these higher concentrations there is much reduced possibility for water–water hydrogen bonds – the water network becomes largely destroyed, so that water behaves more like a simple liquid with many non-bonded molecules, perhaps?

1 A. K. Soper and M. A. Ricci, *Phys. Rev. Lett.*, 2000, **84**, 2881–2884.
2 A. V. Okhulkov, Yu. N. Damianets and Yu. E. Gorbaty, *J. Chem. Phys.*, 1994, **100**, 1578.
3 R. Leberman and A. K. Soper, *Nature*, 1995, **378**, 364–366.
4 A. Botti, F. Bruni, S. Imberti, M. A. Ricci and A. K. Soper, *J. Chem. Phys.*, 2004, **120**, 10154–10162.

Dr Vila Verde asked: You are trying to correlate the changes in the position of the second peak of the RDF at 298 K with changes in freezing temperature of these solutions. It would be useful to repeat your experiments near the freezing temperature at each concentration.

Dr Soper answered: Thank you for raising this important point. We have conducted the neutron diffraction experiments that you refer to and are currently investigating the relevant RDFs and clustering that is found. We expect to report on these new results in the near future.

Dr Vila Verde addressed Mr Towey and Dr Soper :Can you tell us the size of your simulation box? For how many steps did you run the simulation for the final potential?

The reason I ask is that we know that the dynamics of cluster formation and concentration fluctuations are very slow for length-scales larger than 10 Å and, since you fit your simulation potentials to the experimental data also for $L > 10$ Å, it is important to demonstrate that your simulation box is large enough and that the simulations are well-sampled. I note that even if you didn't actually sample these length-scales very well, your results still hold because they depend on the behavior of the $G(r)$ at smaller length scales, which are easier to sample. It would be very useful, however, to publish potentials that lead to the experimentally-measured behavior of solutions at $L > 10$ Å because they could then be used to investigate cluster formation and density fluctuations at these larger lengths-scales, and would likely be more reliable than existing potentials.

Mr Towey answered: The information regarding the size of the simulation boxes can be found in the supporting information of a previously published article.[1] Each of the water–glycerol simulation boxes contains around 14 000 atoms with the number of molecules ranging from 1000 for pure glycerol to 4000 for the most dilute systems. The atomic number density for these systems varies between 0.103 to 0.115 atoms $Å^{-3}$. This leads to cubic simulation boxes with a side length of around 50 Å. The use of an empirical potential drives the simulation to agreement with the experimental data. This yields an ensemble of molecular configurations that are compatible with the neutron diffraction experiments. The average number of configurations that have been sampled for each water–glycerol concentration is over 5000. As far as we can ascertain, running the simulation for longer does not change the overall picture of micro-segregation, suggesting that these simulations are well equilibrated.

The relevance of using the EPSR potential for other simulations where structure refinement is not the primary goal remains an open question. It must be remembered these are purely pair-wise additive, empirical potentials (EP), derived from the scattering data: they have no physical basis in themselves, other than being able to define the distribution of atoms in the simulation box in a manner to reproduce the scattering data. Even the dynamics implied by these potentials is not known, though it would certainly be interesting to find out. Given the complex, probably many-body, forces that are likely to be acting in these systems it is highly doubtful that they can be used for other purposes, and certainly not at thermodynamic state points different from the ones at which they are derived. Note that the potentials used in these simulations are readily available in numerical form from the authors.

1 J. J. Towey, A. K. Soper and L. Dougan, Molecular insight into the hydrogen bonding and micro-segregation of a cryoprotectant molecule, *J. Phys. Chem. B*, 2012, **116**(47), 13898.

Professor Angell said: You refer to the large scattering in certain cases of binary solutions. If this scattering is due to concentration fluctuations, then the value at zero q vector (difficult to obtain) gives information on the thermodynamics of the solution, *via* the Bhatia–Thornton theory. This has often been applied in neutron scattering studies of unmixing liquids, near the consolute point.[1,2] Is there any chance of making this connection in your case, do you think?

1 P. Chieux and P. Damay, On the long wavelength thermodynamic limit of a neutron scattering experiment in the vicinity of a liquid–liquid critical point. Application to the concentration fluctuations in the Li-ND3 system, *Chem. Phys. Lett.*, 1978, **58**, 619–621.
2 G. Chabrier, J. F. Jal, P. Chieux and J. Dupuy, A neutron scattering investigation of the structural order of Rb-RbBr solutions, *Phys. Lett. A*, 1982, **93**, 47–51.

Dr Soper answered: This is an interesting idea, but we are not sure as to its practicality. Currently Bhatia–Thornton theory applies strictly to two component systems, though an extension to calculate the concentration–concentration correlation function for three-component systems has been proposed[1] and used in some recent work.[2] For glycerol–water, treating all C, H and O atoms on the glycerol molecule as the same (which is not strictly valid since the hydroxyl hydrogen atoms behave quite differently from the methyl hydrogen atoms) we already have a five component system. Alternatively, treating this simply as a two component mixture of water and glycerol, there is a need to deal with the orientational correlations, which also do not feature in Bhatia–Thornton. What is also clear is that although we see clustering in the simulations, there is not the slightest hint of enhanced, critical-type, scattering towards $Q = 0$ in either the data or the simulations. The same was true of the previous work on methanol in water.[3] Hence one's suspicion is that the concentration–concentration term, if it could be calculated, would not show anything of great interest - these materials are not de-mixed at larger length scales.

Contrast this with what is seen in *tert*-butanol, where there is a large increase in scattering towards $Q = 0$ - see the paper by Anisimov *et al.*[4] in this Discussion.

1 D. Gazillo, Stability of fluids with more than two components I. General thermodynamic theory and concentration–concentration structure factor, *Mol. Phys.*, 1994, **83**, 1171–1190.
2 A. K. Soper, *J. Phys. Condens. Matter*, 2010, **22**, 404210.
3 L. Dougan, S. P. Bates, R. Hargreaves, J. P. Fox, J. Crain, J. L. Finney, V. Reat and A K Soper, *J. Chem. Phys.*, 2004, **121**, 6456.
4 D. Subramanian, C. T. Boughter, J. B. Klauda, B. Hammoud and M. A. Anisimov, *Faraday Discuss.*, 2013, **167**, DOI: 10.1039/C3FD00070B.

Professor Chandler remarked: Dr. Soper calls our attention to the significant sensitivity of molecular simulation results to the choice of intermolecular potential model. While I agree with this point as far as it goes, it reflects a focus on one particular thermodynamic state. In a broader perspective, most reasonable models behave similarly when examined at the same corresponding state. Demonstrating this fact requires examining each model over a broad range of conditions. This, of course, is a principal point of the Faraday Discussion paper by Limmer and myself.[1]

1 D. T. Limmer and D. Chandler, *Faraday Discuss.*, 2013, **167**, DOI: 10.1039/C3FD00076A.

Professor Molinero continued the discussion of the paper by Prof. Ben-Amotz: You emphasized that the effect of water is to pull the TBA apart, not to bring them together. Molecular simulations of hydrophobic attraction of methane molecules in water reveal that solvent-separated pair configurations are cooperatively favored with respect to contact pair configurations by the clustering of four or more methanes.[1,2] Could you comment on whether the effect you are reporting arises from pairs of TBA or larger clusters?

1 M. Matsumoto, Four-body cooperativity in hydrophobic association of methane. *J. Phys. Chem. Lett.*, 2010, **1**, 1552–1556.
2 L. C. Jacobson, W. Hujo and V. Molinero, Nucleation pathways of clathrate hydrates: effect of guest size and solubility. *J. Phys. Chem. B*, 2010, **114**, 13796–13807.

Professor Ben-Amotz replied: Our results are entirely consistent with what you have suggested, as the results shown in Fig. 4B and 5B of our paper imply that the non-linearity of the concentration dependence of the aggregate population is significantly stronger than would be expected for a dimerization process. Furthermore, we have tried to fit the observed non-linear concentration dependence to higher order aggregation equilibrium expressions and found that the observed non-linearity is approximately consistent with what one would expect if the initially formed (highly hydrated) aggregates each contain approximately four TBA molecules.

Mr Limmer commented: Following the discussion of the percolation and domain formation in aqueous solutions, and their role in cryoprotection I have a comment. In these disordered systems, while there may be a characteristic length scale for domains, that length scale is only the mean in a distribution that in the thermodynamic limit encompasses a broad range of sizes. Therefore attempts to associate the mechanism for cryoprotection to bounding a single size of a single critical nucleus at this characteristic length scale is misguided. Indeed, by simply adapting classical nucleation theory with an additional osmotic pressure term from an ideal solution, Warkentin and coworkers have shown that the critical cooling rates of a wide variety of solutions, across large concentration regimes can be predicted.[1]This simple theory works even though these solutions may percolate at some concentration or even microscopic domains.

1 M. Warkentin, J. P. Sethna, and R. E. Thorne, Critical Droplet Theory Explains the Glass Formability of Aqueous Solutions, *Phys. Rev. Lett.*, 2013, **110**, 015703.

Professor Wang said: Have you measured TBA aggregation at higher temperatures to see if you get closer to random mixing? If so, do you see a gradual transition as you increase the temperature or is there a sharp transition at a certain temperature indicating a possible phase-transition like behavior?

Professor Ben-Amotz responded: Preliminary experiments up to 70 °C show evidence of more direct contact aggregates than at 20 °C. We have not found any evidence of a lower consolute temperature in this system (above which liquid–liquid phase separation would occur).

Dr Sefcik asked: Results in your paper indicate that aggregation of TBA in aqueous solutions starts at concentrations around 0.5 M. Below this concentration TBA molecules appear to be fully hydrated monomers and thus not aggregated (in the sense as this term is used in the paper). However, is it possible that these fully hydrated TBA monomers are themselves associated in clusters in aqueous solutions, analogous to glycine,[1] alanine,[2] urea[3] or methylene glycol[4] mesoscale clusters observed in aqueous solutions? Have there been any experimental measurements done on TBA aqueous solutions to study such clustering effects?

1 A. Jawor-Baczynska, J. Sefcik and B. D. Moore, 250 nm Glycine-Rich Nanodroplets Are Formed on Dissolution of Glycine Crystals But Are Too Small To Provide Productive Nucleation Sites, *Cryst. Growth Des.*, 2013, **13**, 470–478.
2 A. Jawor-Baczynska, B. D. Moore, H. S. Lee, A. V. McCormick and J. Sefcik, Population and size distribution of solute-rich mesospecies within mesostructured aqueous amino acid solutions, *Faraday Discuss.*, 2013, **167**, DOI: 10.1039/C3FD00066D.
3 A. J. Alexander, M. R. Ward, A. D. Ward and S. W. Botchway, Second-harmonic scattering in aqueous urea solutions: evidence for solute clusters?, *Faraday Discuss.*, 2013, **167**, DOI: 10.1039/C3FD00089C.

4 K. Z. Gaca, J. A. Parkinson and J. Sefcik, Molecular Speciation and Mesoscale Clustering in Formaldehyde–Methanol–Water Solutions in the Presence of Sodium Carbonate, *J. Phys. Chem. B*, 2013, **117**, 10548–10555.

Professor Ben-Amotz replied: Our Raman-MCR evidence is consistent with our conclusion that below 0.5 M the hydration-shell of TBA is concentration independent (and we have performed experiments down to 0.1 M). This implies that if there were any sort of aggregation of TBA below 0.5 M then the corresponding aggregates would have to have a structure in which there are at least two water layers between each pair of TBA molecules. In other words, such aggregates would very closely resemble a collection of fully hydrated TBA monomers, and so would probably not be detectable using Raman-MCR technique.

Professor Walker commented: The non-ideal behavior observed in the Raman spectra is assigned to aggregation of two or more TBA solutes. Previous studies reported the onset of aggregation at relatively high concentrations (~1.3 M) but the data reported in this paper show evidence of aggregation at concentrations as low as 0.5 M. Shouldn't this low concentration behavior show up in macroscopic measures of TBA nonideality? Have the activity coefficients of TBA in H_2O been measured and reported in the literature? If so, do the activity coefficients correlate with the reported Raman spectroscopic data and if not, can the authors propose a reason(s) for the discrepancy?

Professor Ben-Amotz responded: This is of course an important issue. There is much previous work on aqueous TBA that we would like to link to our new Raman-MCR measurements. We do hope to include some such comparison in a forthcoming publication. For example, we have performed densitometry and sound velocity measurements, which provide information about the partial molar volume and compressibility of TBA in water. Our results are consistent with our conclusion that subtle changes in the hydration shell structure of TBA begin below 1 M.

Dr Vila Verde asked: What is the minimum lifetime of the aggregates that you would be able to detect?

Professor Ben-Amotz replied: Our spectra contain features arising from species that are present in an equilibrium mixture. There is no lifetime information in the spectra, except that imposed by lifetime broadening (or dephasing), which implies that very short lived species would necessarily produce very broad spectral features. The width of the sharpest (dangling OH) features that we observe imply that the species we have detected each live longer than about 0.2 ps (and perhaps much longer).

Dr Vila Verde opened the discussion of the paper by Professor Drummond: It would be useful if you could investigate the free energy of solvation of small hydrophobic and hydrophilic solutes of different sizes in your solvents, and look at the onset of aggregation for these solutes. This information might be useful to then rationalize the solubility and aggregation behavior of more complex solutes in the same solvents. From a fundamental perspective, it would also be interesting to relate the relationship between the free energy of solvation of your small solutes with the magnitude of density fluctuations of your solvents and their propensity to form molecular clusters of a given size and lifetime.

Professor Drummond replied: We agree that this would be a useful next set of studies. It would be good to obtain these parameters experimentally and then compare them with the results of computational modelling approaches to continue to build our understanding of the link between liquid structure and amphiphile self-assembly.

Professor Angell asked: What would be the effect on your observations if you had a hydroxy group on the end of the sidechain?

Professor Drummond responded: Based on structure–property comparisons obtained to date with both molecular solvents and ionic liquids we would anticipate that the mesostructure of the molecular solvents would be mitigated by the presence of hydroxyls on the side chains, at least for side chains with small hydrocarbon unit lengths.

Dr Sefcik opened the discussion of the paper by Professor Anisimov: In your paper you stated that in order to form mesoscale heterogeneities in aqueous solutions of nonionic hydrotropes, the presence of a third, more hydrophobic, component is needed. This is demonstrated by disappearance of the slow decay mode in the DLS autocorrelation function after filtration of TBA aqueous solutions through 20 nm Anotop filters. We have observed similar phenomena after filtration of glycine[1] and alanine[2] aqueous solutions. However, in both of these cases mesoscale clusters reappeared after prolonged stirring times at similar cluster sizes and number concentrations as before filtration, indicating that there is a significant activation energy barrier associated with their formation. Could there possible be something similar happening in TBA aqueous solutions even without a third hydrophobic component? This is of course not in any contradiction with the concept of mesoscale solubilisation as discussed in the paper.

1 A. Jawor-Baczynska, J. Sefcik and B. D. Moore, 250 nm Glycine-Rich Nanodroplets Are Formed on Dissolution of Glycine Crystals But Are Too Small To Provide Productive Nucleation Sites, *Cryst. Growth Des.*, 2013, **13**, 470–478.
2 A. Jawor-Baczynska, B. D. Moore, H. S. Lee, A. V. McCormick and J. Sefcik, Population and size distribution of solute-rich mesospecies within mesostructured aqueous amino acid solutions, *Faraday Discuss.*, 2013, **167**, DOI: 10.1039/C3FD00066D

Professor Anisimov replied: We have not observed mesoscale inhomogeneities in TBA aqueous solutions after removal of a hydrophobic impurity. We believe that in this particular case the thermodynamically stable mesoscale aggregates exist only in the presence of a hydrophobe. However, it does not mean that such aggregates are not possible in other hydrotropic systems without a hydrophobe, especially if ions are present or if hydroptope molecules have a more complex structure (such as low molecular weight PEG).

Professor Ben-Amotz asked: You refer to small clusters of TBA as "micelle-like", and yet you allude to the fact that these structures are very short lived and remain highly hydrated. I think it would be useful to have a more precise definition of what we mean by terms such as "micelle-like". More specifically, perhaps we should consider restricting the use of the term "micelle" or "micelle-like" to assemblies that not only have a hydrophobic core and hydrophilic surface, but also have a relatively narrow aggregate size distribution, whose width is no larger than its mean, and a relatively long lifetime, relative to the time scale of concentration fluctuations in the corresponding non-aggregated aqueous solution (on a length scale comparable to the mean aggregate size).

Professor Anisimov replied: I agree with your comment. "Micellar-like" is just a slang term that was used to emphasize the fundamental difference between the conventional fluctuations of concentration and loose short-lived molecular clusters in which a hydrophobic part tends to be screened from water molecules by hydroxyl–water hydrogen bonds.

Dr Drummond remarked: Professor Anisimov, it should be possible to measure the Langmuir adsorption isotherm for TBA at the oil–water interface by using the spinning drop technique with TBA serving as a surface active species. This technique would allow the adsorption density as a function of TBA concentration to be determined and help answer a previous question with respect to what is the surface density of the TBA species.

Professor Anisimov responded: It would be useful to measure the interfacial tension between oil-rich and water-rich phases by the spinning-drop technique. Especially interesting would be monitoring the surface activity in real time because of long-time equilibration. However, since the drop will significantly larger than the size of the meso droplets (~100 nm), one will measure the "macroscopic" interfacial tension, and not an effective mesoscale tension which is more relevant to the mesoscale inhomogeneities.

Professor Ben-Amotz asked: I wonder if it would be appropriate to think about the mesoscopic cyclohexane droplets which you observe as analogous to surfactant stabilized oil drops in water, with TBA playing the role of the surfactant? If so, then I wonder if the size of the large aggregates in the TBA–cyclohexane–water system is determined by the optimal density of TBA on the surface of the hydrophobic drops? In other words, if the drops were significantly smaller or larger than the optimal size, then the number of TBA molecules in the system would be either too low or too high to support the optimal TBA surface density.

Professor Anisimov answered: I basically agree. One needs to have certain number of TBA molecules in the intermediate layer to guarantee sufficient osmotic-pressure repulsion between droplets, which would prevent coalescence of the aggregates into larger droplets.

Professor Wang asked: The mesoscale inhomogeneity in the ternary system is very fascinating. When you prepare the ternary system, how do you mix in cyclohexane? Is it possible that the 100 nm scale structure is caused by the mixing procedure?

Professor Anisimov replied: This is correct. The mixing procedure does affect the formation of the mesoscale aggregates. They emerge either very fast or with a significant delay, depending on how we add a drop of cycloxhexane and shake the vial. However, they always present if an oily compound is added.

Professor Mallamace commented: From your data can it be verified if the internal structure of these mesoscale inhomogeneities are homogeneous or not? In some cases amphiphilic water mixtures in the low concentration regime near the "critical micellar concentrations" are just self similar, whereas in the micellar phase they are compact and homogeneous. Please can you comment on that?

Professor Anisimov answered: We speculate that the hydrophobic core of these mesoscale aggregates is mesoscopically homogeneous, while the surface layer is inhomogeneous with a significant gradient of the TBA concentration.

Professor Angell commented: Wouldn't it be interesting to extend the measurements of heat capacity to below 0 °C? We know that heat capacity of water itself becomes very striking at sub-zero temperatures. Speedy[1] showed that the T_s of his power law for conductivity moved to higher values when the water contained a clathrate-forming salt. It implies that certain impurities can make response functions even more dramatic than observed to date.

1 L. H. I. U. Mettananda and R. J. Speedy, Supercooled aqueous tetra-n-butylammonium bromide solutions: conductivity and differential thermal analysis studies, *J. Phys. Chem.*, 1984, **88**, 4163–4166.

Professor Anisimov replied: I fully agree with this comment. Heat capacity measurements of supercolled aqueous solutions could also shed the light on the highly debated water polyamorphism.

Dr Sweatman remarked: We have a poster here that might help explain the physics of some of these aggregating systems. The poster shows it is possible to have micelle-like behaviour, *i.e.* thermodynamically stable clustering, in a pure fluid of spherical particles (a simple fluid) with competing short-range attractive and long-range repulsive interactions. The long-range interactions can be very small in magnitude, but because of their long-range it is possible that they can balance the stronger short-range interactions, and clustering is then expected.

How do you get these effective attractions/repulsions at long range (~10 nm)? There could be a range of mechanisms for generating these long-ranged repulsive interactions. A screened Coulomb interaction is one possibility, another is the long-ranged steric repulsions of well solvated polymer tethered to nanoparticles. The theory takes as input an effective interaction between solutes – the mechanism does not matter within the theory. We have simulation results that validate the theory – essentially, clustering is expected for these kinds of system. Accepting this result, one should now search for mechanisms in each system of interest that generate these kinds of competing effective interactions.

Professor Anisimov replied: Yes, weak long-range repulsion is a possible explanation of the stability of these mesoscale droplets. As these are non-ionic hydrotropes, the repulsion is likely associated with osmotic pressure caused by excess of TBA concentration in the surface layer.

Professor Caupin enquired: Would you tell more about the temperature dependence of the mesoscale inhomogeneities (MIs)? You mention that "as the temperature is increased, this slow mode disappears and reappears as the temperature is lowered". Have you measured the hydrodynamic radius of the MIs with DLS as a function of temperature?

Professor Anisimov responded: The size of the mesoscopic inhomogeneities does not much depend on temperature under the same rate of cooling or heating. The emerging size is usually smaller for fast cooling from 50 °C (no inhomogeneities are observed) to 8–10 °C than for slow cooling, but still of the same order of magnitude. However, the number of the inhomogeneities dramatically increases upon cooling.

Professor Caupin asked: Would you also explain why you are using *cold* filtration to remove the MIs?

Professor Anisimov answered: We believe that the origin of the 100 nm inhomogeneities in TBA–water–cyclohexane solutions is the mesoscale solubilization of cyclohexane by TBA–water clusters. Removing these inhomogeneities at a low temperature by filtration through a 20 nm filter makes the solution practically free of cyclohexane.

Dr Perera asked: In your talk you focus on lifetime of clusters. While it is undeniably important to understand cluster formation in these systems, I would like to point out that even a simple Lennard-Jones (LJ) liquid has inherent cluster formation, depending on how one defines them. Studies of LJ clusters have been conducted in the past.[1,2] Similarly, one could study cluster formation in supercooled liquids. However, there is a fundamental difference between these clusters and those you report in your work: your clusters have an unmistakable static signature in the form of a pre-peak in the partial structure factors $S(q)$, while the other types of clusters do not give such a signature in $S(q)$. In view of this fact, how would you characterize the specificity of the lifetime of your clusters, *versus* the life time of LJ clusters?

1 A. Coniglio and W. Klein, *J. Phys. Math. Gen.*, 1980, **13**, 2775.
2 N. Sator, *Phys. Rep.*, 2003, **376**, 1.

Professor Anisimov answered: I feel there could be some confusion with the terminology. As far as I understand, clustering in LJ fluids are associated with fluctuations of density/concentration. These fluctuations can be characterized by the correlation length which grows from the molecular scale far away from the critical point to a mesoscale in the vicinity of the critical point. Such fluctuational clusters form and dissipate by diffusion dynamics. The correlation length can be obtained by SANS (structure factor) or static light scattering (if the correlation length is large). The diffusion relaxation time can be obtained by dynamic light scattering. In some aqueous solutions of hydrotropes, as evident from MD simulations, there is an alternative kind of molecular clustering: association of water and hydrotrope molecules by strong hydrogen bonds with the formation of loose, dynamic intermolecular structures. Such clusters form and dissipate by a specific rearrangement of hydrogen bonds. Their life time is a thousand times shorter than the relaxation time of conventional concentration fluctuations and thus cannot be detected by dynamic light scattering. Moreover, there is no observable pre-peak in the SANS structure factor associated with these clusters. However, these clusterings cause observable thermodynamic anomalies, such as the heat capacity maxima and the negative excess chemical potential of water. A strong pre-peak in the SANS structure factor at very small wavenumbers and strong light scattering emerge in hydrotrope solutions only upon addition of a hydrophobic impurity (such as cyclohexane). This is still an open question: how could the dynamic structural clustering in binary hydrotrope–water solutions cause the formation of stable mesoscale inhomogeneities in ternary hydrotrope–water–hydrophobe solutions?

Dr Perera remarked: Professor Anisimov, in Fig. 7 of your paper, you report a scattered intensity $I(q)$ that shows a sharp raise in the $q \rightarrow 0$ limit, followed by a plateau at $q = 0.03$ approximatively. The claim that this plateau corresponds to the Ornstein–Zernike (OZ) regime is unsupported by this $I(q)$, since the OZ limit concerns really the region near $q = 0$, hence the sharp peak at $q = 0$. If we admit that both features can be described by OZ Lorentzians such as $A/(q^2 + b^2)$, then the sum of these two contributions leads to a denominator of the Teubner–Strey form $(q^4 + u \times q^2 + v)$, which indicates that the plateau is in fact a pre-peak feature. This pre-peak would correspond to aggregated domain size of 200 Å, consistent with what is reported. Would you agree that, within this interpretation, the small-q raise indicate strong concentration fluctuations of the system upon cluster formation?

Professor Anisimov responded: A strong pre-peak in the SANS structure factor at very small wavenumbers and strong light scattering emerge only upon addition of a hydrophobic impurity (such as cyclohexane). This is still an open question: how could the dynamic structural clustering in binary hydrotrope–water solutions cause the formation of mesoscale inhomogeneities in ternary solutions hydrotrope–water–hydrophobe?

I agree that the structure factor with a sharp peak and a plateau can be described by two OZ Lorentzians. However, to fully understand and correctly interpret this picture, we also need to look at the structure factor of the original binary TBA–water solution and at the results of static and dynamic light scattering in the same system. Without a hydrophobe (cyclohexane) there is no sharp peak in SANS. The remaining “plateau” is consistent with the correlation length of concentration fluctuation of an about 0.5 nm correlation length. On the other hand, upon addition of a hydrophobe, the dynamic light scattering clearly shows the Brownian diffusion of emerging mesoscale droplets (about 100 nm in radius), while the short-range concentration fluctuations remain the same. The fact that the mesoscale droplets are not conventional fluctuations of concentration is also confirmed by confocal microscopy images.

Dr Perera addressed Professor Anisimov and Professor Ben-Amotz: Prof. Ben-Amoz addressed the question of the nature of the micelle-like features reported in the paper of Prof. Anisimov, as compared to what micelles in micro-emulsions are. I would like to point out that, following Prof. Ben-Amoz findings in his paper, the solute clusters of small molecules, such as both authors report independently, are likely to be strongly hydrated. This is consistent with both Prof. Ben-Amoz's findings and those in computer simulations.[1,2] I propose here that the difference between these micelle-like structures, with small solute molecules, and those that are generically considered as micelles, with larger solute molecules, relies on water being expelled from these latter structures. In other words, small micelles would be strongly hydrated while larger ones would be dry to a large extent. Both can be called micelles because they are formed under the same underlying mechanism of water segregating solutes, which is at the origin of micro-heterogeneity in these mixtures.

1 B. Kežić and A. Perera, *J. Chem. Phys.*, 2012, **137**, 014501.
2 R. Gupta and G. N. Patey, *J. Chem. Phys.*, 2012, **137**, 034509.

Professor Ben-Amotz replied: Please see my earlier question addressed to Professor Anisimov, regarding a way in which the term "micelle-like" might be more narrowly defined.

Professor Anisimov added: Your comment on hydration is interesting. However, there is a distinct dynamic difference between the conventional micelles formed by surfactants and "micelle-like" clustering in solutions of non-ionic hydrotropes. Micelles are thermodynamiically stable objects. "Micelle-like" clusters are fluctuations of structure with extremely fast dynamics.

Professor Molinero commented: Your work shows that the formation of mesoscopic aggregates in TBA–water solutions with small amount of organic molecules occurs in the one-phase region, but very close to the boundary of phase segregation of the ternary mixture. Do you expect formation of mesoscopic aggregates in all ternary systems with similar phase diagrams even when the interactions are very disparate from those in the case you studied (*e.g.* metals) or you expect the formation of mesoscale aggregates to be related to the specific nature of water and water-mediated interactions?

Professor Anisimov replied: At least in TBA–water–cyclohexane solutions, mesoscale solubilization occurs only in a specific range of TBA/water ratio, namely around a few mol % TBA. This fact is probably related to the specific nature of water–hydrophobe hydrogen bonding.

Professor Walker remarked: The mesoscopic droplets formed in the ternary solutions of water, hydrotrope and hydrophobe were proposed in the paper as being candidates for new drug delivery systems. An important criterion for drug delivery is that one must be able to load the delivery system with the drug to be delivered.

Is there any evidence (or reason to believe) that pharmaceutical agents (with a distribution of hydrogen bonding and hydrophobic properties) will be solubilized and retained in the mesoscopic droplets described in this work?

Professor Anisimov answered: So far, there has been no experimental study of this important problem.

Dr Drummond remarked: Professor Anisimov, ideally for the application of the proposed mesoscale objects as drug delivery vehicles you need the objects to be either thermodynamically or kinetically stable in excess water. The body is essentially a vast reservoir of water, and you could anticipate that drug delivery vehicles would need to be stable to essentially infinite dilution in order to avoid disassembly on placing in the body reservoir. Is this the situation with these mesoscale objects?

Professor Anisimov answered: Yes, I believe this is the situation with mesoscale inhomogeneities in aqueous solutions of tertiary butanol upon addition of a hydrophobe (such as cyclohexane). The inhomogeneities are "thermodynamically" stable at certain concentration of alcohol and long-lived (at least many hours) upon dilution. The number of the inhomogeneities decreases but their size remains practically unchanged.

Mr Shephard asked: Have you investigated the diffusion of molecules from solution into the hydrotope–water shells of your stable mesoscopic droplets (Fig. 11 of the paper)? To put it another way, at what rate are water and hydrotope molecules in the shells replaced with molecules from the surrounding solution? As you are able to separate the droplets from solution by filtration? An isotopic labeling procedure might enable you to measure the diffusion rate.

Professor Anisimov replied: We have not investigated this issue, but of course it would be interesting.

Professor Angell commented: While we are talking about composition fluctuations stabilized by solvent, what is your opinion about Bubstons (nanobubbles stabilized by ions), as described by Bunkin?[1] Do they exist? They could be the explanation for the new "no-man's land" described by Caupin[2] – an unrecognized universal factor that enables cavitation when the tension on the water is increased beyond 30 MPa.

1 N. F. Bunkin, A. V. Kochergin, A. V. Lobeyev, B. W. Ninham and O. I. Vinogradova, Existence of charged submicrobubble clusters in polar liquids as revealed by correlation between optical cavitation and electrical conductivity, *Colloids Surf., A*, 1996, **110**, 207–212.
2 M. E. Azouzi, C. Ramboz, J.-F. Lenain and F. Caupin, A coherent picture of water at extreme negative pressure, *Nat. Phys.*, 2012, **9**, 38–41.

Professor Anisimov replied: Existence of nanobubles stabilized by ions is quite possible. We have saturated TBA aqueous solutions by nitrogen and observed nanobubbles. They were not thermodynamically stable. However, they could probably be stabilized by ions or surfactants.

Mr Wexler returned to the discussion of the paper by Dr Dougan and communicated: Cryoprotectant induction of pressure and temperature like trends in the gOwOw(r) along with the findings on bi-percolating clusters sheds significant light on the sensitivity and mutability of the aqueous hydrogen bonding environment. Was it also possible to extract *g*OwHw(*r*) and *g*HwHw(*r*) and if so how does this information complete the picture specifically with respect to pressure distorted H-bonds?

Dr Soper communicated in response: It is possible to extract all of the pairwise interactions from the EPSR analysis of the neutron diffraction experiments. This is a key factor when using this technique, although it must be remembered these are only estimates based on the EPSR modelling procedure, since there are not enough isotope contrasts in this case to uniquely identify all the relevant site–site correlation functions. Interestingly, it has been shown that changes in pressure of pure water[1] and changes in glycerol concentration[2] do not initiate large changes in the *g*OwHw(*r*) and *g*HwHw(*r*) radial distribution functions. This suggests that it is not the hydrogen bonds themselves that are modified but rather the structure of the hydrogen bond network. That is, the angle formed by three oxygen atoms joined by two hydrogen bonds is altered readily with pressure or concentration, while the hydrogen bond length itself between neighbouring molecules changes only marginally.

1 A. K. Soper, The radial distribution functions of water and ice from 220 to 673 K and at pressures up to 400 MPa, *Chem. Phys.*, 2000, **258**(2–3), 121–137.
2 J. J. Towey, A. K. Soper and L. Dougan, Molecular insight into the hydrogen bonding and micro-segregation of a cryoprotectant molecule, *J. Phys. Chem. B*, 2012, **116**(47), 13898–13904.

Professor Walker continued the discussion of the paper by Professor Drummond and communicated: Do the amphiphiles reported in this work (Fig. 2) show surface activity when dissolved in the non-aqueous solvents shown in Fig. 1? Do the amphiphiles show typical monolayer formation tendencies and can surface excess concentrations be determined *via* traditional surface tension measurements?

Professor Drummond answered: In molecular solvents that support amphiphile self-assembly, amphiphiles also generally exhibit surface activity at the liquid-air surface. The difference between the liquid–air surface tensions for many of the neat molecular solvents and that of the surface with adsorbed amphiphile is less than that seen in aqueous systems. Nevertheless, we have used the du Nouy ring technique to measure amphiphile adsorption at the liquid–air surface in some of these systems and calculated surface excess concentrations.

Mr Towey continued the discussion of the paper by Professor Ben-Amotz and communicated: In your article you state that your results are "quite insensitive to the precise values of n and V". One might expect that the factors such as steric constraints would contribute to the relative proportions of solute/solvent molecules found in the solvation shell. Previously, steric effects on random mixing have been accounted for by using molecules constructed using only Lennard-Jones potentials.[1,2] The molecular clustering statistics are then compared to those found in a realistic model of the system.

Could you please expand on your statement and what effect steric constraints have on the proportion of molecules found in the solvation shell?

1 J. J. Towey, A. K. Soper and L. Dougan, Preference for isolated water molecules in a concentrated glycerol-water mixture, *J. Phys. Chem. B*, 2011, **115**, 7799.
2 J. J. Towey and L. Dougan, Structural examination of the impact of glycerol on water structure, *J. Phys. Chem. B*, 2012, **116**, 1633.

Professor Ben-Amotz replied: When you say steric constraints, I assume you mean the sort of packing effects that are also present in hard-sphere (or hard-body) fluids. Such effects certainly do lead to structuring as evidenced, for example, in pair distribution functions that are greater than one in the first coordination layer, and may contain subsequent maxima, indicative of longer range ordering. We have here defined an idealized random mixture as one for which the local concentration of a solute in any sub-volume of the system is identical to its bulk concentration. Thus, such idealized random mixture are not the same as hard-sphere fluids. More specifically, packing (or steric) effects in hard-sphere fluids lead to deviations from random mixing predictions. One might describe such deviations as a kind of aggregation, since it produces a first coordination shell concentration that is greater than the bulk concentration. Our Raman-MCR results obtained from aqueous TBA solutions imply that below a TBA concentration of 1 M, the local TBA concentration around each TBA is lower, not higher, than the bulk concentration. Thus, our experiments imply that the local concentration is not only below that which one would expect in a hard-body fluid, but also below that predicted for a hypothetical random mixture.

Since you began your comment be referring to the insensitivity of the random mixture predictions to the precise values of n and V, let me also try to be more clear about that. What we have found is that the predicted aggregate size distributions $P(k)$, obtained for a given total first-coordination shell volume V, are remarkably independent of the number of sub-volumes n into which we subdivided the volume V. However, the random mixture prediction for the average number of TBA molecules in the first coordination shell is linearly proportional to V, since $p = V[c]$, where [c] is the bulk TBA concentration. The value of V depends on how one chooses to define the extent of the coordination shell of interest. When making comparisons with experiments (or simulations) it is clearly important to try to define V in a way that is consistent with the volume that is probed experimentally (or in a simulation). In our case, we have assumed that the experimentally observed changes in the hydration-shell OH band TBA CH frequency all arise from the first coordination layer of TBA molecules around a given TBA, and so we have defined n and V in a way that is consistent with that assumption. Our assumption that there are no significant (measurable) water structure changes beyond the first hydration shell of TBA is also consistent with our observed concentration dependence, as perturbations beyond the first hydration shell would be expected to lead to an earlier onset of concentration dependent changes.

Faraday Discussions RSC Publishing

PAPER

Exploring the behaviour of the hydrated excess proton at hydrophobic interfaces

Revati Kumar,[ab] Chris Knight[ab] and Gregory A. Voth*[ab]

Received 13th May 2013, Accepted 24th June 2013
DOI: 10.1039/c3fd00087g

The affinity of the excess proton for the aqueous solution–hydrophobic interface was examined for two specific examples, the air–water and hydrophobic wall–water cases, using a multiconfigurational molecular dynamics algorithm. The use of a reactive simulation method is important as it allows for a realistic description of the excess proton, namely, its propensity to hop between water molecules *via* the Grotthuss mechanism. The free energy profile reveals a minimum at these interfaces due to a favourable enthalpic term that outweighs the entropic penalty. The key factors that contribute to this enthalpic minimum were examined using a generalization of a scheme that decomposes the interaction energy into separate terms arising from various local environments [Otten *et al.*, *Proc. Natl. Acad. Sci. USA*, 109, 701 (2012)] (coordination shell, bulk, and interface) and the delocalization energy (which allows the proton to hop). For both systems, it was observed that the energetic penalty for loss of coordinating water molecules as the excess proton moves toward the hydrophobic interface is more than compensated by the displacement of unfavourable interfacial water molecules. In addition, the ion becomes more delocalized, more Zundel-like, and therefore possesses a larger effective radius as it moves to the interface. The fluctuations of the instantaneous interface were reduced near the vicinity of the ion, thereby giving rise to an entropic penalty. This paper will discuss the application of energy decomposition schemes to multiconfigurational simulations and the resulting consequences realized for the excess proton at hydrophobic interfaces.

1. Introduction

The presence of hydrated excess protons at the air–water interface has been the subject of intense debate with a renewed enthusiasm towards the topic in recent years.[1–17] The general consensus, based on both experimental and computer simulation studies, is that common cations prefer to remain in the bulk while large "floppy" anions tend towards the interface.[5] However, vibrational sum

[a]Department of Chemistry, James Franck Institute, Computational Institute, The University of Chicago, 5735 South Ellis Avenue, Chicago, IL, USA. E-mail: gavoth@uchicago.edu

[b]Computing, Environment, and Life Sciences Directorate, Argonne National Laboratory, 9700 S. Cass Avenue, Argonne, IL, USA

frequency generation as well as resonance enhanced second harmonic generation spectroscopy, both surface sensitive experiments, suggest the presence of the excess protons and not hydroxide ions at these interfaces.[4,7,10] These latter results have been further validated by heterodyne-detected electronic sum frequency spectroscopy, which suggests that the pH of the surface is lower than that of the bulk by around 1.7.[15] This is to be contrasted with another set of experiments, electrophoretic experiments on water bubbles, which suggest the presence of a negative charge at this interface, possibly arising from the presence of the hydroxide ion.[18] In a recent study Colussi *et al.* looked at the deprotonation of gaseous carboxylic acid (RCOOH) on the air–water interface using electrospray ionization mass spectrometry experiments.[19] The results of their limited quantum calculations on water clusters suggested that the kinetic barrier for the deprotonation is very high unless OH^- was present. They thus concluded that the detection of $RCOO^-$ in the interfacial region was due to the presence of OH^- ions. Despite the general trends that several common ions seem to suggest, the deceptively simple hydrated proton and hydroxide ions appear to create their own set of controversies.

Computer simulations of the hydrated excess proton are complicated by the so called "Grotthuss" shuttling, wherein the proton hops between neighbouring water molecules.[20,21] Conventional molecular dynamics (MD) force fields assume a constant chemical bonding topology throughout a simulation and do not allow for changes in bond topology, which automatically preclude their use in modelling proton transfer reactions and other chemical reactions without some significant modification. More advanced MD simulations, such as those calculating the electronic structure and hence the forces on the nuclei "on-the-fly" as in *ab initio* molecular dynamics (AIMD),[22] do not presume a molecular topology and explicitly treat the electronic degrees of freedom. Thus, they are a natural choice for modelling reactive processes and have been used to investigate the excess proton and hydroxide at the air–water interface.[8,23,24] However, due to their demanding computational cost, these AIMD simulations are currently limited in the amount of statistical sampling that can be achieved with respect to both temporal and spatial scales, as well as by the accuracy of the underlying electronic underlying density functional theory (DFT) and basis sets used in these methods. At a slightly reduced computational cost, hybrid quantum mechanical/molecular mechanics (QM/MM)[25,26] simulations can be used to model the reactive species within a QM region that is embedded within a larger nonreactive MM environment.[14] Owing to the rather high diffusion rate of the excess proton, from the proton hopping along water hydrogen *via* the Grotthuss mechanism, QM/MM methods would be difficult to apply to hydrated protons or hydroxide anions because a fixed QM region is typically assumed throughout the calculation. This limitation of QM/MM calculations has been addressed with the recent development of methods to deal with the swapping of identities of species between QM and MM subsystems, but it remains to be seen if these methods are appropriate for proton transport processes.[27–29]

A computationally efficient method for treating the hydrated proton and hydroxide ions must treat the dynamic bond making and breaking within a region of interest, whilst leaving the remaining "spectator" region modelling with a simpler MM force field. The collection of multiconfigurational MD methods is one class of algorithms that possess these general features.[30–42] These algorithms model a system as a linear combination of physically relevant chemical bonding topologies. For the subset of multiconfigurational MD methods, the bonding topologies are chosen "on-the-fly" over the course of a simulation in response to changes in the environment. Voth and coworkers have successfully used these latter methods to study proton transport in a myriad of condensed phase systems, including aqueous systems and proteins.[31–33,36,37,43] A more detailed review of this reactive simulation methodology is presented elsewhere[36,37,42] and the reader is referred to the appropriate references for further details.

In a previous study using our multi-state empirical valence bond (MS-EVB) version of the multiconfigurational MD methods (specifically the MS-EVB3 model[37]), Iuchi *et al.* calculated the free energy profile for bringing the hydrated excess proton to both the air–water and a hydrophobic wall–water interface.[11] Their results showed a minimum at the interface for both cases, and the subsequent decomposition of the free energy revealed an entropic penalty compensated for by a large favourable enthlapic contribution at the interface. The preference for the hydrated proton at the interface was also observed in path integral MD simulation reported in that work where nuclear quantum effects were explicitly included. However, the underlying mechanism for the hydrated proton's attraction to the interface remains to be analysed through a more detailed theoretical picture. Another MD study using an explicitly polarizable version of the MS-EVB model of the proton did not show a free energy minimum at the air–water interface,[16] while an AIMD study showed a minimum of a depth of around 1.3 kcal mol^{-1} at the interface,[24] in general agreement with the simulations of Iuchi *et al.* Interestingly in the case of the polarizable model mentioned above, a minimum in the free energy was observed at the interface when the polarizability was turned off.[16] (It should be noted that the MS-EVB model is already polarizable in the hydrated proton complex, so the addition of additional electronic polarizability runs the risk of overpolarizing the complex and thus leading to an oversolvation of it by the water solvent around it, as seen in the simulations of Wick. This behaviour affects the propensity for the hydrated proton to be near the interface relative to the bulk.)

A number of simulation studies have also been carried out to determine the presence of simple monoatomic ions at the air–water interface.[3,44–50] Otten *et al.*[51] have shown that in the case of the iodide ion the enthalpic behaviour at the interface is dominated by two competing effects – the penalty imposed by losing energetically favourable solvent molecules when the ion is brought to the surface and the gain due to the displacement of higher energy interfacial water molecules. With an appropriate and non-trivial modification, this type of local energy decomposition scheme can prove particularly useful to examine the case of hydrated excess protons at the interface, but first, the scheme requires a generalization to account for the delocalized nature of the electronic charge defect associated with the hydrated proton. The present work provides such a generalization with an application to the hydrated excess proton. In turn, it is shown to provide a deeper analysis of the origin of the thermodynamic forces driving the hydrated proton to the air–water and hydrophobic interfaces, while also unravelling the competition between enthalpy and entropy at such interfaces. Interestingly, with this generalization the simple picture of Otten *et al.* is still seen to hold, although key differences do exist due to the delocalized nature of the excess proton charge defect.

In the next section, the computational methods employed in this work will be discussed, including a generalization of the energy decomposition scheme to multi-state reactive MD algorithms. The results from the reactive simulations and subsequent energy decomposition will be described in section 3. This is followed by a discussion of the results before concluding with a summary in the last section.

2. Methods and computation

As stated earlier, the modelling of a delocalized charge defect, such as the excess proton, can be challenging to accurately describe in a computationally efficient manner. This is largely due to the fact that the proton can readily shuttle between water molecules that are hydrogen-bonded (H-bond) to one another. Thus, the net positive charge density of the excess proton is actually delocalized over several nearby solvating water molecules instead of being localized at a single point in space. This has the consequence that transport of the excess proton involves the coordinated rearrangement of several water molecules. Additionally, this diffusion process involves principally two components (hopping and vehicular) with separate timescales, both of which are coupled to the dynamics of the surrounding H-bond network.

Multiconfigurational MD methods have been successful in their application to proton (and hydroxide ion) solvation and transport in aqueous, biological, and materials systems. In the MS-EVB class of methods, a quantum-like Hamiltonian matrix is constructed and evaluated at each step in the simulation. The dimension of this Hamiltonian dynamically changes over the course of the simulation depending on how many reactant molecules are proximal to an initial reactive species. The initial reactive molecules and surrounding reactant molecules form the reactive complex, within which, the bonding topology is variable. Outside this reactive complex, the remainder of the system, the environment, is modelled using a single bonding topology. The diagonal elements of this Hamiltonian matrix correspond to distinct bonding topologies resulting from chemical reactions between proximal reactant molecules, such as a proton hopping to a nearby water molecule. The likelihood for a successful reaction is then a function of the off-diagonal couplings between diabatic states and environmental influences that may modulate the reaction barriers. These multistate methodologies are somewhat similar in spirit to the QM/MM class of methods. However, instead of explicitly treating the electronic degrees of freedom using an expensive QM method for the reactive portion of the system, the Hamiltonian matrix elements are typically modelled using MM-type expressions for reasons of computational efficiency. This is not a limitation of the multistate methods and in fact, if one chose to, the electronic degrees of freedom could be explicitly retained and the multistate Hamiltonian matrix elements evaluated using QM approaches. Once the Hamiltonian matrix has been evaluated and diagonalized, the ground state eigenvector is used with the Hellmann–Feynman theorem to obtain atomic forces. Once the total forces on all atoms are known, then an MD simulation would proceed as normal and integrate Newton's equations of motion to propagate the system. Additional details describing these methodologies can be found in ref. 32, 36, 37, and 42.

For the reactive simulations discussed here, the MS-EVB3 model[37] was used to describe the hydrated proton at two types of interfaces: the air–water and the air–hydrophobic wall.[11] This model for the hydrated proton was primarily chosen to facilitate the comparison with previous work, and in fact, has been shown to yield results in good agreement with a similar model explicitly derived from AIMD simulations.[52] The air–water interface was modelled in these simulations as a water slab, with vacuum on either side along the z-axis, consisting of 998 water molecules and an excess proton in a periodic box of dimensions 31.07 × 31.07 × 100 Å. Thus, the pH of the system was effectively equal to 1.3. The water molecules were modelled using the simple point charge flexible (SPC/Fw) water model,[53] as utilized in the MS-EVB3 model. In a second set of simulations having the same number of water molecules, a rigid atomistic hydrophobic wall consisting of carbon atoms arranged in a graphene-like lattice was used with Lennard Jones interactions parameters $\sigma = 3.8$ Å and $\varepsilon = 0.1051$ kcal mol^{-1} defined between wall particles and oxygen atoms. The system dimensions for this system were 31.93 × 31.90 × 100 Å with the walls centred at $z = 0.0$ and $z = 36.0$ Å.

All MS-EVB simulations were calculated using a modified version of the LAMMPS MD code.[54] Unless otherwise stated, the Particle–Particle–Particle Mesh (PPPM) method with a precision of 10^{-5} was used for long-range electrostatic interactions. The equations of motion were integrated with a 1.0 fs timestep in the constant NVT ensemble at 300 K using the Nose–Hoover thermostat with a relaxation time of 0.05 ps. The potential of mean force (PMF) for moving the excess proton towards the interface was calculated using the umbrella sampling[55] method, with a total of 11 windows evenly spaced between 5.0 and 15.0 Å from the centre of the water slab. The harmonic force constant for each of the bias potential in each window was 5 kcal mol^{-1} Å^{-1}. The umbrella sampling collective variable was defined to be the distance, in the z-direction, between the centre of mass of the system and the hydrated proton centre of excess charge (CEC). Since the identity of the excess proton continuously changes over the course of an MS-EVB simulation, the CEC coordinate (r_{cec})[11,37] is a convenient coordinate that allows one to continuously track the reactive species (proton) during the simulation. This coordinate is defined as

$$\boldsymbol{r}_{\text{CEC}} = \sum_i c_i^2 \boldsymbol{r}_{i,\text{COC}} \tag{1}$$

where c_i^2 is the square of the ground state eigenvector coefficient of the ith diabatic state and $\boldsymbol{r}_{i,\text{COC}}$ is the centre of charge coordinate for each diabatic state in the MS-EVB matrix, which mostly corresponds to the hydronium oxygen atom in each state. More specifically, the z component of this CEC coordinate was used as the definition of the collective variable. The potentials of mean force were constructed from the umbrella sampling data using the weighted histogram analysis method (WHAM).[56] In order to decompose the free energy, eqn (2) below, into entropic, $-T\Delta S(z)$, and enthalpic, $\Delta U(z)$, contributions, additional PMFs were calculated at two new temperatures (280 K and 320 K for the air–water case and 320 K and 340 K for the wall–water case). The set of PMFs at three different temperatures were then used in a finite difference approach to extract the two contributions as well as estimate the statistical errors:[11]

$$\Delta F(z) = \Delta U(z) - T\Delta S(z) \tag{2}$$

2.1 Local energy decomposition scheme (enthalpy)

For the case of a single simple monoatomic ion in a slab of water with vacuum on either side, it was recently found that the total energy of the system is well approximated by a sum of interactions that are local in their nature. Using this local decomposition scheme as proposed by Otten *et al.*, the total energy is decomposed into contributions due to solvent molecules that are coordinating the ion, the bulk water molecules, and the interfacial water molecules.[51] If the ion is at a distance z from the interface, the total energy can be approximated as $U_{\text{local}}(z)$, which is defined to be

$$U_{\text{local}}(z) = E_{\text{c}}n_{\text{c}}(z) + E_{\text{b}}n_{\text{b}}(z) + E_{\text{i}}n_{\text{i}}(z), \tag{3}$$

where E_{c} is the average energy of a water molecule coordinating the ion, E_{b} is the average energy of a water molecule in the bulk (but not coordinated to the ion), and E_{i} the average energy of an interfacial water molecule (not coordinated to the ion). The three average interaction energies (E_{b}, E_{c}, and E_{i}) are calculated from molecular simulations when the ion is in the bulk. Each of these energies can be further decomposed into contributions arising from water–water and water–ion interactions. After these energies have been determined, one simply has to count the number of water molecules in the interfacial (n_{i}), bulk (n_{b}), and coordination shell (n_{c}) subsystems and then use eqn (3) to calculate the energy (enthalpy) when the ion is at a distance z from the interface. For the simple case of the iodide ion, Otten *et al.* obtained quantitative agreement between the local approximation to the total energy and the true enthalpy extracted from their simulations. The advantage of this local scheme is that it allows one to study the effect of various interactions on the enthalpy.

For the case of the hydrated excess proton, however, the situation is more complicated because one is no longer dealing with a monoatomic spherical species, but also because the hydrated proton is delocalized charge defect spanning several water molecules. In the original decomposition scheme, the energy quantities E_x were given by the relations

$$E_x = \frac{1}{2}\langle E_{\text{water-water}}\rangle_x + \langle E_{\text{water-ion}}\rangle_x \tag{4}$$

where $E_{\text{water-water}}$ and $E_{\text{ion-water}}$ are the average interaction energies between water–water and water–ion pairs, respectively, in the region x (bulk, coordination shell, or interface). For the case of the hydrated proton as described by MS-EVB simulations, it is necessary to recast each contribution as a weighted averaged over the diabatic states sampled in a reactive simulation so as to incorporate the delocalization of the charged defect. The generalization of these terms is given in eqn (5) as summations over MS-EVB states and the corresponding interaction energy for the ith state, given by

$$\langle E_{\text{ion-water}}\rangle = \sum_{i=1}^{m}\langle c_i^2 E^{\text{i}}_{\text{ion-water}}\rangle \text{ and } \langle E_{\text{water-water}}\rangle = \sum_{i=1}^{m}\langle c_i^2 E^{i}_{\text{water-water}}\rangle \tag{5}$$

The weights in these terms are the ground state MS-EVB eigenvector coefficients, c_i^2. Reasonable convergence of the quantities in eqn (5) does not require a complete summation over MS-EVB states. It was found to be satisfactory for the summation to instead be taken over only those m states with the largest amplitudes with m

chosen to be 4. In bulk water simulations with the MS-EVB3 model for the excess proton, the total number of states at any point in the simulation typically exceeds 20, although usually no more than a few at a given time dominate the behaviour. While most of these MS-EVB states are crucially important to include in the energy/force calculations to properly conserve constants of motion during the simulation, the results of this post-processing procedure are not significantly altered when additional states are included. Physically, these 4 dominant states mostly correspond to a reactive complex involving those water molecules that accept strong H-bonds from the hydronium cation when in Eigen-type (H_9O_4+) configurations and during proton hopping events when the structure of the proton more closely resembles that of the Zundel ion ($H_5O_2^+$).

So far, the discussion has focused only on the contributions to the local energy approximation from the diagonal elements of the MS-EVB Hamiltonian matrix. Since it is only possible in this multistate reactive MD algorithm for chemical reactions to occur when the off-diagonal couplings are nonzero, eqn (3) needs to be supplemented with an appropriate contribution from the off-diagonal matrix elements. This correction takes the form of the average off-diagonal energy, $U_{\text{off-d}}(z)$, when the ion is at a particular distance z from the interface and can be extracted from simulations. The resulting final expression for the local approximation to the ion total energy is

$$U_{\text{local}} = E_c n_c(z) + E_b n_b(z) + E_i n_i(z) + U_{\text{off-d}} \tag{6}$$

2.2 Instantaneous interface surface fluctuations (entropy)

Unlike the case of enthalpy, which can be decomposed into a series of terms accounting for different interactions, it is not possible to straightforwardly decompose the entropic contribution to the free energy in a similar manner. However, the fluctuations of the instantaneous interface can provide useful information to help elucidate the origins of the entropic behavior of ions near the interface. In order to construct this instantaneous interface, the coarse-grained density field approach of Willard and Chandler can be employed.[57] Briefly, at every oxygen position in the simulation cell, a Gaussian function whose width is one water molecular diameter is added to give rise to a smooth "density field". The instantaneous interfacial surface is then defined as the surface for which this coarse-grained density is equal to half of the bulk value. For a more detailed discussion, the reader is directed to the original paper.[57] The quantity of interest to be extracted from this instantaneous surface is the average fluctuation of this surface as a function of radial distance from the ion, when the ion is at the interface, and is defined as follows

$$\partial h(r_p) = h(r_p) - \langle h(r_p) \rangle \tag{7}$$

where $h(r_p)$ is the vertical height of the instantaneous surface at a radial distance, r_p, from the ion. If $\langle \partial h(0)^2 \rangle$ (*i.e.*, average fluctuation at the vicinity of the ion) is smaller than the corresponding value at large r_p, then it implies that the ion is pinning the nearby instantaneous interface, thus giving rise to an entropic penalty. The smaller the value at $r_p = 0$, the greater the pinning, and hence the entropic penalty.

Therefore, the ratio of $\langle \delta_H(r_\mathrm{p})^2 \rangle$ to the value at large r_p, $\langle \partial h(r_\mathrm{L})^2 \rangle$, proves to be a useful measure to study the effect of the ion on the instantaneous interface and provides information regarding entropic contributions to the free energy.

3. Results and discussion

3.1 Free energy profile

It will prove useful in later discussions to partition the water slab structure into distinct regions. Based on the equilibrium structure of the water slab, these regions can be straightforwardly identified upon examination of the average water density as a function of distance from the centre of mass (COM) of the system. This distribution is plotted in the inset of Fig. 1 with the density normalized to the density of bulk water at ambient conditions. From this plot, it is clear that the water slab is mostly bulk-like up to 12.0 Å before density-depletion is observed near the air–water interface. Based on this density profile, the interfacial region can be divided into 3 parts: an inner interface from 11.0 to 12.5 Å with water density close to that of the bulk, a middle region from 12.5 to 14.0 Å where the density first starts to decrease, and an outer region that extends beyond 14.0 Å into the vacuum.

The free energy profile (Fig. 1) for the excess proton shows a well depth of 1.8 kcal mol^{-1} at a separation of around 15.0 Å from the slab COM. The minimum is located just after the onset of the outer interfacial region where the water density has already dropped to less than half of the bulk value. As expected, this result is in agreement with previous, separate reports that used MS-EVB3 to study the water–air interface.[11] By analysing the enthalpic and entropic contributions that arise from the finite difference decomposition of the PMF at 300 K, it is revealed that the enthalpic contribution favors the interface and more than compensates for the entropic penalty. The corresponding maximum and minimum of the individual contributions are located at separations just below 15.0 Å from the slab COM, which is closer to the location at which the slab water

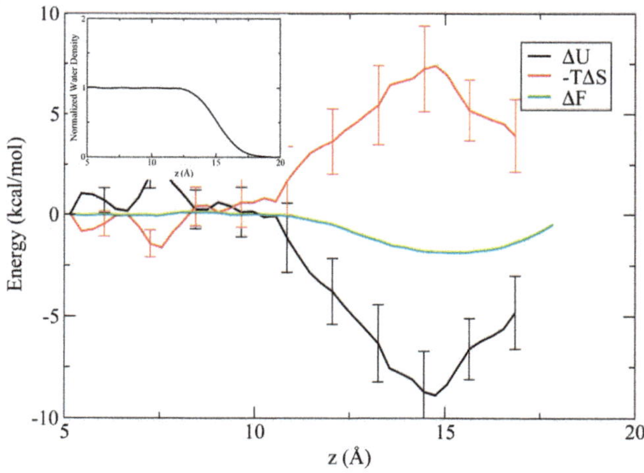

Fig. 1 Free energy profile of the hydrated excess proton for the air–water case along with the decomposition into enthalpic and entropic terms from eqn (2). The inset shows the average water density for the air–water system as a function of distance from the centre of mass.

density is half of the bulk value. These results are again in good agreement with previous work on decomposing the PMF for the proton at the air–water interface,[11] thus confirming the reproducibility of results from this reactive simulation methodology produced using completely separate software packages.

Similar results are also observed for the case of the hydrophobic wall (Fig. 2). With the presence of the wall, the deviations in the water density from bulk values are more pronounced with clear evidence of water layering adjacent to the wall with the two peaks near 11.0 and 14.0 Å. The effects of the hydrophobic wall are even observed at short separations (~5–7 Å) from the slab COM, where no such deviations from the bulk density are apparent. Despite this apparent difference in the interfacial structure of the water slab near the hydrophobic wall compared to that of the air–water interface, there is still a minimum in the hydrated proton PMF located near the position of the first water layer. The depth of this free energy well is 2.0 kcal mol^{-1}, which is slightly larger than the value observed for the air–water interface, but the two values are still within statistical error of one another. The enthalpic and entropic contributions resulting from the decomposition of the PMF are similar in structure to the curves obtained for the air–water interface, through the magnitude of the change is smaller, despite the similarity of the well-depths of the PMFs. For example, the entropic barrier for the air–water system has a maximum near 7.5 kcal mol^{-1} while the same curve for the hydrophobic interface only rises up to 1.5 kcal mol^{-1}. Similar to the air–water interface, there is a pronounced decrease in magnitude for both decomposed curves resulting in the PMF having a value of nearly zero only a few Angströms away from the interface.

3.2 Local energy decomposition

Attention now turns to the main thrust of this paper, which is to address the question of whether a complex charged defect, such as the hydrated excess proton, can be effectively modelled using a short-ranged decomposition of the

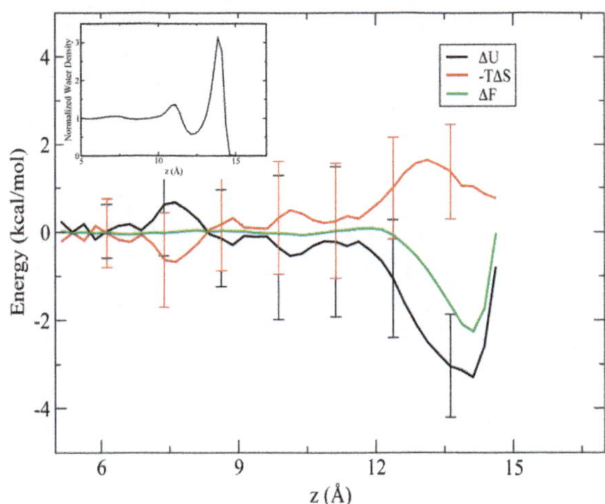

Fig. 2 Free energy profile of the hydrated excess proton for the wall–water case along with the decomposition into enthalpic and entropic terms using eqn (2). Inset shows the average water density for the wall–water system as a function of distance from the centre of mass.

total energy. Using the results from reactive MS-EVB simulations at 300 K in which the ion was constrained to be near the centre of mass of the system, the average water–water and water–ion energies were calculated for both the air–water and wall–water systems using the multistate generalized expressions for the interaction energies, eqn (5). The resulting energy contributions (water–water and ion–water) are shown in Fig. 3 as contour plots for both systems. Along the y-axis is the vertical distance (r_z) of a water molecule from the slab COM, while the x-axis for each contour plot indicates the radial distance from the ion in the plane parallel to the interface. In each plot, the hydrated excess proton is located in the middle of the slab (bulk region) with the interfaces found at the top and bottom of each plot. The water–water and ion–water energies are shown in Fig. 3a and b, respectively, for the case of the air–water interface. It is clear from these figures that there exists an obvious division of the system into a coordination shell region (of about 5 Å) surrounding the ion, a bulk region, and an interfacial region (around $z = \pm 13$). The wall–water interface shows a similar trend (Fig. 3c and d) with essentially the same scale of energies. The average energy in each of the three regions was determined from the MS-EVB simulation.

The average number of water molecules in each region as a function of the distance (z) of the ion from the centre of mass was calculated from the different umbrella sampling simulations, as was the average off-diagonal energy. Using eqn (6) the energy of moving the ion from the bulk to the interface was constructed (Fig. 4). For both systems, a substantial drop in enthalpy is observed with the minimum located at the interface in both instances, albeit at the inner interface rather than the outer interface. For both systems, the decrease in enthalpy is roughly equivalent at 6 kcal mol^{-1}, while the position of the minimum

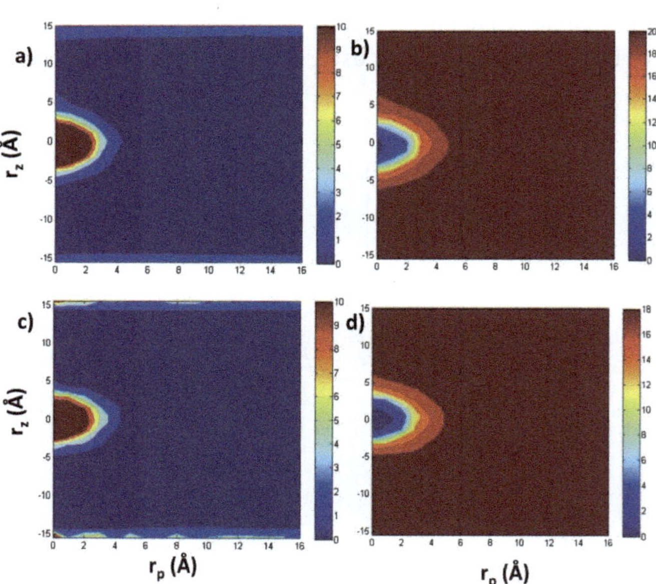

Fig. 3 a) Water–water interaction energy (in kcal mol^{-1}) for the air–water case as a function of radial distance from the ion (r_p) which is at the origin and the vertical distance from the center of mass (r_z). b) Shows the water–ion interaction energy (in kcal mol^{-1}) for the air–water case as a function of r_p and r_z. c) and d) Plots of the water–water and water–ion interaction energies, respectively, for the wall–water case.

would appear to be shifted towards the inner interfacial region. While there is a steep enthalpic penalty for bringing the ion close to the hydrophobic wall near 15.0 Å, similar to the case for the PMF decomposition in Fig. 2, the same curve for the air–water interface is presumably tending towards zero.

The water molecules in the ion coordination shell are energetically more favourable by around 2.5 kcal mol^{-1} compared to those water molecules in the bulk for both the wall and air cases. On the other hand, the interfacial water molecules are energetically less stable than the bulk by approximately 1.5 kcal mol^{-1} for the air case and around 1.0 kcal mol^{-1} for the wall case. As one moves to the interface, energetically favourable coordinating waters are removed, but so are energetically unfavourable interfacial waters. The excess proton retains a large portion of its coordination shell as it moves to the interface, while at the same displacing unfavourable interfacial waters, which then results in an enthalpic minimum (attraction) at the interface.

Interestingly, the average off-diagonal energy is more favourable by 1.5 kcal mol^{-1} at the outer interface compared to bulk for the air–water case. The off-diagonal energy is a measure of the delocalization of the net positive charge defect of the hydrated proton ion and the more negative value at the interface suggests the presence of a more delocalized and hence larger effective ion. In Fig. 5, the distribution of the largest EVB state amplitude,[32] ${c_1}^2$, obtained from the ground state MS-EVB eigenvector at each step in the simulation is plotted for the air–water interface. When this weight is close to 0.5, it indicates that the system is in a predominantly Zundel state, *i.e.* a state in which the excess proton is shared equally between two adjacent water molecules as opposed to the Eigen state around 0.6 in which case the proton is predominantly on the central water (hydronium cation). As the ion moves through the interface, the fraction of Zundel species increases, which is consistent with the more favourable off-diagonal energy.

For the wall–water case (see Fig. 6) the fraction of Zundel species first increases as the ion moves through the interface, but decreases at the outer interface. The comparison of the distribution at the outer interface for the air–water and

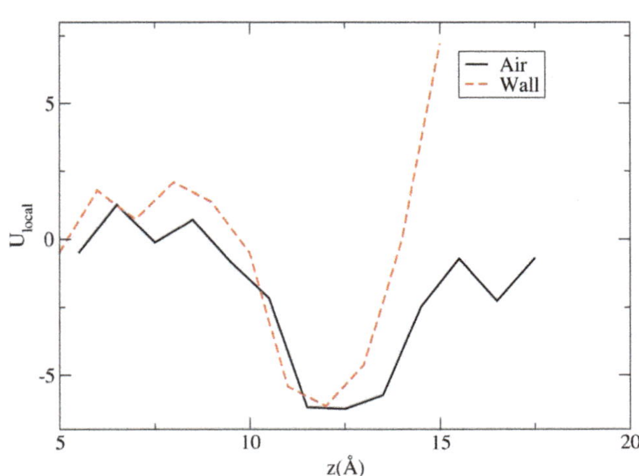

Fig. 4 Local energy for the air and wall interfaces.

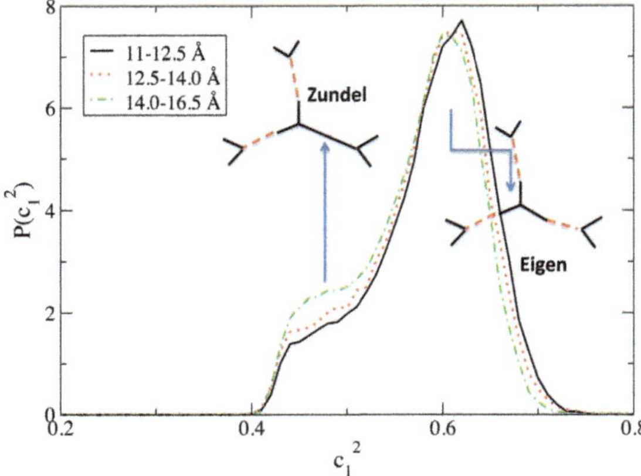

Fig. 5 Normalized probability distribution of the highest MS-EVB coefficient weight for the air–water case for the inner, middle and outer interface.

wall–water cases (see Fig. 7) shows that the fraction of Zundel species is smaller for the wall–water case implying that the hydrated excess proton ion at the wall–water interface is effectively "smaller" than in the air–water case.

The local energy scheme assumes that the radius used to define the coordination shell does not change as it moves to the interface. This is clearly not the case for the hydrated excess proton. Secondly, in the case of a complex ion like the excess proton where specific H-bonds are part of the solvation shell, spherical symmetry is a crude approximation, especially when the ion is at the interface. In addition, only intermolecular interaction energies were considered and the effect of moving the ion to the interface on the intramolecular geometry of the ion as well as the coordinating water molecules was not considered. Lastly, the repulsive effect of the hydrophobic wall on the interaction energy was ignored. Despite these approximations, the local scheme does help identify the key interactions that cause the large enthalpic minimum at the interface.

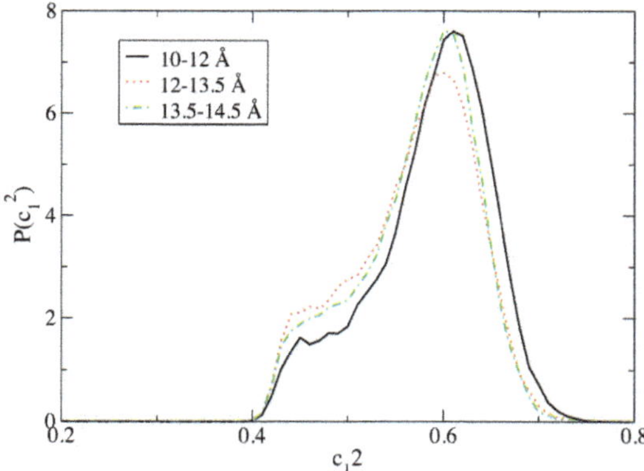

Fig. 6 Normalized probability distribution of the largest MS-EVB coefficient weight for the wall–water interface for the inner, middle, and outer regions of the interface.

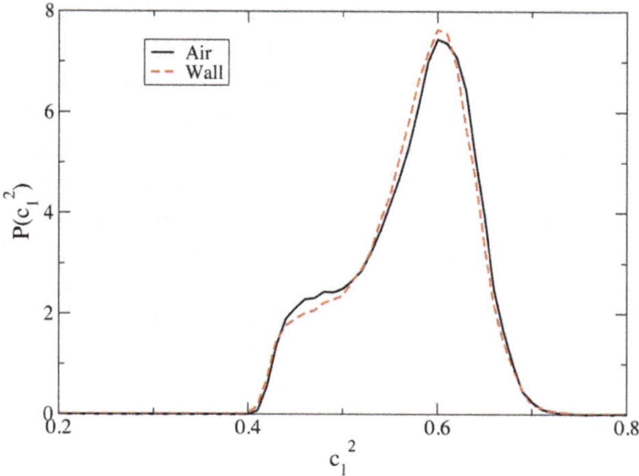

Fig. 7 Normalized probability distribution of the largest MS-EVB coefficient weight for the air–water and wall–water cases at the outer interface.

3.3 Instantaneous interface

The average fluctuation of the instantaneous interface at $z = 14$ Å as a function of radial distance from the ion and normalized to the value far away from the ion is shown in Fig. 8 for both types of interfaces. The fluctuations near the ion are much smaller for the air–water case, indicating the pinning of the interface and hence the high entropic penalty. On the other hand, for the wall case, the fluctuations near the ion are not dampened as much, which is consistent with the smaller value of the unfavourable entropic contribution. The hydrophobic wall dampens overall the fluctuations at the interface. In addition to this, the effective size of the ion in the wall case is smaller and therefore should exhibit less pinning. These two effects clearly contribute to the entropic behaviour near the wall.

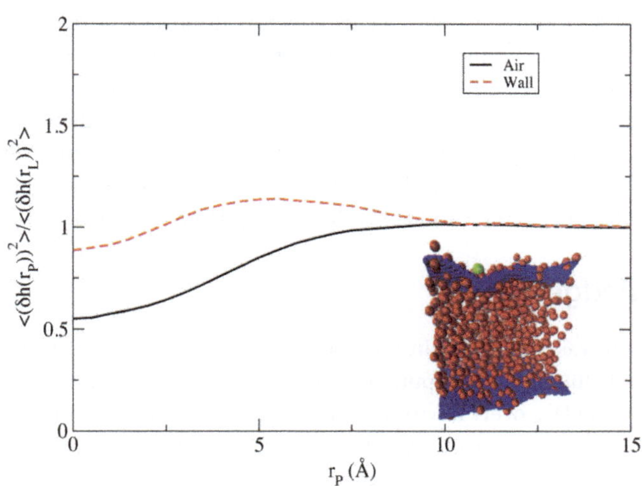

Fig. 8 The fluctuations of the instantaneous interfacial surface as a function of radial distance away from the ion, when the ion is at the outer interface. The inset shows the instantaneous surface (blue) for the air–water case with the ion (green) and the water oxygen atoms (red).

The simple model of two competing terms, the loss of favourable coordinating water molecules *versus* the gain in favourable energy due to displacement of interfacial waters, does explain the overall trend in the enthalpy profile near the interface. However, this analysis is further complicated by the delocalization of the excess proton charge defect and the change in this delocalization as the ion moves to the interface. The hydrated proton ion gets effectively larger as it moves to the hydrophobic interface and this is more pronounced for the air–water case. This does suggest that rather than the sign of the charge, it is instead the ratio of the absolute value of the charge divided by the effective ion radius that governs the presence of ions at the aqueous–hydrophobic interface. The excess proton is effectively a large ion and therefore moves to the interface. This same effectively large ion pins the instantaneous surface near it, resulting in an unfavourable entropic term near the interface.

4. Conclusions

The free energy profile for moving the hydrated excess proton close to two different hydrophobic interfaces, the air–water and hydrophobic wall–water interfaces, was calculated using a reactive MD simulation methodology. For both cases, a minimum in the free energy was found with a depth of approximately 2 kcal mol^{-1}. The enthalpic contribution to the free energy shows a clear minimum at the interface, which more than compensates for the entropic penalty. A scheme for decomposing the total energy into a sum of short-ranged interactions was generalized for the case of MS-EVB simulations. The terms arising in this scheme from the coordination environment of the excess proton, the bulk, and the interface along with a correction for the charge defect delocalization energy yielded a minimum at both interfaces of similar depths. Through analysis of the individual terms in the energy decomposition, it was found that the minimum arises from the displacement of unfavourable interfacial water molecules as the ion moves towards the interface, and this outweighs the energy cost from the loss of favourable coordinating water molecules. In addition, the hydrated proton ion effectively gets larger as it moves towards the interface. The ion pins the water molecules near it, thereby decreasing the fluctuations of the instantaneous interfacial surface in its vicinity, which in turn contributes to the entropic penalty near the interface. This effect is significantly reduced for the hydrophobic wall interface. It will be very interesting to compare these results with those obtained from a similar analysis for the hydroxide ion near interfaces in order to rationalize the various results coming from the experimental studies.

Acknowledgements

This research was supported in part by the National Science Foundation (NSF grant CHE-1214087). The computations in this work utilized the Extreme Science and Engineering Discovery Environment (XSEDE), which is supported by National Science Foundation grant number OCI-1053575, with an allocation of time on TACC computing resources. The computations were also supported in part by a grant of computer time from the DOD High Performance Computing Modernization Program at the Air Force and Navy DOD Supercomputing Resource Centres.

Notes and references

1 M. K. Petersen, S. S. Iyengar, T. J. F. Day and G. A. Voth, *J. Phys. Chem. B*, 2004, **108**, 14804–14806.
2 A. Mucha, T. Frigato, L. M. Levering, H. C. Allen, D. J. Tobias, L. X. Dang and P. Jungwirth, *J. Phys. Chem. B*, 2005, **109**, 7617–7623.
3 P. Jungwirth and D. J. Tobias, *Chem. Rev.*, 2005, **106**, 1259–1281.
4 P. B. Petersen and R. J. Saykally, *J. Phys. Chem. B*, 2005, **109**, 7976–7980.
5 P. B. Petersen and R. J. Saykally, *Annu. Rev. Phys. Chem.*, 2006, **57**, 333–364.
6 L. M. Pegram and M. T. Record, *Proc. Natl. Acad. Sci. U. S. A.*, 2006, **103**, 14278–14281.
7 T. L. Tarbuck, S. T. Ota and G. L. Richmond, *J. Am. Chem. Soc.*, 2006, **128**, 14519–14527.
8 V. Buch, A. Milet, R. Vacha, P. Jungwirth and J. P. Devlin, *Proc. Natl. Acad. Sci. U. S. A.*, 2007, **104**, 7342–7347.
9 R. Vacha, V. Buch, A. Milet, J. P. Devlin and P. Jungwirth, *Phys. Chem. Chem. Phys.*, 2007, **9**, 4736–4747.
10 L. M. Levering, M. R. Sierra-Hernandez and H. C. Allen, *J. Phys. Chem. C*, 2007, **111**, 8814–8826.
11 S. Iuchi, H. Chen, F. Paesani and G. A. Voth, *J. Phys. Chem. B*, 2009, **113**, 4017–4030.
12 A. Gray-Weale and J. K. Beattie, *Phys. Chem. Chem. Phys.*, 2009, **11**, 10994–11005.
13 S. Enami, M. R. Hoffmann and A. J. Colussi, *J. Phys. Chem. Lett.*, 2010, **1**, 1599–1604.
14 H. Takahashi, K. Maruyama, Y. Karino, A. Morita, M. Nakano, P. Jungwirth and N. Matubayasi, *J. Phys. Chem. B*, 2011, **115**, 4745–4751.
15 S. Yamaguchi, A. Kundu, P. Sen and T. Tahara, *J. Chem. Phys.*, 2012, **137**, 151101–151104.
16 C. D. Wick, *J. Phys. Chem. C*, 2012, **116**, 4026–4038.
17 R. J. Saykally, *Nat. Chem.*, 2013, **5**, 82–84.
18 K. G. Marinova, R. G. Alargova, N. D. Denkov, O. D. Velev, D. N. Petsev, I. B. Ivanov and R. P. Borwankar, *Langmuir*, 1996, **12**, 2045–2051.
19 H. Mishra, S. Enami, R. J. Nielsen, L. A. Stewart, M. R. Hoffmann, W. A. Goddard and A. J. Colussi, *Proc. Natl. Acad. Sci. U. S. A.*, 2012.
20 C. J. T. de Grotthuss, *Ann. Chim.*, 1806, **LVIII**, 54–74.
21 N. Agmon, *Chem. Phys. Lett.*, 1995, **244**, 456–462.
22 R. Car and M. Parrinello, *Phys. Rev. Lett.*, 1985, **55**, 2471–2474.
23 C. J. Mundy, I. F. W. Kuo, M. E. Tuckerman, H.-S. Lee and D. J. Tobias, *Chem. Phys. Lett.*, 2009, **481**, 2–8.
24 H.-S. Lee and M. E. Tuckerman, *J. Phys. Chem. A*, 2009, **113**, 2144–2151.
25 A. Warshel and M. Levitt, *J. Mol. Biol.*, 1976, **103**, 227–249.
26 H. Lin and D. Truhlar, *Theor. Chem. Acc.*, 2007, **117**, 185–199.
27 T. Kerdcharoen and K. Morokuma, *Chem. Phys. Lett.*, 2002, **355**, 257–262.
28 A. Heyden, H. Lin and D. G. Truhlar, *J. Phys. Chem. B*, 2007, **111**, 2231–2241.
29 C. N. Rowley and B. Ã. Roux, *J. Chem. Theory Comput.*, 2012, **8**, 3526–3535.
30 I. J. Åqvist and A. Warshel, *J. Biol. Chem.*, 1992, **224**, 7.
31 U. W. Schmitt and G. A. Voth, *J. Phys. Chem. B*, 1998, **102**, 5547–5551.
32 U. W. Schmitt and G. A. Voth, *J. Chem. Phys.*, 1999, **111**, 9361–1981.
33 Y. Wu and G. A. Voth, *Biophys. J.*, 2003, **85**, 864–875.
34 G. A. Voth, *Acc. Chem. Res.*, 2006, **39**, 143–150.
35 C. M. Maupin, K. F. Wong, A. V. Soudackov, S. Kim and G. A. Voth, *J. Phys. Chem. A*, 2006, **110**, 631–639.
36 J. M. J. Swanson, C. M. Maupin, H. Chen, M. K. Petersen, J. Xu, Y. Wu and G. A. Voth, *J. Phys. Chem. B*, 2007, **111**, 4300–4314.
37 Y. Wu, H. Chen, F. Wang, F. Paesani and G. A. Voth, *J. Phys. Chem. B*, 2008, **112**, 467–482.
38 I. S. Ufimtsev, A. G. Kalinichev, T. J. Martinez and R. J. Kirkpatrick, *Phys. Chem. Chem. Phys.*, 2009, **11**, 9420–9430.
39 C. D. Wick and L. X. Dang, *J. Phys. Chem. A*, 2009, **113**, 6356–6364.
40 C. Knight, G. E. Lindberg and G. A. Voth, *J. Chem. Phys.*, 2012, **137**, 22A525.
41 T. Yamashita, Y. Peng, C. Knight and G. A. Voth, *J. Chem. Theory Comput.*, 2012, **8**, 4863–4875.
42 C. Knight and G. A. Voth, *Acc. Chem. Res.*, 2012, **45**, 101–109.
43 H. Chen, Y. Wu and G. A. Voth, *Biophys. J.*, 2006, **90**, L73–L75.
44 L. X. Dang, *J. Phys. Chem. B*, 2002, **106**, 10388–10394.
45 C. Caleman, J. S. Hub, P. J. van Maaren and D. van der Spoel, *Proc. Natl. Acad. Sci. U. S. A.*, 2011, **108**, 6838–6842.
46 B. C. Garrett, *Science*, 2004, **303**, 1146–1147.
47 Y. Levin, A. P. dos Santos and A. Diehl, *Phys. Rev. Lett.*, 2009, **103**, 257802.

48 D. Horinek, A. Herz, L. Vrbka, F. Sedlmeier, S. I. Mamatkulov and R. R. Netz, *Chem. Phys. Lett.*, 2009, **479**, 173–183.
49 R. R. Netz and D. Horinek, *Annu. Rev. Phys. Chem.*, 2012, **63**, 401–418.
50 A. C. Stern, M. D. Baer, C. J. Mundy and D. J. Tobias, *J. Chem. Phys.*, 2013, **138**, 114709–114708.
51 D. E. Otten, P. R. Shaffer, P. L. Geissler and R. J. Saykally, *Proc. Natl. Acad. Sci. U. S. A.*, 2012, **109**, 701–705.
52 C. Knight, C. M. Maupin, S. Izvekov and G. A. Voth, *J. Chem. Theory Comput.*, 2010, **6**, 3223–3232.
53 Y. Wu, H. L. Tepper and G. A. Voth, *J. Chem. Phys.*, 2006, **124**, 024503.
54 S. Plimpton, *J. Comput. Phys.*, 1995, **117**, 1–19.
55 J. Kästner, *Wiley Interdiscip. Rev.: Comput. Mol. Sci.*, 2011, **1**, 932–942.
56 S. Kumar, D. Bouzida, R. H. Swendsen, P. A. Kollman and J. M. Rosenberg, *J. Comput. Chem.*, 1992, **13**, 1011–1021.
57 A. P. Willard and D. Chandler, *J. Phys. Chem. B*, 2010, **114**, 1954–1958.

Faraday Discussions RSC Publishing

PAPER

Monolayer and bilayer structures in ionic liquids and their mixtures confined to nano-films

Alexander M. Smith,[a] Kevin R. J. Lovelock[b] and Susan Perkin*[a]

Received 5th May 2013, Accepted 10th July 2013
DOI: 10.1039/c3fd00075c

The confinement of liquids to thin films can lead to dramatic changes in their structural arrangement and dynamic properties. Ionic liquids display nano-structures in the bulk of the liquid, consisting of polar and non-polar domains, whereas a solid surface can induce layered structures in the near-surface liquid. Here we compare and contrast the layer structures in a series of imidazolium and pyrrolidinium-based ionic liquids upon confinement of the liquids to films of ~0–20 nm between two negatively charged mica surfaces. Using a surface force balance (SFB) we measured the force between the two atomically smooth mica surfaces with ionic liquid between, directly revealing the ion packing and dimensions of layered structures for each liquid. The ionic liquids with shorter alkyl chain substituents form alternating cation–anion monolayer structures on confinement, whilst a longer alkyl chain leads to alignment of the cations in bilayer formation. The crossover from monolayers to bilayers, however, occurs at different alkyl chain lengths for imidazolium- and pyrrolidinium-based ionic liquids with a common anion. In addition, we find that imidazolium cation bilayers are arranged in toe-to-toe orientation, whereas pyrrolidinium cations form bilayers consisting of fully interdigitated alkyl chains. Results for a mixture of monolayer-preferring (*i.e.* short alkyl chain) and bilayer-preferring (*i.e.* long alkyl chain) liquids indicate alkyl chain segregation and bilayer-like structures. We discuss the driving forces for these self-assembly effects, and the contrasting behaviour of the imidazolium and pyrrolidinium-type ionic liquids.

1 Introduction

Inspection of the ion structures of typical ionic liquids – salts which are liquid under ambient conditions – immediately reveals large ions, often irregular in shape, and with delocalised charge.[1] The cations are often amphiphilic: only part of the ion structure holds the charge whilst another part (or parts) of the ion is

[a]*Department of Chemistry, Physical and Theoretical Chemistry Laboratory, University of Oxford, South Parks Road, Oxford OX1 3QZ, UK. E-mail: susan.perkin@chem.ox.ac.uk*
[b]*Department of Chemistry, Imperial College London, London SW7 2AZ, UK*

relatively uncharged. This asymmetry results in more complex interactions than when compared with high temperature molten salts; certain preferred orientations of an ion relative to its neighbours cause meso-structure and order in the liquid, both in the bulk and at interfaces.[2–8] This complexity introduced by multiple and directional interactions – such as electrostatics, van der Waals, dipole interactions and hydrogen bonding – means that subtle changes in the ion structure or external environment can lead to dramatic switches in the preferred ion ordering in the liquid. The mesostructure in ionic liquids is in some ways akin to the self-assembly observed for amphiphilic solutes in water, such as the aggregation of surfactants into micelles or lipids into bilayers. There is, however, an important difference: ionic liquids have the additional constraint that the whole of the volume of the liquid must be taken up by the ions themselves, with no solvent to 'fill the spaces'. The result must be a balance between the optimised inter-ion interactions and the space-filling requirement.[2–5]

Confinement of ionic liquids by surfaces or to a thin film leads to a template effect whereby the charge and geometry of the surface kicks-off a layer structure which can extend ~5 nm into the liquid or double that distance for liquid confined between two surfaces.[7,9,10] The liquid layers alternate between positive and negative charge due to ion correlations which cause overscreening.

Several applications of ionic liquids take advantage of their interfacial and confinement properties, for example their use as lubricants, electrolytes, and colloidal dispersion media.[11–13] Of particular interest for these applications has so-far been the highly stable and hydrophobic anions such as bis[(trifluoromethane)sulfonyl]imide, [NTf_2], and cations based on imidazolium and pyrrolidinium groups. Pyrrolidinium-based ionic liquids have similar melting points and viscosities compared to their imidazolium-based analogues.[14] The major driving force for their study so far has been for electrochemical applications, due to their large electrochemical windows[15,16] as compared to other ionic liquids. Recently, there has been much interest in using mixtures of ionic liquids as a route to further tune the liquid properties and achieve greater flexibility with a smaller range of components.[17,18]

Here we compare and contrast layered ion structures detected for two homologous series of ionic liquids (imidazolium- and pyrrolidinium-based with a common anion) as well as their mixtures, confined between macroscopic, charged, atomically smooth surfaces. Analysis of the ion geometries in terms of a 'packing parameter' allows rationalisation of the different self-assembled structures observed. New results presented here for a mixture of short- and long-chain pyrrolidinium-based ionic liquids reveal how bilayers dominate in the mixed system. The results allow new insight into the driving forces for meso-structuring in confined ionic liquids.

2 Experimental procedure

2.1 Surface force measurements

The SFB technique (shown schematically in Fig. 1) used in this study to measure normal interaction forces between mica surfaces across liquids with sub-molecular resolution in film thickness has been described in detail elsewhere.[19] Two large (~cm^2) mica pieces from the same facet – hence guaranteed to be atomically smooth and free of steps over the whole area – of 1.5–2.5 μm thickness are silvered

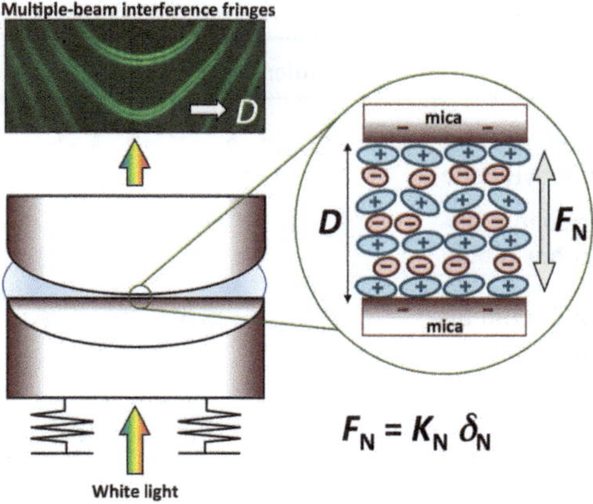

Fig. 1 Schematic of the surface force balance (SFB) used to measure the normal force, F_N, across molecularly confined liquid films as a function of surface separation, D.

on their backside and glued to cylindrical silica lenses. All surface preparation is carried out in a laminar flow hood to avoid any particulate contamination. The lenses are then mounted in the apparatus in a crossed-cylinder configuration, and in this arrangement the mica surfaces have a point of closest approach geometrically equivalent to a sphere near a flat surface. The separation between the surfaces, D, is determined using white-light multiple beam interferometry with Ångstrom resolution by means of constructive interference fringes of equal chromatic order (FECO).[20] Using a piezoelectric tube upon which one of the lenses is mounted, the mica surfaces are brought carefully to molecular contact in dry air to determine mica thickness through interferometry, after which they are widely separated ($\approx$1 mm) and a droplet of ionic liquid (50–100 μL) injected between them. Care is taken to keep the atmosphere dry inside the SFB chamber by purging with dry, reduced hydrocarbon nitrogen before sealing, and a vial containing P_2O_5 placed inside the apparatus to absorb any surrounding residual water vapour.

The deflection of a horizontal leaf spring (of known spring constant) is measured *via* interferometry to determine the normal surface force, F_N, with resolution better than 10^{-7} N. F_N is normalised by the local radius of curvature, R, of the glued mica surfaces ($R \approx 1$ cm), which is determined directly from the shape of the FECO fringes viewed through the spectrometer. Usefully, F_N/R is proportional to the interaction energy per unit area between parallel plates, allowing quantitative comparison of force magnitudes both between different experiments and between different contact areas within each experiment.

2.2 Materials

Mica was highest grade of the ruby muscovite variety, supplied by S&J Trading Inc. The ion structures of ionic liquids investigated in this study are shown in Table 1, and include 1-alkyl-3-methylimidazolium bis[(trifluoromethane)sulfonyl]imide, $[C_nC_1Im][NTf_2]$ with $n = 4$ and 6, and 1-alkyl-1-methylpyrrolidinium bis[(trifluoromethane)sulfonyl]imide, $[C_nC_1Pyrr][NTf_2]$ with $n = 4$, 6, 8, and 10. All ionic liquids investigated here were kindly provided by Prof. Tom Welton (Imperial

Table 1 A summary of the ionic liquids investigated in this study

Name	Chemical structure	Abbreviation
1-Butyl-3-methylimidazolium bis-[(trifluoromethane)sulfonyl]imide	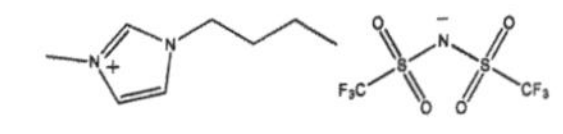	$[C_4C_1Im]$ $[NTf_2]$
1-Hexyl-3-methylimidazolium bis-[(trifluoromethane)sulfonyl]imide	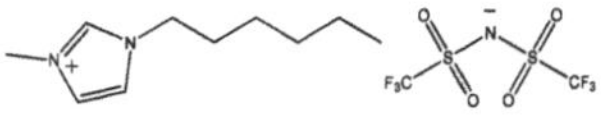	$[C_6C_1Im]$ $[NTf_2]$
1-Butyl-1-methylpyrrolidinium bis-[(trifluoromethane)sulfonyl]imide	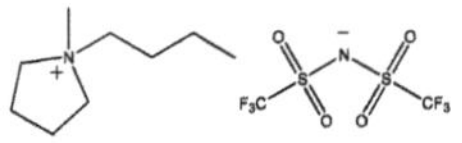	$[C_4C_1Pyrr]$ $[NTf_2]$
1-Hexyl-1-methylpyrrolidinium bis-[(trifluoromethane)sulfonyl]imide	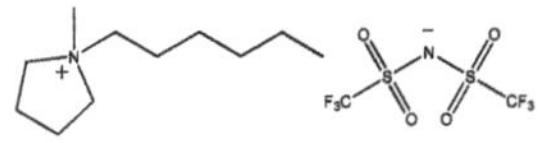	$[C_6C_1Pyrr]$ $[NTf_2]$
1-Octyl-1-methylpyrrolidinium bis-[(trifluoromethane)sulfonyl]imide	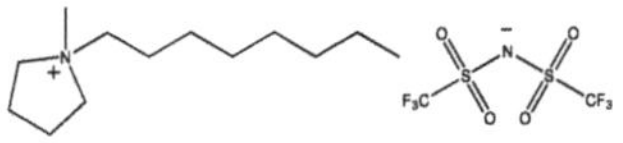	$[C_8C_1Pyrr]$ $[NTf_2]$
1-Decyl-1-methylpyrrolidinium bis-[(trifluoromethane)sulfonyl]imide	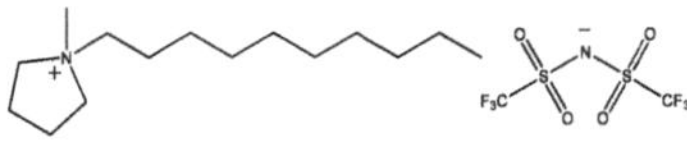	$[C_{10}C_1Pyrr]$ $[NTf_2]$

College London) and Prof. Peter Licence (University of Nottingham), and were synthesised using modifications of existing literature methods[21,22] followed by characterisation using NMR spectroscopy, electrospray ionisation mass spectrometry and Karl Fischer titration. We find that an exceptional level of ionic liquid purity is required for successful SFB experiments, as the presence of even minute amounts of any surface active contamination dramatically affects reproducibility of the measured forces. They were not treated by column chromatography, since this has been shown to introduce particulate impurities.[23] All ionic liquids were thoroughly washed with water (>10 times) to remove any unreacted Li $[NTf_2]$ from ion-exchange synthetic procedures, which has previously been shown to be hard to remove from $[C_nC_1Pyrr][NTf_2]$.[24] After drying *in vacuo* (10^{-2} mbar, 70 °C) for at least 24 h, Karl Fischer titration showed the water levels were <50 ppm. However, the injection process between the surfaces in the apparatus is likely to introduce a small amount of water resulting in a final estimated water content of <200 ppm.

3 Results

3.1 Imidazolium-based ionic liquids

The interaction force as a function of separation distance between atomically smooth mica surfaces immersed in the imidazolium-based ionic liquids[25] $[C_4C_1Im][NTf_2]$ and $[C_6C_1Im][NTf_2]$ are shown in Fig. 2. For both liquids the normal interaction force, F_N, between the surfaces oscillates between repulsive walls and attractive minima for surface separations below ~10–15 nm. Forces were measured from a mica–mica separation distance of >300 nm, however no force was detected (above the resolution of our measurement) until the surfaces approached to the range shown in the figure.

It is now well established[26,27] that these oscillatory forces arise from the structure in the liquid near solid surfaces, as follows. The liquid is arranged in

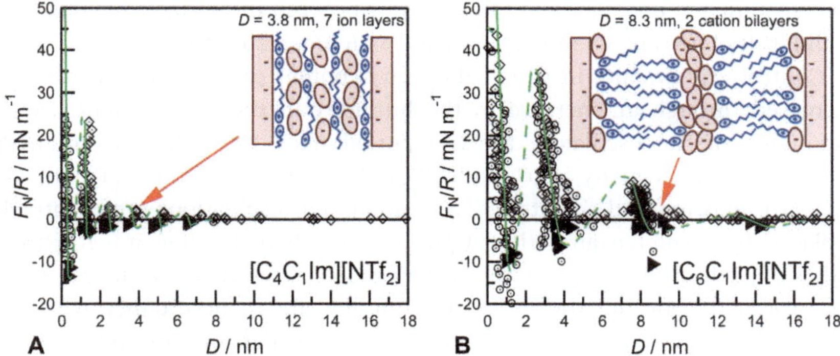

Fig. 2 Measured normal force, F_N, normalised by radius of curvature, R, between mica surfaces across (A) $[C_4C_1Im][NTf_2]$ and (B) $[C_6C_1Im][NTf_2]$, as a function of surface separation, D. Open diamond data points were measured on approach and open circles on retraction of the surfaces. Jumps inwards occur during attractive regions due to a spring instability when $dF_N/dD > K_N$ shown as dashed lines, at which point ionic liquid layers are squeezed out of the film and the surfaces approach the next closest stable film configuration. Energy minima are similarly detected as jumps outwards to large separations and are shown by filled triangles. Schematics indicate the likely ion layering structures, showing example film configurations for the repulsive walls indicated consisting of alternating cation/anion monolayers for $[C_4C_1Im][NTf_2]$ and toe-to-toe cation bilayers for $[C_6C_1Im][NTf_2]$.

layers between the surfaces, so when the surfaces are at a distance corresponding to a stable layered film then a certain load (or normal force, F_N) can be applied without squeezing out the liquid and this is observed as a (near-)vertical 'wall' in the force profile. Increasing F_N at some point leads to a whole layer being squeezed out of the film, at which point the surfaces jump together to their next stable distance with fewer layers between. Thus, the D-values corresponding to these walls in the oscillatory profile can be used to deduce the thickness of the stable layer-structures in the film between the surfaces.

Whilst uncharged molecular liquids can give rise to structural forces with oscillatory period (or inter-wall distance) close to one molecular diameter due to squeeze-out of one molecular layer at a time,[28] ionic liquids have an additional constraint: in order to maintain electroneutrality equal number of cations and anions must be squeezed out each time the film decreases in thickness.

From the oscillation period in Fig. 2A, which is close to the dimension of one cation–anion pair, we can infer that the ions are arranged in layers of cations and layers of anions. If the anions and cations were co-planar there would be no requirement for the repeat distance to be as great as a full anion–cation pair size. This structural interpretation of the observed layer dimensions is shown schematically in the inset to Fig. 2A for one particular stable film thickness, and is in line with many other interpretations of ionic liquid interfacial and thin-film structure.[6,9,29–31]

The interaction force between the mica surfaces as a function of their separation distance across $[C_6C_1Im][NTf_2]$ is shown in Fig. 2B. Although the alkyl chain substituent is only incrementally increased relative to the $[C_4C_1Im][NTf_2]$, the oscillations in force are dramatically different indicating an entirely different structure. The observed layer dimensions in this case are no longer consistent with alternating cation and anion monolayers, and instead indicate bilayers are the preferred structure as shown in the inset to Fig. 2B. Repulsive walls corresponding to one, two, and three bilayers between the surfaces are observed at film thicknesses of 3.4 nm, 8.3 nm, and 14.6 nm, respectively.

Control experiments in which the water-content of the ionic liquids was increased through an increase in the humidity in the environment, led to expunging of the oscillatory forces reported here. The effect of gradually increasing water activity is to gradually reduce the magnitudes of both the repulsive maxima and the attractive minima, and oscillations are eventually replaced by a sudden attraction. Upon contact there is often much greater adhesion, reflecting the presence of condensed water at the surfaces. These observations regarding the effect of humidity are in accordance with the destabilisation effect of water on nanoparticle dispersions in similar ionic liquids.[32] With precautions taken as described in the Experimental procedure section, we observe no change in forces during the first 5–10 h of each experiment, indicating no substantial increase in water content in the ionic liquid samples during this time.

3.2 Pyrrolidinium-based ionic liquids

Fig. 3 shows F_N/R *vs.* D profiles for a homologous series of pyrrolidinium-based ionic liquids with the same anion, $[C_nC_1Pyrr][NTf_2]$, and again for small D values the forces are oscillatory in nature.[33] Alkyl chain lengths of 4, 6 and 8 carbons show very similar film thicknesses for stable layer structures in the film (Fig. 3A–C). Although the first barrier at $\sim$1.4 nm could not be surmounted to reduce the film thickness to that of only one layer of cations, we can see that the layered structure in this case is consistent with alternating layers of cations and ions described above for $[C_4C_1Im][NTf_2]$. In fact, the ΔD-values are so similar for alkyl chains of 4, 6, and 8 carbons that we can infer a tendency for alkyl chains to lie more or less parallel to surface (rather than normal to the surfaces), *i.e.* in-plane orientation, even for hexyl and octyl cations as shown in the inset cartoons in Fig. 3A–C. In-plane orientation of the alkyl chains would lead to a decrease in the number density of cations within each layer as the alkyl chain increases in length. This leads to a gradually increasing mismatch in the maximum possible ion concentration between the anion and cation layers, and so might be expected to 'frustrate' the overscreening alternating anion–cation layer structure. We propose that these may be the reasons for the substantially lower magnitude of oscillatory forces for $[C_8C_1Pyrr][NTf_2]$ compared to $[C_4C_1Pyrr][NTf_2]$: mismatch of the anion and cation in-plane area for $[C_8C_1Pyrr][NTf_2]$ prevents ideal packing into pure anion layers leading to greater disorder and lower forces required to disrupt the layers. Relatedly, the increasing 2-dimensional 'footprint' of the cation leads to greater disparity between the area per negative charge on the mica and the cation footprint, and so a reduction in the overscreening of the mica charge by one cation layer.

Moving now to look at the $[C_{10}C_1Pyrr][NTf_2]$ in Fig. 3D we see substantially different oscillatory forces compared to $[C_8C_1Pyrr][NTf_2]$. This is attributed to a flip from monolayer to bilayer structure, as explained above for $[C_6C_1Im][NTf_2]$. However a closer inspection of the bilayer repeat-spacing reveals differences in the bilayer structure of $[C_{10}C_1Pyrr][NTf_2]$ compared to $[C_6C_1Im][NTf_2]$: the chains are this time interdigitated and the anions must reside between the headgroups in order for the bilayer thickness to match the measured value. The origin of this

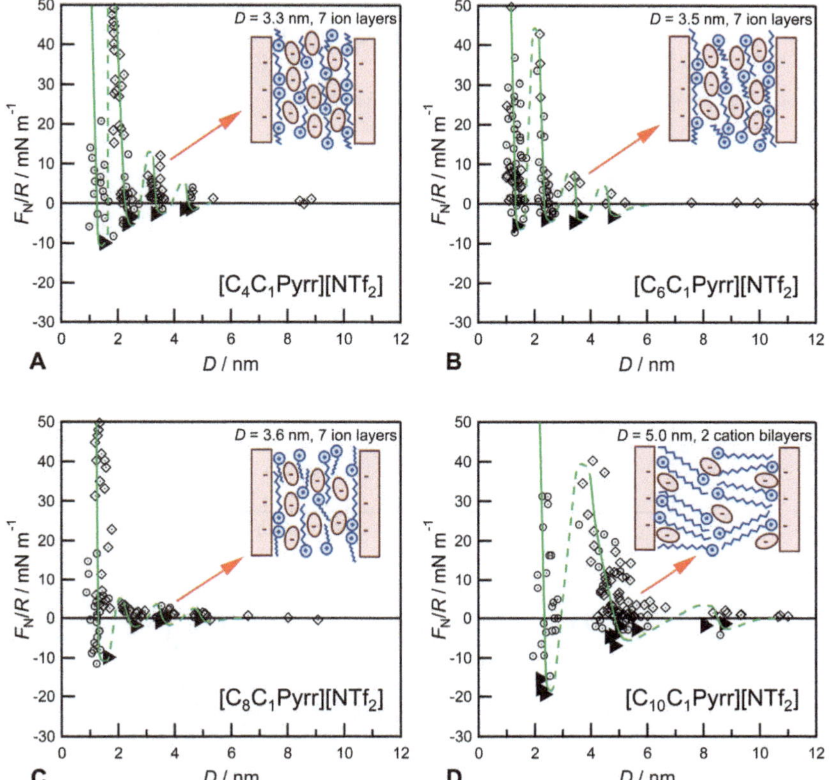

Fig. 3 Measured normal force, F_N, normalised by radius of curvature, R, between mica surfaces across (A) $[C_4C_1Pyrr][NTf_2]$, (B) $[C_6C_1Pyrr][NTf_2]$, (C) $[C_8C_1Pyrr][NTf_2]$ and (D) $[C_{10}C_1Pyrr][NTf_2]$ as a function of surface separation, D. Schematics indicate the likely ion layering structures, showing example film configurations consisting of alternating cation/anion monolayers for $[C_nC_1Pyrr][NTf_2]$, where $n = 4, 6, 8$, and interdigitated cation bilayers for $[C_{10}C_1Pyrr][NTf_2]$.

contrasting structure, which is shown schematically in the inset to Fig. 3D, is discussed in Section 4.

3.3 Equimolar binary mixture of ionic liquids

Fig. 4 presents the interaction force between mica surfaces across the equimolar binary mixture $[C_4C_1Pyrr]_{0.5}[C_{10}C_1Pyrr]_{0.5}[NTf_2]$. Oscillatory structural forces are clearly evident, yet are different to either of the simple ionic liquids $[C_4C_1Pyrr][NTf_2]$ or $[C_{10}C_1Pyrr][NTf_2]$. The layer thickness of the mixture differs appreciably from those of alternating ion layers (monolayers) and are more similar to those of the bilayers as seen for simple $[C_{10}C_1Pyrr][NTf_2]$ (data also plotted on Fig. 4 for comparison). The similarity in the D-values suggests alkyl chain segregation and bilayer-like structures in the $[C_4C_1Pyrr]_{0.5}[C_{10}C_1Pyrr]_{0.5}[NTf_2]$. However the layers are slightly thinner than for simple $[C_{10}C_1Pyrr][NTf_2]$ and the force required to disrupt the bilayer is substantially lower. The shallow slope of the F_N/R *vs.* D for the 2-bilayer film (around 4 nm) indicates that the bilayers are more easily compressed, although the extent to which they can be compressed (difference in the layer thickness between its force minimum and maximum) is similar to simple $[C_{10}C_1Pyrr][NTf_2]$.

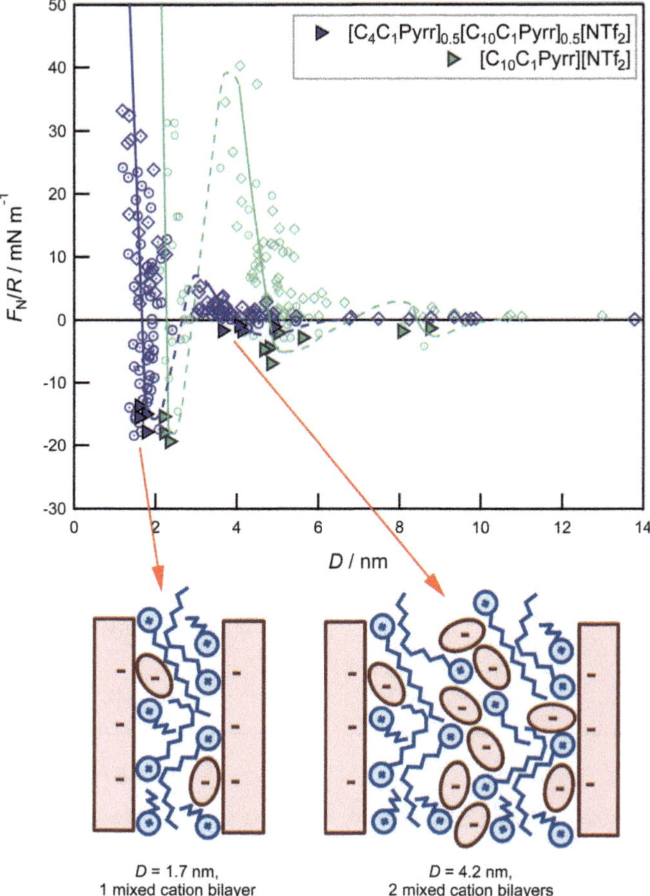

Fig. 4 Measured normal force, F_N, normalised by radius of curvature, R, between mica surfaces across a binary equimolar mixture of $[C_4C_1Pyrr][NTf_2]$ and $[C_{10}C_1Pyrr][NTf_2]$. Also shown for comparison are results for simple $[C_{10}C_1Pyrr][NTf_2]$. Schematics indicate possible layering structures for the mixture, showing bilayer-like structures.

4 Discussion

4.1 Surface charge induced structuring

Two general features of all of the force profiles here are worth noting: (a) for a given liquid, the *period* of oscillation is somewhat larger for thicker films, indicating ion packing that is less ordered (or layers which are more dynamic or disordered) with increasing distance from the surface. The electrostatic interactions result in ions nearest to the surface being held more tightly, with ions further away exhibiting more fluid-like behaviour within a layer. This is likely to be critical to lubrication mechanisms of ionic liquid films.[34] Interestingly, even neutral non-polar molecules behave in a similar manner when confined between mica,[28] suggesting that the liquid molecules simply pack more efficiently adjacent to a surface irrespective of charge. (b) The oscillatory profiles show greatest magnitude of force for small D-values and decreasing *amplitude* as the D increases. Thus, a greater force is required to disrupt the layering for thinner films, and this is likely to be related to decreasing layer-purity – excess of anions or cations – in the layers further from the surfaces.

Is the structure observed in confined liquids induced by the confinement itself, or is the order present at each individual surface independently of the other? Although the results of these oscillatory force profiles seem qualitatively similar to those of spherical non-polar liquids[28,35] and even linear alkanes,[36] the origin of structuring is very different. For non-polar molecular liquids, layered structures upon confinement to thin films are only observed for molecules with a high degree of symmetry which are able to pack together efficiently at film thicknesses corresponding to an integral number of molecular layers.[28,35,36] Importantly, high forces are not required for squeezing out successive layers, suggesting that this liquid to solid-like phase transition is due to geometric constraints alone.[37] Not unexpectedly, mixtures of liquids with differing molecular dimensions such as the addition of cyclohexane to octamethylcyclotetrasiloxane (OMCTS)[38] leads to a reduction in the range and magnitude of the structural force with increasing mole fraction.

Ionic liquids, with their asymmetric ion shapes and frequent mismatch in size between cations and anions, would not be expected to form such rich, layered structures induced by confinement alone. Indeed, experiments using a similar apparatus with ionic liquids between ultrasmooth methyl-terminated surfaces (SAM-coated mica) exhibit no pronounced layering.[30] A poorer surface state compared to mica is however ruled out by the detection of OMCTS solvation forces between the same surfaces. X-ray reflectivity data has drawn similar conclusions about ionic liquid structure at a single, charged surface such as mica and at uncharged graphene.[39]

Instead, the layering observed in ionic liquids is charge-templated and has been observed in many systems as long as the surfaces hold a net charge.[10,40] In the case of the mica surfaces used here, the origin of the negative charge is dissolution of K^+ ions from the (001) plane leaving a negative charge shared between oxygen atoms on the underlying aluminosilicate layer. The solubility of K^+ in these ionic liquids is high,[41] and the ion-exchange of K^+ for imidazolium on mica[42] appears to be favourable, together providing the driving force for removal of the K^+. The maximum density of negative charge on the mica surface, 1 e per 0.48 nm^2 (arising if every K^+ ion is dissociated), is close to half the positive charge density on a full monolayer of [C_4C_1Im] cations.[29] For all the ionic liquids studied here the positive charge density of a cation monolayer is greater than the mica negative density. Thus a full cation monolayer which is attracted to the negative surface overscreens the surface charge and kicks-off an oscillation between positive and negative overscreening, as has been well documented recently.[10,43] Thus the force measured between two mica surfaces, each dressed in its own ion layers, is in fact the result of interference between the two ordered structures emanating from each surface.

4.2 Architecture of bilayer structures

The monolayer-to-bilayer transition and the resulting bilayer architecture for imidazolium-based and pyrrolidinium-based ionic liquids were seen to differ in two ways and in this section we attempt to account qualitatively for these differences.

1. Crossover from monolayer to bilayer behaviour occurs at different chain lengths for imidazolium- and pyrrolidinium-based ionic liquids with the same anion.

For imidazolium cations the crossover in behaviour occurs between alkyl chain lengths of 4 and 6 carbons, whilst for pyrrolidinium cations the crossover is between 8 and 10 carbons. This is likely due to more favourable cation–cation interactions in the case of imidazolium cations: the planar delocalised rings can interact favourably through π–π stacking,[44] allowing close approach of the alkyl chains and favourable dispersion interactions in the non-polar regions. Pyrrolidinium cations, on the other hand, cannot stack at such close spacing and so require longer alkyl chains to drive bilayer formation. Atomistic molecular dynamics simulations of the same imidazolium-based ionic liquids studied here have revealed an equivalent crossover in orientational ordering at a mica surface:[45] imidazolium cations with ethyl or butyl groups were found to lie with their alkyl tails parallel to the surface, while cations with hexyl or octyl tails were oriented perpendicular to it. Cation orientation at a silica surface seems to similarly depend on alkyl chain length, regardless of anion.[46] Bulk nanostructure appears at equivalently longer alkyl chain lengths for pyrrolidinium compared to imidazolium ionic liquids.[8,47]

Weaker force magnitudes in [C_8C_1Pyrr][NTf_2] are another hint as to the origin of the structural transition: longer alkyl chains mean that the ion occupies more area within the layer for monolayer structures. This anion–cation area mismatch eventually frustrates the monolayer structure and favours bilayers.

2. The layer thickness of the bilayers indicates qualitatively different architectures for imidazolium- and pyrrolidinium-based bilayers.

The layer thickness of the bilayer structures in Fig. 2B and 3D indicate qualitatively different bilayer architectures for the imidazolium and pyrrolidinium ionic liquids in two important ways. First: comparing the measured bilayer thickness dimensions of the extended ions suggests *toe-to-toe* bilayers for [C_6C_1Im][NTf_2], but for [$C_{10}C_1$Pyrr][NTf_2] the alkyl chains are significantly *interdigitated* (crumpled or tilted chains are also compatible with the measured thickness). Second: the [NTf_2] anion is sitting on top of the cation headgroup in the case of [C_6C_1Im][NTf_2] bilayers, whereas the [NTf_2] anions are located between the cation headgroups – in the same plane – in the case of [$C_{10}C_1$Pyrr][NTf_2] bilayers. These features are represented schematically in the cartoons in Fig. 2B and 3D.

What is the driving force for the different bilayer architectures? We might start to rationalise these differences by considering the possibility of π–π stacking for the imidazolium cations leading to close approach of the cations and exclusion of the [NTf_2] to a separate layer above and below. Despite their positive charge, favourable π–π stacking structures are expected between imidazolium cations.[44] In contrast, the pyrrolidinium ionic liquids have less favourable headgroup interactions (no π–π stacking is possible) which also encourages the anions to intercalate between the cation headgroups and screen their repulsive electrostatic interaction. The inclusion of the anions within the bilayer for [$C_{10}C_1$Pyrr][NTf_2] leads to some contact between the [NTf_2] anion and the non-polar tails of the cation; this is not unexpected since this polarisable anion can well engage in dispersion interactions. As a result of the inclusion of [NTf_2] anions between the cations of the pyrrolidinium bilayer, the cross-sectional area of each ion pair is very much larger than the cross-section of the hydrocarbon tail. This leads to the interdigitation of the hydrocarbon chains on opposite sides of the bilayer leaflet in order to optimally fill the space in the film, and thus smaller bilayer thickness compared to imidazolium. This difference is in accordance with small and wide angle scattering experiments on bulk ionic liquids where the smaller non-polar domains detected for piperidinium ionic liquids (in comparison to imidazolium analogues) were attributed to the bulky and non-aromatic headgroup.[48]

Having made such observations, we suggest that the bilayer structures could be rationalised using a single geometrical factor or packing parameter, P, as is commonly employed to explain self-assembled architectures in water:

$$P = v_c/al_c$$

Where v_c is the volume of the alkyl chain, l_c is the length of the alkyl chain, and a is the headgroup area including cation and anion. Although our data suggests that the anions are between the headgroups for the $[C_{10}C_1Pyrr][NTf_2]$ and outside of the cation bilayer for $[C_6C_1Im][NTf_2]$ there remain several possibilities for their relative positions which lead to a range of possible values for the headgroup area, a.

Nonetheless, by considering the possible positions which give rise to the maximum and minimum possible headgroup areas, we can reach limiting values of P for each ionic liquid. These orientations and the resulting packing parameters are shown in Table 2. For $[C_6C_1Im][NTf_2]$ three possible anion–cation positions are shown: A and B are most likely, and lead to $P \sim 0.8$–1.0 in line with the observed toe-to-toe bilayer structure. For $[C_{10}C_1Pyrr][NTf_2]$ the most likely of the two orientations shown is D which leads to $P \sim 0.4$; again in line with the observed interdigitation in the bilayer. Although this packing parameter analysis was merely demonstrated to be reasonable in this case – since we used knowledge of the structures from our measurements to predict the orientations – we suggest that it may be usefully applied to other ionic liquid structures in rationalising or predicting self-assembled structure.

Table 2 Possible cation–anion arrangements for the two bilayer-forming ionic liquids, $[C_6C_1Im][NTf_2]$ and $[C_{10}C_1Pyrr][NTf_2]$, and the resulting packing parameters, P, for each

Bilayer-forming ionic liquids	$[C_6C_1Im][NTf_2]$			$[C_{10}C_1Pyrr][NTf_2]$	
Possible ion orientations	A	B	C	D	E
v_c/nm^3	0.2	0.3	0.2	0.3	0.3
a/nm^3	0.2	0.5	0.5	0.4	0.7
l_c/nm^3	0.9	0.9	0.9	1.5	1.5
Packing parameter	1.0	0.8	0.4	0.4	0.3

4.3 Structure of the binary mixture $[C_4C_1Pyrr]_{0.5}[C_{10}C_1Pyrr]_{0.5}[NTf_2]$

It is immediately clear that the structural forces for the binary mixture are different to either of the simple liquids: this result implies that both $[C_4C_1Pyrr]^+$ and $[C_{10}C_1Pyrr]^+$ cations are present in the confined film, rather than one component being substantially more surface active than the other and thus segregating from the bulk mixture in the thin film. This is in agreement with angle-resolved X-ray photoelectron spectroscopy (ARXPS) of the mixture $[C_2C_1Im]_{0.9}[C_{10}C_1Im]_{0.1}[NTf_2]$ showing[49] homogeneous distribution and no specific enrichment of longer chain cations at the ionic liquid–vacuum surface.

The oscillatory structural forces show repulsive walls at distances too great to be explained by monolayer arrangements and similar to those observed for the simple $[C_{10}C_1Pyrr][NTf_2]$ bilayers; thus suggesting bilayers are present in the mixture. Although the bilayers have similar thickness to the simple $[C_{10}C_1Pyrr]$-$[NTf_2]$ bilayers (or slightly thinner), they are substantially more compressible and require lower force to rupture or squeeze-out the bilayers.

Why does the equimolar mixture take on the structure of the longer-chain cation rather than the shorter-chain cation? We might have expected that the 'average' chain length would determine the structure, since this gives the fraction of the liquid taken up by non-polar groups. However the mean alkyl chain length for this mixture is 7 carbons per ion pair, whereas we have shown above that simple $[C_8C_1Pyrr][NTf_2]$ still forms monolayers. So it appears that the longer of the chains dictates the structure in this case. This is reminiscent of the effect of mixtures on bulk nanostructure, where the longer chain liquid is again predominant in determining the size of the domains[50] (see also the subsequent discussion of this work[51]). It will be interesting to see in future how the concentration of $[C_{10}C_1Pyrr][NTf_2]$ in $[C_4C_1Pyrr][NTf_2]$ alters the structure: how little $[C_{10}C_1Pyrr][NTf_2]$ need one to add to $[C_4C_1Pyrr][NTf_2]$ in order to induce it to flip from monolayer to bilayer structure?

5 Conclusions

The ionic liquids investigated in this study all demonstrate oscillatory structural forces upon confinement to thin films between atomically smooth and charged surfaces. These alternating forces of repulsion and attraction are attributed to sequential squeeze-out of charge-induced layers of ions between the surfaces, originating from the overscreening phenomenon. Consequently, the dominant structure (for most ionic liquids with short alkyl chain substituents) at charged surfaces is one of alternating layers of cations and anions. Increasing the alkyl chain length and hence amphiphilicity of the cation, however, can result in cation bilayer formation reminiscent of surfactant self-assembly in water and domain formation in bulk ionic liquids. In this case the non-polar parts of the cation segregate together away from the polar, charged moieties.

Here we have compared and contrasted layered structures of imidazolium and pyrrolidinium-based ionic liquids with the same anion. The chemistry of the cation headgroup plays a pivotal role in determining both the chain length required for this behaviour crossover and the resultant bilayer architecture. We employ packing parameter arguments to rationalise the different architectures, and propose that the orientation of the $[NTf_2]^-$ anions in relation to the cation headgroups is key in determining whether or not the alkyl segregation involves interdigitation, and hence the dimensions of the segregated domains. A mixture of short and long chain pyrrolidinium-based ionic liquids – components which individually would favour different structures – give rise to oscillatory periods indicative of alkyl chain segregation, revealing how bilayer-like structures are favoured in the mixed system.

These results have direct relevance to applications of ionic liquids at solid surfaces in confined mesoscopic geometries; including electrochemical applications (where the electrolyte is confined within porous electrodes, the use of ionic liquids as lubricants and for nanoparticle stabilisation. Furthermore they illustrate the remarkable self-assembly effects that can occur in ionic liquids at solid interfaces.

Acknowledgements

This work was supported by The Leverhulme Trust (F/07 134/DK and F/07 134/DN), Taiho Kogyo Tribology Research Foundation, Weizmann UK, Infineum UK, The Office of Naval Research (N00014-10-1-0096) and the EPSRC (EP/J015202/1). We are grateful to Patricia Hunt and Tamar Greaves for useful discussions.

References

1 J. P. Hallett and T. Welton, *Chem. Rev.*, 2011, **111**, 3508–3576.
2 Y. Wang and G. A. Voth, *J. Am. Chem. Soc.*, 2005, **127**, 12192–12193.
3 A. Triolo, O. Russina, H.-J. Bleif and E. Di Cola, *J. Phys. Chem. B*, 2007, **111**, 4641–4644.
4 J. N. A. Canongia Lopes and A. A. H. Pádua, *J. Phys. Chem. B*, 2006, **110**, 3330–3335.
5 B. L. Bhargava, R. Devane, M. L. Klein and S. Balasubramanian, *Soft Matter*, 2007, **3**, 1395–1400.
6 R. Hayes, S. Z. El Abedin and R. Atkin, *J. Phys. Chem. B*, 2009, **113**, 7049–7052.
7 S. Perkin, *Phys. Chem. Chem. Phys.*, 2012, **14**, 5052–5062.
8 S. Li, J. L. Bañuelos, J. Guo, L. Anovitz, G. Rother, R. W. Shaw, P. C. Hillesheim, S. Dai, G. A. Baker and P. T. Cummings, *J. Phys. Chem. Lett.*, 2011, **3**, 125–130.
9 M. Mezger, H. Schröder, H. Reichert, S. Schramm, J. S. Okasinski, S. Schöder, V. Honkumaki, M. Deutsch, B. M. Ocko, J. Ralston, M. Rohwerder, M. Stratmann and H. Dosch, *Science*, 2008, **322**, 424–428.
10 R. M. Lynden-Bell, A. I. Frolov and M. V. Fedorov, *Phys. Chem. Chem. Phys.*, 2012, **14**, 2693–2701.
11 F. Zhou, Y. Liang and W. Liu, *Chem. Soc. Rev.*, 2009, **38**, 2590–2599.
12 M. Armand, F. Endres, D. R. MacFarlane, H. Ohno and B. Scrosati, *Nat. Mater.*, 2009, **8**, 621–629.
13 J. Dupont and J. D. Scholten, *Chem. Soc. Rev.*, 2010, **39**, 1780–1804.
14 S. A. Forsyth, S. R. Batten, Q. Dai and D. R. MacFarlane, *Aust. J. Chem.*, 2004, **57**, 121–124.
15 S. P. Ong, O. Andreussi, Y. B. Wu, N. Marzari and G. Ceder, *Chem. Mater.*, 2011, **23**, 2979–2986.
16 R. Wibowo, S. E. W. Jones and R. G. Compton, *J. Chem. Eng. Data*, 2010, **55**, 1374–1376.
17 H. Niedermeyer, J. P. Hallett, I. J. Villar-Garcia, P. A. Hunt and T. Welton, *Chem. Soc. Rev.*, 2012, **41**, 7780–7802.
18 E. T. Fox, J. E. F. Weaver and W. A. Henderson, *J. Phys. Chem. C*, 2012, **116**, 5271–5275.
19 S. Perkin, L. Chai, N. Kampf, U. Raviv, W. Briscoe, I. Dunlop, S. Titmuss, M. Seo, E. Kumacheva and J. Klein, *Langmuir*, 2006, **22**, 6142–6152.
20 J. Israelachvili, *J. Colloid Interface Sci.*, 1973, **44**, 259–272.
21 S. Men, K. R. J. Lovelock and P. Licence, *Phys. Chem. Chem. Phys.*, 2011, **13**, 15244–15255.
22 M. A. Ab Rani, A. Brant, L. Crowhurst, A. Dolan, M. Lui, N. H. Hassan, J. P. Hallett, P. A. Hunt, H. Niedermeyer, J. M. Perez-Arlandis, M. Schrems, T. Welton and R. Wilding, *Phys. Chem. Chem. Phys.*, 2011, **13**, 16831–16840.
23 F. Endres, S. Z. El Abedin and N. Borissenko, *Z. Phys. Chem.*, 2006, **220**, 1377–1394.
24 S. Men, B. B. Hurisso, K. R. J. Lovelock and P. Licence, *Phys. Chem. Chem. Phys.*, 2012, **14**, 5229–5238.

25 S. Perkin, L. Crowhurst, H. Niedermeyer, T. Welton, N. N. Gosvami and A. M. Smith, *Chem. Commun.*, 2011, **47**, 6572–6574.
26 R. G. Horn, D. F. Evans and B. Ninham, *J. Phys. Chem.*, 1988, **92**, 3531–3537.
27 R. G. Horn and J. N. Israelachvili, *J. Chem. Phys.*, 1981, **75**, 1400–1411.
28 R. Horn and J. Israelachvili, *J. Chem. Phys.*, 1981, **75**, 1400–1411.
29 S. Perkin, T. Albrecht and J. Klein, *Phys. Chem. Chem. Phys.*, 2010, **12**, 1243–1247.
30 I. Bou-Malham and L. Bureau, *Soft Matter*, 2010, **6**, 4062–4065.
31 K. Ueno, M. Kasuya, M. Watanabe, M. Mizukami and K. Kurihara, *Phys. Chem. Chem. Phys.*, 2010, **12**, 4066–4071.
32 E. Vanecht, K. Binnemans, S. Patskovsky, M. Meunier, J. W. Seo, L. Stappers and J. Fransaer, *Phys. Chem. Chem. Phys.*, 2012, **14**, 5662–5671.
33 A. M. Smith, K. R. J. Lovelock, N. N. Gosvami, P. Licence, A. Dolan, T. Welton and S. Perkin, *J. Phys. Chem. Lett.*, 2013, 378–382.
34 A. M. Smith, K. R. J. Lovelock, N. N. Gosvami, T. Welton and S. Perkin, *Phys. Chem. Chem. Phys.*, 2013, **15**, 15317.
35 J. Klein and E. Kumacheva, *J. Chem. Phys.*, 1998, **108**, 6996–7008.
36 H. K. Christenson, D. W. R. Gruen, R. G. Horn and J. N. Israelachvili, *J. Chem. Phys.*, 1987, **87**, 1834–1841.
37 J. Klein and E. Kumacheva, *Science*, 1995, **269**, 816–819.
38 H. K. Christenson, *Chem. Phys. Lett.*, 1985, **118**, 455–458.
39 H. Zhou, M. Rouha, G. Feng, S. S. Lee, H. Docherty, P. Fenter, P. T. Cummings, P. F. Fulvio, S. Dai, J. McDonough, V. Presser and Y. Gogotsi, *ACS Nano*, 2012, **6**(11), 9818–9827.
40 R. Hayes, G. G. Warr and R. Atkin, *Phys. Chem. Chem. Phys.*, 2010, **12**, 1709–1723.
41 R. Wibowo, L. Aldous, S. E. W. Jones and R. G. Compton, *Chem. Phys. Lett.*, 2010, **492**, 276–280.
42 B. H. Cipriano, S. R. Raghavan and P. M. McGuiggan, *Colloids Surf., A*, 2005, **262**, 8–13.
43 M. V. Fedorov and A. A. Kornyshev, *J. Phys. Chem. B*, 2008, **112**, 11868–11872.
44 P. A. Hunt and R. P. Matthews, submitted.
45 R. S. Payal and S. Balasubramanian, *ChemPhysChem*, 2012, **13**, 1764–1771.
46 J. B. Rollins, B. D. Fitchett and J. C. Conboy, *J. Phys. Chem. B*, 2007, **111**, 4990–4999.
47 C. S. Santos, N. S. Murthy, G. A. Baker and E. W. Castner, *Journal of Chemical Physics*, 2011, **134**, 4.
48 A. Triolo, O. Russina, B. Fazio, G. B. Appetecchi, M. Carewska and S. Passerini, *J. Chem. Phys.*, 2009, **130**, 164521–164526.
49 F. Maier, T. Cremer, C. Kolbeck, K. R. J. Lovelock, N. Paape, P. S. Schulz, P. Wasserscheidc and H. P. Steinruck, *Phys. Chem. Chem. Phys.*, 2010, **12**, 1905–1915.
50 O. Russina and A. Triolo, *Faraday Discuss.*, 2012, **154**, 97–109.
51 *Faraday Discussions*, 2012, **154**, 189–220.

Faraday Discussions RSCPublishing

PAPER

A comparative study on bulk and nanoconfined water by time-resolved optical Kerr effect spectroscopy

Andrea Taschin,[a] Paolo Bartolini,[a] Agnese Marcelli,[a] Roberto Righini[ab] and Renato Torre*[ac]

Received 24th April 2013, Accepted 28th May 2013
DOI: 10.1039/c3fd00060e

The low frequency (ν < 500 cm^{-1}) vibrational spectra of hydrated porous silica are specifically sensitive to the hydrogen bond interactions and provide a wealth of information on the structural and dynamical properties of the water contained in the pores of the matrix. We investigate systematically this spectral region for a series of Vycor porous silica samples (pore size ≈ 4 nm) at different levels of hydration, from the dry matrix to completely filled pores. The spectra are obtained as the Fourier transforms of time-resolved heterodyne detected optical Kerr effect (HD-OKE) measurements. The comparison of these spectra with that of bulk water enables us to separately extract and analyze the spectral contributions of the first and second hydration layers, as well as that of bulk-like inner water. We conclude that the extra water entering the pores above ≈10% water/silica weight ratio behaves very similarly to bulk water. At lower levels of hydration, corresponding to two complete superficial water layers or less, the H-bond bending and stretching bands, characteristic of the tetrahedral coordination of water in the bulk phase, progressively disappear: clearly in these conditions the H-bond connectivity is very different from that of liquid water. A similar behavior is observed for the structural relaxation times measured from the decay of the time-dependent HD-OKE signal. The value for the inner water is very similar to that of the bulk liquid; that of the first two water layers is definitely longer by a factor ≈4. These findings should be carefully taken into account when employing pore confinement to extend towards lower temperatures the accessible temperature range of supercooled water.

1 Introduction

The surface interactions between a liquid and a solid produce local modification on both material properties. In particular, the liquid layers at the interface show structural and dynamic alterations that turn into non-trivial modification of the

[a]European lab. for Non-Linear Spectroscopy (LENS), Univ. di Firenze, via N. Carrara 1, I-50019 Sesto Fiorentino, Firenze, Italy. E-mail: torre@lens.unifi.it
[b]Dip. di Chimica, Univ. di Firenze, via Della Lastruccia 13, I-50019 Sesto Fiorentino, Firenze, Italy
[c]Dip. di Fisica e Astronomia, Univ. di Firenze, via Sansone 1, I-50019 Sesto Fiorentino, Firenze, Italy

fundamental chemical–physical properties of the liquid. Thus, in case of complete spatial confinement the behaviour of the liquid is determined by its interfacial properties.[1]

The investigation and study of confined liquids is relevant to a broad variety of scientific topics, spanning from biology to geophysics. Recently, studies aimed to understand water anomalies have boosted a large interest on the proprieties of water confined in silica nanopores. It is well known that a number of chemical–physical properties of bulk water differ from those of the other liquids.[2,3] In particular the temperature dependence of several experimental observables have unexpected and counter-intuitive behaviours. For example, some thermodynamic quantities (*e.g.* isothermal compressibility, isobaric heat capacity) show a sharp increase upon cooling and as well, some dynamic features (*e.g.* viscosity and structural relaxation time) present a critical slowing down when temperature is lowered. Many experimental and simulation works suggested the existence of a water singularity temperature, T_s, around 223–228 K at atmospheric pressure, which should reflect the existence of a critical process taking place in the supercooled water phase (*i.e.* the metastable non-equilibrium phase that water enters when it is cooled below the melting point and crystallization does not take place). The nature of this critical phenomenon is still largely debated: some authors postulate the existence of a liquid–liquid phase transition,[4,5] another approach sees the water anomalies as non-equilibrium phenomena precursory of the liquid–crystal transition.[6] Moreover, the relevance of ice spontaneous nucleation has been recognized.[7,8] A relevant experimental drawback is that the critical evidence is expected to occur at temperature/pressure values not directly accessible in bulk samples, and this prevents a direct experimental solution of the problem; homogeneous nucleation fixes the lowest reachable temperature at about 231 K for bulk water. Actually, in a macroscopic bulk sample the crystallization phenomena and the experimental difficulties limit the minimum temperatures around 243 K.

Nanoconfinement has the advantage of preventing water freezing and of enabling the experimental investigation of the supercooled phase down to very low temperatures. Nevertheless, the water crystallization doesn't take place only if the confinement is very tight, reaching the nanometric length scale. Already the confinement of water in silica nanopores of 10 nm diameter produces a lowering of the crystallization temperature of about 15 degrees. In order to maintain the water liquid approaching T_s a confinement of about 2.5 nm is required,[9] in pore of diameter $\leq$2 nm water remains in a liquid-like phase also below that singularity temperature.[10]

Here comes the basic question: to what extent do the interactions with the pore surfaces modify the water properties?

The debate on this issue is open.[11–22] All the researchers agree that some structural and dynamic properties of water at the silica interface are modified from those of the bulk liquid, but there is no agreement about if these modifications involve also the water fraction situated in the inner part of the pore, not in direct contact with the surface.

In order to give a contribution to the understanding of this problem, we undertook a comparative investigation of bulk and nanoconfined water by means of time-resolved optical Kerr effect spectroscopy.

Because of the intrinsic experimental difficulties, time-resolved non-linear spectroscopy[23] has been adopted in a limited number of cases to investigate the dynamics of nanoconfined liquids. Optical Kerr effect experiments (OKE) proved to be capable of measuring the sub-picosecond dynamics of liquids inside nanopores;[24] the transient grating technique has been utilized to study slower dynamical processes in liquid filled nanoporous glasses.[11,25–29] In a previous study[30] the effect of nanoconfinement on the relaxation dynamics of liquid water has been investigated by OKE experiments.

Here we present an experimental investigation of nanoconfined water by heterodyne detected optical Kerr effect spectroscopy. By using an improved experimental set-up,[31,32] we were able to measure the entire water dynamics, including the vibrational components not previously investigated. Moreover, the precise determination of the instrumental function, the very good signal-to-noise level of the acquired data and the wide dynamic range investigated enabled us to Fourier transform the measured data into the frequency domain with very high accuracy.

2 Optical Kerr effect experiments

In an optical Kerr effect (OKE) experiment[33,34] a linear-polarized short laser pulse (the pump) induces a transient birefringence in an optically transparent medium. The induced birefringence is measured by monitoring the polarization changes produced in second laser pulse (the probe), which is spatially superimposed with the pump pulse into the sample. The time behaviour of the induced birefringence is reconstructed by changing the time delay between the pump and probe.

The resulting signal reflects the relaxation and vibrational response of the molecules in the sample. The method enables to measure the fast relaxation processes and the low frequency vibrational correlation functions of the system by employing femtosecond laser pulses. OKE techniques enable measurements in a wide time window (from tens of femtoseconds to hundreds of picoseconds) and is able to reveal very different dynamic regimes, from that typical of simple liquids to that characterizing supercooled liquids and glass formers.[35–39]

In our case, the laser system is a self-mode-locked Ti:sapphire laser producing pulse of 20 fs duration with energy of 3 nJ. A detailed description of the optical set-up and of the experimental apparatus is reported in ref. 31, 32. We adopted a peculiar configuration for the heterodyne detection,[40] with a circularly polarized probe beam and differential acquisition of two opposite-phase signals on a balanced double photodiode. Thus, the measured OKE signal is automatically heterodyned and free from possible spurious signals. Finally, the photodiode output signal is processed by a lock-in amplifier phase locked to the reference frequency at which the pump beam is chopped.

With optical heterodyne detection, the OKE signal is directly proportional to the material response function, $R(t)$, convoluted with the instrumental function, $G(t)$. The response $R(t)$ is directly connected to the time derivative of the time-dependent correlation function of the dielectric susceptibility:[33,34]

$$R(t) \propto -\frac{\partial}{\partial t}\langle \chi(t)\chi(0) \rangle \tag{1}$$

The Fourier transform of $R(t)$ corresponds to the frequency-dependent response measured in depolarized light scattering (DLS) experiments.[33,34]

In order to get access to the OKE response, the knowledge of the correct instrumental function $G(t)$ is fundamental, especially for the short delay time range.

This is particularly critical for weak signals characterized by complex relaxation dynamics, as in the case of bulk water or confined water. In short, the critical point in the acquisition of the instrumental function is that any minor, apparently negligible, adjustment of the optical set-up when passing from recording the instrumental function to measuring the water signal, prevents the extraction of the correct instrumental function. The resulting inaccurate deconvolution of the water signal degrades the quality of the overall fitting. To minimize these effects, $G(t)$ was obtained in our experiment by measuring the OKE signal of a CaF_2 plate of the same thickness as the Vycor sample placed side by side in the same cell. The switching from the water measurement configuration to the instrumental one was achieved by simply translating the cell perpendicularly to the optical axis of the experiment without even touching the rest of optical set-up.[32]

Our samples are Vycor slabs (code 7930 by Corning Company) of $8 \times 8 \times 2\ mm^3$ dimensions, with porosity 28% of volume, internal surface area $S = 250\ m^2\ g^{-1}$, average pore diameter of 4 nm and density of 1.5 g cm^{-3} (in dry condition), according to the manufacturer technical data sheet. The samples were cleaned with 35% hydrogen peroxide solution (heating up to 90 °C for 2 h) and washed in distilled water. They were then stored in P_2O_5 (phosphoric anhydride) until use. All the measurements have been carried out at room temperature.

The HD-OKE experiments require a very good optical quality of the samples investigated. Vycor porous glass is among the few porous glasses having a solid structure that enables preparation of macroscopic samples with polished surfaces of optical quality.

In spite of the good surface quality, the Vycor samples have, especially at low hydration level, an intrinsic scattering ability, probably due to the presence of some heterogeneities whose size is comparable to the laser wavelength. Part of the scattered pump light propagates collinearly to the probe and interferes with it on the photodetector, giving rise to an unwanted signal. This spurious contribution appears only during the time superposition of the two pulses and degrades the quality of the OKE data at very short delay time. This disturbing signal was removed by inserting a piezo-driven vibrating mirror in the optical path of the pump arm. We set the amplitude of the mirror stroke and the frequency of the sinusoidal driving signal in a way that the induced frequency modulation transfers the spectral content of the spurious signal out of the acceptance band of the lock-in amplifier (the bandwidth is determined by the chopper frequency and by the bandwidth of the notch-filter at the lock-in amplifier input). This device allowed us to almost completely remove the unwanted contribution.

3 Results on bulk liquid water

The hydrogen bond network largely determines the intra- and inter-molecular vibrational spectrum of liquid water; thus the experimental investigation of the spectral features gives precious information on the local H-bond network structures. The investigation of the high frequency vibrational spectrum, 1000–4000 cm^{-1}, gives access to the intramolecular vibrational dynamics. The low frequency spectrum of water, $\nu < 1000\ cm^{-1}$, reflects its intermolecular dynamics and so it is particularly sensitive to local molecular structures. This frequency range of the water spectrum has been investigated in several light scattering studies, see for example ref. 41–47. Two main broad peaks are observed around

50 and 175 cm^{-1} at room temperature, generally attributed to the hydrogen bond "bending" and "stretching" vibrations, respectively. The spectra show also a very low frequency wing, $\nu < 10$ cm^{-1}, which has been attributed to "relaxation" processes.[42] The light scattering data were measured at relatively high temperatures and have rather poor signal-to-noise ratio due to the very weak water signal. These drawbacks did not enable identification of all the possible vibrational modes possibly contributing to the water spectrum in this region. Time-resolved non-linear spectroscopy has been shown in recent years to be a very useful tool for the investigation of complex liquid dynamics,[23,33] and in particular of liquid water.[32,38,48–51] Recent heterodyne-detected optical Kerr effect (HD-OKE) investigations of liquid water[32,38] at temperature below the melting point (*i.e.* in the supercooled phase) report interesting evidence of critical phenomena.

In Fig. 1 we report the HD-OKE data of liquid water at 293 K: the data clearly show the signature of fast vibrational dynamics at short times, extending up to 1 ps, followed by a slower monotonic relaxation. The initial oscillatory component provides information about the intermolecular hydrogen-bond dynamics, and can be directly compared to the low-frequency Raman spectrum.[48] At longer times the data display a monotonic decay due to structural relaxation. The relaxation features presented by water are very similar to those of the glass-former liquids,[36] and the slow dynamics are found to be in agreement with the main mode-coupling theory predictions.[38] For the analysis of the whole time-dependent correlation function measured in optical experiments a model is required that enables an operative parametrization of the vibrational modes present in the water dynamics. This turns out to be a complex task, and different attempts have been pursued.[32,49–51] We decided to use a relatively simple approach, which allows the comparative analysis of HD-OKE data obtained both in bulk and nanoconfined water.

The data are fitted in the time scale according to the following expressions:

$$S(t) = \int[k\delta(t - t') + R(t - t')]G(t')\mathrm{d}t' \tag{2}$$

$$R(t) = B\frac{\mathrm{d}}{\mathrm{d}t}\exp\left[-\left(\frac{t}{\tau}\right)^{\beta}\right] + \sum_i C_i\exp(-\gamma_i^2 t^2)\sin(\omega_i t) \tag{3}$$

Eqn (2) describes the convolution of the response function, $R(t)$, with the instrumental function, $G(t)$, where the δ-function reproduces the instantaneous electronic response.[52] Eqn (3) gives the response function simulating the material dynamics. The liquid water dynamics is described by the sum of a relaxation function in the form of a stretched exponential,[38] and of a few damped oscillators (DO). We are aware that this is a simple fitting function that turns out to be inappropriate to reproduce the supercooled water HD-OKE data at low temperatures,[32] nevertheless it is sufficient to describe the liquid water at relative high temperature or/and in nanoconfinement. The fitting parameters of the model are: the structural relaxation time τ, the stretching factor β, the frequency ω_i and the damping constants γ_i of the DOs.

In the upper panel of Fig. 1 we show the fit, obtained with eqn (2) and (3), of the measured HD-OKE signal: the fitting function is able to reproduce correctly the experimental data over the whole time window. The parameters used to fit the data are: structural relaxation: $\tau = 0.35$ ps and $\beta = 0.6$; vibrational dynamics: the bending mode at about 50 cm^{-1} is described by 2 DOs ($\omega_1 = 47$ cm^{-1}, $\omega_2 = 100$ cm^{-1}) and the stretching mode around 175 cm^{-1} needs 2 DOs ($\omega_3 =$ 173 cm^{-1} and $\omega_4 = 238$ cm^{-1}) to be properly reproduced.

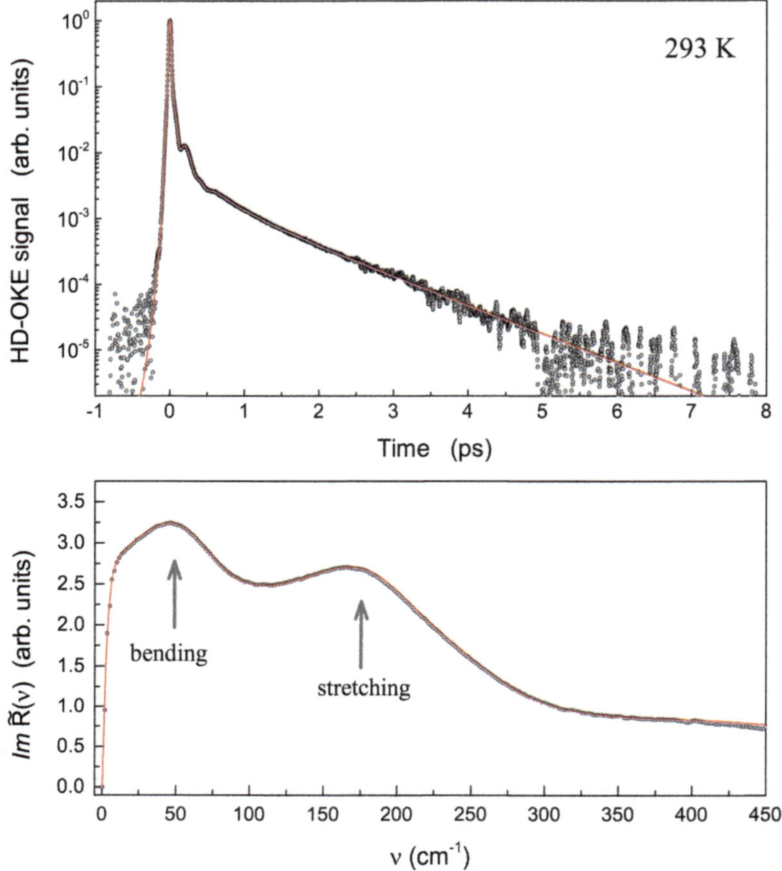

Fig. 1 We report here a HD-OKE data of bulk water at 293 K. In the upper panel, the experimental data are reported in the time scale with the best fit result (red line). The data are reported on a log–linear scale. In the lower panel, we show the HD-OKE response function Fourier transform in the frequency domain. The water spectrum displays the two main vibrational bands: bending and stretching located around 50 and 175 cm^{-1}, respectively.

The HD-OKE data, collected in time domain, can be Fourier transformed in the frequency domain to yield the corresponding spectra. This enables a more immediate visualization of the different vibrational modes present in the water dynamics and a direct comparison with the depolarized light scattering experiments.[23,32,33] In order to get the spectra of the HD-OKE response function, we need to: i) Fourier transform the measured data, ii) deconvolute them from the instrumental response and iii) retain the imaginary part of it, i.e. $\text{Im}[\tilde{R}(\nu)] \propto \text{Im}\{\text{FT}[S(t)]/\text{FT}[G(t)]\}$. The details about these procedures are reported by Taschin *et al.*[32]

The $\text{Im}[\tilde{R}(\nu)]$ obtained from the HD-OKE data and the corresponding fits are reported in the lower panel of Fig. 1. The frequency response shows clearly the main vibrational features present in the liquid water spectrum: the two bands around 50 and 175 cm^{-1}, the "bending" and "stretching" modes. These vibrations are characteristic of the intermolecular dynamics of the first neighbour molecular cage,[53–55] that for liquid water are largely determined by the hydrogen bonds. The shoulder appearing at very low frequency, $\nu < 10$ cm^{-1}, is due to the relaxation processes. As expected, this feature is hardly detectable in the frequency domain, whereas the vibrational modes are clearly visible.

4 Results on nanoconfined water

4.1 Hydration control and FT-IR spectroscopy

We studied Vycor glasses characterized by variable water filling, from "dry" to "fully hydrated". The hydrophilic properties of porous silica samples depend on the degree of surface hydroxylation, *i.e.* on the amount of Si–OH groups. This quantity determines the potential hydration rate of the surface and may influence also the structure of the first water layer, in which the molecules interact with the silanol groups. In the literature, silanols are often indicated as chemical (or chemisorbed) water, as they are the result of dissociative chemisorption of water molecules onto the silica surface, while the molecular water interacting with the surface is referred to as physical (or physisorbed) water.[56,57]

In spite of the amount of work done on this subject, water filling in silica nanopores is still a debated issue.[13,58] Basically, two different hypothesis have been formulated. According to the first model, water fills the pore following a layer by layer process until the full hydration level is reached.[58] A second different filling process has been described in a recent study:[13] at low hydration level water fills the pore forming the first layer, as in the previous model, but at higher hydration water starts to form liquid plugs connecting the pore surfaces that coexist with the surface-adsorbed water.

Fourier Transform Infrared (FT-IR) spectroscopy is a valid tool to qualitatively and quantitatively characterize the hydration process of porous glass. Spectra were recorded by an ALPHA FT-IR spectrometer (Bruker Optics) in the range of 4200–5500 cm^{-1}, where two absorption bands can be well-distinguished at $\approx$4550 and $\approx$5260 cm^{-1} arising from combination of stretching and bending modes of silanol and water, respectively.[59]

Dry samples were obtained by heating Vycor at 400 °C for 10 h, while samples at different hydration levels were prepared by exposing dry Vycor slabs to a water atmosphere for 12 h in closed vials, previously purged with nitrogen. Samples with different amounts of water filled nanopores were obtained *via* a vapor phase exothermic transfer from the bulk.[60] The water concentration in each sample was adjusted adding different volumes of bulk water into the vial containing Vycor by means of a microsyringe. After this procedure, all the samples were weighed to check the water content and immediately closed into a hermetic cell for spectroscopic measurements. No significant variation of the sample mass was found after the OKE measurements.

The hydration level can be quantify using the Filling Fraction parameter: $f = H_2O$ (g) / Vycor (g). This is defined as the ratio between the weight of water contained into Vycor glass and the weight of the borosilicate glass composing the Vycor matrix. Even if the heating temperature of 400 °C is above the temperature needed to remove the absorbed physical water, some re-absorption occurs during the cooling process and the weighing procedure. Indeed, in the spectrum of the Vycor sample recorded after heating (upper panel of Fig. 2) the absorption band of water at 5260 cm^{-1} has very low intensity. From the area of this band we can estimate the actual amount of physical water left in the Vycor glass. From the molar integrated absorption coefficient, $\varepsilon_{H_2O} = 0.22$ (cm μmol^{-1}),[59] we estimated

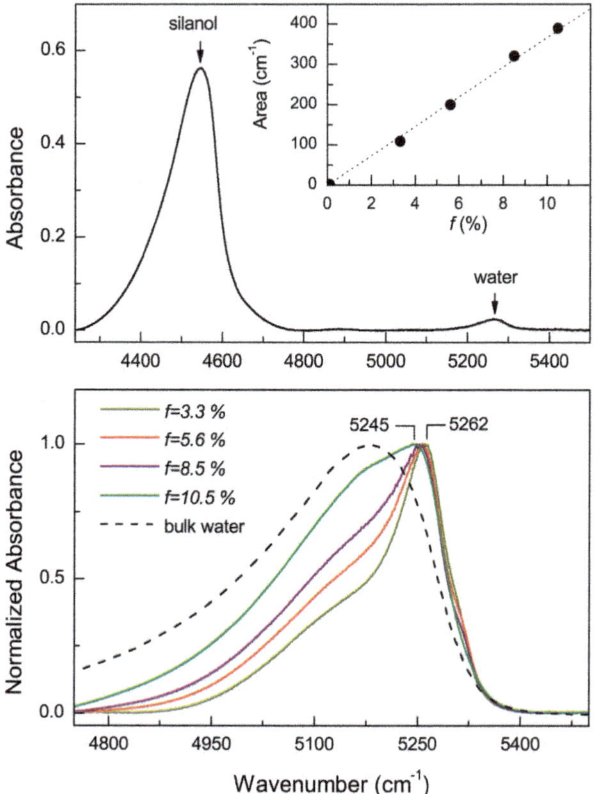

Fig. 2 Upper panel: FTIR spectrum of Vycor after heating the sample up to 400 °C for 10 h. The absorption bands of silanol and water are indicated. In the inset, we show the variation of the area value of the water band at ≈5260 cm^{-1} as a function of hydration level. This is measured by the filling fraction, *f*, defined as the ratio between water and silica weights. Lower panel: The absorption band of water on partially filled Vycor glass at variable hydration levels, *f*. The absorption band recorded in bulk water[47] is also reported.

that the concentration of water in the sample was equal to $f = 0.1\%$. Thus we obtained a good measure of the mass of the completely dehydrated slab.

In summary, we prepared samples at 6 hydration levels, whose water content was accurately determined from their weights: $f = 0.10 \pm 0.02\%$, $f = 3.3 \pm 0.2\%$, $f = 5.6 \pm 0.3\%$, $f = 8.5 \pm 0.4\%$, $f = 10.5 \pm 0.5\%$ and $f = 24.3 \pm 0.9\%$. We can consider the first sample as substantially "dry" Vycor, in fact at this hydration the number of water molecules inside the pores is much less than the number of silanol groups present on the pore surface. On account of the porous glass properties, the hydration of 5.6% ensures enough water inside pores to generate the "first layer", *i.e.* the full coverage of the pore surface with a mono-molecular water layer. The sample with 10.5% hydration contains about "two layers", whereas the sample with 24.3% hydration is in the "full hydration" condition.[60] We verified that for the partially hydrated samples the integrated absorption area linearly increases with the absorbed water concentration (see the inset of Fig. 2). The spectrum of "dry" Vycor also shows a prominent band at 4550 cm^{-1} due to silanol absorption. Its integrated area value allowed us to evaluate the OH surface concentration, in accordance with the following relationship:

$$n_{OH} = \frac{\text{Area}}{\varepsilon_{OH} l} \times \frac{1}{\rho} \times \frac{1}{S} \times 10^{-24} \times N_A \quad (4)$$

where n_{OH} indicates the number of OH groups per nm^2, ε_{OH} is the molar integrated absorption coefficient (0.16 cm μmol^{-1}),[59] l, ρ and S are the thickness (0.2 cm), the density (1.5 g cm^{-3}) and the internal surface area (250 m^2 g^{-1}), respectively, and N_A is Avogadro's number. We obtained a value of 5 OH per nm^2, in agreement with the results obtained by Zhuravlev[56] on silica surface of amorphous materials. The determination based on the integrated absorption coefficient of the silanol band has the advantage that it does not significantly vary in the presence of H-bonding with surrounding water molecules.

Moreover, the comparison between the absorption spectra of physical water, reported in the lower panel of Fig. 2, clearly indicates that the structure of water drastically changes as a function of water content. The band maximum is red-shifted with increasing water concentration, f, while a broad absorption grows below 5200 cm^{-1}, gradually approaching the spectral profile of bulk water. The sharp peak at ≈5260 cm^{-1} is reasonably assigned to vibrations of monomeric water, the H-bonding interactions can reasonably lead to the broad red-shifted absorption. The latter band is present even at the lowest water concentration (<5%), showing that water aggregation occurs before full surface coverage is attained.

4.2 HD-OKE data

We measured the HD-OKE response at room temperature, 293 K, of the six Vycor samples at different hydration levels (f = 0.1%, 3.3%, 5.6%, 8.5%, 10.5% and 24.3%) described in the previous section. The experimental results are collected in Fig. 3. The upper panel shows a log–linear plot of the data in the short time range; the results for longer delay times are shown in the lower panel in a linear–linear plot. Just a simple look at the data reveals several differences between the water-in-Vycor and that of bulk water shown in Fig. 1. The data measured at different hydration levels enable disentanglement of the nanoconfined water signal from the contribution of the dry Vycor matrix. As shown in Fig. 3, the signal of dry Vycor (f = 0.1%) displays a relatively slow oscillation extending in the picosecond time-scale. According to our fit, it corresponds to an oscillatory mode characterized by a frequency of $\nu \approx 5$ cm^{-1} and it likely due to an acoustic-like vibration localized on the pore surface.[61] The presence of liquid water adds a monotonic decay that becomes the dominant feature at full hydration; this contribution is attributed to the relaxation processes of nanoconfined water.

The fast vibrational components visible in the upper panel of Fig. 3 are rather complex: their characteristic features become clearer when transformed into the frequency domain. In order to fit the HD-OKE signal we use the equation introduced in the previous section, see eqn (2) and (3). The response function must be completed to include the dry Vycor matrix response, $R_{dry}(t)$. So the complete response becomes:

$$R(t) = AR_{dry}(t) + B\frac{d}{dt}exp\left[-\left(\frac{t}{\tau}\right)^{\beta}\right] + \sum_i C_i exp(-\gamma_i^2 t^2)\sin(\omega_i t) \quad (5)$$

We simulated the $R_{dry}(t)$ response as the sum of a series of damped oscillators and exponential functions, whose characteristic parameters were determined by fitting of the HD-OKE signal of the dry sample (f = 0.1%).

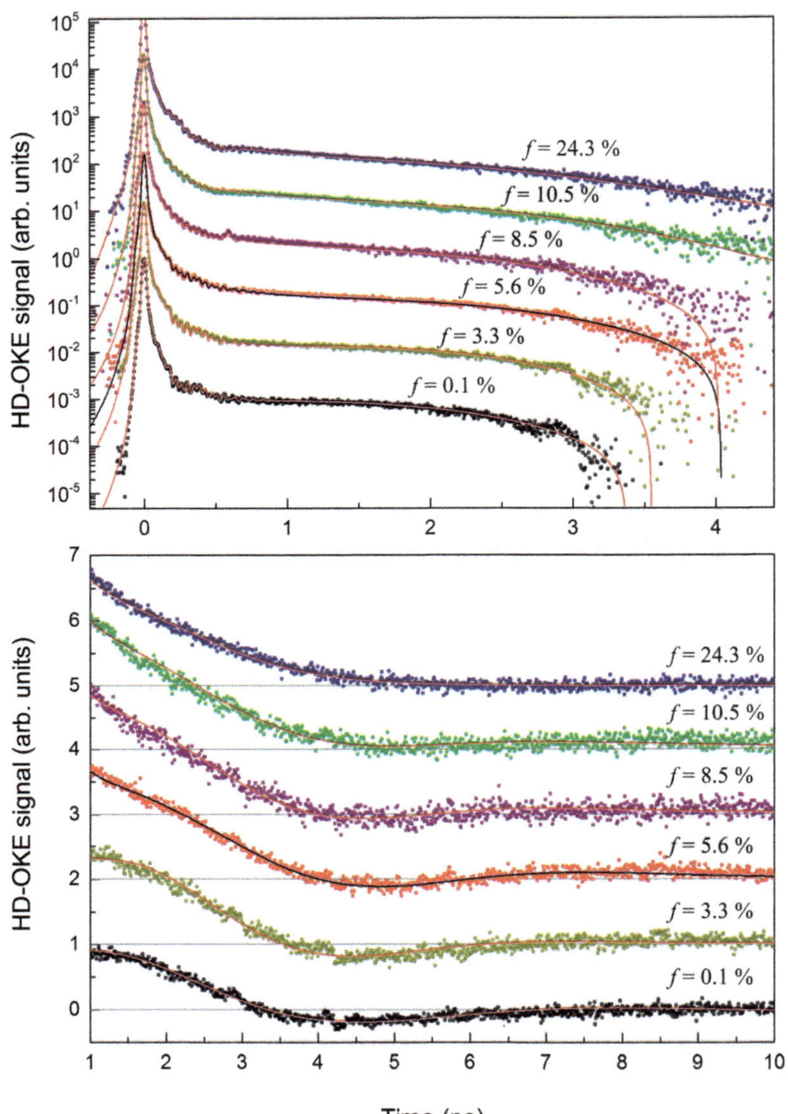

Fig. 3 HD-OKE data of nanoconfined water at 293 K and at variable hydration levels, from dry condition, $f = 0.1\%$, to full hydration, $f = 24.3\%$. Full circles: experimental data; continuous line: best fit results. In the upper panel the data are reported in a log–linear plot. In the lower panel the signal decays at longer times are shown in linear–linear plot. The data have been vertically shifted to make all kinetics clearly visible.

The best fit procedure enables one to get the structural relaxation times of nanoconfined water and these are reported in Table 1. As expected, the amplitude of the structural relaxation component decreases with the hydration level, so that at $f = 3.3\%$ it becomes unmeasurable. Moreover, it is impossible to extract a reliable value for the stretching exponent; we then fixed it to the bulk value $\beta = 0.6$. The results collected in Table 1 show that the structural relaxation time does not vary appreciably on going from hydration $f = 5.6\%$ to 10.5%, whereas the $f = 24.3\%$ sample is characterized by a faster relaxation process.

Table 1 Structural relaxation parameters obtained by the best fitting of HD-OKE data. f is the filling fraction, τ the relaxation time and β the stretching parameter

f (%)	τ (ps)	β
5.6	1.6 ± 0.3	0.6
8.5	1.4 ± 0.3	0.6
10.5	1.6 ± 0.3	0.6
24.3	0.45 ± 0.10	0.6

The structural relaxation time of nanoconfined water at full hydration is indeed comparable to the bulk water time, $\tau = 0.35$ ps. On the contrary, when the degree of hydration is less than two water layers, the structural relaxation experiences a clear slowing down, with an increase of the characteristic time scale of about a factor of 4. In principle, the structural relaxation of fully hydrated Vycor results from the superposition of the contributions of the inner water (faster) and of the surface layers (slower). Actually, our data analysis is not able to detect any slow decaying component in the OKE results for the $f = 24.3\%$ sample. The direct measurement of the distinct relaxation times of inner and surface water in fully hydrated samples is in fact a very arduous experimental task, which is only viable when the characteristic time constants are very different. The structural relaxation time measured in the full hydrated sample corresponds probably to a weighted average of the interfacial and inner dynamics. We notice also that in weakly hydrated pores the surface water faces silica on one side and air on the other one, while in fully hydrated samples the outer layers are in contact with silica and with the inner water: their dynamics might not be the same in the two cases. In this respect, computer simulations are of great help, as they allow disentangling of the structural and dynamical features of the interfacial water; the differences in the two cases mentioned above are shown to be minor.[58]

In summary, our results suggest that the collective rearrangements of interfacial and inner water are characterized by similar structural processes (both require a stretched exponential function to be properly reproduced), but with clearly different characteristic time scales. The dynamics of the inner water does not differ noticeably from that of bulk water. These findings are in agreement with several computer simulations[19,58,62,63] and experiments;[64–67] but in partial contradiction with the results reported by Scodinu and Fourkas[30] on the basis of an OKE investigation of water confined in a similar, but not identical, nanoporous glass. According to this work in fact, the structural relaxation of the inner water in fully hydrated samples is slower than that of bulk water.

In Fig. 4 we report the Fourier transform of the measured response function after deconvolution from the instrumental response, using the procedure summarized in the previous section and described in detail in ref. 32. In the main panel of Fig. 4 we show the spectra obtained by Fourier transformation of the HD/OKE data of Fig. 3. The dry Vycor sample, $f = 0.1\%$, shows a frequency spectrum extending up to 500 cm^{-1} characterized by a broad peak at about 100 cm^{-1}. We notice also a narrow peak at about 5 cm^{-1}, hardly visible in the figure, whose nature was previously discussed in connection to the description of the time domain HD-OKE data, and a peak at 800 cm^{-1}, not shown in the figure. Apart from the 5 cm^{-1} band, the dry Vycor spectra resemble closely the spectrum of amorphous SiO_2, as measured by depolarized Raman experiments.[68] The

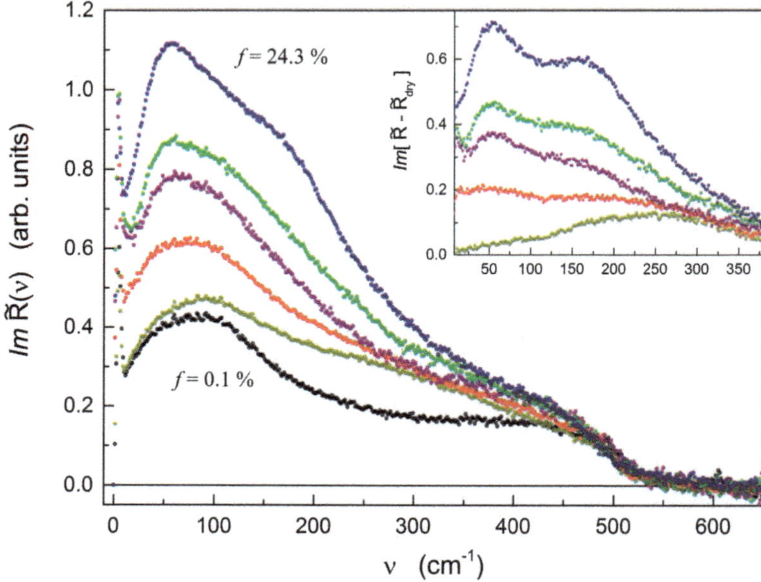

Fig. 4 Fourier transforms of the HD-OKE data, deconvoluted from the instrumental response, at different filling fractions: from dry condition, $f = 0.1\%$, to full hydration, $f = 24.3\%$. In the inset we report the same data after subtraction of the dry Vycor signal; The spectra give the contribution of nano-confined water.

presence of liquid water inside the Vycor pores causes an evident increase of the vibrational intensity in the frequency range from 20–350 cm^{-1}. In order to extract the contribution of nanoconfined water, we subtract from the data the dry Vycor spectrum, *i.e.* that for $f = 0.1\%$; the result is reported in the inset of Fig. 4. At 3.3% hydration the spectrum of nanoconfined water is featureless; there is no signature of characteristic bending and stretching water peaks. At higher hydration, $f = 5.6\%$, the spectrum is unchanged in the high frequency part, $\nu > 200$ cm^{-1}, while on the low frequency side a weak signature of the bending peak appears. The spectra at the two highest levels of hydration show a growing presence of the bending and stretching features. These results are in agreement with a light scattering investigation of water confined in a similar nanoporous glass.[65]

The present data suggest the following scenario for the dynamics of nano-confined water.

Up to $f = 5.6\%$, corresponding to one water layer, only "interfacial water" exists inside the Vycor pores. This water is in direct contact with the surfaces, it is linked by the H-bonds formed with the silanol groups present on the surface. The water H-bonding network of water is strongly modified by the surface interactions, these modifications affect the intermolecular vibration spectrum. Our data show that this spectrum doesn't show any specific signature, suggesting a large network distortion characterized by a wide distribution of angles and distances. Moreover, the structural relaxation is much slower than in bulk water suggesting a strong hindrance of the collective rearrangement and/or diffusion process.

Starting from $f = 10.5\%$, *i.e.* two water layers or pore plugs, some of the water molecules can form H-bonds with other water molecules without having a direct link to the surface silanol groups. In other words, we have some "inner water".

This water presents a liquid structure that resembles bulk water. The vibrational part of our data partially supports this point of view, even if the measured spectrum still presents several differences from that of bulk water. The measured intermolecular vibrations indicate that an H-bond network of nanoconfined water is characterized by deformations, which are larger and more numerous than in free bulk water. Also, the structural phenomena are strongly affected by the surface interactions, as proved by the slowing down of the structural relaxation.

Under full hydration conditions, $f = 24.3\%$, both the slow structural processes and the fast vibrational dynamics become similar to those of bulk water. This suggests that in this case the inner water shows several dynamic characteristics reminiscent of the bulk liquid water.

4.3 Comparison of bulk and nano-confined water dynamics

The direct comparison of nanoconfined and bulk water spectra, see the upper panel of Fig. 5, makes differences and similarities immediately evident. Clearly, both spectra display the bending and stretching vibrational bands.

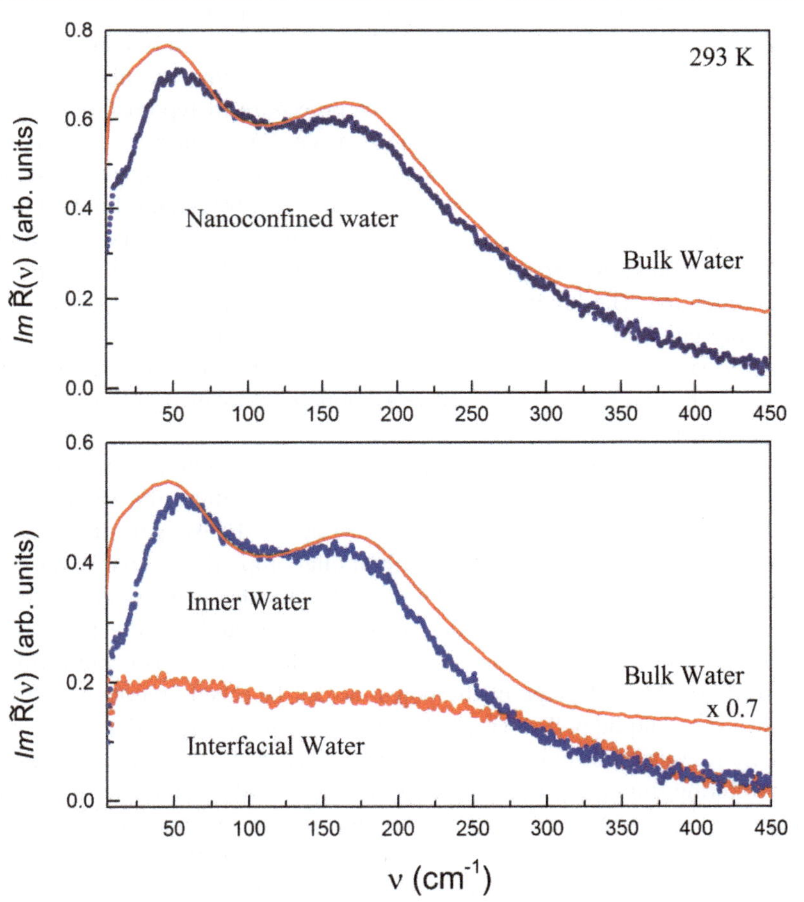

Fig. 5 Upper panel: comparison of the HD-OKE response function in the frequency of bulk water (red line) and nanoconfined water (blue circles). Lower panel: decomposition of the total HD-OKE spectrum of nanoconfined water in the interfacial (red circles) and inner (blue circles) contributions.

The low frequency part of the spectra, $\nu < 50\ \mathrm{cm}^{-1}$, is mainly characterized by the large intensity difference of the band attributed to structural relaxation; this spectral component is much weaker in nanoconfined water, and this causes a steeper rise of the low frequency side of the bending band. At higher frequencies, the bending and stretching vibrational bands appear in both cases, but they are less definite in nanoconfined water then in bulk water. This is probably ascribable to a larger inhomogeneous broadening of those modes, due to the contribution of water in contact with the silica surfaces. At room temperature the water spectrum above $300\ \mathrm{cm}^{-1}$ is generally attributed to librational modes;[54] in nanoconfined water we observe a depletion of the librational components in this spectral region. In other words, it seems that nanoconfinement hinders the orientational oscillatory dynamics.

Assuming that progressive hydration simply adds contributions to the HD-OKE signal, we applied a decomposition procedure in order to recover separately the contributions of interfacial and inner water dynamics.

The spectra of interfacial water is immediately obtained by subtracting the $f = 0.1\%$ spectrum, assumed to represent that of the dry Vycor matrix, from the spectrum of the $f = 5.6\%$ sample. The difference spectrum represents the contribution of the first water layer. By subtracting this interfacial water spectrum and that of the dry Vycor matrix from the spectrum measured for the fully hydrated sample ($f = 24.3\%$) we finally extracted the spectral contribution of the inner water. The results of this decomposition procedure are reported in the lower panel of Fig. 5. The interfacial water shows a very broad spectrum lacking the spectral features characteristic of the bulk phase. This unambiguous evidence proves that the H-bond network of interfacial water is substantially different from that of bulk water. Not surprisingly, the spectrum of inner water is largely similar to that of the bulk; in particular, the bending and stretching bands are present and well defined. The librational features of the spectrum above $300\ \mathrm{cm}^{-1}$ are weakly present both in interfacial and inner water. In contrast, in the low frequency part ($\nu < 50\ \mathrm{cm}^{-1}$) of the inner water spectrum we notice the same intensity reduction of the structural relaxation component observed in nanoconfined water.

As a final conclusion, we can try to give an answer to the initial question asking to what extent the interaction with the pore surfaces modifies the water properties. According to our results for hydration levels $f \geq 10.5\%$, part of the nanoconfined water presents structural and vibrational properties similar to those of bulk water. In other words, the liquid water added inside the pores above the $f = 10.5\%$ amount has large similarities with that characterizing the bulk water structure. At this level it is not possible to establish whether this is indicative of the presence of liquid plugs in the pores, or can be taken as evidence of the existence of real liquid polls. Nevertheless, we would like to stress that the second water layer, even if not in direct contact with silica surface, has dynamic properties differing from those of bulk water. The van der Waals diameter of a water molecule is of the order of 0.3 nm,[47] so the thickness of the first two water layers covers about 1.2 nm of the pore diameter. If the pore diameter is 4 nm, as it is in Vycor porous glasses, there are about 4–5 layers of inner water; but if the pore diameter reduces below 1.5 nm there is no space available for inner water layer and all the water molecules are in the first or second layer. This suggests that when water is confined in a hydrophilic narrow pore, practically all the water molecules interact effectively with the surfaces, either directly (*i.e* first interfacial layer) or indirectly by water–water H-bonds (*i.e* second layer).

We believe that the findings reported in this work should be carefully taken into account when employing pore confinement in attempts to approach experimentally the putative singularity temperature T_s around 225 K in the supercooled phase of water.

Acknowledgements

The research has been performed at LENS. This work was supported by Regione Toscana POR-CRO-FSE 2007–2013 by EC COST Action MP0902-COINAPO. We acknowledge M. De Pas and A. Montori for providing their continuous assistance in the set-up of the electronics; R. Ballerini and A. Hajeb for the mechanical realizations; M. Pucci for the optical treatments of Vycor glasses.

References

1 I. Brovchenko and A. Oleinikova, *Interfacial and Confined Water*, Elsevier, 2008.
2 P. Debenedetti and H. E. Stanley, *Phys. Today*, 2003, **56**, 40.
3 P. Debenedetti, *J. Phys.: Condens. Matter*, 2003, **15**, R1669–R1726.
4 P. H. Poole, F. Sciortino, U. Essmann and H. E. Stanley, *Nature*, 1992, **360**, 324–328.
5 P. H. Poole, R. K. Bowles, I. Saika-Voivod and F. Sciortino, *J. Chem. Phys.*, 2013, **138**, 034505.
6 D. T. Limmer and D. Chandler, *J. Chem. Phys.*, 2011, **135**, 134503.
7 E. B. Moore and V. Molinero, *J. Chem. Phys.*, 2009, **130**, 244505.
8 E. B. Moore and V. Molinero, *Nature*, 2011, **479**, 506.
9 S. Jähnert, F. Vaca Chávez, G. E. Schaumann, A. Schreiber, M. Schönhoff and G. H. Findenegg, *Phys. Chem. Chem. Phys.*, 2008, **10**, 6039.
10 M. Oguni, Y. Kanke, A. Nagoe and S. Namba, *J. Phys. Chem. B*, 2011, **115**, 14023–9.
11 A. Taschin, R. Cucini, P. Bartolini and R. Torre, *Europhys. Lett.*, 2010, **92**, 26005.
12 P. Gallo, M. Rovere and S.-H. Chen, *J. Phys. Chem. Lett.*, 2010, **1**, 729.
13 E. de la Llave, V. Molinero and D. A. Scherlis, *J. Chem. Phys.*, 2010, **133**, 034513.
14 S. Capaccioli, K. Ngai, S. Ancherbak, P. Rolla and N. Shinyashiki, *J. Non-Cryst. Solids*, 2011, **357**, 641654.
15 S. L. Caer, S. Pin, S. Esnouf, Q. Raffy, J. P. Renault, J.-B. Brubach, G. Creff and P. Roy, *Phys. Chem. Chem. Phys.*, 2011, **13**, 17658.
16 X. Limei and M. Valeria, *J. Phys. Chem. B*, 2011, **115**, 14210.
17 D. Banerjee, S. N. Bhat, S. V. Bhat and D. Leporini, *PLoS One*, 2012, **7**, e44382.
18 D. T. Limmer and D. Chandler, *J. Chem. Phys.*, 2012, **137**, 044509.
19 A. A. Milischuk, V. Krewald and B. M. Ladanyi, *J. Chem. Phys.*, 2012, **136**, 224704.
20 F. G. Alabarse, J. Haines, O. Cambon, C. Levelut, D. Bourgogne, A. Haidoux, D. Granier and B. Coasne, *Phys. Rev. Lett.*, 2012, **109**, 035701.
21 N. Giovambattista, P. Rossky and P. Debenedetti, *Annu. Rev. Phys. Chem.*, 2012, **63**, 179.
22 C. E. Bertrand, Y. Zhang and S.-H. Chen, *Phys. Chem. Chem. Phys.*, 2013, **15**, 721–45.
23 R. Torre, *Time-Resolved Spectroscopy in Complex Liquids, an experimental perspective*, Springer, New York, 2008.
24 R. A. Farrer and J. T. Fourkas, *Acc. Chem. Res.*, 2003, **36**, 605–12.
25 R. Cucini, A. Taschin, C. Ziparo, P. Bartolini and R. Torre, *Eur. Phys. J. Spec. Top.*, 2007, **141**, 133–136.
26 A. Taschin, R. Cucini, C. Ziparo and R. Torre, *Philos. Mag.*, 2007, **87**, 715–722.
27 A. Taschin, R. Cucini, P. Bartolini and R. Torre, *Europhys. Lett.*, 2008, **81**, 58003.
28 R. Cucini, A. Taschin, P. Bartolini and R. Torre, *J. Mech. Phys. Solids*, 2010, **58**, 1302–1317.
29 R. Cucini, A. Taschin, P. Bartolini and R. Torre, *J. Phys.: Conf. Ser.*, 2010, **214**, 012032.
30 A. Scodinu and J. T. Fourkas, *J. Phys. Chem. B*, 2002, **106**, 10292–10295.
31 P. Bartolini, A. Taschin, R. Eramo, R. Righini and R. Torre, *J. Phys.: Conf. Ser.*, 2009, **177**, 012009.
32 A. Taschin, P. Bartolini, R. Eramo, R. Righini and R. Torre, *Nat. Commun.*, 2013, 2401, DOI: 10.1038/ncomms3401.

33 N. T. Hunt, A. A. Jaye and S. R. Meech, *Phys. Chem. Chem. Phys.*, 2007, **9**, 2167–2180.
34 P. Bartolini, A. Taschin, R. Eramo and R. Torre, in *Optical Kerr Effect Experiments on Complex Liquids, A Direct Access to Fast Dynamic Processes*, ed. R. Torre, Springer, New York, 2008, ch. 2, pp. 73–127.
35 R. Torre, P. Bartolini and R. Pick, *Phys. Rev. E: Stat. Phys., Plasmas, Fluids, Relat. Interdiscip. Top.*, 1998, **57**, 1912–1920.
36 R. Torre, P. Bartolini, M. Ricci and R. Pick, *Europhys. Lett.*, 2000, **52**, 324–329.
37 M. Ricci, P. Bartolini and R. Torre, *Philos. Mag. B*, 2002, **82**, 541–551.
38 R. Torre, P. Bartolini and R. Righini, *Nature*, 2004, **428**, 296–298.
39 M. Ricci, S. Wiebel, P. Bartolini, A. Taschin and R. Torre, *Philos. Mag.*, 2004, **84**, 1491–1499.
40 G. Giraud, C. Gordon, I. Dunkin and K. Wynne, *J. Chem. Phys.*, 2003, **119**, 464–477.
41 S. Krishnamurthy, R. Bansil and J. Wiafe-Akenten, *J. Chem. Phys.*, 1983, **79**, 5863–5870.
42 F. Aliotta, C. Vasi, G. Maisano, D. Majolino, F. Mallamace and P. Migliardo, *J. Chem. Phys.*, 1986, **84**, 4731.
43 G. Walrafen, M. Fisher, M. Hokmabadi and W.-H. Yang, *J. Chem. Phys.*, 1986, **85**, 6970.
44 J. L. Rousset, E. Duval and A. Boukenter, *J. Chem. Phys.*, 1990, **92**, 2150.
45 K. Mizoguchi, Y. Hori and Y. Tominaga, *J. Chem. Phys.*, 1992, **97**, 1961.
46 A. Sokolov, J. Hurst and D. Quitmann, *Phys. Rev. B: Condens. Matter*, 1995, **51**, 12865.
47 M. Chaplin, http://www.lsbu.ac.uk/water/, 2013.
48 E. Castner, Y. Chang, Y. Chu and G. Walrafen, *J. Chem. Phys.*, 1995, **102**, 653–659.
49 S. Palese, S. Mukamel, R. Miller and W. Lotshaw, *J. Phys. Chem.*, 1996, **100**, 10380.
50 K. Winkler, J. Lindner and P. Vohringer, *Phys. Chem. Chem. Phys.*, 2002, **4**, 2144–2155.
51 B. Ratajska-Gadomska, B. Biakowski, W. Gadomski and C. Radzewicz, *Chem. Phys. Lett.*, 2006, **429**, 575–580.
52 R. W. Hellwarth, *Prog. Quantum Electron.*, 1977, **5**, 1–68.
53 M. Skaf and M. Sonoda, *Phys. Rev. Lett.*, 2005, **94**, 137802.
54 A. DeSantis, A. Ercoli and D. Rocca, *J. Chem. Phys.*, 2004, **120**, 1657.
55 J. A. Padró and J. Martí, *J. Chem. Phys.*, 2004, **120**, 1659.
56 L. T. Zhuravlev, *Colloids Surf., A*, 2000, **173**, 1–38.
57 A. Burneau, J. Lepage and G. Maurice, *J. Non-Cryst. Solids*, 1997, **217**, 1–10.
58 P. Gallo, M. Ricci and M. Rovere, *J. Chem. Phys.*, 2002, **116**, 342.
59 J.-P. Gallas, J.-M. Goupil, A. Vimont, J.-C. Lavalley, B. Gil, J.-P. Gilson and O. Miserque, *Langmuir*, 2009, **25**, 5825–5834.
60 E. Tombari, G. Salvetti, C. Ferrari and G. P. Johari, *Phys. Chem. Chem. Phys.*, 2005, **7**, 3407–3411.
61 T. Woignier, J. Sauvajol, J. Pelous and R. Vacher, *J. Non-Cryst. Solids*, 1990, **121**, 206–210.
62 P. Gallo, M. Rovere and E. Spohr, *J. Chem. Phys.*, 2000, **113**, 11324.
63 A. a. Milischuk and B. M. Ladanyi, *J. Chem. Phys.*, 2011, **135**, 174709.
64 J. Zanotti, M. Bellissent-Funel and S. Chen, *Phys. Rev. E: Stat. Phys., Plasmas, Fluids, Relat. Interdiscip. Top.*, 1999, **59**, 3084–3093.
65 V. Crupi, A. Dianoux, D. Majolino, P. Migliardo and V. Venuti, *Phys. Chem. Chem. Phys.*, 2002, **4**, 2768–2773.
66 V. Crupi, D. Majolino, P. Migliardo, V. Venuti and M. C. Bellissent-Funel, *Mol. Phys.*, 2003, **101**, 3323–3333.
67 F. Mallamace, C. Corsaro, S.-H. Chen and H. E. Stanley, in *Liquid Polymorphism*, ed. H. E. Stanley, John Wiley & Sons, 2013, vol. 152, ch. 10, p. 203.
68 O. Pilla, A. Fontana, S. Caponi, F. Rossi, G. Viliani, M. Gonzalez, E. Fabiani and C. Varsamis, *J. Non-Cryst. Solids*, 2003, **322**, 53.

Faraday Discussions RSC Publishing

PAPER

Liquid organization and solvation properties at polar solid/liquid interfaces

Eric A. Gobrogge,†[a] B. Lauren Woods†[a] and Robert A. Walker*[b]

Received 2nd May 2013, Accepted 30th May 2013
DOI: 10.1039/c3fd00071k

Second order nonlinear optical spectroscopy has been employed to examine the organization of four different liquids at the hydrophilic silica/liquid interface. The liquids – cyclohexane, methylcyclohexane, 1-propanol, and 2-propanol – were chosen to isolate how intermolecular forces between the liquid and the substrate competed with steric effects to control liquid structure and solvating properties across the interfacial region. Vibrational sum frequency generation (VSFG) data showed that cyclohexane structure at the silica/liquid cyclohexane interface closely resembled the structure of a cyclohexane monolayer adsorbed to the silica/vapor interface. Methylcyclohexane, however, showed evidence of large structural reorganization between the silica/liquid and silica/monolayer/vapor interfaces. 1-Propanol at a silica/vapor interface formed a well-ordered, Langmuir-like monolayer due to strong hydrogen bonding with the surface silanols and cohesive van der Waals interactions between carbon chains. 1-Propanol at the silica/liquid interface retained the same ordered structure. In contrast, 2-propanol adopted different structures adsorbed to the solid/vapor and at the solid/liquid interfaces. Specifically, the plane defined by 2-propanol's three carbon atoms changed orientation from being perpendicular to the surface (silica/vapor) to parallel to the surface (silica/liquid). Surface mediated liquid structure affected the solvation of adsorbed solutes. Resonance enhanced second harmonic generation (SHG) data showed that silica/alkane interfaces were significantly more polar than would be expected based on a solute's bulk solution solvatochromic behavior. Both silica/alcohol interfaces exhibited alkane-like polarity, a result that was interpreted in terms of a reduction in hydrogen bonding opportunities for adsorbed solutes.

1. Introduction

Solid surfaces force adjacent liquids to adopt structures not found in bulk solution. Liquid organization at a solid/liquid interface will depend sensitively on

[a]*Department of Chemistry & Biochemistry, Bozeman, Mt 59717, USA. E-mail: eric.gobrogge@gmail.com; woodslauren88@gmail.com; Tel: +01 406 994 6739*

[b]*Department of Chemistry & Biochemistry, Office 057, Bozeman, Mt 59717, USA. E-mail: rawalker@chemistry.montana.edu; Fax: +01 406 994 5407; Tel: +01 406 994 7928*

† Both authors made equivalent contributions to this work.

both geometric and dipolar considerations as well as differences between monomer–monomer and monomer–substrate affinities. These effects can be subtle; small differences between these intra- and inter-phase interactions can have pronounced consequences that control how far into solution interfacial anisotropy extends and how properties of the interfacial liquid change from bulk limits. Surface effects on liquid structure can be inferred from macroscopic measurements, where tuning liquid–substrate interactions and measuring solid/liquid contact angles provide insight into the role played by dipolar and dispersive forces in solvent wetting.[1–7] However, molecular insight requires that these interfaces be examined by methods that have intrinsic surface and molecular specificity. Studies described in this work use 2nd order nonlinear optical spectroscopy to examine how polar, hydrophilic silica surfaces affect liquid organization across the silica/liquid interface. Liquids have been chosen to explore how competition between liquid/substrate association and steric effects between liquid monomers at the surface impact monomer structure and orientation. Additional NLO experiments explore how this surface mediated liquid structure affects interfacial solvation by recording effective excitation spectra of adsorbed solutes. These latter experiments highlight how a solvent's dielectric properties at interfaces can differ substantially from what might be inferred from additive or mean-field models.

In the most general sense, liquids at solid/liquid interfaces can be categorized as being either weakly associating or strongly associating.[8–10] Weakly associating liquids will interact with a solid substrate through induced dipole or dispersion forces. While weakly associating liquids may wet a polar substrate, they tend to do so incompletely and form films with measurable contact angles.[11–13] Strongly associating liquids, in contrast, will completely wet a polar substrate due to hydrogen bonding or more general dipolar forces that promote liquid spreading. From these macroscopic observations and empirical parameterization,[14,15] one is left to infer details about liquid structure.

From a molecular perspective, questions about liquid structure and organization at a solid/liquid interface can be nuanced. For example, X-ray scattering experiments performed by Doerr *et al.* reported that the silicon oxide/liquid cyclohexane interface was characterized by a dense, solid-like cyclohexane layer in direct contact with the substrate followed by a reduced density region that extended ~3–4 nm into the bulk.[16] *n*-Decane density at the same solid surface approached bulk values in only one solvent layer and showed no further anomalies as a function of distance away from the solid surface. In contrast to both cyclohexane and *n*-decane, *n*-hexane experienced significant solvent depletion within the first solvent layer and this low-density region persisted ~3 nm into bulk solution. These results were interpreted in terms of differences between liquid packing densities (ρ^*)[16] with cyclohexane being able to form much denser films than either decane or hexane. Data from surface force apparatus measurements also showed that *n*-alkanes ($n \geq 8$) formed layered structures when confined between silica and mica surfaces whereas thin liquid films of branched alkanes were much more unstable and were readily expelled from between the two surfaces.[16–20] Again, data were interpreted with "geometric" explanations that considered solvent packing and organization as the primary contributors to liquid behavior at an interface.

While liquid organization at solid/liquid alkane interfaces appears to depend sensitively on molecular shape, strongly associating liquids such as alcohols, amines, and nitriles will exploit dipole–dipole interactions and hydrogen bonds between the liquid and the substrate to create regions having structural and dynamic properties that differ significantly from bulk liquid limits. AFM measurements of alcohol solutions at the silica surface report that n-alcohols with $n > 3$ stand upright and form hydrophobic films at the solid/liquid interface.[21,22] At the related silica/methanol interface, surface specific vibrational spectra imply that methanol forms a tightly coordinated bilayer structure where methyl groups from the first, hydrogen bonded solvent layer point away from the silica surface and methyl groups from the second solvent layer are directed towards the silica surface. Effects from the silica surface on methanol structure are assumed to extend no further than two solvent layers into bulk solution.[23] This picture contrasts with combined experimental and theoretical studies of acetonitrile at the silica/acetonitrile solid/liquid interface.[24] Experimental and simulation data both showed that the first layer of acetonitrile accepted strong hydrogen bonds from the surface silanol groups, and this first layer served to template repeating bilayer structure in the liquid that extended multiple layers into the bulk.

Liquid structure that has been altered by a surface will have different solvating properties than in the bulk where solvent species are free to organize without anisotropic constraints. In this context, solvation describes the noncovalent interactions a solute has with its surroundings. Numerous spectroscopic studies and simulations report that local dielectric properties, hydrogen bonding opportunities and reorientation dynamics sampled by solutes adsorbed to silica/liquid interfaces differ significantly from bulk solution limits.[10,25–32] Many of the experimental results, however, are rationalized in terms of assumed or anticipated solvent organization rather than directly measured structure. For example, slow solute reorientation at the silica/butanol[33] and sapphire/butanol[34] interfaces has been attributed to a restrictive environment created by strong hydrogen bonding between the liquid and the solid substrate, but solvent structure has not been measured directly. Similarly, several studies have examined solvatochromic shifts in a solute's excitation wavelength to infer details about the local polarity at and across solid/liquid interfaces. Again, however, conclusions about why a given solid/liquid interface is more or less polar than a bulk solution limit, rely upon indirect evidence from bulk liquid studies or simulations.

Findings reported in this work couple spectroscopic studies of liquid structure and organization at the silica/liquid interface with findings that describe the interfacial dielectric environment sampled by adsorbed solutes. The liquids chosen in this work include two strongly associating solvents, 1- and 2-propanol and two weakly associating liquids, cyclohexane and methylcyclohexane (Fig. 1A). Each pair of liquids were chosen because differences in intermolecular steric interactions were expected to influence liquid monomer organization and solvation at silanol-terminated, polar silica surfaces. While surface specific vibrational sum frequency generation (VSFG) experiments show that a liquid's molecular structure and intraphase forces play the strongest role in determining if monomolecular films adsorbed to the solid/vapor interface retain their structure when a saturated vapor is replaced by the neat liquid, resonance enhanced second harmonic generation (SHG) data from coumarin 152 (C152) (Fig. 1B) provide clear evidence that local polarity across a solid/liquid interface depends

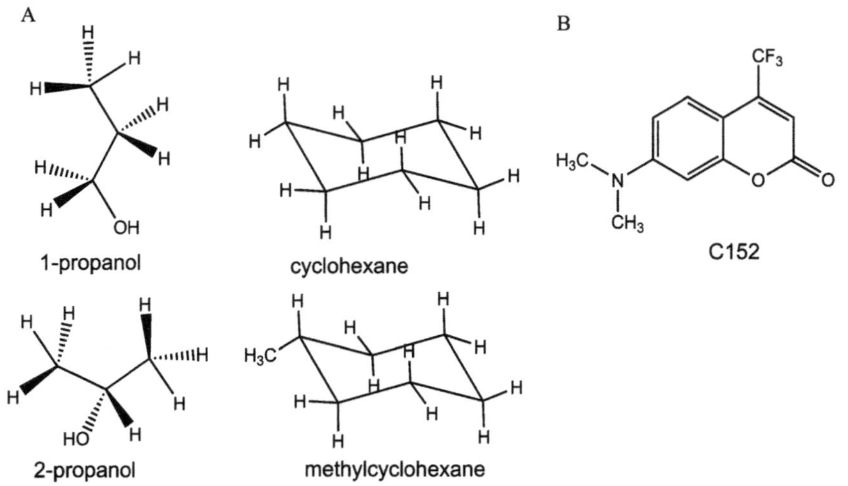

Fig. 1 Solvents studied in this work are shown in (A); solvation effects were studied using 7-dimethylamino-4-(trifluoromethyl)coumarin (C152) shown in (B).

primarily on interactions between the liquid and the substrate. If the silica–liquid interactions are strong, then the solute will sample a polarity that is sensitive to the anisotropic solvent structure induced by the silica substrate. If, however, silica–solute interactions are stronger than those between the silica and the liquid, then a solute's interfacial polarity will be dominated by the surface silanol groups.

2. Experimental

A. Materials

Anhydrous methylcyclohexane was purchased from Sigma Aldrich, >99% cyclohexane was purchased from Acros, HPLC grade 2-propanol was received from EMP, and 1-propanol was purchased from Fisher. Laser grade coumarin 151 and coumarin 152 were obtained from Exciton. All liquids and solutes were used as received. Silica slides from SPI Inc. were cleaned using a 50/50 (by volume) sulfuric/nitric acid mixture and rinsed thoroughly with deionized water (Milipore, 18.2 MΩ). Slides were then affixed in the sample cell in direct contact with either a vapor phase saturated with the appropriate liquid vapor or a liquid phase containing a pure liquid or a solution with a given coumarin solute.

B. Apparatus

VSFG and SHG experiments employ a Libra-HE Ti:sapphire laser (Coherent, 3.3W 85 fs pulse duration, 1kHz repetition rate) coupled to a visible optical parametric amplifier (Coherent OPerA Solo) to generate visible and IR light. VSFG experiments were performed with co-propagating IR and visible fields passing through the 0.5 mm silica slide. The signal was detected in the reflected direction. The IR wavelength was tuned from 3.2 to 3.7 μm in 0.05 μm increments and the IR field was focused onto the sample at an angle of 73° with respect to normal. The visible beam was spectrally stretched and sliced using an 1800 g mm^{-1} grating and variable width slits resulting in a spectrally narrowed visible beam (20 cm^{-1}). After passing through two different delay stages, this beam was focused onto the surface at an angle of 67° with respect to normal. When the IR and visible fields were spatially and temporally overlapped, sum frequency signal was generated and directed into a monochromator (SpectraPro-300i, Acton Research Corporation) where it was dispersed onto a 1340 × 100 pixel CCD (PIXIS100B, Princeton Instruments). SFG spectra were combined and normalized to a non-resonant gold system response using homemade routines written in Igor Pro (v.6). Resonance-enhanced SHG signal was collected using a PMT and photon counting electronics. Incident power of the visible light before the sample ranged from 0.5 mW to 3.5 mW and for the solute used in these studies, C152, SHG experiments covered a SH wavelength range of 350–408 nm. Additional details about the SHG assembly can be found in previous reports.[35,36]

C. Nonlinear spectroscopy

VSFG is a 2nd order nonlinear optical process that measures vibrational spectra of molecules in environments lacking inversion symmetry. In a VSFG experiment, two high intensity fields (one visible and one infrared) induce a coherent, nonlinear polarization in molecules at a surface or interface. This polarization oscillates at a frequency equal to the sum of the incident visible and IR fields, and is responsible for the detected signal. When the IR frequency is resonant with a vibrational mode of a surface molecule, the sum frequency (SF) response experiences resonance enhancement. The intensity of this SF response is dependent on two different frequencies according to eqn (1):

$$I(\omega_{\text{sum}}) \propto |P^{(2)}|^2 = |\varepsilon_0\chi^{(2)}(E_1\cos\omega_1 t + E_2\cos\omega_2 t)^2|^2 \propto |\chi^{(2)}|^2 I(\omega_1)I(\omega_2) \tag{1}$$

Expanding this expression shows that the non-linear polarization is responsible not only for the SF response, but it also produces a DC field (no frequency dependence), second harmonic generation for both frequencies, and difference frequency generation.[37] Furthermore, the second order susceptibility tensor can be separated into both a resonant (${\chi_R}^{(2)}$) and non-resonant (${\chi_{NR}}^{(2)}$) part. The resonant contribution is described by:

$${\chi_R}^{(2)} = \frac{N\langle\beta_{ijk}\rangle}{\varepsilon_0},\ \beta = \frac{A_K M_{IJ}}{\omega_\nu - \omega_{IR} - i\Gamma_\nu} \tag{2}$$

where β is the molecular hyperpolarizability, A_K and M_{IJ} are the IR and Raman transition moment integrals, respectively, ω_ν is the resonant mode frequency, and Γ is related to the inhomogeneous linewidth.[38] Since signal intensity is proportional to the square of the magnitude of the non-linear polarization, when the IR frequency matches a vibrational mode, the resonant second order susceptibility increases, causing the polarization and therefore the intensity to increase. For dielectric systems without formal charges, ${\chi_{NR}}^{(2)}$ is often quite small compared to the resonant contribution.

By altering the polarization of the light incident on the sample and the polarization at which the SF beam is collected, orientation data from detected surface species can also be obtained by solving for the four independent non-zero elements of the 27 element second order susceptibility tensor, (${\chi_{zzz}}^{(2)}$, ${\chi_{yyz}}^{(2)} = {\chi_{xxz}}^{(2)}$, ${\chi_{xzx}}^{(2)} = {\chi_{yzy}}^{(2)}$, ${\chi_{zxx}}^{(2)} = {\chi_{zyy}}^{(2)}$). The individual polarizations in such a VSFG spectrum are identified by a three-letter symbol (*e.g. ssp* or *sps*) where *s* indicates the electric field vector of the light is parallel to the surface, and *p* indicates the electric field vector is perpendicular to the surface. These three letters represent the polarization of the three frequencies in order of decreasing energy (*i.e.* SF, visible, IR).[38]

VSFG spectra were corrected first for IR power using the non-resonant SFG response across the frequency region of interest from a gold-coated silicon wafer. Spectra were then normalized by the SFG signal from a clean, hydrophilic silica/vapor interface acquired under the appropriate polarization condition. This procedure was followed in order to avoid artificial enhancements of SFG intensity at the high and low frequency spectral boundaries. While effective for its intended application, normalizing to the clean silica/air interface can introduce small artifacts into the SFG spectra. One example of this effect *might* be the small, negative-going feature at 2830 cm^{-1} in the *sps* spectra of silica/liquid interfaces.[39] This frequency corresponds to the symmetric stretch of methoxy groups bound to silica surfaces. While the spectrum of the clean silica/vapor interface does not show clear intensity in this region under *sps* (or any other polarization) condition, a small, systematic contribution to the reference can appear as a contaminant in the referenced spectrum of interest. Regardless of its origin, this behavior in the *sps* spectra is very weak and spectrally isolated from other bands of interest and will not be considered further in the discussion below.

All spectra were then calibrated using the 2910 cm^{-1} methyl symmetric stretch of DMSO at the liquid/vapor interface.[40] All interfaces were probed using two different polarization combinations – *ssp* and *sps* – and vibrational modes were subsequently assigned using a host of both bulk and surface specific literature values. Differences in absolute intensities between solid/vapor and solid/liquid spectra have been shown to depend on differences in adjacent phase refractive indices (assuming oscillator strength, surface coverage, and orientation are all equal).[41,42]

Resonance enhanced SHG is another popular 2nd order NLO method used to study adsorption and interfacial solvation,[43–45] and has been used to identify how surfaces change solvation environments from bulk solution limits.[45–47] Similar to VSFG, the strength of the solute's SHG signal is related to the population and average orientation of adsorbed solutes. The intensity of the SHG response, $I(2\omega)$ depends directly on the intensity of the incident light $I(\omega)$ through the induced nonlinear polarization and is given by:

$$I(2\omega) \propto |P^{(2)}(2\omega)|^2 = |\chi^{(2)}{:}E(\omega)E(\omega)|^2 = |\chi^{(2)}|^2{:}I(\omega)^2 \tag{3}$$

The nonlinear susceptibility can again be separated into resonant and non-resonant portions. The non-resonant susceptibility is intrinsic to any boundary where inversion symmetry is broken. ${\chi_{NR}}^{(2)}$ is often assumed to be single valued, but can also be fitted to a linear function across a broader spectral range.[48,49] The resonant portion is the same as eqn (2), only the hyperpolarizability is now given by:

$$\beta = \frac{A}{\omega_0 - \omega - i\Gamma} \tag{4}$$

where A is a constant associated with SH amplitude, ω_0 is the resonant frequency of the transition, and ω is the frequency of the incident light, and Γ is the transition linewidth.

3. Results and discussion

A. Neat solvents

VSFG was used to examine the structure of the four different liquids at the polar, silica surface. Of particular interest was how well the structure of a thin film formed from a saturated vapor described the structure of the interfacial liquid at a buried solid/liquid interface. Films that retained the same structure when the vapor was replaced by a liquid were assumed to enjoy strong lateral interactions between adsorbed monomers. Films that underwent significant reorganization at the solid/liquid interface were subject to weaker monomer–monomer and correspondingly stronger monomer–substrate interactions at the interface. By examining the polarization dependent VSFG band intensities, each liquid's interfacial organization at the silica surface was determined.

i. 1- and 2-propanol. Fig. 2 shows VSFG spectra acquired from 1-propanol at both silica/vapor and silica/liquid interfaces using two different polarization combinations.

Four vibrations are apparent in the 1-propanol silica/vapor *ssp* spectrum at 2853 cm^{-1}, 2876 cm^{-1}, 2910 cm^{-1}, and 2941 cm^{-1}. The two largest features are

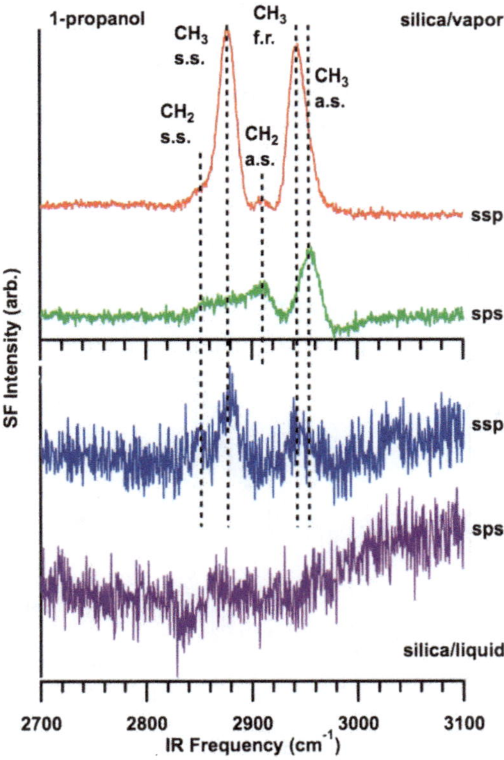

Fig. 2 SFG spectra of 1-propanol at the silica/vapor (top) and silica/liquid (bottom) interface in both the *ssp* and *sps* polarization combinations.

assigned to the CH_3 symmetric stretch (2876 cm^{-1}) and a CH_3 Fermi resonance (2941 cm^{-1}).[50–52] These two assignments leave the remaining two peaks (2853 cm^{-1}, and 2910 cm^{-1}) to be assigned to the CH_2 symmetric and antisymmetric stretch, respectively.[50] The solid/vapor *sps* spectrum has two main features at 2910 cm^{-1} and 2956 cm^{-1}. These two features are assigned to the CH_2 antisymmetric stretch and the CH_3 antisymmetric stretch, respectively. A broad feature at 2854–2894 cm^{-1} is also apparent, and may contain contributions from both the CH_2 and CH_3 symmetric stretches. The appearance of these bands show that the 1-propanol is standing near vertical with the C_3 axis of the terminal methyl group aligned close to the surface normal, similar to the structure first reported by Barnette and Kim.[53,54] While significantly attenuated, a similar pattern appears in the solid/liquid *ssp* spectrum, leading us to believe that 1-propanol structure observed at the solid/vapor interface persists at the solid/liquid interface.

Equivalent measurements were made with 2-propanol. These spectra are shown in Fig. 3.

Four assignments can be made for the *ssp* silica/2-propanol vapor spectrum. The methyl symmetric stretch at 2876 cm^{-1} and the Fermi resonance at 2945 cm^{-1} both match corresponding functional group assignments from the 1-propanol spectrum. Additionally, the 2-propanol *ssp* silica/vapor spectrum shows a feature at 2919 cm^{-1} corresponding to the isolated CH methine

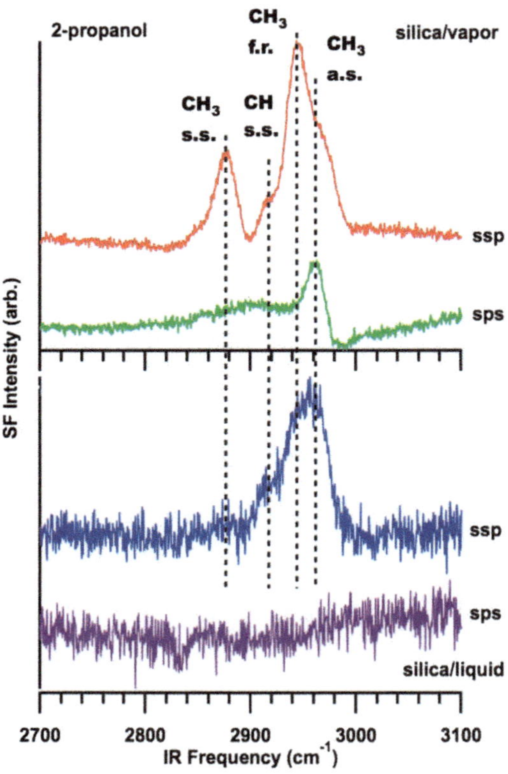

Fig. 3 SFG spectra of 2-propanol at the silica/vapor (top) and silica/liquid (bottom) interface in both the *ssp* and *sps* polarization combinations.

oscillator, and the methyl antisymmetric stretch at 2962 cm^{-1}. Wang and co-workers proposed that the methine stretch is shifted from its normal frequency of ~2900 cm^{-1} due to its close proximity to 2-propanol's alcohol group.[50] Similar effects have been reported for the CH_2 groups in ethylene glycol.[55] The *sps* solid/vapor spectrum shows only the methyl antisymmetric stretch at 2961 cm^{-1} with a broad feature covering both the methyl symmetric stretching and methine stretching regions. Taken together, features in the *ssp* and *sps* silica/vapor spectra indicate that the methyl C_3 axes of 2-propanol adsorbed from the vapor are directed at an angle relative to the surface normal so that both the symmetric and antisymmetric methyl stretches appear. This orientation is in contrast to how 2-propanol positions itself at the silica/liquid interface. Fig. 3 shows that all the features in *sps* polarization disappear, and in the *ssp* spectrum, the CH_3 symmetric stretch is significantly attenuated while the relative intensities are shifted in the three mode combination peak comprised of the CH stretch (2918 cm^{-1}), CH_3 Fermi resonance (2946 cm^{-1}), and the CH_3 antisymmetric stretch (2965 cm^{-1}). (We note that this broad feature may also contain several different contributions assigned nominally to a non-degenerate methyl antisymmetric stretch and not to a methyl Fermi resonance.) The shifting intensities show that the 2918 cm^{-1} mode has grown with respect to the other two high frequency modes, implying that the CH bond has aligned itself to be more perpendicular to the surface. This pattern of spectral intensities suggests that at

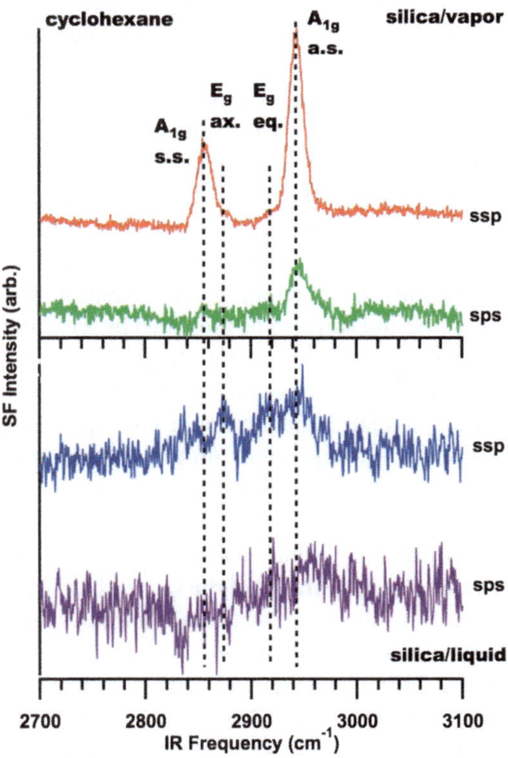

Fig. 4 SFG spectra of cyclohexane at the silica/vapor (top) and silica/liquid (bottom) interface in both the *ssp* and *sps* polarization combinations. Silica/liquid spectra have been smoothed using a 2-point box average.

the silica/liquid interface, 2-propanol molecules lie at an angle with their methyl C_3 axes parallel to the silica surface and with the lone C–H bond pointing almost normal to the surface.

ii. Cyclohexane and methylcyclohexane. Fig. 4 shows VSFG spectra of cyclohexane adsorbed to the silica/vapor interface and at the silica/liquid cyclohexane interface. One might be surprised that cyclohexane gives rise to an SFG signal at all. Previous studies have shown that at room temperature, >99% of all cyclohexane monomers are in the chair conformation.[56,57] Because the more energetically stable chair conformer has a center of inversion (D_{3d} point group symmetry), mutual exclusion considerations require that modes can not be both Raman and IR active and therefore cyclohexane should not give rise to a SFG spectrum.

The significant SFG signal seen in Fig. 4 shows clearly that molecular point group symmetry must be broken at the interface. This phenomenon has also been observed with other centrosymmetric molecules and most notably with benzene.[58,59] Recent work reported by Morita and co-workers used molecular dynamics simulations and quantum chemical calculations to calculate the origin of benzene's strong SFG response. The authors concluded that SF activity was due to a breakdown of the normal mode description of benzene's –CH stretching vibrations. In the local mode picture that emerged, benzene's SF activity arose from a mixing of IR and Raman active vibrations to produce vibrational motions with nonzero changes in benzene's molecular dipole and polarizability.

While we cannot discount the fact that the cyclohexane's SF activity of cyclohexane might also arise from a breakdown of cyclohexane's normal mode vibrational structure, we also cannot deduce, *a priori*, what coordinates will be involved in the new local mode vibrations. Consequently, SF active vibrational modes for silica/cyclohexane vapor and silica/cyclohexane liquid are assigned based on experiments and computational calculations by Wiberg *et al.*,[60–62] Keefe *et al.*,[63] and others.[41,64,65] In the silica/cyclohexane vapor *ssp* spectrum, two main features are evident, with two smaller features appearing as shoulders. The two main peaks at 2856 cm^{-1} and 2943 cm^{-1} have been assigned previously to the CH_2 symmetric and antisymmetric stretching modes, respectively. Both of these modes have A_{1g} symmetry, and can also be modelled by recasting cyclohexane C–H vibrations into C–H_{axial} vibrations and C–$H_{equatorial}$ vibrations. Using this model, the low frequency mode corresponds predominantly to the stretching of the axial protons while the high frequency mode corresponds predominantly to the vibrational motion of the equatorial protons. Using the axial/equatorial model, the two shoulders at 2871 cm^{-1} and 2927 cm^{-1} are assigned to two doubly degenerate modes with E_g symmetry.

The two A_{1g} modes are also visible in the *sps* spectrum albeit with weaker intensity and less resolution. Appearance of the same vibrational features in both *ssp* and *sps* spectra indicates that the interfacial molecules are aligned with the relevant IR transition moments projected at some intermediate angle between the surface normal and the surface plane. Assigning the low frequency modes in the VSFG spectrum to symmetric methylene vibrations and the high frequency modes to anitsymmetric methylene modes, we use methods developed by Wang and coworkers to estimate the average orientation of cyclohexane monomers adsorbed to the surface.[55] This analysis suggests that cyclohexane's C_3 axis is canted ~50–70° away from the surface normal. At the silica/liquid interface, the same four vibrational features are visible in the *ssp* spectrum although the relative intensities of specific bands change. A change in the relative intensities suggests some degree of structural reorganization at the solid/liquid interface, but again the vibrations assigned to the equatorial CH bonds are stronger, implying that cyclohexane's organization at the silica surface has changed very little. The *sps* spectrum shows no obvious features. This picture of cyclohexane structure at the silica surface is consistent with expectations based on previously mentioned X-ray scattering data from Doerr *et al.*[16] These studies concluded that at the silicon oxide/liquid cyclohexane interface, the first layer of cyclohexane formed a very dense film that was followed by a region of depleted solvent density. If adjacent cyclohexane layers at the silica surface associate with each other only weakly, then the structure adopted by cyclohexane at the solid/vapor interface is likely to persist when the vapor phase is replaced by the neat liquid.

As with cyclohexane, methylcyclohexane is most stable in its chair conformation. The methyl group can therefore be in either an axial or an equatorial position.[66] Many sources have both experimentally and computationally found methylcyclohexane's axial–equatorial conformational energy difference to be $\sim$7–8 kJ mol^{-1}.[67–71] From this value, calculations show that at room temperature in solution, most ($\sim$95%) methylcyclohexane monomers will have the methyl group in an equatorial position. One important distinction between methylcyclohexane and cyclohexane is that the methyl group of methylcyclohexane

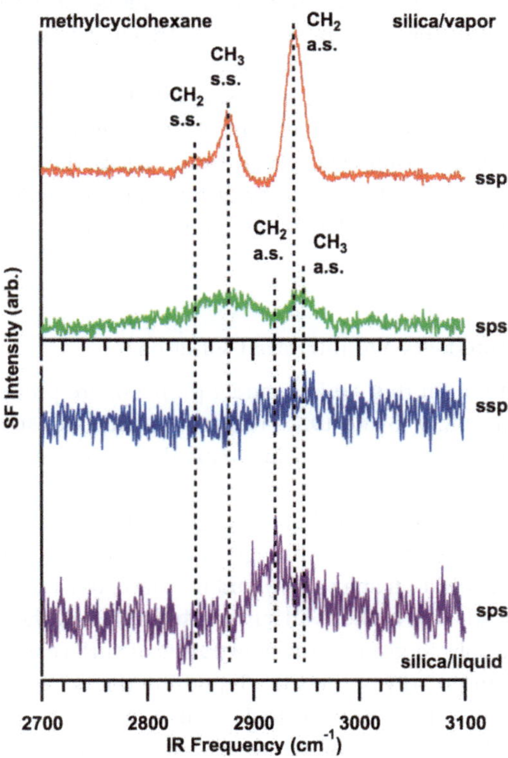

Fig. 5 SFG spectra of methylcyclohexane at the silica/vapor (top) and silica/liquid (bottom) interface in both the *ssp* and *sps* polarization combinations. Silica/liquid spectra have been smoothed using a 2-point box average.

reduces the symmetry of this molecule from D_{3d} to C_s and all molecular vibrations become SFG symmetry allowed according to eqn (2).

The silica/methylcyclohexane spectra shown in Fig. 5 have three distinctive features in the *ssp* polarization combination, and two broad features in the *sps* combination that can be attributed to multiple vibrational modes. Assignment of vibrational features in methylcylohexane's SFG spectra are based on bulk vibrational experiments and computer simulations run by Gardiner *et al.*[69] and Durig *et al.*[66] Bands in the *ssp* spectrum at 2845 cm^{-1}, 2876 cm^{-1}, and 2938 cm^{-1} are assigned to symmetric CH_2 stretching, symmetric CH_3 stretching, and antisymmetric CH_2 stretching, respectively. The *sps* spectrum is more difficult to interpret. Both CH_2 symmetric (2854 cm^{-1}) and CH_3 symmetric (2878 cm^{-1}) stretching may contribute to the broad, low frequency feature. The higher frequency feature contains both the CH_2 antisymmetric mode (2940 cm^{-1}) and CH_3 antisymmetric mode (2951 cm^{-1}). Our approach to determine methylcyclohexane orientation at the silica/vapor surface relies on the orientation of the methyl symmetric and antisymmetric IR transition dipoles and the polarization combinations that enable these features to appear in spectra. Since the antisymmetric stretch appears predominantly in the *sps* polarization and the symmetric stretch appears primarily in the *ssp* polarization, we conclude that the equatorial methyl group's C_3 axis is directed almost normal to the surface meaning the methylcyclohexane is standing on end. This result contrasts with the orientation of methylcyclohexane at the silica/liquid interface. In the liquid *ssp* spectrum the only feature that appears is very weak and is located 2952 cm^{-1}. Based on previous work, this feature is assigned to the CH_3 antisymmetric stretch. The *sps* spectrum on the other hand shows one distinct feature, a strong CH_2 antisymmetric stretch at 2920 cm^{-1}. The *sps* spectrum also contains a hint of intensity at 2866 cm^{-1} that could be due to either the CH_3 symmetric or CH_2 symmetric stretch, although this assignment should be considered tenuous. Nevertheless, the features observed in the solid/liquid spectra imply that methylcyclohexane molecules are lying flat at the silica/liquid interface with the methyl group's C_3 axis aligned parallel to the surface.

B. Solid/liquid interfaces and interfacial solvation

Of the liquids chosen in this work, 1- and 2-propanol are protic and polar with dielectric constants of 20.45 and 19.92, respectively. In contrast, cyclohexane and methylcyclohexane are nonpolar, having static dielectric constants of 2.02. The polarity of the propanol isomers derives from the alcohol functional group and correspondingly large molecular dipole. Both cyclohexane and methylcyclohexane can only interact with solutes through van der Waals and dispersive forces. When considering solvation at a solid/liquid interface, one needs to consider not only how a solute adsorbs to the solid surface, but also how the solute associates with surrounding solvent given to the constraints of surface anisotropy.

To probe solvent polarity at the silica/liquid interfaces described above, SHG spectra of C152 (Fig. 1) were measured to determine the effective excitation wavelengths. C152 belongs to the family of 7-aminocoumarins, a collection of solutes whose bulk solution photophysical properties have been extensively documented.[31,72,74] C152 undergoes a +4 D change in molecular dipole upon photoexcitation meaning that this solute exhibits a bathochromic solvatochromic

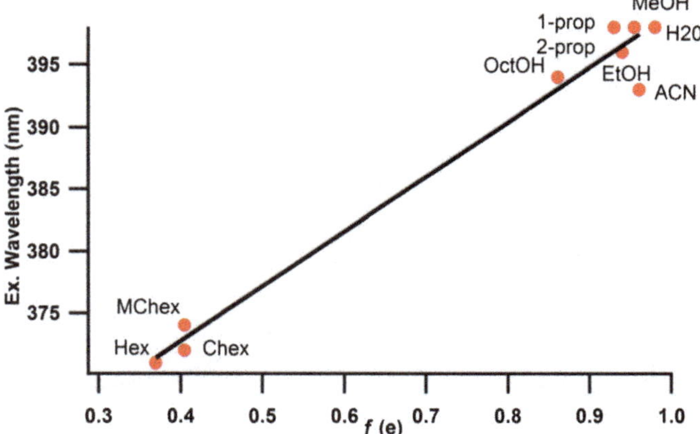

Fig. 6 Solvatochromic behavior of C152 in bulk solutions, solvents are organized by the Onsager polarity function $f(\varepsilon)$. Solvent abbreviations are as follows: hexane (hex), cyclohexane (chex), methylcyclohexane (mchex), octanol (OctOH), ethanol (EtOH), methanol (MeOH), and acetonitrile (ACN), 1-propanol (1-prop), 2-propanol (2-prop), and water (H_2O). Excitation wavelengths were collected in our laboratory and/or reported by Pal *et.al.*[72]

behavior or a red-shift in its excitation spectra with increasing solvent polarity. This behavior is illustrated in Fig. 6 using excitation wavelengths measured in our laboratory and also reported by Pal *et al.* (Fig. 6).[72,74]

SHG spectra of C152 adsorbed to the silica/1- and 2-propanol interfaces are shown in Fig. 7. Some of the noise in the data between 355–365 nm arose from a necessary switch between filter sets. SHG data in this region were all normalized to the non-resonance response of a clean gold surface. The data show clear resonance enhancement at shorter wavelengths. Fitting the spectra to eqn (3) and 4 results in excitation wavelengths of 356 ± 2 and 352 ± 2 nm for 1-propanol and 2-propanol, respectively. For reference the excitation wavelength of C152 in bulk 1- and 2-propanol is 398 nm. In fact, the SHG data show that C152 adsorbed to the silica/1- and 2-propanol interfaces samples an environment that is even less polar than bulk alkane solvents.

Such behavior is not unprecedented. Earlier studies of solvent polarity across strongly associating silica/*n*-alcohol and aqueous/*n*-alcohol interfaces reported the existence of extremely nonpolar environments for $n \geq 8$. Experiments using variable length surfactants that systematically varied solute location across the interface suggested that the width of this nonpolar region extended approximately one solvent length away from the surface. The origin of this low dielectric region had been attributed to a paucity of hydrogen bonding across the interfacial region. If the propanol isomers hydrogen bond strongly to the silica surface forming a nonpolar film, then any adsorbed solutes will experience an environment that is significantly less polar than the bulk liquid. Furthermore, if solvent density experiences partial depletion or conformational restrictions relative to bulk limits, then the local dielectric environment can be even rarer than in solutions of alkanes. This picture is consistent with the proposed organization of 1- and 2-propanol at the silica/liquid interface deduced from VSFG data presented above.

Interfacial polarity at the silica/cyclohexane and silica/methylcyclohexane interfaces are radically different from the silica/propanol systems (Fig. 8). SHG

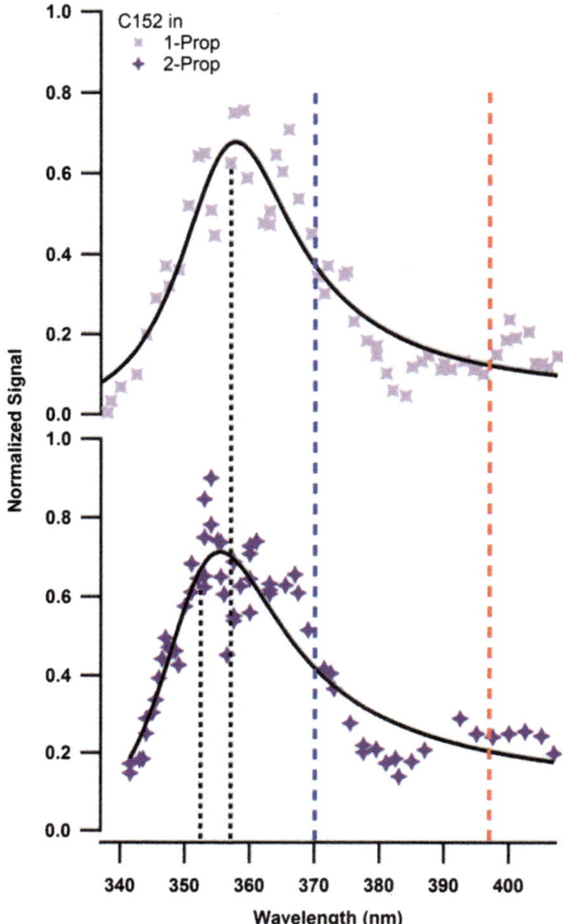

Fig. 7 SHG data of C152 in polar solvents (1-prop and 2-prop). The peak wavelengths were found to be 356 nm and 352 nm for 1- and 2-propanol, respectively. The resonance wavelengths fall out of the range of the solvatochromatic window, shown here with the blue (alkane limit) and red (polar limit) dashed lines. See text for details and explanation.

spectra of C152 adsorbed to both of these silica/alkane interfaces report excitation wavelengths of 399 ± 2 nm, data that fall near the long-wavelength, polar edge of C152's solvatochromic window. Unlike cyclohexane and methylcyclohexane, C152 can associate with surface silanol groups through dipole–dipole interactions. The solvatochromic behavior of C152 does not allow us to discern whether or not dipolar association between C152 and the surface involves hydrogen bond donation or acceptance from the surface silanol groups. In bulk polar media, C152's excitation wavelength is ~400 nm regardless of whether the solvent is a strong hydrogen bond donor (such as H_2O) or acceptor (such as DMSO). Instead, all one can deduce from the SHG spectra and C152's bulk solution behavior is that C152 solvation at the silica/cyclohexane and silica/methylcyclohexane interfaces is dominated by a distinctly polar environment likely created by the high density of surface dipoles. If hydrogen bonding interactions between the surface silanols and C152's carbonyl (in the 2-position) and/or tertiary amine (in the 7-position)

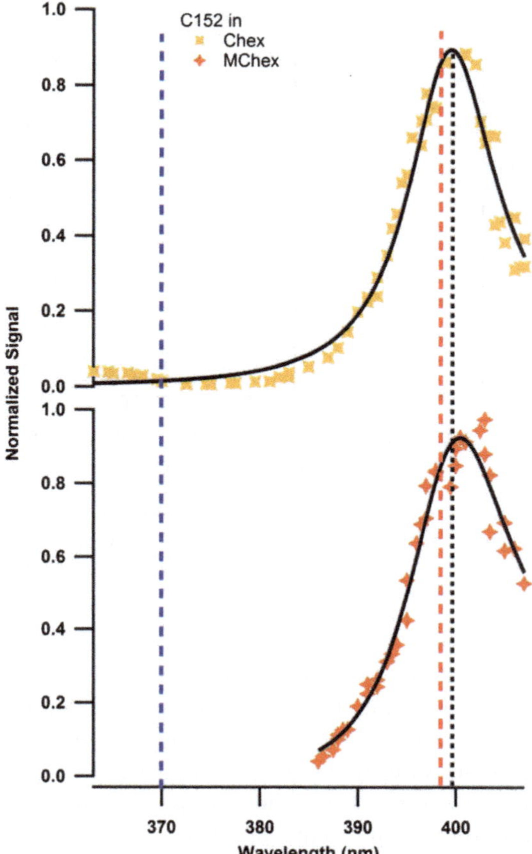

Fig. 8 SHG data of C152 in nonpolar solvents (chex and mchex). The peak wavelengths were found to be 400 nm for both solvents. The resonance wavelength was near the polar limit, C152 appears to be sampling an environment similar to bulk methanol. The solvatochromatic window is shown here with the blue (alkane limit) and red (polar limit) dashed lines.

are strong enough to displace interfacial solvent species, then the extremely polar environment reported by C152 adsorbed to these silica/alkane interfaces can be can assigned to strong substrate/solute interactions.

An alternative explanation for polar environment sampled by C152 at the silica/alkane interfaces is that the C152 is simply solvated by thin layer of water bound to the silica surface itself. (For reference, the excitation wavelength of C152 in aqueous solution is 398 nm.) Thin water films have been reported by Schlangen and others and we cannot discount the possibility that adventitious water might be present at the silica/alkane interfaces sampled in this work.[41,42,73] Several independent observations, however, lead us to believe that trace interfacial water is not playing a significant role in either liquid organization or interfacial solvation. First, water adsorbed to a silica surface shows a strong broad SFG signal centered at 3140 cm^{-1}.[75–77] This feature is assigned to tetrahedrally coordinated water in a strong hydrogen bonding environment. SFG spectra reported in Fig. 4 and 5 show little evidence of a strong feature at high frequencies.

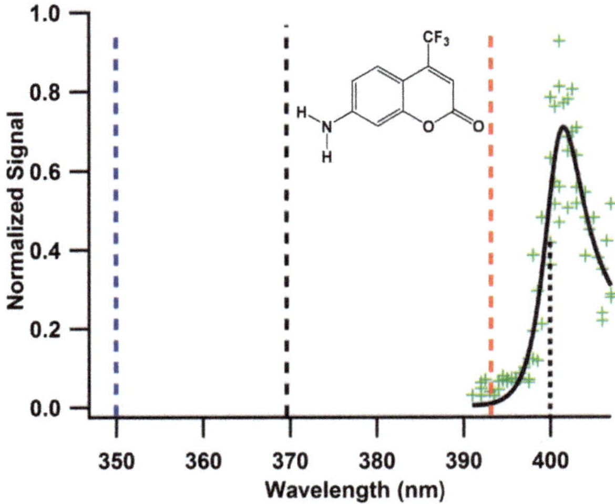

Fig. 9 SHG spectral data of 4 μM coumarin 151 (shown as inset) in cyclohexane. The resonance wavelength (400 ± 1 nm) is red shifted from the polar limit (red dashed line). The solvatochromatic behavior of C151 in nonpolar solvents is shown by the blue dashed line. For reference, the excitation wavelength of C151 in water is included at 369nm.

Second, C152's solvatochromic behavior in polar solvents is sensitive primarily to polarity (a nonspecific solvation mechanism), functional-group specific solvation interactions such as hydrogen bond donation or acceptance do not significantly affect excitation wavelengths. In contrast, coumarin 151 (C151), the 1° amine analog to C152, is extremely sensitive to specific solvation effects. In polar solvents such as acetonitrile and methanol, C151's λ_{exc} is ~380–390 nm. In strong hydrogen bond accepting solvents such as DMSO, λ_{exc} for C151 is red shifted to 412 nm and in solvents capable of donating strong hydrogen bonds such as water, λ_{exc} shifts to shorter wavelengths (369 nm). Such effects have been explored previously and can be stood in terms of hydrogen bonding at the amine position. In the current context, these differences in solvatochromic behavior can serve as a sensitive indicator of hydrogen bonding opportunities for the adsorbed solute.

Fig. 9 shows the SHG spectrum of C151 adsorbed to the silica/cyclohexane interface. Like C152, C151's SHG spectrum is shifted to longer wavelengths and is remarkably narrow. This result implies that C151 is sensitive to the high dipole density presented by the silica surface and is likely to be donating hydrogen bonds to surface silanols. Were significant water present at the silica/cyclohexane interface, we would expect C151 to show an SHG resonance closer to the 369 nm aqueous limit.

Table 1 SHG spectral data collected for Coumarin 151 and 152

Solvent	λ_{max} (nm)	Γ (nm)	[C152] (μM)
Chex	399 ± 1	11 ± 2	14.00
Mchex	399 ± 1	13 ± 2	31.10
1-prop	356 ± 2	22 ± 3	279.90
2-prop	352 ± 2	23 ± 3	108.86

Table 1 lists the SHG data for C152 adsorbed to all for silica/liquid interfaces. Included in the table are SHG resonance wavelengths and linewidths. Also included are the concentrations of the bulk solutions.

Included in this table are the resonance wavelength peak (λ_{max}), the linewidth (Γ), and the concentration of coumarin 152 in solution. The solvents of study are cyclohexane (chex), methylcyclohexane (mchex), 1-propanol (1-prop), and 2-propanol (2-prop).

Several observations warrant mentioning. First, the interfacial polarity reported by adsorbed C152 tracks inversely with bulk solvent polarity: silica/propanol interfaces appear extremely nonpolar to adsorbed C152 solutes while silica/cyclohexane interfaces provide a local polarity consistent with what one would expect of high dielectric solvents. Second, linewidths are closely correlated with SHG resonance wavelengths. SHG linewidths at the nonpolar, silica/propanol interfaces are ~20 nm (FWHM), a result that is more narrow than excitation linewidths in bulk solution (~40 nm) but consistent with previously reported SHG spectra. At the polar silica/alkane interfaces, however, the SHG linewidths are remarkably narrow at ~12 nm. The origin of this effect is uncertain, but if C152 adsorbs to the silica surface by accepting hydrogen bonds through both the 2-position carbonyl and 7-position amine *and* if the surrounding solvent monomers are restricted in their motion, then the narrow linewidths may reflect a very low degree of inhomogeneous broadening. Finally, we note that the bulk solution C152 concentrations required to create a surface excess sufficient to measure with SHG were very different between the propanols and cyclohexanes. In both instances, these concentrations corresponded to what was needed to form full monolayer coverage.

4. Conclusions

Experiments described in this letter examine the structure of four different liquids at the silica/liquid interface. Data from surface specific nonlinear vibrational spectra are used to determine the structure and organization of cyclohexane, methylcyclohexane, 1- and 2-propanol, and resonance enhanced second harmonic spectra measure how a liquid's solvating properties at the interface differs from bulk solution limits. Two important findings emerge from these studies.

1. High registry between liquid monomers at the silica surface leads to a retention of film structure of monolayers adsorbed to the silica/vapor interface when the saturated vapor is replaced by the neat liquid. Conversely, liquid monomers that can not pack together efficiently undergo large structural reorganization when the saturated vapor is replaced by the neat liquid.

2. Interfacial solvation depends less on solvent monomer structure and more upon the noncovalent associations between the interfacial liquid and the adjacent substrate. As a result, the least polar liquids studied in this work, cyclohexane and methylcyclohexane, create surprisingly polar environments at the silica/alkane interface, whereas 1- and 2-propanol form interfacial regions having interfacial polarity less than that of what one observes in bulk alkanes.

Since both 1- and 2- propanol –OH dipoles hydrogen bond strongly to the substrate, adsorbed C152 molecules experience a nonpolar environment and have limited hydrogen bonding opportunities. This conclusion is supported by the surface vibrational spectra. Cyclohexane and methylcyclohexane associate only weakly with the silica surface through van der Waals and dispersion forces. As a result, adsorbed C152 can associate with the silica surface through stronger dipole/dipole interactions and sample a much more polar environment than it would were it solvated in bulk alkane solution.

References

1 D. Beaglehole and H. K. Christenson, *J. Phys. Chem.*, 1992, **96**, 3395–3403.
2 L. J. M. Schlangen, L. K. Koopal, M. A. C. Stuart, J. Lyklema, M. Robin and H. Toulhoat, *Langmuir*, 1995, **11**, 1701–1710.
3 L. K. Koopal, *Adv. Colloid Interface Sci.*, 2012, **179**, 29–42.
4 P. Horng, M. R. Brindza, R. A. Walker and J. T. Fourkas, *J. Phys. Chem. C*, 2010, **114**, 394–402.
5 H. Tavana and A. W. Neumann, *Adv. Colloid Interface Sci.*, 2007, **132**, 1–32.
6 R. Vazquez, R. Nogueira, S. Busquets, J. L. Mata and B. Saramago, *J. Colloid Interface Sci.*, 2005, **284**, 652–657.
7 F. M. Fowkes, *J. Phys. Chem.*, 1962, **66**, 382.
8 J. C. Bonnet and F. P. Pike, *J. Chem. Eng. Data*, 1972, **17**, 145–150.
9 P. Suppan, *J. Photochem. Photobiol., A*, 1990, **50**, 293–330.
10 M. R. Brindza and R. A. Walker, *J. Am. Chem. Soc.*, 2009, **131**, 6207–6214.
11 E. Chibowski and L. Holysz, *Langmuir*, 1992, **8**, 710–716.
12 E. Chibowski and L. Holysz, *J. Adhes. Sci. Technol.*, 1997, **11**, 1289–1301.
13 T. J. Horr, J. Ralston and R. S. Smart, *Colloids Surf., A*, 1995, **97**, 183–196.
14 R. Correa and B. Saramago, *J. Colloid Interface Sci.*, 2004, **270**, 426–435.
15 P. Levinson, M. P. Valignat, N. Fraysse, A. M. Cazabat and F. Heslot, *Thin Solid Films*, 1993, **234**, 482–485.
16 A. K. Doerr, M. Tolan, J. P. Schlomaka and W. Press, *Europhys. Lett.*, 2000, **52**, 330–336.
17 H. K. Christenson, D. W. R. Gruen and R. G. Horn, *J. Chem. Phys.*, 1987, **87**, 1834–1841.
18 J. P. Gao, W. D. Luedtke and U. Landman, *J. Chem. Phys.*, 1997, **106**, 4309–4318.
19 H. K. Christenson, *J. Dispersion Sci. Technol.*, 2006, **27**, 617–624.
20 Y. Wang, K. Hill and J. G. Harris, *Langmuir*, 1993, **9**, 1983–1985.
21 P. T. Mikulski, L. A. Herman and J. A. Harrison, *Langmuir*, 2005, **21**, 12197–12206.
22 Y. Kanda, S. Iwasaki and K. Higashitani, *J. Colloid Interface Sci.*, 1999, **216**, 394–400.
23 W. T. Liu, L. N. Zjang and Y. R. Shen, *Chem. Phys. Lett.*, 2005, **412**, 206–209.
24 F. Ding, Z. H. Hu, Q. Zhong, K. Manfred, R. R. Gattass, M. R. Brindza, J. T. Fourkas, R. A. Walker and J. D. Weeks, *J. Phys. Chem. C*, 2010, **114**, 17651–17659.
25 W. H. Steel, Y. Y. Lau, C. L. Beildeck and R. A. Walker, *J. Phys. Chem. B*, 2004, **108**, 13370–13378.
26 X. Y. Zhang, M. M. Cunningham and R. A. Walker, *J. Phys. Chem. B*, 2003, **107**, 3183–3195.
27 I. Benjamin, *Chem. Rev.*, 1996, **96**, 1449–1475.
28 C. J. Yu, G. Evmenenko, A. G. Richter, A. Datta, J. Kmetko and P. Dutta, *Appl. Surf. Sci.*, 2001, **182**, 231–235.
29 K. Bhattacharyya, E. V. Sitzmann and K. B. Eisenthal, *J. Chem. Phys.*, 1987, **87**, 1442–1443.
30 S. W. Ong, X. L. Zhao and K. B. Eisenthal, *Chem. Phys. Lett.*, 1992, **191**, 327–335.
31 D. Roy, S. Piontek and R. A. Walker, *Phys. Chem. Chem. Phys.*, 2011, **13**, 14758–14766.
32 S. Ishizaka, H. B. Kim and N. Kitamura, *Anal. Chem.*, 2001, **73**, 2421–2428.
33 X. M. Shang, A. V. Benderskii and K. B. Eisenthal, *J. Phys. Chem. B*, 2001, **105**, 11578–11585.
34 M. Yanagimachi, N. Tamai and H. Masuhara, *Chem. Phys. Lett.*, 1992, **200**, 469–474.
35 B. L. Woods and R. A. Walker, *J. Phys. Chem. A*, 2013, **117**(29), 6224–6233.
36 A. R. Siler, M. R. Brindza and R. A. Walker, *Anal. Bioanal. Chem.*, 2009, **395**, 1063–1073.
37 A. G. Lambert, P. B. Davies and D. J. Nelvandt, *Appl. Spectrosc. Rev.*, 2005, **40**, 103–145.
38 G. L. Richmond, *Chem. Rev.*, 2002, **102**, 2693–2724.
39 C. Y. Wang, H. Groenzin and M. J. Shultz, *J. Phys. Chem. B*, 2004, **108**, 265–272.
40 F. Ding, Q. Zhong, M. R. Brindza, J. T. Fourkas and R. A. Walker, *Opt. Express*, 2009, **17**, 14665–14675.
41 A. M. Buchbinder, E. Weitz and F. M. Geiger, *J. Phys. Chem. C*, 2010, **114**, 554–566.
42 X. Zhuang, P. B. Miranda, D. Kim and Y. R. Shen, *Phys. Rev. B: Condens. Matter*, 1999, **59**, 12632–12640.
43 N. Bloembergen, *Opt. Acta*, 1966, **13**, 311–322.
44 X. Y. Zhang, M. M. Cunningham and R. A. Walker, *J. Phys. Chem. B*, 2003, **107**, 3183–3195.

45 A. R. Siler and R. A. Walker, *J. Phys. Chem. C*, 2011, **115**, 9637–9643.
46 G. Marowsky, R. Steinhoff, L. F. Chi, J. Hutter and G. Wagniere, *Phys. Rev. B*, 1988, **38**, 6274–6278.
47 K. B. Eisenthal, *Chem. Rev.*, 2006, **106**, 1462–1477.
48 B. Busson and A. Tadjeddine, *J. Phys. Chem. C*, 2009, **113**, 21895–21902.
49 S. Z. Can, C. F. Chang and R. A. Walker, *Biochim. Biophys. Acta, Biomembr.*, 2008, **1778**, 2368–2377.
50 R. Lu, W. Gan, B. H. Wu, Z. Shang, Y. Guo and H. F. Wang, *J. Phys. Chem. B*, 2005, **109**, 14118–14129.
51 C. D. Stanners, Q. Du, R. P. Chin, P. Cremer, G. A. Somorjai and Y. R. Shen, *Chem. Phys. Lett.*, 1995, **232**, 407–413.
52 P. Guyotsionnest, J. H. Hunt and Y. R. Shen, *Phys. Rev. Lett.*, 1987, **59**, 1597–1600.
53 A. L. Barnette and S. H. Kim, *J. Phys. Chem. C*, 2012, **116**, 9909–9916.
54 O. Esenturk and R. A. Walker, *J. Chem. Phys.*, 2006, **125**, 174701.
55 R. Lu, W. Gan, B. H. Wu, H. Chen and H. F. Wang, *J. Phys. Chem. B*, 2004, **108**, 7297–7306.
56 G. Gill, D. M. Pawar and E. A. Noe, *J. Org. Chem.*, 2005, **70**, 10726–10731.
57 M. Squillacote, R. S. Sheridan, O. L. Chapman, *97*, 1975, **11**.
58 E. L. Hommel and H. C. Allen, *Analyst*, 2003, **128**, 750–755.
59 T. Kawaguchi, K. Shiratori, Y. Henmi, T. Ishiyama and A. Morita, *J. Phys. Chem. C*, 2012, **116**, 13169–13182.
60 K. B. Wiberg and A. Shrake, *Spectrochim. Acta, Part A*, 1970, **27**, 1139–1151.
61 K. B. Wiberg and A. Shrake, *Spectrochim. Acta, Part A*, 1973, **29**, 583–594.
62 K. B. Wiberg, V. A. Walters and W. P. Dailey, *J. Am. Chem. Soc.*, 1985, **107**, 4860–4867.
63 C. D. Keefe and J. E. Pickup, *Spectrochim. Acta, Part A*, 2009, **72**, 947–953.
64 R. G. Snyder and J. H. Schachts, *Spectrochim. Acta*, 1965, **21**, 169.
65 J. R. Durig, C. Zheng, D. El and M. Ahmed, *J. Raman Spectrosc.*, 2009, **40**, 197–204.
66 J. R. Durig, R. M. Ward, G. A. Guirgis and T. K. Gounev, *J. Raman Spectrosc.*, 2009, **40**, 1919–1930.
67 K. B. Wiberg, J. D. Hammer, H. Castejon, W. F. Bailey, E. L. DeLeon and R. M. Jarret, *J. Org. Chem.*, 1999, **64**, 2085–2095.
68 F. Anet, C. H. Bradley and G. W. Buchanan, *J. Am. Chem. Soc.*, 1971, **93**, 258.
69 D. J. Gardiner, G. S. Bassi and G. D. Galvin, *Appl. Spectrosc.*, 1984, **38**, 313–317.
70 Y. N. Huang and J. Leech, *J. Phys. Chem. A*, 2001, **105**, 6965–6970.
71 O. A. Subbotin and N. M. Sergeyev, *J. Chem. Soc., Chem. Commun.*, 1976, 141–142.
72 S. Nad, M. Kumbhakar and H. Pal, *J. Phys. Chem. A*, 2003, **107**, 4808–4816.
73 L. J. M. Schlangen, L. K. Koopal, M. Stuart, J. Lyklema, M. Robin and H. Toulhoat, *Langmuir*, 1995, **11**(5), 1701–1710.
74 S. Nad and H. Pal, *J. Phys. Chem. A*, 2001, **105**, 1097–1106.
75 L. Zhang, S. Singh, C. Tian, Y. R. Shen, Y. Wu, M. A. Shannon and C. J. Brinker, *J. Chem. Phys.*, 2009, **130**, 154702.
76 O. Isaienko, S. Nihonyanagi, D. Sil and E. Borguet, *J. Phys. Chem. Lett.*, 2013, **4**, 531–535.
77 Y. R. Shen and V. Ostroverkhov, *Chem. Rev.*, 2006, **106**, 1140–1154.

Faraday Discussions RSC Publishing

PAPER

Dynamics and vibrational spectroscopy of water at hydroxylated silica surfaces†

Prashant Kumar Gupta and Markus Meuwly*

Received 28th May 2013, Accepted 4th July 2013
DOI: 10.1039/c3fd00096f

In the present study, the structural and dynamical properties of water at hydroxylated silica surfaces are investigated with classical molecular dynamics simulations. Depending on the nature of the interface, water molecules are observed to have well defined ordering and slower dynamics compared to bulk water. These properties include the orientation of water near the surface, reorientational relaxation times, translational diffusion coefficients, planar density distribution and vibrational spectroscopic features. The dynamical and structural features are affected up to ≈6–10 Å away from the surface, depending on the properties considered (lateral diffusion coefficient, charge density profile, rotational orientational time). Water molecules at the silica surface show a marked decrease in the diffusion coefficient and an increase in rotational correlation times. The presence of the polar –OH group on the hydroxylated silica surface provides adsorption sites for water with preferred orientations. In addition to the known broadening of the water spectrum in the bonded OH-stretch region for first-layer water molecules, the present simulations find a characteristic band at 1150 cm^{-1}, which is assigned to the HO(water)–H(SiOH) bending vibration. Because this signal only occurs for water molecules in the first layer, it should be experimentally accessible through surface sensitive techniques such as vibrational sum frequency generation spectroscopy (VSFG).

1. Introduction

Water molecules at the solid–liquid interface play important roles in surface science (adsorption/desorption), geology, material science, diffusion of ions in nanopores, biological membranes and in interfacial chemistry.[1–6] A number of experimental[7–12] and theoretical studies[13–20] have provided deeper insight into the molecular level behavior of water itself and at interfaces. It has been found that properties of water at surfaces can be strongly affected by the existence and the nature of the solid substrate and is due to the heterogeneous environment

Department of Chemistry, University of Basel, Klingelbergstrasse 80, CH-4056 Basel, Switzerland. E-mail: m.meuwly@unibas.ch

† Electronic supplementary information (ESI) available: planar density distribution, vibrational spectra of water with the simple charge model (TIP3P). Vibrational spectra for isotopically substituted water molecules (HOD and D_2O) with MTP water model. See DOI: 10.1039/c3fd00096f

generated by the surface itself. The resulting effects on water are observed through changes in the spectroscopic response, pronounced density layers, the change in the local dielectric constant and in different structural and dynamical properties of water at these interfaces.[21,22] Such phenomena are also important technologically in the context of electrolytes near a charged surface[23] and in chromatographic processes.[24,25] As to the latter, various theoretical studies involving classical molecular dynamics, *ab initio* molecular dynamics and experimental techniques like non-linear optical spectroscopy have been used to characterize silica–water interfaces.[26,27] Computer simulations considered interfacial water at quartz,[27–34] metal[35,36] and other mineral surfaces.[37,38] These investigations clearly show how the solid substrate perturbs the local structure and hydrogen bond dynamics of water up to 10–15 Å away from the surface.

Water has been investigated particularly well at the water–air interface.[26,27,39–47] These studies found spectroscopic features that extend from around 3200 to 3700 cm^{-1}. The analysis of the spectroscopic signatures in terms of their origin and the local orientation of the water molecules has lead to a number of, sometimes conflicting, interpretations.[39,47–49] Also, water at mineral surfaces has been investigated by spectroscopic means. For water at silica surfaces the pH-dependent VSFG spectra were recorded at a quartz(0001) surface. A prominent two-peak structure with spectroscopic signals around 3200 cm^{-1} and 3400 cm^{-1} was found at high pH (SiO^-) whereas at low pH (SiOH) the double-peak was less prominent.[50]

The hydrogen bond and orientational dynamics of water in contact with nanoscopic environments has also attracted considerable attention. Of particular relevance are experimental and simulation studies of water in reverse micelles, which provide a well defined chemical environment to study water dynamics in contact with a surface.[51–54] The reorientational dynamics were found to contain multiple time scales, extending from the sub-picosecond librational dynamics to several ten picoseconds involving the reorganization of the water-hydrogen-bonded network. Comparing experimental and simulation results it was found that the nonexponential decay of the orientational diffusion is comparable although they do not agree quantitatively.[54]

In the context of chromatography, the role of water molecules has been recognized for quite some time. The existence of more or less ordered water networks has been postulated or found in chromatographic columns. Already 30 years ago, Scott and co-workers have suggested that individual water molecules can bind to the –OH groups of a silica surface and form a thermodynamically stable (fractional) monolayer.[55] These water molecules, thought to be present up to temperatures of several hundred Kelvin, can serve as H-bond donors to additional water molecules, which eventually form filaments between the stationary (silica surface) and the mobile (solvent mixture) phase. The existence, morphology, and stability of such water filaments has been recently investigated by using atomistic simulations.[25] These studies also corroborated the notion of water layers with differing stability depending on their separation from the stationary phase in chromatographic systems. The existence of surface-bound water molecules in chromatographic systems has been also established from atomistic simulations using a range of force fields and simulation strategies.[25,56,57]

In the present study, classical molecular dynamics simulations are used to provide insight into atomistic details of water at the silica–water interface relevant to chromatographic studies. We report on the structural and dynamical properties of the solvent including the effect of multipolar interactions in the electrostatics on vibrational properties of water and provide an atomistic interpretation of the observed data. To this end, two different water models are used in the simulations – the widely used TIP3P parametrization[58] and a flexible, multipolar water model. First, the computational realization of the system and the analysis of the data is described. Then, static and dynamical properties of water at a silica (101) face at low pH are reported and finally, the infrared spectroscopy is discussed.

2. Computational methods

2.1. Description of the system

The model silica layer was constructed by slicing two 8.75 Å thick segments of the (101) face of a quartz crystalline lattice with dimensions of 36 × 41 Å. This resulted in two surfaces with 64 silica hydroxy groups. The distance between the two silica layers was 30 Å, which resulted in a unit cell with dimensions 36 × 41 × 47.5 $Å^3$. A 30 Å wide water box was superimposed and all water molecules overlapping with the silica layers were removed. Fig. 1 shows a schematic representation of the system, which has also been used previously in the study of chromatographic interfaces.[25,56,59,60]

In the present work, two water models were employed. One is the TIP3P water model[58] and the second one is a flexible model based on the parametrization by Kumagai, Kawamura and Yokokawa (KKY),[61] which was used with a quadrupolar charge model (see below). The functional form of the KKY potential for the stretching and bending energies is

$$\begin{aligned} E_{\mathrm{str}} &= D\{1 - \exp[-\beta(r - r_0)]\}^2 \\ E_{\mathrm{bend}} &= 2f_{\mathrm{k}}k_1k_2\sin^2(\theta - \theta_0) \end{aligned} \tag{1}$$

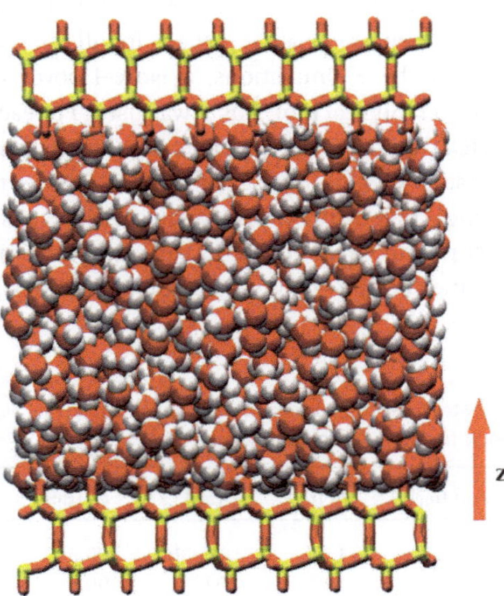

Fig. 1 The solid silica layer with water. Water molecules are in CPK representation, the silica layer is in licorice representation. Color code: oxygen atoms of water and the silica layer are shown in red, silica atoms are shown in yellow and hydrogen atoms are shown in white. The *z*-axis is normal to silica surface.

where

$$k_i = 1/\{\exp[g_r(r_i - r_m)] + 1\}$$

in which r_i is the equilibrium distance of the O–H bond of the water molecule and g_r and r_m are reported in Table 1. For the electrostatics, a monopolar (*i.e.* TIP3P) and a multipolar charge model with a geometry independent quadrupole on the oxygen atom and up to monopoles on the hydrogens was used.[62] The electrostatic parameters for the multipolar model were obtained from *ab initio* calculations at the MP2/aug-cc-pVQZ level of theory and are described in Table 2. The Lennard-Jones (LJ) parameter for this flexible water model was the same as for the TIP3P water model.[58] The parameters for the silica surface were taken from our previous work.[56,59,60]

With the original parametrization of the KKY potential[61] for water the stretching frequencies obtained were $\nu_1 = 3720$ cm^{-1}, $\nu_2 = 1525 \times 10^4$ cm^{-1} and $\nu_3 = 3763$ cm^{-1}, which do not yield correct frequencies for the water bending mode. In the present study a reparametrization was used, which gives fundamentals at $\nu_1 = 3689$ cm^{-1}, $\nu_2 = 1595$ cm^{-1} and $\nu_3 = 3724$ cm^{-1}.[63]

2.2. Molecular dynamics simulations

All MD simulations were carried out with the CHARMM program[64] with provisions for multipolar interactions.[65] The number of water molecules in the system was 1310. Simulations for H_2O, HOD and D_2O followed identical protocols as described below. All simulations were carried out with periodic boundary conditions (PBC) and nonbonded interactions (electrostatic and Lennard-Jones) were truncated at a distance of 12 Å and switched between 10 and 12 Å. The atomic position of the bulk quartz surfaces with the exception of the exposed hydrogen atoms of the silanol groups (1664 atoms) were held fixed during the simulations. Initially the system was heated and equilibrated for 20 ps followed by 100 ps of production at 300 K in the NVT ensemble. The time step used was $\Delta t = 0.4$ fs for all the simulations with the flexible KKY model. For these simulations, a Nose–Hoover thermostat with a thermal piston mass of $Q = 50$ kcal mol^{-1} ps^{-2} was used. For each of these systems with H_2O, D_2O and HOD, multiple independent simulations, each of 100 ps, were carried out and the results were averaged. For the TIP3P water model two simulations were run, one single trajectory of 5 ns length with $\Delta t = 1$ fs using SHAKE and one trajectory for 100 ps with $\Delta t = 0.4$ fs, to calculate the vibrational spectra of water. However for some specific analysis (exchange dynamics), simulations with

Table 1 Parameters for water potential reported in the work of Kumagai *et al.* and used in this work, together with the harmonic frequencies computed from them[a]

	D (E_h)	β (a_0^{-1})	r_0 (a_0)	f_k (E_h)	θ_0 (deg)	r_m (a_0)	g_r (a_0^{-1})
KKY[b]	0.120	1.45	1.55	2.5×10^6	99.5	2.65	3.7
This work	0.120	1.44	1.81	0.0388	99.5	2.65	3.7

[a] Atomic units are used for D, β, r_0, f_k, r_m and g_r. θ_0 is given in degrees. [b] The literature data were converted to atomic units from the following parameters: $D = 75.0$ kcal mol^{-1}, $\beta = 2.74$ Å^{-1}, $r_0 =$ Å, $f_k = 1.1 \times 10^{-1}$ J, $r_m = 1.40$ Å and $g_r = 7.0$ Å^{-1}.

Table 2 Multipole moments of H_2O used in this work. MP2/aug-cc-pVQZ calculations were used for the parametrization. Each multipole component is static, *i.e.*, it does not depend on the geometry of the molecule[a]

	Q^{O}_{00} (e)	Q^{O}_{10} (ea_0)	Q^{O}_{20} ($ea_0{}^2$)	Q^{O}_{22c} ($ea_0{}^2$)	Q^{H}_{00} (e)
H_2O	−0.595922	−0.065831	−0.041591	−1.125991	0.297961

[a] In the multipole component Q^{x}_{u}, u represents angular momentum labels (00, 10, 11c, 11s, 20, 21c, 21s, 22c, and 22s) and x represents the atom type (O or H). The units of Q^{x}_{u} are given in parentheses in the table. Only nonzero components are shown.

the MTP water model are run for several nanoseconds. The MTP water model with inclusion of quadrupole is a computationally expensive water model and this is the reason for carrying out multiple short simulations (100 ps). It had been observed from other simulations performed on silica surfaces[66,67] that the water dynamics (translational diffusion, orientational relaxation time) converges well within 10–30 ps. Furthermore, convergence tests for calculated IR spectra of single spectroscopic probes suggest that 500 ps simulations are appropriate for this property.[68]

Exchange dynamics. For the analysis of the water dynamics, the system was divided into layers of thickness 3 Å each, parallel to the Si-surface and commensurate with the typical size of a water molecule. The first layer (layer I) includes water molecules which are within 3 Å of the oxygen atoms of the silica layer, which serve as the reference $z = 0$. The second layer (layer II) extends from 3 to 6 Å, and the third layer (layer III) from 6 to 9 Å. The exchange of water molecules is quantified by considering water molecules which are in layer I at $t = 0$ and move to layer II during the simulation and *vice versa*.

Infrared spectra. The infrared (IR) spectrum is calculated from the Fourier transform of the dipole moment autocorrelation function $C(t) = \langle \vec{\mu}(0)\vec{\mu}(t)\rangle$, which is accumulated over 2^n time origins, where n is an integer such that 2^n corresponds to between 1/3 and 1/2 of the trajectory, with the time origins separated by 0.8 fs. $\hat{C}(\omega)$, which is the Fourier transform of $C(t)$, is computed using a fast Fourier transform (FFT) with a Blackman filter to reduce noise. The final infrared absorption spectrum $A(\omega)$ is then calculated from

$$A(\omega) = \omega\{1 - \exp[-\omega/(k_B T)]\}\hat{C}(\omega) \tag{2}$$

where k_B is the Boltzmann constant and T is the absolute temperature in Kelvin.

3. Results and discussion

With a fully hydroxylated surface (SiOH), the simulations in the present work correspond to low-pH conditions in actual experiments. First, static properties are discussed, after which the translational, rotational and exchange dynamics are considered and the infrared spectra are presented.

3.1. Static properties

Charge density profile. The charge density profile of water and deuterated water as a function of the distance away from the silica surface is shown in Fig. 2, with the silanol oxygen atom defining $z = 0$. Fig. 2 shows that the presence of the surface perturbs the interfacial water structure up to ≈5 Å away from the silica surface. A similar behavior is observed for HOD and D_2O. The data in Fig. 2 is consistent with an orientation of the interfacial water where the water-H atom points towards the surface. The positive charge excess at $z = 0$ is due to the water oxygen H-bonded to the hydrogen of silanol group and the water-H atom pointing downwards between two adjacent silanol groups and was also found in previous studies.[29,69,70] The excess of negative charge at $z = 4$ is due to the second layer of water molecules interacting through hydrogen bonding with the first layer, as shown in the inset. These charge density profiles are expected to affect the distribution of charged species at solid–water interfaces, possibly contributing to the adsorption of lactates[71] and amino acid[72] analogues on the hydrated silica surface.

Orientational behavior of water. The presence of the surface is expected to perturb the local structure of water.[32,33,38] The orientation of water molecules near the surface can be quantified by considering the water-OH vector relative to the space fixed z-axis. For this, the scalar product $\cos\alpha = \frac{\vec{r}_{OH}\vec{e}_z}{|\vec{r}_{OH}||\vec{e}_z|}$ is computed for water in different layers. Fig. 3 reports the orientational probability $p(\alpha)$ for the OH-vectors involving both hydrogen atoms, H1 and H2, of the water molecules. For bulk water the water-OH vectors are expected to exhibit quite a uniform distribution as shown in Fig. 3A, whereas water molecules near the surface should have clearly preferred orientations. This is indeed observed for the three orientations I to III in Fig. 3B. For orientation I (H_w–O_w points towards the surface-O atom, *i.e.* "H_w-down" configuration) the scalar product is ≈0, whereas for orientation II (H_w–O_w points towards the water above) it is $-1.0 \leq \cos(\alpha) \leq -0.8$ or $140° \leq \theta \leq 180°$, see Fig. 3B. For orientation III both water-H_w–O_w vectors form an equal angle with the SiOH-hydrogen atom and yield $0.4 \leq \cos(\alpha) \leq 0.6$ as shown in Fig. 3B. Comparing this result with the orientation of the water TIP3P model (see Fig. 3C), we note that orientation III is considerably less populated

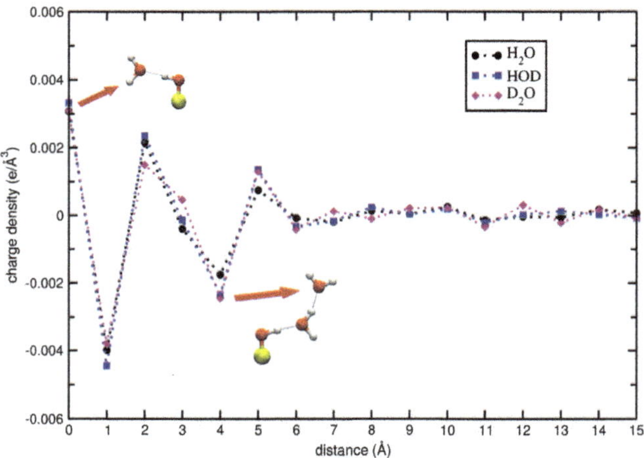

Fig. 2 Charge density profile of water as a function of the vertical distance z from the silica layer where $z = 0$ corresponds to the oxygen atom of the silanol group. This data is plotted for H_2O, HOD, D_2O with the quadrupole moment on oxygen atoms. Color code: black for H_2O, blue for HOD, magenta for D_2O.

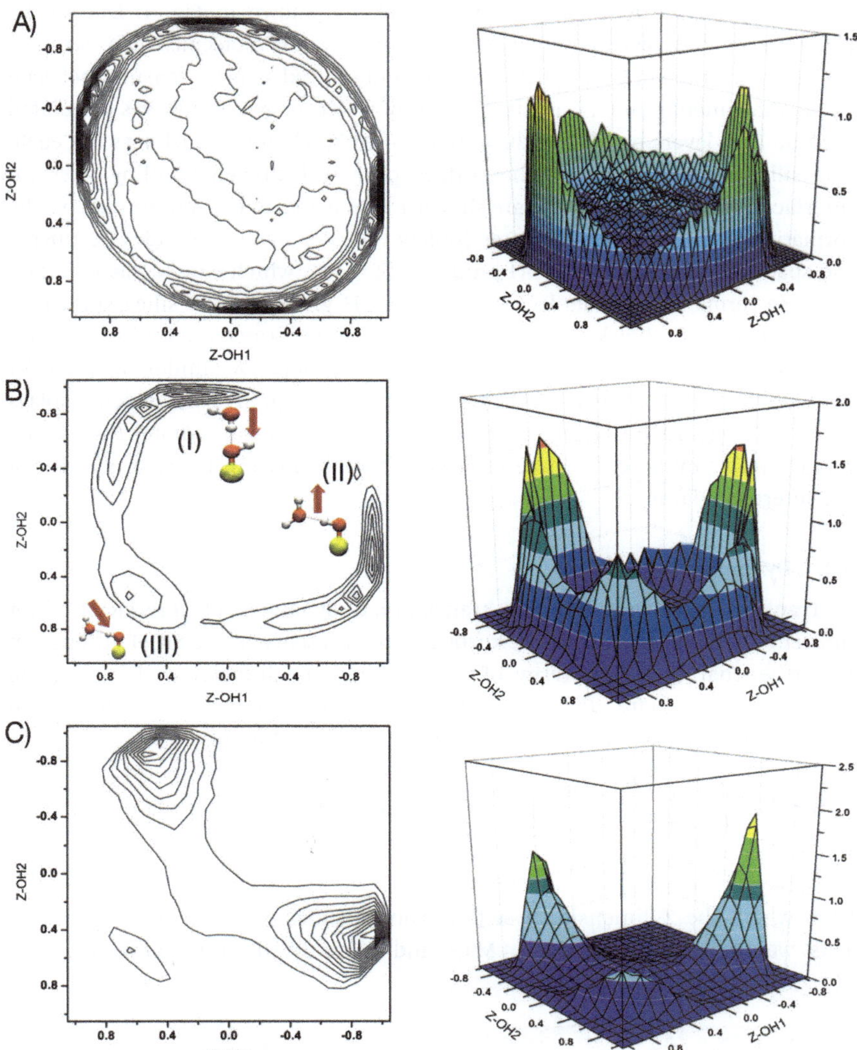

Fig. 3 Contour plots (left) and normalized 3D probability distributions (right) for the orientation of water molecules in the system. A) Orientation of water molecules in the bulk with the MTP model. B) The three preferred orientations of MTP water molecules in the first hydration layer. State I corresponds to the OH-down configuration while state II corresponds to OH-up. In state III the water-OH vector is in a sideways orientation relative to the surface. C) Water orientation in the first hydration layer from simulations with the TIP3P water model.

when using a TIP3P water model. This can be understood in terms of a favourable and stabilizing multipole-charge interaction between the water molecules and the surface-silanol hydrogen atoms compared to charge–charge interaction in the TIP3P water model. The increased population of state III should also affect other properties of the system, including the diffusional behaviour and the spectroscopy of surface-near water molecules.

Planar density distribution. The simulations allow to characterize the distribution and ordering of surface-bound water molecules. The planar density distribution of the first layer water molecules is shown in Fig. 4. Due to the presence of the SiOH-groups, surface-bound water molecules can strongly interact with the surface, which leads to pronounced density maxima between these attachment points. The distribution of water oxygen atoms was calculated separately in layers parallel to the surface (x–y-plane) and of thickness 3 Å each. The silica oxygen atoms on the surface provide hydrogen bonding sites for interfacial water molecules and are shown as white dots in Fig. 4A. We observe the formation of closed ring structures (yellow trace) of water molecules occupying the space around a single Si-OH group, see Fig. 4A, which can be easily understood in terms of water molecules around –OH groups of the silica surface as shown in Fig. 4B (black dotted ring). Consequently, water molecules optimize their orientation relative to the surface SiOH groups. A similar analysis for simulations with the TIP3P water model is shown in Fig. S1, ESI,† which displays a less ordered structuring for surface bound water. This corroborates the finding that including multipoles in intermolecular interactions affects the structuring of the interfacial layer.

3.2. Dynamical properties

Translational dynamics. The translational dynamics of interfacial water molecules close to the surface perturbed by the presence of the SiOH groups can be further characterized in terms of their translational diffusion coefficients. The in-plane mean-square displacement $MSD_{||}$ for each layer is obtained from following the motion of the water-oxygen atoms according to

$$\langle D\rangle = \lim_{t\to\infty}\frac{1}{4t}\left\langle \sum_{i\in \text{layer}}[(x_i(t)-x_i(t=0))^2+(y_i(t)-y_i(t=0))^2]\right\rangle = \lim_{t\to\infty}\frac{1}{4t}\langle\Delta r^2(t)\rangle \quad (3)$$

from which the 2-dimensional self-diffusion coefficient D can be determined. This was done for the TIP3P, MTP and a modified MTP parametrization

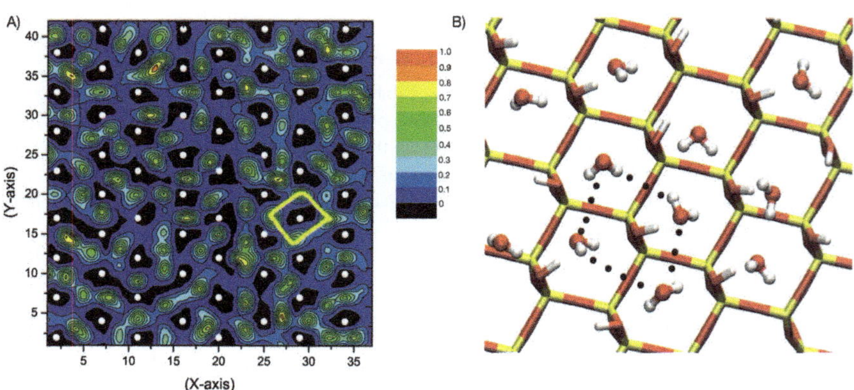

Fig. 4 A) Surface water density distribution parallel to the surface (x–y-plane) for water molecules in the first hydration layer. The white dots in panel A correspond to the oxygen atoms of the silica surface. "0" and "1" correspond to the minimum and maximum occupation. Relative water occupations are color coded from black (0.0) to red (1.0). B) Top view for distribution of water molecules adsorbed on the silica layer due to hydrogen bonding with –OH group of silica layer. Water molecule is shown in CPK and hydroxylated silica surface in licorice representation.

(see below) for the water molecules. As expected, the translational mobility of water molecules away from the interface is more rapid compared to layers close to the interface as shown in Fig. 5. The in-plane diffusion coefficient $D_{\|}$ for different regions were determined by fitting the 4 to 20 ps portion of $\langle \Delta r^2(t) \rangle$ to eqn (3). With the MTP water model the in-plane diffusion coefficient for layer I (0.004 $Å^2$ ps^{-1}) is 35 times smaller than that in the middle of the column which is $D_{\|} = 0.14$ $Å^2$ ps^{-1}. This demonstrates that motion of water molecules adsorbed at the surface is slowed down due to strong electrostatic interactions with the surface. For the TIP3P model a similar observation is made although quantitatively, the numbers differ somewhat. They are 0.007 and 0.21 $Å^2$ ps^{-1}. For layer III the diffusion coefficients are 0.06 and 0.08 $Å^2$ ps^{-1}, respectively, for the MTP and TIP3P model.

In order to put these results into perspective, the (3-dimensional) diffusion coefficients for the MTP and TIP3P water models were also determined from 100 ps simulations of pure water (without silica surfaces) in a $30 \times 30 \times 30$ $Å^3$ box with 1000 water molecules. They were found to be 0.052 $Å^2$ ps^{-1} and 0.32 $Å^2$ ps^{-1} for the MTP and TIP3P models, respectively, which compares with 0.22 $Å^2$ ps^{-1} for the experimentally determined[73] diffusion coefficient. Hence, the quadrupolar model considerably underestimates the experimental value. This is related to the fact that the TIP3P van der Waals ranges were used in the simulations with the MTP electrostatic model. Repeating the MTP simulations with slightly increased (+2%) van der Waals ranges on the oxygen atom yields diffusion coefficients in close agreement with the experimental value. Using these scaled van der Waals ranges for the MTP model in simulations of water in contact with the silica surface, the corresponding 2-dimensional diffusion coefficients for layer I (0.008 $Å^2$ ps^{-1}) is 28 times smaller than that in the middle (0.28 $Å^2$ ps^{-1}). A similar behavior was already observed for water diffusion on hydroxylated aluminum[37] and other polar surfaces.[38]

Reorientational relaxation. The orientational behavior of water is highly dependent upon the surrounding environment and can result in slower dynamics close to the interface compared to bulk water. To characterize this we computed the

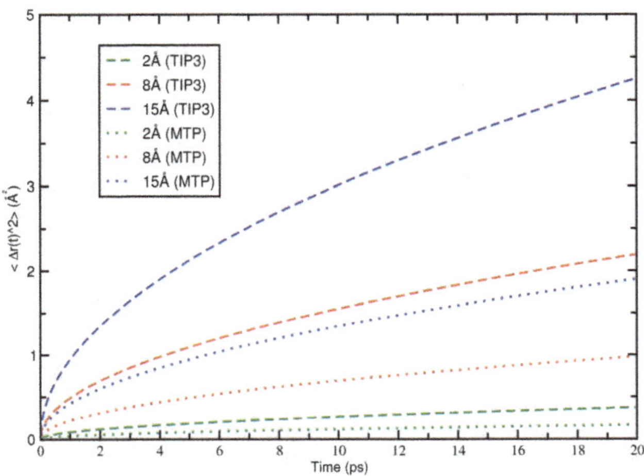

Fig. 5 Mean square displacement for MTP and TIP3P water molecules in three different layers.

reorientational autocorrelation functions for the water dipole moment vector ($\vec{\mu}$) and the H–H vector ($\vec{r}_{HH}$). The autocorrelation function used for this analysis is

$$C_{\hat{\rho}}(t) = \frac{\left\langle \hat{\rho}(0)\hat{\rho}(t) \right\rangle}{\left\langle \hat{\rho}(0)\hat{\rho}(0) \right\rangle} \tag{4}$$

where $\hat{\rho}$ represents either the unit dipole moment vector $\vec{\mu}$ or the H–H vector $\vec{r}_{HH}$, respectively. The data for H_2O in different layers is reported in Fig. 6A. The reorientational relaxation times can be obtained from $C(\tau)$. For water molecules in different layers we find that in the time regime from 0–30 ps the curves can be reasonably well described by triple exponentials $A_1 \exp(-t/\tau_1) + A_2 \exp(-t/\tau_2) + (1 - A_1 - A_2) \exp(-t/\tau_3)$, with decay constants τ_1, τ_2 and τ_3. Furthermore, a

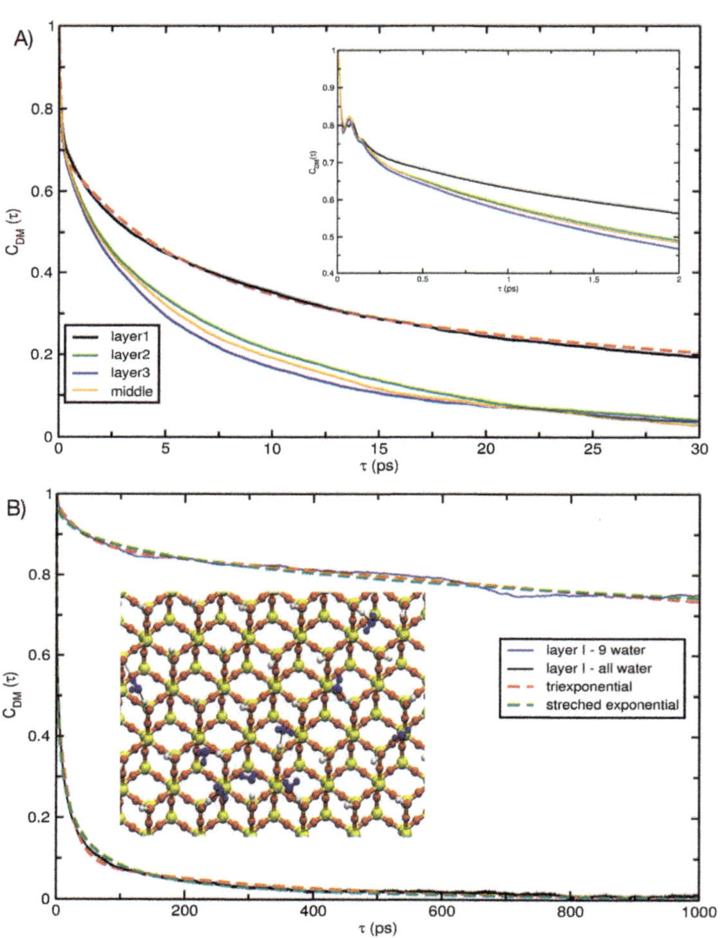

Fig. 6 The dipole moment auto-correlation function for H_2O in different layers. A) Correlation function for $\vec{\mu}$ for 4 different layers with the fitted data. Color code: layer I (black), layer II (green) and layer III (blue), middle (orange) and fitted data (red). Only the fit to layer I is shown. All other fits are of similar quality. The inset shows the correlation function on short times. B) Correlation function for the 9 water molecules which remain adsorbed to the surface for the entire 2.5 ns trajectory (blue) and for all water molecules in layer I (black).

stretched exponential exp $(-t/\tau)^{\beta}$ was considered and was found to equally well describe the data. The reorientational relaxation within the hydration layer (from $z = 0$ to 3 Å, layer I) is significantly slower than away from the surface (from $z = 6$ to 9 Å, layer III) and in the bulk (from $z = 12$ to 15 Å, middle). For all water molecules, irrespective of the layer they are in, the fastest decay occurs on a sub-picosecond time scale with $\tau_1 = 0.5$–0.8 ps. The second decay constant ranges from 2–5 ps with faster time scales for water in the middle of the system (*i.e.* "bulk") and longer time constants for water molecules in layer I. The third decay constant has values between 11 ps and 52 ps, for water molecules in the bulk and in layer I, respectively. When fitting $C(\tau)$ with a stretched exponential, β assumes values between 0.38 and 0.53.

For the dynamics on the nanosecond time scale, water molecules adsorbed for the entire simulation were analyzed separately. There were 9 molecules, *i.e.* 10% of the total number of water molecules in layer I (see Fig. 6B), which never exchange with neighboring layers. The corresponding rotational correlation function can be fit with time scales $\tau_1 = 1.6$ ps, $\tau_2 = 42$ ps and $\tau_3 = 6.0$ ns. When all water molecules in layer I are analyzed together, $\tau_1 = 0.6$ ps, $\tau_2 = 19$ ps and $\tau_3 = 307$ ps are obtained. Hence, the dynamics of the permanently adsorbed water molecules are distinctly different from those of the layer I molecules on average, which also contains molecules that exchange with subsequent layers. It is also interesting to note that a stretched exponential fit yields $\beta = 0.33$ for the 9 permanently adsorbed water molecules, which is indicative of collective and frustrated behaviour as also found in glassy systems. This observation is consistent with rotational relaxations of water on silica or other surfaces, where molecules at the surface generally reorient more slowly than in the bulk.[37,38,74] Such long time decays have also been observed for water molecules at the surface of reverse micelles and hydrocarbons[54,75] and also measured by NMR experiments for water molecules at silica surfaces.[74] We further note that for $\vec{\mu}$, the relaxation in layer II (from $z = 9$ to 12 Å) (Fig. 6) is very similar to bulk water, demonstrating that at a distance of 6 Å from the surface the rotational dynamics of water approaches that in the bulk. Another quantity, related to proton-NMR investigations, is the time scale of the rotational dynamics of the H–H vector ($\vec{r}_{HH}$). Fitting the $\vec{r}_{HH}$ correlation function to a stretched exponential gives $\beta = 0.34$ and $\tau = 238$ ps for layer I water molecules and $\beta = 0.49$ and $\tau = 172$ ps for water molecules in the middle. This is consistent with the heterogeneous dynamics found from the dipole moment correlation functions.

Exchange dynamics. In chromatographic systems facile water exchange between neighbouring water layers within the alkyl chains was observed.[60] The water dynamics within the hydrophobic chains[25] can lead to formation of water filaments, which allows intercalation of analyte molecules within hydrophobic regions.[60] It is therefore also of interest to consider the water exchange dynamics in the present system. As was mentioned in the previous section, on a time scale of 2.5 ns about 10 of the water molecules do not desorb from the surface. The absolute number of water molecules exchanged between the first and the second layer from the silica layer for the MTP water model, together with average number of water molecules in layer I and layer II are reported in Fig. 7. Data from Fig. 7 for individual layers suggest that layer I contains around 104 water molecules, which is more than the number of silanol groups present at the surface. This is due to water molecules occupying interstitial sites between the silanol groups on the

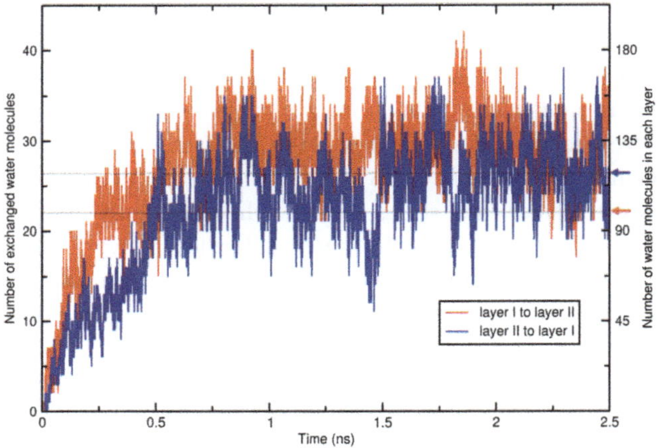

Fig. 7 Exchange of water molecules between layers I and II. The red trace corresponds to water molecules which were initially in layer I and exchanged to layer II and the blue trace represents water molecules initially in layer II and exchanged to layer I. Each layer has a width of 3.0 Å. The two arrows on the right hand side correspond to the average number of water molecules in layers I and II which are 95 and 124, respectively.

surface. Simulations with the MTP model find that the water flux equilibrates on the ns time scale and at equilibrium about 30 water molecules exchange between the two layers closest to the silica surface. However, it is interesting to note that even without the alkyl chains present, "avalanches" of water molecules can desorb or adsorb onto the silica surface, see *e.g.* the rapid change in the blue trace in Fig. 7 around 1.5 ns, which also suggests collective behaviour.

3.3. Vibrational spectroscopy

Fig. 8A gives an overview of the IR spectrum from simulations with the MTP model computed for all water molecules. Also the layer-specific spectra for layers I to IV are shown. By comparing the spectrum of layer I with those from layers II to IV it is evident that surface-bound water molecules give rise to considerably different spectroscopic signals. In particular the bonded –OH stretching region (3000 to 3750 cm^{-1}) differs in its width and around 1150 cm^{-1} a new band, labelled α in Fig. 8A, arises which is absent for water molecules in layers II and beyond. Water molecules in layer I have a more heterogeneous environment compared to water molecules in layers II and beyond as they are surrounded by the silica layer and water molecules. Contrary to that, water molecules in layers away from the interface have a significantly more homogeneous environment. Experimentally, vibrational spectroscopy is a sensitive and powerful means to characterize structurally disordered condensed phase systems. In particular, surface sensitive SFG spectroscopy together with atomistic simulations[34,38] can provide detailed information about the relationship of structure and spectroscopy at interfaces.

In order to better understand the spectroscopic features of water molecules in the first layer, power spectra for different coordinates involving water molecules were also determined. In Fig. 8B, the water-IR spectrum for first layer molecules is enlarged. Fig. 8C reports power spectra for coordinates including the water-HOH bend, the water-OH and the O_w–H_{SiOH} stretch, and the HO_w–H_{SiOH} bend modes.

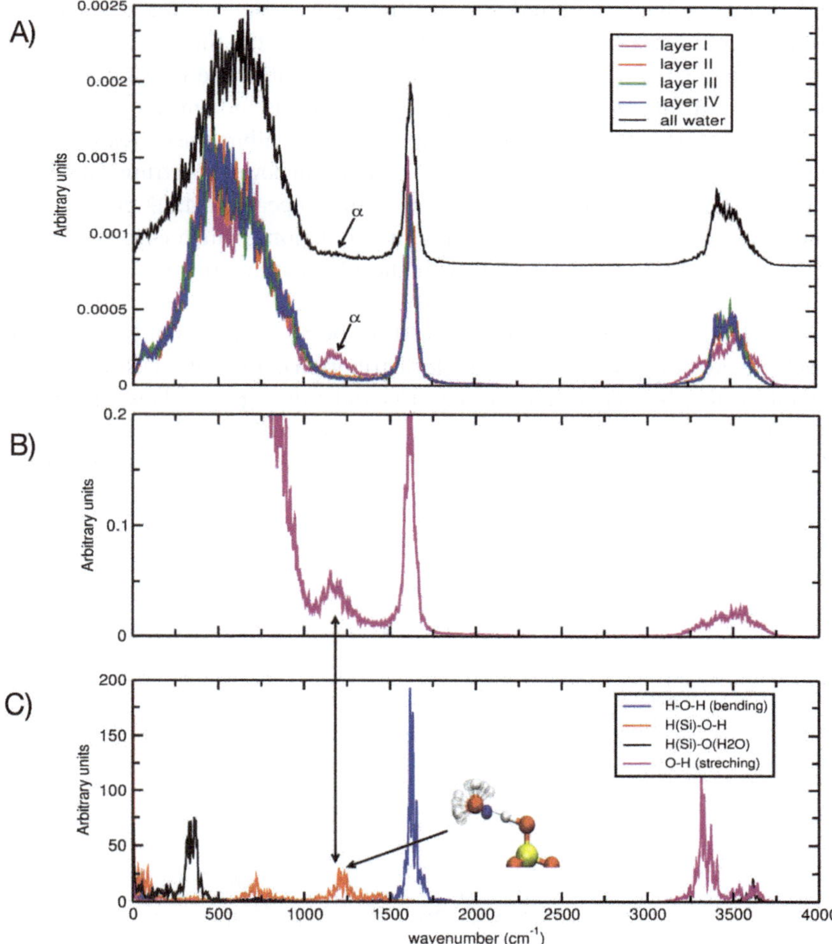

Fig. 8 A) Vibrational spectra of all water molecules (black) with the KKY plus quadrupolar water model together with layer-specific spectra for 4 different layers where each layer has a width of 3 Å. The arrow indicates the shoulder near the H(Si)–O(W)–H(W) bending mode. B) The enlarged IR spectra for water molecules in layer I. C) Power spectra corresponding to different modes as shown in the inset. The black arrow shown in the figure correspond to the mode which gives rise to the new peak in the spectra of water in the first hydration layer. A ball-and-stick model illustrating the H(Si)–O(W)–H(W) bending modes is drawn. Blue spheres indicate the oxygen lone pairs.

Comparison with the IR spectrum identifies the HO_w–H_{SiOH} bend as being responsible for the IR peak at $\approx$1150 cm^{-1}. The SiOH-hydrogen atom and water-oxygen atom give rise to strong hydrogen bonding, which leads to a libration mode as indicated in the inset of Fig. 8C. With the quadrupolar model for water, the tetrahedral geometry of the water molecules is captured more realistically. Simulations with the TIP3P water model without using SHAKE leads to a spectroscopic signal around 1150 cm^{-1} but the power spectrum for the HO_w–H_{SiOH} bend is featureless and extends from a few hundred up to 2000 cm^{-1} in stark contrast with simulations using the MTP model.

In order to study the isotope dependence, the same simulations were repeated for HOD and D_2O with the MTP model. The overall IR spectra are reported in

Fig. 9. The water-bending vibration shifts to the red as expected due to the larger mass in HOD and D_2O compared to H_2O. Concomitantly, the 1150 cm^{-1} band shifts also to the red by -15 cm^{-1} and -35 cm^{-1} for HOD and D_2O, respectively. In the IR spectrum for D_2O, the DO_{DOD}–H_{SiOH} band is hidden under the DOD bending mode. This is shown in Fig. S5, ESI,† which exhibits the correspondence between the spectra for water (D_2O) in layer I and the power spectrum calculated for different modes. Experimentally,[76] this band appears at 1209 cm^{-1} which compares with 1180 cm^{-1} from the present simulations, which emphasizes the quality of the MTP model together with the KKY parametrization for spectroscopic applications.

The orientation of water molecules with respect to the silanol group of the silica layer can be determined from angle distribution of the water-OH vector. The probability distribution is shown in Fig. 10A and 10B. Fig. 10A shows the distribution from simulations with the MTP model, whereas 10B reports that from using the TIP3P model. While the multipolar water model yields a narrow

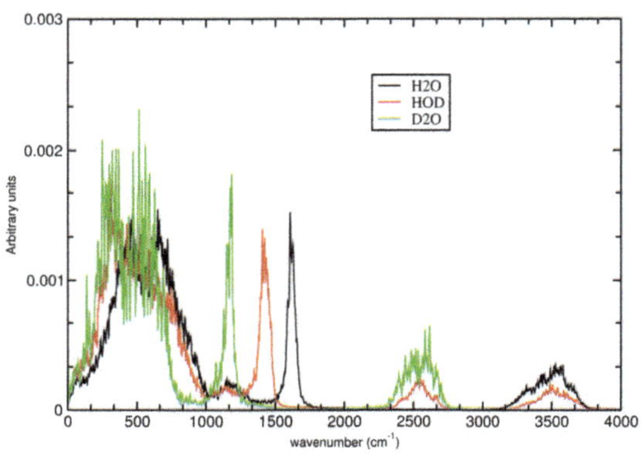

Fig. 9 Vibrational spectra of H_2O, HOD and D_2O from simulations with the KKY model with multipolar electrostatics. The isotope effect on the vibrational spectra is evident. Specifically, the band at 1150 cm^{-1} for H_2O shifts to the red upon isotopic substitution.

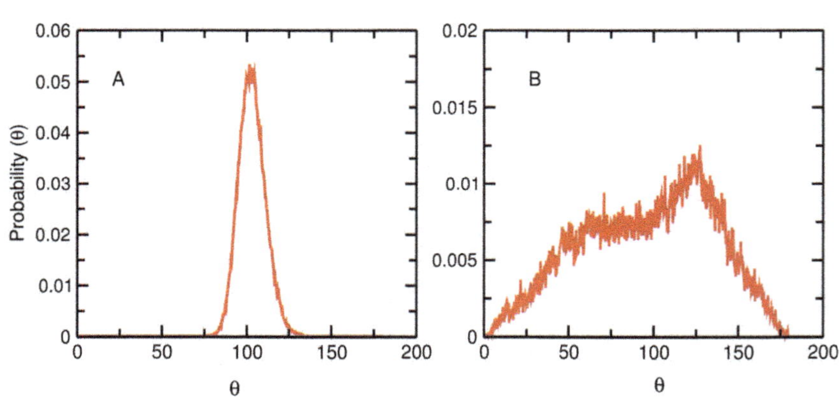

Fig. 10 Distribution for the H(SiOH)–OH(water) angle with the KKY model and multipoles (panel A) and the TIP3P water model (panel B).

distribution peaked around 100°, indicative of a tetrahedral environment of the surface-bound water molecules, the distribution for the TIP3P model is very broad and bimodal. A similar bimodal distribution is observed for TIP3P water molecules in contact with nanoporous silica layers in which the study was focused on the orientation of water depending upon the adsorbed amount of water[69] in the silica pore. This is due to the increased orientational freedom of water molecules described by simple point charge models – such as TIP3P – compared to parametrizations which account for higher electrostatic moments.

4. Discussion and conclusion

The present work reports all-atom molecular dynamics simulations of the dynamics and spectroscopy of water at hydroxylated silica surfaces. For this, two different water models were used: a point charge-only model and one multipolar, flexible water model. Specifically, the local charge density, in-plane mobility, rotational correlation time, exchange dynamics and vibrational spectra and effect of isotopic substitution on vibrational spectra of water molecules at the interface and in the bulk were considered. Our results suggest that the presence of the solid substrate affects both structural properties of water and their dynamical behaviour. In particular, the simulations provide layer-specific information which will be of interest for future surface-sensitive experiments. The in-plane diffusion of water at the surface is considerably slowed down through H-bonding interactions and the dipole moment ($\vec{\mu}$) and the HH-vector ($\vec{r}_{HH}$) correlation functions show exponents very different from 1, which suggests strongly inhomogeneous, collective behaviour. The fitting of rotational correlation functions with a sum of exponentials (3 in the present case) or with a stretched exponential is a matter of some debate. Although a 3 exponential function may describe the correlation function well, it is often difficult to associate the time scales with specific processes due to the inhomogeneous nature of the system. The orientation of water molecules at the interface is studied in terms of "OH-up" and "OH-down" configurations. The presence of the hydroxyl group on the silica surface, characteristic for silica surfaces at low pH, provides a preferential adsorption site for water molecules. Such a situation is encountered in chromatographic systems, where the functionalized and partially hydroxylated silica surface acts as a solid support for the stationary phase.

The infrared spectrum of first-layer H_2O at the SiOH interface features a band around 1150 cm^{-1} which was assigned to the OH(SiOH)–O(W) bending vibration. This band is absent for water molecules in subsequent layers and could be an interesting marker band in future experimental studies. The bonded-OH stretch region for first-layer water molecules is considerably broadened compared to that for water molecules in layers away from the interface and extends from 3200 to 3700 cm^{-1}. Specifically, the first-layer spectrum has a contribution to the red side of the bonded-OH stretch compared to second- and higher-layer water molecules. This is in qualitative agreement with experimental studies.[50] However, for a more detailed discussion of this spectral range the vibrational sum-frequency spectrum would have to be determined, which is outside the scope of the present work.[46,77] One of the differences between the present work and previous experiments is the face of silica that was used in the simulations. Following our previous work on chromatographic systems, the (101) face was employed here whereas the experiments were carried out with quartz(0001). The orientational probability functions of first-layer water molecules show three different kinds of orientations of water molecules relative to the SiOH surface. The two major populations have the water-OH bond predominantly oriented "up" and "down" relative to the SiO bond whereas a minor component has the OH(SiOH) vector pointing along the water-HOH-bisector. This last configuration is only occasionally populated when a TIP3P model is used for water whereas the quadrupolar model allows to align the water molecules along the oxygen-lone pairs. This conformation is also responsible for the predicted spectroscopic band at 1150 cm^{-1}.

Commensurate with the present applications – linear spectroscopy, diffusion and rotational motion – the force constants in the KKY model realistically describe the water monomer vibrations and the multipoles correctly capture the electrostatic potential around a water monomer. The diffusion coefficient computed with the present model is in reasonable agreement with experiment but no other thermodynamic observables were determined and compared with reference data. Contrary to that, the recently parametrized E3B model contains explicit many body contributions but it is a rigid water model.[78] However, we do not expect that the primary conclusions of the present work are affected by the limited comparison with thermodynamic data because all quantities of interest are of spectroscopic nature. An advantage of the "KKY plus MTP" model may be that it is a flexible model, hence no maps are required for the OH-stretch vibrations, and the electrostatics is based on a multipolar expansion, which makes it also suitable for simulations in heterogeneous environments such as the one in the present work. For such studies, water models for pure water simulations or water at the water–air interface may be less well suited.

The most notable findings of the present work concern the prediction of a spectroscopic feature around 1150 cm^{-1} involving the surface-bound water molecules and the dramatic slowdown of translational and rotational dynamics of the first- and second-layer water molecules near an SiOH surface. The spectroscopic signal is assigned to the HO_w–H_{SiOH} bending vibration. Such quantities may be observable experimentally and provide meaningful atomistic information about surfaces relevant for chromatographic systems and the behaviour of water at interfaces in general.

Acknowledgements

This work has been supported by the Swiss National Science Foundation through the NCCR-MUST and grant 200021-117810. We thank Dr Myung Won Lee and Dr Tibor Nagy (University of Basel) for insightful discussions.

References

1 Y. L. A. Jaing, J. Chen, M. Cadene, B. Chait and R. Mackinnon, *Nature*, 2002, **417**, 515–522.
2 F. Fornasiero, H. G. Park, J. K. Holt, M. Stadermann, C. P. Grigoropoulos, A. Noy and O. Bakajin, *Proc. Natl. Acad. Sci. U. S. A.*, 2008, **105**, 17250–17255.
3 D. Argyris, D. R. Cole and A., *Proc. Natl. Acad. Sci. U. S. A.*, 2008, **105**, 17250–17255.
4 J. Kim, W. Lu, W. Qiu, L. Wang, M. Caffrey and D. Zhong, *J. Phys. Chem. B*, 2006, **110**, 21994.
5 D. Chandler, *Nature*, 2005, **437**, 640.
6 C. Tanford, *Science*, 1978, **200**, 1012.
7 P. M. Dove and S. F. Elston, *Geochim. Cosmochim. Acta*, 1992, **56**, 4147–4156.

8 P. M. Dove, *Am. J. Sci.*, 1994, **294**, 665–712.
9 P. M. Dove, *Geochim. Cosmochim. Acta*, 1999, **63**, 3715–3727.
10 B. R. Bickmore, J. C. Wheeler, B. Bates, K. L. Nagy and D. L. Eggett, *Geochim. Cosmochim. Acta*, 2008, **72**, 4521–4536.
11 P. M. Dove, N. Z. Han and J. J. De Yoreo, *Proc. Natl. Acad. Sci. U. S. A.*, 2005, **102**, 15357–15362.
12 S. Roe-Garrett, F. Perakis, F. Rao and P. Hamm, *J. Phys. Chem. B*, 2011, **115**, 6976.
13 A. Pelmenschikov, H. Strandh, L. Petterson and J. Leszczynski, *J. Phys. Chem. B*, 2000, **104**, 5779–5683.
14 A. Pelmenschikov, J. Leszczynski and L. G. M. Petterson, *J. Phys. Chem. A*, 2001, **105**, 9528–9532.
15 L. J. Criscenti, J. D. Kubicki and S. L. Brantley, *J. Phys. Chem. A*, 2006, **110**, 198–206.
16 Y. T. Xiao and A. C. Lasaga, *Geochim. Cosmochim. Acta*, 1996, **60**, 2283–2295.
17 Y. T. Xiao and A. C. Lasaga, *Geochim. Cosmochim. Acta*, 1994, **58**, 5379–5400.
18 S. Nangia and B. J. Garrison, *J. Phys. Chem. A*, 2008, **112**, 2027–2033.
19 S. Nangia and B. J. Garrison, *J. Mol. Phys.*, 2009, **107**, 831–843.
20 F. Rao, S. Roe-Garrett and P. Hamm, *J. Phys. Chem. B*, 2010, **114**, 15598.
21 A. Striolo, A. A. Chialvo, P. T. Cummings and K. E. Gubbins, *Langmuir*, 2003, **19**, 8583.
22 J. Janecek and R. R. Netz, *Langmuir*, 2007, **23**, 8417.
23 I. C. Bourg and G. Sposito, *J. Colloid Interface Sci.*, 2011, **360**, 701.
24 S. M. Melnikov, A. Holtzel, A. Seidel-Morgenstern and U. Tallarek, *J. Phys. Chem. C*, 2009, **113**, 9230.
25 M. Orzechowski and M. Meuwly, *J. Phys. Chem. B*, 2010, **114**, 12203.
26 L. Zhang, S., S., C. Tian, Y. R. Shen, Y. Wu, M. A. Shannon and J. C. Brinker, *J. Chem. Phys.*, 2009, **130**, 154702–154710.
27 Y. Zheng, L. Qifeng, R. G. Murray and C. C. Keng, *Langmuir*, 2010, **26**, 16397–16400.
28 J. Yang, S. Meng, L. Xu and E. G. Wang, *Phys. Rev. B*, 2005, **71**, 35413.
29 J. Yang and E. G. Wang, *Phys. Rev. B*, 2006, **73**, 35406.
30 A. A. Hassanali and S. J. Singer, *J. Phys. Chem. B*, 2007, **111**, 11181.
31 T. S. Mahadevan and S. H. Garofalini, *J. Phys. Chem. C*, 2008, **112**, 1507–1515.
32 A. A. Skelton, P. Fenter, J. D. Kubicki, D. J. Wesolowski and P. T. Cummings, *J. Phys. Chem. C*, 2011, **115**, 2076.
33 A. A. Skelton, D. J. Wesolowski and P. T. Cummings, *Langmuir*, 2011, **27**, 8700.
34 M. Sulpizi, M.-P. Gaigeot and M. Sprik, *J. Chem. Theory Comput.*, 2012, **3**, 1037.
35 S. Meng, E. G. Wang and S. Gao, *Phys. Rev. B*, 2004, **69**, 195404.
36 D. N. Denzler, C. Hess, R. Dudek, S. Wagner, C. Frischkorn, M. Wolf and G. Ertl, *Chem. Phys. Lett.*, 2003, **376**, 618.
37 D. Argyris, T. Ho, D. R. Cole and A. Striolo, *J. Phys. Chem. C*, 2011, **115**, 2038.
38 O. Z. Tan, K. H. Tsai, M. C. H. Wu and J.-L. Kuo, *J. Phys. Chem. C*, 2011, **115**, 22444.
39 Q. Du, R. Superfine, E. Freysz and Y. R. Shen, *Phys. Rev. Lett.*, 1993, **70**, 2313–2316.
40 Q. Du, E. Freysz and Y. R. Shen, *Science*, 1994, **264**, 826–828.
41 E. A. Raymond and G. L. Richmond, *J. Phys. Chem. B*, 2004, **108**, 5051–5059.
42 D. S. Walker, D. K. Hore and G. L. Richmond, *J. Phys. Chem. B*, 2006, **110**, 20451–20459.
43 D. E. Gragson and G. L. Richmond, *J. Phys. Chem. B*, 1998, **102**, 3847–3861.
44 M. Sovago, R. K. Campen, H. J. Bakker and M., B., *Chem. Phys. Lett.*, 2009, **470**, 7–12.
45 L. F. Scatena, *Science*, 2001, **292**, 908–912.
46 P. A. Pieniazek, C. J. Tainter and J. L. Skinner, *J. Chem. Phys.*, 2011, **135**, 44701–44712.
47 Y. R. Shen, *Annu. Rev. Phys. Chem.*, 2013, **64**, 129.
48 M. Sovago, R. K. Campen, G. W. H. Wurpel, M. Muller, H. J. Bakker and M. Bonn, *Phys. Rev. Lett.*, 2008, **100**, 173901.
49 P. A. Pieniazek, C. J. Tainter and J. L. Skinner, *J. Chem. Phys.*, 2011, **135**, 044701.
50 V. Ostroverkhov, G. Waychunas and Y. Shen, *Phys. Rev. Lett.*, 2005, **94**, 046102.
51 J. Faeder and B. Ladanyi, *J. Phys. Chem. B*, 2000, **104**, 1033–1046.
52 H. Tan, I. Piletic, R. Riter, N. Levinger and M. Fayer, *Phys. Rev. Lett.*, 2005, **94**, 057405.
53 H. Tan, I. Piletic and M. Fayer, *J. Chem. Phys.*, 2005, **122**, 174501.
54 P. A. Pieniazek, Y.-S. Lin, J. Chowdhary, B. M. Ladanyi and J. L. Skinner, *J. Phys. Chem. B*, 2009, **113**, 15017–15028.
55 R. P. W. Scott, *Faraday Symp. Chem. Soc.*, 1980, **15**, 49–68.
56 A. Fouqueau and M. Meuwly, *J. Phys. Chem. B*, 2007, **111**, 10208.
57 J. L. Rafferty, J. I. Siepmann and M. R. Schure, *J. Chromatogr., A*, 2011, **1218**, 2203–2213.
58 W. Jorgensen, J. Chandrashekhar, J. Madura, R. Impey and M. Klein, *J. Chem. Phys.*, 1983, **79**, 926.
59 J. Braun, A. Fouqueau, J. Bemish and M. Meuwly, *Phys. Chem. Chem. Phys.*, 2008, **10**, 4765.
60 P. K. Gupta and M. Meuwly, *J. Phys. Chem. B*, 2012, **116**, 10951.
61 N. Kumagai, K. Kawamura and T. Yokokawa, *Mol. Simul.*, 1994, **12**, 177–186.

62 N. Plattner and M. Meuwly, *J. Mol. Model.*, 2009, **15**, 687.
63 M. W. Lee and M. Meuwly, *J. Phys. Chem. A*, 2011, **115**, 5053–5061.
64 B. R. Brooks, R. E. Bruccoleri, B. D. Olafson, D. J. States, S. Swaminathan and M. Karplus, *J. Comput. Chem.*, 1983, **4**, 187–217.
65 N. Plattner and M. Meuwly, *Biophys. J.*, 2008, **94**, 2505–2515.
66 S. N. Lee and P. J. Rossky, *J. Chem. Phys.*, 1994, **100**, 3334–3344.
67 A. A. Milischuk and B. M. Ladanyi, *J. Chem. Phys.*, 2011, **135**, 174709–10.
68 M. W. Lee and M. Meuwly, *J. Phys. Chem. B*, 2012, **116**, 4154–4162.
69 P. A. Bonnaud, B. Coasne and R. J.-M. Pellenq, *J. Phys.: Condens. Matter*, 2010, **22**, 284110.
70 D. Argyris, D. R. Cole and A. Striolo, *J. Phys. Chem. C*, 2009, **113**, 19591–19600.
71 J. Ha, T. Hyun Yoon, Y. Wang, C. B. Musgrave and J. G. E. Brown, *Langmuir*, 2008, **24**, 6683.
72 R. Notman and T. R. Walsh, *Langmuir*, 2009, **25**, 1683.
73 W. S. Price, H. Ide and Y. Arata, *J. Phys. Chem.*, 1999, **103**, 448.
74 M. R. Warne, N. L. Allan and T. Cosgrove, *Phys. Chem. Chem. Phys.*, 2000, **2**, 3663.
75 J. Chowdhary and B. M. Ladanyi, *J. Phys. Chem. B*, 2009, **113**, 4045–4053.
76 C. Chapados and J. J. Max, *J. Chem. Phys.*, 2009, **131**, 184505.
77 A. Morita and J. Hynes, *J. Phys. Chem. B*, 2002, **106**, 673–685.
78 C. J. Tainter, P. A. Pieniazek, Y. S. Lin and J. L. Skinner, *J. Chem. Phys.*, 2011, **134**, 184501–10.

Faraday Discussions RSC Publishing

DISCUSSIONS

General discussion

DOI: 10.1039/c3fd90038j

Dr Henchman opened the discussion of the paper by Professor Voth: How does the hydrated proton sit on the surface? Does it lie flat?

Professor Voth replied: For the most part the each of the H atoms of the hydronium species faces the bulk while the O atom is towards the vapor (see Fig. 1).

Dr Henchman asked: Along the lines of your energy argument, can the greater stability of the hydrated proton at the interface be explained by its lower coordination number than the hydroxide ion? This could be more readily accommodated with the vacuum on one side.

Professor Voth answered: The iodide ion has a larger coordination number than the Cl^- ion. However the iodide ion is seen at the interface and not the Cl^- ion. Although the hydronium (H_3O^+) has a smaller coordination number than the

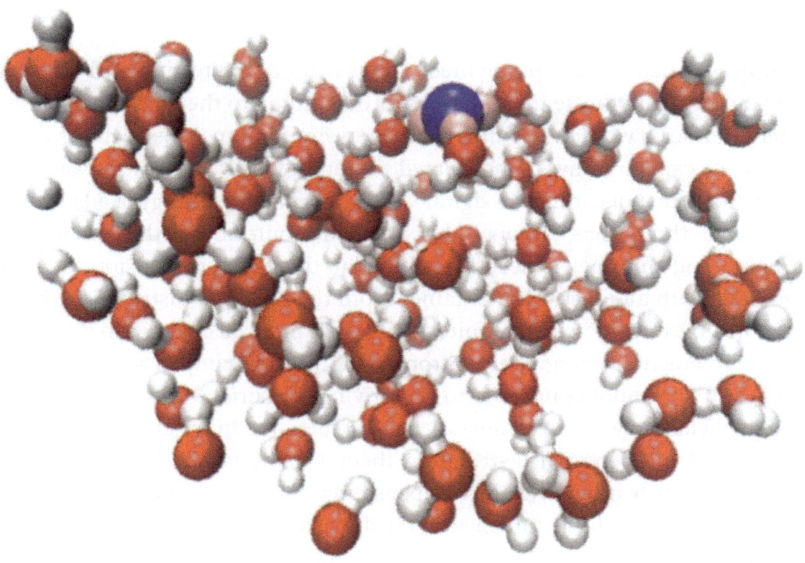

Fig. 1 Snapshot of the hydronium (H_3O^+, O in blue and H in pink) at the air–vapor interface.

OH^- ion, the fact is that the charge defect associated with the excess proton is more delocalized at the interface and is therefore larger than a simple H_3O^+ species. On the other hand, the OH^- shows much smaller delocalization in general. Therefore, it seems that the |charge/effective ion radius| ratio is a much better indicator of whether ions go to the interface. The smaller the value of this ratio the greater the likelihood of the ion going to the interface. Hydronium (hydrated proton) appears to be the only "simple" cation that satisfies this criterion.

Dr Perkin commented: You have demonstrated nicely the entropy and enthalpy contributions associated with moving a proton through water towards the interface. Somewhere in the water (real water) there must be a counterion (e.g. OH^-) for this proton, and the action of moving the proton towards the surface may therefore involve charge separation if the counterion remains in the bulk. On the other hand, the anion might also move with the H^+ towards the interface. Is it possible to incorporate a counterion in the simulation and so include this energy in the total?

Professor Voth responded: In dilute solutions a counterion like the Cl^- ion tends not to form a contact ion pair with the proton.[1] In previous simulations of the air–water interface with both an excess proton and Cl^- present, the chloride tended to remain in the bulk region and did not really track the proton (which of course went to the interface) [Petersen et al., J. Phys. Chem. B, Vol. 108, pp 14804-14806 (2004)] . SFG spectra on the water-vapor interface also show the low preference of Cl- ions for the interface.[2,3] We therefore decided to ignore the Cl^- ion., although it is possible to include it in our description and we have done so in our past studies.

1 Wang *et al.*, *J. Chem. Phys.*, 2005, **122**, 144105.
2 T. Tarbuck *et al.*, *J. Am. Chem. Soc.*, 2006, **128**, 14519.
3 L. M. Levering *et al. J. Phys. Chem. C*, 2007, **111**, 8814.

Professor Ben-Amotz commented: I find it quite interesting and significant that your results indicate that the transfer of a proton to the air–water and solid–water interface have a quite similar free energy minimum. This suggests that the pinning of capillary waves (and related proton-induced changes in water–water interactions) do not contribute significantly to the net free energy of adsorption of a proton on the surface, although they clearly do contribute to the corresponding enthalpy and entropy. It seems to me that the latter conclusion is entirely consistent with an exact result obtained from the potential distribution theorem (in its inverse form)[1,2] which requires that solute induced changes in the solvent–solvent interaction energy and entropy must exactly compensate (cancel each other) and so cannot contribute to the driving for surface adsorption, which is entirely dictated by direct solute–solvent interactions. In other words, I am suggesting that the potential distribution theorem may imply that capillary wave pinning primarily influences the compensating water–water interaction enthalpy and energy, and has little influence on the direct ion–water interaction energy and its fluctuation, which give rise to the non-compensating energetic and entropic contributions to the mean force potential associated with moving a proton from the bulk water to the interface. This suggestion could be quantitatively tested by comparing the proton–water interaction energy change associated with the transfer of a proton from the bulk to either the air–water or solid–water interface. If the latter two energies are similar, then that (combined with the similar free energy minimum) would require that capillary wave pinning primarily influences the compensating water–water interaction energy and entropy changes.

1 B. Widom, *J. Phys. Chem.*, 1982, **86**, 869–872.
2 D. Ben-Amotz, F. Raineri and G. Stell, *J. Phys. Chem. B*, 2005, **109**, 6866–6878.

Professor Voth responded: We generally agree with Professor Ben-Amotz's comment on this point. However, it will be important to confirm this idea in the future.

Professor Chandler addressed Professor Ben-Amotz and Professor Voth: Professor Ben-Amotz is referring to the well-known identity, $\exp(-\beta\Delta\mu) = \langle\exp[-\beta\Delta U(X)]\rangle$, where $1/\beta$ is temperature times Boltzmann's constant, and $\Delta\mu$ is the solvation free energy for a solute. The solute couples to the solvent in configuration X with the potential energy $\Delta U(X)$. The angle brackets stand for the average over X with the statistical weight of the pure equilibrium solvent. That weight, of course, depends crucially upon the nature of the solvent. The solvents intermolecular interactions, its thermodynamic state (*e.g.*, whether it exhibits an interface separating coexisting phases), and so forth are all significant in determining the value of the average.

The factor being averaged, $\exp[-\beta\Delta U(X)]$, is generally small in typical configurations of the pure solvent. It usually becomes appreciable only in configurations that are highly unlikely in the pure solvent. For example, pinned interfaces can contribute substantially to the solvation of charged particles even though such structures do not normally appear in the pure solvent. I believe Professor Ben-Amotz overlooks the important role of solvent–solvent interactions because he overlooks the crucial role of these interactions implicit in the averaging. Professor Phillip Geissler made this point at an earlier Faraday Discussion.[1]

1 P. L. Geissler, *Faraday Discuss.*, 2013, **160**, 103–120.

Professor Ben-Amotz replied: The identity to which I refer is not exactly the one that you have written, but rather the following expression obtained from the inverse form the same potential distribution theorem[1] $\exp[+\beta\mu^{\times}] = \langle\exp[+\beta\Psi]\rangle$ in which the average is now performed over the equilibrium configurations of the fully-coupled system containing the solute (ion) and solvent (water). The excess chemical potential of the ion $\mu^{\times} = \mu - \mu^{\mathrm{IG}}$ is that relative to an isolated (ideal gas) ion at the same number density, and the energy Ψ is that associated with turning on ion–water interactions in a system containing a fixed configuration of water molecules (sampled from the equilibrium distribution pertaining to the fully coupled system). One may further express the latter energy in terms of its mean value $\Psi = \langle\Psi\rangle + (\Psi - \langle\Psi\rangle) = \langle\Psi\rangle + \delta\Psi$ $\exp[+\beta\mu^{\times}] = \langle\exp[+\beta\Psi]\exp[+\beta\delta\Psi]\rangle$ which is equivalent to the following expression for the ions excess chemical potential: $\beta\mu^{\times} = \beta\langle\Psi\rangle + \ln\langle\exp[+\beta\delta\Psi]\rangle$.
Thus $\mu^{\times}$ may be obtained entirely from the average value of the ion–water interaction energy $\langle\Psi\rangle$ and the average of the above exponential involving

fluctuations in the ion-water interaction energy $\delta\Psi$. Although it is certainly true that ion-induced changes in the structure of water indirectly influence these averages, the above expression also clearly indicates that one need not calculate the changes in the direct water–water interaction energy in order to obtain the ion's chemical potential. On the other hand, the ion's total (experimental) hydration entropy and energy (or enthalpy) do certainly include contributions from the latter changes in the direct water–water interactions,[2] which can in fact be quite substantial for an ion hydration process.[3] Thus, it is also clear that the latter contributions to the energy and entropy must exactly cancel each other when evaluating the ion's hydration free energy, as further discussed in the Faraday Discussion 160, as well as the following publications (and references therein).[3,4]

1 B. Widom, *J. Phys. Chem.*, 1982, **86**, 869–872.
2 D. Ben-Amotz, F. Raineri and G. Stell, *J. Phys. Chem. B*, 2005, **109**, 6866–6878.
3 D. Ben-Amotz and R. Underwood, *Acc. Chem. Res.*, 2008, **41**, 957–967.
4 F. O. Raineri, G. Stell and D. Ben-Amotz, *Mol. Phys.*, 2005, **103**, 3209–3221.

Professor Voth responded: We generally agree with Professor Chandler's comment on this point. However, it will be important to confirm this idea in the future.

Professor Meuwly remarked: In the discussion you mentioned a coordination number of 4 for OH^- in water. This compares with a number closer to 5 from experiment.[1] Can you comment on these findings?

1 Botti *et al.*, *J. Chem. Phys.*, 2003, **119**, 5001–5004.

Professor Voth noted: In the paper by Botti *et al.*, they conclude that the fifth water molecule bonds weakly to the $(H_9O_5)^-$ (OH^- with four coordinating waters). When they integrated the first peak of their hydroxyl O^- water O $g(r)$, the coordination number they got was 3.7 ± 0.3. The hydroxyl H $g(r)$ shows a broad distribution, which is pushed towards longer distances compared to the corresponding water H–water O $g(r)$. Therefore, 4 water molecules coordinate strongly to the O of the hydroxide ion and a fifth water coordinates weakly to the H of the hydroxide ion.

Dr Perkin remarked: In your profile of energy and entropy as a function of H^+ distance from the interface (Fig. 1 in your paper) there is an energy decrease and $-T\Delta S$ increase in the first layer; but also apparently a smaller effect in the opposite direction (*i.e.* ΔU increase and $-T\Delta S$ decrease) in the next layer away from the surface. Can you comment on the origin of this, and might it lead to an activation barrier for access to the first surface layer (in particular for other ions which cannot infiltrate via the Grotthuss mechanism)?

Professor Voth answered: Given the error bars in the $-T\Delta S$ and energy terms in that region, I am not convinced that the second layer ordering is real. It is certainly plausible though that there might be some sort of small barrier in those results to get to the surface.

Professor Meuwly commented: Your finding of entropy–enthalpy compensation for the solvation of the surface-bound proton could suggest experimental ways for clarifying the situation. As the entropic contribution depends on temperature, are there experiments which do capitalize on the fact that at lower temperature the surface-bound state should be stabilized?

Professor Voth answered: Recently Yamaguchi *et al.*[1] have estimated the pH of the water surface using heterodyne-detected electronic sum frequency generation. It might be possible to determine this as a function of temperature. Also, there are experiments from Pacific Northwest National Laboratory in the US by Cowin and co-workers[2] on hydronium deposited on ice layers on metal electrode surfaces. They see evidence that the hydrated proton remains at the surface of the ice with a larger barrier to move to the bulk. Ice is of course not water, but it is colder and our predictions, as you note, suggest that the entropic effect should go down with temperature while the energy minimum at the surface should remain the same or get even deeper.

1 Yamaguchi *et al.*, *J. Chem. Phys.*, 2013, **137**, 151101.
2 Cowin *et al.*, *Nature*, 1991, **398**, 405.

Professor Mount inquired: I am interested in how control, establishment or application of a potential difference between bulk and interface would affect the amounts of protons and hydroxide ions at the interface, whether it is clear that some of the conflicting experimental measurements of these amounts previously made and referenced in this paper could not have significant interfacial potential differences which might explain the observed variations in these interfacial concentrations and whether the author has investigated or is planning to investigate such effects.

Professor Voth replied: We are currently studying the proton at a metal electrode–water interface with and without the application of an applied voltage. Preliminary results suggest that without the application of a potential difference the proton shows two minima close to the electrode–water interface. Simulations with an applied potential difference are underway.

Ms Zhuang asked: In the Hofmeister series, the larger ions like iodide and bromide are found to be enriched at the water–air interface, while smaller ions like fluoride and chloride are not. In your paper, hydrogen ions, though small in size (at least the bare ion is small in size), are enriched at the interface. Do you think hydrogen ions involve very different physics from other ions, or is the physics is similar, but we should be considering hydrogen ions as hydronium ions (or some aggregates that are even bigger)?

Professor Voth replied: The hydrated excess proton is delocalized over a number of solvating water molecules and therefore exists as a larger ion than the bare proton or even H_3O^+ (hydronium). It is this large solvated ion that goes to the air–water interface. Therefore this is consistent with the idea that large ions, irrespective of the sign of their charge, go to the interface. Interestingly the hydrated excess proton is likely the only simple cation with the propensity for the water–vapor interface.

Dr Henchman commented: The negative entropy of surface association for the hydrated proton indicates that the association would become stronger at lower temperatures.

Professor Voth added: Our simulations show that the depth of the free energy minimum increases with lowering of temperature.

Mr Wexler communicated: Recent experimental research on electro-hydrodynamic bridges produced from water has revealed a mid-infrared emission feature at 2150 cm^{-1} which can be linked to enhanced proton mobility.[1] The energetic favorability of the hydrated excess proton to reside at the surface along with increased ion radius reported in your paper supports the proposed mechanisms for the IR emission feature. However, the entropic penalty due to surface pining would appear to work against such increased ion mobility. What is unclear is how the addition of a moderate potential (1–3 MV m^{-1}) across the hydrophobic interface will affect your simulation results. Given such a case would you expect a significant change in the proton kinetics observed in your model? Specifically how would the effective ion size change, or conversely, where would the free energy minimum move with respect to the interface?

1 E. C. Fuchs, A. Cherukupally, A. H. Paulitsch-Fuchs, L. L. F. Agostinho, A. D. Wexler, J. Woisetschläger and F.T. Freund. Investigation of the Mid-Infrared Emission of a Floating Water Bridge, *J. Phys. D: Appl. Phys.*, 2012, **45**, 475401.

Professor Voth communicated in reply: This is an interesting question, but we have not performed these calculations so any answer from me at this point would be speculation.

Professor Seddon opened the discussion of the paper by Professor Perkin: What is the water content of your ionic liquids? I would be interested in discussing how sensitive your measurements are to even small levels of water. It was clear from the manuscript that you have been scrupulously careful, but I wanted you to emphasise why this is so important.

Dr Perkin replied: Professor Seddon, this is a really important question because even small quantities of water can have dramatic effects on the surface interactions and properties of ionic liquids. The ionic liquids we use are purified as described in the paper then re-dried immediately before use. Water content after drying has been measured on some of our samples using a Karl-Fisher Coulometer – typically revealing <200ppm water and in some cases substantially lower – however we suspect the liquids inside the SFB experiments are dryer than this because of the more rigorously dry environment of the SFB compared to the procedure of using the coulometer itself. Importantly, we performed control experiments allowing gradual humidification of the liquid, to observe in a controlled way the effect of increasing water content on the surface forces. In extreme cases this can be dramatic: water which is soluble in the bulk IL can phase separate at the surface and give rise to a capillary force between the two mica sheets and so a strong attraction between them, obscuring all the subtle oscillatory forces shown in our paper. In general, as the ionic liquid becomes wet the amplitude of the oscillatory forces decreases gradually. Being aware of these observations as a 'signature' of the possible water effect, we are able to make our measurements over a period of time (typically up to 10 hours) during which there is no observable change in the forces.

Dr Billard commented: Are the differences evidenced in your experiments between bulk ILs and confined ILs expected to induce differences in various properties of these liquids, for example: melting point, solubilities, transport properties (conductivity, diffusion of species *etc.*) as compared to the bulk ILs?

Dr Perkin replied: Indeed our experiments show that the structure of ionic liquids in the near-surface region – extending typically ~3–5 nm from each surface – is ordered in layers and so is different to the bulk structure. The bulk and surface structures are still related, though; for example the nano-scale segregation emerging in the bulk as alkyl chain length increases is related to the bilayer structures we observe for longer alkyl chain ionic liquids. The properties you mention – such as solubilities and transport properties – are bulk properties and hard to define in few-nm surface regions, although we can make some measurements which hint at how the changed structure leads to alteration of related properties. (i) We perform shear experiments on the confined films[1,2] which reveal that the layered film has a non-zero yield stress, i.e. a deviation from purely viscous fluid behaviour. Molecular diffusion and conductivity, however, we do not measure directly in these experiments. (ii) Variation in solubility in the interfacial region could also be considered as surface activity of the solute; and this is indeed observed for some amphiphilic solutes in ionic liquids as well as for water which can phase-separate at the surface due to preferential wetting of the surface.

1 S. Perkin, T. Albrecht, and J. Klein, *Phys. Chem. Chem. Phys.*, 2010, **12**, 1243.
2 A. M. Smith *et al.*, *Phys. Chem. Chem. Phys.*, 2013, **15**, 15317.

Professor Chandler commented: Your measurements are extremely interesting and seemingly important, but I wonder if your interpretations are overly simplified. I accept that oscillations in surface force measurements manifest sizes of molecules. In the case of a liquid mixture of similar-sized molecules, there is one important molecular length scale, and this length will be reflected in the oscillations of the measured forces. On the other hand, liquid matter is disordered and ergodic, which makes me suspect that the idealized layering depicted in your cartoons are inaccurate. Is there direct experimental evidence for anions and cations to segregate perfectly into separate macroscopic layers as you suggest? For sure, molecular simulations of ionic melts[1] suggest a richer and more disorganized picture.

1 For example, O. J. Lanning and P. A. Madden, *J. Phys. Chem. B*, 2004, **108**, 11069.

Dr Perkin answered: Although the inset diagrams in our Fig. 2–4 show only pure cation and anion layers for clarity, we have described in the text of the manuscript (section 4.1) how the "layer-purity", *i.e.* the excess of anions/cations or the net charge density within any particular layer, decreases with increasing distance from the charged surface. This is reflected in the damping of the oscillatory solvation force, which has a length scale corresponding to dying-away of the surface-imposed ordering as mentioned by Professor Bob Evans in an earlier question. The extent to which the first layer adjacent to a charged surface is in fact "pure" (*i.e.* with charge density equal to the maximal possible charge density for a densely packed layer of counterions) depends on several factors including charge density on the solid surface, ion footprint-area, and molecular interactions (*e.g.* dispersion interactions) between like ions or between one ion or other with the surface. For the particular case of two mica surfaces sandwiching between them a single layer of imidazolium-based ionic liquid, the layer is likely composed almost exclusively of cations as the quantity required to charge-compensate the mica create a fully packed monolayer.

You mention the length scale in a mixture of uncharged similar-sized molecules; *i.e.* the molecular size. An important point of comparison here is that the period of force oscillations measured in our (and other) experiments with ionic liquids corresponds to the size of a cation and anion *together*; there is of course some molecular conformation freedom which makes this quantity hard to define, as Professor Seddon has earlier mentioned, but it is clear that our lengthscale matches much better the ion pair size than either individual ion size. This implies that the layer structure involves oscillation of charge density, because if the ions were ordered with precisely equal numbers in-plane (and thus no charge oscillation) then the force (or energy) oscillation would have the period equal to the mean cation/anion size rather than the sum. This conclusion that there exist layers of alternating charge density is supported by a number of other experimental studies such as the X-ray reflectivity experiments of Mezger *et al.*,[1]and the AFM experiments of Atkin *et al.*,[2]and the simulation studies – as you note – which give insights into subtle variation in structure away from surfaces of varying charge with pioneering work in this area carried out by Lanning and Madden.[3]

1 M. Mezger *et al.*, *Science*, 2008, **322**, 424.
2 R. Atkin, and G. G. Warr, *J. Phys. Chem. C*, 2007, **111**, 5162.
3 O. J. Lanning, and P. A. Madden, *J. Phys. Chem. B*, 2004, **108**, 11069.

Dr Limmer inquired: Is it possible to do these experiments under an applied voltage? From recent work,[1] I suspect that significant correlations between ions could emerge under such conditions, leading possibly to an as yet undetected surface phase transition.

1 D. T. Limmer, C. Merlet, M. Salanne, D. Chandler, P. A. Madden, R. van Roij, B. Rotenberg, Charge fluctuations in nano-scale capacitors, *Phys. Rev. Lett.*, 2013, **111**, 106102.1–5.

Dr Perkin replied: Yes, it is indeed possible to apply voltage to one or both surfaces: similar experiments carried out using and AFM (*i.e.* with a sharp tip near a planar surface, rather than the two planar surfaces in our experiment) have done exactly that, showing several interesting effects.[1] Using apparatus similar to ours (surface force apparatus/balance) there have been other approaches to applying voltage directly to one or both of the confining surfaces, for example ref. 2.

1 H. Li, F. Endres, and R. Atkin, *Phys. Chem. Chem. Phys.*, 2013, **15**, 14624–14633.
2 L. Chai, and J. Klein, *Langmuir*, 2009, **25**, 11533.

Dr Soper asked: The results you describe are quite intriguing. Do these materials intercalate into clay minerals? Clay minerals form atomistically smooth surfaces and, like mica, are charged, but less so usually. If these ionic liquids go into mica they might go into other clays? If you can do this you can do scattering experiments and so get independent verification of the orientational order near the surface.

Dr Perkin answered: I think this is a good idea, and the results would be interesting not only to us but in many directions. Dynamic studies of that system would also be of interest. However the ionic liquids we studied until now do not actually swell mica – this is part of the reason we use mica – we are putting them in contact with a mica interface which is pre-cleaved. We would have to find clays which swell in ILs naturally.

Professor Seddon added: Prof. Soper asked (me specifically) whether there had been any structural studies on ionic liquids in structured solids, such as clays. I am not aware of any, and I would be surprised if any have been undertaken, as the water content of the clay would interfere with the measurements.

Professor Seddon commented: Ordering has been seen on silica by X-ray reflectivity[1] supporting your model.

1 A. J. Carmichael, C. Hardacre, J. D. Holbrey, M. Nieuwenhuyzen and K. R. Seddon, *Mol. Phys.*, 2001, **99**, 795–800.

Professor Seddon then went on to ask: How did you judge the size of the anion? You mentioned that the spacing observed was the sum of the size of the anion and the cation. Bistriflamide can be in a *cis-* or *trans-* conformation, and either vertical or horizontal to the surface. So how did you choose a conformation? Maybe Raman experiments would tell you about the structure?

The implication of the question is that the error in the size estimation would be enough to mask interdigitation/penetration of the layers, so that the "ideal" cationic and anionic layers which have been the subject of other questions might actually be interpenetrating.

Dr Perkin replied: Your description of the conformational possibilities for the anion are of course right (and indeed some freedom also exists for the cations), and we cannot distinguish between these in our experiment. However when we consider the range of possible structures we find that (i) for the imidazolium-based ILs, the bilayer repeat distance is consistent with either toe-to-toe (*i.e.* no interpenetration of the alkyl tails) or only a small amount of interpenetration. Further, the anions are most likely in a separate layer from the cations. In contrast, (ii) for the pyrrolidinium-based ILs the bilayer repeat thickness is consistent with strongly interpenetrated cation alkyl tails and anions lying (all or mostly) in-plane with the cations.

Professor Chandler addressed Dr Perkin and Professor Seddon: I do not believe that X-ray scattering can generally distinguish anions from cations. For sure, X-ray scattering can detect layering of molecules, but that is not the issue I raise with Dr Perkin. My question concerns whether layering can so perfectly segregate anions from cations. I do not think such perfection is possible in liquid matter.

Dr Perkin replied: The X-ray scattering experiments by Mezger *et al.* referred to previously demonstrate a layer spacing which they propose is best fit by a double-layer (*i.e.* alternating cation layer/anion layer) structure, rather than a "checker-board" structure. The high counterion (cation) density in the first layer adjacent to the (negative) solid surface proposed by both Mezger *et al.* and in our experiments can interpreted in terms of overscreening effects expected in highly correlated dense Coulomb systems.[1] This can kick-off charge density oscillations in some situations (but not all – *e.g.* see recent work of Kirchner *et al.*[2]); however entropy of course will not allow such an ordered structure to persist far from the surface – the structure becomes less ordered with distance – and the decay length of any (charge or density) oscillations was discussed already by Professor Evans.

1 Y. Levin, *Rep. Prog. Phys.*, 2002, **65**, 1577.
2 K. Kirchner *et al.*, *Electrochim. Acta.*, 2013, **110**, 762.

Professor Seddon clarified: I was referring to ref. 1 in my previous comment.

Professor Evans said: This question is concerned with the decay of the oscillations measured in the normal force as D, the distance between surfaces, increases. For a one-component neutral fluid (atomic or a simple molecule such as OMCTS) one can show that the oscillatory part of the solvation force (excess pressure) $f_s(D)$, for two identical surfaces, decays as $\exp(-a_0D)\cos(a_1D + c)$ as D approaches infinity. Here c is a constant phase and the inverse lengths are given by the complex poles of the structure factor of the bulk liquid at the same chemical potential and temperature, *i.e.* the roots of $1 - c(q) = 0$ with $q = ia_0+a_1$, where $c(q)$ is the Fourier transform of the pair direct correlation function.[1] Thus the thermodynamic quantity $f_s(D)$ decays in the same fashion as does the structural quantity, the pair correlation function $r\text{h}(r)$, in the bulk . For simple ionic liquids (NaCl-like) much is known about how the partial pair correlation functions $r\text{h}_{ij}(r)$, with i, j =cation, anion, decay in bulk.[2] In other words charge ordering is well-understood.

My question is whether anyone has analyzed the corresponding decay of the solvation force. It might be revealing to ascertain just what features of charge ordering determine the decay of oscillations in $f_s(D)$ for a simple ionic system. Clearly your ionic liquids are not simple and the thrust of your work is in understanding ordering at small D but I think the general question of what determines the asymptotic decay could be interesting.

1 R. Evans *et al.*, *Molec. Phys.*, 1993, **80**, 755.
2 R. J. F. Leote de Carvalho and R. Evans, *Molec. Phys.*, 1994, **83**, 619.

Dr Perkin replied: It is certainly interesting and pertinent to ask what features of charge ordering determine the observed oscillatory solvation force such as seen in our experiments. Although I am not aware of any analysis like this having been carried out until now, the bulk experiments required to determine a_0 and a_1 (as defined in your question) are available so the analysis could be done. This could be revealing indeed: for example, the effect of a cross-over from charge oscillations to density oscillations (as you and others have described in terms of $rh(r)$ for molten salts) on the solvation force. An additional parameter in the ionic case is the surface charge density; or more usefully the ratio of surface charge density to the theoretical maximum density of counterions in one layer (κ in the model of Fedorov *et al.*[1]). There, it was shown that charge oscillations in the liquid have greatest magnitude when κ is small, and are expunged when $\kappa = 1$. In order to probe the effect of κ experimentally we need to tune the surface charge density (or the ion foot-print size, although this has more limited scope); several groups are working in this direction so these regimes may be accessed in future.

1 K. Kirchner *et al.*, *Electrochim. Acta.*, 2013, **110**, 762.

Dr Drummond said:My recollection is that there has been other surface force apparatus work reported on high concentrations of ions in aqueous electrolyte solutions where oscillatory force-distance profiles were also observed. How do these observations relate to your work?

Dr Perkin answered: Indeed oscillations in force, as a function of film thickness, have been seen in the past for aqueous electrolytes at various concentrations.[1,2] In each of these cases the oscillations are attributed to water layers squeezing out (the period oscillation is close to the water molecule dimension), and so these can be understood in a manner similar to the structural (oscillatory) forces seen for non polar liquids. In ionic liquids, oscillations observed when there is a charge on the confining surface have an oscillatory period similar to the cation + anion pair size, indicating a structure involving double layers. That is not to say structure is absent in the near-surface region when there is no surface charge, only that it is not detected in this type of experiment.[3]

1 R. Pashley, and J. Israelachvili, *J. Col. Int. Sci.*, 1984, **101**, 511.
2 P. M. McGuiggan, and R. Pashley, *J. Phys. Chem.*, 1988, **92**, 1235.
3 I. Bou-Malham, and L. Bureau, *Soft Matter*, 2010, **6**, 4062.

Professor Mallamace opened the discussion of the paper by Renato Torre: Your experimental technique represents a nice and appropriate approach to study fast dynamics. Are there some connections between the OKE and the opportunity to study the system vibrational density of states? As you know Raman and Neutron spectroscopies are used for this, but there is an open question about the best technique to study this complex dynamics. Please can you comment on that?

Dr Torre answered: The extraction of the vibrational density of states from the optical spectroscopic investigations is a dated problem. Indeed, the OKE response and the depolarized Raman response are equivalent, see for example Chapter 2 of ref. 23 from my paper. The difference between these response functions is that the first is defined and measured in the time domain and the second in the frequency domain. So the wide literature on the connection between the molecular observable, single and/or collective, and the collective polarizability apply also to the definition of the OKE response. We must consider that having the response in the time domain can be in some cases an advantage.

Professor Meuwly commented: In fitting your data you kept the β-exponent constant at $\beta = 0.6$. This appears questionable as for low water content one would expect a more highly frustrated system with β closer to 0.3 or 0.4. Can you elaborate more on why it turns out to be difficult to let this parameter change in the fits?

Dr Torre responded: In the slow part of the HD-OKE signal from Vycor samples there are two main contributions: a monotonic decay that can be fitted by a stretched exponential and a relatively slow oscillation extending in the picosecond time-scale. To our present understanding, the oscillating signal is an acoustic-like vibration localized on the pore surface. The amplitudes of these two contributions are strongly dependent on the hydration level. At high hydrations, the stretched exponential decay increases substantially, enabling the extraction of the relaxation times of nano-confined water. Unfortunately, the stretching parameter, β, can't be extracted reliably and we fixed it to the bulk water value.

Professor Meuwly remarked: One possibility to do your fits would be a multivariate ansatz which is also being used for fitting data from time resolved X-ray diffraction.

Dr Torre agreed: A multivariate analysis of the HD-OKE data would be very interesting, we could attempt a self-modeling curve resolution similar to the procedure used by Prof. Ben-Amotz to analyze the Raman data.[1]

1 D. S. Wilcox, B. M. Rankin and D. Ben-Amotz, *Faraday Discuss.*, 2013, **167**, DOI: 10.1039/C3FD00086A.

Professor Caupin asked: Based on ref. 13 and 58 in your paper, you mention that "water filling in silica nanopores is still a debated issue", with two mechanisms being proposed for the pore filling: layer by layer or by formation of plugs. It seems to me that a theory of capillary condensation based on first principles has been given by M. W. Cole and W.F. Saam.[1,2] Due to the attraction of the fluid to the wall, at low vapor pressure a film forms on the pore surface. The thickness of the film increases, and it becomes metastable compared to the filled state. However, total pore filling does not occur immediately, because, the energy barrier to overcome between the film and filled states is large. Therefore the film continues to grow, up to a thickness close to a critical value at which it becomes unstable (if the pore radius is larger than this thickness). The pore is then filled, and the curvature of the liquid–vapor meniscus is set by the fluid activity. At higher hydration, the curvature decreases until the liquid–vapor interface becomes flat at saturated vapor pressure. This is at the origin of the capillary condensation hysteresis. If the fluid activity is reduced, the liquid–vapor interface starts to increase its curvature, and eventually recedes in the pore. This occurs at coexistence with the film state, at a lower vapor pressure than the filling. Of course this scenario neglects several effects (pore interconnectivity, roughness...). The general trend has been confirmed by molecular dynamics simulations for a model of water in Vycor, see work by Puibasset and Pellenq.[3,4]

1 M. W. Cole and W. F. Saam, *Phys. Rev. Lett.*, 1974, **32**, 985.
2 W. F. Saam and M. W. Cole, *Phys. Rev. B*, 1975, **11**, 1086.

3 J. Puibasset and R. J.-M. Pellenq, *Phys. Chem. Chem. Phys.*, 2004, **6**, 1933.
4 J. Puibasset and R. J.-M. Pellenq, *J. Chem. Phys.*, 2006, **122**, 094704.

Dr Limmer said: I have one comment to Dr Torre and one following up on a comment by Prof. Caupin. First, Dr Torre has presented his study on the short time dynamics of water in partially filled porous silica. These results are in excellent agreement with theoretical predictions made by Prof. Chandler and myself[1] regarding the lengthscales over which water is bulk-like in confinement. Prof. Caupin asked whether or not the system studied by Dr Torre underwent capillary condensation, which would be a way of determining pore size and the mechanism by which the pore fills. It is important to note that under strong confinement, *i.e.* the limit of microscopically small pore diameters, fluctuation effects will necessarily destroy macroscopic signatures of condensation. Moreover, in that limit the system becomes one-dimensional, and there can be no long range correlations. Such exponentially decaying correlations necessarily result in plug-like configurations, though they may be preceded by thin or thick film growth on the surface.

1 D. T. Limmer, and D. Chandler, "Phase diagram of supercooled water conned to hydrophilic nanopores, *J. Chem. Phys.*, 2012, **137**, 044509.

Dr Torre answered: As you nicely summarized in your question the porous glasses present the interesting phenomena named "capillary condensation". This physical process has been known for many years and there are several theoretical and phenomenological models explaining its main features. The water adsorption in silica nanopores, Vycor in particular, has many strong relationships with the "classic theory" of capillary condensation, nevertheless several differences must be considered. The peculiar structural and topological features of Vycor nanopores, the presence of hydroxyl groups on the silica surfaces forming hydrogen bonds with the water molecules, as well as the nanometric dimension of the pore diameters does not enable to use the simpler approaches. As it has been correctly pointed out by Dr Limmer, the nanometric diameters of the pores increase the importance of fluctuation phenomena. In the simulation work from E. de la Llave *et al.* (ref. 13 of our paper) is shown how the water density fluctuations towards the nanopore center are strongly dependent on the pore dimensions. These fluctuations define the nucleation point of the liquid plug, increasing the pore diameters increase the water filling percent needed to onset the plugs. So, for relatively large nanopores, a layer-by-layer filling process cannot be excluded. Our experimental investigation is not precisely focused on the water filling mechanism but mainly on the dynamics of the nano-confined water. Nevertheless we can attempt to extract some indirect information on the nature of the filling. The structural relaxation times remain constant up to $f = 10.5\%$, see Table 1 in the paper; and the vibrational features show the gradual set-up of the bending and stretching bands, see inset of Fig. 4 in the paper. This suggests a layer-by-layer growth because the formation of plugs would correspond to dynamic features more similar to the bulk water. Nevertheless, it can be either that water forms the plugs at higher hydration level or the filling proceeds layer-by-layer up to the full hydration condition.

Professor Caupin asked: Did you measure the adsorption isotherm of water in your sample?

Dr Torre replied: No we did not, but the sample we used Vycor (code 7930 by Corning Company) has been investigated in many research papers. I would like to quote a very recent paper from Ogawa and Nakamura,[1] where the adsorption–desorption isotherm of water in Vycor 7930 is measured together with a light scattering investigation.

1 S. Ogawa and J. Nakamura, *J. Opt. Soc. Am. A*, 2013, **30**, 2079.

Professor Caupin commented: It seems to me that a diameter of 4 nm is large enough for capillary condensation to occur as described previously. This would occur at a vapor pressure lower than the saturated vapor pressure, and could lead to a liquid–vapor meniscus with a radius of curvature R close to that of the pore. Based on Laplace pressure, one would then expect a pressure in the liquid of $-2\ \sigma/R$, that is *ca.* -70 MPa. This is a rough estimate, since the attraction from the wall, the non-isotropy of pressure in the confined fluid, or the small radius of curvature, may make the Laplace formula inaccurate. Yet it should give a correct order of magnitude for 4 nm pores, taking into account the Tolman length less than -0.05nm.[1] This negative pressure would then correspond to a liquid at lower density than the bulk. This effect might have be to considered when comparing the results with bulk water, and related to the difference observed in your Fig. 5.

1 M. El Mekki-Azouzi *et al.*, *Nature Phys.*, 2013, **9**, 38.

Dr Torre answered: As reported previously in the answer to your previous question, our data does not seem to support the plug formation. So we attributed the dynamic modifications measured in the nano-confined water at low hydration levels to the interactions with the nanopore surfaces. These interactions produce strong effects that extend over two water layers. Whereas, the inner water has dynamics very similar to the bulk water. Alternatively we could suppose the formation of plugs and the water dynamic modifications induced by the negative pressure effects. Nevertheless, in this case there are three hydration regimes. In fact, the water filling of the nanopores always starts with a single layer that grows towards multiple layers that eventually form water plugs; at higher hydration levels the pores will be filled completely with water.

So also the water dynamics should reflect these three steps. Our OKE measurements do not evidence the intermediate regime.

Professor Wynne commented: In our recent OKE experiments on concentrated aqueous LiCl solutions,[1,2] we observed very little change in the part of the spectrum normally associated with the hydrogen-bond bend and stretch modes and the water liberation, even though there where as few as 7 water molecules per LiCl ion pair. In your experiments, you vary the hydration levels to obtain a single layer or a double layer of water molecules on the surface of the nano pores. Hence, the bulk phase is missing. As the water molecules in this case are unable to form four hydrogen bonds in an approximate tetrahedral pattern, it is perhaps not surprising that you find in your paper that the spectrum changes dramatically compared to the bulk. Then the question arises whether your results are pertinent to the case when the pores are fully filled. In other words, does the interfacial water layer have the same structure and dynamics at low hydration levels and when the pores are fully filled?

1 D. A. Turton, C. Corsaro, M. Candelaresi, A. Brownlie, K. R. Seddon, F. Mallamace and K. Wynne, *Faraday Discuss.*, 2011, **150**, 493–504.
2 D. A. Turton, C. Corsaro, D. F. Martin, F. Mallamace and K. Wynne, *Phys. Chem. Chem. Phys.*, 2012, **14**, 8067–8073.

Dr Torre answered: The dynamics of interfacial water (*i.e.* water layers in contact with the porous silica at low hydration levels) can be different from the outer water (*i.e.* water layers in contact with the porous silica but at full hydration). Unfortunately, OKE experiments cannot measure selectively the outer water dynamics. I believe that any experimental investigation can hardly select it. The computer simulation can do it. According to the simulation reported in ref. 58 the characteristics of interfacial and outer water are very similar. This can be explained by the presence of numerous hydroxyl groups on the pore silica surfaces that give a support to the H-bond water network. Nevertheless these surface H-bonds force the interfacial water network into a distorted pattern. This surface induced pattern persist also at full hydration, so that the outer water has structural and dynamics features very similar to the interfacial water.

Dr Royall commented: In your paper, Fig. 3, you show the time-correlation functions. I understand these are fitted with eqn 3, which is the sum of a stretched exponential and an oscillatory term. Your resulting timescales (from the stretched exponential component) show that the timescale is shorter in the case of more complete absorption. Yet in Fig. 3 it appears that these time-correlation functions do not decay, whereas those at longer absorption do. Can you provide a simple physical explanation of this effect – I presume it is related to the relative contributions of the oscillatory and exponential terms in eqn 3?

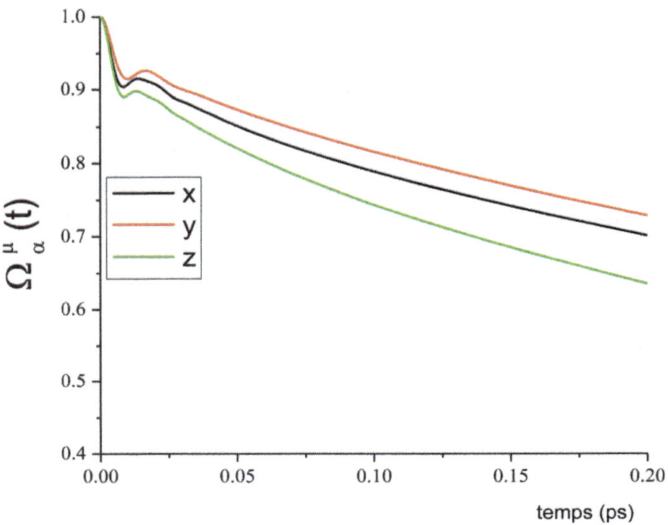

Fig. 2 Reorientation dynamics of water along the molecular axes.

Dr Torre answered: As you already suggests this is an interplay between the amplitudes of the oscillatory contributions and the stretched exponential decay. The slow part of the HD-OKE signal from Vycor samples is characterized by the a relatively slow oscillation extending in the picosecond time-scale. To our present understanding, this is an acoustic-like vibration localized on the pore surface. Nevertheless, in the HD-OKE signal there is a monotonic decay that can be fitted by a stretched exponential. At the lower hydration conditions, $f = 0.1\%$ and $f = 3.3\%$, the slow oscillation dominate the signal and hide the stretched exponential decay. At higher hydrations, $f > 5.6\%$, the amplitude of the stretched exponential decay increase substantially, enabling the extraction of the relaxation times of nano-confined water. Unfortunately, the stretching parameter, beta, can't be extracted reliably and we fixed it to the bulk water value.

Professor Idrissi asked: Due to the anisotropy of the topology of the Vycor, is there any signature of this anisotropy in the dynamics of water?

Dr Torre responded: In the Vycor sample the nanopores are curved and randomly distributed in a complex pattern, let me say in a spaghetti-like structure. So we cannot experimentally select the water–silica surface orientation.

Professor Ben-Amotz commented: The results in Fig. 5 of your paper suggest that the "inner water" differs from the "bulk water" in a way that is consistent with a weakening of water H-bonds in the "inner water" (relative to "bulk water"). In other words, weakening of H-bonds would be expected to shift the H-bond stretch to lower frequency and the H-bond bend to higher frequency (because of the lower effective mass of the bending vibration), which is consistent with your observations.

Dr Torre replied: I agree that the nano-confined water spectrum suggests this picture. Nevertheless we cannot exclude that the frequency shifts of the bending and stretching bands are only apparent due to a reduction, with respect to the bulk water,[1] in the amplitude of the other vibrational and relaxational components present in the spectrum.

1 Taschin *et al.*, *Nat. Commun.*, 2013, **4**, 2401.

Professor Angell noted: I'm puzzled by these statements on the physical effect of confinement. At the recent Relaxation meeting in Barcelona, Lars Petersson from Sweden described the condition of water confined between BaF2 plates. He made a convincing case for the notion that the layer between the plates was under, or behaved as if it was under, a high pressure?

Professor Caupin replied: I think that you refer to a recent work on a water film (up to 2.4 monolayers) on top of a *single* BaF2 plate, in contact with water vapor at 1.5 Torr, rather than confined between two plates.[1] Nevertheless, you raise an interesting point. It is true that, in this case of a flat film, the Laplace equation would predict a liquid pressure equal to the vapor pressure, 1.5 Torr. This would not be consistent with the density higher than bulk that is reported, which would suggest a large positive pressure. However, a water molecule close to a wall can experience a very large attraction due to the van der Waals interaction. In this situation, the pressure cannot be defined as an isotropic scalar quantity, but should be described by an anisotropic tensor, which depends on the distance to the wall. The Laplace law will fail in this situation. Furthermore, layering can occur in a confined liquid, with the molecules concentrated around planes at specific distances from the wall. The density in each plane can be different.

However, far enough from the wall, one expects to recover a more homogeneous liquid, with a more isotropic pressure. The question is therefore about the lengthscale above which macroscopic laws such as the Laplace equation can be used, and below which they fail. We can use as a reference the imbibition experiments by S. Gruener *et al.*[2] They used two porous Vycor glass monoliths with two different mean pore diameters: 7 and 10 nm. They concluded that the results "can be described in a reasonable way by macroscopic hydrodynamics provided a sticky preadsorbed boundary layer of about two monolayers of water molecules is considered". It is interesting to note that this "sticky" layer has the same thickness as the slowly relaxing outer water reported in Dr Torre's paper. The macroscopic hydrodynamics law mentioned is the Lucas–Washburn equation, derived using Laplace and Poiseuille equations, with the bulk surface tension and viscosity. This supports the idea that Laplace equation can be used as a reasonable guide to estimate the pressure experienced by a water molecule near the center of the pore, for diameters down to 7 nm. The pores used Dr Torre's paper are smaller – 4 nm in diameter – and one might wonder about the validity of the Laplace equation. For water in MCM-41 silica (pores with a diameter of 2.5 nm), analysis of neutron diffraction data suggested[3] that there were two layers of water next to the pore wall (one "overlap" layer with low density, and an "interfacial" layer with high density), and liquid water at uniform density in the core region. Molecular dynamics simulations of water in Vycor with pores 4 nm in diameter[4] are also consistent with a double layer of water at high density near the wall, and a uniform water density in the core region. I believe that, for pores large enough to have a core region with uniform density, and equilibrated with a vapour below the saturated vapour pressure, the pressure at the center might well be negative, and the liquid density below the bulk value.

1 S. Kaya *et al.*, *Sci. Rep.*, 2013, **3**, 1074.
2 S. Gruener *et al.*, *Phys. Rev. E*, 2009, **79**, 067301.
3 A. Soper, *J. Phys.: Condens. Matter*, 2012, **24** 064107.
4 P. Gallo, M. A. Ricci, and M. Rovere, *J. Chem. Phys.*, 2002, **116**, 342.

Dr Perera addressed Dr Torre and Professor Tanaka and Professor Molinero: In connection with the bond-ordered domains proposed by Prof. Tanaka, could the origin of the water-plug you report be due to the appearance of such a single bond-ordered domain? More generally, would water under confinement be mostly bond-ordered domains, hence acting like if it were under pseudo super-cooled conditions, compatible with the proposal of Prof. Tanaka?

Professor Tanaka answered: Surface field can in general enhance bond orientational order near the surface, which can be interpreted as a sort of surface wetting effect. We observed such effects for supercooled hard sphere liquids[1,2] and also for the liquid–liquid transition of triphenyl phosphite.[3] In the case of water, a hydrophilic surface may enhance ice-like structural order near the surface. We note that in hydrogen bonding liquids, local structural ordering involves both bond orientational and translational ordering, since there is strong constraint on both direction and length of bonds. Under confinement in a pore, not only arrangement of hydrogen bonding sites but also the surface curvature can be important factors controlling the degree of surface-induced ordering. Such ordering near surfaces leads to spatial heterogeneity not only in structure but also in dynamics in a pore.[1] As suggested by Prof. Perera, thus, we may argue that water under confinement is effectively similar to that at a lower temperature, but non-locality (spatial heterogeneity) should also play a crucial role, which makes a simple temperature mapping difficult.

1 K. Watanabe, T. Kawasaki and H. Tanaka, Structural origin of enhanced slow dynamics near a wall in glass-forming systems, *Nature Mater.*, 2011, **10**, 512.
2 H. Tanaka, Bond orientational order in liquids: Towards a unified description of water-like anomalies, liquid–liquid transition, glass transition, and crystallization, *Eur. Phys. J. E*, 2012, **35**, 1.
3 K. Murata, and H. Tanaka, Surface-wetting effects on the liquid–liquid transition of a single-component molecular liquid, *Nat. Commun.*, 2010, **1**, 16.

Dr Torre added: Honestly I can't answer your question, but let me comment on it. To my understanding, the bond-order domains proposed by Prof. Tanaka is formulated for bulk liquid phase based on the expansion of the free energy on local order parameters. In some respects, this is similar to the Landau–De Gennes theory for isotropic–nematic transitions in liquid crystals. In order to apply this picture to the confined liquids the interaction with the surfaces should be included. Indeed Prof. Tanaka *et al.* discussed the problem in a recent paper,[1]for confined colloidal systems.

The sample investigated in our experiment is characterized by strong hydrogen bond interactions with the silanol groups present on the silica pore surfaces that can induce relevant modification on water local network structures. Any theoretical approach should take into account this issue. Overall it seems difficult to relate the theoretical approach based on bond-order domains with the formation of a water plug inside the pores.

1 K. Watanabe, T. Kawasaki and H. Tanaka, *Nature Mater.*, 2011, **10**, 512.

Professor Molinero continued: The formation of a liquid plug in nanopores is a manifestation of a first order transition in nanoconfined water. The liquid phase nucleates from a supersaturated surface-adsorbed water phase, and under conditions of constant water content forms a liquid plug that does not fully fill the pore.[1,2] Further addition of water at constant chemical potential increases the amount of liquid in the pore (see, for example, Fig. 1 of ref. 1). Except for the outermost layer in contact with the pore surface, the simulations indicate that confined liquid water has a local order and density that is similar to that of bulk liquid water at the same temperature. It should also be noted that confinement results in a decrease of the melting point of ice,[3] therefore the temperatures at which water confined in nanopores is actually supercooled (below its respective melting point) are lower than for bulk water.

1 E. De La Llave, V. Molinero and D. A. Scherlis, Water filling of hydrophilic nanopores, *J. Chem. Phys.*, 2010, **133**, 034513.

2 E. De La Llave, V. Molinero and D. A. Scherlis, Role of confinement and surface affinity on filling mechanisms and sorption hysteresis of water in nanopores, *J. Phys. Chem. C*, 2012, **116**, 1833–1840.
3 S. Jaehnert, *et al.*, Melting and freezing of water in cylindrical silica nanopores, *Phys. Chem. Chem. Phys.*, 2008, **10**, 6039–6051.

Dr Lovelock opened the discussion of the paper by Professor Walker: How do you know the thickness of your monolayer? How do you know it is indeed a monolayer? Is the monolayer consistently reproduced?

Professor Walker responded: Without additional context, the term "monolayer" can have several different meanings. If by "monolayer" the question refers to the film formed by the saturated vapor formed at the silica surface, then the SFG data show that such a film is reproduced consistently. The thickness of the film is unknown – our efforts to use ellipsometry to quantify this thickness have been unsuccessful – but given the similarity between spectra reported in the top panel of Fig. 2 and 3 in my paper monolayers of alcohol surfactants on aqueous-vapor interfaces, we assume that the films at the silica vapor interface have thicknesses less than the extended length of adsorbed monomers.

We do not know the thickness of alkane films formed at the silica–vapor interface. (Top panels of Fig. 4 and 5) but given the differences between the silica–vapor interface and the silica–liquid interface, we expect that the silica–vapor films have thicknesses corresponding to a single adsorbed monomer.

If the "monolayer" corresponds to the coumarin adsorbates probed by SHG experiments (Fig. 7 and 8), then the question of monolayer thickness is harder to answer. What we can say with confidence is that the adsorbed solutes are located within a region that is sensitive to the anisotropy intrinsic to the solid–liquid interface. How far away from the surface this region extends is unknown, but simulations predict that depending on solvent structure, surface effects typically persist for no more than 2–3 solvent lengths.

Dr Meech asked: In Fig. 2 in your paper, the differences in signal-to-noise between the silica–vapour and silica–liquid interfaces is quite dramatic. What is the origin of this?

Professor Walker responded: The very high S/N observed at silica–vapor interfaces where the vapor phase consists of air saturated with a given solvent's vapor is due to a well ordered monolayer of 1-propanol adsorbates. (A similar response is appropriate for all of the silica–vapor spectra reported in Fig. 2–5.) In addition, the critical angle for the silica–vapor interface is ~44° meaning that for the silica–vapor SFG experiments, all data are collected under conditions of total internal reflection.

At the silica–liquid interface, all SFG spectra suffer from low S/N. Several sources can be responsible. First, if adjacent solvent layers adopt a structure that introduces local inversion symmetry within the interfacial region, the second order susceptibility will become very small and the SFG response will suffer. Second, given that the refractive indices of propanol and cyclohexane are 1.39 and 1.43, respectively, the critical angles for silica–solvent systems are 73° away from surface normal (propanol) and 80° (cyclohexane). Consequently, experiments are performed in an external reflection geometry and some of the SFG signal will necessarily be transmitted into the liquid phase. (Simple Fresnel considerations place an upper limit on this "lost" or transmitted signal of = 20% in the case of cyclohexane.) A third reason why SFG signals at the solid–liquid interface are much lower than at the solid–vapor interface can be due to rapid dephasing or solvent reorientation dynamics.[1] While we expect that dynamic effects should be slowed at a silica–liquid interface, this mechanism is still one that can not be discounted *a priori*.

1 See, for example, J. T. Fourkas *et al.*, *J. Phys. Chem. C*, 2007, **111**, 8902–8915.

Professor Meuwly inquired: In your Fig. 2 the SFG spectra at the silica–vapor interface in the ssp polarization decays to a flat line to both sides of the CH_x signals. This is not what is found at the silica–liquid interface. Can you comment? Also, what would be the effect of doing the experiments in D_2O instead of H_2O?

Professor Walker responded: Comparing differences in baseline behavior between solid–vapor (s–v) and solid–liquid (s–l) interfaces can be deceptive. Based simply on signal to noise considerations, one sees that the response from the s–v interface is considerably stronger than from the s–l interface. In fact, for equivalent acquisition times, the difference in signal magnitude is ~40-fold. The difference in baseline behavior noted in the question is not likely to reflect meaningful scientific differences in terms of molecular structure and organization.

Instances where discrepancies in baselines are important will arise when the detected SFG signal contains interferences between the resonant and non-resonant parts of the second order susceptibility or between adjacent resonant contributions from different vibrational transitions. Such interference can lead to line asymmetry and derivative line shapes if the magnitude of the non resonant contribution to the second order susceptibility approaches that of the resonant contribution.

Examples abound, but one that is particularly relevant to the question posed previously is a study of a cationic surfactant (cetyltrimethylammonium bromide) adsorbed to aqueous silica interfaces.[1] SFG spectra in this work show clear evidence of these interferences both for individual vibrational transitions as well as for resonant and non-resonant pieces. Swapping H_2O for D_2O would shift the large SFG water intensity from higher frequencies (above those of –CH stretching frequencies) to lower frequencies (below those of –CH stretching frequencies) and likely change the shape of any observed interferences.

1 E. Tyrode *et al.*, *J. Am. Chem. Soc.*, 2008, **130**, 17434–17445.

Dr Vila Verde continued: Would it be interesting to look at the OH frequency, or the OD one if that is more convenient in your setup, to gain further insight into the orientation of the two types of propanol at the surface?

Professor Walker responded: Examining the –OH and/or –OD frequencies may be helpful to discern whether or not adventitious water is present at the solid–liquid interfaces studied in this work. In the case of the silica–1-propanol and silica–2-propanol interfaces, efforts to observe –OH frequencies may not prove instructive given that a) strong hydrogen bonding between the silica substrate and the adjacent solvent is likely to create very broad features and b) hydrogen bonding between the surface silanol and solvent –OH will lead to interfacial structure that has approximate inversion symmetry and thus a small 2nd order susceptibility. In the case of (b), the SFG response from surface –OH will be vanishingly small.

At the silica–alkane interface, the free –OH stretching frequency of surface silanol groups might prove very instructive. While a search of the literature turned up no examples of silanol –OH stretching at silica–liquid interfaces, studies of liquid–liquid and liquid vapor interfaces[1] show that the free –OH stretch shifts ~30 cm^{-1} depending on the polarizability and hydrogen bonding ability of the adjacent phase. If this correlation remains true at the silica–cyclohexane and silica–methylcyclohexane interfaces studied in this work, then any shifts observed in the free –OH frequency might contain information about the strength of solvent–substrate interactions.

1 G. L. Richmond *et al.*, *Science*, 2001, **292**, 908–912.

Dr Meech asked: Your conclusion 2, that the nonpolar liquid provides a polar interface and *vice versa* for the polar medium is really quite counter-intuitive. Can you speculate on the origin of such remarkable differences to the bulk solution behaviour.

Professor Walker replied: The conclusions that the nonpolar liquids create a polar interface at the silica–liquid boundary and polar alcohols create a nonpolar interface are based on the resonance enhanced SHG spectra presented in Fig. 7 and 8. The data fit reasonably well to the traditional expressions for the second order susceptibility and resonance wavelengths can be extracted with high accuracy.

The SHG experiments are sensitive to the local solvation environment surrounding the adsorbed solute and this environment will be controlled by the interfacial anisotropy created by solvent-substrate associations.

The nonpolar environment at the silica–propanol interfaces may be easiest to understand. Strong hydrogen bonding between the silica and solvent will reduce hydrogen bonding opportunities for adsorbed solutes. Furthermore, we proposed that order imposed on hydrogen bonded propanol will force the interfacial solvent to adopt a long range structure reminiscent of Langmuir films with a hydrophobic region created by the propanol tails immediately adjacent to the silica surface. Reduced hydrogen bonding and a thin hydrophobic sheet can create the nonpolar solvation observed in Fig. 7. We propose that the polar environment created at the silica–alkane interfaces reflects relatively weak association between the substrate and solvent. Weak interactions will allow interfacial solvent to be easily displaced by solutes capable of hydrogen bonding with the surface silanol groups. The high density of SiOH groups can create the polar environment sampled by the adsorbed Coumarin solutes.

Professor Wang opened the discussion of the paper by Markus Meuwly: It is commonly accepted in recent years that damping of short range electrostatics is important in multipole based force fields. Does the KYY model explicitly damp the short-range electrostatics interactions?

Professor Meuwly replied: It should be clarified that we only use the bonded interactions of the KKY model whereas the non-bonded interactions are multipolar terms augmented by van der Waals interactions. As in all of our multipole work[1–8] we do not damp the electrostatic terms at short range but rather employ suitable van der Waals interactions.

1 N. Plattner and M. Meuwly, The Role of Higher CO-Multipole Moments in Understanding the Dynamics of Photodissociated Carbonmonoxide in Myoglobin, *Biophys. J.*, 2008, **94**, 2505.
2 N. Plattner and M. Meuwly, Higher order multipole moments for MD simulations, *J. Mol. Mod.*, 2009, **15**, 687.
3 M. W. Lee and M. Meuwly, On the Role of Nonbonded Interactions in Vibrational Energy Relaxation of Cyanide in Water, *J. Phys. Chem. A*, 2011, **115**, 5053.
4 C. Kramer, P. Gedeck and M. Meuwly, Atomic Multipoles: Electrostatic Potential Fit, Local Reference Axis Systems and Conformational Dependence, *J. Comp. Chem.*, 2012, **33**, 1673.
5 C. Kramer, P. Gedeck, and M. Meuwly, Multipole-based Force Fields from *ab initio* Interaction Energies and the Need for Jointly Refitting all Intermolecular Parameters, *J. Chem. Theo. Chem.*, 2013, **9**, 1499.
6 T. Bereau, C. Kramer, F. W. Monnard, E. S. Nogueira, T. R. Ward and M. Meuwly, Scoring multipole electrostatics in condensed-phase atomistic simulations, *J. Phys. Chem. B*, 2013, **117**, 5460.
7 M. W. Lee, J. K. Carr, M. Goellner, P. Hamm and M. Meuwly, 2D IR Spectra of Cyanide in Water Investigated by Molecular Dynamics Simulations, *J. Chem. Phys.*, 2013, **139**, 054506.
8 M. W. Lee and M. Meuwly, Hydration Free Energies of Cyanide and Hydroxide Ions from Molecular Dynamics Simulations with Accurate Force Fields, *Phys. Chem. Chem. Phys.*, 2013, **15**, 20303.

Professor Wang asked: The intramolecular bending term in the KKY model couples the bend with bond stretch. What is the physical motivation behind this particular energy expression?

Professor Meuwly answered: Such an energy expression correctly reproduces the spectroscopy in using a flexible water model, as reported in ref. 1.

1 J. J. Szymczak, F. Hofmann, and M. Meuwly, Structure and Dynamics of Solvent Shells around Photoexcited Metal Complexes, *Phys. Chem. Chem. Phys.*, 2013, **15**, 6268.

Professor Idrissi remarked: The spectral contribution that it is observed in the spectra may be associated with the enhancement of the local anisotropy in the water dynamics (translation, rotation and reorientation) as a result of the anisotropy of the water–silica interface. In the following, I show the results of molecular dynamics simulations using the SPC water potential model. Furthermore, in order to analyze the local anisotropy, I defined autocorrelation correlation (ACF) of the dynamical property **A** projected onto the principal axes of the molecule $\mathbf{e}_\infty(t=0)$ (for $\alpha = x, y, z$), the z-axis is perpendicular to the (x, y) plane of the water molecule: $\Omega_\alpha{}^{\mathbf{A}}(\mathrm{t}) = \langle \mathbf{A}(t)\cdot\mathbf{e}_\alpha(t)\mathbf{A}(0)\cdot\mathbf{e}_\alpha(0)\rangle$. We have calculated the dipole ($\mathbf{A} = \mathbf{u}$) autocorrelation functions which are drawn in the following figures.

Fig. 2 displays the reorientation autocorrelation function about the water molecular axes and clearly shows the anisotropy in the reorientation dynamics. This anisotropy may be enhanced when the water molecule at the silica interface.Finally, I want to point out that as the topology of the silica–water interface is not isotropic, the study of the dynamics using an isotropic correlation function may be inadequate. It is perhaps more helpful to analyze the dynamics along the axes associated with the silica layer, the z-axis being the normal to the silica surface.[1,2]

1 D. Dumont, D. Bougeard, *Zeolites*, 1995, **15**, 650–655.
2 B. Smit, T. L. M. Maesen, Molecular Simulations of Zeolites: Adsorption, Diffusion, and Shape Selectivity, *Chem. Rev.*, 2008. **108**, 4125–4184.

Professor Meuwly replied: This observation seems to be in line with power spectra along the three Cartesian directions for first-layer water molecules at the water–CO_2 interface[1] where also slight differences were found for different polarization directions.

1 M. W. Lee, N. Plattner and M. Meuwly, Structure, Spectroscopy and Dynamics of layered H_2O and CO_2 ices, *Phys. Chem. Chem. Phys.*, 2012, **14**, 15464.

Dr Meech asked: Is the low frequency (*ca.* 800 wavenumber) component in Fig. 8 due to the water librational mode? If so it is surprising that it is only weakly perturbed in the first layer. Can you comment?

Professor Meuwly replied: The water librational mode with multipolar interactions is rather around 600 wavenumbers.[1–3] The feature at approx. 800 wavenumbers in Fig. 8C rather involves the H(Si)–O–H bending.

1 M. W. Lee and M. Meuwly, On the Role of Nonbonded Interactions in Vibrational Energy Relaxation of Cyanide in Water, *J. Phys. Chem. A* 2011, **115**, 5053.
2 M. W. Lee, N. Plattner and M. Meuwly, Structure, Spectroscopy and Dynamics of layered H_2O and CO_2 ices, *Phys. Chem. Chem. Phys.*, 2012, **14**, 15464.
3 J. J. Szymczak, F. Hofmann, and M. Meuwly, Structure and Dynamics of Solvent Shells around Photoexcited Metal Complexes, *Phys. Chem. Chem. Phys.*, 2013, **15**, 6268.

Professor Wang commented: When an in plane diffusion constant in a layer is determined, there are two possible methods for selecting water molecules included for the calculations. The first method only counts water molecules that stayed in the same layer during the entire duration of the mean square displacement (MSD) calculation; the second method also keeps statistics from water molecules that leave the layer during the MSD calculation. Our calculation of normal diffusion constant indicates the first method can result in a systematic error. For the in-plain diffusion constant calculation, which of the two methods did you use and did you notice a difference between the two approaches beyond statistical uncertainty?

Professor Meuwly responded: Only the water molecules located in the first layer were considered. In other words, if a water molecule left the first layer it was disregarded from the average diffusion coefficient and was replaced by the water molecule that exchanged coming from the layer above.

Dr Henchman commented: TIP3P water is known to be poor at treating diffusion. Have you compared with other water models that reproduce diffusion constants more accurately?

Professor Meuwly answered: The essential parts of our contribution were carried out with a multipolar water model which reproduces experimental spectra. For this, it was shown that the diffusion coefficient depends sensitively on the van der Waals parametrization. For a more comprehensive comparison see, *e.g.*, ref. 1.

1 P. Mark and L. Nilsson, *J. Phys. Chem. A*, 2001, **105**, 9954–9960.

Ms Zhuang said: As the water molecules in bulk and at the Si–water interface experience different reactions fields, interfacial water and bulk water are likely to have different dipole moments. Has a polarizable MD simulation been considered?

Professor Meuwly answered: No, a polarizable simulation has not been carried out so far. We agree that such a simulation would be of interest *per se* if there was suitable experimental data to be compared with.

Professor Wales said: TIP3P is well known to give unphysical geometries for small water clusters. Could you relax the configurations you already have and do some local equilibration with a better potential to see if the structure is recovered?

Professor Meuwly responded: It would certainly be possible to relax such water structures with an improved potential, such as the KKY with multipoles. However, the use of TIP3P in the present work is rather to illustrate its limitations compared with a more physically motivated potential including multipolar interactions. An interesting simulation would be one with an extended point charge model, such as TIP5P or SPC/E.

Faraday Discussions RSC Publishing

PAPER

Low-density liquid water is the mother of ice: on the relation between mesostructure, thermodynamics and ice crystallization in solutions

Griffin Bullock and Valeria Molinero*

Received 11th May 2013, Accepted 24th May 2013
DOI: 10.1039/c3fd00085k

Predicting the temperature and extent of ice freezing in aqueous solutions is crucial for areas as diverse as cryobiology and materials design. It has long been recognized that the thermodynamics of liquid water controls the temperature and kinetics of ice crystallization. Parameterizations of the freezing temperatures in terms of the water activity of the solution have been successfully established, but the fundamental origin of the thermodynamic control of the non-equilibrium crystallization of ice has remained elusive. Here we use large-scale molecular simulations to elucidate the relationship between the structure, thermodynamics, and ice crystallization temperatures for solutions of mW water and a strongly hydrophilic solute that mimics LiCl ions. Fast cooling of solutions with up to 20 mol% ions results in the formation of nanosegregated glasses with domains of low-density amorphous ice and an ion-rich vitrified solution. Slow cooling of the mixtures results in nucleation and growth of ice within the domains of four-coordinated liquid water. The temperature of crystallization T_X coincides with the temperature of appearance of nanoscopic domains of four-coordinated liquid water in the mixture, T_L. We use the insight provided by the simulations to derive a thermodynamic expression for the crystallization temperature as a function of the water activity, $T_X(a_W)$, analogous to the dependence of the melting temperature, $T_m(a_W)$. The simple expression derived in this work provides a good account of the experimental freezing temperatures of water and the well-known steepest dependence of T_X on solute concentration compared to that of T_m.

1. Introduction

Water is a ubiquitous component of biological, geological and synthetic materials. The state of water – liquid, glass or ice – strongly impacts the function, physical properties and chemical stability of materials. Ice formation, in particular, has a deleterious effect in the stability of micro- and nanostructured

Department of Chemistry, The University of Utah, 315 South 1400 East, Salt Lake City, UT 84112-0850, USA. E-mail: Valeria.Molinero@utah.edu

systems, from living cells[1–3] to membranes for fuel cells,[4,5] and much effort is directed towards the development of new materials and processes in which water is unable to crystallize (*i.e.* vitrifies) at low temperatures. It is critical, therefore, to understand what controls the lowest temperature at which liquid water can be equilibrated before freezing to ice – and whether the liquid will form ice at all or produce an amorphous solid, a glass.

There has been a long quest for understanding what determines the homogeneous freezing temperature of water in solutions. In 1977 Kanno and Angell reported the temperatures of melting (T_m) and homogeneous ice nucleation (T_H) for solutions of LiCl, NaCl and KCl as a function of concentration and pressure, and found them to be strongly dependent on the water to ion ratio in the mixture, but essentially independent of the cation in the salt.[6] The similarity in T_H for these salts prompted Kanno and Angell to suggest that the depression in the freezing point might be a colligative property for dilute solutions. They noted that the experimental temperature of homogeneous ice nucleation T_H decreased with increasing pressure or solute concentration with a steeper slope than the equilibrium melting line and suggested that it might be due to the smaller value of the enthalpy of crystallization of ice at the (non-equilibrium) freezing temperature than at the (equilibrium) melting temperature.[6] The same year, MacKenzie presented a study of the non-equilibrium freezing of aqueous solutions of polymers, soluble organic molecules and electrolytes in which he reported that the ratio K between the depression in the homogeneous freezing temperature ΔT_H and the depression in the melting temperature ΔT_m, measured with respect to the values for pure water, $K = \Delta T_H/\Delta T_m$, was a constant for each solute.[7] MacKenzie found that the values of K for small solutes, either electrolytes or organic molecules, were all clustered around 2.0 ± 0.2. More recent analyses of experimental freezing temperatures of various solutes render K from 1.4 to 2.2 for small molecules and ions and larger values for polymers.[7–9] The first theoretical attempt to interpret these results came from Rasmussen, who developed a non-classical theory of nucleation that established, for transitions with a critical point, a linear relationship between a thermodynamic limit of supercooling of a parent phase and the equilibrium phase transition temperature.[10] The theory predicts the existence of a "physical spinodal", defined as the temperature at which the free energy of formation of the critical nucleus vanishes. The size of the critical cluster of the daughter phase was predicted to contain about 20 monomers at the physical spinodal temperature, which Rasmussen assumed to be the same as the experimental temperatures of homogeneous nucleation of the new phase. Although the liquid–ice transition does not end in a critical point and the expression $K = \Delta T_H/\Delta T_m$ does not satisfy the form predicted in the theory, Rasmussen considered the proportionality between depression of freezing and melting points to support the existence of a thermodynamic lower limit of stability at the experimental freezing temperature of water.[11]

Koop and coworkers analyzed the freezing temperature of water as a function of solute concentration and pressure and found that the freezing temperature can be universally represented as a function of the difference $\Delta a_W = 0.305$ (later re-parameterized to $\Delta a_W = 0.313^9$) between the water activity in the solution at the freezing point and in equilibrium with ice, independent of the nature of the solute.[12] While Rasmussen and MacKenzie's approach was based on evaluating the decrease in the freezing and melting temperatures at fixed solute concentration, the analysis of Koop *et al.* focused on the evaluation of Δa_W at constant temperature. DeMott compared several parameterizations of ice crystallization rates, and showed that the rates predicted with the water activity criterion of ref. 12 closely follow those predicted with Rasmussen's approach using $K = 1.7$.[8] The correlations derived by Koop *et al.* reinforced the points, to our knowledge first made by Kanno and Angell, that the freezing temperature behaves as a colligative property of the solution, and that pressure and solutes have analogous effects on the freezing temperatures because they produce equivalent changes in the structure of the mother liquid from which ice nucleates.[6,12]

The results discussed above indicate that the equilibrium thermodynamics of supercooled water controls the kinetics of non-equilibrium freezing of ice. Liquid water presents several thermodynamic anomalies in the supercooled region.[13] For example, its density decreases and the heat capacity and compressibility increase on cooling.[14–16] It has been hypothesized that these anomalies are associated to the existence of a first order liquid–liquid phase transition in supercooled water,[17] although a first order transition is not needed to explain the anomalous thermodynamics of liquid water.[16,18,19] Mishima noticed that the hypothesized locus of the liquid–liquid transformation line and its supercritical continuation are almost indistinguishable from the temperature of homogeneous nucleation of ice, below which supercooled liquid water cannot be prevented from crystallizing in the observational time scales available to state of the art experiments.[20] He thoroughly characterized the thermal and volumetric signatures on decompression induced crystallization of LiCl–water glasses with up to 15.4 mol% ions and concluded that the crystallization of ice occurred in two steps: first low-density amorphous ice (LDA) formed within the previously homogeneous glasses, then ice crystallized in the samples.[20–22] Mishima hypothesized that ice nucleated in the LDA-like domains, a conjecture supported by early simulations of nucleation of ice from supercooled liquid water by Matsumoto and coworkers.[23] Moore and Molinero investigated the relationship between the structure and thermodynamics of supercooled water and the rate of crystallization of ice using molecular simulations with the mW model of water.[14] They concluded that a structural transformation that sharply but continuously increases the fraction of four-coordinated molecules in the liquid controls the crystallization temperature and rate of formation of ice.[14] The continuous transformation of water into a low-density liquid is also responsible for the anomalous increase in the heat capacity and compressibility of supercooled liquid water.[14–16] The simulations with mW water predicted that the temperature of maximum crystallization rate T_x coincides with the liquid transformation temperature T_L, for which the rate of change of the density and structure of supercooled liquid water with temperature, $\alpha_p = (d\rho/dT)_p$ and $(df_4/dT)_p$, and the constant pressure heat capacity of the liquid, C_p, are also maxima.[14] Crystallization times as a function of temperature computed using classical nucleation theory with experimental data for water agree with the predictions of the simulations with mW water and predict a maximum crystallization rate at $T_X = 224$ K.[14] This temperature is just a few degrees lower than the experimental temperature of homogeneous nucleation of ice, T_H (231 to 235 K, depending on the cooling rate or experimental observation time). As they are so similar, T_X can be used to predict T_H. We note that, different from T_H, the temperature of the maximum crystallization rate T_X is an intrinsic property of the liquid, independent of the cooling rate.

The temperature of maximum crystallization rate T_X predicted using classical nucleation theory and experimental data for water is essentially the same as the temperature T_s at which a power law fit of the experimental heat capacity and compressibility of liquid water were predicted to diverge.[24–26] Within the liquid–liquid critical point scenario, T_s has been interpreted to belong to the supercritical continuation of the liquid–liquid line, also known as the Widom line.[27] However, mW water does not present a first-order liquid–liquid transition at any pressure,[28] but nevertheless reproduces the anomalies and known phase behavior of liquid water and ice.[14–16,29–32]

In this work we use molecular simulations to investigate the crystallization of aqueous solutions of mW water and a solute that mimics the LiCl ions.[33] Ice crystallization and vitrification have been extensively studied for water–LiCl mixtures,[6,20–22,34–44] allowing a comparison of the predictions of the simulation to the results of experiments. Aqueous solutions of LiCl with up to 20 mol% ions produce ice on slow cooling and vitrify to glasses that present two distinct glass transition temperatures on hyperquenching.[34,45] The lowest glass transition has been assigned to a concentrated water–salt glass[34] and the higher one to low density amorphous ice (LDA, the glass obtained by hyperquenching or vapor deposition of pure water), which readily crystallizes as it softens.[45] Raman spectroscopy of aqueous glasses of LiCl, NaCl and KCl provided further confirmation of the identities of these phases and demonstrated that the fraction of water in the LDA phase depends on the concentration of ions but is quite insensitive to the identity of the cation,[40] a result supported by simulations with the mW water model.[33] While calorimetry and spectroscopy point to the existence of two distinct phases in the vitrified solutions, the glasses are clear and do not present signs of macroscopic phase segregation. Le and Molinero used large-scale molecular dynamics simulations to investigate the microscopic evolution of the structure of the mixtures from the liquid to the glass states.[33] The simulations reveal that solutions with up to 20 mol% ions produce nanophase segregated glasses with characteristic dimensions of phase segregation of about 5 nm. The nanophases in the glass were pure LDA water glass and water–ion glass, as previously inferred from the experimental glass transitions and Raman spectra. The fraction of water in the LDA nanophase as a function of concentration predicted by the simulations is in excellent agreement with the results from Raman spectroscopy in the glasses of LiCl–water.[33,40] The results of the simulations and experiments indicate that pronounced concentration fluctuations and nanophase segregation are common in deeply supercooled solutions of aqueous solutions of very hydrophilic solutes.

In this work we investigate the interplay between mesoscopic concentration fluctuations in the liquid solution and the crystallization of water. We find that, in agreement with Mishima's interpretation, ice nucleates within domains of four-coordinated pure liquid water in the mixture. Assuming that crystallization of ice occurs at the temperature at which nanoscopic domains of low-density liquid water appear in the solutions, we derive an expression for the depression of the freezing point as a function of water activity as a colligative property. We conclude that the steeper dependence of the freezing temperature with water activity (or solute concentration) originates in a smaller enthalpy associated to the formation of LDA from liquid water as compared with the enthalpy of formation of ice from liquid.

2. Model and methods

Models

Water was modeled with the monatomic water model mW which represents each molecule as a single particle that interacts through anisotropic short-ranged potentials that favor "hydrogen-bonded" water structures.[29] The mW model has been previously validated for the study of liquid, amorphous and crystalline water.[14–16,28–33,46–57] Each ion was modeled as the single particle solute S, which presents strong short-range attraction to water, but no long-range (electrostatic) interactions. The parameters of the ion–ion and ion–water interactions are those of ref. 33, and have been shown to reproduce the structural properties of aqueous LiCl solutions.[33] The mild attraction between S ions ensures that their interactions are dominated by water and the ions do not demix. Although S was not parameterized to reproduce the property of a particular salt, water–S mixtures correctly predict several properties of water–LiCl solutions and that the dimensions of phase segregation and the fraction of water in each phase are not very sensitive to the details of the water–solute interaction.[33] As the effect of S on the structure of water and the phase segregation is equivalent to that of a monovalent salt in water, through the text we use the term 'ions' to refer to the coarse-grained solute S of this study. Compositions are indicated as mol% of solute, $x_S = 100 \times n_S/(n_S + n_W)$, where n_S and n_W are the number of molecules of solute and water in the mixture, respectively.

Simulations

Molecular dynamics (MD) simulations were carried out with LAMMPS.[58] The equations of motion were integrated using the velocity Verlet algorithm with a time step of 10 fs. Simulations were performed in the *NpT* ensemble using the Nosé–Hoover thermostat and barostat with damping constants of 1 and 5 ps, respectively. The pressure was 1 atm in all simulations. Periodic boundary conditions were used in the three Cartesian directions.

Analysis

The equilibrium melting temperatures T_m of bulk mixtures was determined from the coexistence of ice and solution. We constructed simulation cells with 9216 particles divided into two blocks: one with the structure of ice I_h containing 4447 mW water molecules and a solution containing 4769 particles with 233, 471 or 797 ions. The two-phase systems were evolved for 8 ns under *NpT* conditions at various temperatures: 265 K for the for the system with 233 ions, 245, 250 and 255 K for the system with 471 ions, 235, 230 and 225 K for the system with 797 ions. The ice front advanced or receded at each temperature, until a new composition of the liquid phase – the concentration x_S in equilibrium with ice at the selected temperature – was attained. Changes in the number of water molecules in the ice and liquid phases were not greater than 1500 particles for any simulation. All ions remained in the liquid phase. The fraction of ice in the simulation cells was determined with the CHILL algorithm, which is based on bond-order correlations between water molecules.[32]

One-phase cubic simulation cells containing 32,768 molecules (linear dimensions ~10 nm) with 5%, 10%, and 15% molar percent of S were prepared as in ref. 33, starting from equilibrated liquid water configurations at 300 K, randomly replacing water molecules with solutes and equilibrating them for 5 ns at 300 K. The temperatures of liquid transformation T_L correspond to the inflection points in the fraction of four-coordinated water molecules of the mixtures cooled at a rate of 10 K ns^{-1}, for which ice did not form.[33] The temperatures of crystallization T_X correspond to the inflection points in the fraction of ice of the mixtures cooled at a rate of 1 K ns^{-1} from 260 to 150 K. The CHILL algorithm was used to compute the fraction of ice and to identify molecules with the order of the cubic and hexagonal ice polymorphs. Simulation cells containing 110592 particles (linear dimensions ~15 nm) were used for the study of the nanophase segregation in the glasses.[33]

3. Results and discussion

A. Equilibrium melting temperature of ice in the solutions

We start by computing the equilibrium melting temperatures of ice in the solution and comparing them with the experimental phase diagram of LiCl–water mixtures to assess the accuracy of the coarse-grained model. The addition of salts to water decreases the chemical potential of water in the solution, resulting in a depression of the melting temperature of ice. The dependence of the ice melting temperature T_m on the molar percent of solute x_S can be obtained by equating the chemical potential of water in the pure ice phase and the solution. Assuming a negligible temperature dependence of the enthalpy of melting of pure ice, the temperature of melting of the solution is given by

$$T_m(x_S) = \left(\frac{1}{T_m^*} - \frac{R(\ln a_W)}{\Delta H_m^*}\right)^{-1} = \left(\frac{1}{T_m^*} - \frac{R(\ln(1 - x_S/100) + \ln(\gamma_W))}{\Delta H_m^*}\right)^{-1} \quad (1)$$

where T_m^* is the melting temperature of ice in pure water (273.15 K in experiments, 274 K for mW water[29,59]), ΔH_m^* is the enthalpy of melting of pure ice (6.0 kJ mol^{-1} in experiments,[60] 5.3 kJ mol^{-1} for mW water[29]), R is the gas constant, and a_W is the activity of water in the solution with x_S molar percent of solute. The activity is the product of the molar fraction of water and the activity coefficient γ_W of water in the solution. Fig. 1 shows the equilibrium melting temperature of the solutions measured in equilibrium simulations, along with the experimental curve for LiCl solutions from ref. 61. The melting temperatures of mW ice as a function of concentration of S solute are in quantitative agreement with the experimental $T_m(x_s)$ for ice in LiCl–water mixtures for solute content up to 15 mol%. At higher temperatures the simulations slightly underestimate the depression in the melting point of ice. Good agreement between experiments and simulations may be expected for an ideal solution in view of the values of T_m^* and ΔH_m^* of the mW water model. The solutions in this study, however, are strongly non-ideal; therefore the extent of agreement between simulations and experiments implies that the non-ideal coefficients γ_W are also well represented by the coarse-grained model.

B. Nucleation of ice is preceded by the formation of domains of four-coordinated liquid water in aqueous solution

Aqueous solutions can be supercooled to temperatures well below their equilibrium melting points, because homogeneous nucleation of ice is a rare, activated

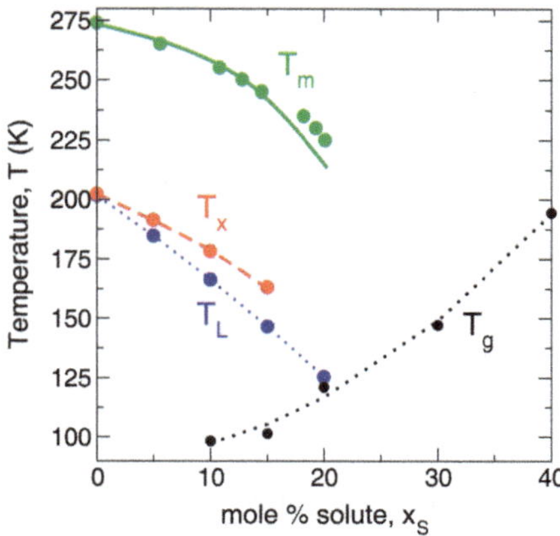

Fig. 1 Supplemented phase diagram. Equilibrium melting temperature of ice T_m (green circles), temperature of maximum crystallization rate of ice T_X (red circles), temperature of structural transformation of liquid water in the solution T_L (blue circles) and glass transition temperature of the mixture T_g corresponding to a relaxation time of 0.1 μs (black circles) as a function of the molar percent of ions in the mixtures. T_L and T_g from ref. 33. Cooling of mixtures with more than 20% ions results in a homogeneous glass,[33] in agreement with the experimental results for LiCl solutions.[62] The experimental melting temperature of ice in LiCl solutions[61] (solid green line) is shown for reference. The dotted black and blue lines are guides for the eye. The dashed red lines correspond to the freezing temperatures predicted with eqn (2) using $\Delta H_L^* = 2.0$ kJ mol^{-1}, $T_X^* = 202$ K and assuming that the activity coefficients γ_W computed from the melting line are independent of temperature.

process. Fast cooling of the aqueous solutions leads to vitrification, while slow cooling results in the formation of ice crystals (Fig. 2). At a cooling rate of 10 K ns^{-1}, water–S solutions with less than 20% ions yield nanosegregated glasses containing domains of pure low-density amorphous ice and a solute-enriched amorphous aqueous mixture with characteristic dimensions of phase segregation in the glasses of about 5 nm.[33] The results from the simulations explain the experimental observation of two distinct glass transition temperatures in glasses with up to 20% ions, although the glasses did not present signs of phase segregation for characteristic dimensions down to 20 nm.[34,45,63] The fraction of water in the LDA nanophase as a function of solute content predicted by simulations[33] is in quantitative agreement with the values derived from Raman spectroscopy.[40]

Cooling at a rate of 1 K ns^{-1}, results also in a nanosegregated system with pure crystalline ice in coexistence with a solute-enriched vitrified solution. The ice formed from the solutions is a hybrid ice I with cubic (C) and hexagonal (H) ice layers in a ratio of about 2C : 1H (Fig. 3). The hexagonal layers occur mostly as growth faults (single H flanked by C, CHC) and deformation faults (CHHC), as reported for ice homogeneously nucleated in bulk, nanopores and nanoparticles.[31,32,50,52,64,65] Molecular simulations and modeling of experimental spectra show that the hybrid ice I produces diffraction patterns which only display the peaks common to cubic ice and hexagonal ice, plus a peak or shoulder in a position close to that of the 100 peak of hexagonal ice.[31,64–67] The structure of the ice formed in the simulations upon slow cooling of the solutions is consistent with the structure of the ice obtained from LiCl solutions in experiments, which displays the diffraction pattern of cubic ice with hexagonal stacking faults.[39,44]

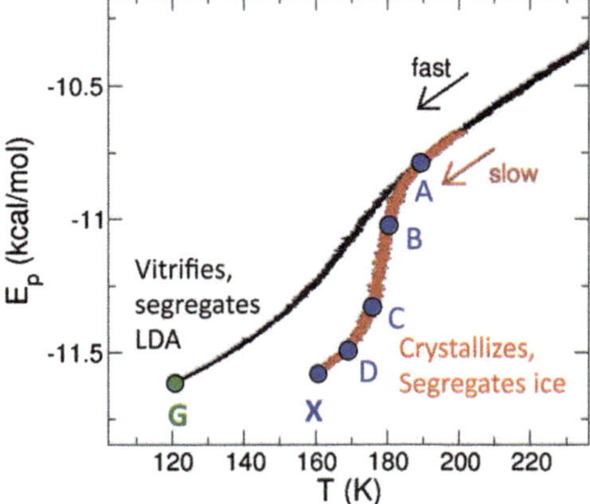

G nanosegregated glass

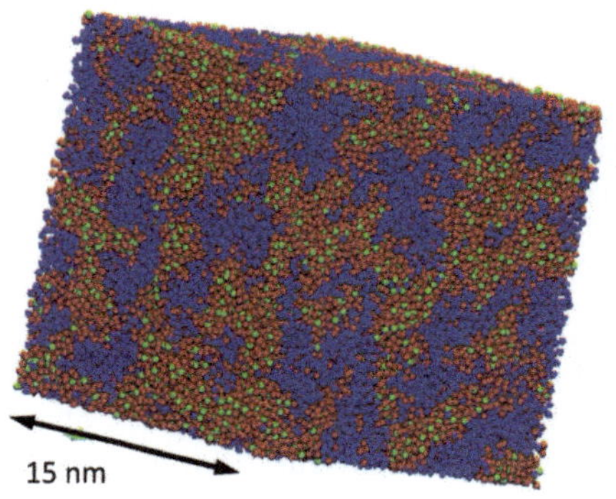

X crystallized; 55% ice

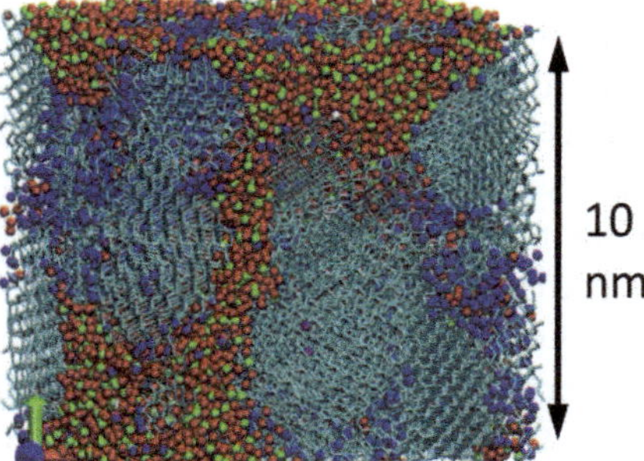

Fig. 2 The fate of the mixture on cooling, vitrification or crystallization, is determined by the cooling rate. Upper panel: potential energy of a solution with 10% ions in water as a function of temperature as the mixture is cooled at two rates q. The black curve corresponds to $q = 10$ K ns^{-1} and results in vitrification of the mixture into a nanophase segregated glass **G**. The red curve corresponds to $q = 1$ K ns^{-1}, the faster cooling rate that results in ice formation, and results in a system with ice in coexistence with a solute-rich glass, **X**. Middle panel: snapshot of the 10% nanosegregated glass (marked **G** in upper panel) obtained by cooling a solution containing 110592 molecules (from ref. 33). The glass contains domains of pure low-density amorphous ice (blue balls, 46% of the water in the system) and a glassy mixture of high-coordinated water (red balls) with ions (green balls). The dimensions of phase segregation of the

The temperature T_X at which ice crystallizes from the solution decreases with increasing solute concentration and closely tracks the temperature T_L (Fig. 1) that signals the inflection point in the fraction of four-coordinated water molecules in the liquid. The temperature of the maximum crystallization rate of pure mW water occurs at 202 K,[14] approximately 30 K below the temperature of homogeneous nucleation of ice in experiments. The shift in the temperature of crystallization accompanies an identical shift in the temperature of the maximum density and heat capacity anomaly in mW water with respect to real water.[14,15,29] At temperatures below T_X, in what is known as "water's no-man's land", pure liquid mW water cannot be equilibrated because the time scale of equilibration of the liquid becomes comparable or shorter than the time scale of crystallization.[14,30]

Koop and coworkers demonstrated that the experimental nonequilibrium freezing temperature of aqueous solutions can be predicted from the water activity of the mixture.[12] The process of incipient phase segregation and crystallization from the low-density liquid phase provides a microscopic foundation for this puzzling observation. The simulations reveal that the formation of nanoscopic patches of four-coordinated molecules precedes the crystallization, and that ice forms within these mesoscopic domains, as seen in Fig. 4. The results of this work indicate that the patches of four-coordinated liquid water are the birthplace of the ice embryos, as proposed by Mishima[20] and in agreement with the results of simulations of crystallization of pure water.[14,23]

If we assume that the domains of low-density liquid water are in equilibrium with the aqueous solution rich in ions,[33] then we can derive an expression analogous to that relating the melting point to the activity of water (eqn (1)) for the coexistence temperature between a hypothetical low-density liquid state and the aqueous solution. Low-density liquid mW water cannot be equilibrated before if forms ice,[14,30] hence the locus of equilibrium between mixture and the four-coordinated liquid corresponds to the lowest temperature T_X at which the supercooled aqueous mixture can be equilibrated without crystallization,

$$T_X(a_W) = \left(\frac{1}{T_X^*} - \frac{R\ln(a_W)}{\Delta H_L^*}\right)^{-1}, \tag{2}$$

where T_X^* is the temperature of freezing of pure water and ΔH_L^* the enthalpy difference between a hypothetical four-coordinated liquid and pure liquid water at a temperature just above freezing. In deriving eqn (2) we assumed that ΔH_L^* is independent of temperature. This approximation can be removed using the actual temperature dependence of the excess enthalpy of the liquid with respect to the ice.[14,16] We note that the T_L values of Fig. 1 were determined by cooling the solution at 10 K ns^{-1}, and they become increasingly lower than the T_X measured at 1 K ns^{-1} as the temperature of the structural transformation of the liquid approaches the glass transition line of the mixture: it takes a finite time to shovel away the ions to grow the pure water nanophase. We consider T_X to be representative of the locus of formation of four-coordinated domains large enough to allow for the nucleation of ice. As the radius of the critical nuclei in pure water is

glass are 5 nm.[33] Lower panel: structure of crystallized solution (marked **X** in upper panel) in which 55% of the water is ice (cyan lines), 9% is four-coordinated amorphous water (blue balls), and the remaining 36% (red balls) vitrified with the ions (green balls).

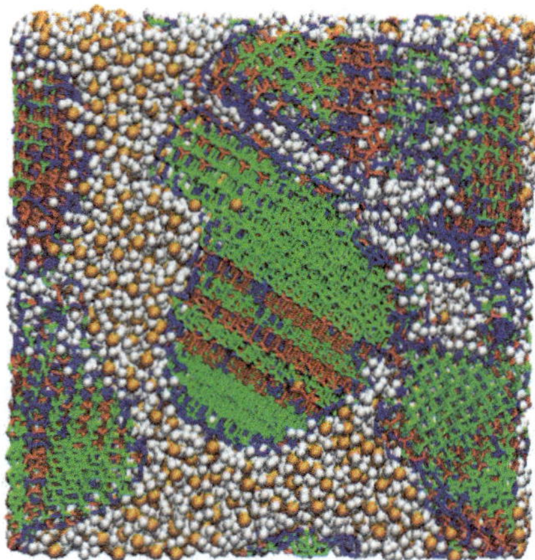

Fig. 3 Ice crystallized from the mixture is a hybrid of ice I with cubic and hexagonal layers. Structure obtained by cooling the 10% ion solution containing 32,768 particles at a rate of 1 K ns^{-1} (Fig. 2). The color coding corresponds to ice (red: hexagonal, green: cubic, blue: interfacial), liquid water (white) and ions (orange). There is a clear segregation of ice and ions. The ice contains hexagonal and cubic layers. The hexagonal stacks are 1 to 3 layers thick, while the cubic regions range from 1–5 layers. The ratio of cubic to hexagonal layers is about 2 : 1.

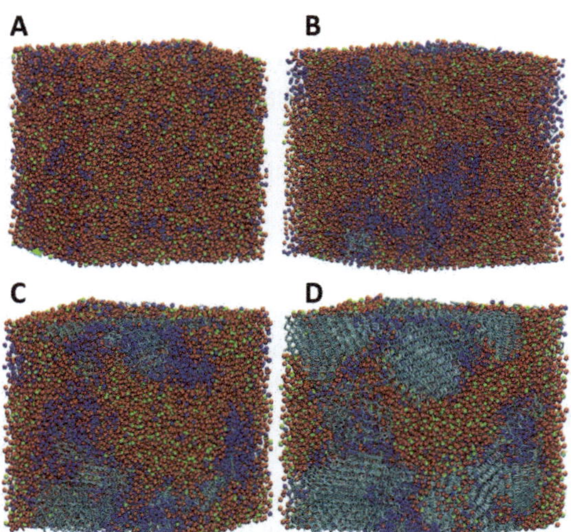

Fig. 4 Evolution the structure of the solution with 10% ions cooled at the fastest rate that results in ice crystallization. The four panels correspond to the system at the temperatures marked as A, B, C and D in Fig. 2. Ice (cyan lines) forms within the domains of pure four-coordinated liquid water (blue balls). The ions (green balls) are mixed with water molecules that have coordination larger than four in their first solvation shell (red balls). The time scale of ice growth within the four-coordinated liquid domains is shorter than the time scale required for coarsening of the structure of the phase separated mixture, resulting in systems with nanoscopic ice crystallites at the end of the ~50 ns simulations. The ice phase comprises 55% of the water at the end of the crystallization process in the 10% solute system, compared with 46% of water in the LDA phase obtained by faster cooling for the same solution. The higher fraction of water as ice (or the lower water content in the solution in equilibrium with it) reflects the lower chemical potential of the crystal with respect to LDA.[14]

about 1 nm,[14] the characteristic dimensions of the domains that nucleate ice must be at least 2 nm. The results presented in section C support this estimation.

Different from the first-order liquid-ice transition, the transformation of mW water into a low-density four-coordinated liquid (and the concomitant change in enthalpy) in the absence of the solute is steep, but continuous, with inflection temperature $T_L^* = T_X^* = 202$ K.[14,15,29] As it is not straightforward to determine ΔH_L^* from simulations of pure water, we computed it from eqn (2), assuming that the water activity coefficients γ_W determined at T_m do not change with temperature (the same assumption made in formulating the thermodynamic relations for ice nucleation in ref. 12). Eqn (2) with $\Delta H_L^* = 2.0$ kJ mol^{-1} (dashed red line in Fig. 1) reproduces the freezing temperatures $T_X(a_W)$ of the simulations, which encompass water activities ranging from 1 to 0.75. Remarkably, we derive $\Delta H_L^* = 2.2 \pm 0.2$ kJ mol^{-1} from the analysis of the experimental freezing temperatures as a function of water activity of ref. 12 using the experimental temperature of homogeneous nucleation of pure water in lieu of T_X^*. We considered experimental data with $a_W > 0.8$ because the approximation of temperature independent ΔH_L^* worsens at low temperatures and also the experimental freezing temperatures of solutions are highly dispersed in the low water activity region.

The experimental determination of T_X is only possible when the fastest crystallization time is slower than the observation time in the experiments. MacFarlane, Kadiyala and Angell exploited the closeness of the glass transition line to the freezing line in LiCl solutions approaching 20% ion content (well reproduced by the simulations, see Fig. 1) to determine the time–temperature transformation (TTT) curves for LiCl solutions with concentration between 17% and 19.8% ions.[36,38] Theirs was, in fact, the first determination of TTT curves for the crystallization of a supercooled liquid. We find excellent agreement between the experimental temperatures of maximum crystallization rate measured for water–LiCl solutions and the ones derived from the simulations with the mW–S mixture, after adding the 22 K that correspond to the difference between the $T_X^* = 202$ K for mW and the one predicted using classical nucleation theory and experimental data for water, $T_X^* = 224$ K. For example, for 17% ions $T_X^{\text{water–LiCl}} = 171$ K *vs* $T_X^{\text{mW–S}} + 22$ K $= 178$ K and for 18.5% ions, $T_X^{\text{water–LiCl}} = 160$ K *vs* $T_X^{\text{mW–S}} + 22$ K $= 171$ K. Considering the approximations involved in the extrapolation of eqn (2) to very low temperatures, the agreement between simulations and experiments is quite remarkable.

The lower value of ΔH_L^* with respect to ΔH_m^* explains the experimental observation that the depression of the temperature of homogeneous nucleation of ice is more pronounced than the depression of the equilibrium melting temperature as a function of solute concentration, *i.e.* $K = \Delta T_X/\Delta T_m > 1$. We used eqn (1) and 2, with $\Delta H_m^* = 6.0$ kJ mol^{-1}, $\Delta H_L^* = 2.2 \pm 0.2$ kJ mol^{-1}, $T_m^* = 273.15$ K and $T_X^* \approx T_H^* = 232$ K, to compute λ. The calculations show that K is not truly a constant, but decreases with decreasing water activity (*e.g.* from 1.96 to 1.74 as a_W decreases from 0.99 to 0.75) even if ΔH_L^* and ΔH_m^* are assumed to be temperature independent. A careful look at the published experimental data sets also evidences a drift of K towards lower values with increasing solute content.[8] The larger the ΔH_L^*, the lower the value of K. $\Delta H_L^* = 2.2 \pm 0.2$ kJ mol^{-1} results in $K = 1.83 \pm 0.15$ for $a_W = 0.9$, in good agreement with the $K = 1.7$ that was found[8] to closely represent the parameterization of the freezing rates in terms of Δa_W of ref. 12.

The ΔH_L^* computed from the freezing temperatures represents the abrupt – but continuous – jump in the enthalpy of pure liquid water around T_L.[14,16] We evaluated the enthalpy gap for mW water at $T_X^* = 202$ K as the difference between the tangents to the enthalpy of mW water at 150 K and 250 K and obtained $\Delta H = 2$ kJ mol^{-1}. This value is lower than the excess enthalpy of liquid water with respect to ice at T_X^*, 3.5 kJ mol^{-1}: the freezing temperatures are not controlled by the gap between the liquid and crystal but the change in enthalpy within the liquid state. We note that while the anomalous thermodynamics of liquid water in experiments[19] and in simulations with the mW model[16] are well reproduced assuming water is an athermal mixture of two "components" (a four-coordinated liquid and a liquid with the structure of water at 350 K), the difference in enthalpy between these two "components", 3.54 kJ mol^{-1}, is larger than the ΔH_L^* that determines the temperature dependence of the freezing temperature. This indicates that ΔH_L^* measures the difference between the enthalpy of the four-coordinated liquid and a supercooled liquid water state that, from the point of view of the two-state model, is already a mixed state. The value of ΔH_L^* that controls the crystallization of water is 37% of the enthalpy of melting of ice both for the data collected in simulations with mW water and in experiments.

C. The extent of ion segregation modulates the temperature of freezing and the fraction of water crystallized

The results presented in the previous section indicate that domains of four-coordinated liquid water are the birthplace of ice. In this section we investigate how the dimensions of these domains of low-density liquid water affect the temperature of freezing and the amount of water that crystallizes. To this end, we investigated to which extent the distribution of the ions in the mixture, evaluated at constant concentration, impacts the homogeneous nucleation and growth of ice. We prepared four model systems that contain the same concentration of ions, 10%, with different degrees of spatial segregation. We selected four configurations along a crystallization simulation of the 10% ion solution cooled at 1 K ns^{-1}: configurations A, B, C and D of Fig. 2 and 4. In the simulations that follow, the ions were held fixed at their initial positions in A, B, C and D (*i.e.* their equations of motion were not integrated) while the water was evolved under *NpT* conditions. By fixing the positions of the ions in space we treat the degree of segregation and concentration of ions as independent variables.

Fig. 5 shows the progression in fraction of ice along these five simulation trajectories corresponding to a cooling rate 1 K ns^{-1}: the original unconstrained trajectory, in which all particles move freely, and the four constrained trajectories. The fraction of crystallized water along the unconstrained trajectory was 0.5% in A, 4.4% in B, 22.2% in C and 48.6% in D. These four configurations were evolved with the ions fixed for 0.8 ns at 260 K to melt all ice, and subsequently cooled down to 160 K at 1 K ns^{-1}. The degree of ion segregation controls the maximum amount of ice that can be formed and the temperature of freezing (Fig. 5). We discuss these two aspects below.

The crystallization temperature increases with the degree of ion segregation, approaching the one of pure water for configurations with large water domains. The temperatures at which ice forms at the highest rate in the ion-fixed cooling simulations (*i.e.* the inflection points of the curves in Fig. 5) are 194 K for D_{fix},

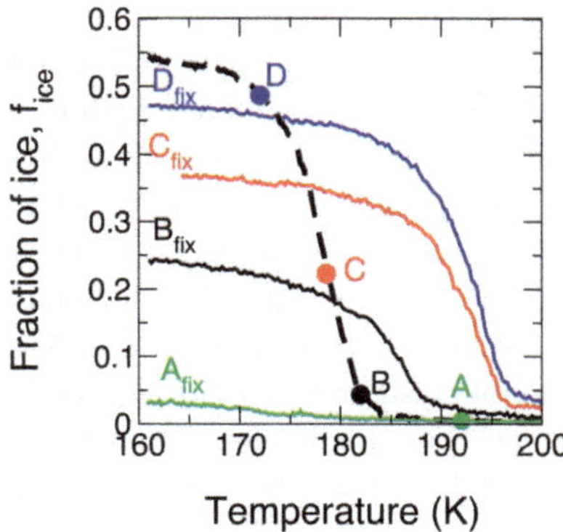

Fig. 5 Effect of ionic distribution on the fraction of ice and freezing temperature. Simulations of 10% solute mixtures cooled at 1 K ns^{-1}. Four simulations were evolved from the liquid state with the ions constrained in the positions they have in the unconstrained (dashed black) trajectory: A (A_{fix}, green), B (B_{fix}, solid black), C (C_{fix}, red) and D (D_{fix}, blue). The ice yield for D_{fix} is slightly less ice than the free-ion ramp. By comparing the locus of ice in the original D configurations and at the end of the D_{fix} quenching simulation, we determined that the lower ice yield in D_{fix} is due to the restriction the fixed ions impose on the growth of ice.

193 K for C_{fix}, 186 K for B_{fix}. These freezing temperatures are higher than for the ion-free (*i.e.* unrestrained) simulation cooled at the same rate, 178 K, implying that ion segregation within the solution precedes the formation of ice in the unrestrained simulation. The fraction of water that can be crystallized from the solution increases with the degree of ion segregation. The more segregated ion configurations produce more ice because they contain large domains of pure, ion-free water. The fraction of four-coordinated liquid water in these ion-fixed simulations increases before ice nucleates and peaks at the point where ice begins to form at a significant rate, supporting the conclusion that low-density liquid water is the mother of ice.

Very little ice is formed in the A_{fix} simulations because the fixed ions are uniformly scattered through the water severely limiting the size of domains of pure water. The ions significantly affect water molecules up to the second solvation shell, decreasing the tetrahedral order of water.[33,42,68] The number of free water molecules (defined as those at least 0.5 nm away from any ion) increases as the ions become more segregated. The number of free water molecules in the liquid also tracks the fraction of water molecules involved in the ice phase at the end the crystallization. We conclude that the decrease in crystallization temperatures in configurations with lower degrees of ion segregation is due to a lower availability of ion-free water to form the large clusters of four-coordinated water molecules that nucleate ice.

To further analyze the effect of ion distribution on ice crystallization we modified the fixed-ion protocol described above to investigate the nucleation and growth of ice at constant temperature for the fixed-ion configurations. We first melted the ice in the fixed-ion configurations of Fig. 4, and then evolved the fixed-ion simulations at constant temperature for 50 ns. Fig. 6 shows the fraction of ice as a function of time along the six isothermal simulations for B_{fix}, at temperatures from 205 K to 180 K, every 5 K. The crystallization times display a non monotonic temperature dependence, same as previously found in simulations of pure mW water in bulk, nanodroplets and in nanopores[14,30,32,52] and in experiments for LiCl solutions.[36,38] The temperature of maximum crystallization rate T_X for the "gel" with 10% ions fixed at the positions of the B configurations is between 185 and 190 K, in good agreement with the temperature of maximum crystallization rate determined from the 1 K ns^{-1} cooling ramp of Fig. 5, 186 K. Increasing the degree of ion segregation, *i.e.* the size of the domains of pure water, facilitates the nucleation of ice. Configuration D, for example, which has the greatest degree of segregation among those studied here, formed ice at 205 K within 25 ns, while all other configurations did not crystallize in 50 ns simulations at 205 K.

The fixed ions in the "gel" confine the domains of pure water. Confinement is known to decrease the stability of ice, because of the significant effect of the unfavorable interfacial free energy on small crystallites. The equilibrium melting point of bulk ice in equilibrium with the 10% solution is 254 K (Fig. 1). We determined the melting temperature of ice in the systems crystallized under different ion-segregation conditions. The crystallized fixed-ion systems obtained at the end of the cooling ramps of Fig. 5 were heated at a constant rate of 1 K ns^{-1} from 200 K to 270 K, and the fraction of ice was monitored. Slow heating allows for the determination of the equilibrium melting temperatures, because ice with open interfaces does not superheat. The melting points of ice, determined here as the temperatures at which half of the ice has melted (Fig. 7), increase with ion segregation: 229 ± 5 K for B_{fix}, 242 ± 5 K for C_{fix} and 246 ± 5 K for D_{fix}. The increase in the melting temperatures reflects an increase in the ratio of volume to area for the larger crystals. The ice and the glassy mixture nanophases in the 10% solute systems form two interpenetrated percolated networks, the same as the LDA and mixture nanophases in the vitrified solution.[33] Unconnected crystallites would produce well-defined steps in the loss of ice on heating, not observed in Fig. 7.

Although the ice crystallites are not spherical (see, for example, Fig. 4), we compute an effective radius of the crystallites in the ion-fixed simulations using the Gibbs–Thomson equation derived for the melting of spherical ice nanoparticles.[52] The melting temperatures correspond to effective ice crystal radii of

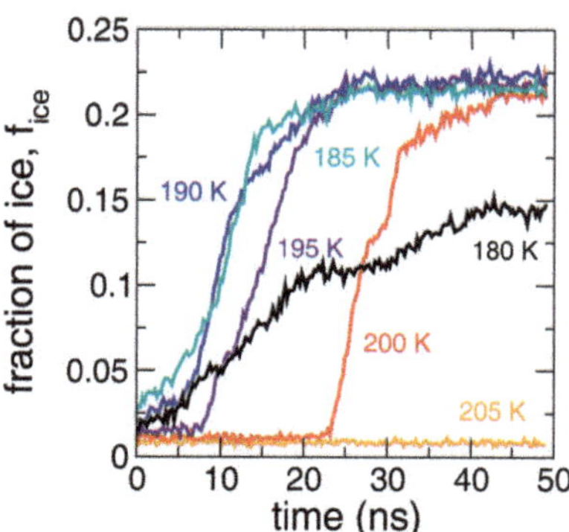

Fig. 6 Isothermal crystallization of ice in mixtures with 10% ions fixed at their positions in configuration B of Fig. 5. The crystallization rate is the highest between 185 and 190 K, in agreement with the temperature of maximum crystallization rate T_X deduced from the B_{fix} quenching trajectory in Fig. 5. Crystallization at temperatures above T_X is characterized by long induction times followed by fast growth of ice, indicating that the formation of ice at $T > T_X$ is limited by nucleation. At temperatures below T_X, on the other hand, ice nucleation is fast and the crystallization is limited by growth.

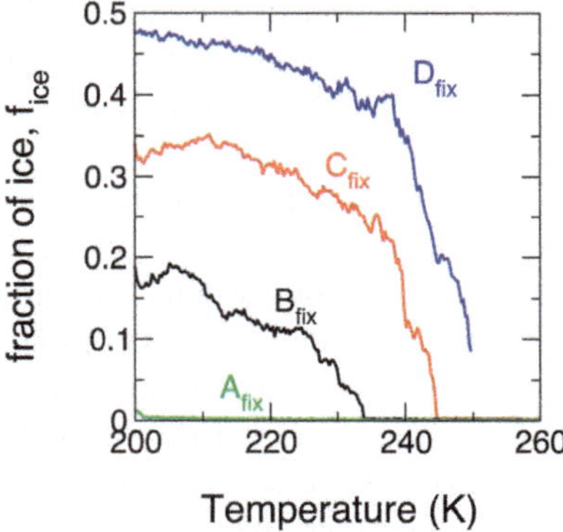

Fig. 7 Melting of the ice formed in systems crystallized with the same fraction of ions, 10%, at fixed positions. The crystallized systems correspond to the final configurations of Fig. 5. The melting points, determined here as the temperatures at which half of the ice has melted, are 229 ± 5 for B_{fix}, 242 ± 5 for C_{fix} and 246 ± 5 K for D_{fix}.

2.1 ± 0.2 nm for B_{fix}, 2.8 ± 0.2 nm for C_{fix} and 3.2 ± 0.2 nm K for D_{fix}. As the ions are fixed and ice does not solvate the ions, the size of the crystallites can be considered a good proxy for the size of the domains of pure water in these ion-fixed systems. Simulations[14] and experiments[69] indicate that the radius of the critical ice nuclei at the freezing temperature is about 1 nm. We conclude that ice nucleation will be strongly hindered in systems where the pure water domains are restricted because of the topology of the solute distribution (*e.g.* a gel, a polymer) to spaces with dimensions comparable or smaller than 2 nm.

4. Conclusions

We used large-scale molecular dynamics simulations to investigate the interplay between mesostructure and thermodynamics of the solution and crystallization of ice in mixtures of mW water and a solute that mimics LiCl ions. Fast cooling of solutions with up to 20 mol% of ions results in vitrification to produce a nanosegregated glass with domains of pure low-density amorphous ice and an amorphous mixture of water and ions.[33] The formation of a nanosegregated glass is consistent with the experimental observation of two glass transitions and spectroscopic signals for LDA and a water–LiCl glass for solutions with up to 20% ions.[34,40,45] The phase segregation in the glass is driven by the competition between two incompatible local orders for the water molecules: tetrahedrally coordinated order that optimizes the hydrogen bonds between water molecules and high density ordering of the water molecules in the hydration shell of the strongly attractive solutes.

It is known that mW water does not present a first order liquid–liquid transition at ambient pressure[15] (or at any other pressure[28]), nevertheless a distinct LDA nanophase forms when the mixture is hyperquenched at 1 atm. Chatterjee and Debenedetti predicted that the addition of strongly interacting solutes to water-like models with and without a liquid–liquid critical point could lead to first order liquid–liquid phase transitions.[70] These predictions concerned equilibrium phase transitions and did not account for crystalline phases. While the presence of ions induce nanophase segregation in the supercooled aqueous solutions, it should be noted that low-density liquid mW water cannot be equilibrated because its characteristic times of crystallization and equilibration are comparable at T_L.[14,28]

The four-coordinated amorphous water nanophase can only persist for long periods out of equilibrium as a glass. Slow cooling of the mixtures results in the crystallization of ice from the domains of low-density liquid water. The ice formed from the solution is a hybrid ice I with short stacks of cubic and hexagonal layers. As the stacks of hexagonal ice comprise very few layers, the diffraction pattern of the hybrid ice does not present the characteristic 10*n* peaks of hexagonal ice but only those that are common to the cubic and hexagonal ice polymorphs,[31] as seen in the diffraction patterns measured for ice crystallized from LiCl–water mixtures.[39]

The spatial distribution of the ions, and not just their global concentration, impacts the rate and temperature of crystallization of ice and the amount of ice formed. Nucleation of ice occurs when the characteristic size of the four-coordinated liquid water domains exceeds the size of the critical nucleus, about 2 nm diameter at the temperature of homogeneous nucleation. We investigated the crystallization of water in "gels" made of solutions for which the ions were kept at fixed positions that reflected various degrees of segregation. The freezing temperatures increase with the degree of segregation and they correspond well with the freezing temperatures of free water nanodroplets[52] with radii equivalent to the effective size (as measured by their corresponding melting temperatures) of the pure water domains in the "gels".

An increase in the concentration of ions results in less water available to form pure water domains and a higher viscosity of the solution. The temperature of the liquid transformation T_L that measures the inflection point of the growth of four-coordinated water molecules and the glass transition temperature T_g of the solution cross over at a solute content just above 20% ions. Cooling of solutions with larger solute concentrations results in the formation of a homogeneous glass both in the simulations[33] and experiments.[40,45,62] Ice crystallization is not observed for these very concentrated solutions even at very slow cooling rates, as the temperature of homogeneous nucleation of ice drops below the T_g of the solution.[62]

The freezing temperature of ice in aqueous solutions can be predicted entirely from the activity of water.[12] We explain this behavior in terms of the structural (thermodynamic) transformation of liquid water into a four-coordinated liquid that, we show, is the mother of ice. The crystallization temperature T_X of the solutions decreases with increase in solute content and closely tracks the temperature of structural transformation of water into a mostly four-coordinated liquid, T_L. We derived an expression for T_X as a function of water activity a_W, assuming that the crystallization temperature T_X corresponds to the onset of formation of a pure four-coordinated water phase in equilibrium with the mixture. The expression for the freezing temperature $T_X(a_W)$ is analogous to the one that describes the colligative effects in the melting temperature of ice, $T_m(a_W)$, and reproduces the experimental data compiled in ref. 12. Moreover, the T_X predicted from the simulation results are in excellent agreement with the experimental determinations of T_X for concentrated aqueous solutions of LiCl by MacFarlane *et al.*,[38] after correcting for the difference in the T_X^* of pure water in mW and water. The enthalpy gap that controls the crystallization, ΔH_L^* reflects a change in enthalpy within the liquid phase and is about one third of the enthalpy of melting of ice, both for the simulations and the experiments. The lower value of ΔH_L^* with respect to ΔH_m^* explains why the ice freezing temperature has a steeper dependence on concentration than the melting temperature.

This work presents a framework for interpreting and predicting the temperatures of homogeneous nucleation of ice from solutions based on the structural transformation and anomalous thermodynamics of liquid water. Koop and coworkers conjectured that "solutions and pure water under pressure melt and freeze when their hydrogen bonding network is the same".[12] The present study indicates that the solutions of strongly hydrophilic solutes such as salts develop significant composition fluctuations on approaching the temperature of homogeneous nucleation. These fluctuations originate on incompatible competing low-energy orderings for the water molecules and – if the solute content is not too high – result in the formation of nanoscopic patches of pure four-coordinated water molecules that are the birthplace of the ice nuclei.

Acknowledgements

We gratefully acknowledge support by the National Science Foundation through awards CHE-1012651 and CHE-1125235, the Arnold and Mabel Beckman Foundation through a Young Investigator Award, and the Camille and Henry Dreyfus Foundation through a Teacher-Scholar Award. We thank the Center of High Performance Computing of The University of Utah for technical support and allocation of computing time.

References

1 P. Mazur, *Science*, 1970, **168**, 939–949.
2 P. Mazur, *The American journal of physiology*, 1984, **247**, C125–142.
3 W. F. Rall and G. M. Fahy, *Nature*, 1985, **313**, 573–575.
4 Y. Ishikawa, H. Hamada, M. Uehara and M. Shiozawa, *J. Power Sources*, 2008, **179**, 547–552.
5 H. Mendil-Jakani, R. Davies, E. Dubard, A. Guillermo and G. Gebel, *J. Membr. Sci.*, 2010, **369**, 148–154.
6 H. Kanno and C. A. Angell, *J. Phys. Chem.*, 1977, **81**, 2639–2643.
7 A. P. MacKenzie, *Philos. Trans. R. Soc. London, Ser. B*, 1977, **278**, 167–189.
8 P. J. DeMott, in *Cirrus*, ed. D. K. Lynch, K. Sassen, D. O. C. Starr and G. Stephens, Oxford University Press, New York2002.
9 T. Koop and B. Zobrist, *Phys. Chem. Chem. Phys.*, 2009, **11**, 10839–10850.
10 D. H. Rasmussen, *J. Cryst. Growth*, 1982, **56**, 45–55.
11 D. H. Rasmussen, *J. Cryst. Growth*, 1982, **56**, 56–66.
12 T. Koop, B. P. Luo, A. Tsias and T. Peter, *Nature*, 2000, **406**, 611–614.
13 P. G. Debenedetti, *J. Phys.: Condens. Matter*, 2003, **15**, R1669–R1726.
14 E. B. Moore and V. Molinero, *Nature*, 2011, **479**, 506–508.
15 E. B. Moore and V. Molinero, *J. Chem. Phys.*, 2009, **130**, 244505–244512.
16 V. Holten, D. T. Limmer, V. Molinero and M. A. Anisimov, *J. Chem. Phys.*, 2013, **138**, 174501.
17 P. H. Poole, F. Sciortino, U. Essmann and H. E. Stanley, *Nature*, 1992, **360**, 324–328.
18 S. Sastry, P. G. Debenedetti and F. Sciortino, *Phys. Rev. E: Stat. Phys., Plasmas, Fluids, Relat. Interdiscip. Top.*, 1996, **53**, 6144–6154.
19 V. Holten and M. A. Anisimov, *Sci. Rep.*, 2012, **2**, 713.
20 O. Mishima, *J. Chem. Phys.*, 2005, **123**, 154506.
21 O. Mishima, *J. Chem. Phys.*, 2007, **126**, 244507.
22 O. Mishima, *J. Phys. Chem. B*, 2011, **115**, 14064–14067.
23 M. Matsumoto, S. Saito and I. Ohmine, *Nature*, 2002, **416**, 409–413.

24 R. J. Speedy and C. A. Angell, *J. Chem. Phys.*, 1976, **65**, 851–858.
25 D. H. Rasmussen, A. P. MacKenzie, C. A. Angell and J. C. Tucker, *Science*, 1973, **181**, 342–344.
26 E. Tombari, C. Ferrari and G. Salvetti, *Chem. Phys. Lett.*, 1999, **300**, 749–751.
27 L. M. Xu, P. Kumar, S. V. Buldyrev, S. H. Chen, P. H. Poole, F. Sciortino and H. E. Stanley, *Proc. Natl. Acad. Sci. U. S. A.*, 2005, **102**, 16558–16562.
28 D. T. Limmer and D. Chandler, *J. Chem. Phys.*, 2011, **135**, 134503.
29 V. Molinero and E. B. Moore, *J. Phys. Chem. B*, 2009, **113**, 4008–4016.
30 E. B. Moore and V. Molinero, *J. Chem. Phys.*, 2010, **132**, 244504.
31 E. B. Moore and V. Molinero, *Phys. Chem. Chem. Phys.*, 2011, **13**, 20008–20016.
32 E. B. Moore, E. de la Llave, K. Welke, D. A. Scherlis and V. Molinero, *Phys. Chem. Chem. Phys.*, 2010, **12**, 4124–4134.
33 L. Le and V. Molinero, *J. Phys. Chem. A*, 2011, **115**, 5900–5907.
34 C. A. Angell and E. J. Sare, *J. Chem. Phys.*, 1968, **49**, 4713.
35 C. A. Angell and E. I. Sare, *J. Chem. Phys.*, 1970, **52**, 1058.
36 D. R. MacFarlane, R. K. Kadiyala and C. A. Angell, *J. Phys. Chem.*, 1983, **87**, 1094–1095.
37 D. R. MacFarlane, R. K. Kadiyala and C. A. Angell, *J. Phys. Chem.*, 1983, **87**, 235–238.
38 D. R. MacFarlane, R. K. Kadiyala and C. A. Angell, *J. Chem. Phys.*, 1983, **79**, 3921–3927.
39 A. Elarby-Aouizerat, J.-F. Jal, P. Chieux, J. M. Letoffé, P. Claudy and J. Dupuy, *J. Non-Cryst. Solids*, 1988, **104**, 203–210.
40 Y. Suzuki and O. Mishima, *Phys. Rev. Lett.*, 2000, **85**, 1322–1325.
41 R. Souda, *J. Phys. Chem. B*, 2007, **111**, 5628–5634.
42 K. Winkel, M. Seidl, L. T, L. E. Bove, S. Imberti, V. Molinero, F. Bruni, R. Mancinelli and M. A. Ricci, *The Journal of Chemical Physics*, 2011, 134.
43 Y. Suzuki and O. Mishima, *J. Chem. Phys.*, 2002, **117**, 1673–1676.
44 A. Elarby-Aouizerat, J.-F. Jal, J. Dupuy, H. Schildberg and P. Chieux, *J. Phys.*, 1987, **48**, 465–470.
45 H. Kanno, *J. Phys. Chem.*, 1987, **91**, 1967–1971.
46 J. C. Johnston, N. Kastelowitz and V. Molinero, *Journal of Chemical Physics*, 2010, in press.
47 N. Kastelowitz, J. C. Johnston and V. Molinero, *J. Chem. Phys.*, 2010, **132**, 124511.
48 T. Li, D. Donadio, G. Russo and G. Galli, *Phys. Chem. Chem. Phys.*, 2011, **13**, 19807–19813.
49 L. Xu and V. Molinero, *J. Phys. Chem. B*, 2010, **114**, 7320–7328.
50 E. G. Solveyra, E. de la Llave, D. A. Scherlis and V. Molinero, *J. Phys. Chem. B*, 2011, **115**, 14196–14204.
51 E. B. Moore, J. T. Allen and V. Molinero, *J. Phys. Chem. C*, 2012, **116**, 7507–7514.
52 J. C. Johnston and V. Molinero, *J. Am. Chem. Soc.*, 2012, **134**, 6650–6659.
53 E. de La Llave, V. Molinero and D. A. Scherlis, *J. Phys. Chem. C*, 2012, **116**, 1833–1840.
54 T. D. Shepherd, M. A. Koc and V. Molinero, *J. Phys. Chem. C*, 2012, **116**, 12172–12180.
55 D. T. Limmer and D. Chandler, *J. Chem. Phys.*, 2012, **137**, 044509.
56 A. Reinhardt and J. P. K. Doye, *J. Chem. Phys.*, 2012, **136**, 054501.
57 B. Shadrack Jabes, D. Nayar, D. Dhabal, V. Molinero and C. Chakravarty, *J. Phys.: Condens. Matter*, 2012, **24**, 284116.
58 S. J. Plimpton, *J. Comput. Phys.*, 1995, **117**, 1–19.
59 L. C. Jacobson, W. Hujo and V. Molinero, *J. Phys. Chem. B*, 2009, **113**, 10298–10307.
60 CRC-Handbook, *Handbook of Chemistry and Physics*, CRC, Boca Raton, 2000–2001.
61 C. Monnin, M. Dubois, N. Papaiconomou and J.-P. Simonin, *J. Chem. Eng. Data*, 2002, **47**, 1331–1336.
62 C. A. Angell, E. J. Sare, J. Donnella and D. R. Macfarlane, *J. Phys. Chem.*, 1981, **85**, 1461–1464.
63 J. Dupuy, J. Jal, C. Ferradou, P. Chieux, A. Wright, R. Calemczuk and C. Angell, *Nature*, 1982, **296**, 138–140.
64 T. L. Malkin, B. J. Murray, A. V. Brukhno, J. Anwar and C. G. Salzmann, *Proc. Natl. Acad. Sci. U. S. A.*, 2012, **109**, 1041–1045.
65 W. F. Kuhs, C. Sippel, A. Falenty and T. C. Hansen, *Proc. Natl. Acad. Sci. U. S. A.*, 2012, **109**, 21259–21264.
66 T. C. Hansen, M. M. Koza and W. F. Kuhs, *J. Phys.: Condens. Matter*, 2008, **20**, 285104.
67 T. C. Hansen, M. M. Koza, P. Lindner and W. F. Kuhs, *J. Phys.: Condens. Matter*, 2008, **20**, 285105.
68 R. Mancinelli, A. Botti, F. Bruni, M. A. Ricci and A. K. Soper, *Phys. Chem. Chem. Phys.*, 2007, **9**, 2959–2967.
69 J. Liu, C. E. Nicholson and S. J. Cooper, *Langmuir*, 2007, **23**, 7286–7292.
70 S. Chatterjee and P. G. Debenedetti, *J. Chem. Phys.*, 2006, **124**, 154503.

Faraday Discussions RSCPublishing

PAPER

The microscopic features of heterogeneous ice nucleation may affect the macroscopic morphology of atmospheric ice crystals†

Stephen J. Cox,[ab] Zamaan Raza,[a] Shawn M. Kathmann,[c] Ben Slater[a] and Angelos Michaelides[ab]

Received 24th April 2013, Accepted 13th May 2013
DOI: 10.1039/c3fd00059a

It is surprisingly difficult to freeze water. Almost all ice that forms under "mild" conditions (temperatures > −40 °C) requires the presence of a nucleating agent – a solid particle that facilitates the freezing process – such as clay mineral dust, soot or bacteria. In a computer simulation, the presence of such ice nucleating agents does not necessarily alleviate the difficulties associated with forming ice on accessible timescales. Nevertheless, in this work we present results from molecular dynamics simulations in which we systematically compare homogeneous and heterogeneous ice nucleation, using the atmospherically important clay mineral kaolinite as our model ice nucleating agent. From our simulations, we do indeed find that kaolinite is an excellent ice nucleating agent but that contrary to conventional thought, non-basal faces of ice can nucleate at the basal face of kaolinite. We see that in the liquid phase, the kaolinite surface has a drastic effect on the density profile of water, with water forming a dense, tightly bound first contact layer. Monitoring the time evolution of the water density reveals that changes away from the interface may play an important role in the nucleation mechanism. The findings from this work suggest that heterogeneous ice nucleating agents may not only enhance the ice nucleation rate, but also alter the macroscopic structure of the ice crystals that form.

I. Introduction

Ice formation is a process important in numerous fields, ranging from microbiology[1-3] to understanding and predicting chemical processes in the atmosphere where, for example, it is known that ice particles in the polar stratospheric regions catalyse the formation of radicals responsible for ozone depletion.[4] Almost all ice

[a]Thomas Young Centre and Department of Chemistry, University College London, London, WC1E 6BT, UK
[b]London Centre for Nanotechnology, University College London, 17–19 Gordon Street, London, WC1H 0AH
[c]Physical Sciences Division, Pacific Northwest National Laboratory, Richland, Washington 99352, United States

† Electronic supplementary information (ESI) available. See DOI: 10.1039/c3fd00059a

formation is facilitated by the presence of a (solid) foreign body, in a process known as *heterogeneous nucleation*, and it is well reported that different materials affect the rate of ice formation to different extents.[5–9] However, despite the wide ranging consequences of ice formation, little is understood about how the surface properties of a foreign body affect its ice nucleating ability. By furthering our knowledge of the microscopic details of heterogeneous nucleation, it is possible that new pathways to the rational design of materials that either inhibit or enhance ice formation can be explored, with implications for the atmospheric[10–16] and climate sciences,[17,18] along with the food and transport industries.

Whereas experiments aimed at measuring the ice nucleating ability of different materials relevant to the atmosphere provide useful information for global climate models, as well as telling us which materials actually make good ice nucleating agents, most of our molecular level understanding of water–surface interactions are a result of detailed surface science studies (for an overview see ref. 19–25). For example, through the combined use of density functional theory (DFT) calculations and experiments such as scanning tunnelling microscopy and infrared spectroscopy, the structures of the first wetting layer at Pt(111)[26,27] and sub-monolayer chains at Cu(110)[28] have been elucidated. Such studies are, however, unable to provide the simultaneous spatial and temporal resolution required to probe the heterogeneous ice nucleation mechanism, making computer simulation an appropriate tool to study such a process. Despite a number of computer simulation studies of homogeneous nucleation,[29–37] there have been very few that have directly probed heterogeneous nucleation. Yan and Patey[38,39] have performed an excellent set of molecular dynamics (MD) simulations aimed at investigating the effect of strong electric fields on ice nucleation, finding that ferroelectric cubic ice forms in the region exposed to the electric field. Although this provides some insight into the role of electric fields on nucleation, the fields used are relatively smooth, whereas those exerted by real surfaces are likely to greatly vary on molecular length scales. Solveyra *et al.*[40] have also looked at the effect of confinement on ice nucleation in both hydrophilic and hydrophobic nanopores, using the single-site mW water model.[41] Use of such coarse grained force fields to describe the molecular interactions has the distinct advantage of being able to simulate large length- and time-scales at reasonable computational cost, but would unfortunately be inappropriate for the current study, where the electrostatic interactions between the surface and water are significant. The work presented here is unique in that, to our knowledge, it will be the first to directly simulate the dynamical process of heterogeneous nucleation where the atomic structure of both water and the substrate is taken into account.

As with homogeneous nucleation, there are many computational techniques at our disposal for looking at heterogeneous nucleation. One possible route is to use a free-energy based method such as metadynamics[42] or umbrella sampling[43] (for applications of these methods to homogeneous ice nucleation see ref. 35–37). The advantage of methods such as these is that one is able to obtain free energy barriers to nucleation along a specified reaction coordinate, but with the drawback that the system has to be driven along a predetermined set of collective variables, with no guarantee that the 'true' reaction pathway is being sampled. Another approach is to perform a number of unbiased MD simulations, starting with water in the supercooled liquid state, over suitably long time-scales until the nucleation event is observed. Although adopting such an approach may be seen as computationally inefficient, with recent advances in computer technology and software, the timescales involved are realisable at a reasonable computational cost for small to medium system sizes. Furthermore, by only performing unbiased MD simulations, we are no longer imposing *a priori* the reaction coordinate that the system must traverse. This direct approach has been used to seemingly good effect to study homogeneous ice nucleation, first by Matsumoto *et al.*[29] and subsequently by Jungwirth and co-workers.[30–32]

With our aim of understanding heterogeneous ice nucleation, we have opted to explore the clay mineral kaolinite as our model ice nucleating agent. Each year, as much as 3000 Tg of mineral dust (naturally occurring crystalline solid compounds) is transported into the troposphere from desert regions[44] where it catalyses the formation of ice.[5,9] The composition of mineral dust is diverse with quartz, feldspar, calcite and clays all present in significant proportions in typical atmospheric dust samples. Clays are the most frequently observed group in atmospheric mineral dust, of which kaolinite forms a substantial fraction.[9] Apart from being a known effective ice nucleating agent,[7,8,45] the binding of water to the pristine hydroxyl-terminated (001) face has been well characterised theoretically,[46,47] which aids in the analysis of our nucleation simulations.

Kaolinite is a layered silicate mineral with chemical composition $Al_2Si_2O_5(OH)_4$. Each layer consists of a tetrahedral silica sheet alternating with an octahedral alumina sheet, terminated with hydroxyl groups (see Fig. 1). In the bulk, these layers are bound by hydrogen bonds between the hydroxyl-terminated face and the silica-terminated face, giving rise to facile cleavage along the (001) plane, exposing the hydroxyl- and silicate-terminated faces. It is believed that the hydrophilic hydroxyl-terminated face is the origin of the ice nucleating efficacy of kaolinite, with the textbook explanation being that the pseudo-hexagonal arrangement of –OH groups acts as a template upon which the basal face of ice I_h can grow.[5] Despite its attractive simplicity, the validity of this explanation has been questioned; a series of DFT calculations by Hu and Michaelides[46–48] indicate

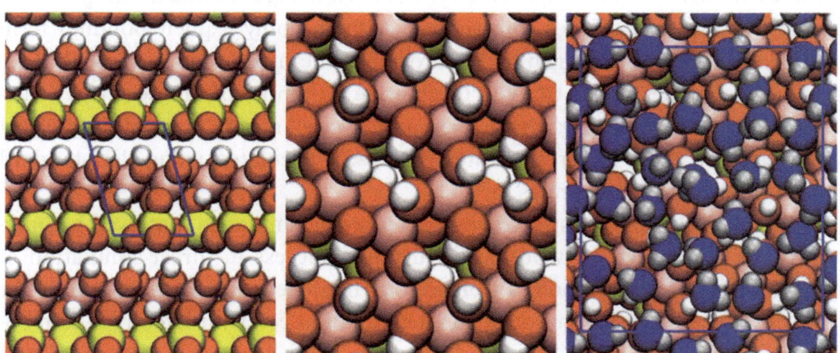

Fig. 1 Structure of kaolinite. On the left we show the layered bulk structure of kaolinite. As the layers are bound by hydrogen bonds between the hydroxyl-terminated and silicate-terminated faces, facile cleavage is observed along in the (001) plane. The middle panel shows the hydroxyl-terminated (001) face. DFT calculations[46,47] show that upon cleavage, 1/3 of the OH groups rotate into the plane of the surface, making it amphoteric *i.e.* able to both accept and donate hydrogen bonds with water. On the right is a snapshot from one of the MD simulations showing the first contact layer of supercooled water. We can see that the water molecules are densely packed and disordered. The colour scheme is: Si, yellow; Al, pink; O, red; and H, white. Water molecules in the first contact layer are shown in blue.

that the most stable ice-like bilayer at the kaolinite surface is actually hydrophobic with respect to growth of further layers of ice, a property attributed to the amphoteric nature of the hydroxyl-terminated surface; whilst grand canonical Monte Carlo simulations by Croteau *et al.*[49–51] have shown that only small regions of hexagonal motifs form in the first water overlayer and that these are somewhat stretched relative to bulk ice. In this work, we will directly probe the ice nucleation mechanism at the kaolinite (001) surface using MD simulations in a bid to shed further light onto the process of ice formation in the presence of this important mineral, as well as make inroads into understanding heterogeneous ice nucleation in general.

In what follows, we will see that, rather than the basal face, we exclusively see the prism face growing from the kaolinite surface. We will show that density fluctuations in the supercooled water away from the kaolinite slab play an important role in the heterogeneous nucleation mechanism. We will also discuss the role finite size effects play in our simulations before concluding and discussing the implications for the macroscopic crystal structure of our findings.

II. Methods

As we wish to simulate many molecules over long time scales it will be necessary to use classical force fields as opposed to quantum mechanical methods such as DFT. To this end, we employ the TIP4P/2005 water model[52] and the CLAYFF potential of Cygan *et al.*[53] to describe the kaolinite. TIP4P/2005, a rigid point charge water model, has been shown to replicate the phase diagram of water qualitatively well along with the transport properties of bulk water, even though it predicts the melting point of ice I_h to be *ca.* 252 K. It also reproduces the experimental bulk densities of liquid water, hexagonal and cubic ice very well, making it a suitable choice for modelling ice nucleation. The CLAYFF potential has been widely used for studying water at various clay mineral interfaces,[54–57] and in particular for the study of ice nucleation at kaolinite by grand canonical Monte Carlo methods.[49–51] In this approach, the clay atoms are treated as simple point charges with Lennard-Jones interactions, with the only explicit bonding term occurring between the oxygen and hydrogen of the hydroxyl groups. Such flexibility in the model allows CLAYFF to describe a number of different clay structures and phases satisfactorily, as well as the swelling of clays with increased water content.[53] The water–clay interaction was calculated using the standard Lorentz–Berthelot mixing rules.[58,59]

For the MD simulations, we followed a protocol similar to that used by Jungwirth and co-workers,[30–32] who have had much success in direct simulation of homogeneous ice nucleation. To create our homogeneous systems, 192 water molecules were placed in an orthogonal simulation cell with lateral (xy) dimensions of *ca.* 13.2 × 15.6 Å^2.[60] Due to the small x- and y-dimensions, a small cutoff of 6.5 Å was employed. Electrostatic interactions were calculated using the smooth particle mesh Ewald method, with a pseudo 2D correction[61] for the slab geometry, giving an effective z-dimension of at least 100 Å. The geometry of this system can thus be best described as an infinite slab with two liquid–vapour interfaces. For the heterogeneous system, the kaolinite was modelled as a single slab and 192 water molecules were placed on the hydroxyl-terminated (001) face, creating a solid–liquid interface, whilst leaving a liquid–vapour interface. Due to the presence of the kaolinite substrate, the lateral dimensions were *ca.* 15.5 × 17.9 Å^2, slightly larger than in the homogeneous case. To ensure that the kaolinite slab did not drift, one of the silicon atoms was fixed throughout the simulations.

To propagate the dynamics, the velocity Verlet algorithm was used with a timestep of 2 fs. Simulations were performed in the canonical ensemble and the temperature was controlled using a Nosé–Hoover chain of length 10 and a temperature coupling constant of 0.5 ps. Both systems were equilibrated at 300 K for 2 ns from which initial configurations for the production runs were sampled. For the production simulations, the systems were quenched to 220 K (*i.e.* approximately 30 K supercooled) and ran for the order of 1 μs or until nucleation was observed. The water geometry was maintained using the SETTLES[62] algorithm, whereas the P-LINCS[63] algorithm was used to constrain the O–H bond in kaolinite. All simulations were performed using the GROMACS 4.5 simulation package.[64]

III. Results and discussion

In total we ran 27 heterogeneous simulations, observing 10 nucleation events and 30 homogeneous simulations, observing 9 nucleation events. Movies of some of these are available in the supporting information (ESI†). Before doing any detailed analysis, one trend was immediately clear: on the kaolinite we exclusively formed hexagonal ice whereas in the homogeneous simulations we generally observed a mixture of hexagonal and cubic stacking patterns. In all but one of the homogeneous simulations, at least half of the ice formed consisted of cubic sequences with the bilayers parallel to the liquid–vapour boundary. Only one simulation resulted in solely hexagonal ice. This is qualitatively consistent with X-ray diffraction data and Monte Carlo simulations performed by Malkin *et al.*, which demonstrated that the homogeneous nucleating phase is stacking disordered (denoted as ice I_{sd}), consisting of roughly equal numbers of cubic and hexagonal sequences.[65,66] This mixture of cubic and hexagonal layers is also consistent with previous simulation studies.[30,67] Furthermore, when ice forms homogeneously we see a variety of crystal orientations within the simulation cell whereas when ice forms on the kaolinite, we always see growth along the prism face of ice and not the basal face *i.e.* the ice bilayers grow perpendicular to the kaolinite slab. In Fig. 2 we show diagrams of the basal and prism faces at the kaolinite surface. The observation that the prism face nucleates at kaolinite is interesting, as it means that the pseudo-hexagonal arrangement of –OH groups at the kaolinite surface are not acting as a template for the basal face of ice.

From visual inspection of the ice-forming trajectories it was noticed that, during the nucleation event, considerable rearrangement of the water molecules always seemed to occur in the second water layer above the kaolinite surface (note that this statement does not preclude any rearrangement occurring in the first or third layers). To provide evidence for this observation we measured how the density of water varies with height along the *z*-direction during the transition. In Fig. 3 we present this analysis for a single heterogeneous and homogeneous simulation along with the corresponding snapshots. First of all, we can see that the supercooled liquid (shown at 55 ns) has an extremely sharp and intense density peak at the kaolinite surface, as well as a pronounced, but broader, second peak.[68] After 61.5 ns the nucleation event has occurred and we can see the intensity of the first peak has decreased slightly, although it still remains much

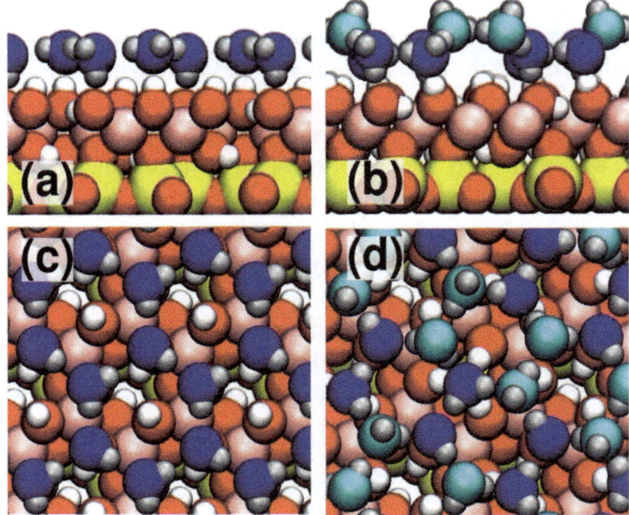

Fig. 2 Diagram of ice-like structures at the kaolinite surface. In panel (a) we show a side view of the basal face of ice bound to kaolinite in the "H-down bilayer" configuration.[47] All water molecules bind with similar heights from the surface. Panel (b) shows a side view of the prism face bound to kaolinite. In this structure, the water molecules come in high-lying (light blue) and low-lying (dark blue) pairs. Note that the prism face structure donates hydrogen bonds to the surface, as well as having 'dangling' hydrogen bonds pointing away from the surface (these dangling hydrogen bonds are absent in the basal face structure). Panels (c) and (d) show top views of the basal and prism face structures, respectively.

higher than anywhere else in the system. We also see that the second peak has started to split (highlighted in yellow), indicative of an ice-like layer forming. It is only after this change in density in the second layer that we see the first layer transform fully to ice. We can compare this to the homogeneous case, where the density in the supercooled liquid is essentially uniform and the nucleation event seems to occur by two or three layers concurrently forming ice. In both the homogeneous and heterogeneous scenarios, once the initial nucleation event has occurred the growth of ice then proceeds, with a quasi liquid-like layer remaining at the water–vapour interface, consistent with previous simulation studies.[30,67,69,70] We note that the observed changes away from the surface have striking similarity with the previously reported 'collective mechanism' for ice growth along non-basal faces at temperatures below 240 K.[71,72]

To investigate these structural changes away from the surface further, for each heterogeneous nucleation event observed we have computed the density difference:

$$\Delta\rho(z) = \rho(z) - \langle\rho_{\text{liq}}(z)\rangle \tag{1}$$

where $\rho(z)$ is the instantaneous water density at a height z and $\langle\rho_{\text{liq}}(z)\rangle$ is the water density at a height z averaged over supercooled liquid configurations. The results are presented in Fig. 4 (for reference, panel (b) corresponds to the heterogeneous nucleation event presented in Fig. 3). We can clearly see that in all instances, just after the onset of nucleation, there is a change in the density of the second (and often the third) water layer that is comparable to the changes observed in the first

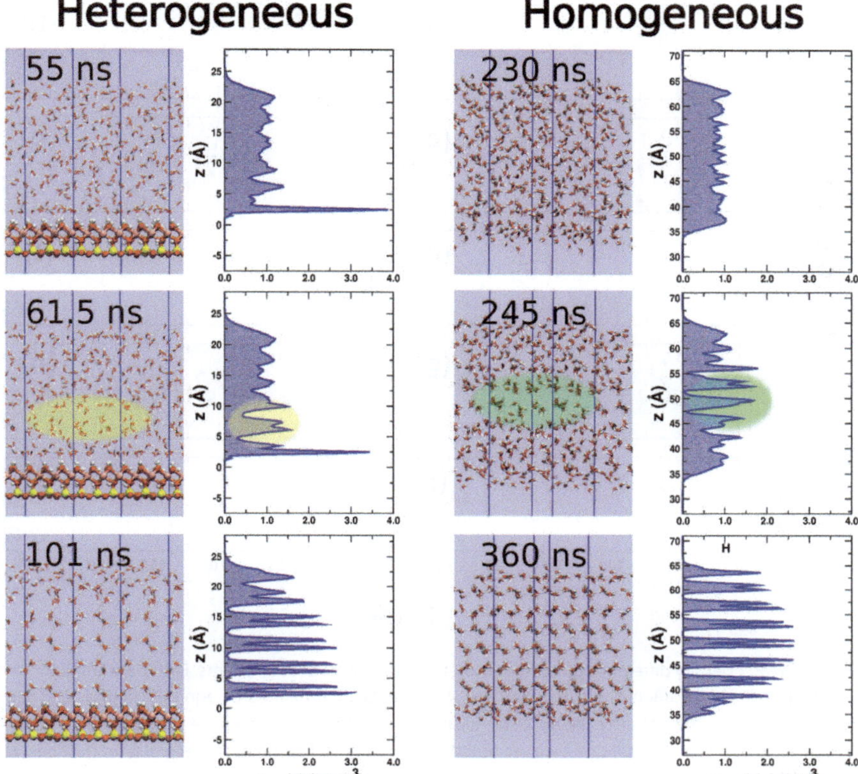

Fig. 3 Snapshots and water density profiles for a homogeneous (right) and heterogeneous (left) nucleation event. In the presence of kaolinite, the supercooled water (55 ns) has a high density peak corresponding to the first contact layer. There is also a noticeable second peak, but this is far less intense and much broader. After 61.5 ns, nucleation has started. There is a slight reduction in density of the first density peak, but this is still much higher than anywhere else in the system. Rearrangement of water molecules in the second layer associated with a split in the density peak (highlighted in yellow), is also seen and is indicative of an ice-like layer forming. By 101 ns the first contact layer has fully transformed to ice and the density is similar to that observed in the rest of the system. Note that it is the prism face of ice exposed to the kaolinite surface and that we only observe hexagonal ice. In contrast, for the homogeneous slab we see a fairly uniform density profile in the supercooled regime (230 ns). We also see a mixture of hexagonal and cubic stacking. In this instance, the initial nucleation event (highlighted in green) leads to a cubic stacking arrangement. The densities are averages over a 2.5 ns interval centred at the specified time. The colour scheme is the same as Fig. 1.

layer. In none of the simulations do we observe the first layer fully transform to ice without this signature splitting of the second layer density.

It is important to consider how significant the changes in density away from the surface are in the ice nucleation mechanism; after all, one may argue that these are just a consequence of the initial changes seen in the first layer and are merely indicative of ice growth rather than playing a role in the nucleation mechanism itself. We have therefore performed a committor analysis on the heterogeneous trajectory presented in Fig. 3 (and panel (b) in Fig. 4), using the CHILL algorithm of Moore *et al.*[73] to monitor ice formation. This was done by choosing different configurations along this trajectory and starting 10 new trajectories with random velocities drawn from the Maxwell–Boltzmann

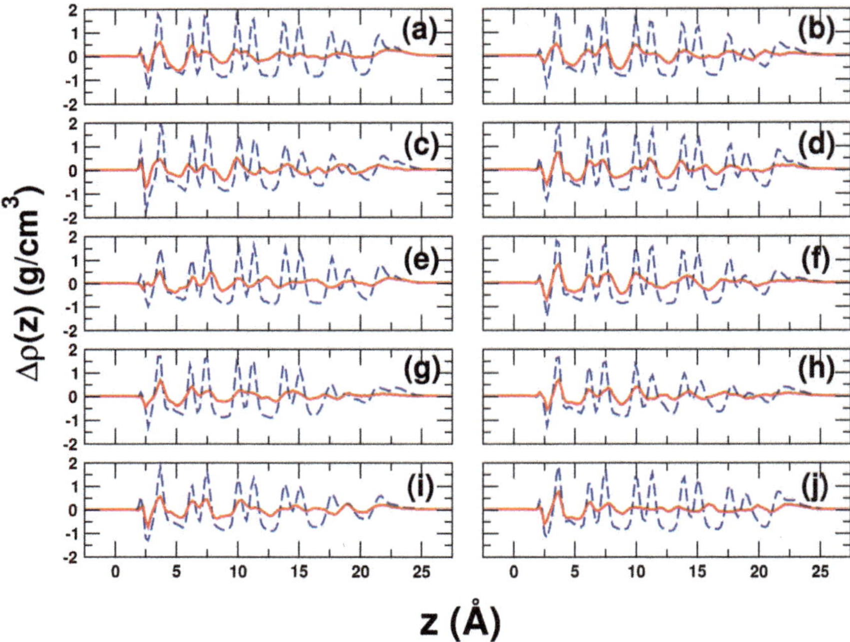

Fig. 4 Water density difference profiles for all heterogeneous nucleation events. Each panel (a–j) shows an independent nucleation event. The quantity plotted is $\Delta\rho(z)$ as defined by eqn (1). The red solid line shows $\Delta\rho(z)$ at a time just after the onset of nucleation and the blue dashed line shows $\Delta\rho(z)$ at a later time after ice has grown. In all cases, we see that there are density changes in the second layer (just below 7.5 Å) of a similar size to those in the first layer, before ice goes on to form fully (noticeable changes in the third layer are also often observed). In the case of (b), we know from a committor analysis that the red line corresponds to a pre-critical configuration. The displayed densities are averages over 2.5 ns.

distribution. Results from three starting configurations are presented in Fig. 5, where we can clearly define a pre- and post-critical region (initial configurations from 62.5 ns and 70.0 ns of the initial trajectory, respectively). In between these two regimes, however, we do not see an expected 50 : 50 split of trajectories going on to reach the liquid and ice states, rather we see some that definitely go to ice, some that definitely go to liquid but some trajectories that stay somewhere in between, even over fairly long timescales (*ca.* 50 ns). As the cost of this committor analysis is high, we have not attempted to refine our search further and remain satisfied that the configuration sampled at 65.0 ns is a reasonable representation of the 'transition region'. What is relevant to our discussion regarding the density changes in the second layer is that $\Delta\rho(z)$ shown by the red line shown in Fig. 4(b) corresponds to the configuration sampled at 62.5 ns *i.e.* the splitting in the second peak for this trajectory occurs in the pre-critical regime, indicating that these structural changes are part of the nucleation mechanism rather than a feature of growth. We have also performed a similar analysis for the trajectory in Fig. 4(d), which we present in the supporting information (ESI†), along with movies showing how $\Delta\rho(z)$ varies during the nucleation event.

It is interesting to attempt to explain some of these observations. To help understand why we see the formation of ice with its prism rather than basal face exposed to the kaolinite surface, we have investigated how the adsorption energy of ice changes with the number of ice-like layers, for both the prism and basal

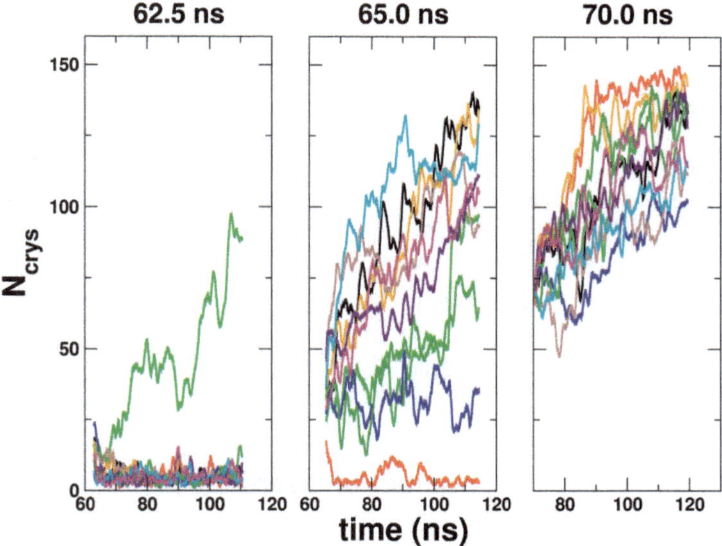

Fig. 5 Committor analysis from one of the heterogeneous ice nucleation trajectories. Results here are shown for initial configurations sampled at 62.5 ns, 65.0 ns and 70.0 ns from the initial ice forming trajectory. 10 independent trajectories were started from each configuration by giving the particles random velocities sampled from a Maxwell–Boltzmann distribution. By monitoring the number of water molecules defined as being ice by the CHILL algorithm,[73] N_{crys}, we are able to determine whether or not ice forms. We can clearly see that at 62.5 ns we are in a pre-critical regime and by 70 ns all trajectories continue to form ice. At 65.0 ns we do not see all trajectories form ice or liquid, but some stay somewhere in between the two states, even over the *ca.* 50 ns timescale. Results are presented as running averages over a 1 ns interval.

faces bound to the kaolinite (details of these calculations are given in the supporting information, ESI†). The results of this analysis are presented in Fig. 6, where we present the data in terms of adsorption energy per water molecule and adsorption energy per conventional unit cell of kaolinite. When only the first contact layer is present, the basal face of ice is more strongly bound than the prism face by approximately 15 meV/H_2O. As soon as we go beyond the first layer, however, the prism face becomes more stable, with the difference becoming more pronounced as more layers are added. The prism face also binds with a higher coverage than the basal face (5.33 *vs.* 4 H_2O per conventional unit cell) meaning that the prism face is more stable per unit cell of kaolinite independent of the number of ice-like layers. To understand these differences, it is useful to examine the structure of the ice-like layers when binding through the prism and basal faces, which we show in Fig. 2. Here it can be seen that the water molecules in the basal face structure bind with similar heights from the kaolinite, with half the molecules donating one hydrogen bond to the surface and the other half accepting a hydrogen bond from the kaolinite whilst donating two hydrogen bonds to other water molecules (the "H-down bilayer" structure as described in ref. 47).[74] By adopting this structure, the water molecules maximise their bonding to the kaolinite and maintain good hydrogen bonding between each other, giving a large overall adsorption energy for the first layer. This structure, however, saturates all hydrogen bonds leaving no 'dangling' hydrogen bonds that water molecules in above layers can bind to and consequently, the adsorption energy

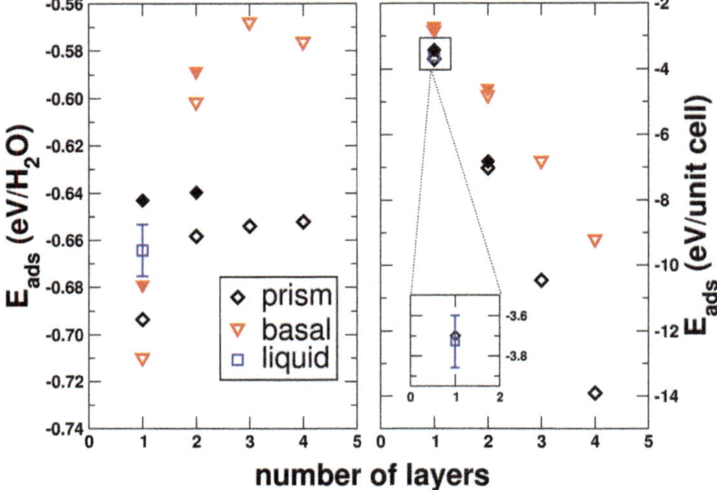

Fig. 6 Variation of the adsorption energy (E_{ads}) of ice to kaolinite as the number of ice layers changes. The black diamonds show results for ice binding to kaolinite through its prism face, whilst the red triangles show results for ice binding through its basal face. Filled symbols show results from DFT calculations. The left panel shows the adsorption energy calculated per water molecule, whereas the right panel shows the adsorption energy per conventional unit cell of kaolinite. For the first contact layer on its own, the adsorption energy per water molecule is stronger for the basal face than the prism face, but upon adsorption of other layers, the prism face structure becomes significantly more stable. When ice binds through the prism face, the coverage of water molecules is higher than when it binds through the basal face, meaning that the adsorption energy per unit cell of kaolinite is more stable for the prism face independent of the number of adsorbed layers. Data for the first liquid layer is also shown (the bars indicate estimates of the thermal fluctuations). On a per molecule basis, this is less stable than the ice-like structures, but is more stable per unit cell of kaolinite.

rapidly becomes less exothermic as other layers are added. This finding is consistent with previous findings from a DFT study,[47] as well as the experimental observation that the availability of dangling hydrogen bonds determines the multilayer wetting behaviour of water on metal and metal oxide surfaces.[75] On the other hand, the prism face binds with a somewhat more corrugated configuration, with the water molecules coming in high-lying and low-lying pairs. One of the molecules in the low-lying pair donates one hydrogen bond to the kaolinite whilst its partner accepts hydrogen bonds from the kaolinite. The high-lying pairs bridge the low-lying pairs through hydrogen bonds, with the important feature that one of these high-lying molecules has an OH bond directed away from the surface *i.e.* the prism face exhibits dangling hydrogen bonds. The fact that half of the molecules come in high-lying pairs means that the adsorption energy per water molecule of the first layer is less for the prism face than it is for the basal face, but the ability of the prism face to donate and accept hydrogen bonds to both the surface and the above water layers means that it becomes more stable as the number of water layers increases. We have also computed the adsorption energies of the first and second layers with DFT using the Perdew–Burke–Ernzerhof (PBE) exchange–correlation functional[76] (full details of these calculations are given in the supporting information, ESI†). Although agreement is not exact between our force field setup and PBE (which should not be taken as a benchmark) the trend that the prism face becomes more stable than the H-down bilayer upon adsorption of a second layer of ice is still seen. This suggests that this observation is not an artifact of our choice of force field.

In Fig. 6 we also show the average adsorption energy of the first layer from 25 configurations selected from the supercooled liquid.[77] On a per molecule basis, the liquid layer is less stable than either of the ice-like structures, but from the right hand panel of Fig. 6 we can see that per unit cell of kaolinite, the liquid layer is slightly more stable. This result may help us explain the observed density changes away from the surface during the nucleation process. If we draw an analogy to the grand canonical ensemble, we may consider the first water layer as a subsystem that is able exchange heat and particles with the bulk liquid above.[78] In the supercooled state, therefore, there will be some pseudo-equilibrium number of water molecules in the first layer, which we have measured to be 5.61 H_2O/unit cell (*cf.* 5.33 H_2O/unit cell for the prism face). Thus, although the adsorption energy per water molecule is stronger for the first layer of ice, on average more water molecules are present in the first liquid layer leading to an overall stabilisation. For ice to form and persist at the surface, it is therefore required that the average number of water molecules at the surface decreases. In keeping with the analogy to the grand canonical ensemble, this amounts to a need for a change in the chemical potential of the reservoir of water molecules above the first layer, which manifests itself as the structural changes away from the surface discussed previously. We can also see from the right panel of Fig. 6 that the adsorption energy of the prism face per unit cell is within our estimate of the thermal fluctuations from the average liquid value. This may be one of the reasons for kaolinite's good ice nucleating ability.

Finally, it is important that we mention the role of finite size effects in this work. We have attempted to perform these simulations with system sizes doubled in the lateral dimensions (768 water molecules, cutoff for interactions extended to 9 Å) but no nucleation was observed in a total simulation length of 15.5 μs over a temperature range spanning of 190–220 K. We also simulated 2 μs at 240 K using the TIP4P/ice water model,[79] which has a melting point similar to experiment, but still no nucleation was observed. This discrepancy can be explained by the fact that in the small systems, there is a self interaction of the growing ice nucleus with its periodic images that lowers the interfacial free energy cost of nucleation. We have also performed 14 homogeneous simulations in the same cell used for the heterogeneous simulations (*i.e.* lateral dimensions of *ca.* 15.5 × 17.9 Å without the kaolinite slab). No nucleation events were observed. Although it would have been desirable to have observed nucleation in these simulations, so that we could have compared homogeneous and heterogeneous rates, one pleasing aspect of this last null result is that it means that the heterogeneous nucleation results presented earlier are not completely dominated by finite size effects. As we are able to routinely observe nucleation in this cell when the kaolinite slab is present, but not homogeneously, we are left to conclude that kaolinite significantly enhances the rate. We are not currently in a position, however, to go beyond this qualitative level.

As a final test of the finite size effects, we doubled the lateral cell dimensions and used a configuration from one of our heterogeneous simulations, replicated in both dimensions to fill the larger cell, as an initial configuration. Taking the configuration that we have determined to be representative of the transition region from the committor analysis as a 'seed' configuration for the larger cell, we still see ice growth in the same manner as the small cells. The fact that we see growth and not a collapse of the crystal suggests that the prism face is stable on the hydroxyl-terminated (001) kaolinite face and is not solely stabilised by periodic boundary effects present in the small cells.

IV. Conclusions

We have investigated ice nucleation in thin water films, both homogeneously and heterogeneously in the presence of a kaolinite slab, using regular molecular dynamics simulation. We have performed many simulations on the order of one microsecond, observing many nucleation events. In agreement with previous simulation studies and recent experiments, in the case of homogeneous nucleation we see a mixture of cubic and hexagonal arrangements. Contrary to expectation, at the kaolinite surface we always see growth along the prism face of ice, suggesting that the source of kaolinite's good ice nucleating ability does not lie with its good epitaxial match with the basal face of ice. By monitoring the density of water above the kaolinite slab during the nucleation event, we see that changes in the second water layer appear crucial to the nucleation mechanism. The growth of the prism face rather than the basal face is due to the ability of the former to bind favourably to both the surface and water layers above, as well as having a higher coverage. The observed structural changes away from the surface have been explained as allowing the average number of water molecules in the first layer to decrease, which subsequently allows the remaining water molecules to form the favoured ice-like structure. We have, however, seen that finite size effects are non-negligible in these simulations, with no nucleation observed upon moving to bigger cells. Nevertheless, the fact that we do not observe homogeneous nucleation in the cell size used for the heterogeneous nucleation simulations suggests that the results on kaolinite are not entirely dominated by the finite size effects. This result also shows that kaolinite is a potent ice nucleating agent.

Given the finite size effects, it would be highly desirable to implement a free energy method that could definitively probe the heterogeneous nucleation mechanism proposed here. Even with the current state-of-the-art in free energy methods and advanced sampling techniques, freezing water is still likely to be difficult. The reason for this is that slow dynamics offered by the hydrogen bonding network present in supercooled water makes it very difficult for methods such as umbrella sampling and metadynamics to equilibrate the system as it is pushed along the chosen order parameter.[80] Furthermore, advanced sampling techniques that exploit natural dynamics, such as transition path sampling[81] or forward flux sampling[82] are likely to suffer as the actual transition time is relatively long (tens of nanoseconds, as seen in Fig. 5), which may make sampling computationally prohibitive. One way to circumvent this problem is through the use of a coarse grained potential such as the mW model[41] which, by treating hydrogen bonding in a mean-field sense, reduces the complexity of the underlying potential energy surface and results in faster dynamics. This has already been used to good effect with both direct molecular dynamics (see *e.g.* ref. 83) and forward flux sampling[84] for homogeneous nucleation. Such methods could be used to verify previous homogeneous simulations that suffer from similar finite size effects.[30–32] This approach is unlikely to work in the case of heterogeneous nucleation on substrates such as clays, however, where electrostatics are dominant. How to proceed in such cases is at present unclear, but given the industrial and environmental implications of ice formation, the topic deserves a major research effort.

Finally, the fact that the pristine kaolinite surface promotes the growth of the prism face over the basal face may have consequences for the macroscopic crystal structure of ice that forms. Ice exhibits a complex habit diagram[85] and as the surface cleavage energies of the prism and basal faces are very similar[86] it is possible that different heterogeneous ice nucleating agents could tip the balance to favour different ice habits under the same conditions. As the macroscopic structure of an ice crystal can affect its light scattering properties, understanding the effect of ice nucleating agents may be important for global climate models. Future calculations will probe the influence of other ice nucleating agents on nucleation and growth processes with the aspiration of comparing with measurements from cloud chamber experiments.

Acknowledgements

The authors would like to thank Dr. Xiao-Liang Hu for many helpful discussions when commencing this work and for providing Fig. S2.† This research used resources of the National Energy Research Scientific Computing Center, which is supported by the Office of Science of the U.S. Department of Energy under Contract No. DE-AC02-05CH11231. We are grateful to the London Centre for Nanotechnology and UCL Research Computing for computational resources. *Via* our membership of the UK's HPC Materials Chemistry Consortium, which is funded by EPSRC (EP/F067496), this work made use of the facilities of HECToR, the UK's national high-performance computing service, which is provided by UoE HPCx Ltd. at the University of Edinburgh, Cray Inc., and NAG Ltd., and funded by EPSRC's High End Computing Programme. The authors also acknowledge the use of the UCL *Legion* High Performance Computing Facility, and associated support services, in the completion of this work. S.M.K. was supported fully by the U.S. Department of Energy, Office of Basic Energy Sciences (BES), Division of Chemical Sciences, Geosciences, and Biosciences. S.J.C. was supported by a student fellowship funded jointly by UCL and BES. A.M. is supported by the European Research Council and the Royal Society through a Royal Society Wolfson Research Merit Award.

References

1 K. A. Sharp, *Proc. Natl. Acad. Sci. U. S. A.*, 2011, **108**, 7281.
2 Y. Liou, A. Tocilj, P. Davies and Z. Jia, *Nature*, 2000, **406**, 322.
3 B. C. Christner, C. E. Morris, C. M. Foreman, R. Cai and D. C. Sands, *Science*, 2008, **319**, 1214.
4 J. P. D. Abbatt, *Chem. Rev.*, 2003, **103**, 4783.
5 H. R. Pruppacher and J. D. Klett, *Microphysics of Clouds and Precipitation - Second Revised and Enlarged Edition with an Introduction to Cloud Chemistry and Cloud Electricity*, Kluwer Academic Publishers, Dordrecht, The Netherlands, 1997.
6 W. Cantrell and A. Heymsfield, *Bull. Am. Meteorol. Soc.*, 2005, **86**, 795.
7 F. Zimmermann, S. Weinbruch, L. Schuetz, H. Hofmann, M. Ebert, K. Kandler and A. Worringen, *J. Geophys. Res.: Atmos*, 2008, **113**, D23204.
8 M. L. Eastwood, S. Cremel, C. Gehrke, E. Girard and A. K. Bertram, *J. Geophys. Res.: Atmos*, 2008, **113**, D22203.
9 B. J. Murray, D. O'Sullivan, J. D. Atkinson and M. E. Webb, *Chem. Soc. Rev.*, 2012, **41**, 6519.
10 P. DeMott, D. Rogers and S. Kreidenweis, *J. Geophys. Res.: Atmos*, 1997, **102**, 19575.

11 D. Rogers, P. DeMott, S. Kreidenweis and Y. Chen, *Geophys. Res. Lett.*, 1998, **25**, 1383.
12 R. W. Saunders, O. Möhler, M. Schnaiter, S. Benz, R. Wagner, H. Saathoff, P. J. Connolly, R. Burgess, B. J. Murray, M. Gallagher, R. Wills and J. M. C. Plane, *Atmos. Chem. Phys.*, 2010, **10**, 1227.
13 M. B. Baker, *Science*, 1997, **276**, 1072.
14 J. Verlinde, J. Y. Harrington, G. M. McFarquhar, V. T. Yannuzzi, A. Avramov, S. Greenberg, N. Johnson, G. Zhang, M. R. Poellot, J. H. Mather, D. D. Turner, E. W. Eloranta, B. D. Zak, A. J. Prenni, J. S. Daniel, G. L. Kok, D. C. Tobin, R. Holz, K. Sassen, D. Spangenberg, P. Minnis, T. P. Tooman, M. D. Ivey, S. J. Richardson, C. P. Bahrmann, M. Shupe, P. J. DeMott, A. J. Heymsfield and R. Schofield, *Bull. Am. Meteorol. Soc.*, 2007, **88**, 205.
15 G. M. McFarquhar, G. Zhang, M. R. Poellot, G. L. Kok, R. McCoy, T. Tooman, A. Fridlind and A. J. Heymsfield, *J. Geophys. Res.: Atmos*, 2007, **112**, D24201.
16 B. J. Murray, T. W. Wilson, S. Dobbie, Z. Cui, S. M. R. K. Al-Jumur, O. Möhler, M. Schnaiter, R. Wagner, S. Benz, M. Niemand, H. Saathoff, V. Ebert, S. Wagner and B. Kaercher, *Nat. Geosci.*, 2010, **3**, 233.
17 Y.-S. Choi and C.-H. Ho, *Geophys. Res. Lett.*, 2006, **33**, L21811.
18 J. Lee, P. Yang, A. E. Dessler, B.-C. Gao and S. Platnick, *J. Atmos. Sci.*, 2009, **66**, 3721.
19 P. A. Thiel and T. E. Madey, *Surf. Sci. Rep.*, 1987, **7**, 211.
20 M. Henderson, *Surf. Sci. Rep.*, 2002, **46**, 1.
21 A. Hodgson and S. Haq, *Surf. Sci. Rep.*, 2009, **64**, 381.
22 A. Verdaguer, G. Sacha, H. Bluhm and M. Salmeron, *Chem. Rev.*, 2006, **106**, 1478–1510.
23 G. Ewing, *Chem. Rev.*, 2006, **106**, 1511–1526.
24 J. Carrasco, A. Hodgson and A. Michaelides, *Nat. Mater.*, 2012, **11**, 667.
25 A. Michaelides, *Faraday Discuss.*, 2007, **136**, 287.
26 S. Nie, P. J. Feibelman, N. C. Bartelt and K. Thürmer, *Phys. Rev. Lett.*, 2010, **105**, 026102.
27 P. J. Feibelman, N. C. Bartelt, S. Nie and K. Thürmer, *J. Chem. Phys.*, 2010, **133**, 154703.
28 J. Carrasco, A. Michaelides, M. Forster, S. Haq, R. Raval and A. Hodgson, *Nat. Mater.*, 2009, **8**, 427.
29 M. Matsumoto, S. Saito and I. Ohmine, *Nature*, 2002, **416**, 409.
30 L. Vrbka and P. Jungwirth, *J. Phys. Chem. B*, 2006, **110**, 18126.
31 E. Pluhařová, L. Vrbka and P. Jungwirth, *J. Phys. Chem. C*, 2010, **114**, 7831.
32 S. Bauerecker, P. Ulbig, V. Buch, L. Vrbka and P. Jungwirth, *J. Phys. Chem. C*, 2008, **112**, 7631.
33 E. B. Moore and V. Molinero, *Phys. Chem. Chem. Phys.*, 2011, **13**, 20008.
34 E. B. Moore and V. Molinero, *J. Chem. Phys.*, 2010, **132**, 244504.
35 A. Reinhardt and J. P. K. Doye, *J. Chem. Phys.*, 2012, **136**, 054501.
36 R. Radhakrishnan and B. Trout, *J. Am. Chem. Soc.*, 2003, **125**, 7743.
37 D. Quigley and P. M. Rodger, *J. Chem. Phys.*, 2008, **128**, 154518.
38 J. Y. Yan and G. N. Patey, *J. Phys. Chem. Lett.*, 2011, **2**, 2555.
39 J. Y. Yan and G. N. Patey, *J. Phys. Chem. A*, 2012, **116**, 7057.
40 E. González Solveyra, E. de la Llave, D. A. Scherlis and V. Molinero, *J. Phys. Chem. B*, 2011, **115**, 14196.
41 V. Molinero and E. B. Moore, *J. Phys. Chem. B*, 2009, **113**, 4008.
42 A. Laio and M. Parrinello, *Proc. Natl. Acad. Sci. U. S. A.*, 2002, **99**, 12562.
43 G. Torrie and J. Valleau, *J. Comput. Phys.*, 1977, **23**, 187.
44 P. A. Smithson, *Int. J. Climatol.*, 2002, **22**, 1144.
45 B. J. Murray, S. L. Broadley, T. W. Wilson, J. D. Atkinson and R. H. Wills, *Atmos. Chem. Phys.*, 2011, **11**, 4191.
46 X. L. Hu and A. Michaelides, *Surf. Sci.*, 2007, **601**, 5378.
47 X. L. Hu and A. Michaelides, *Surf. Sci.*, 2008, **602**, 960.
48 X. L. Hu and A. Michaelides, *Surf. Sci.*, 2010, **604**, 111.
49 T. Croteau, A. K. Bertram and G. N. Patey, *J. Phys. Chem. A*, 2008, **112**, 10708.
50 T. Croteau, A. K. Bertram and G. N. Patey, *J. Phys. Chem. A*, 2009, **113**, 7826.
51 T. Croteau, A. K. Bertram and G. N. Patey, *J. Phys. Chem. A*, 2010, **114**, 2171.
52 J. L. F. Abascal and C. Vega, *J. Chem. Phys.*, 2005, **123**, 234505.
53 R. Cygan, J. Liang and A. Kalinichev, *J. Phys. Chem. B*, 2004, **108**, 1255.
54 J. P. Larentzos, J. A. Greathouse and R. T. Cygan, *J. Phys. Chem. C*, 2007, **111**, 12752.
55 J. Wang, A. G. Kalinichev, R. J. Kirkpatrick and R. T. Cygan, *J. Phys. Chem. B*, 2005, **109**, 15893.
56 D. Argyris, D. R. Cole and A. Striolo, *ACS Nano*, 2010, **4**, 2035.
57 B. Rotenberg, A. J. Patel and D. Chandler, *J. Am. Chem. Soc.*, 2011, **133**, 20521.
58 H. A. Lorentz, *Ann. Phys. Chem.*, 1881, **248**, 127.
59 D. Berthelot and C. R. Hebd, *Séanc. Acad. Sci.*, 1898, **126**, 1703.

60 The lateral cell dimensions for the homogeneous simulations were obtained from a 0.01 K NPT simulation of a proton ordered configuration of hexagonal ice. In the case of the heterogeneous simulations, the lateral cell dimensions are constrained to be commensurate with the kaolinite slab. The kaolinite structure used was based upon the experimental structure of Bish,[87] with the α and γ angles altered slightly to make the cell orthogonal.
61 I. Yeh and M. Berkowitz, *J. Chem. Phys.*, 1999, **111**, 3155.
62 S. Miyamoto and P. A. Kollman, *J. Comput. Chem.*, 1992, **13**, 952.
63 B. Hess, H. Bekker, H. J. C. Berendsen and J. G. E. M. Fraaije, *J. Comput. Chem.*, 1997, **18**, 1463.
64 B. Hess, C. Kutzner, D. van der Spoel and E. Lindahl, *J. Chem. Theory Comput.*, 2008, **4**, 435.
65 T. L. Malkin, B. J. Murray, A. V. Brukhno, J. Anwar and C. G. Salzmann, *Proc. Natl. Acad. Sci. U. S. A.*, 2012, **109**, 1041.
66 W. F. Kuhs, C. Sippel, A. Falenty and T. C. Hansen, *Proc. Natl. Acad. Sci. U. S. A.*, 2012, **109**, 21259.
67 M. A. Carignano, *J. Phys. Chem. C*, 2007, **111**, 501.
68 This high density peak is also a feature of water at kaolinite at 300 K, which we show in the supporting information (ESI†), along with a density profile from a DFT-MD simulation at 330 K that also exhibits such a peak.
69 D. Pan, L.-M. Liu, B. Slater, A. Michaelides and E. Wang, *ACS Nano*, 2011, **5**, 4562.
70 C. L. Bishop, D. Pan, L. M. Liu, G. A. Tribello, A. Michaelides, E. G. Wang and B. Slater, *Faraday Discuss.*, 2009, **141**, 277.
71 H. Nada and Y. Furukawa, *J. Cryst. Growth*, 1996, **169**, 587.
72 D. Rozmanov and P. G. Kusalik, *J. Chem. Phys.*, 2012, **137**, 094702.
73 E. B. Moore, E. de la Llave, K. Welke, D. A. Scherlis and V. Molinero, *Phys. Chem. Chem. Phys.*, 2010, **12**, 4124.
74 The "H-up bilayer" structure was also tested, but this was always less stable than the "H-down" structure and so has been omitted for clarity.
75 M. Salmeron, H. Bluhm, M. Tatarkhanov, G. Ketteler, T. K. Shimizu, A. Mugarza, X. Deng, T. Herranz, S. Yamamoto and A. Nilsson, *Faraday Discuss.*, 2009, **141**, 221–229.
76 J. P. Perdew, K. Burke and M. Ernzerhof, *Phys. Rev. Lett.*, 1996, **77**, 3865–3868.
77 Water molecules are defined as being in the first layer if their oxygen atoms are within 5 Å of the average height of the kaolinite surface oxygen atoms. This corresponds to a cutoff that is approximately halfway between the first and second ice layers.
78 We must emphasise that this is an analogy to the grand canonical ensemble and that the first water layer and the above liquid are strongly coupled.
79 J. L. F. Abascal, E. Sanz, R. G. Fernández and C. Vega, *J. Chem. Phys.*, 2005, **122**, 234511.
80 A. Reinhardt, J. P. K. Doye, E. G. Noya and C. Vega, *J. Chem. Phys.*, 2012, **137**, 194504.
81 P. G. Bolhuis, D. Chandler, C. Dellago and P. L. Geissler, *Annu. Rev. Phys. Chem.*, 2002, **53**, 291–318.
82 R. J. Allen, C. Valeriani and P. R. ten Wolde, *J. Phys.: Condens. Matter*, 2009, **21**, 463102.
83 E. B. Moore and V. Molinero, *Nature*, 2011, **479**, 506.
84 T. Li, D. Donadio, G. Russo and G. Galli, *Phys. Chem. Chem. Phys.*, 2011, **13**, 19807.
85 M. P. Bailey and J. Hallett, *J. Atmos. Sci.*, 2009, **66**, 2888.
86 D. Pan, L.-M. Liu, G. A. Tribello, B. Slater, A. Michaelides and E. Wang, *J. Phys.: Condens. Matter*, 2010, **22**, 074209.
87 D. L. Bish, *Clays Clay Miner.*, 1993, **41**, 738.

Faraday Discussions RSC Publishing

PAPER

Lifetimes and lengthscales of structural motifs in a model glassformer

Alex Malins,[ab] Jens Eggers,[c] Hajime Tanaka[d] and C. Patrick Royall[aef]

Received 7th May 2013, Accepted 7th June 2013
DOI: 10.1039/c3fd00078h

We use a newly-developed method to identify local structural motifs in a popular model glassformer, the Kob–Andersen binary Lennard-Jones mixture. By measuring the lifetimes of a zoo of clusters, we find that 11-membered bicapped square antiprisms, denoted as 11A, have longer lifetimes on average than other structures considered. Other long-lived clusters are similar in structure to the 11A cluster. These clusters group into ramified networks that are correlated with slow particles and act to retard the motion of neighbouring particles. The structural lengthscale associated with these networks does not grow as fast as the dynamical lengthscale ξ_4 as the system is cooled, in the range of temperatures our molecular dynamics simulations access. Thus we find a strong, but indirect, correlation between static structural ordering and slow dynamics.

1 Introduction

The nature of the rapid increase in viscosity as liquids are cooled toward the glass transition is the subject of many theoretical approaches, however there is no consensus on its fundamental mechanism.[1–4] One plausible scenario is the emergence of self-induced memory effects upon supercooling of liquids, which causes slow dynamics.[5] However the recent discovery of dynamic heterogeneities, *i.e.*, spatial heterogeneities in the relaxation dynamics that emerge on supercooling,[6–9] is suggestive of the importance of a growing dynamic length scale in the slowing down approaching the glass transition.

In addition to these dynamical phenomenona, the idea of a structural change leading to vitrification has a long history.[10] Sir Charles Frank suggested that upon supercooling liquids would form polyhedral motifs such as icosahedra that do not fill space. A related approach, geometric frustration,[11] suggests that the glass transition can be thought of as a manifestation of a crystallisation-like transition

[a]*School of Chemistry, Cantock's Close, Bristol, UK, BS8 1TS*
[b]*Bristol Centre for Complexity Science, University of Bristol, Bristol, UK, BS8 1TS*
[c]*School of Mathematics, University Walk, Bristol, UK, BS8 1TW*
[d]*Institute for Industrial Science, The University of Tokyo, Komaba-4-6-1, Meguro-ku, Tokyo, 153-8505, Japan*
[e]*HH Wills Physics of Laboratory, Tyndall Avenue, Bristol, UK, BS8 1TL*
[f]*Centre for Nanoscience of Quantum Information, Tyndall Avenue, Bristol, UK, BS8 1FD*

that would occur in curved space (where the structural motifs tessellate[12]), but that growth of the "crystal nuclei" of polyhedral motifs is frustrated in Euclidean space. Another approach based on frustration against crystallization (due to random disorder effects or competing orderings)[13,14] addresses the origin of slow dynamics and the physical factors controlling glass-forming ability within the same framework.

It has become clear that a range of glass formers exhibit a change in structure upon the emergence of slow dynamics,[15,16] and it is debated as to whether there exists a static lengthscale that underlies the growing lengthscale for the dynamical correlations. Broadly speaking two types of structure have been identified: spatially extendable crystal-like ordering,[15,17–19] and non-extendable polyhedral ordering.[20–28] For the former it has been suggested that critical-like fluctuations of crystalline order are the origin of dynamic heterogeneities in certain classes of supercooled liquid.[17] The latter concerns particles organised into polyhedra that cannot tile Euclidean space due to geometrical constraints.[10,11] Instead they form ramified structures with a fractal dimension that is less that the dimensionality of the system, *i.e.* non-extendable ordering. Some metallic glasses have been shown to exhibit this second type of ordering.[29,30] The exact relationship of the polyhedral order to the dynamic heterogeneities is unclear, however measurements have shown that the polyhedral domains are slow to relax.[16] Although the two types of orderings have a different nature, we note that both are induced to lower the free energy locally in the situation where its global minimization (crystallization) is prohibited.[14] In relation to this, we note that even polyhedral ordering often has some connection to crystalline order, for example icosahedra in quasicrystal approximants.[26] Furthermore, even in the case where extendable order dominates slow dynamics, non-extendable polyhedral order competes with extendable crystal-like order in systems such as 2D spin liquids[15] and polydisperse hard spheres.[19]

One of the main difficulties with identifying structural correlations in supercooled liquids is that it is not known *a priori* which (if any) type of static order is important for the dynamic slowdown. Frequently studies have employed structure detection methods, such as Voronoi face analysis,[31] common neighbour analysis,[32] bond orientational order analysis,[33] or the topological cluster classification employed here[34–36] that search for predefined types of structural "motif" and investigate the change in population as a function of temperature.[20,22,28,35,37] However recently a number of "order-agnostic" schemes have been devised to identify structural correlations without first having to define what structures will be searched for.[38–44] In relation to this, it has recently been pointed out that to access such hidden structural ordering in apparently random structures it is crucial to focus on many-body correlations.[45,46]

To strengthen the link between structure and glassy behaviour three types of evidence have been presented. Firstly, many-body structural correlation functions have been presented that clearly show structural changes occur on supercooling towards the glass transition.[16,26,46] Secondly, dynamically slow regions have been correlated with different types of local ordering.[16,24] Thirdly, the presence of structural and dynamic length scales that grow similarly has been sought.[15,17,18] This is motivated by strong evidence that, for sufficient cooling (which leads to relaxation timescales that may or may not be accessible to computer simulations as we employ here), an increasing dynamical lengthscale necessitates an increasing lower bound to a structural lengthscale.[47] The case of growing structural and dynamic length scales is controversial. One of us has identified a direct correspondence between the growing dynamical and structural lengthscales in polydisperse systems displaying crystal-like ordering,[17,18,45] while others have claimed that structural lengthscales are decorrelated from the dynamic lengthscales and only grow weakly on cooling.[48–50]

Local structural ordering has yet to be found in some glassforming systems. For these systems no one-to-one correspondence between an order-agnostic structural lengthscale and the dynamical lengthscale has been found,[41,44,48,51] but static perturbation analysis has suggested a growing structural lengthscale.[52]

The Kob–Andersen binary Lennard-Jones system[53] of interest here is known to display polyhedral ordering.[16] In this system, weakly growing structural lengthscales have been identified by the order-agnostic "point-to-set" analysis,[43] while other approaches using static perturbation of inherent structures[39] and finite size scaling[54] find a stronger increase in static lengthscales. The polyhedral ordering in the Kob–Andersen (KA) mixture,[53] take the form of the bicapped square antiprism ["11A" in the topological cluster classification (TCC) nomenclature[34–36,55] owing to its original identification as minimum energy cluster of the Morse potential[56,57]]. This structure has been linked to slow dynamics[16] and frustrates crystallisation, which in the KA mixture occurs spontaneously by phase separation into two face-centred cubic lattices.[58] We note that crystals based on 11A could in principle coexist with other structures [the stoichimetry of the 11A crystal with a small particle in the centre (see Section 4.3 and Fig. 5) is not compatible with the KA mixture[59]]. Another possibility is four-fold symmetric crystals which have been predicted as low-lying energy minima for the KA mixture.[60] An order parameter associated with the formation of 11A clusters has been shown to control the dynamical phase transition in trajectory-space, a hallmark of dynamical facilitation theory for the glass transition,[61] indicating that the transition has both structural and dynamical character.[62] Here we study the spatial correlations between the domains of 11A in the supercooled liquid.

We consider the lifetimes of a multitude of structures in the supercooled liquid using the TCC algorithm. We detail our simulation protocol in section 2, and briefly review the KA model's dynamical behaviour in section 3. We use the TCC to identify any structural changes that occur in these mixtures on cooling towards the glass transition in section 4. We then study in section 4.2 the lifetimes of the clusters that are found at deeply supercooled state points in order to gauge which structures are likely candidates to be associated with slow domains of dynamic heterogeneities. Finally we analyse how correlation lengths for the domains of structured particles are related to the growing dynamic lengthscale in section 5 before concluding.

2 Model and simulation details

The Kob–Andersen (KA) binary mixture is composed of 80% large (A) and 20% small (B) particles of the same mass m.[53] The nonadditive Lennard-Jones interactions between each species, and the cross interaction, are given by $\sigma_{AA} = \sigma$, $\sigma_{AB} = 0.8\sigma$, $\sigma_{BB} = 0.88\sigma$, $\varepsilon_{AA} = \varepsilon$, $\varepsilon_{AB} = 1.5\varepsilon$, and $\varepsilon_{BB} = 0.5\varepsilon$. The results are quoted in reduced units with respect to the A particles, *i.e.* we measure length in units of σ, energy in units of ε, time in units of $\sqrt{m\sigma^2/\varepsilon}$, and set Boltzmann's constant k_B to unity. The interactions are truncated and smoothed using the Stoddard–Ford method.[63] The truncation lengths are in proportion to the interaction lengths,[53] *i.e.* $r_{\rm tr}^{\rm AA} = 2.5$, $r_{\rm tr}^{\rm AB} = 2.0$ and $r_{\rm tr}^{\rm BB} = 2.2$. The simulations consist of $N = 10\,976$ particles in 3D with periodic boundary conditions such that $N_{\rm A} = 8781$ and density $\rho = 1.2$. The α-relaxation time $\tau_\alpha^{\rm A}$ for each state point is defined by fitting the Kohlrausch–Williams–Watts stretched exponential to the decay of the intermediate scattering function (ISF) of the A-type particles.

Equilibrated samples at each temperature were prepared by simulating for $100\tau_\alpha^{\rm A}$ in the canonical *NVT*-ensemble using the Nosé–Poincaré thermostat with coupling parameter 1.0.[64] The thermostat was switched off, then further equilibration was performed in the microcanonical *NVE*-ensemble using the velocity Verlet algorithm for $1000\tau_\alpha^{\rm A}$. On completion of the equilibration process, trajectories of length $300\tau_\alpha^{\rm A}$ were sampled for analysis. Colder state points were obtained in a step-wise fashion by quenching instantaneously from an equilibrated configuration of the previous higher temperature state point. The stability of the deep quenches was checked by ensuring there was no time evolution in the ISF, the partial radial distribution functions $g_{\rm AA}(r)$, $g_{\rm AB}(r)$ and $g_{\rm BB}(r)$, or the number of clusters detected by the TCC algorithm across the trajectories. Crystallisation was not seen in any of the simulations.

3 Dynamical behaviour

In Fig. 1 the relaxation time $\tau_\alpha^{\rm A}$ as a function of inverse temperature is plotted. For the equilibrium liquid state points at high temperatures, the relaxation times are well fitted by an Arrhenius function. As the temperature is lowered a cross-over in the dynamical behaviour occurs and the relaxation times increase faster than predicted by the Arrhenius equation.[2,65,66]

We fit the two regimes for the relaxation time delimited by an onset temperature for slow dynamics T^*.[16] For $T > T^*$ an Arrhenius form is used, while for lower temperatures the Vogel–Fulcher–Tammann (VFT) equation is fitted.[67–69]

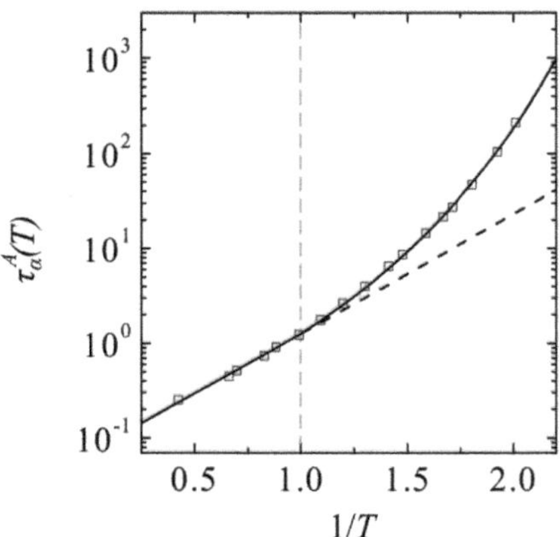

Fig. 1 Increase in the alpha relaxation time, $\tau_\alpha^{\rm A}$, on cooling. The data are fitted with a hybrid Arrhenius-VFT fit (solid lines). The dotted lines indicate the relaxation times predicted by the high-*T* Arrhenius fit. Orange dashed line indicates the crossover from Arrhenius to super-Arrhenius (T^*).

$$\tau_\alpha^{A} = \begin{cases} \tau_\infty \exp(E_\infty/T) & \text{for } T \geq T^*, \\ \tau_\infty{}' \exp\left(\dfrac{DT_0}{T - T_0}\right) & \text{for } T < T^*. \end{cases} \quad (1)$$

We set $T^* = 1.00$ and fit the Arrhenius equation finding $\tau_\infty = 0.0693$ and $E_\infty = 2.91$. For the VFT fit the fragility parameter is found to be $D = 7.48$, the VFT temperature is $T_0 = 0.325$ and $\tau_\infty{}'$ is set to ensure continuity of the fit at T^*. While it is possible to use other fitting forms,[70,71] and although VFT may be physically reasonable,[1] there remains no clear consensus as to which form best describes data such as those plotted in Fig. 1.[72] Nonetheless, following our previous study,[28] here we choose this fitting procedure to reflect the onset of slow dynamics (super-Arrhenius) for $T < 1$[2,66] and, over the range of temperatures we consider, find good agreement with the assumption of a crossover to a VFT regime.

4 Structural analysis

4.1 Fraction of particles participating within clusters

We analyse how the particles in the supercooled liquids are structured using the topological cluster classification algorithm. This algorithm identifies a number of local structures as shown in Fig. 2, including those which are the minimum energy clusters for $m = 5$ to 13KA particles in isolation. The first stage of the TCC algorithm is to identify the bonds between neighbouring particles. The bonds are detected using a modified Voronoi method with a maximum bond length cut-off of $r_c = 2.0$ for all types of interaction (AA, AB and BB).[34,36,55] A parameter which controls identification of four- as opposed to three-membered rings f_c is set to unity thus yielding the direct neighbours of the standard Voronoi method.[36,55,73–75]

In Fig. 3 we plot the fraction of particles detected within each type of cluster, N_C/N. The onset temperature for slow dynamics is indicated by the orange dotted line on each of the plots. It is clear from these order parameters that the liquid sees continuous changes in local structure as it is cooled. The majority of clusters see an increase in their numbers upon cooling. All of the particles are identified within certain simple structures such as the 5A triangular bipyramid irrespective of temperature, while other more complicated clusters, such are 10K, 11B and HCP, are almost never seen.

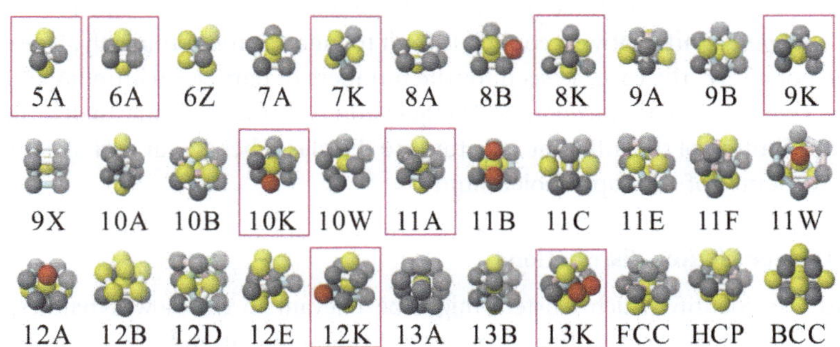

Fig. 2 Clusters detected by the topological cluster classification. Highlighted are minimum energy clusters for the Kob–Andersen system. The colours of the particles and the bonds are pertinent to the detection method of the clusters.[36,55]

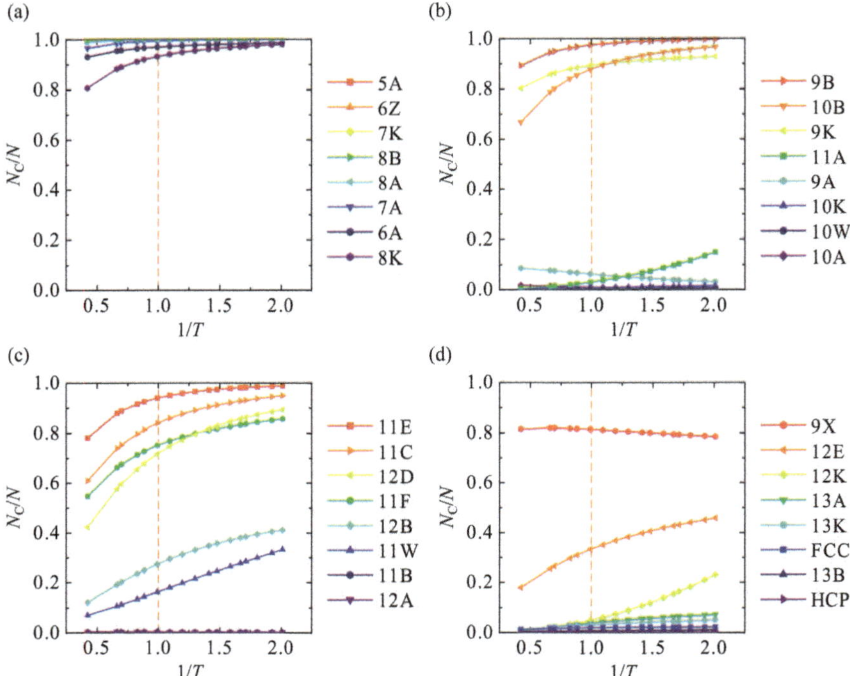

Fig. 3 The fraction of particles participating in each cluster type for the KA mixture. The dotted orange lines mark the onset temperature of slow dynamics T^*. (a) Clusters 5A to 8K, (b) 9A to 11A, (c) 11B to 12D, (d) 12E to 13K and the crystal clusters.

Given the variety and range of structural changes that occur, it is not clear *a priori* which of the structures, if any, are important for the formation of dynamic heterogeneities on cooling. It should not be assumed that just because structural changes occur within the supercooled regime that they are responsible for the formation of dynamic heterogeneities, as structural changes also occur in the Arrhenius regime which is not characterised by glassy behaviour. The numbers of particles within 11A, 11W, 12B, 12E and 12K clusters show the largest relative increases in the super-Arrhenius regime. Moreover, the rate of increase in these clusters grows upon further supercooling.

We also find clusters in which almost all particles participate, while other clusters are only seen in trace quantities. In general the trend for clusters where N_C/N changes significantly on cooling is for it to increase monotonically. However this is not always the case, as seen for the numbers of particles within 9A and 9X clusters which decrease on cooling. The question remains as to how to determine the contribution of the different structures to the glassy behaviour and dynamical heterogeneities of the supercooled liquid.

4.2 Cluster lifetime distributions

In order to identify which clusters might be relevant to the slow dynamics, we employ the *dynamic* topological cluster classification algorithm[28,36] to measure the lifetimes of the different TCC clusters at the lowest temperature state point. A lifetime τ_ℓ is assigned to each "instance" of a cluster, where an instance is defined by the unique indices of the particles within the cluster and the type of TCC cluster. Each

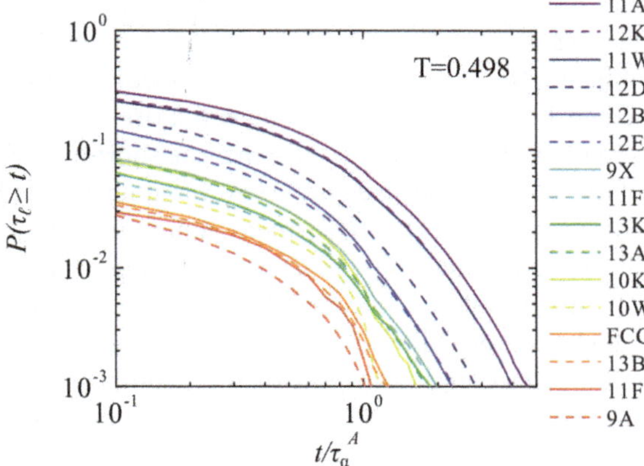

Fig. 4 Lifetime autocorrelation functions for the clusters $P(\tau_\ell \geq t)$ for the lowest temperature state point $T = 0.498$.

instance cluster occurs between two frames in the trajectory and the lifetime is the time difference between these frames. Any periods where the instance is not detected by the TCC algorithm are shorter than τ_α^A in length, and no subset of the particles becomes un-bonded from the others during the lifetime of the instance.

The measurement of lifetimes for all the instances of clusters in these $N = 10976$ simulations is intensive in terms of the quantity of memory required to store the instances, and the number of searches through the memory required by the algorithm each time an instance of a cluster is found to see if it existed earlier in the trajectory. Therefore we do not measure lifetimes for the clusters where $N_C/N > 0.8$, since the vast majority of particles are found within such clusters and it is not immediately clear how dynamic heterogeneities could be related to structures that are pervasive throughout the whole liquid.

In Fig. 4 we plot the lifetime autocorrelation function $P(\tau_\ell \geq t)$ for $T = 0.498$. Fig. 4 clearly shows that the most persistent or the longest lived of the different types of clusters are the 11A bicapped square antiprisms. All other clusters display lifetime autocorrelation functions that decay more quickly than 11A. The long-time tail of the 11A autocorrelation function indicates that some of these clusters preserve their local structure on timescales far longer than τ_α^A. As we shall see below, this effect is enhanced when the 11A group into domains.

The two next-slowest decaying clusters are 11W and 12K. 12K is a KA minimum energy cluster, formed by bonding an additional particle to 11A. There is a high degree of overlap between the 11W clusters and the 11A clusters as their bonding is similar, and only small fluctuations are required for an 11A to be reclassified as an 11W. However the faster-decaying clusters also contain KA minimum energy clusters, for example 13K and 10K. Moreover the lifetimes of all cluster types hold no simple relationship to their size and frequency of occurrence. For example the $n = 11$ particle 11F cluster is much more numerous than 11A, yet displays far quicker decay of $P(\tau_\ell \geq t)$. There is also no monotonic trend in the lifetime of the ground state clusters with the cluster size, as the 10K and 13K decay faster than the 12K and 11A. These results clearly demonstrate that the average lifetime of each type of cluster is a property of the local ordering of the particles rather than the size of the cluster or its pervasiveness.

The fast initial drops of $P(\tau_\ell \geq t)$ reflect the existence of large numbers of clusters with lifetimes $\tau_\ell \ll \tau_\alpha^A$. The lifetimes of these clusters are comparable to the timescale for beta-relaxation where the particles fluctuate within their cage of neighbours. It could be argued that these clusters arise spuriously due to the microscopic fluctuations within the cage, and that the short-lived clusters are not representative of the actual liquid structure. However almost no 11A are found at higher temperatures, *cf.* Fig. 3(b), where microscopic fluctuations in the beta-regime also occur. We have not yet found a way to distinguish between the short and long-lived 11A structurally, so we conclude that the measured distribution of 11A lifetimes, which includes short-lived clusters, is representative of the true lifetime distribution. However, given the structural similarity of 11A and 11W, small fluctuations leading to reclassification could contribute to the drop at short times. We will see below that as 11A overlap, one particle may be a member of multiple clusters and that the majority of particles found in short-lived 11A also participate in longer-lived 11A. In other words short- and long-lived 11A mainly lie in the same regions of the liquid.

Another interpretation for the initial drop in $P(\tau_\ell > t)$ of the clusters is that our intuition that there is constant local structure in the beta-regime, which then relaxes on the timescale of the alpha-regime, is incorrect. It may be the case that microscopic ballistic and "cage-rattling" motions are enough to reorder local structures without relying on the "cage-hopping" motions of the alpha-regime. It has been seen in previous studies that deeply supercooled liquids can crystallise on a timescale before the diffusive range of the mean squared displacement is reached, and occurs with most particles moving by less than one diameter.[76–79] Those results demonstrate that significant changes in local structure (*i.e.* liquid-like to crystalline) are possible with only small movements of the particles, which could explain the initial drops of $P(\tau_\ell \geq t)$ that occur on a timescale $\simeq 0.1\tau_\alpha^A$.

4.3 Composition and dynamics of particles in long-lived clusters

The structural analysis above was performed by treating all particles identically with the TCC algorithm. Here we examine the composition of the 11A bicapped square antiprisms in terms of A- and B-species. The 11A cluster consists of a central particle surrounded by 10 outer (or shell) particles. We find that the central particle is a B-species in >99% of all instances of 11A clusters. This is a different composition than the crystal structures found to be low-lying energy minima for the KA mixture,[60] and may be related to the stoichiometry of the system. Thus the 11A we find are unrelated to any underlying crystal.

In Fig. 5 we plot the compositions of the shell particles of the 11A clusters. The majority of 11A clusters have $m_A = 10$ A-species in the shell of the cluster. We note that this arrangement maximises the number of AB bonds for the central B-particle, which is energetically favourable for the central B-particle with the KA Lennard-Jones interactions.

In Fig. 6 we examine how the dynamics of the 11A clusters translates into the dynamics of individual particles. We show in Fig. 6(a) that the number of particles within 11A clusters as a function of the cluster lifetime. Although there is a fast initial drop in the lifetime autocorrelation function of 11A clusters on the beta-relaxation timescale (Fig. 4, solid purple line), as the 11A overlap there remains a significant fraction of the particles these clusters with lifetimes comparable to the

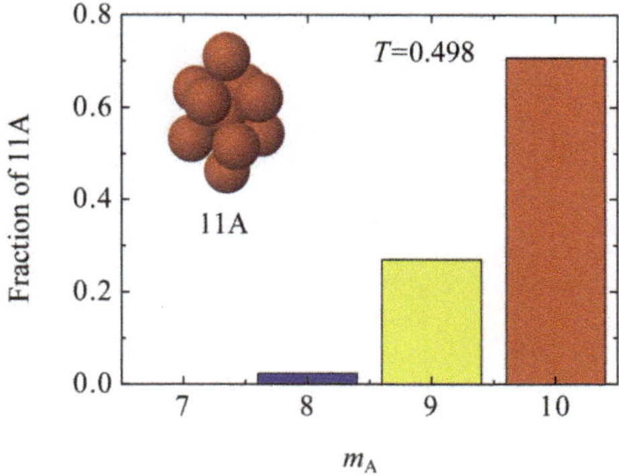

Fig. 5 Composition of the 11A clusters at $T = 0.498$. Almost all 11A have a small B-particle at the centre. m_A is the number of A-species in the shell of the cluster. The height of the bars show the relative proportions that each of the compositions occurs.

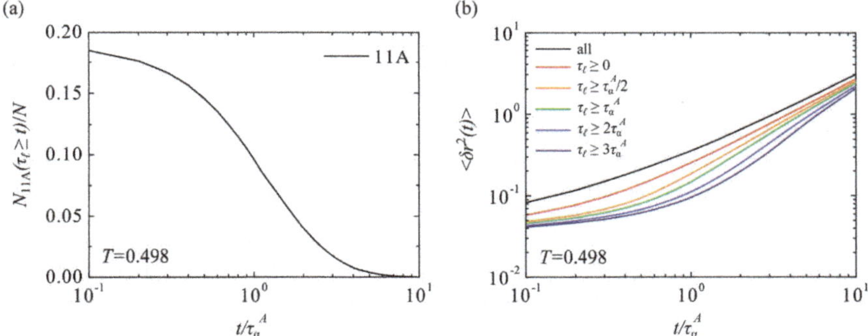

Fig. 6 Dynamics of the particles within 11A clusters in the KA mixture. (a) The fraction of particles participating in 11A clusters with lifetime $\tau_\ell > t$. $N_{11A}(\tau_\ell \geq 0)/N = 0.24$. (b) The mean squared displacement of particles identified initially within 11A polyhedra of various lifetimes.

dynamic heterogeneities $\approx \tau_\alpha^A$. The difference between $N_{11A}(\tau_\ell \geq 0.1\tau_\alpha^A)/N$ and $N_{11A}(\tau_\ell \geq 0)/N$ indicates that only 6% of the particles are members of 11A with $\tau_\ell < 0.1\tau_\alpha^A$ and *not* a member of an 11A with a longer lifetime as well.

Fig. 6(b) shows the mean squared displacement (MSD) of the particles identified initially within 11A clusters (coloured lines) and compares this to the system-wide MSD (black line). The MSD is defined as the ensemble average $\langle \delta r^2(t) \rangle = \langle |\mathbf{r}_i(t + t_0) - \mathbf{r}_i(t_0)|^2 \rangle$ for the subset of particles of interest (indexed by i). All of the particles within 11A relax more slowly than the system-wide average (black line), and the time they take to attain diffusive motion increases as the lifetime of the 11A in which they participate in at t_0 increases. In other words the longer the lifetime of the 11A cluster, the slower the particles become. Since some 11A last for very long times [Fig. 4], it is expected that these particles may exhibit very low mobilities as they maintain some of their nearest neighbours throughout (*e.g.* the central particle in an 11A cluster will always have the same shell particles as its nearest neighbours). For the longest lived 11A (blue lines) there appears to be a super-diffusive regime after the initial sub-diffusive regime, indicating that the particles in these clusters may be hopping out of their cage of neighbours as the 11A structure relaxes.

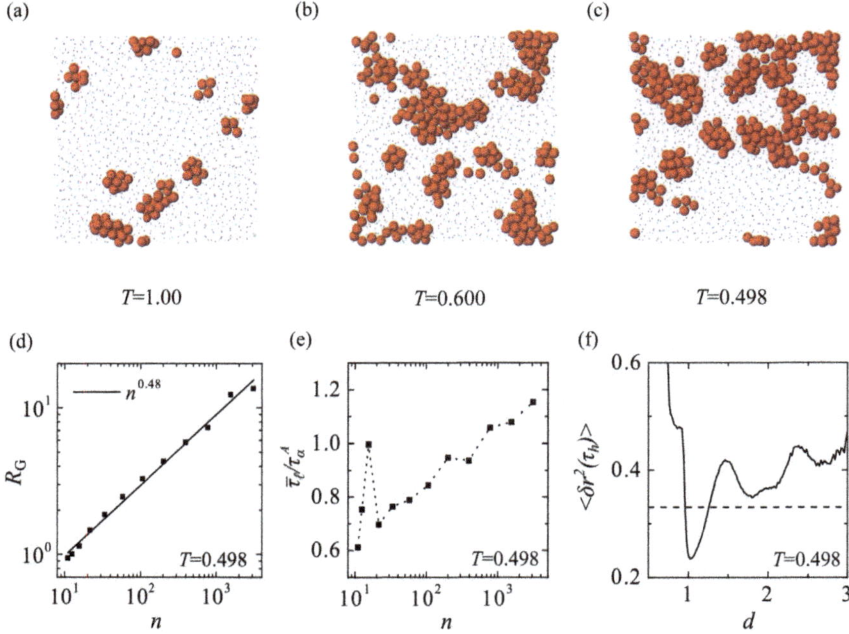

Fig. 7 Analysis of the domains of 11A clusters. (a)–(c) Domains form on cooling from high to low temperature (slices through 3D simulation box). Particles in 11A clusters are shown full size in red, other particles are blue dots. (d) The radius of gyration R_G of the domains *versus* the number of particles in the domain n for $T = 0.498$. R_G is well fitted by $n^{0.48}$ indicating the domains have a fractal dimension $d_f \simeq 2$. (e) The mean lifetime of 11A clusters $\bar{\tau}_\ell$ *versus* the domain size n. (f) 11A domains affect the motion of neighbouring particles. The MSD $\langle \delta r^2(\tau_h) \rangle$ of non-11A particles as a function of distance from 11A domains d (solid line). The dotted line is the MSD over τ_h of all particles not in 11A clusters and independent of the distance from an 11A domain (d).

4.4 Analysis of structured domains

On cooling, the number of 11A clusters in the KA mixture increases. At high temperatures the clusters are generally isolated from one another [Fig. 7(a)]. As the temperature is lowered and the number of clusters increases, domains of clustered particles form. We now analyse the character of these domains of clustered particles and determine the effect the domains have on individual particle dynamics.

The domains of 11A that form on cooling in the KA mixture are shown in Fig. 7(a)–(c). At high temperature 11A are predominantly isolated [Fig. 7(a)]. Upon cooling, the 11A overlap and join together [Fig. 7(b)] to form (transient) networks at low temperatures [Fig. 7(c)].

In order to investigate the structure of the domains, we calculate their radius of gyration,

$$R_G = \frac{1}{2n^2} \sum_{i,j} \left(\mathbf{r}_i - \mathbf{r}_j \right)^2, \quad (2)$$

where n is the number of particles in the domain and the double sum extends over all pairs of particles in the domain. For configurations where the domains of 11A percolate throughout the simulation box R_G cannot be defined. These configurations are rare at $T = 0.498$ and are excluded from the analysis. The consequences of percolating 11A domains are discussed further below.

The radius of gyration shows a power law growth in the size of the domain n with an exponent of 0.48 [Fig. 7(d)]. In other words the domains have a fractal dimension $d_f \simeq 2$, indicating that they are not space-filling. The individual 11A have enhanced stability as the size of the 11A domains grow [Fig. 7(e)]. The time $\bar{\tau}_\ell$ is the mean lifetime of 11A clusters constituting a domain of size n in a configuration. For $T = 0.498$ the general trend is that the mean 11A lifetime increases with the size of the 11A domains. Fig. 7(e) indicates that there is particularly stable arrangement of two overlapping 11A clusters ($n \simeq 16$) relative to domains from similar size. The mean lifetime $\bar{\tau}_\ell$ doubles between isolated 11A ($n = 11$) and extended domains ($n \approx 1000$).

We now consider the effect the domains have on the remainder of the system. In Fig. 7(f), we plot the MSD of the particles not in 11A domains $\langle \delta r^2(\tau_h) \rangle$ against the distance d from the nearest 11A particle at time $t = t_0$. The time $\tau_h \simeq \tau_\alpha^A$ is the time of the maximum in the dynamic susceptibility $\chi_4(t)$ defined below.[80] For distances $d < 0.96$ the non-11A particles are mainly B-species due to the nonadditive nature of the KA Lennard-Jones interactions. These particles are more mobile than the majority A-species, independent of d, due to their smaller size. The number of particles with $d < 0.96$ of a 11A domain is around 10% of the system, *i.e.* half of all the B-species.

Moving further away from the 11A domains, there is then a region of particles with $d \simeq 1$ with reduced mobility compared to the average for non-11A particles [dotted line in Fig. 7(f)]. Around 37% of the particles are found in this region. Subsequent neighbours of the 11A domains for $d > 1.26$ have increased mobility relative to the average. Therefore, excluding the minority B-species, the first nearest neighbours of the domains have suppressed mobility compared to the average, indicating coupling between the structured domains of 11A clusters and the dynamics of the neighbouring particles. Correspondingly the second and third shells of neighbours to the 11A domains have higher mobility compared to the average, indicating a hierarchy of spatial dynamics related to the domains of 11A clusters.

5 Correlation lengths

5.1 Dynamic correlation lengths

Finally we consider whether the structured domains of particles are related to the increasing dynamic correlation lengths in supercooled liquids. In order to do this, we calculate the dynamic correlation length ξ_4, following Lačević *et al.*[80] Details of this procedure can be found elsewhere.[28,80,81] ξ_4 has been previously calculated for the Kob–Andersen model and our values correspond closely to those in the literature.[81] The dynamical correlation length ξ_4 is obtained by analogy to critical phenomena.[80] A (four-point) dynamical susceptibility is calculated as

$$\chi_4(t) = \frac{V}{N^2 k_B T}\left[\langle Q(t)^2 \rangle - \langle Q(t) \rangle^2\right], \tag{3}$$

where

$$Q(t) = \frac{1}{N}\sum_{j=1}^{N}\sum_{l=1}^{N} w\left(\left|\mathbf{r}_j(t+t_0) - \mathbf{r}_l(t_0)\right|\right). \tag{4}$$

The overlap function $w(|\mathbf{r}_j(t+t_0)-\mathbf{r}_l(t_0)|)$ is defined to be unity if $|\mathbf{r}_j(t+t_0)-\mathbf{r}_l(t_0)|\leq a$, 0 otherwise, where $a = 0.3$. The dynamic susceptibility $\chi_4(t)$ exhibits a peak at $t = \tau_h$, which corresponds to the timescale of maximal correlation in the dynamics of the particles. We then construct the four-point dynamic structure factor $S_4(\mathbf{k}, t)$:

$$S_4(\mathbf{k},t)=\frac{1}{N\rho}\left\langle\sum_{jl}\exp[-i\mathbf{k}\cdot\mathbf{r}_l(t_0)]w(|\mathbf{r}_j(t+t_0)-\mathbf{r}_l(t_0)|)\right.$$
$$\left.\times\sum_{mn}\exp[i\mathbf{k}\cdot\mathbf{r}_n(t_0)]w(|\mathbf{r}_m(t+t_0)-\mathbf{r}_n(t_0)|)\right\rangle,$$

where j, l, m, n are particle indices and $\mathbf{k}$ is the wavevector. For time τ_h, the angularly averaged version is $S_4(k, \tau_h)$. The dynamic correlation length ξ_4 is then calculated by fitting the Ornstein–Zernike (OZ) function to $S_4(k, \tau_h)$, as if the system were exhibiting critical-like spatio-temporal density fluctuations,

$$S_4(k,\tau_h)=\frac{S_4(0,\tau_h)}{\left(1+(k\xi_4(\tau_h))^2\right)}, \quad (5)$$

to $S_4(k, \tau_h)$ for $k < 2$.[80] The resulting ξ_4 are plotted in Fig. 8.

We carried out an unconstrained fit to the ξ_4 data, according to $\xi_4(T) = \xi_4^0(T-T_C)^{-\nu}$. The line is plotted in Fig. 8. We find the "critical exponent" is $\nu = 0.588 \pm 0.02$, the "critical temperature" is $T_C = 0.471 \pm 0.002$ and the prefactor is $\xi_4^0 = 0.59 \pm 0.02$. Under the caveat that obtaining ξ_4 from fitting S_4 in limited size simulations is notoriously problematic[81,82] and thus any numerical values should be treated with caution, we observe that the value of T_C is not hugely different to the glass transition temperature found by fitting Mode-Coupling theory to this system, around 0.435.[53,83] We also note that $\nu = 0.588$ lies between mean field ($\nu = 0.5$) and 3D Ising ($\nu = 0.63$) criticality. Here it is worth noting that Onuki and his coworkers pointed out that the dynamical correlation length estimated from four-point density correlations may be affected seriously

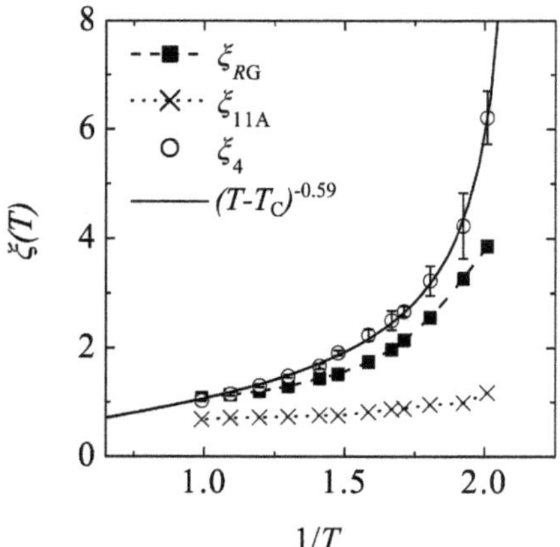

Fig. 8 Comparison of static and dynamic correlation lengths. Structural ξ_{RG} (squares) and ξ_{SC} (crosses), and dynamical ξ_4 (circles) correlation lengths. ξ_4 is fitted with a power law (solid line), which diverges at a value $T_C = 0.47$.

by thermal low-frequency vibration modes, which may lead to a strong system-size dependence.[84] They also showed that a bond-breakage correlation length is free from these vibrational modes, and thus a suitable measure of dynamical coherence.

5.2 Static correlation lengths

We consider two static correlation lengths for the domains of particles in 11A clusters. The first method allows for direct comparison with the dynamic lengthscale ξ_4. We define a structure factor restricted to the particles identified within 11A:

$$S_{11A}(\mathbf{k}) = \frac{1}{N\rho}\left\langle \sum_{j=1}^{N_{11A}} \sum_{l=1}^{N_{11A}} \exp\left[-i\mathbf{k}\cdot\mathbf{r}_j(t_0)\right]\exp\left[i\mathbf{k}\cdot\mathbf{r}_l(t_0)\right]\right\rangle, \tag{6}$$

where N_{11A} is the number of particles in 11A clusters. We then fit the Ornstein–Zernike equation (eqn (5)) to the low-k behaviour of the angularly-averaged $S_{11A}(k)$ in order to extract a structural correlation length ξ_{11A}. This is plotted in Fig. 8. This procedure is akin to the calculation of the dynamic lengthscale ξ_4: first a structure factor is calculated from a selected fraction of the particles (either immobile or structured), and the Ornstein–Zernike expression used to extract a correlation length.

The second lengthscale we consider for the structured particles is derived from the radius of gyration of the domains of clusters. We define

$$\xi_{RG} = R_G^{11A}(\langle n\rangle/m)^{1/d_f}, \tag{7}$$

where ${R_G}^{11A}$ is the radius of gyration of a single cluster, $\langle n\rangle$ is the ensemble average of the domain size, m is the number of particles in the cluster, and $1/d_f$ is the exponent of the power law fitted to R_G *versus* n. This correlation length does not probe the correlations between the domains, as per ξ_{11A}, rather it characterises the growth in size of the domains on cooling until a percolation transition is reached.

The temperature behaviour of the different correlation lengths is shown in Fig. 8. All three correlation lengths increase on cooling, however the manner in which each of the lengths increases is quite different. The main result is that the growth in the dynamic correlation length ξ_4 is not matched by the growth in the structural correlation length ξ_{11A}. Thus we do not find one-to-one correspondence between the behaviour of structural and dynamic correlation lengths.

5.3 Discussion

The fact that different behaviour between the dynamic and static lengths is found is in agreement with some recent studies on 2D and 3D systems,[43,44,48–50] however we note that a one-to-one correspondence in the growth in a lengthscale relating to static crystalline order and the dynamic correlation length for polydisperse quasi-hard sphere systems has been found.[15,17,18,85]

The structural order we find is distinct from the crystalline structure and is thought to frustrate crystallisation. We note that in the studies on 3D systems that have found one-to-one correspondence between the structural and dynamical lengthscales,[17,85] the relative increase in the static lengthscales going from state points in the Arrhenius regime to into the supercooled regime is no more than a factor of 2. Increases of comparable magnitude have been found in studies using point-to-set measures,[43] and more indirect methods[39,52] rather than focusing on explicit structures. In Fig. 8 our static length ξ_{11A} shows an increase of the same order between Arrhenius and supercooled state points, however the increase in ξ_4 is relatively much greater (factor of ≈ 6 or more between high and low T). We note a recent 2D study that suggests one-to-one correspondence in lengthscales for crystalline order and dynamic heterogeneities on quasi-hard spheres breaks down with the addition of attractions between the particles.[86] However, we also note that 3D polydisperse hard-sphere and Lennard-Jones systems have the same link between the correlation length of crystal-like order and slow dynamics, albeit over a limited range of correlation length.[17] These points need to be clarified in the future.

The structured domains form rarefied networks with $d_f \simeq 2$. These networks do not fill space, whereas it is thought that any critical-like nature of the dynamic heterogeneities would imply that the domains have $d_f \simeq 2.5$.[87] Therefore geometrically it appears that the domains of slow and structured particles may be different, at least at the state points we have accessed. Moreover the number of particles in the domains is unlikely to coincide across a range of temperatures, as the N_{11A} for the structured particles increase monotonically on approaching the glass transition, while the number of particles selected by $Q(t)$ is relatively constant as a function of temperature.

Here we note that Mosayebi *et al.*[39,52] found behaviour consistent with critical-like static correlations in the same system as ours, which implies $d_f \sim 2.5$. This seems to suggest that 11A together with other clusters or kinetically pinned particles may correspond to their static structures with solid-like nature. Now the temperature range over which we were able to equilibrate our simulations is less than that for which Mosayebi *et al.*[39,52] present data. Indeed over our range of temperature, the relative increase in static lengthscale they measure is consistent with ours. However, since our $d_f \sim 2.0$, our findings are not consistent with Ising-like criticality, although critical-like behaviour at lower temperatures than we have been able to access cannot be ruled out. We also note that the structural motifs investigated here are of intrinsically of discrete nature, whereas the structural measures showing critical-like behaviours ($d_f \sim 2.5$), such as bond orientational order[14,17] and static structures[39,52] have a continuous nature.

The MSD of the 11A particles indicates that there is not a one-to-one correspondence between the particles selected in S_4 and S_{11A}. The particles selected by w for S_4 all have $\delta r^2(\tau_h) \leq 0.09$ strictly (by the definition of w). However, as can been seen from the red line in Fig. 6(b), the MSD of the 11A particles over the same timescale is $\langle \delta r^2(\tau_h) \rangle \approx 0.1$. Only the particles in the longest-lived 11A clusters [blue lines in Fig. 6(b)] have squared displacements over τ_h comparable to the immobile particles on which S_4 is measured. Thus on this basis direct correspondence between ξ_4 and ξ_{11A} should not necessarily be expected. We note that this might also be related to the effects of low-frequency vibrational modes.[84]

Note χ_4 has been shown to exhibit dependence upon system size for $N \lesssim 1000$.[88] We expect that such effects are reasonably small here, and have taken care to only consider temperatures where all our measured lengths are smaller than the system size. While system size effects cannot be ruled out, we do not believe these make a significant impact on the conclusions we draw.

The lengthscale ξ_{RG} indicates how the size of the domains grows on cooling. In our related study on the Wahnström binary Lennard-Jones glass former,[28] where the concentration of clusters was higher, a percolating network of icosahedra was formed. In a study on the Kob–Andersen model considered here,[62] some of us found evidence for an increase in the population of 11A clusters at lower temperatures than we have been able to access here. Thus if the 11A clusters percolate, this would suggest a diverging structural lengthscale at a temperature higher than either the VFT or MCT temperatures ($\sim$0.325 and $\sim$0.435 respectively). In any case, percolating domains of 11A clusters do not imply structural arrest since each cluster has a finite lifetime. This scenario contrasts with colloidal gels where a percolating network of local structures leads to dynamic arrest.[35]

We also note that Fig. 7(f) strongly indicates that the effect of the domains of clustered particles on the surrounding liquid extends around one particle diameter from the domains. This shows that the dynamical effect of the structured particles is hierarchical and not solely limited to the structured domains themselves. However the correlation lengths we measured from a structure factor of the domains and their first nearest neighbours was no greater than the lengthscale ξ_{11A} for the domains themselves.

Thus the question remains as to the most appropriate structural (and dynamical) correlation length to explain the viscous slowing down in supercooled liquids, and how order-specific correlation lengths, such as ξ_{RG} and ξ_{11A}, are related to order-agnostic structural correlation lengths.[38–40,44,52,89] We conclude this discussion with the following observations.

1. Static correlation lengths have been measured in a variety of systems. Most seem to grow less strongly than dynamic correlation lengths in the regime accessible to computer simulation (and real-space colloid experiments).[28,38–40,43,44,52,89] Those that are observed to grow in a way comparable to the dynamic correlation length ξ_4 are often related to crystalline order.[15,17,90] Here we note that a bond orientational order parameter is a continuous variable, whereas structural motifs are discrete by the definition. There might also be some effects of low-frequency vibrational modes on the estimation of the dynamical correlation length in our system.[84] These points need further investigation.

2. We note that geometric frustration suggests a term in R^5 to be included in a classical nucleation theory like equation, where R is the size of a growing domain of the preferred structure (11A here).[11] This R^5 term strongly suppresses growth of such domains, consistent with ramified structures as we (and others[16,24]) find. Now the arguments supporting geometric frustration tended to focus on icosahedra, not 11A bicapped square antiprisms as we find here. However, in our study of the Wahnström model (whose local structural motif is the icosahedron),[28] we found very similar behaviour to that reported here. In any case, crystals can be formed of 11A, though not for the 80–20 composition of the KA mixture,[59] and icosahedra (with the inclusion of Frank–Kasper bonds).[26] Assuming crystallisation is avoided, given the presence of such an R^5 term, one might enquire as to why growth of polyhedral domains is expected in the first place.

3. Apart from ξ_4, dynamic correlation lengths have also been measured by other means.[41,43] In particular, evidence has been found for non-monotonic behaviour of such dynamic correlation lengths around the Mode-Coupling transition.[41] While ξ_4 itself does not exhibit non-monotonic behaviour as such, there

is evidence that a different scaling is followed for quenches below the Mode-Coupling transition.[91] Moreover, that our fit of ξ_4 implies divergence at $T = 0.47$ suggests that some other scaling might be found at lower temperatures than we have accessed. Another way to approach the discrepancy in Fig. 8 is to enquire whether ξ_4 is the "right" dynamical correlation length.[92] One could even speculate that dynamical correlations are enhanced around the Mode-Coupling transition (as our results imply, along with those of Kob *et al.*[41]), and that at least the lengthscale does not grow significantly at lower temperatures. This would rationalise the estimates of dynamic correlation lengths close to the *molecular* glass transition (some 10 orders of magnitude slower in relaxation time than our simulations are able to access) which suggest correlation lengths only of a few molecular diameters.[93,94] However, we should note that there is a possible deficiency of the standard dynamical correlation length ξ_4.[84] We point out that the wavenumber dependence of the transport coefficient, viscosity, is a physically appealing method for estimating the intrinsic dynamical correlation length.[95–97]

6 Summary and conclusions

We have demonstrated that by studying the lifetimes of different structural orderings within the Kob–Andersen supercooled liquid, the relatively stable orderings of particles can be detected unambiguously. This method alleviates some of the difficulties in identifying structural correlations relevant to glassy behaviour from the temperature dependency of the number of particles participating in clusters. The most stable cluster found in the Kob–Andersen supercooled liquid is the 11A bicapped square antiprism. The relaxation of particles within these clusters proceeds more slowly as the lifetime of the cluster increases. This is consistent with previous work based on Voronoi polyhedra.[16]

The long-lived clusters form rarefied domains on cooling with a fractal dimension $d_f \simeq 2$, *i.e.* the structured domains are non-space-filling, at least in the regime we have accessed. The lifetime of the 11A clusters increases markedly with the size of the domains of these clusters. The non-11A particles neighbouring these domains have reduced mobility compared to particles further from the domains, suggesting a link between structure and dynamic heterogeneity at this level of detail. In other words, the network of 11A clusters acts to "pin" its neighbours.

We examined the relationship between the structured domains and the dynamic heterogeneities by considering static and dynamic correlation lengths of structured/slow particles. A static correlation length calculated in a like-for-like manner with the dynamic correlation length was found to grow moderately on cooling, however its increase was outmatched by the growth in the dynamic correlation length ξ_4. The difference in behaviour of the correlation lengths was rationalised by noting that the structured domains grow in a non-space filling manner, and that the correlation between the structured and slow particles is not perfect. The relationship between the our static and dynamic correlation lengths, and other lengthscales for static order, remains an open question, (see section 5.3).

Finally we consider a possible direction for future study. It has been shown recently that an order parameter associated with the population of 11A clusters can be used to drive a first order transition in an ensemble of trajectories.[62] The susceptibility of the transition to the field coupled to the structural order parameter was found to be higher than when biasing with a field coupled to the dynamical activity, which is the usual method that the transition is accessed. Furthermore a recent study by Singh *et al.* has shown that "ultra-stable" KA glasses prepared by a vapour deposition technique have high numbers of clusters equivalent to 11A polyhedra.[98] Together these results provide further evidence for a connection between the atomic level structure, most easily accessed with high-order structural correlation functions, and the glass transition. The biasing of fields coupled to structural order parameters in trajectory space, and possibly in configuration space as well, thus opens up a new route for the preparation of ultra-stable glassy states[98] pertinent to temperatures well below the Mode-Coupling temperature. The relaxation times inferred for these states are many orders of magnitude higher than those which can currently be prepared with conventional simulations. Study and characterisation of the properties of these states will shed further light on the nature and role of local structure in the glass transition. In other words, while Fig. 8 (like much of the recent literature[28,39,41,43,44,52]) indicates a decoupling between structural and dynamical lengthscales, other static and/or dynamical measures might finally lead to coupling of structure and dynamics.

Acknowledgements

We gratefully acknowledge stimulating discussions with Rob P. Jack, Thomas Speck and Stephen Williams. A. M. is funded by EPSRC grant code EP/E501214/1. C. P. R. thanks the Royal Society for funding. H. T. acknowledges support from a grant-in-aid from the Ministry of Education, Culture, Sports, Science and Technology, Japan and the Aihara Project, the FIRST program from JSPS, initiated by CSTP. This work was carried out using the computational facilities of the Advanced Computing Research Centre, University of Bristol.

References

1 A. Cavagna, *Phys. Rep.*, 2009, **476**, 51–124.
2 L. Berthier and G. Biroli, *Rev. Mod. Phys.*, 2011, **83**, 587–645.
3 F. H. Stillinger and P. G. Debenedetti, *Annu. Rev. Condens. Matter Phys.*, 2013, **4**, 263–285.
4 G. Biroli and J. P. Garrahan, *J. Chem. Phys.*, 2013, **138**, 12A301.
5 W. Götze, *Complex Dynamics of Glass-Forming Liquids: A Mode-Coupling Theory*, Oxford University Press, Oxford, 2008.
6 M. M. Hurley and P. Harrowell, *Phys. Rev. E: Stat. Phys., Plasmas, Fluids, Relat. Interdiscip. Top.*, 1995, **52**, 1694–1698.
7 R. Yamamoto and A. Onuki, *J. Phys. Soc. Jpn.*, 1997, **66**, 2545–2548.
8 M. Ediger, *Annu. Rev. Phys. Chem.*, 2000, **51**, 99–128.
9 L. Berthier, G. Biroli, J.-P. Bouchaud, L. Cipelletti and W. van Saarloos, *Dynamical Heterogeneities in Glasses, Colloids, and Granular Media*, Oxford University Press, Oxford, 1st edn, 2011.
10 F. C. Frank, *Proc. R. Soc. London, Ser. A*, 1952, **215**, 43–46.
11 G. Tarjus, S. A. Kivelson, Z. Nussinov and P. Viot, *J. Phys.: Condens. Matter*, 2005, **17**, R1143–R1182.
12 D. R. Nelson, *Defects and Geometry in Condensed Matter Physics*, Cambridge University Press, 2002, p. 392.
13 H. Tanaka, *J. Phys.: Condens. Matter*, 1998, **10**, L207–L214.
14 H. Tanaka, *Eur. Phys. J. E*, 2012, **35**, 113.
15 H. Shintani and H. Tanaka, *Nat. Phys.*, 2006, **2**, 200–206.
16 D. Coslovich and G. Pastore, *J. Chem. Phys.*, 2007, **127**, 124504.
17 H. Tanaka, T. Kawasaki, H. Shintani and K. Watanabe, *Nat. Mater.*, 2010, **9**, 324–31.

18 F. Sausset and G. Tarjus, *Phys. Rev. Lett.*, 2010, **104**, 065701.
19 M. Leocmach and H. Tanaka, *Nat. Commun.*, 2012, **3**, 974.
20 H. Jónsson and H. Andersen, *Phys. Rev. Lett.*, 1988, **60**, 2295–2298.
21 T. Kondo and K. Tsumuraya, *J. Chem. Phys.*, 1991, **94**, 8220.
22 T. Tomida and T. Egami, *Phys. Rev. B: Condens. Matter*, 1995, **52**, 3290–3308.
23 R. Jullien, P. Jund, D. Caprion and D. Quitmann, *Phys. Rev. E: Stat. Phys., Plasmas, Fluids, Relat. Interdiscip. Top.*, 1996, **54**, 6035–6041.
24 M. Dzugutov, S. I. Simdyankin and F. H. M. Zetterling, *Phys. Rev. Lett.*, 2002, **89**, 195701.
25 E. Lerner, I. Procaccia and J. Zylberg, *Phys. Rev. Lett.*, 2009, **102**, 125701.
26 U. R. Pedersen, T. B. Schrøder, J. C. Dyre and P. Harrowell, *Phys. Rev. Lett.*, 2010, **104**, 1–4.
27 D. Coslovich, *Phys. Rev. E: Stat., Nonlinear, Soft Matter Phys.*, 2011, **83**, 8.
28 A. Malins, J. Eggers, C. P. Royall, S. R. Williams and H. Tanaka, *J. Chem. Phys.*, 2013, **138**, 12A535.
29 T. Schenk, D. Holland-Moritz, V. Simonet, R. Bellissent and D. M. Herlach, *Phys. Rev. Lett.*, 2002, **89**, 75507.
30 D. B. Miracle, *Nat. Mater.*, 2004, **3**, 697–702.
31 M. Tanemura, Y. Hiwatari, H. Matsuda, T. Ogawa, N. Ogita and A. Ueda, *Prog. Theor. Phys.*, 1977, **58**, 1079–1095.
32 J. D. Honeycutt and H. C. Andersen, *J. Phys. Chem.*, 1987, **91**, 4950–4963.
33 P. J. Steinhardt, D. R. Nelson and M. Ronchetti, *Phys. Rev. B*, 1983, **28**, 784–805.
34 S. R. Williams, arXiv:0705.0203, 2007.
35 C. P. Royall, S. R. Williams, T. Ohtsuka and H. Tanaka, *Nat. Mater.*, 2008, **7**, 556–61.
36 A. Malins, J. Eggers and C. P. Royall, 2013, in preparation.
37 R. Della Valle, D. Gazzillo, R. Frattini and G. Pastore, *Phys. Rev. B: Condens. Matter*, 1994, **49**, 12625–12632.
38 G. Biroli, J. P. Bouchaud, A. Cavagna, T. S. Grigera and P. Verrochio, *Nat. Phys.*, 2008, **4**, 771–775.
39 M. Mosayebi, E. Del Gado, P. Ilg and H. C. Öttinger, *Phys. Rev. Lett.*, 2010, **104**, 205704.
40 F. Sausset and D. Levine, *Phys. Rev. Lett.*, 2011, **107**, 045501.
41 W. Kob, S. Roldán-Vargas and L. Berthier, *Nat. Phys.*, 2011, **8**, 164–167.
42 C. Cammarota and G. Biroli, *Europhys. Lett.*, 2012, **98**, 36005.
43 G. M. Hocky, T. E. Markland and D. R. Reichman, *Phys. Rev. Lett.*, 2012, **108**, 225506.
44 A. J. Dunleavy, K. Wiesner and R. C. P., *Phys. Rev. E: Stat., Nonlinear, Soft Matter Phys.*, 2012, **86**, 041505.
45 M. Leocmach, R. J and H. Tanaka, *J. Chem. Phys.*, 2013, **138**, 12A536.
46 D. Coslovich, *J. Chem. Phys.*, 2013, **138**, 12A539.
47 A. Montanari and G. Semerjian, *J. Stat. Phys.*, 2006, **125**, 23–54.
48 B. Charbonneau, P. Charbonneau and G. Tarjus, *Phys. Rev. Lett.*, 2012, **108**, 4.
49 B. Charbonneau, P. Charbonneau and G. Tarjus, *J. Chem. Phys.*, 2013, **138**, 12A515.
50 P. Charbonneau and G. Tarjus, *Phys. Rev. E: Stat., Nonlinear, Soft Matter Phys.*, 2013, **87**, 042305.
51 A. Widmer-Cooper and P. Harrowell, *Phys. Rev. Lett.*, 2006, **96**, 185701.
52 M. Mosayebi, E. Del Gado, P. Ilg and H. C. Öttinger, *J. Chem. Phys.*, 2012, **137**, 024504.
53 W. Kob and H. C. Andersen, *Phys. Rev. E: Stat. Phys., Plasmas, Fluids, Relat. Interdiscip. Top.*, 1995, **51**, 4626–4641.
54 S. Karmakar and I. Procaccia, arXiv:1204.6634, 2012, p. 6.
55 A. Malins, PhD thesis, University of Bristol, 2013.
56 J. P. K. Doye, D. J. Wales and R. S. Berry, *J. Chem. Phys.*, 1995, **103**, 4234–4249.
57 J. P. K. Doye and D. J. Wales, *J. Chem. Soc., Faraday Trans.*, 1997, **93**, 4233–4243.
58 S. Toxvaerd, T. B. Schrøder and J. C. Dyre, arXiv:0712.0377, 2007.
59 J. R. Fernandez and P. Harrowell, *Phys. Rev. E: Stat. Phys., Plasmas, Fluids, Relat. Interdiscip. Top.*, 2003, **67**, 011403.
60 T. Middleton, J. Hernandez-Rojas, P. N. Mortenson and D. J. Wales, *Phys. Rev. B: Condens. Matter*, 2001, **64**, 184201.
61 D. Chandler and J. P. Garrahan, *Annu. Rev. Phys. Chem.*, 2010, **61**, 191–217.
62 T. Speck, A. Malins and C. P. Royall, *Phys. Rev. Lett.*, 2012, **109**, 195703.
63 S. Stoddard and J. Ford, *Phys. Rev. A: At., Mol., Opt. Phys.*, 1973, **8**, 1504.
64 S. Nose, *J. Phys. Soc. Jpn.*, 2001, **70**, 75.
65 M. Goldstein, *J. Chem. Phys.*, 1969, **51**, 3728–3739.
66 L. Berthier and J. P. Garrahan, *Phys. Rev. E: Stat. Phys., Plasmas, Fluids, Relat. Interdiscip. Top.*, 2003, **68**, 041201.
67 H. Vogel, *Phys. Z.*, 1921, **22**, 645.
68 G. S. Fulcher, *J. Am. Ceram. Soc.*, 1925, **8**, 339–355.
69 G. Tammann and W. Hesse, *Z. Anorg. Allg. Chem.*, 1926, **156**, 245–257.

70 V. K. De Souza and D. J. Wales, *Phys. Rev. B: Condens. Matter Mater. Phys.*, 2006, **74**, 134202.
71 H. Tanaka, *J. Non-Cryst. Solids*, 2005, **351**, 3385–3395.
72 T. Hecksler, A. I. Nielsen, N. Boye Olsen and J. C. Dyre, *Nat. Phys.*, 2008, **4**, 737–741.
73 J. L. Meijering, *Philips Res. Rep.*, 1953, **8**, 270.
74 W. Brostow, J.-P. Dussault and B. L. Fox, *J. Comput. Phys.*, 1978, **29**, 81–92.
75 N. Medvedev, *J. Comput. Phys.*, 1986, **67**, 223–229.
76 E. Zaccarelli, C. Valeriani, E. Sanz, W. C. K. Poon, M. E. Cates and P. N. Pusey, *Phys. Rev. Lett.*, 2009, **103**, 135704.
77 I. Saika-Voivod, R. Bowles and P. H. Poole, *Phys. Rev. Lett.*, 2009, **103**, 225701.
78 E. Sanz, C. Valeriani, E. Zaccarelli, W. C. K. Poon, P. N. Pusey and M. E. Cates, *Phys. Rev. Lett.*, 2011, **106**, 215701.
79 J. Taffs, S. R. Williams, H. Tanaka and C. P. Royall, *Soft Matter*, 2013, **9**, 297–305.
80 N. Lačević, F. W. Starr, T. B. Schrųder and S. C. Glotzer, *J. Chem. Phys.*, 2003, **119**, 7372.
81 E. Flenner and G. Szamel, *Phys. Rev. E: Stat., Nonlinear, Soft Matter Phys.*, 2009, **79**, 051502.
82 E. Flenner and G. Szamel, *J. Phys.: Condens. Matter*, 2007, **19**, 205125.
83 W. Kob and H. Andersen, *Phys. Rev. E: Stat. Phys., Plasmas, Fluids, Relat. Interdiscip. Top.*, 1995, **52**, 4134–4153.
84 H. Shiba, T. Kawasaki and A. Onuki, *Phys. Rev. E: Stat., Nonlinear, Soft Matter Phys.*, 2012, **86**, 041504.
85 T. Kawasaki and H. Tanaka, *J. Phys.: Condens. Matter*, 2010, **22**, 232102.
86 W.-S. Xu, Z.-Y. Sun and L.-J. An, *Phys. Rev. E: Stat., Nonlinear, Soft Matter Phys.*, 2012, **86**, 041506.
87 A. Onuki, *Phase Transition Dynamics*, Cambridge University Press, Cambridge, 2002.
88 S. Karmakar, C. Dasgupta and S. Sastry, *Proc. Natl. Acad. Sci. U. S. A.*, 2009, **106**, 3675.
89 C. Cammarota and G. Biroli, *Proc. Natl. Acad. Sci. U. S. A.*, 2012, **109**, 8850–5.
90 T. Kawasaki, T. Araki and H. Tanaka, *Phys. Rev. Lett.*, 2007, **99**, 2–5.
91 E. Flenner and G. Szamel, *J. Chem. Phys.*, 2013, **138**, 12A523.
92 P. Harrowell, *Dynamical Heterogeneities in Glasses, Colloids and Granular Materials*, OUP, Oxford, 2010.
93 L. Berthier, G. Biroli, J.-P. Bouchaud, L. Cipelletti, D. El Masri, D. L'Hote, F. Ladieu and M. Pierno, *Science*, 2005, **310**, 1797–1800.
94 D. Fragiadakis, R. Casalini and R. C. M., *Phys. Rev. E: Stat., Nonlinear, Soft Matter Phys.*, 2011, **84**, 042501.
95 A. Furukawa and H. Tanaka, *Phys. Rev. Lett.*, 2009, **103**, 135703.
96 A. Furukawa and H. Tanaka, *Phys. Rev. E: Stat., Nonlinear, Soft Matter Phys.*, 2011, **84**, 061503.
97 A. Furukawa and H. Tanaka, *Phys. Rev. E: Stat., Nonlinear, Soft Matter Phys.*, 2012, **86**, 030501.
98 S. Singh, M. D. Ediger and J. J. de Pablo, *Nat. Mater.*, 2013, **12**, 139–44.

Faraday Discussions RSCPublishing

PAPER

Population and size distribution of solute-rich mesospecies within mesostructured aqueous amino acid solutions

Anna Jawor-Baczynska,[a] Barry D. Moore,*[b] Han Seung Lee,[c] Alon V. McCormick[c] and Jan Sefcik*[a]

Received 26th April 2013, Accepted 23rd May 2013
DOI: 10.1039/c3fd00066d

Aqueous solutions of highly soluble substances such as small amino acids are usually assumed to be essentially homogenous systems with some degree of short range local structuring due to specific interactions on the sub-nanometre scale (*e.g.* molecular clusters, hydration shells), usually not exceeding several solute molecules. However, recent theoretical and experimental studies have indicated the presence of much larger supramolecular assemblies or mesospecies in solutions of small organic and inorganic molecules as well as proteins. We investigated both supersaturated and undersaturated aqueous solutions of two simple amino acids (glycine and DL-alanine) using Dynamic Light Scattering (DLS), Brownian Microscopy/Nanoparticles Tracking Analysis (NTA) and Cryogenic Transmission Electron Microscopy (Cryo-TEM). Colloidal scale mesospecies (nanodroplets) were previously reported in supersaturated solutions of these amino acids and were implicated as intermediate species on non-classical crystallization pathways. Surprisingly, we have found that the mesospecies are also present in significant numbers in undersaturated solutions even when the solute concentration is well below the solid–liquid equilibrium concentration (saturation limit). Thus, mesopecies can be observed with mean diameters ranging from 100 to 300 nm and a size distribution that broadens towards larger size with increasing solute concentration. We note that the mesospecies are not a separate phase and the system is better described as a thermodynamically stable mesostructured liquid containing solute-rich domains dispersed within bulk solute solution. At a given temperature, solute molecules in such a mesostructured liquid phase are subject to equilibrium distribution between solute-rich mesospecies and the surrounding bulk solution.

[a]*EPSRC Centre for Innovative Manufacturing in Continuous Manufacturing and Crystallisation, Department of Chemical and Process Engineering, University of Strathclyde, 75 Montrose Street, Glasgow, G1 1XJ, UK. E-mail: anna.jawor-baczynska@strath.ac.uk; jan.sefcik@strath.ac.uk; Fax: +44 (0)141 5482539; Tel: +44 (0)141 548 2837*

[b]*WestCHEM, Department of Pure and Applied Chemistry, University of Strathclyde, 295 Cathedral Street, Glasgow, G1 1XL, UK. E-mail: b.d.moore@strath.ac.uk; Fax: +44 (0)141 548 4822; Tel: +44 (0)141 548 2301*

[c]*Department of Chemical Engineering and Materials Science, University of Minnesota, 421 Washington Ave. SE., Minneapolis, MN, 55455-0132, US. E-mail: leex3870@umn.edu; mccormic@umn.edu; Fax: +(001)612-626-7246; Tel: +(001)612-625-1313*

Introduction

Aqueous solutions of simple highly soluble molecules at thermodynamic equilibrium are usually assumed to be homogeneous systems with local microscopic structuring on the sub-nanometre scale due to system-specific interactions typically involving not more than several solute molecules.[1–3] These small molecular-scale structures (molecular clusters) are commonly observed in aqueous solutions of many organic and inorganic molecules.[4–7] Recent theoretical and experimental studies of supersaturated aqueous solutions for a wide range of solutes have also revealed presence of submicron liquid-like or amorphous solid-like mesospecies.[8–19] These mesospecies have been generally seen as metastable and often considered as a separate phase with respect to the bulk multicomponent solution (assuming the Gibbs phase rule is satisfied), composed of solute-rich droplets well-dispersed in the surrounding solution (*e.g.*, quasi-emulsion).[20–22] However, if there are droplets of a separate phase present in a solution they would be expected to undergo coalescence sooner or later, depending on their size and number concentration, unless there is some colloidal stabilisation mechanism which could slow down the coalescence process considerably. Coalescence would eventually result in two macroscopic liquid phases appearing (oiling out).[14,18]

It is presently not fully understood what role if any is played by molecular clusters and/or solution mesospecies in nucleation of solid crystalline forms from supersaturated solutions. There have been numerous investigations of possible links between the assembly of molecular clusters in supersaturated solutions and the solid form which preferentially nucleates from such solutions, but these have been difficult to establish unequivocally.[5,6,23–26] While molecular clusters are understood to be in equilibrium with the surrounding solution, it is not clear if this might be the case with mesospecies as well. The presence of mesoscale domains in supersaturated solutions has been associated with two step nucleation mechanisms[27,28] where these domains were seen as a metastable phase where molecular organisation of crystal nuclei would be preferentially formed. A two-step nucleation mechanism was proposed by Wolde and Frenkel who theoretically studied homogenous nucleation using a Monte Carlo technique.[29] Further theoretical investigations of a two step nucleation mechanism were presented by Kashchiev *et al.*, where droplets of dense liquid formed first, followed by crystallisation within the droplets upon ordering of a critical number of molecules.[30]

The contact of solute rich domains with various solid surfaces could also have an impact on the kinetics of crystal nucleation and polymorphism selection at solid–liquid interfaces.[31–33]

Microscopic liquid–liquid separation as an intermediate step in the crystal nucleation process was first experimentally studied in the crystallisation of large molecules such as proteins.[11,34–36] It was found that the existence of a separate liquid phase with a high protein concentration strongly affects the nucleation kinetics.[30] Interestingly, mesoscale structures have also been reported in undersaturated protein solutions.[37] Crystallisation of calcium carbonate was reported to start with the formation of prenucleation clusters, subject to aggregation into initially amorphous nanoparticles with diameters of around 30 nm. After reaching a critical size, the nanoparticles underwent a solid transformation resulting in crystalline domains inside an amorphous matrix.[38]

Further reports referred to supersaturated solutions of small organic molecules where liquid-like or amorphous solid-like mesospecies have been found.[39–42] Crystal nucleation of a small organic molecule, DL-alanine, from aqueous solution in the absence of any additives was studied by Cölfen and coworkers.[40,41] Dynamic light scattering experiments showed colloidal-scale mesospecies with hydrodynamic radii around 40–100nm, which were presumably initially amorphous or liquid-like structures which aligned *via* directed aggregation and at some point self-organised into a mesocrystalline form.[43–46] Myerson *et al.* studied the nucleation of glycine crystals during cooling crystallisation using Small Angle X-ray Scattering (SAXS).[39] The authors proposed that the first step of glycine crystallization involves the formation of liquid-like clusters of solute molecules with a size of a few hundreds of nanometers, while the second involves the reorganization of such clusters into an ordered crystalline structure.[39] Another interesting study showed that it is possible to crystallise glycine from undersaturated glycine–D_2O solution by irradiating the solution with a focused continuous wave near-infrared laser. The authors claimed that liquid-like clusters could be optically trapped and subjected to molecular alignment, leading to nucleation.[47] Recently we investigated the nucleation of glycine crystals in slightly supersaturated solutions at a constant temperature and found that glycine-rich nanodroplets or mesospecies with a mean diameter of about 250 nm were present in glycine aqueous solution, in equilibrium with glycine crystals.[48] Furthermore, we found that whilst these mesospecies did not lead to productive nucleation, if their coalescence was induced by gentle tumbling it resulted in much larger nanodroplets, with diameters over 750 nm, and lead to very significant acceleration of crystal formation.

Previous experimental investigations indicated the existence of sub-micron mesospecies in various solutions even at concentrations well below the solubility limit.[49–55] Investigations of undersaturated solutions of substances such as sodium chlorate, citric acid, glucose and urea using static and dynamic light scattering showed mesoscale structures with broad size distributions within the range of several hundred nanometers. Detailed light scattering studies showed that they can be characterized as discrete domains of a higher solute density in a less dense solution with the number of solute molecules varying between 10^3 and 10^8.[50–52]

In this work we investigated both super- and undersaturated aqueous solutions of two simple amino acids (glycine and DL-alanine) using Dynamic Light Scattering (DLS), Brownian Microscopy/Nanoparticles Tracking Analysis (NTA) and Cryogenic Transmission Electron Microscopy (Cryo-TEM). Solute-rich mesospecies were found to be present in the aqueous solutions of both amino acids at concentrations well below the solubility limit of the crystalline solid phases. The number and size distribution of the mesospecies varied with the nature and concentration of the solute and when removed by filtration they reappeared, indicating a mesostructured liquid phase is the equilibrium position for these solutes.

Experimental procedure

Sample preparation

Aqueous solutions of glycine and DL-alanine with range of concentrations between 10–270 mg ml^{-1} and 2–240 mg ml^{-1}, respectively, were prepared by combining solid glycine (≥99% NT from Fluka) or DL-alanine (≥99% TLC from Sigma) and deionised water (from an in-house Millipore Water System, 18 MΩ cm^{-1}) in glass vials with a screw-on cap. The solute was completely dissolved under stirring at 55 °C in an incubator. Prior to analysis (DLS, NTA) some samples were filtered at 55 °C using either PTFE (hydrophobic) or alumina (hydrophilic) 100 nm syringe filters and control water samples were also run. The filters used in the study were PTFE membrane Puradisc 13 Syringe filters, 100 nm, Whatman, cat. No. 6784-1301 and alumina membrane filers Anotop 10 Syringe filters, 100 nm, Whatman, cat. no. 6809-1012. The practical performance of these filters was assessed using standard solid polystyrene latex microspheres (100 and 200 nm diameter) from Thermo Scientific (3000 Series Nanospheres Size Standards). All syringes, filters, cuvettes and tubes were preheated in an incubator at 55 °C to avoid premature cooling of the supersaturated solutions during filtering and transfers. NTA and DLS sample cells were preheated to 50 °C before the solution was introduced and then cooled down and equilibrated at 25 °C for 5 min prior to the start of data collection. Tumbling experiments were carried out in glass HPLC vial (volume 1.5 ml from Chromacol) with a PTFE-coated stirrer bar (12 mm × 4.5 mm; with an octagonal ring around its axis) at constant temperature 25 °C. Continuous inversion of the vials (tumbling) was carried out using a blood tube rotator (SB2 Stuart with a constant speed of 20 rpm, fixed mixing angle of 65° and tube holder SB3/1).

Dynamic light scattering (DLS)

DLS measurements were carried out using an ALV/LSE- 5004 instrument, equipped with temperature control, using photon correlation spectroscopy, at a scattering angle of $\theta = 90°$ and laser light wavelength $\lambda = 632.8$ nm. DLS is a well established experimental technique for studying nanoscale particles in dispersions. By measuring the time-dependent fluctuations of the scattered light intensity arising from Brownian motion, average diffusion coefficients and the corresponding mean hydrodynamic radii can be inferred. From the analysis of a measured autocorrelation function, the average hydrodynamic diameter of the mesospecies was estimated using the cumulant method.[56]

Nanoparticle tracking analysis (NTA)

Mesospecies size distribution and estimated concentrations were determined with a Nanosight LM10 instrument with a temperature control unit. Nanosight NTA2.1 software was used to analyze the videos and calculate the size and concentration of the nanodroplets. The camera setting of the instrument was set using the 'Autosettings' option on the software to allow the software to optimize the shutter and gain settings and all capture, pre-processing and analysis settings were kept the same for all measurements. The sample was introduced into the viewing unit, and an image of the particles' scattering of the laser light was captured by a CCD camera attached to a microscope. A video of the sample was recorded and processed, with each observed individual particle 'tracked' by the nanoparticle tracking analysis (NTA) software. Each video was recorded for 90 s and the processing parameters of brightness and gain were optimized by the software. From the Brownian motion analysis the particle size was calculated for each individual particle tracked. The diffusion coefficient from the mean squared displacement of the tracked particle was calculated, and substituted into the Stokes–Einstein equation to obtain the hydrodynamic diameter. Estimation of the particle concentration was based on the particle count in the illuminated volume calculated from the dimensions of the field of view (at a given magnification) and the dimension of the laser beam.[57] The total particle concentration can be subject to a variation of up to 25% between identical samples. Particle size distributions for each sample were measured at least 4 times and averaged results were reported.

Cryogenic transmission electron microscopy (Cryo-TEM)

A lacy carbon Cu grid (01881, 200-mesh, Ted Pella, Ltd., Redding, CA) was washed with 200 proof absolute ethyl alcohol (Aaper Alcohol and Chemical Co., Shelbyville, KY) for 5 s and was dried on a No. 1 Whatman filter paper (Whatman International Ltd, Maidstone, England). The washed TEM grids were glow discharged in a vacuum evaporator at 65–70 mTorr (DV-502A, Denton Vacuum Moorestown, NJ) for 30 s. A 2 μL aliquot of fresh sample was pipetted onto the carbon coated side of the grid at 22 °C in a Mark III Vitrobot chamber (FEI Company, Hillsboro, OR) with a relative humidity of 100%. After the application, the sample was relaxed for 15 s and the excess sample was removed with 595 filter papers (Ted Pella, Ltd., Redding, CA) with a −1.5 mm offset parameter for 5 s. After blotting, the sample was relaxed for 2 s before being plunged into liquid ethane. The sample, physically fixed by vitrification, was transferred to a Gatan 626 cryo-transfer unit (Gatan, Pleasanton, CA) at −194 °C and characterized at −178 °C in the microscope. A 120 kV FEI Tecnai Spirit BioTWIN was used and images were taken digitally with an Eagle™ 2k CCD camera (FEI Company, Hillsboro, OR) in a low-dose mode. The images were processed with TEM Imaging and Analysis (Version 4.2 SP1 build 816, FEI Company, Hillsboro, OR) software.

Results

Mesospecies populations and sizes: DLS and NTA

The dissolution of glycine and DL-alanine crystals in pure water produced an optically clear solution for all solute concentrations investigated. However, using Dynamic Light Scattering (DLS) and Nanoparticle Tracking Analysis (NTA) it was observed that a small population of the nanoscale scattering species were present in both systems, with the size and number varying with the chemical nature and concentration of the solute. Fig. 1 shows an example of typical DLS autocorrelation functions and the corresponding decay time distributions for aqueous solutions of DL-alanine and glycine. The results from DLS measurements clearly show two characteristic decay times corresponding to two typical sizes of nanostructures present in the solutions. The first decay, with a characteristic decay time of around 0.005 ms, corresponds to freely diffusing objects with a mean hydrodynamic diameter of up to 1 nm (molecular clusters) with a more clearly distinguishable distribution peak in glycine solutions. Small molecular clusters in aqueous amino acid solutions have been previously observed and reported.[5,39] The second decay, with a characteristic decay time of around 2 ms, corresponds to a mean hydrodynamic diameter of around 300 nm showing that mesospecies are also present in these solutions.

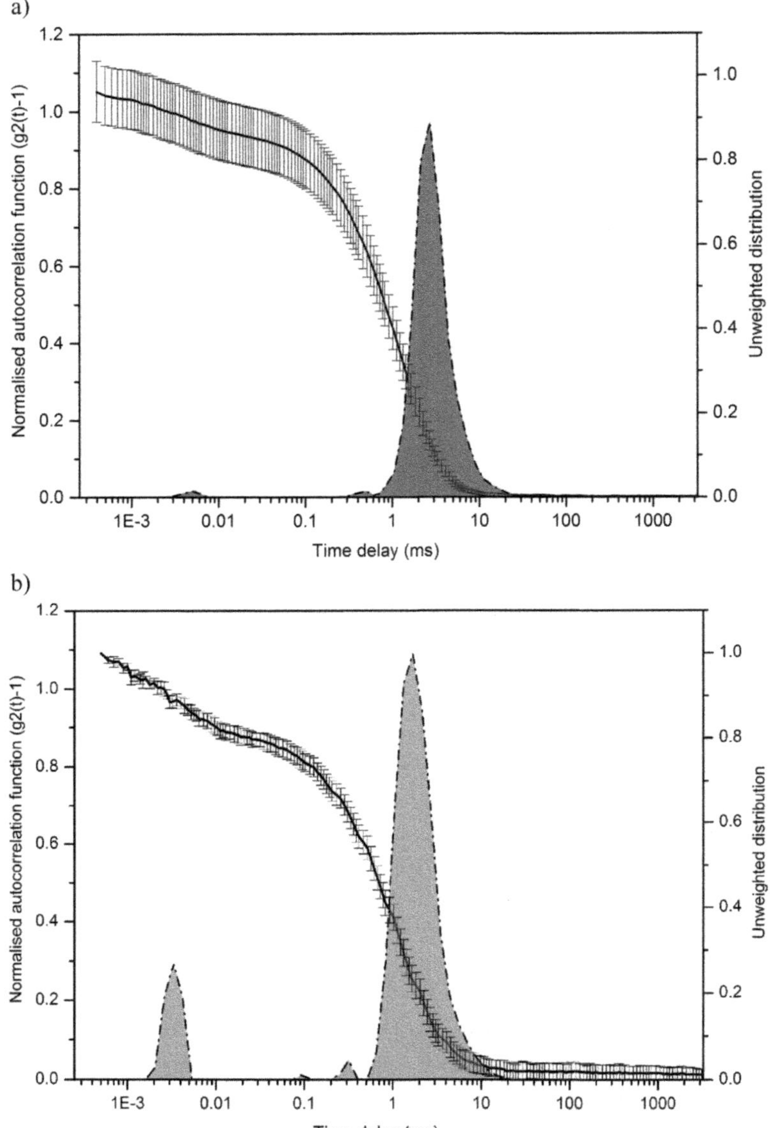

Fig. 1 Dynamic light scattering measurements of DL-alanine and glycine aqueous solutions at 25 °C. Example of an autocorrelation function (solid line) and the corresponding decay time distribution (dotted line), (a) 125 mg ml^{-1} ($S = 0.75$) aqueous solution of DL-alanine not filtered, (b) 180 mg ml^{-1} ($S = 0.75$) aqueous solution of glycine filtered using a 0.1μm PTFE filter.

While DLS can reliably provide information on ensemble averaged mean hydrodynamic diameters, assuming that the autocorrelation function decay is caused by free diffusion movement of the scattering species, Brownian microscopy/NTA measurements were used to estimate the concentration of scattering mesospecies as well as their hydrodynamic diameter distributions. We note that DLS provides an intensity weighted particle size average, which is more sensitive toward larger sizes, while Brownian microscopy directly provides number based particle size distributions and corresponding averages, as well as an estimate of particle number concentrations.[56–60] The range of the sizes, which could be analysed using NTA, is reported to be from around 20 nm up to 1 μm (depending on particle material and solvent type),[57] precluding detection of small molecular clusters. However, it has to be noted that detection of particles with different optical contrasts in the same sample may be hampered by imaging and image processing artifacts (see below).

In Fig. 2 we plotted the number based size distributions of mesospecies obtained from under- and supersaturated aqueous solutions of the two amino

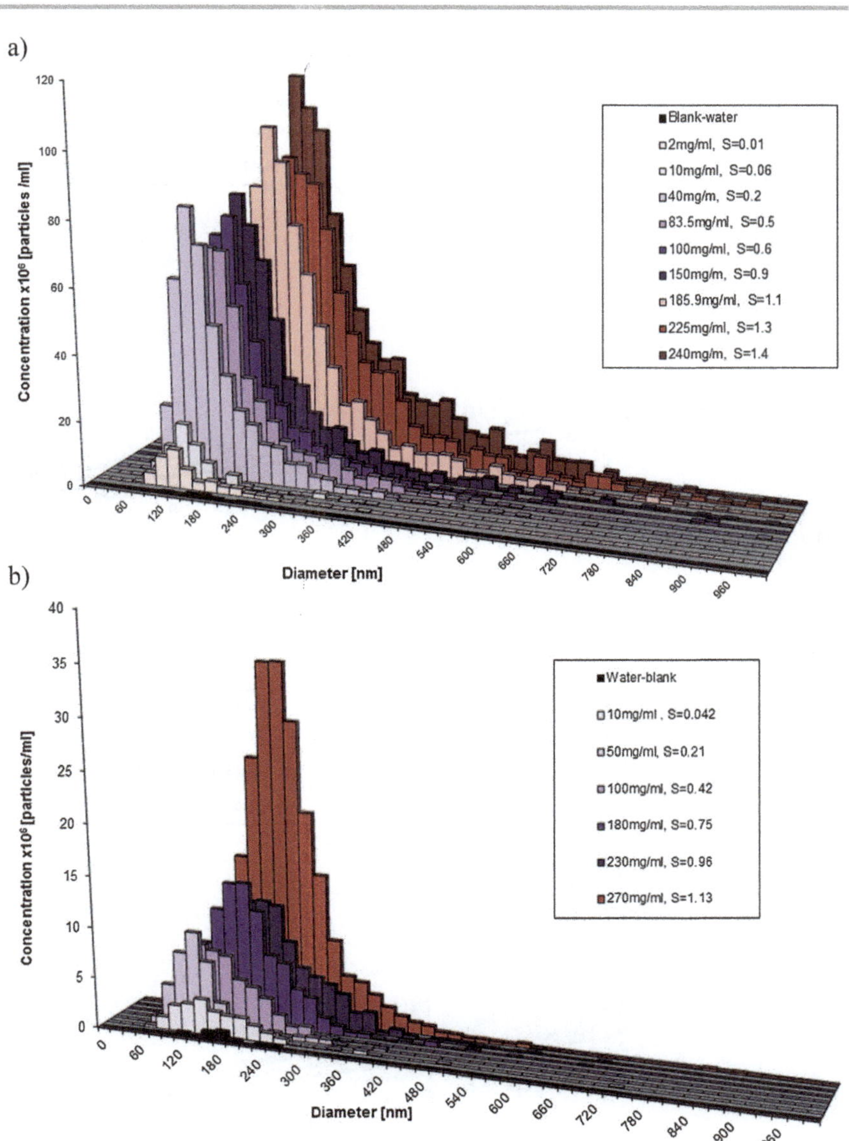

Fig. 2 Number based hydrodynamic diameter distributions of mesospecies (from NTA) at 25 °C, (a) aqueous solutions of DL-alanine together with blank-water (not filtered), (b) aqueous solutions of glycine and blank-water, filtered using a 100 nm PTFE filter.

acids investigated. It is clear that increasing the solute concentration leads to the formation of broader size distributions of scattering mesospecies with the tail extending towards larger sizes (Fig. 2). Multiple recrystallizations of the solutes from a range of solvents had no effect on the size or number of mesospecies produced.[48]

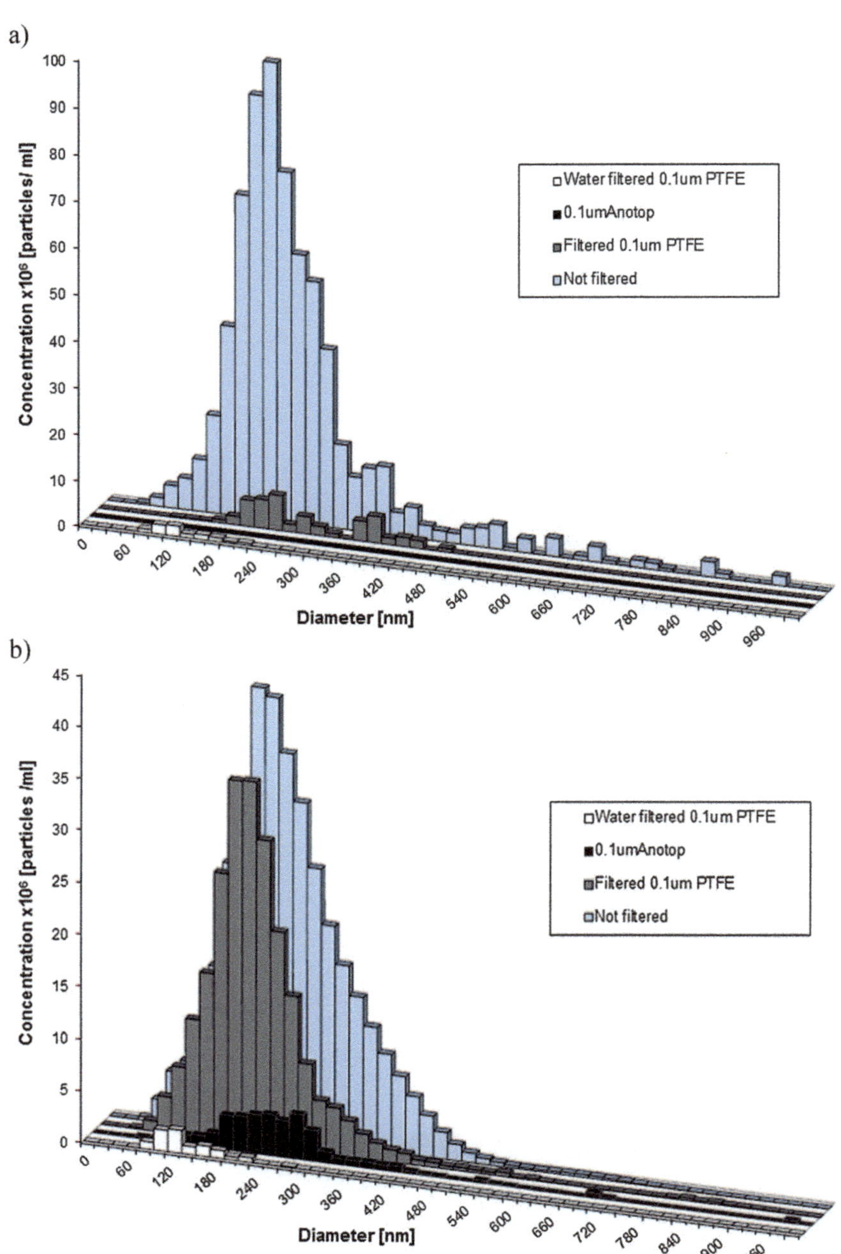

Fig. 3 Number based hydrodynamic diameter distribution of mesospecies (from NTA) at 25 °C, not filtered, filtered using 100 nm PTFE filter or 100 nm Anotop filter, aqueous solution of (a) DL-alanine, 225 mg ml^{-1} ($S = 1.3$), (b) glycine, 270 mg ml^{-1} ($S = 1.13$).

We attempted to remove the scattering mesospecies from amino acid solutions by filtration through 100 nm PTFE (hydrophobic) and 100 nm Anotop (hydrophilic aluminum oxide) filters. As reported previously,[48] in the case of glycine solutions, using a 100 nm PTFE filter did not significantly change either the number or the size of the scattering mesospecies (Fig. 3b). Because mesospecies with sizes larger than 100 nm passed through the 100 nm PTFE filter, it was determined that the mesospecies could not be solid nanoparticles; they were therefore identified as liquid-like nanodroplets. On applying the 100 nm Anotop filter, however, it was possible to lower the population of mesospecies present by an order of magnitude (Fig. 3b), and removal of nanodroplets was also accompanied by a clear increase in backpressure, indicating that the liquid-like mesospecies were retained on this particular filter; perhaps due to their affinity for the hydrophilic filter membrane material. As a control the performance of the 100 nm PTFE filters was tested using commercial solid polystyrene latex particle standards with diameters of 100 and 200 nm. The filters preformed as expected and quantitatively removed both 100 and 200 nm solid nanoparticles from the suspensions being filtered. Interestingly, applying the same filtration protocol to DL-alanine solutions reduced the number of mesospecies to barely detectable limits with both types of filters (Fig. 3a). Although this would seem to suggest that mesospecies in DL-alanine solutions may be more solid-like, this cannot be established unequivocally. Since the mesospecies in glycine solutions passed through one filter and not the other it is possible that more hydrophobic (even liquid-like) mesospecies in DL-alanine solutions could bind to both types of filter.

As was reported previously, even after removal by filtration with 100 nm Anotop filters, the liquid-like mesospecies reappeared in the supersaturated glycine solutions if gentle mechanical action (tumbling) was applied to the filtered solutions.[48] A similar effect was investigated here for under- and supersaturated DL-alanine solutions. Filtration of DL-alanine solutions using 100 nm PTFE filters, resulted in removal of the previously observed mesospecies and mechanical agitation was then applied by placing solutions in glass HPLC vials with a small PTFE coated bar at constant temperature (25 °C) with continuous inversion of the vials (tumbling), carried out using a blood tube rotator. Similar to the glycine system, the mesospecies reappeared. Fig. 4 shows the mean diameters and concentrations observed in DL-alanine solutions (under- and supersaturated) after 120 h of tumbling. The results show that the same number and mean size of mesospecies were observed in the undersaturated DL-alanine aqueous solutions as before filtration. This suggests that a mesostructured solution is actually the equilibrium position and contains a fixed concentration and size of mesospecies. The mean size of the mesospecies observed after tumbling the filtered supersaturated solutions was similar to that seen in undersaturated solutions but lower than in the initial unfiltered supersaturated solutions. This indicates that the size distribution of metastable mesospecies in supersaturated solutions depends on the history of the sample preparation, probably because the system is not at equilibrium.

Fig. 5a and b show the mean size of mesospecies measured by DLS and NTA and the total number concentrations (from NTA) and overall scattered intensity (from DLS) for DL-alanine solutions. Larger values of the mean diameter were

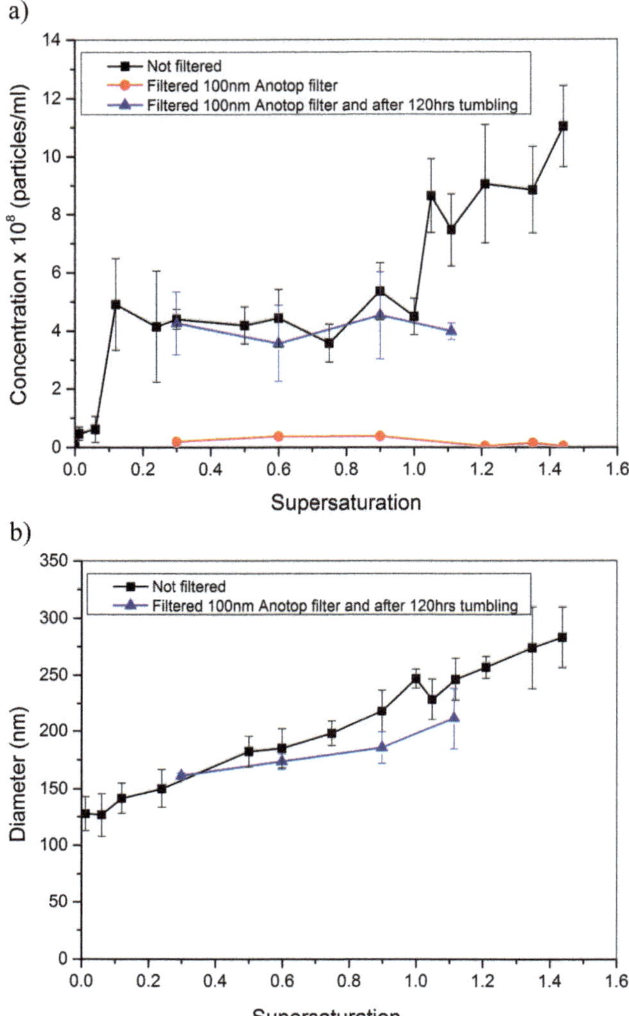

Fig. 4 Estimated concentration (a) and mean number averaged hydrodynamic diameter (b) of mesospecies in DL-alanine aqueous solutions obtained using NTA at 25 °C, not filtered (■), filtered using a 100 nm Anotop filter (●), filtered using a 100 nm Anotop filter and subsequently subjected to 120 h tumbling (▲).

obtained using DLS compared to NTA, which is not unexpected, since DLS provides intensity weighted size averages which are more sensitive toward larger species.[56–58]

It can be seen from Fig. 5b that two important solute concentration boundaries exist in aqueous DL-alanine solutions: the concentration required for equilibrium mesospecies formation and the concentration required for equilibrium solid formation (*i.e.*, solubility, where the relative supersaturation $S = 1$).

Upon passing through each of these concentration boundaries, the number concentrations of the mesospecies were found to increase significantly. In aqueous solutions of DL-alanine the critical solute concentration for mesospecies formation is around 17 mg ml^{-1} (relative supersaturation S around 0.1). Once the

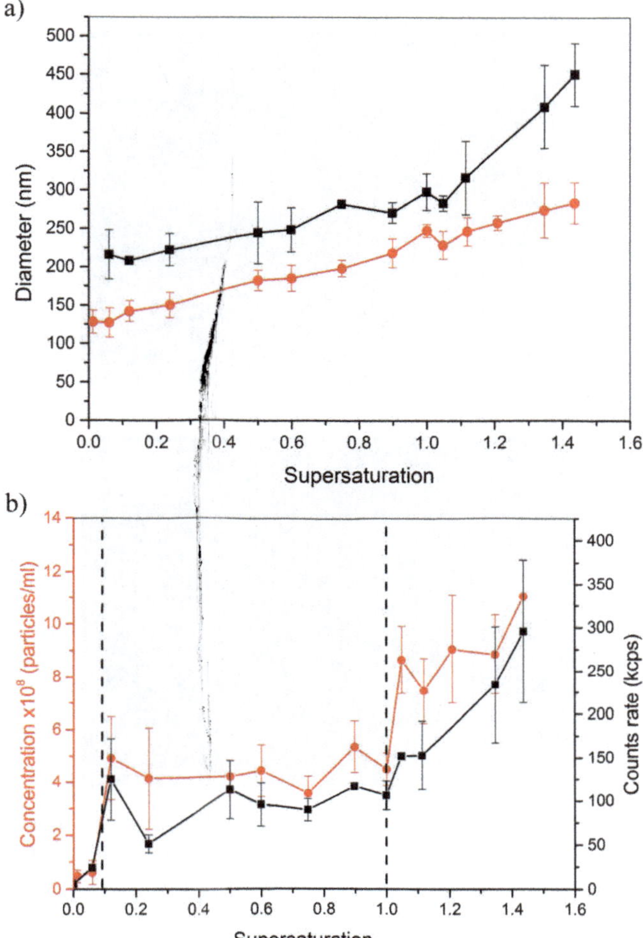

Fig. 5 NTA (●) and DLS (■) data for DL-alanine aqueous solutions (not filtered), (a) mean hydrodynamic diameter *vs.* solute supersaturation, (b) number concentration and scattered intensity (counts rate) *vs.* solute supersaturation for mesospecies at 25 °C.

overall solute concentration in the solution increases above this value, the number concentration of mesospecies is approximately constant until the solute concentration reaches the solubility limit, while the mean size gradually increases with solute concentration. In supersaturated DL-alanine solutions the number concentration of the mesospecies increased gradually with overall solution concentration over the region that could be investigated; above $S = 1.44$ the measurement was disturbed by the nucleation of crystals during the time scale of the measurements.

We note that the total concentration of mesospecies in the glycine solutions seems to be much lower compared to the DL-alanine solutions (Fig. 2), although this could be affected by differences in the optical properties of the scattering species. The mesospecies observed in the DL-alanine system seem to be significantly brighter (*i.e.*, have higher optical contrast) than those observed in the glycine solutions (Fig. 6) and thus may remain detectable at a greater distance

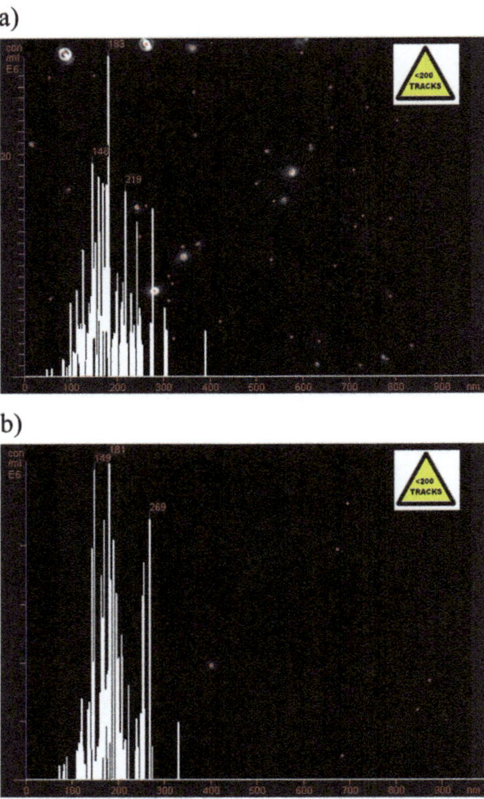

Fig. 6 Still images of software 'tracking' particles using NTA at 25 °C, (a) DL-alanine aqueous solutions, 150 mg ml^{-1} (not filtered), (b) glycine aqueous solution, 230 mg ml^{-1} (filtered using a 100 nm PTFE filter).

from the beam centre than would those of lower optical contrast. Mesospecies with higher optical contrast may contain higher solute concentrations and/or form denser, more compact, or more solid-like, structures. Differences in refractive index were reported amongst the parameters which may significantly affect the accuracy of particle concentration determination using NTA.[57,60]

Cryogenic transmission electron microscopy (Cryo-TEM)

High-resolution Cryo-TEM studies of fresh glycine and DL-alanine undersaturated aqueous solutions with concentrations of 200 and 150 mg ml^{-1}, respectively, showed colloidal scale objects, which were not seen in blank-water solutions with dimensions of around 300–400nm in both investigated systems (Fig. 7). These objects are likely to correspond to the mesospecies observed by DLS and NTA measurements. Using high resolution Cryo-TEM therefore allowed us to visualize mesospecies in undersaturated amino acid solutions. Additionally, by obtaining electron diffraction patterns, we were able to determine that they were amorphous structures. As can be seen from the images, these mesospecies are fairly compact domains with highly irregular surfaces, which is consistent with the *in situ* SAXS measurements on nanodroplets in supersaturated glycine aqueous solutions that were reported previously.[48]

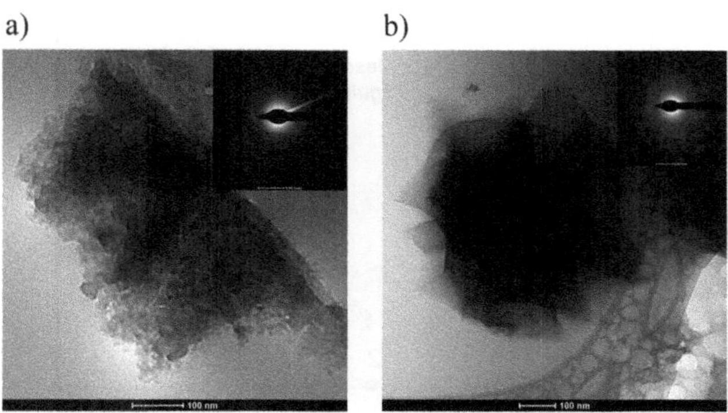

Fig. 7 Cryo-TEM images of mesospecies in undersaturated aqueous solutions (not filtered) together with electron diffraction patterns (a) DL-alanine (150 mg ml^{-1}) and (b) glycine (200 mg ml^{-1}).

Discussion

Using both DLS and NTA, the size distributions and number concentrations of the mesospecies in aqueous solutions of glycine and DL-alanine at 25 °C were investigated. In both systems the mean sizes of mesospecies increased approximately linearly with solute concentration. After passing the solubility limit this trend continued further even under non-equilibrium conditions. The increasing mean size was accompanied by broadening size distributions with the tail at the large sizes clearly developing. Using NTA, we were able to estimate number concentrations of the mesospecies. This clearly showed that for DL-alanine solutions there are two important solute concentration boundaries. The first, above which significant numbers of mesospecies are present in equilibrium with the bulk solution, was found to be around 17 mg ml^{-1} for DL-alanine. The second, above which the crystalline solid phase is present in equilibrium, corresponds to the usual crystalline solid phase solubility. Interestingly, while the number concentration of the mesospecies was approximately constant in undersaturated DL-alanine solutions, from the critical concentration for mesospecies formation to the saturated concentration the number concentration then strongly increased with solute concentration in supersaturated solutions prepared by the direct dissolution of solids (Fig. 5).

A proposed phase diagram for these amino acids in aqueous solutions is illustrated schematically in Fig. 8. Passing along an isotherm indicated by the dotted horizontal line a–b, at the lowest solute concentrations (left), solutions are essentially devoid of mesospecies and solutes are present in the form of solvated molecules accompanied by some molecular clusters. Upon reaching a critical concentration for mesospecies formation (around 17 mg ml^{-1} for DL-alanine at 25 °C), the most stable state for the solutions is mesostructured, consisting of a dispersion of solute-rich mesospecies suspended within the solute-poorer bulk solution. The formation of mesospecies in solutions is a spontaneous process although there is a kinetic barrier for their formation after being removed by filtration. As solute concentration increases, the microheterogeneous mesostructured liquid phase remains thermodynamically stable, with additional solute distributed between both the solute rich and solute poor domains, until an upper composition limit, corresponding to crystalline solid solubility, is reached

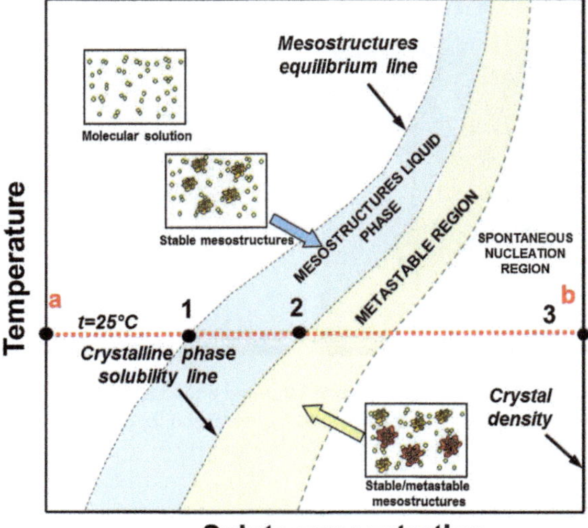

Fig. 8 A proposed phase diagram of amino acid (glycine, DL-alanine) solutions in water.

at point (2). Above this concentration any additional solute will be thermodynamically more stable in the crystalline phase than in any part of the mesostructured liquid phase.

The presence of mesospecies in undersaturated solutions could be taken to mean that these structures are not necessarily essential for crystal nucleation, unless they undergo some further transformation in terms of composition, size or internal structure. Increasing solute concentration leads to the formation of larger sized mesospecies (widening the size distribution) and this could indicate that the size of the mesospecies may be linked to their potential role in a possible nucleation event. Indeed when metastable supersaturated solutions are prepared by cooling, mesospecies with larger sizes are formed. Evidently the whole mesostructured supersaturated solution will be metastable with respect to the crystalline state but crystals might be expected to nucleate with higher likelihood within these solute-rich mesospecies.

Conclusions

Our results show that the dissolution of two simple amino acids, glycine and DL-alanine, in water invariably leads to the formation of a mesostructured solution containing submicron size mesospecies, with their size distribution and number concentration dependent on concentration and chemical properties of the solute. While glycine mesospecies show liquid-like properties, DL-alanine mesospecies may be denser and more solid-like. The mesostructured solution, consisting of mesospecies dispersed within the bulk solution, forms a single liquid phase at thermodynamic equilibrium. The concentration of solute in the mesospecies is higher than in the bulk solution so when the overall solution composition crosses the saturation limit, crystalline solid nucleation is expected to be more likely to occur within them. However, as was demonstrated previously with glycine[48] there is likely to be a critical size of mesospecies required for a productive nucleation process to take place.

Acknowledgements

This project was supported by the EPSRC Centre for Innovative Manufacturing in Continuous Manufacturing and Crystallisation, grant number EP/I033459/1.

Notes and references

1 M. K. Cerreta and K. A. Berglund, *J. Cryst. Growth*, 1987, **84**, 577–588.
2 I. T. Rusli, G. L. Schrader and M. A. Larson, *J. Cryst. Growth*, 1989, **97**, 345–351.
3 S. A. Hassan, *Phys. Rev. E: Stat., Nonlinear, Soft Matter Phys.*, 2008, **77**, 031501.
4 S. Parveen, R. J. Davey, G. Dent and R. G. Pritchard, *Chem. Commun.*, 2005, 1531–1533.
5 D. Erdemir, S. Chattopadhyay, L. Guo, J. Ilavsky, H. Amenitsch, C. U. Segre and A. S. Myerson, *Phys. Rev. Lett.*, 2007, **99**, 115702.
6 S. Hamad, C. E. Hughes, C. R. A. Catlow and K. D. M. Harris, *J. Phys. Chem. B*, 2008, **112**, 7280–7288.
7 R. A. Chiarella, A. L. Gillon, R. C. Burton, R. J. Davey, G. Sadiq, A. Auffret, M. Cioffi and C. A. Hunter, *Faraday Discuss.*, 2007, **136**, 179–193.
8 J. F. Lutsko and G. Nicolis, *Phys. Rev. Lett.*, 2006, **96**, 046102.
9 L. F. Filobelo, O. Galkin and P. G. Vekilov, *J. Chem. Phys.*, 2005, **123**.
10 T. O. Drews, M. A. Katsoulakis and M. Tsapatsis, *J. Phys. Chem. B*, 2005, **109**, 23879–23887.
11 O. Galkin and P. G. Vekilov, *J. Am. Chem. Soc.*, 2000, **122**, 156–163.
12 M. C. R. Heijna, W. J. P. van Enckevort and E. Vlieg, *Phys. Rev. E*, 2007, **76**, 011604.
13 K. Onuma and N. Kanzaki, *J. Cryst. Growth*, 2007, **304**, 452–459.
14 S. Grouazel, J. Perez, J. P. Astier, F. Bonnete and S. Veesler, *Acta Crystallogr., Sect. D: Biol. Crystallogr.*, 2002, **58**, 1560–1563.
15 L. Lafferrere, C. Hoff and S. Veesler, *J. Cryst. Growth*, 2004, **269**, 550–557.
16 S. Veesler, L. Lafferrere, E. Garcia and C. Hoff, *Org. Process Res. Dev.*, 2003, **7**, 983–989.
17 S. Veesler, E. Revalor, O. Bottini and C. Hoff, *Org. Process Res. Dev.*, 2006, **10**, 841–845.
18 P. E. Bonnett, K. J. Carpenter, S. Dawson and R. J. Davey, *Chem. Commun.*, 2003, 698–699.
19 T. J. Sorensen, P. C. Sontum, J. Samseth, G. Thorsen and D. Malthe-Sorenssen, *Chem. Eng. Technol.*, 2003, **26**, 307–312.
20 X. Wang, J. M. Gillian and D. J. Kirwan, *Cryst. Growth Des.*, 2006, **6**, 2214–2227.
21 X. Wang and D. J. Kirwan, *Cryst. Growth Des.*, 2006, **6**, 2228–2240.
22 K. Waizumi and T. Eguchi, *Chem. Lett.*, 2005, **34**, 1654–1655.
23 J. Huang, T. C. Stringfellow and L. Yu, *J. Am. Chem. Soc.*, 2008, **130**, 13973–13980.
24 R. J. Davey, W. Liu, M. J. Quayle and G. J. T. Tiddy, *Cryst. Growth Des.*, 2002, **2**, 269–272.
25 D. Gidalevitz, R. Feidenhansl, S. Matlis, D.-M. Smilgies, M. J. Christensen and L. Leiserowitz, *Angew. Chem., Int. Ed. Engl.*, 1997, **36**, 955–959.
26 R. J. Davey, S. L. M. Schroeder and J. H. ter Horst, *Angew. Chem., Int. Ed.*, 2013, **52**, 2166–2179.
27 D. Erdemir, A. Y. Lee and A. S. Myerson, *Acc. Chem. Res.*, 2009, **42**, 621–629.
28 P. G. Vekilov, *Nanoscale*, 2010, **2**, 2346–2357.
29 P. R. t. Wolde and D. Frenkel, *Science*, 1997, **277**, 1975–1978.
30 D. Kashchiev, P. G. Vekilov and A. B. Kolomeisky, *J. Chem. Phys.*, 2005, **122**, 244706.
31 A. Singh, I. S. Lee, K. Kim and A. S. Myerson, *CrystEngComm*, 2011, **13**, 24–32.
32 V. López-Mejías, J. L. Knight, C. L. Brooks and A. J. Matzger, *Langmuir*, 2011, **27**, 7575–7579.
33 Y. Diao, T. Harada, A. S. Myerson, T. A. Hatton and B. L. Trout, *Nat. Mater.*, 2011, **10**, 867–871.
34 O. Galkin and P. G. Vekilov, *Proc. Natl. Acad. Sci. U. S. A.*, 2000, **97**, 6277–6281.
35 S. T. Yau and P. G. Vekilov, *Nature*, 2000, **406**, 494–497.
36 O. Galkin, K. Chen, R. L. Nagel, R. E. Hirsch and P. G. Vekilov, *Proc. Natl. Acad. Sci. U. S. A.*, 2002, **99**, 8479–8483.
37 O. Gliko, W. Pan, P. Katsonis, N. Neumaier, O. Galkin, S. Weinkauf and P. G. Vekilov, *J. Phys. Chem. B*, 2007, **111**, 3106–3114.
38 E. M. Pouget, P. H. H. Bomans, J. A. C. M. Goos, P. M. Frederik, G. de With and N. A. J. M. Sommerdijk, *Science*, 2009, **323**, 1455–1458.
39 S. Chattopadhyay, D. Erdemir, J. M. B. Evans, J. Ilavsky, H. Amenitsch, C. U. Segre and A. S. Myerson, *Cryst. Growth Des.*, 2005, **5**, 523–527.
40 Y. Ma, H. Cölfen and M. Antonietti, *J. Phys. Chem. B*, 2006, **110**, 10822–10828.
41 D. Schwahn, Y. Ma and H. Cölfen, *J. Phys. Chem. C*, 2007, **111**, 3224–3227.
42 D. D. Medina and Y. Mastai, *Cryst. Growth Des.*, 2008, **8**, 3646–3651.

43 H. Cölfen and S. Mann, *Angew. Chem., Int. Ed.*, 2003, **42**, 2350–2365.
44 M. Niederberger and H. Colfen, *Phys. Chem. Chem. Phys.*, 2006, **8**, 3271–3287.
45 S. Mann, B. R. Heywood, S. Rajam and J. D. Birchall, *Nature*, 1988, **334**, 692–695.
46 M. Xu and K. D. M. Harris, *J. Phys. Chem. B*, 2007, **111**, 8705–8707.
47 T. Rungsimanon, K.-i. Yuyama, T. Sugiyama and H. Masuhara, *Cryst. Growth Des.*, 2010, **10**, 4686–4688.
48 A. Jawor-Baczynska, J. Sefcik and B. D. Moore, *Cryst. Growth Des.*, 2013, **13**, 470–478.
49 Y. Georgalis, A. M. Kierzek and W. Saenger, *J. Phys. Chem. B*, 2000, **104**, 3405–3406.
50 M. Sedlak, *J. Phys. Chem. B*, 2006, **110**, 4329–4338.
51 M. Sedlak, *J. Phys. Chem. B*, 2006, **110**, 4339–4345.
52 M. Sedlak, *J. Phys. Chem. B*, 2006, **110**, 13976–13984.
53 M. Sedlak and D. Rak, *J. Phys. Chem. B.*, 2013, **117**, 2495–2504.
54 D. Hagmeyer, J. Ruesing, T. Fenske, H. W. Klein, C. Schmuck, W. Schrader, M. E. M. da Piedade and M. Epple, *RSC Advances*, 2012, **2**, 4690–4696.
55 K. Z. Gaca, J. A. Parkinson and J. Sefcik, *J. Phys. Chem. B.*, 2013, **117**, 10548–10555.
56 P. Linder, T. Zember and P. N. Pusey, *Neutron, X-rays and light scattering methods to soft condensed matter. Introduction to Scattering experiments*, Elsevier Science B.V., 2002.
57 *NanoSight LM10&NTA 2.0 Analytical Software*, 2009.
58 R. Finsy, *Adv. Colloid Interface Sci.*, 1994, **52**, 79–143.
59 R. Finsy, L. Deriemaeker, E. Geladé and J. Joosten, *J. Colloid Interface Sci.*, 1992, **153**, 337–354.
60 J. A. Gallego-Urrea, J. Tuoriniemi and M. Hassellöv, *TrAC, Trends Anal. Chem.*, 2011, **30**, 473–483.

Faraday Discussions

RSC Publishing

PAPER

Second-harmonic scattering in aqueous urea solutions: evidence for solute clusters?†

Martin R. Ward,[a] Stanley W. Botchway,[b] Andrew D. Ward[b] and Andrew J. Alexander*[a]

Received 14th May 2013, Accepted 17th June 2013

DOI: 10.1039/c3fd00089c

Measurements of second-harmonic scattering (SHS) from concentrated aqueous solutions of urea are reported for the first time using scanning microscopy. SHS signal was measured as a function of solution concentration (C) over a range of saturation conditions from undersaturated ($S = 0.15$) to supersaturated ($S = 1.86$), where $S = C/C_{sat}$ and C_{sat} is the saturation concentration. The results show a non-linear increase in SHS signal against concentration, with local maxima near $S = 0.95$ and 1.75 suggesting a change in solution structure near these points. Rayleigh scattering images indicate the presence of particles in nearly saturated ($S = 0.95$) urea solutions. Time-dependent SHS measurements indicate that signals originate from individual events encountered during scanning of the focal volume through the solution, consistent with second harmonic generation (SHG) from particles. SHG from aqueous dispersions of barium titanate ($BaTiO_3$) nanoparticles with diameters <200 nm, showed signals ~20 times larger than urea solutions. The results suggest the existence of a population of semi-ordered clusters of urea that changes with solution concentration.

1. Introduction

Nucleation, the process of formation of a new phase from an existing phase, is of fundamental scientific interest. It is also tremendously important economically because it is used, for example, in the production of agrochemicals, cosmetics, foods, and pharmaceuticals. Spontaneous nucleation is often referred to as primary nucleation, in contrast to secondary nucleation which requires a seed of the new phase.[1] Nucleation can also be classed as homogeneous or heterogeneous depending on whether the new phase nucleates alone or on the surface of a third body, respectively. Understanding the mechanisms of nucleation in detail would give us greater scope to explore new phases of materials with unusual properties. A major factor that hinders us from studying nucleation is its stochastic nature;

[a]*School of Chemistry, University of Edinburgh, Edinburgh, EH9 3JJ, Scotland, UK. E-mail: andrew.alexander@ed.ac.uk*

[b]*STFC, Rutherford Appleton Laboratory, R92, Harwell-Oxford, Oxfordshire, OX11 OQX, UK*

† Celebrating 300 years of Chemistry at Edinburgh.

we generally can't predict where or when nucleation will take place. This hindrance is further exacerbated by the small length and time-scales involved.

The longest-standing theoretical model of nucleation is classical nucleation theory.[1,2] This model considers the free-energy change for formation of a cluster, comprised of surface energy and bulk energy terms. The theory predicts a critical cluster size beyond which spontaneous growth can be sustained. The solution can be viewed as a dynamic system where clusters continually grow and dissolve, unit by unit, until chance creates a critical cluster. Classical nucleation theory does a remarkably good job; there are several criticisms of the theory, however. One major criticism is the use of bulk thermodynamic parameters, such as interfacial tension, to describe nanoscale clusters. Another problem is that the theory does not explicitly consider variations in cluster structure.[3]

A more promising approach to nucleation theory is the two-step mechanism.[2,3] The first step involves formation of dense regions of solute or metastable clusters, which are considered to be liquid-like and may include solvent; the second step involves structural organisation or nucleation within these dense regions. A key feature of the mechanism is the incorporation of both concentration and structure co-ordinates in the free-energy pathway to nucleation. The mechanism is much more in accord with experiments on protein nucleation.[4–6] Current experimental and computational evidence suggest that the two-step nucleation mechanism is also applicable to small-molecule systems.[7]

Both the classical and two-step theories suggest a population of clusters of solute exist in solution prior to nucleation. There is a lot of circumstantial evidence to support the existence of these clusters for small-molecule systems, such as diffusivity measurements,[8] Raman spectroscopy,[9] small-angle neutron scattering,[10] static (Rayleigh) light scattering,[11,12] and dynamic light scattering.[13–15] Other experiments, however, suggest that clustering is not at all prevalent.[10,16]

Recently there has been much interest in the use of laser light to induce nucleation.[17–21] A comprehensive review of the methods and systems explored lies outside the scope of the present article, and the reader is directed elsewhere for more details.[20,22] The method of non-photochemical laser-induced nucleation (NPLIN) is particularly interesting because it is believed to proceed without photomechanical or photochemical damage to the system.[17] NPLIN involves exposing supersaturated solutions or supercooled liquids to nanosecond-duration pulses of visible or near-infrared laser light.[23–25] The exact mechanism for NPLIN has not been resolved. The favoured theory is that the intense electric field ($\sim 10^7$ V m^{-1}) at the peak of the laser pulse interacts with the polarizability of clusters of solute molecules, causing them to re-structure to become critical nuclei.[17,19,26] NPLIN was discovered by accident: to quote Garetz *et al.*,[17] "Recently, while attempting to observe second harmonic generation in supersaturated solutions of urea in water, we have noticed that pulses from a Q-switched Nd:YAG laser can induce nucleation in such solutions." The observation of nucleation under those conditions was remarkable, but distracted attention from the original aim of the work.

Second-harmonic scattering (SHS) is a non-linear optical process that doubles the frequency of light.[27] In the electric-dipole approximation, bulk second-harmonic generation (SHG) requires the incoming and outgoing photons to be phase-matched in a material with a non-zero second-order susceptibility tensor $\chi^{(2)}$. This material is typically a single crystal from a class that is non-centrosymmetric, *e.g.*, potassium titanyl phosphate (KTP, commonly used to convert near-infrared diode laser light to visible light in green laser pointers). Surface SHG is possible, even at the surfaces of centrosymmetric materials, due to the breaking of inversion symmetry at the interface. The disordered, homogeneous nature of an aqueous solution would be expected to preclude bulk SHG. However, SHS may also be observed in liquids due to hyper-Rayleigh scattering (HRS), which is caused by local density and orientation fluctuations.[28,29]

There is only one known crystal polymorph of urea at ambient pressures.[30] The crystallographic space group is $P\bar{4}2_1m$, which is non-centrosymmetric and also second-harmonic active.[31,32] If solute clusters do exist in solution, what is the structure of a typical cluster? Do the clusters have any ordered (*e.g.*, crystalline) component, or are they disordered? In the case of urea, if these clusters have a crystalline component, it is very likely to be that of the known solid phase, and therefore second-harmonic active. SHS has recently been applied to monitor nucleation and growth of second-harmonic active materials.[33,34] In order to test for the presence of clusters in aqueous (metastable) supersaturated urea solutions, we have re-visited the experiments of Garetz *et al.* to attempt to observe second-harmonic generation.

2. Experimental methods

The SHS microscope used was based on a fluorescence-lifetime imaging microscopy (FLIM) setup that has been described in detail previously.[35] The light source was a diode-pumped Ti:sapphire laser (Coherent, Verdi V18 and Mira F900) operating at 800 nm to produce 180 fs pulses at 75 MHz. The 800 nm pump beam was focussed through an inverted water-immersion microscope objective ($\times$60, NA 1.2) onto the sample, which was mounted on a motor-controlled *XY* translational stage (Märzhäuzer Wetzlar GmbH & Co. KG, SCAN IM120 $\times$ 100). Photon emission was collected through the same objective and detected using an external fast micro-channel plate photomultiplier tube (Hamamatsu, R3809U-50). Emission was recorded using time-correlated single photon counting (TCSPC) through a computer interface (Becker and Hickl, SPC830). Optical filters were employed to select the emission wavelength range detected. A visible band-pass filter (Schott glass, BG39) was used to block out the source light, while allowing detection of wavelengths in the range 340–610 nm. In addition, optional interference filters could be used to pass narrow bands at 400 $\pm$ 40, 450 $\pm$ 40, and 458 $\pm$ 10 nm (Comar Optics 400IU25, 450IU25 and 458IL25, respectively). The emission was detected without selection of polarization. Collection of emission in the forward direction was investigated using a second objective. It was found that the signals obtained were fractionally lower. The reduced signal may be attributed in some part to the difficulty in aligning un-matched objectives. Laser powers reported here were those measured before the objective: 1 mW corresponds to a single pulse peak power density of ~50 GW cm^{-2}.

Sample vessels were prepared by gluing glass rings (17 mm diameter $\times$ 4 mm tall) onto coverslips (No. 1.5) using optical glue (Norland, Optical Adhesive 61). Two samples of urea were used, one from VWR (BDH, AnalR grade) and the other from Sigma Aldrich (puriss p.a. ACS grade, 99.8%). The experiments were conducted at 21 °C: the saturation concentration of urea at this temperature was calculated to be C_{sat} = 18.14 mol kg^{-1}. Urea solution was hot-filtered through

0.22-μm syringe filters (Millex GS) and placed as a droplet onto the coverslip (volume ~0.02 cm^3) inside the glass ring; the sample vessel was sealed with another coverslip. A range of samples with supersaturations $S = C/C_{sat}$ from 0.15 to 1.86 at 21 °C were tested.

For comparative experiments, barium titanate nanoparticles ($BaTiO_3$, 99.9%, tetragonal phase) were purchased from Nanostructured and Amorphous Materials, Inc. (Houston, TX). The particles were approximately spherical, with mean diameter, $d = 200$ nm. The crystal space group of the tetragonal phase of $BaTiO_3$ is *P4mm*, which is non-centrosymmetric and second-harmonic active.[32,36] To disperse the particles, 8 mg of $BaTiO_3$ powder was added to 11.6 g of urea solution (2.72 mol kg^{-1}) followed by shaking and sonication. 1.00 g of this crude stock dispersion was filtered through a 0.22-μm syringe filter into 8.50 g of urea solution (1.02 mol kg^{-1}). The resulting solution contained $BaTiO_3$ nanoparticles ($d <$ 200 nm) in 1.19 mol urea kg^{-1}. Successful dispersion was verified using the microscope; no further analysis was carried out.

The data collection involved integrating the time-dependent TCSPC photon counts over a fixed period. In order to measure the photon signal over a large volume of the solution, the sample was placed on a motorised *XY* translational stage to allow scanning. A typical scan consisted of a series of translational steps: along the *X* direction for 1 mm followed by a 2 μm step in the *Y* direction, and so on. Each 1 mm step in the *X* direction took ~2.25 s. The depth of the focal volume into the solution was typically 100 μm beyond the coverslip. The photon count was continually integrated during the scanning procedure.

3. Results

The microscope setup was optimized and tested using a sample of crushed, crystalline potassium dihydrogen phosphate (KDP), which gives a strong SHG response.[37] The TCSPC time-dependent photon count is shown in Fig. 1. The trace shows a very sharp response at time, t = 1 ns, which is the 400-nm second harmonic of the input pump light at 800 nm, and represents the instrument response function. The peak can be fitted using a single Gaussian function with full-width at half-maximum (FWHM) of 55 ps.

Time-dependent photon counts obtained from water and from slightly undersaturated urea solution (S = 0.98) are shown in Fig. 2. The water trace (Fig. 2c) was obtained using the visible band-pass filter and shows no clear peak at t = 1 ns. The urea trace obtained using a narrow-pass filter at 400 nm (Fig. 2a) shows a sharp peak. The peak was fitted using a single Gaussian function with FWHM of 43 ps, which is close to the value obtained for KDP (Fig. 1). The urea trace taken using a narrow-pass filter at 458 nm (Fig. 2b) shows no peak, supporting the assignment of the emission as SHS from the urea solution. The results in Fig. 2 demonstrate that, under similar conditions, a second-harmonic signal was obtained from urea solution but not from water.

We conducted several tests to verify that the signal observed was consistent with SHS from the solution, and not from other sources. We verified that the total SHS signals ($I_{2\omega}$) from both water and from urea solution (S = 0.98) showed a quadratic dependence on incident intensity (I_0), $I_{2\omega}^{(SHS)} \propto {I_0}^2$. One potential artefact could be multiphoton-induced breakdown of the liquid, resulting in plasma emission at the focal volume. We would expect the plasma emission to have a

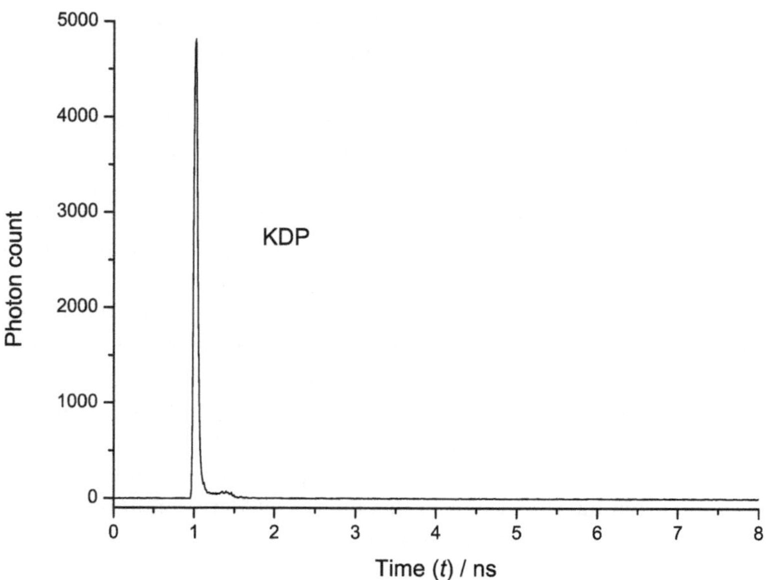

Fig. 1 Time-dependent photon count for solid potassium dihydrogen phosphate crystals. The peak is attributed to second-harmonic generation at 400 nm using an 800 nm pump-pulse train (laser power, $P = 0.7$ mW, 10 s integration). The shape of the peak represents the instrument response function, and can be fitted using a single Gaussian function with full-width at half-maximum (FWHM) of 55 ps.

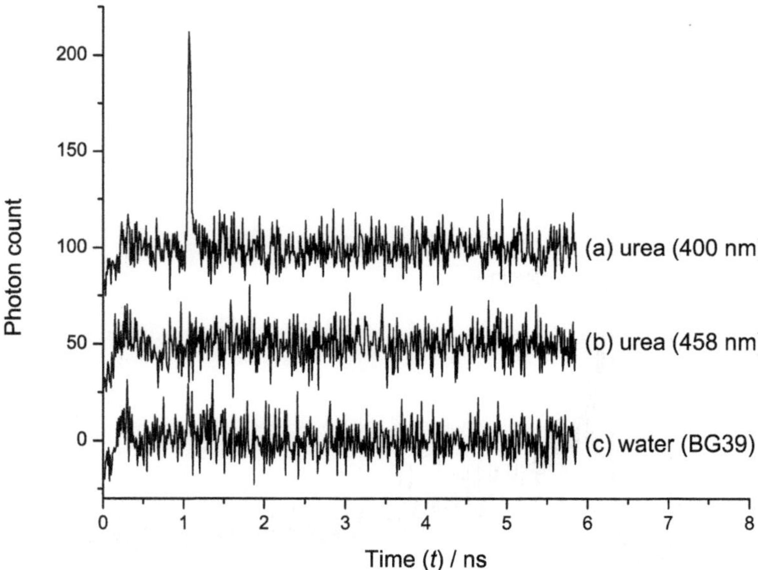

Fig. 2 Time-dependent photon counts for undersaturated urea solutions ($S = 0.98$) and water. All traces were integrated for 180 s while scanning the beam through the droplet using identical pump laser powers ($P = 3.8$ mW). The background level of each trace has been offset to aid comparison. Trace (c), for water, was acquired using a visible band-pass filter (BG39). The top two traces were acquired for urea solution using (a) 400 ± 40 nm and (b) 458 ± 10 nm interference filters. The peak in (a) can be fitted using a Gaussian function with FWHM of 43 ps.

similar temporal profile to SHS. We note, however, that the power densities employed here are orders of magnitude lower than would be required to cause breakdown.[38] Plasma emission was ruled out by using the interference filters to show that the emission contained a very narrow range of wavelengths at 400 nm rather than a broad band in the UV–blue region of the spectrum.

Other potential artefacts include extrinsic sources of SHG, *e.g.*, from the coverslip, or at the coverslip–droplet or droplet–vapour interfaces. At the laser powers ($P = 3.8$ mW) used to obtain the traces in Fig. 2, a small signal was observed using only water when the excitation beam was focussed directly onto the coverslip–droplet interface. Moving the focal volume into the liquid by ~20 μm was sufficient to remove this signal. Raising the laser power to $P = 10$ mW a small signal was obtained even at ~20 μm into the droplet. The signal was found to decrease with increasing depth into the droplet, and therefore we attribute it to surface SHG from the droplet–coverslip interface. All of the measurements on urea reported here were carried out at sufficient powers ($P < 4.0$ mW) and depths (~100 μm) to avoid interference from this interfacial SHG; conditions were verified periodically throughout the experiments.

We attempted to detect SHS from both glycine and sucrose solutions. In each case a long decay signal was observed, masking any SHS, as shown in Fig. 3. The cause of this 2-photon fluorescence is not clear, since there are no electronic states of glycine at the right energy.[39] It may be possible that fluorescence arises due to glycine aggregation at these high concentrations. We did not pursue further analysis of these systems for the present work, and in the absence of further data we attribute the emission to trace impurities in the sample used.

In Fig. 4 we show results for detection of SHS from urea solutions with supersaturations ranging from $S = 0.15$ (undersaturated) to $S = 1.86$ (highly

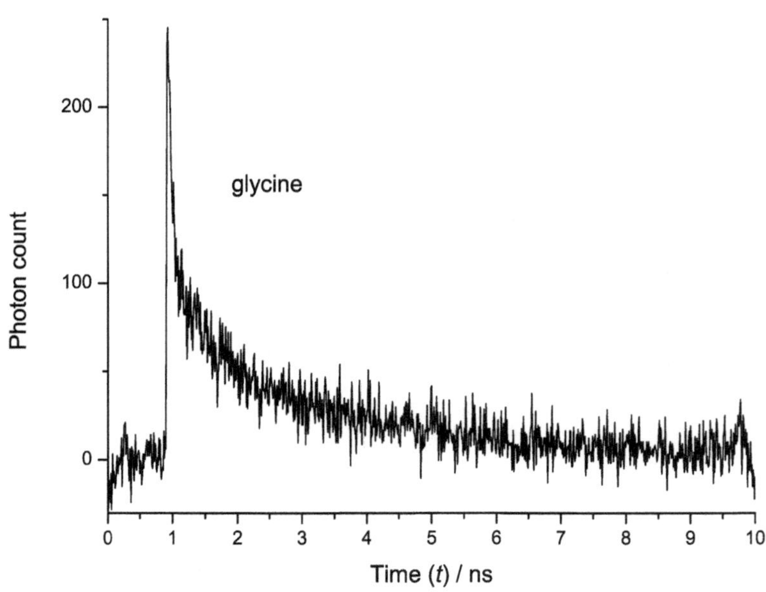

Fig. 3 Time-dependent photon count for undersaturated glycine solution ($S = 0.95$). The pump-laser power was $P = 4.0$ mW and signal was integrated for 180 s while scanning through solution. The decay of the trace can be fitted with two exponential functions (lifetimes, $\tau_1 = 0.16$ ns and $\tau_2 = 2.16$ ns).

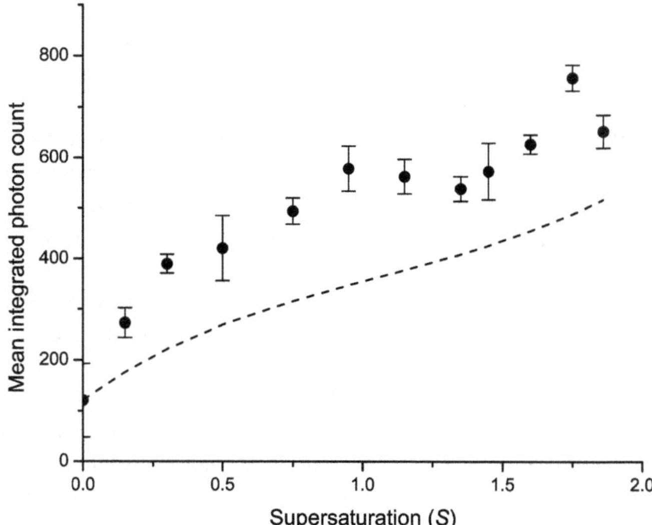

Fig. 4 Concentration dependence of SHS signal from urea solutions. The pump-laser power was $P =$ 3.8 mW, and the depth was ~100 μm. Emission was detected while scanning the focal volume through the solution, integrating for 180 s per trace. For each trace, a baseline signal was calculated by averaging data from 2 to 3 ns (*cf.* Fig. 2) and this was subtracted from the data. The total photon count under the peak was integrated. The mean value of these signal counts, calculated from repeat measurements, is plotted on the vertical axis. Error bars represent 95% confidence intervals from the repeat measurements. The dashed line represents model calculations of hyper-Rayleigh scattering, which have been scaled by a constant factor to match the experimental data at the point $S = 0$ (water).

supersaturated). The signal is seen to increase with concentration up to the point of saturation ($S = 0.95$) and then decrease up to $S = 1.35$; the signal then increases again up to $S = 1.75$, and drops again at the highest supersaturation ($S = 1.86$). Repeat measurements were made with samples tested in a random order. There were no signs of nucleation or damage to samples. The dependence of the signal on concentration does not increase monotonically, as would be expected for hyper-Rayleigh scattering,[40] and is strong evidence that the signal that we have measured is intrinsic, *i.e.*, it originates from the structure of the urea solution. The most remarkable feature is the apparent turnover near saturation ($S = 1$): this is the cross-over point where the solution becomes metastable, where we might expect the structure of the solution to change. The results accord well with Guoy interferometry measurements of Sorell and Myerson, who saw a sharp decrease in urea diffusivity above saturation, which they attributed to the formation of larger urea clusters.[8]

The influence of hyper-Rayleigh scattering was determined as follows. Scattered HRS intensity can be written as $I_{2\omega}^{(\mathrm{HRS})} = GI_0{}^2B^2$, where ($I_0$) is the incident intensity and G is a constant containing geometrical and electric-field factors.[40] The HRS scattering term B^2 can be expanded as

$$B^2 = N_{\mathrm{urea}}\langle\beta_{\mathrm{urea}}{}^2\rangle + N_{\mathrm{water}}\langle\beta_{\mathrm{water}}{}^2\rangle \quad (1)$$

where N_{urea} and N_{water} are the molecule number densities of urea and water, respectively. The space-averaged first-order hyperpolarizabilities for each molecule are given by $\langle\beta^2\rangle = \langle\beta_{ZZZ}{}^2\rangle + \langle\beta_{XXZ}{}^2\rangle$, which represent laboratory-frame averages over the molecule-frame hyperpolarizabilities.[41] Using theoretical values of molecule-frame hyperpolarizabilities (β_{xxz}, β_{yyz}, β_{zzz}) for water[42] and for urea[43] (both have C_{2v} symmetry) we calculated $\langle\beta_{urea}{}^2\rangle = 0.14 \times 10^{-30}$ esu and $\langle\beta_{water}{}^2\rangle = 0.011 \times 10^{-30}$ esu. We used these values to calculate the concentration dependence of B^2, taking into account the change in solution density with concentration.[44] The results are shown as the dashed curve in Fig. 4; the curve has been scaled by a constant factor to match the experimental results at $S = 0$ (water). Ignoring the arbitrary scaling factor, the overall trend in the HRS scattering model matches the increasing signal with concentration observed in the experimental data. However, at higher concentrations, the experimental SHS deviates significantly from the model.

Rayleigh scattering measurements, similar to those of Li and Ogawa on KDP and KCl solutions,[12] reveal a population of particles in urea solutions: see Fig. 5.[45] To determine if the features in the experimental data shown in Fig. 4 could be attributed to SHG from floating particles, we lowered the data-acquisition integration time to 0.2 s, to acquire a sequence of traces while scanning through the solution. This integration time corresponds to an X displacement of 90 μm. In Fig. 6(a) we show the results from a sequence of 2000 traces for urea ($S = 0.98$). Each point in this figure represents the integrated photon count from the emission peak. In Fig. 6(b) we show the total time-dependent photon-count obtained by summing up all traces. The total trace shows a clear peak. Fig. 6(a) demonstrates that the SHS signal we have observed is caused by individual events encountered while scanning through the solution. This further strengthens the case for the signal as being intrinsic to the solution, and not an artefact from the coverslip or elsewhere in the system.

The scan for the results in Fig. 6 took approximately 400 s to complete, and the effective distance travelled was ~0.18 m. If we assume a focal-volume width of 400 nm and depth of 600 nm, the total volume probed during the scan was 4.3×10^{-14} m^3. Setting an arbitrary threshold, we count 7 peaks in Fig. 5(a). This equates to roughly 1.6×10^8 particles cm^{-3}. This estimate is comparable to that obtained by Lian *et al.*, who measured 1.3×10^6 particles cm^{-3} for supersaturated KDP solution ($S = 1.001$) using Rayleigh scattering.[46] Jawor-Baczynska *et al.* report

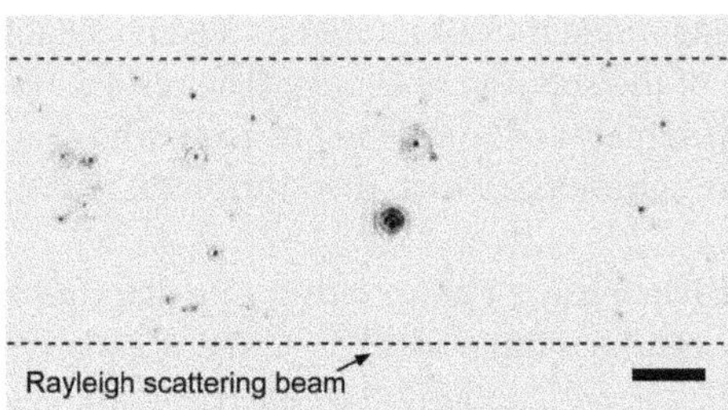

Fig. 5 Rayleigh scattering image of undersaturated urea solution ($S = 0.95$). The 488 nm beam (P = 55 mW) was focussed into solution using a lens (focal length = +200 mm) and the Rayleigh scattering imaged at right angles through an objective (×50). Because the dimensions of the particles are much smaller than the wavelength of scattered light, the particles can be considered as point sources and appear in the image as concentric rings (point spread function). The scale bar (bottom right) represents 10 μm.

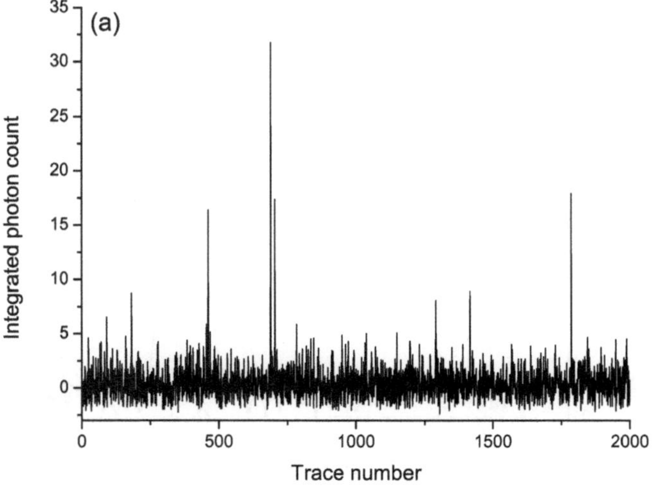

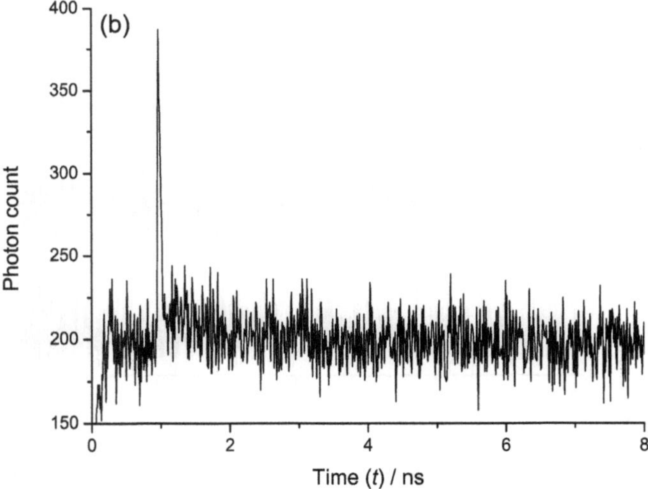

Fig. 6 (a) Integrated peak photon counts from a series of traces taken while scanning through urea solutions ($S = 0.98$). The integration time for each trace was 0.2 s. The signal under the peak for each trace was integrated as detailed in Fig. 4. The pump-laser power was $P = 3.4$ mW. The total signal from summing over all 2000 traces is shown in (b). The peak at ~1 ns is the SHS signal. The results suggest that the total signal originates from individual peak events detected while scanning through the solution.

estimates of 10^9 particles cm^{-3} in their studies of supersaturated ($S = 1.08$) glycine solutions.[15] Our Rayleigh scattering measurements suggest densities of 2.9×10^8 particles cm^{-3},[45] close to the value estimated here.

To make comparisons with particulate material that is known to be SHG-active, we carried out scanning measurements on dispersions of $BaTiO_3$ nanoparticles.[47] In Fig. 7(a) we show a direct comparison between integrated peak photon counts for $BaTiO_3$ and for urea solution ($S = 0.98$). The integration time for each trace was 10 s. Two consecutive traces taken for the urea sample are shown in Fig. 7(b) to illustrate the contribution that a single event makes to the

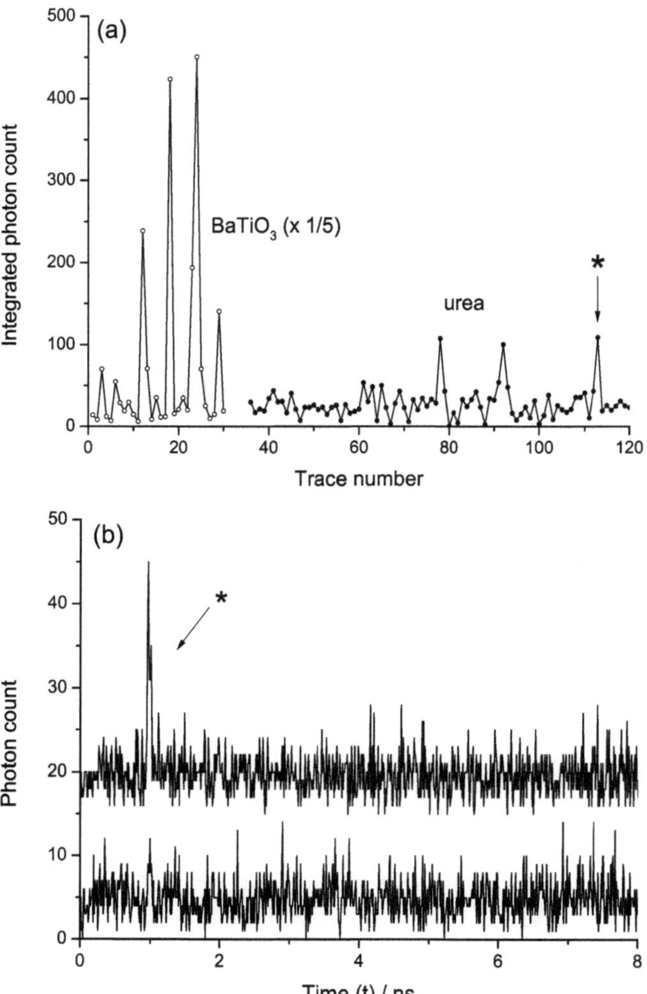

Fig. 7 (a) Integrated photon counts from a series of traces taken while scanning through the solutions; these are total counts obtained by integrating under the peak for each trace, as described in Fig. 4. The integration time for each trace was 10 s. The first set of 30 points (open circles) were obtained using a dispersion of $BaTiO_3$ nanoparticles with diameters $d < 200$ nm; laser power $P = 3.8$ mW. The $BaTiO_3$ data have been multiplied by a factor of 1/5 to plot them on the same scale. The second set of points (closed circles) from 36–120 were taken using urea solution ($S = 0.98$); laser power $P = 3.4$ mW. In (b) we show individual traces, numbers 113 (marked with an asterisk *) and 114; the top trace has been offset (by 15 counts) to aid comparison.

overall signal. Similar scans for water samples did not show spikes like those seen for $BaTiO_3$ and urea samples. The peaks from the $BaTiO_3$ scan are approximately 20 times larger than those from urea.

4. Discussion

The results shown in Fig. 4–7 suggest that the SHS signal we observe is dominated by clusters floating in solution. At supersaturation ($S = 1.0$), the mole fraction of urea is $x = 0.33$, *i.e.*, nearly one-third of molecules is solute; at $S = 1.86$, we have $x = 0.61$.

Clusters are liable to form due to the competition between solute–solute interactions and solute–solvent (and also solvent–solvent). As the concentration increases, the persistence (lifetime) and size of these clusters would be expected to increase.

The bulk SHG efficiency (η) of urea relative to quartz ($\eta = 400$) is similar to that of $BaTiO_3$ ($\eta = 130$).[37] Taking the efficiency factor into account, the signal from $BaTiO_3$ nanoparticles ($d < 200$ nm), is a factor of 60 larger than signals for nearly saturated urea solution. If clusters of urea do have a crystalline structure, this factor could simply be due to the smaller size of the urea particles. For bulk SHG, considering the dependence of SHG conversion efficiency on the interaction length (L), and taking into account the cross-sectional area of the particles illuminated (σ), we calculate $\eta_{2\omega}^{(SHG)} \propto L^2\sigma$.[48] For a particle of diameter d, $\eta_{2\omega}^{(SHG)} \sim d^4$, and we estimate that the particles of urea would be 2.8 times smaller than the equivalent $BaTiO_3$ particles. This is a very crude estimate, however.

The mechanism for second harmonic generation from nanoparticles is distinct from bulk SHG. Since the particle is very much smaller than the wavelength of the exciting radiation, the mechanism is analogous to HRS. The incident electric field induces local dipoles (and multipoles) in and around the nanoparticle. The contributions from the induced dipoles combine coherently to produce the resulting SHG.[49] SHG can therefore be considered as a signature of locked-in order, *e.g.*, in particles, as compared to statistical fluctuations that result in HRS.

Bulk and surface SHG generate similar lifetimes and we are unable to assign whether SHG originates from native urea clusters or an unknown impurity. We consider each of the possible sources in turn.

(1) Bulk

For nanoparticles of non-centrosymmetric materials, bulk SHG has been demonstrated to be very efficient.[47,50] The peak signals that we measured for urea are only 1–2 orders of magnitude lower than for SHG-active $BaTiO_3$ nanoparticles. As outlined above, this reduction may be due to crystalline clusters of urea that are smaller than the $BaTiO_3$ nanoparticles. We note that full crystalline order is not required for bulk SHG: all that is required is some non-centrosymmetric orientational order from the urea molecules at these length scales.[49] Electronic structure calculations on small urea clusters (up to the septimer) show near-linear additivity in the hyperpolarizability, and also highlight the importance of contributions from hydrogen bonding.[51] The changes in SHS with concentration, shown in Fig. 4, could originate from changes in the internal structures or number densities of semi-ordered clusters.

(2) Surface

If the core of a urea cluster is completely centrosymmetric (*e.g.*, disordered), the bulk contribution to SHG vanishes. SHG can then take place due to the breaking of inversion symmetry at the cluster–solution interface. For a spherical cluster, the integrated surface contribution vanishes. In the present case, surface SHG could occur in two ways: (i) from non-spherical clusters or (ii) from clusters that are spherical but with an outer shell of molecules having some degree of orientational order.[52]

(3) Impurities

SHG due to solid impurity particles essentially follows the mechanisms (1) and (2) above. It is possible that the impurity happens to be SHG active. Alternatively, even for spherical centrosymmteric impurities, the orientational order of urea molecules adsorbed onto the particle could produce surface SHG.[52] At the high urea concentrations employed, we would expect the surface coverage of the particles to be saturated, even at $S = 0.15$. Assuming that any impurities come from the urea, the SHG for either case (bulk or surface) would be expected to increase linearly with concentration, contrary to our observations (Fig. 4). It is always possible, however, that the impurity particles induce some order in the surrounding solution. We note that no impurities were detected in water samples prepared in an identical way.

The sources of SHG listed above share a common feature, *viz.* that increasing orientational order of the solute molecules increases the second harmonic scattering. At the present time it is not clear why the SHS signal (Fig. 4) appears to dip in the region above $S = 1.0$ and again at $S = 1.7$. One explanation is that we are measuring changes in the structure within clusters and the number density of clusters, both of which change with concentration. At the point of supersaturation ($S = 1.0$), we might expect a large number of small clusters. Our signal would then be higher because the many small clusters are more frequently encountered during the scanning of the focal volume through the solution. As concentration is increased, larger clusters are formed, and these are encountered less often during the scan. Another way of looking at this is to consider that the changes at $S = 1.0$ and 1.7 signal spinodal decomposition: effectively the appearance of a distinct phase that is metastable and disordered, as water becomes the minority component in the system.

Further work is required to support the preliminary results presented here. This could include correlating SHS with Rayleigh scattering measurements to demonstrate that the peak signals come from long-lived particles. Perhaps more challenging will be to measure angle-dependent scattering, to separate coherent SHG scattering from incoherent HRS, and thereby to determine more details of the structure of the scattering particles.

5. Conclusions

In summary, we have demonstrated measurements of second-harmonic scattering (SHS) from aqueous solutions of urea using scanning microscopy. SHS was measured as a function of solute concentration ranging from $S = 0.15$ (undersaturated) to $S = 1.86$ (supersaturated). The results show an overall increase in the SHS signal with concentration; local maxima near $S = 0.95$ and 1.75 suggest significant changes in the solution structure at these concentrations. Rayleigh scattering experiments indicate a population of free particles in aqueous urea solutions. Time-dependent SHS experiments revealed distinct peaks in SHS signal while scanning the focal volume through solution. These peaks are consistent with SHG from particles; similar peaks were observed when scanning through aqueous dispersions of $BaTiO_3$ nanoparticles that are known to exhibit bulk second harmonic generation (SHG). SHG intensities from the $BaTiO_3$ nanoparticles (diameters, $d < 200$ nm) were ~20 times greater than from urea solution

($S = 0.98$). Our experiments suggest that semi-ordered urea clusters exist in aqueous urea solutions, and that the structures and number densities of these clusters change with concentration.

Acknowledgements

We thank Dr Philip Camp (Edinburgh) for useful discussions. We wish to acknowledge the Science and Technology Facilities Council (STFC) and the Engineering and Physical Sciences Research Council (EPSRC) for supporting this work (EP/G067546/1), and the Royal Society (London) for a research grant.

References

1 J. W. Mullin, *Crystallization*, Butterworth-Heinemann, Oxford, 2001.
2 D. Erdemir, A. Y. Lee and A. S. Myerson, *Acc. Chem. Res.*, 2009, **42**, 621–629.
3 P. G. Vekilov, *Cryst. Growth Des.*, 2010, **10**, 5007–5019.
4 D. Vivares, E. W. Kaler and A. M. Lenhoff, *Acta Cryst. D*, 2005, **61**, 819–825.
5 P. R. ten Wolde and D. Frenkel, *Science*, 1997, **277**, 1975–1978.
6 O. Galkin, K. Chen, R. L. Nagel, R. E. Hirsch and P. G. Vekilov, *Proc. Natl. Acad. Sci. U. S. A.*, 2002, **99**, 8479–8483.
7 D. Zahn, *Phys. Rev. Lett.*, 2004, **92**, 040801.
8 L. S. Sorell and A. S. Myerson, *AIChE J.*, 1982, **28**, 772–779.
9 R. L. Frost and D. W. James, *J. Chem. Soc. Faraday Trans.*, 1982, **78**, 3223–3234.
10 P. E. Mason, G. W. Neilson, S. R. Kline, C. E. Dempsey and J. W. Brady, *J. Phys. Chem. B*, 2006, **110**, 13477–13483.
11 L. Lian, T. J. Lu, Y. Zaizu and T. Ogawa, *J. Mater. Res.*, 1996, **11**, 387–390.
12 L. Li and T. Ogawa, *J. Cryst. Growth*, 2000, **211**, 286–289.
13 Y. Georgalis, A. M. Kierzek and W. Saenger, *J. Phys. Chem. B*, 2000, **104**, 3405–3406.
14 M. Sedlák, *J. Phys. Chem. B*, 2006, **110**, 4329–4338.
15 A. Jawor-Baczynska, J. Sefcik and B. D. Moore, *Cryst. Growth Des.*, 2013, **13**, 470–478.
16 J. Huang, T. C. Stringfellow and L. Yu, *J. Am. Chem. Soc.*, 2008, **130**, 13973–13980.
17 B. A. Garetz, J. E. Aber, N. L. Goddard, R. G. Young and A. S. Myerson, *Phys. Rev. Lett.*, 1996, **77**, 3475–3476.
18 B. A. Garetz, J. Matic and A. S. Myerson, *Phys. Rev. Lett.*, 2002, **89**, 175501.
19 A. J. Alexander and P. J. Camp, *Cryst. Growth Des.*, 2009, **9**, 958–963.
20 T. Sugiyama, K.-I. Yuyama and H. Masuhara, *Acc. Chem. Res.*, 2012, **45**, 1946–1954.
21 S. Nakayama, H. Y. Yoshikawa, R. Murai, M. Kurata, M. Maruyama, S. Sugiyama, Y. Aoki, Y. Takahashi, M. Yoshimura, S. Nakabayashi, H. Adachi, H. Matsumura, T. Inoue, K. Takano, S. Murakami and Y. Mori, *Cryst. Growth Des.*, 2013, **13**, 1491–1496.
22 M. R. Ward and A. J. Alexander, *Cryst. Growth Des.*, 2012, **12**, 4554–4561.
23 M. R. Ward, G. W. Copeland and A. J. Alexander, *J. Chem. Phys.*, 2011, **135**, 114508.
24 M. R. Ward, S. McHugh and A. J. Alexander, *Phys. Chem. Chem. Phys.*, 2012, **14**, 90–93.
25 C. Duffus, P. J. Camp and A. J. Alexander, *J. Am. Chem. Soc.*, 2009, **131**, 11676–11677.
26 M. R. Ward, I. Ballingall, M. L. Costen, K. G. McKendrick and A. J. Alexander, *Chem. Phys.Lett.*, 2009, **481**, 25–28.
27 R. W. Boyd, *Nonlinear optics*, Academic Press, London, 2003.
28 R. W. Terhune, P. D. Maker and C. M. Savage, *Phys. Rev. Lett.*, 1965, **14**, 681–684.
29 P. D. Maker, *Phys. Rev. A*, 1970, **1**, 923–951.
30 P. W. Bridgman, *Proc. Am. Acad. Arts Sci.*, 1916, **52**, 91–187.
31 S. Swaminathan, B. M. Craven and R. K. McMullan, *Acta Cryst. B*, 1984, **40**, 300–306.
32 *International tables for crystallography, Volume A*, ed. T. Hahn, D. Reidel, Dordrecht, 1983.
33 M. C. D'Arrigo, F. R. Cruickshank, D. Pugh, J. N. Sherwood, J. D. Wallis, C. Mackenzie and D. Hayward, *Phys. Chem. Chem. Phys.*, 2006, **8**, 3761–3766.
34 D. Segets, L. Martinez Tomalino, J. Gradl and W. Peukert, *J. Phys. Chem. C*, 2009, **113**, 11995–12001.
35 S. W. Botchway, A. W. Parker, R. H. Bisby and A. G. Crisostomo, *Microsc. Res. Tech.*, 2008, **71**, 267–273.
36 J. P. Dougherty and S. K. Kurtz, *J. Appl. Crystallogr.*, 1976, **9**, 145–158.
37 S. K. Kurtz and T. T. Perry, *J. Appl. Phys.*, 1968, **39**, 3798–3813.
38 A. Vogel, J. Noack, G. Hüttman and G. Paltauf, *Appl. Phys. B*, 2005, **81**, 1015–1047.

39 R. Maul, M. Preuss, F. Ortmann, K. Hannewald and F. Bechstedt, *J. Phys. Chem. A*, 2007, **111**, 4370–4377.

40 K. Clays and A. Persoons, *Phys. Rev. Lett.*, 1991, **66**, 2980–2983.

41 S. J. Cyvin, J. E. Rauch and J. C. Decius, *J. Chem. Phys.*, 1965, **43**, 4083–4095.

42 A. V. Gubskaya and P. G. Kusalik, *Mol. Phys.*, 2001, **99**, 1107–1120.

43 T. Pluta and A. J. Sadlej, *J. Chem. Phys.*, 2001, **114**, 136–146.

44 *CRC Handbook of Chemistry and Physics*, ed. D. Lide, 86th edn, CRC Press, Boca Raton, FL, 2005.

45 M. R. Ward, A. D. Ward and A. J. Alexander, unpublished work.

46 L. Lian, L. Taijing, K. Sakai and T. Ogawa, *J. Mater. Res.*, 1992, **7**, 3275–3279.

47 C.-L. Hsieh, Y. Pu, R. Grange and D. Psaltis, *Opt. Express*, 2010, **18**, 11917–11932.

48 B. E. A. Saleh and M. C. Teich, *Fundamentals of photonics*, John Wiley & Sons, Inc., Hoboken, NJ, 1991.

49 S. Wunderlich, B. Schürer, C. Sauerbeck, W. Peukert and U. Peschel, *Phys. Rev. B*, 2011, **84**, 235403.

50 L. Le Xuan, C. Zhou, A. Slablab, D. Chauvat, C. Tard, S. Perruchas, T. Gacoin, P. Villeval and J.-F. Roch, *Small*, 2008, **4**, 1332–1336.

51 K. Wu, J. G. Snijders and C. Lin, *J. Phys. Chem. B*, 2002, **106**, 8954–8958.

52 S.-H. Jen, G. Gonella and H.-L. Dai, *J. Phys. Chem. A*, 2009, **113**, 4758–4762.

DISCUSSIONS

General discussion

DOI: 10.1039/c3fd90039h

Professor Chandler opened the discussion of the paper by Professor Molinero: Your paper describes molecular simulation results illustrating how low-density liquid water is an unstable fluctuation or transient state appearing in the formation of ice from cold liquid water. In this sense, your results are consistent with conclusions that Dr Limmer and I have drawn from simulations[1,2] and analysis of experiments.[3,4] Nevertheless, as we have heard in Professor Tanaka's introductory lecture, some persist in picturing low-density liquid water as a distinct meta-stable phase that can reversibly transition between and coexist with a higher-density liquid phase. Yet for all the models of water exhibiting ice-like phases that we have examined, Dr Limmer and I find reversible free energy surfaces possessing only one liquid basin. This finding, of course, relates to models consistent with experimental behavior of water. It does not rule out a two-liquid picture for liquids that do not freeze into ice-like solids.

You refer to models that exhibit liquid-liquid transitions? Do these models exhibit crystal states like those of ice?

1 D. T. Limmer, and D. Chandler, *J. Chem. Phys.*, 2011, **135**, 134503.
2 D. T. Limmer, and D. Chandler, *J. Chem. Phys.*, 2013, **138**, 214504.
3 D. T. Limmer, and D. Chandler, *J. Chem. Phys.*, 2012, **137**, 045509.
4 D. T. Limmer, and D. Chandler, *Faraday Discuss.*, 2013, **167**, DOI: 10.1039/C3FD00076A.

Professor Molinero responded: The models of Chatterjee and Debenedetti[1] that exhibit liquid–liquid transitions induced by solutes are van der Waals models with an added orientation-dependent hydrogen-bonding contribution. The equations of state for these models derived by Chatterjee and Debenedetti[1] and Truskett *et al.*[2] describe only fluid states, without the possibility of competing crystallization.

1 S. Chatterjee and P. G. Debenedetti, Fluid-phase behavior of binary mixtures in which one component can have two critical points, *J. Chem. Phys.*, 2006, **124**, 154503.
2 T. M. Truskett, P. G. Debenedetti, S. Sastry, and S. Torquato, A single-bond approach to orientation-dependent interactions and its implications for liquid water. *J. Chem. Phys.*, 1999, **111**, 2647.

Dr Royall enquired: Is it reasonable to think of an analogy between the phase separation to a bicontinuous network that you see, and colloidal gelation? There the network is driven by demixing into two amorphous phases, such that there is a critical (effective) temperature below which gelation occurs. Of course your system breaks symmetry by crystallising, but I wonder if (1) you see fluctuations of your "red domains" above $T = 200$ K, and (2) have you looked at the coarsening behaviour of these domains?

Professor Molinero replied: Fluctuations in concentration and in local order of the water molecules are already noticeable above the temperature of maximum crystallization rate of the solutions. The fraction of four-coordinated water molecules and the intensity of the diffraction peak associated with the nanophase segregation increase together as a function of supercooling, as seen in Fig. 6 of a previous paper.[1] Large order fluctuations before crystallization are also observed in the simulations of pure supercooled water.[2] These are equilibrium fluctuations and can be studied in the temperature region where water is metastable with respect to the crystal. Below the temperature of maximum crystallization rate the time scale of crystallization and relaxation of mW water are comparable and the liquid is below its metastability limit. In that temperature region, crystallization is faster than the coarsening of the concentration fluctuations, so the latter cannot be studied.

1 L. Le, and V. Molinero, Nanophase segregation in supercooled aqueous solutions and their glasses driven by the polyamorphism of water, *J. Phys. Chem. A*, 2011, **115**, 5900–5907.
2 E. B. Moore, and V. Molinero, Growing correlation length in supercooled water, *J. Chem. Phys.*, 2009, **130**, 244505.

Dr Limmer said: Prof. Molinero has presented her results for a theory of freezing in aqueous solutions and in particular for the temperature of maximum freezing rate as a function of solute concentration. The temperature of this maximal rate reflects the balance between the monotonically decreasing free energy for nucleation and the monotonically increasing free energy for growth. However, the theory that Prof. Molinero has presented neglects any dependence of the crystal growth rate on solute concentration. Is this assumption validated by numerical calculations of the solution diffusivity or viscosity?

Professor Molinero replied: The effect of the solute on the viscosity of the solution can be neglected when the glass transition temperature is far below the temperature of ice crystallization. This is always the case for dilute solutions of small molecules. In the case of soluble polymers or high concentration solutions the effect of the solute on the viscosity should be taken into account. As the effect of solutes on the viscosity is not a universal function of the water activity, a scatter of the freezing temperatures as a function of water activity may be expected in highly concentrated solutions. This is actually the case, as seen in Fig.1 in paper by Koop *et al.*[1] Water activity is not sufficient to describe the freezing temperature of very concentrated solutions, dense gels and other systems in which the solute has high concentration and very low mobility. We investigated an exaggerated, cartoonish version of the latter in section C of our paper for this Discussion[2] and concluded that "ice nucleation will be strongly hindered in systems where pure water domains are restricted, because of the topology of the solute distribution (*e.g.* a gel, a polymer) to spaces with characteristic dimensions comparable or smaller than 2 nm."

1 Koop, T., Luo, B., Tsias, A. and Peter, T. Water activity as the determinant for homogeneous ice nucleation in aqueous solutions. Nature 406, 611??"614 (2000).
2 G. Bullock and V. Molinero, *Faraday Discuss.*, 2013, **167**, 10.1039/C3FD00085K.

Dr Limmer continued: In a related question, a key parameter in Prof. Molinero's theory of the temperature of maximal crystallization rate, $T_x(c)$, as a function of solute concentration, c, is the enthalpy difference between the liquid and a hypothetical ideal, low density liquid. Within her theory, this parameter is necessary for explaining the larger slope of $T_x(c)$ relative to the melting temperature, $T_m(c)$, that she observes in molecular dynamics simulations. In calculations for a two component lattice gas model with one component undergoing an ordering transition and the other component undergoing diffusion, such a trend can be similarly seen. However, in this case there is no intermediate structured state as assumed in Prof. Molinero's theory. Does this mean that the two observations need distinct explanations, or is it sufficient to include only timescales associated with composition fluctuations to understand $T_x(c)$?

Professor Molinero responded: The composition fluctuations control $T_x(c)$. I expect that this may be true for the crystallization of a pure component from any mixture. What is distinct for water is that the locus of crystallization, which according to our simulations coincides with the development of nanoscopic domains of pure four-coordinated water, can be predicted from the chemical potential of water, without resorting to solute properties (at least for solutions with water activity above 0.75). It would be interesting to investigate whether the lattice gas model can reproduce the observed collapse of the freezing temperature with solvent chemical potential observed in the experiments for water. The regime for which the freezing temperature approaches the glass transition of the solution is, unsurprisingly, extremely challenging for the study with molecular simulations. Simplified models such as the one you presented at this conference could provide important insight on the interplay between viscosity and freezing in these solutions.

Professor Angell asked: What is your opinion about describing the transient water nanophase you describe as an "Ostwald stage" in the spirit of the venerable Ostwald "rule of stages", which states that often a final equilibrium state is achieved not by going directly to it, but instead through several states of decreasing free energy, each kinetically facile compared to the direct transition? Your nanophase is clearly transient, but has it a substantial existence in time? Low temperature phase-separated water seems to be an example of an Ostwald stage on the way to crystallization of water. Put another way, do you see any objection to an amorphous phase serving as a Ostwald stage on the way to a stable crystalline state? No one doubts that the phases Ostwald described are *bona fide* metastable phases.

Professor Molinero answered: Homogeneous nucleation usually happens only at high driving forces, when there is a significant free energy gap between liquid and crystal. Under those conditions, a myriad of phases, including disordered ones, could have free energy intermediate between the liquid and the stable crystal state. The phenomenon of two-step nucleation through amorphous precursors is well documented in experiments,[1,2] simulations,[3–7] and theory[7,8] for a variety of systems. The pathway for nucleation, however, is not always controlled by the free energy[9] and in many of the examples of two-step nucleation of crystals, an amorphous phase can assist the nucleation even under conditions for which that phase is not stable (*e.g.* systems above a consolute or critical point[10]). In the case of the simulations presented in our study of ice crystallization from water–salt solutions,[11] the nucleation of ice can be considered to proceed in two stages, mediated by the formation of the domains of pure four-coordinated water molecules. However, it should be noted that the pure water domains have a short lifetime, comparable to the ice crystallization times, and do not form a true metastable phase.

1 P. G. Vekilov, The two-step mechanism of nucleation of crystals in solution, *Nanoscale*, 2010, **2**, 2346–2357.
2 D. Erdemir, A. Y. Lee and A. S. Myerson, Nucleation of Crystals from Solution: Classical and Two-Step Models, *Acc. Chem. Res.*, 2009, **42**, 621–629.
3 P. Wolde and D. Frenkel, Enhancement of protein crystal nucleation by critical density fluctuations, *Science*, 1997, **277**, 1975.
4 N. Duff and B. Peters, Nucleation in a Potts lattice gas model of crystallization from solution, *J. Chem. Phys.*, 2009, **131**, 184101.
5 L. C. Jacobson, W. Hujo and V. Molinero, Amorphous precursors in the nucleation of clathrate hydrates, *J. Am. Chem. Soc.*, 2010, **132**, 11806–11811.
6 L. Xu, S. V. Buldyrev, H. E. Stanley and G. Franzese, Homogeneous Crystal Nucleation Near a Metastable Fluid–Fluid Phase Transition, *Phys. Rev. Lett.*, 2012, 1–6.
7 L. Hedges and S. Whitelam, Limit of validity of Ostwald's rule of stages in a statistical mechanical model of crystallization, *J. Chem. Phys.*, 2011, **135**, 164902.
8 S. Whitelam, Nonclassical assembly pathways of anisotropic particles. *J. Chem. Phys.*, 2010, **132**, 194901.
9 B. Peters, Competing nucleation pathways in a mixture of oppositely charged colloids: Out-of-equilibrium nucleation revisited, *J. Chem. Phys.*, 2009, **131**, 244103.
10 H. Liu, S. K. Kumar and J. F. Douglas, Self-Assembly-Induced Protein Crystallization, *Phys. Rev. Lett.*, 2009, **103**, 018101.
11 G. Bullock and V. Molinero, Low-density liquid water is the mother of ice: on the relation between mesostructure, thermodynamics and ice crystallization in solutions, *Faraday Discuss.*, 2013, 167, DOI: 10.1039/C3FD00085K.

Professor Wang opened the discussion of the paper by Mr Cox: During the movie of the nucleation on kaolinite, you mentioned that the second layer of ice formed first and the first layer remained as a liquid longer. Do you know if the first layer is actually a liquid or a transient ice with more fluctuations than the second layer? Did you measure the lateral diffusion constant for the first layer of water? Did it actually move around?

Mr Cox replied: I must first of all clarify that in the particular movie that I showed (movie 1 of the ESI), I meant to convey the fact that the water molecules in the second layer occupied positions close to the final ice structure before those in the first. To answer fully whether or not the first layer is a "transient ice" during the transition period, I would need a working definition of such a material. As the system is undergoing a phase transition, I imagine that this would be a difficult property to define and obtain meaningful time averages of.

The question of diffusion in the first layer compared to the rest of the water is interesting, and in Fig. 1 below, I show the mean square displacement (MSD) of the water molecules in the first layer compared to those of the water above, in the supercooled state. The figure clearly shows that diffusion is slower in the first layer than in the water above (referred to as "bulk", even though the vapour interface is present). I also show a plot (Fig. 2) showing the fraction of water molecules that stay in the first layer as a function of time. It is clear that there is exchange of water molecules between the first layer and those above. As such, the diffusion for individual layers is difficult to compute for long times.

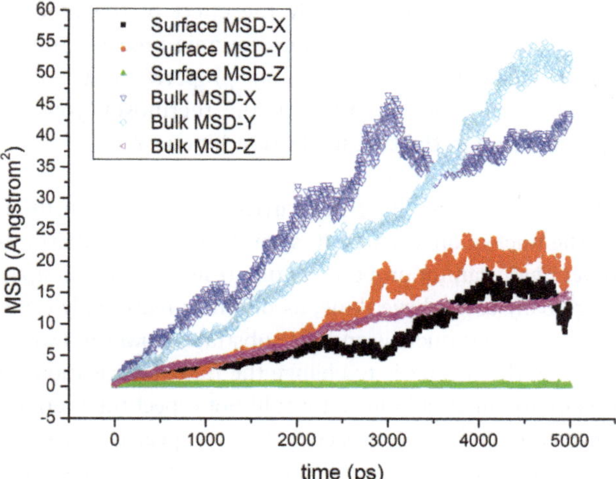

Fig. 1 Mean squared displacement (MSD) for bulk (open symbols) and first layer (filled symbols) water molecules. Note that "bulk" includes contributions from the water molecules at the liquid/vapour interface. The MSD in each Cartesian direction has been calculated separately. We can clearly see that diffusion is slower in the first layer compared to the rest of the system. If a molecule joins or leaves the first layer (see Fig. 2), it is excluded from making any further contribution to the calculation. These results are averages over 5 trajectories, each of 5 ns in length.

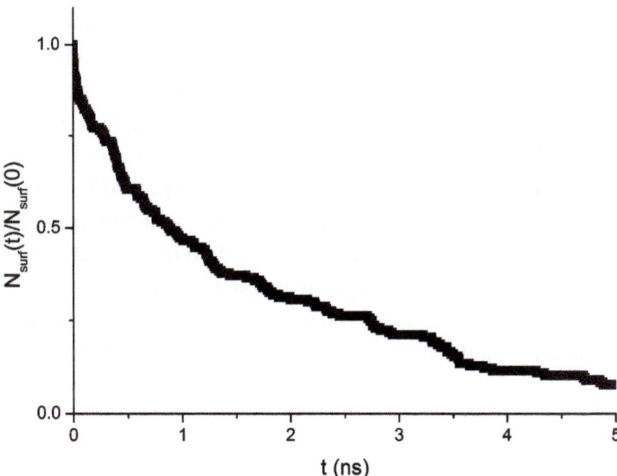

Fig. 2 Survival probability of water molecules in the first layer. $N_{surf}(0)$ is the number of water molecules in the first water layer. $N_{surf}(t)$ is the number of water molecules in the first layer at time t that have never left the first layer. Clearly there is exchange of water molecules between the first layer and those above, making the diffusion of individual water layers difficult to compute. These results are averages over 5 trajectories, each of 5 ns in length.

To compute the MSD for the first layer, as soon as a water molecule leaves the first layer it is excluded from making any further contribution to the calculation. Similarly, a water molecule that joins the first layer from the bulk no longer contributes to the bulk MSD. The data presented in the plots here are averages over 5 trajectories of 5 ns each.

Professor Wang asked: Did you study the effect of the thermostat on the nucleation kinetics? Ice formation is an exothermic process. Standard thermostats, such as the Nose–Hoover chain cannot properly capture the true energy dissipation kinetics. Did you investigate if the thermostat you are using is a sufficient approximation to the true nucleation kinetics?

Mr Cox responded: No, we have not carried out a study on the effect of the thermostat on the nucleation kinetics. It is correct to say that thermostats such as the Nose–Hoover chain cannot properly capture true energetic dissipation kinetics when applied globally as we have done, as one may reasonably expect heat flow between the embryonic ice nucleus and the substrate. However, I would expect the biggest effect of the thermostat is to change the rate of nucleation, which we are unable to measure in our simulations. I would not expect the thermostat to affect the main conclusions presented, especially the unexpected observation of prism rather than basal face growth when nucleation is induced on kaolinite.

Dr Henchman enquired: Do you expect defects on the surface of kaolinite play an important role in ice crystallisation?

Mr Cox answered: Yes, I would expect defects on the kaolinite surface to play an important role in ice nucleation. Certainly, kaolinite particles can be highly defective (see, for example, the AFM images of Bickmore *et al.*[1]) and the presence of defects at the surface is often said to be one of the requirements for a particle to be a good ice nucleating agent (see *e.g.* ref. 5 and 9 of the main article). The exact mechanisms by which surface defects enhance ice nucleation is, however, unclear and warrants further study. From a computational point of view, there are two main challenges that need to be overcome, on top of those already mentioned in the paper pertaining to the proper sampling of ice nucleation: (i) characterisation of the surface defects and (ii) describing the water–defect interaction. Point (i) could be achieved through well controlled surface science experiments or through structure prediction methods (or a combination of the two). Point (ii) will depend on the type of defects that form, but it is unclear whether or not simple point charge force fields, such as those used in this study, will be adequate. It is likely that an *ab initio* method would be required, but this even with these methods challenges still exist, as we have discussed previously.[2,3]

1 B. Bickmore *et al.*, *Am. Mineral.*, 2002, **87**, 780.
2 J. Carrasco, A. Hodgson and A. Michaelides, *Nature Mater.*, 2012, **11**, 667.
3 J. Klimeš and A. Michaelides, *J. Chem. Phys.*, 2012, **137**, 120901.

Professor Molinero remarked: Large ice crystals grown from deeply supercooled water are usually rich in stacking faults.[1–3] If stacking faults were also pervasive at the temperatures for which kaolinite nucleates ice, would they impair the effect that a small ice nuclei have on the control of the morphology of macroscopic ice crystals?

1 E. B. Moore and V. Molinero, Is it cubic? Ice crystallization from deeply supercooled water, *Phys. Chem. Chem. Phys.*, 2011, **13**, 20008–20016.
2 T. L. Malkin, B. J. Murray, A. V. Brukhno, J. Anwar and C. G. Salzmann, Structure of ice crystallized from supercooled water, *Proc. Natl. Acad. Sci. U. S. A.*, 2012, **109**, 1041–1045.
3 W. F. Kuhs, C. Sippel, A. Falenty and T. C. Hansen, Extent and relevance of stacking disorder in "ice I(c)", *Proc. Natl. Acad. Sci. U. S. A.*, 2012, **109**, 21259–21264.

Mr Cox replied: The stacking faults discussed in ref. 1–3 occur exclusively along the *c*-axis of ice, resulting in a mixture of hexagonal and cubic stacking arrangements (referred to as "ice I_{SD}" in reference 2). The result of the basal face of kaolinite studied in our simulations is to give a preferential growth along the prism face of ice, which is orthogonal to the *c*-axis. I would therefore not expect stacking faults to impair the effect of small ice nuclei on the macroscopic crystal morphology. The question of what effect changes in structure of a small ice nucleus can have on the macroscopic morphology is not fully known, and is a speculation based on the findings of our work. It is worth noting, however, that the experiments of Evans[4] suggest that β-AgI is able to affect the *phase* of ice that forms at different pressures, which can be seen as an extreme example of the effect that we suggest ice nucleating agents can have on the macroscopic structure of ice.

1 E. B. Moore and V. Molinero, Is it cubic? Ice crystallization from deeply supercooled water, *Phys. Chem. Chem. Phys.*, 2011, **13**, 20008–20016.
2 T. L. Malkin, B. J. Murray, A. V. Brukhno, J. Anwar and C. G. Salzmann, Structure of ice crystallized from supercooled water, *Proc. Natl. Acad. Sci. U. S. A.*, 2012, **109**, 1041–1045.
3 W. F. Kuhs, C. Sippel, A. Falenty and T. C. Hansen, Extent and relevance of stacking disorder in "ice I(c)", *Proc. Natl. Acad. Sci. U. S. A.*, 2012, **109**, 21259–21264.
4 L. Evans, *Nature*, 1965, **206**, 822.

Mr Limmer said: Mr. Cox has shown how correlations within the water on the kaolinite surface can affect the preferred nucleation pathways for ice crystallization away from the surface. Platinum, which is similar to kaolinite in forming a water adlayer that is hydrophobic, is also regarded as a good surface for growing single ice crystals. Are there reasons to believe that the formation of a hydrophobic interface is related to these surfaces particular ice nucleating abilities? What does this imply for ice nucleation at a liquid–vapor interface?

Mr Cox responded: The question of how a surface's ice nucleating ability is related to the hydrophobicity of its adlayer is interesting, but as we have only been able to nucleate ice in the presence of kaolinite, we are unable to determine with any certainty what this relationship is (*i.e.* a study in which ice nucleation at a variety of surfaces that form different types of adlayer is required). See also comment 514 from Prof. Molinero.

Professor Molinero continued: We studied the heterogeneous nucleation of ice on a variety of carbon surfaces and found that, similar to what was reported by Mr. Cox for kaolinite, liquid water layers at the surface of graphitic surfaces. We found a correlation between increased layering of liquid water at a carbon surface and the surface ability to promote ice nucleation.[1] It is an open question whether layering of liquid water occurs on all surfaces that promote heterogeneous ice nucleation.

1 L. Lupi, A. Hudait and V. Molinero, Heterogeneous nucleation of ice on carbon surfaces, under review.

Dr Royall commented: You mention in your paper and at the end of your talk some concerns about finite size effects. In fact, you say that homogenous nucleation is not observed the case of the larger system of 768 molecules, but that it is in the smaller system of 192, suggesting that finite size effects through the periodic boundaries are influencing nucleation. Can you comment on whether it is possible to, for example, go to deeper quenches or even shallower quenches where the dynamics are faster?

Mr Cox responded: This is a reasonable suggestion and one that we have investigated. We mention in the paper that we tried a temperature range of 190–220 K for TIP4P/2005 water (15.5 μs total trajectory time). We also tried using the TIP4P/ice water model at 240 K (2 μs total trajectory time) but we did not see any nucleation. As the computational expense of these calculations is quite large, we have not investigated a wider temperature range.

Mr Durham asked: Does your simulation allow you to include traces (around 100 μmol per mol) of a non-water molecule such as CO_2, to see if a clathrate or hydrate could exist as part of the hexagonal matrix?

The question derives from Poster 35 (The CO_2–water equilibrium at partial pressures below 101.325 kPa: eighty years of measurements), which seeks an explanation for unexpectedly high CO_2 concentrations in UK rainwater and snowfall.

Mr Cox answered: Our ice nucleation simulations would not allow us to include trace gases at the concentrations that you specify. So far, we have only been able to observe ice nucleation in systems containing 192 water molecules, periodically repeated in x and y. Including just one gas molecule would therefore lead to a concentration of approximately 5 mmol per mol. Even in the larger systems that we investigated (768 water molecules, where ice nucleation was not observed) the concentration would still be a little over 1 mmol per mol. To reach a low concentration of 100 μmols per mol, 10 000 water molecules would be required for every gas molecule. To my knowledge, no heterogeneous ice nucleation simulations involving atomospherically relevant minerals have been performed on such large systems.

Mr Durham communicated in reply: Your oral response tallies with Dr Salzmann's view afterwards, *i.e.* "space in ice"[1] is confined to the hexagonal matrix illustrated in your simulation, and despite its 10% additional molar volume gained at freezing, ice would only be capable of transporting a species as small as helium. This view accords with my own analyses of snowfall and of ice created in CO_2 solutions, but fortuitously this conference illustrated the potential for space in liquid water. An upward deflection in the slope of the molar volume of water begins around 25 °C, whereafter between zero and −34 °C the volume increases by 2.5%.[2,3] Could this offer solvent "space" that is less constrained than the ice matrix? The smooth curve of CO_2 solubility in liquid water at partial pressure of 101.3 kPa is not visibly deflected.[4] However at a partial pressure of 45 Pa of CO_2 (still half again higher than cloud partial pressure), the solute gas would be present at a much lower concentration, and a 2.5% increase in molar volume of the solvent might be relatively more available to the solute molecule. Some encouragement is provided by additional post-conference values for Fig. 5 of Poster 35, showing solubility inverse to temperature (see Fig. 3). If this curve continues to track molar volume into the supercooled zone, cloud water would have a significant dissolved content at the temperature of ice nucleation.

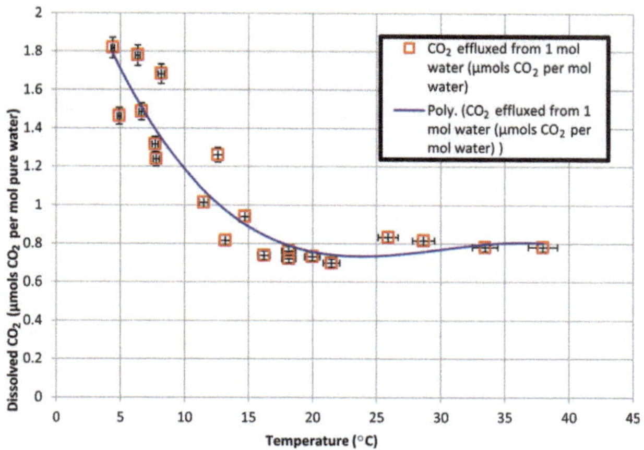

Fig. 3 Aqueous solubility of carbon dioxide at a partial pressure of *c.* 45 Pa of the gas, in the temperature range 4–38 °C. In the lower part of this range solubility is showing as inverse to temperature. Values are RMS averages of 60 readings at Li-Cor's 3% confidence level.

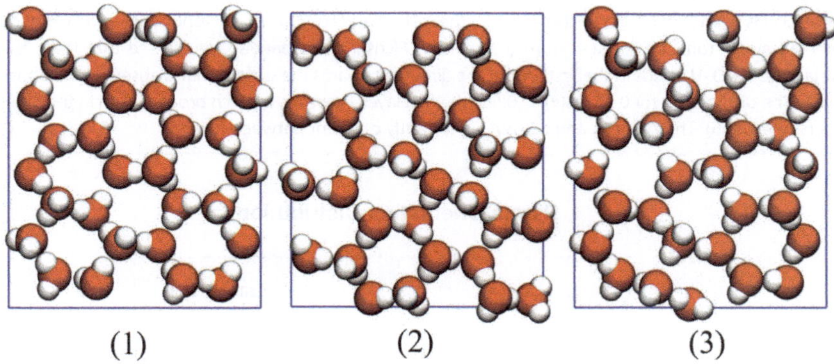

Fig. 4 Snapshots of ice that forms in the first (1), second (2) and third (3) layers in one of our simulations. Here we can see that the ice that forms is not a proton ordered phase.

1 W. H. Bragg, The crystal structure of ice, *Proc. Phys. Soc. London*, 1921, **34**, 98.
2 D. E. Hare and C. M. Sorensen, The density of supercooled water. II. Bulk samples cooled to the homogeneous nucleation limit, *J. Chem. Phys.*, 1987, **87**, 4840.
3 V. Holten and M. A. Anisimov, Entropy-driven liquid–liquid separation in supercooled water, *Sci. Rep.*, 2012, **2**, 1–7, http://dx.doi.org/10.1038/srep00713.
4 P. Scharlin, *IUPAC Solubility data series v 62*, Oxford, 1996.

Professor Wales queried: I wonder if there might be additional proton ordering of the ice 1h oxygen framework near to the surface? There could be some interesting electrostatic effects here that might reflect additional structure.

Mr Cox communicated in reply: This is an interesting question and something we had not looked into. To answer this, I have included snapshots from the 1st, 2nd and 3rd layers of ice that form in one of our simulations (Fig. 4(1), (2) and (3), respectively). From these pictures, we can clearly see that we are not forming a distinct proton ordered phase of ice. However, we have investigated this in more detail, and I include two other figures. Fig. 5 shows the zenith and azimuth O-H

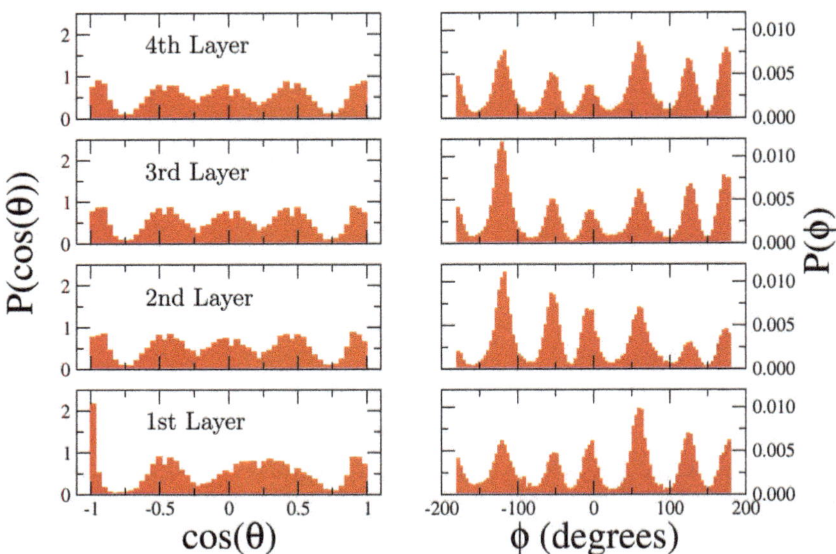

Fig. 5 Zenith (θ) and azimuth (ϕ) angle distributions of O–H bonds. For the zenith angle, $\cos(\theta) = -1$ corresponds to an O–H bond directed towards the surface and $\cos(\theta) = +1$ corresponds to and O–H bond directed away from the surface. From the plots of $P(\cos(\theta))$ (left panels) we can see that there is a preference for O–H bonds in the first layer to be directed towards the surface that is absent in the above layers. The distribution for $0 \leqslant \cos(\theta) \leqslant 0.5$ in the first layer, however, is much broader than in the other layers (see Fig. 6). The azimuth angle is not significantly different between layers.

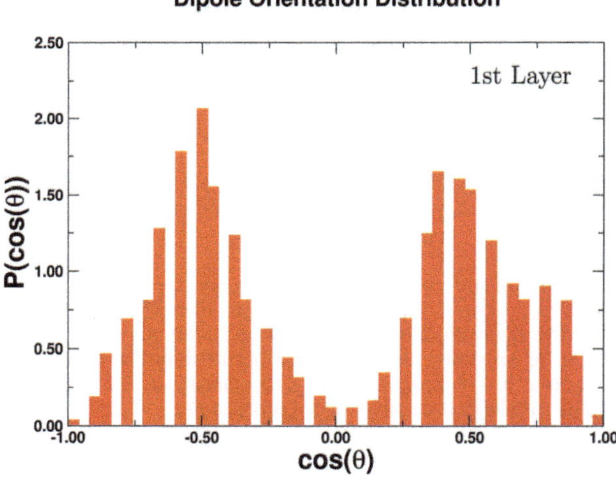

Fig. 6 Zenith angle distribution of water dipole moments in the first layer. For dipoles oriented towards the surface, $\cos(\theta) = -1$. Clearly the distribution is bimodal (average value of $\cos(\theta) = 0.007$, standard deviation $= 0.551$).

bond angle distributions for the 1st to 4th layers of ice (averaged over 44ns). The zenith angle (θ) is defined such that $\cos(\theta) = -1$ corresponds to an O–H bond directed towards the kaolinite surface, and $\cos(\theta) = +1$ corresponds to an O–H bond directed away from the kaolinite. From the distribution of $\cos(\theta)$ ($P(\cos(\theta))$ plots in Fig. 5, we can clearly see a peak at $\cos(\theta) = -1$ in the first layer that is absent in the other layers. This means the surface is inducing a preference for the water molecules to direct an O–H bond directly towards the surface. This is consistent with previous density functional theory calculations.[1,2] However, the distribution for $0 \leq \cos(\theta) \leq 0.5$ in the first layer is broadened compared to the other layers and this is reflected in the dipole orientation distribution in the first layer (Fig. 6). Here we can clearly see a bimodal distribution centred around zero (the average value of $\cos(\theta) = 0.007$, standard deviation $= 0.551$). If we look at the azimuth angle (φ) distributions in Fig. 5 we can also see that the kaolinite does not induce any significant changes in the OH orientations in the plane of the surface (the plots are slightly noisy due to the small lateral dimension used). The azimuth angle is defined such that $\varphi = 0$ corresponds to an O–H bond laying parallel to the *a*-axis of the kaolinite conventional unit cell. To summarise, we do not form a fully proton-ordered phase, but the kaolinite does induce a preference for O–H bonds to be directed towards the surface.

1 X. L. Hu and A. Michaelides, *Surf. Sci.*, 2007, **601**, 5378.
2 X. L. Hu and A. Michaelides, *Surf. Sci.*, 2008, **602**, 960.

Dr Haji-Akbari enquired: Have you studied homogeneous nucleation in the bulk?

Mr Cox replied: No, we have only studied the thin films. This provided the easiest way by which we could compare homogeneous and heterogeneous ice nucleation.

Dr Haji-Akbari asked: Based on your successful trajectories that correspond to homogeneous nucleation of ice in a free-standing thin film, do you have any statistics on the likelihood of crystallization occurring in the subsurface in comparison to bulk crystallization, *i.e.* crystallization initiating in the center of the film? This question is due to conflicting results in the literature on this matter. For instance, Vrbka and Jungwirth[1] study thin films of pure water; they use a six-site potential and observe a higher probability of sub-surface crystallization due to the violation of electrostatic neutrality in the surface. On the other hand, Lu *et al.* study thin films of mW and observe a higher probability of bulk (center) nucleation compared to subsurface nucleation.

1 L. Vrbka and P. Jungwirth, *J. Phys. Chem B*, 2006, **110**, 18126.
2 Y. Lu *et al.*, *J. Phys. Chem. B*, 2013, **117**, 10241.

Mr Cox responded: We have not investigated whether homogeneous ice nucleation occurs in the subsurface or bulk. As we have only investigated 30 Å thick films, our simulations offer little in the way of trying to resolve this issue.

Dr Haji-Akbari asked: How thick is the film in which homogeneous nucleation is studied?

Mr Cox answered: The homogeneous films are approximately 30 Å thick.

Professor Wynne continued the discussion of the paper by Professor Molinero: In your simulations,[1] the ions were modelled as effective single particles rather than as ion pairs. However, experimental work by the Bakker group[2] has shown cooperative effects where water molecules form short chains between cation and ions, and our own work[3] has shown the presence of tetrahedrally hydrogen-bonded "Walrafen" pentamers between Li^+ and Cl^-. Therefore, the question is whether the single-particle representation of the ions is an accurate representation of the relevant physics.

1 G. Bullock and V. Molinero, *Faraday Discuss.*, 2013, **167**, DOI: 10.1039/C3FD00085K.
2 K. J. Tielrooij, N. Garcia-Araez, M. Bonn and H. J. Bakker, *Science*, 2010, 328, 1006–1009.
3. D. A. Turton, C. Corsaro, D. F. Martin, F. Mallamace and K. Wynne, Phys. Chem. Chem. Phys., 2012, **14**, 8067–8073.

Professor Molinero responded: According to our analysis, ions have a colligative effect on the freezing temperature of water. Colligative properties depend on properties of the solvent, but not of the solute. In a colligative property, the only property of the solute that matters is its ability to tune the chemical potential of the solvent (*i.e.* the water activity). Fig. 1 of our paper[1] shows quantitative agreement of the melting temperature of ice in water–LiCl mixtures (experimental) and mW–S mixtures (simulations) for concentrations up to 15 mol% of ions. As the temperature of melting and enthalpy of melting of pure mW water are in good agreement with the experiments, the agreement on the melting temperatures implies a good description of the water activity as a function of solute content. These results indicate that the simulation model, while surely too simple for the study of the hydration structure, captures the relevant physics for the study of the crystallization of solutions.

1 G. Bullock and V. Molinero, *Faraday Discuss.*, 2013, **167**, DOI: 10.1039/C3FD00085K.

Professor Chandler commented: Professor Molinero's paper presents a number of interesting results showing how crystallization follows from fluctuations into low-density liquid congfiurations of water. She also shows how the amorphous solids appear when rapid cooling arrests these fluctuations. As I have already noted, her results are consistent with independent theoretical results that Dr. Limmer and I have described elsewhere.[1–3] Some have questioned whether these results are peculiar to the mW model of water. I believe that any reasonable model one for which the liquid prefers local tetrahedral structure and for which the liquid crystallizes into an ice-like structure will exhibit the same general behaviors shown to us by Professor Molinero. Different models will produce qualitatively different behaviors only to the extent that they are examined at different corresponding states. The usual models applied in molecular simulation studies of water all behave similarly when put at the same corresponding states.[2,4]

1 D. T. Limmer, and D. Chandler, *J. Chem. Phys.*, 2012, **137**, 045509.
2 D. T. Limmer, and D. Chandler, *J. Chem. Phys.*, 2013, **138**, 214504.
3 D. T. Limmer, and D. Chandler, 2013, arXiv:1306.4728.
4 D. T. Limmer, and D. Chandler, *Faraday Discuss.*, 2013, **167**, DOI: 10.1039/C3FD00076A.

Professor Tanaka responded: We agree that the link of locally favoured or mesoscopic structures formed in a liquid with the equilibrium crystalline structure[1] is a very interesting issue. Recently we studied this problem for hard spheres, soft spheres, and mW water. In the case of hard spheres, we have two important structural features: crystal-like bond orientational ordering (fcc- and hcp-type) and icosahedral ordering.[2] We found that the presence of mesoscopic ordering, whose rotational symmetry is consistent with the equilibrium crystal, leads to the lowering of the nucleation barrier[3,4] and the pre-existing order selects crystal polymorphs to be formed.[4,5] We found a similar link between the pre-existing order and the crystal to be formed in soft Gaussian core liquids and mW water. In the case of water, we also found that the locally favoured structures characterized by the translational order in the second shell have a structural feature consistent with the ice structure as well as that inconsistent with it,[6] which leads to an interesting effect on ice crystallization.[7]

1 H. Tanaka, Bond orientational order in liquids: Towards a unified description of water-like anomalies, liquid–liquid transition, glass transition, and crystallization, *Eur. Phys. J.*, 2012, **E35**, 1.
2 M. Leocmach and H. Tanaka, Roles of icosahedral and crystal-like order in hard spheres glass transition, *Nat. Comm.*, 2012, **3**, 974, DOI: 10.1038/ncomms1974.
3 T. Kawasaki and H. Tanaka, Formation of crystal nucleus from liquid, *Proc. Natl. Acad. Sci. U. S. A*, 2010, **107**, 14036–14041.
4 J. Russo and H. Tanaka, The microscopic pathway to crystallization in supercooled liquids, *Sci. Rep.*, 2012, **2**, 505, DOI: 10.1038/srep00505.
5 J. Russo and H. Tanaka, Selection mechanism of polymorphs in the crystal nucleation of the Gaussian core model, *Soft Matter*, 2012, **8**, 4206–4215.
6 J. Russo and H. Tanaka, Understanding water's anomalies with locally favored structures, 2013, arXiv:1308.4231.
7 J. Russo, F. Romano and H. Tanaka, unpublished.

Professor Wang remarked: We have a water model (presented in poster 7) that seems to indicate a more stable transient LDL conformation.[1,2] The neighbor averaged $q6$ order parameter for the metastable LDL configurations calculated with our model is very different from that of ice-Ih even for the first solvation shell waters. If a substantial reorientation is required to transform LDL to ice-Ih, do you anticipate the mW model will underestimate the transition barrier?

1 E. Pinnick, S. Erramilli and F. Wang, Predicing the melting temperature of ice-Ih with only the electronic structure information as input, *J. Chem. Phys*, 2012, **137**, 014510.
2 Y. Li, J. Li and F. Wang, Liquid–liquid transition in supercooled water suggested by microsecond simulations, *Proc. Natl. Acad. Sci. U. S. A.*, 2013, **110**, 12209–12212.

Professor Molinero responded: Within the framework of Classical Nucleation Theory (CNT), the barrier for crystal nucleation depends only on thermodynamic quantities that reflect the competition of the higher stability of the bulk crystal and the free energy cost of the interface between crystal nucleus and liquid mother phase. These quantities are the density, the excess free energy of the liquid with respect to the crystal and the liquid-ice surface tension. These properties are all well reproduced by the mW model, down to the lowest temperatures for which experimental data is available. This has been elaborated upon and discussed in detail in the supporting information a previous paper.[1] However, as the free energy barrier in CNT depends on the cube of the surface tension, a quantitative assessment of the agreement would require data on the dependence of the liquid–ice surface tension with temperature and radius of the crystallites, which to my knowledge is not available for mW nor experimental water.

1 E. B. Moore and V. Molinero, Structural transformation in supercooled water controls the crystallization rate of ice, *Nature*, 2011, **479**, 506–508.

Professor Anisimov enquired: Could the low-density water be also the mother of molecular clustering in solutions of hydrotropes at low temperatures?

Professor Molinero replied: Low-density water is, generally, solutophobic although it is a better solvent for small hydrophobic species than high temperature water.[1] It would be interesting to investigate whether hydrotopes can be dissolved in low-density amorphous ice. The study of Ben-Amotz and coworkers in this Faraday Discussion[2] suggests that the structure of the clusters of TBA and water may resemble the one of "blobs", clusters of solutes at solvent-separated distances, which are precursors for the homogeneous nucleation of clathrate hydrates.[3] While water in clathrates is, like in ice, tetrahedrally coordinated, the local order of water in blobs is different from that in low-density water.

1 D. Paschek, How the liquid–liquid transition affects hydrophobic hydration in deeply supercooled water, *Phys. Rev. Lett.*, 2005, **94**, 217802.
2 D. Ben-Amotz *et al.*, Faraday Discuss., 2013, **167**, DOI: C3FD00086a.
3 L. C. Jacobson, W. Hujo and V. Molinero, Amorphous precursors in the nucleation of clathrate hydrates, *J. Am. Chem. Soc.*, 2010, **132**, 11806–11811.

Professor Mallamace asked: In your work, the water activity is a central quantity; what is the correct experimental way to have a proper measure of it?

Professor Molinero answered: The water activity is defined as the ratio between the vapor pressure of water in the mixture and the vapor pressure of pure water at the same temperature. That definition provides a way to measure water activity by determining the vapor pressure of water. Alternatively, the water activity can be determined from the equilibrium melting temperature using eqn 1 in our paper[1] or a more accurate integration that accounts for the temperature dependence of the enthalpy of melting. We took the second route for the determination of the water activity in the simulations.

1 G. Bullock and V. Molinero, *Faraday Discuss.*, 2013, **167**, DOI: 10.1039/C3FD00085K.

Professor Chandler opened the discussion of the paper by Dr Royall: I wonder whether your focus on specific static structural features in a glass forming material can be valid. As you know, an order parameter distinguishing a glass from its melt must measure deviations from equilibrium ergodic behavior. In particular, a measure of local structure at a given time, call it $a(t)$, has the time average

$$\bar{a} = (1/t_{\text{obs}}) \sum_{t=1}^{t_{\text{obs}}} a(t).$$

Taking the averaging time, t_{obs}, to be large, the quantity $\bar{a}$ is the equilibrium average, $\langle a \rangle$, when the system is at equilibrium. A glass transition occurs at a stage where $\bar{a}$ changes notably from $\langle a \rangle$. In other words, a glass is characterized by an arrested fluctuation that is far from equilibrium, a behavior for which $\bar{a} - \langle a \rangle$ serves as an appropriate order parameter.

This observation is the basis for the so-called s-ensemble method for preparing glassy states in molecular simulation. Ref. 1–4 provide illustrations. In contrast to $\bar{a} - \langle a \rangle$, the behavior of $a(t)$ at one point in time does not distinguish equilibrium from non-equilibrium, which brings me to my question: to what extent can you justify characterizing glass, as you attempt in your paper, through a specific choice of $a(t)$ (like the instantaneous density of a certain type of cluster) rather than through a time average of that choice?

1 L. O. Hedges, R. L. Jack, J. P. Garrahan and D. Chandler, *Science*, 2009, **323**, 1309–1313.
2 R. L. Jack, L. O. Hedges, J. P. Garrahan and D. Chandler, *Phys. Rev. Lett.*, 2011, **107**, 275702.
3 T. Speck, and D. Chandler, *J. Chem. Phys.*, 2012, **136**, 184509.
4 T. Speck, A. Malins and C. P. Royall, *Phys. Rev. Lett.*, 2012, **109**, 195703.

Dr Royall replied: Your points about the s- and mu-ensembles are quite right in that context, and in the latter case for sure, when the time-averaging was taken to one frame, the transition disappeared. Certainly, as these and other papers demonstrate, a transition exists in trajectory space. However, this does not in my view exclude the possibility of a random-first-order transition[1] behaviour of cooperatively re-arranging regions, or indeed concepts such as geometric frustration.[2] Such scenarios are based on static quantities. Now our data, like many others, show a disparity between the static and dynamic correlation lengths. I believe that this may be related to an increase in the dynamic correlation length near the mode-coupling transition.[3] What is clear is that at deeper supercoolings, the increase of the dynamic correlation length in the T > TMCT regime is not maintained.[4,5] I thus expect a coincidence of structural and dynamical lengths at the T_g.

1 V. Lubchenko and P. Wolynes, *Ann. Rev. Phys. Chem.*, 2007, **58**, 235–266.
2 G. Tarjus, *et al.*, *J. Phys.: Condens. Matter*, 2005, **17**, R1143–R1182.
3 P. Charbonneau and G. Tarjus, *Phys. Rev. E.*, 2013, **87**, 042305.
4 M. T. Cicerone and M. D. J. Ediger, *Chem. Phys.*, 1995, **103**, 5684–5892.
5 L. Berthier, G. Biroli, J.-P. Bouchaud, L. Cipelletti, D. El Masri, D. L'Hote, F. Ladieu and M. Pierno, *Science*, 2005, **310**, 1797–1800.

Professor Chandler remarked: There is a general structural feature that distinguishes glass from melt. It is the distribution of distances between excitations. Excitations in glass and glass-forming materials are localized to regions where particle configurations can reorganize.[1] These regions are soft spots in an otherwise rigid material. Aging of the glass, or relaxation of the melt, follows from the dynamics of those excitations or soft spots.[2] At equilibrium, *i.e.*, in the melt, the spatial distribution of soft spots is like that of particles in a dilute gas.[1] In contrast, when the melt is driven out of equilibrium to form glass, a non-equilibrium correlation length, ℓ_{ne}, emerges. The distribution of distances between excitations changes from exponential in the melt to non-exponential in the glass. ℓ_{ne} is the most probable distance in the non-exponential distribution. This length can be large. The more stable the glass the larger it is.[3] It may be that one can relate the equilibrium concentration of excitations to average densities of locally ordered clusters, but it is the arrangement of those excitations (not simply their concentration) that distinguishes non-equilibrium glass from its equilibrium melt. Stillinger and Torquato have also noted that a large correlation length distinguishes a melt from its glass.[4]

1 A. S. Keys, L.O. Hedges, J. P. Garrahan, S. C. Glotzer, and D. Chandler, *Phys. Rev.*, 2011, **X1**, 021013.
2 D. Chandler and J. P. Garrahan, *Annu. Rev. Phys. Chem.*, 2010, **61**, 191–217.

3 A. S. Keys, J. P. Garrahan and D. Chandler, *Proc. Natl. Acad. Sci. U. S. A.*, 2013, **110**, 4482–4487.
4 E. Marcotte, F. H. Stillinger, and S. Torquato, *J. Chem. Phys.*, 2013, **138**, 12A508.

Dr Royall replied: Thank you for this point. I agree that excitations can provide a mechanism for relaxation and that their distribution – and the "surging" events between excitations – may form a key structural aspect to relaxation upon deep supercooling. I should like to add that we expect structural signatures of "fast" regions also to be accessible to our techniques, which may correspond to these excitations.

Professor Barrat asked: Do you have any idea of the local mechanism that governs the dissolution or disappearence of long lived clusters?

Dr Royall replied: The clusters come and go through thermal fluctuation. We define them through a bond network, and can have redefinitions of the bond network on timescales shorter than the α-relaxation time. However we ignore these "breakages" of a cluster if the same particles are again found in the cluster within the α-relaxation time.

We believe that the locally favoured structure (LFS), the longest-lived cluster, is a geometrically stable arrangement. For example, the 11A bicapped square anti-prism which is the LFS for the Kob–Andersen mixture has a full shell around the centre particle. Within the Kob–Andersen stochiometry, we almost always find a small particle at the centre of the 11A cluster, and this can act to minimise strain.

Conversely, in the Wahnstrom binary Lennard-Jones mixture, the LFS is the icosahedron.[1] Again the shell is full, but here the stochimetry is 50–50 such that the shell is around 50–50 large and small particles, while in the Kob–Andersen system the 80–20 stoichiometry means the shell is predominantly large particles.

In hard spheres, the LFS is a cluster we term 10B.[2] These are icosahedra missing three particles. However, we have some evidence that, at deeper super-cooling, more icosahedra form and 10B are "broken icosahedral fragments".

In short, we think the long-lived structures are mechanically robust. This is implied by a filled shell and low strain. These properties will be satisfied by different geometries depending on the system under consideration.

1 A. Malins *et al.*, *J. Chem. Phys.*, 2013, **138**, 12A535.
2 J. Taffs *et al.*, *Soft Matter*, **9**, 297.

Dr Perera commented: There are currently three competing theories of the glass transition, from what is reported in a recent review (random first order transition, frustration and facilitation).[1] The RFOT predicts the hypothetical existence of an ideal glass, made of mosaic of domains; hence suggesting that glasses would be a mosaic of domains. To what extent do you think that the sort of elementary clusters that you report could be used to support the picture provided by RFOT?

1 L. Berthier and G. Biroli, *Rev. Mod. Phys.*, 2011, **83**, 587.

Dr Royall replied: Random first-order theory is proposed to become dominant in the temperature range between the mode-coupling transition and a lower temperature transition to an "ideal glass". At the moment, this regime is hard to access with computer simulation (or other techniques that provide a similar degree of detail such as particle-resolved studies of colloids). I believe that the reason that our measures of structural and dynamic lengths do not coincide may reflect the mode-coupling transition somehow leading to the former increasing strongly. What is clear is that the dynamic correlation length seems unlikely to maintain its rate of increase in the temperature regime above the mode-coupling transition. This is because it has already increased to around 10 particle diameters (see our paper and many others), yet, indirect measurements on molecules at the molecular glass transition[1–3] show a similar length-scale, despite the fact that the molecular glass transition corresponds to a relaxation time around ten orders of magnitude greater than the mode-coupling transition. This view is further supported by recent computer simulations which suggest a change in the growth of the dynamic correlation length around the mode-coupling transition.[4,5]

I would therefore like to think that, in the temperature regime below the mode-coupling transition, the structural and dynamic correlation lengths coincide and fast-moving cooperatively re-arranging regions correspond to clusters distinct from the bicapped square antiprisms, but the bicapped square antiprisms are found in dynamically slow regions. Thus, at lower temperatures than we can access with our simulations, our structural and dynamic lengthscales will coincide, and the mosaic state consists of slow bicapped square anti-prism-based CRRs and fast CRRs of different local structure.

1 M. T. Cicerone and M. D. Ediger, *J. Chem. Phys.*, 1995, **103**, 5684–5892.
2 S. Tatsumi, S. Aso and O. Yamamuro, *Phys. Rev. Lett.*, 2012, **109**, 045701.
3 L. Berthier, *et al.*, *Science*, 2005, **310**, 1797–1800.
4 W. Kob, S. Roldán-Vargas and L. Berthier, *Nature Phys.*, 2011, **8**, 164–167.
5 E. Flenner and G. Szamel, *J. Chem. Phys.*, 2013, **138**, 12A523.

Professor Angell asked: Would it be reasonable to see the glass formed in the laboratory (*i.e.* one that has lost its entropy by cooling at rates that are quite unachievable in computer simulation) as being not merely a percolated structure, but a totally jammed structure made of units like the ones you describe? Glass transition would then correspond to the unjamming, by local excitations, occurring under completely solid-like diffusivity conditions ($D \approx$ 10–22 m^2s^{-1}; the glass transition occurs on time scales many orders of magnitude below the limits of present day calculations. Would that be a viable way of looking at the declustering during heating?

Dr Royall responded: I think what you suggest is a most attractive scenario. Of course, since the dynamical regime accessible to simulation is far from the molecular glass transition, inferences from numerical data are necessarily speculative. However, we[1] and others[2] have devised means of preparing simulation data which by some measures appears closer to the molecular glass transition. These papers suggest a saturation in the number of clusters. We have studied the break-up of the clusters upon effective heating (specifically, by removing the biasing field that generated the clusters). Our results suggest a melting transition concurrent with disintegration of the clusters. A more detailed analysis might well reveal the kind of behaviour you suggest and we will look into it.

1 T. Speck *et al.*, *Phys. Rev. Lett.*, 2012, **109**, 195703.
2 S. Singh *et al.*, *Nature Mater.*, 2013, **12**, 139–144.

Professor Anisimov opened the discussion of the paper by Dr Sefcik: What makes your system different from the TBA solution where the mesoscale inhomogeneities originate from hydrophobic impurities? Are your aggregates closely related to the existence of the solid phase?

Dr Sefcik communicated in reply: We investigated aqueous solutions of glycine and DL-alanine. These are the two simplest amino acids, where for glycine there is no side group present that would provide a hydrophobic character and DL-alanine has a single methyl side group. So one clear difference between these small amino acid molecules and TBA is an absence of hydrophobic functional groups. This would be expected to limit propensity of these amino acids to associate with hydrophobic impurities and/or stabilise corresponding nanoemulsions present in TBA aqueous solutions.[1] We believe that mesoscale clusters observed in aqueous amino acid solutions are not directly related to the existence of solid phases, but rather are thermodynamically stable features of liquid phases well below saturation concentrations (*i.e.*, well above solubility temperatures) with respect to the corresponding solid phase.

1 D. Subramanian *et al.*, *Faraday Discuss.*, 2013, **167**, DOI: 10.1039/C3FD00070b.

Professor Wynne remarked: The data in the first panel of Fig. 4 in your paper[1] shows very distinct jumps at a supersaturation of 0.1 and 1, respectively. Are these jumps associated with some type of phase transition, for example, something akin to a liquid–gas transition involving the molecular cluster you have spoken about? If these are phase transitions, why are there two of them and why are they occurring at these particular concentrations?

1 A. Jawor-Baczynska, B. D. Moore, H. S. Lee, A. V. McCormick and J. Sefcik, *Faraday Discuss.*, 2013, **167**, DOI: 10.1039/C3FD00066D.

Dr Sefcik replied: There are two distinct jumps in mesospecies concentrations in DL-alanine solutions as seen in Fig. 4. We believe that the first jump at superstauration of about 0.1 is due to a phase transition between a molecular fluid (including small molecular clusters with hydrodynamic diameters up to 1 nm) and a cluster fluid. Above this concentration (a critical concentration for cluster formation), the thermodynamically stable state of the solution is a mesostructured liquid containing solute-rich domains (*i.e.*, mesospecies). Why the critical concentration is at this particular value, we cannot explain at present, but it would presumably depend on details of molecular interactions in a given system. We note that small molecular clusters were found to be present in amino acid solutions under all conditions investigated here and their formation is apparently not subject to a significant energy barrier. However, there appears to be a considerable energy barrier for the formation of mesoscale clusters.

Regarding the second jump at supersaturation of 1, we believe that this is in fact related to the presence of metastable conditions in solutions supersaturated with respect to the solid crystalline amino acid and therefore it is not an indication of a phase transition in the liquid phase itself. This is supported by observations of filtered supersaturated solutions after subsequent tumbling (Fig. 4a in our paper here and further unpublished data), where the number concentrations of mesospecies obtained were similar to those for undersaturated solutions at comparable tumbling times, but significantly lower than those observed in unfiltered solutions prepared by direct dissolution of crystalline solids. This shows that mesospecies populations in supersaturated solutions are pathway dependent and thus not in full thermodynamic equilibrium (unlike those in undersaturated solutions).

Dr Alexander asked: Can you go through, in terms of the clusters reappearing, what you think about the effects of walls, and how the energy of agitation affects the system? Where are the clusters hiding?

Dr Sefcik replied: Mesoscale clusters observed in both glycine and DL-alanine solutions can be removed by nanofiltration with suitable filters. When such filtered solutions are mechanically agitated for extended periods of time, the mesospecies population reappears with the same mean size and number concentration as before. This indicates that while the mesospecies appear to be an equilibrium feature of the solution, there is a substantial energy barrier to their formation. Since agitation including gentle mechanical contact with glass container walls appears to enhance cluster reformation, we speculate that formation of clusters (*e.g.*, through a nucleation process) might be more favourable at container walls (through heterogeneous nucleation). Resulting clusters then may be detached from walls through mechanical contact and appear in the bulk solutions where they can be detected by light scattering and Brownian microscopy (NTA).

Dr Vila Verde enquired: Do you know the composition of your droplets?

Dr Sefcik communicated in reply: We have not yet been able to quantify the composition of nanodroplet mesospecies observed in aqueous amino acid solutions. Based on observations from glycine and DL-alanine aqueous solutions[1,2] we believe that mesospecies in these two-component solutions correspond to solute-rich liquid cluster domains with solute concentrations higher than those in surrounding bulk solution.

1 A. Jawor-Baczynska, B. D. Moore, H. S. Lee, A. V. McCormick and J. Sefcik, *Faraday Discuss.*, 2013, **167**, DOI: 10.1039/C3FD00066D.
2 A. Jawor-Baczynska, J. Sefcik and B. D. Moore, *Cryst. Growth Des.*, 2013, **13**, 470–478.

Professor Angell commented: The clusters you describe, delicate assemblies that can be easily destroyed by shear stresses, as in filtration or ultrasonic agitation, but then reappear after a certain time, remind me a lot of the controversial "Fischer clusters", discussed a lot by polymer and glassformer liquid scientists over the years. Would it not be possible to examine the reassembly process more quantitatively to see if they follow the equations of nucleation and growth? There are a variety of experiments that can be carried out to quantify this process. For instance one can carry out combinations of probes to separate nucleation kinetics from nucleation + growth combinations.

Dr Sefcik communicated in reply: It would be certainly very interesting to perform kinetic investigations of nucleation and growth of thermodynamically stable mesoscale clusters observed in aqueous amino acid solutions. In this context it is also worth highlighting experimental work by Sedlak on kinetics of formation of mesoscale clusters in a range of solution systems[1] as well as recent investigations to distinguish mesoscale liquid clusters from nanobubbles.[2]

1 M. Sedlak, *J. Phys. Chem. B*, 2006, **110**, 4339–4345.
2 M. Sedlak and D. Rak, *J. Phys. Chem. B*, 2013, **117**, 2495–2504.

Dr Vila Verde opened the discussion of the paper by Dr Alexander: In Fig. 4, the signal decreases between $S = 0.95$ and $S = 1.35$. You state that this decrease is significant; however, the decrease is small and the uncertainty associated with each point is large, so one could just as easily interpret the data to mean that the signal stays approximately constant in that range. Can you elaborate on why you are confident that the observed decrease in signal is significant? Would it be feasible to perform more experiments so that the statistical uncertainty associated with each point could be reduced?

Dr Alexander responded: In our paper we have tried to be reasonably careful with the language used, so we refer to the changes as apparent rather than definite. As you rightly point out, the difference in the means between the points at $S = 0.95$ and 1.35 is not statistically significant (2-tailed *t*-test, $P = 0.15$); the difference between the points at $S = 1.75$ and 1.86 is significant ($P < 10^{-4}$). More measurements in the region of $S = 1$ are desirable. This should be done by improving detection of the SHS signal from the particles relative to the HRS background. For example, simply increasing the laser power would favour collection of the HRS signal, which is monotonic (dashed line in Fig. 4 of our paper). One possible strategy would be to use optical trapping of these particles, which we have demonstrated in principle but as yet have not combined with the SHS experiment.

Professor Ben-Amotz asked: How many times were each of the experiments repeated in order to obtain the error bars on each of the points in Fig. 4 of your paper? Note that it is always worrisome when removing one point from a set of data points is sufficient to make an apparent feature disappear.

Dr Alexander replied: For the data in Fig. 4 of our paper the number of repeat measurements per point (n) for $S < 0.5$ was $n = 5$, and for $S > 0.6$ was $n = 8$ to 14. The error bounds shown represent 95% confidence intervals (2σ). The repeat experiments at different concentrations were conducted in no particular order to avoid systematic errors.

Professor Wales asked: What are the error bars for the supersaturation values?

Dr Alexander replied: We believe the largest source of uncertainty in our supersaturation values stems from the accuracy of the solubility data rather than experimental uncertainties from preparation of samples. A rudimentary analysis comparing solubility data from several sources suggests that the uncertainties in our supersaturations is <1%, certainly within the boundaries of the symbols used in Fig. 4 of our paper.

Professor Walker commented: The data in Fig. 4 show an increase in integrated signal as a function of concentration. Larger signals might result from either more small particles generating the second harmonic response or a smaller number of particles that grow in size.

Are the authors able to assess which mechanism is operative? (Or another mechanism altogether?) Have the authors attempted to measure SH signals using different polarization combinations in order to perhaps learn more about the structures that are responsible for the observed signals?

Along those lines, the authors acknowledged that particle growth may be occurring around trace contaminants in the solution. One way to test this hypothesis might be to intentionally seed their solutions with a chiral contaminant and then look for chiral SH signal. (Garth Simpson at Purdue has been one of the most active scientists in this area; see, for example ref. 1). Finally, work by Ken Eisenthal and coworkers might be relevant to the authors' results. Specifically, Eisenthal noticed that spherical particles in bulk solution could generate a surprisingly large SH response. A number of papers have been published over the past 10 years looking at nanoparticles, styrene beads and vesicles (see ref. 2 for an example).

1 G. Simpson *et al.*, *Cryst. Growth and Des.*, 2008, **8**(8), 2589–2594, DOI: 10.1021/cg700732n.
2 K. Eisenthal *et al*, *J. Phys. Chem. C*, 2008, **112**(40), 15809–15812, DOI: 10.1021/jp8047168.

Dr Alexander replied: The possible mechanisms for change in SH signal that we outline in our paper are about as far as we can go with speculation for now. Measurements of the angular dependence of the scattering, and the dependence on polarization as Professor Walker suggests would be very useful. The idea for the chiral experiment is particularly innovative. We have not yet attempted such measurements.

Professor Molinero asked: Is it possible to distinguish between liquid and crystal states of the mesoscopic urea–water aggregates using second-harmonic scattering? Do you observe evidence of crystallization of the mesoscopic aggregates in the experiments at high supersaturation?

Dr Alexander replied: We did not see any evidence of crystal nucleation in our samples, which we found can remain metastable for months. In our paper[1] we demonstrate that the signals come from localised particles, floating in solution, but we are not able to say whether these are crystalline or not.

1 M. R. Ward, S. W. Botchway, A. D. Ward and A. J. Alexander, *Faraday Discuss.*, 2013, **167**, DOI: 10.1039/C3FD00089C.

Professor Wynne enquired:In the trapping experiment you have described,[1] would it be possible to send one of the trapping lasers, after it has passed through the sample, to a spectrometer to perform a Raman spectroscopy experiment? This would allow you to ascertain that the scattering centres are indeed urea.

1 M. R. Ward, S. W. Botchway, A. D. Ward and A. J. Alexander, *Faraday Discuss.*, 2013, **167**, DOI: 10.1039/C3FD00089C.

Dr Alexander answered: The image in Fig. 5 of our paper was taken while the particle at the centre was being held in a dual optical trap; full results of these experiments have yet to be published. We have looked at urea and several other solutes at high concentrations using Raman microscopy without trapping, and we found that the signal is dominated by the background solute. A combined trapping and Raman spectroscopy experiment may indeed help to unravel the nature of these particles.

Professor Wynne asked: You mention experiments on vanillin solutions in water, however, this system is know to display a liquid–liquid phase separation.[1] Do you believe that supersaturated aqueous urea solutions also crystallise through a liquid–liquid phase separation process along the lines of the theory proposed for protein crystallisation[2] by Frenkel?

1 M. Svard, S. Gracin and A. C. Rasmuson, *J. Pharm. Sci.*, 2007, **96**, 2390–2398.
2 P. Tenwolde and D. Frenkel, *Science*, 1997, **277**, 1975–1978.

Dr Alexander responded: The analogy of the liquid–liquid phase separation is very tempting. There is now quite a lot of circumstantial evidence for objects floating around in two-component ionic and small-molecule systems. These objects are intrinsic and not foreign to the system; let's call them clusters. I would say that these clusters are not liquid in the sense that they have restricted internal degrees of freedom; I would view them more as disordered liquid-crystalline or glassy objects, which may include molecules of solvent.

Dr Perera commented: In our simulations of aqueous urea, we found that urea aggregates into domains, just like other hydrophilic molecules like DMSO or formamide. However, the experimental observation of the concentration fluctuations, as measured by the Kirkwood–Buff integrals, are surprisingly ideal looking.[1] It seems that the urea clusters are formed while the mixture exhibit very little concentration fluctuation. In view of this, you may want confirm this behaviour by studying DMSO or formamide aqueous mixtures, which are soluble over the entire concentration range.

1 E. Matteoli and L. Lepori, *J. Chem. Phys.*, 1984, **80**, 2856.

Dr Alexander communicated in reply: We agree.

Professor Angell asked: Did I miss something? Did you show us a phase diagram for the urea–water system? Can you tell us what it looks like? Does it have binary compounds or is it just a simple eutectic?

Dr Alexander communicated in reply: The urea–water system exhibits simple eutectic behaviour.[1] There is only one known crystal form of urea at ambient pressures.[2]

1 M. Egal, T. Budtova and P. Navard, *Cellulose*, 2008, **15**, 361–370.
2 F. J. Lamelas, Z. A. Dreger and Y. M. Gupta, *J. Phys. Chem. B*, 2005, **109**, 8206–8215.

Professor Angell communicated: If the latter, it may possible that there are complex crystal structures hiding just below the eutectic, as there are in the case of the Au–Si system that is a simple, but enormously deep, eutectic (at half the T_m of gold, the lower melting component).

Professor Wynne opened a general discussion of Professor Anisimov, Dr Alexander and Dr Sefcik's papers: A number of papers discussed in this meeting so far[1–3] appear to show the existence of very large molecular clusters in solution. My gut feeling is that these huge clusters should be thermodynamically unstable, however, the experiments appear to demonstrate that they are stable and reform after filtering. Non-classical nucleation theories have been proposed—to understand crystal nucleation of calcium carbonate—that involve stable pre-nucleation clusters.[4–6] Could it be that the clusters discussed here are related to pre-nucleation clusters?

1 D. Subramanian, C. T. Boughter, J. B. Klauda, B. Hammouda and M. A. Anisimov, *Faraday Discuss.*, 2013, **167**, DOI: 10.1039/C3FD00070B.
2 A. Jawor-Baczynska, B. D. Moore, H. S. Lee, A. V. McCormick and J. Sefcik, *Faraday Discuss.*, 2013, **167**, DOI: 10.1039/C3FD00066D.
4 M. R. Ward, S. W. Botchway, A. D. Ward and A. J. Alexander, *Faraday Discuss.*, 2013, **167**, DOI: 10.1039/C3FD00089C.
4 D. Gebauer, A. Voelkel and H. Coelfen, *Science*, 2008, **322**, 1819–1822.
5 A. Dey, P. H. H. Bomans, F. A. Mueller, J. Will, P. M. Frederik, G. De With and N. a. J. M. Sommerdijk, *Nat. Mater.*, 2010, **9**, 1010–1014.
6 D. Gebauer and H. Coelfen, Nano Today, 2011, 6, 564-584.

Professor Anisimov replied: I feel that mesoscale aggregates reported in the literature may belong to different physical phenomena. In some systems they reappear after filtration. In other systems (such as TBA–water) they emerge only in the presence of a hydrophobic impurity.

Dr Alexander responsed: These may be stable clusters, or they may be impurity particles that have passed the filtering. We note that our estimates of the number density of these objects given in our paper are very similar to those measured by Jawor-Baczynska *et al.*,[1] who demonstrated that these objects can be reformed after removal by filtering.

1 A. Jawor-Baczynska, B. D. Moore, H. S. Lee, A. V. McCormick and J. Sefcik, *Faraday Discuss.*, 2013, **167**, DOI: 10.1039/C3FD00066D.

Dr Sefcik added: Indeed we believe that it is likely that mesoscale clusters observed in amino acid aqueous solutions are involved in non-classical nucleation pathways to formation of solid crystalline phases. It may well be the case for other solutions, such as calcium carbonate, as well. However, these non-classical nucleation pathways might be more complex that involving just a single pre-nucleation cluster "species". For example, there might be a distribution of cluster sizes and their densities and internal arrangements in a mutual equilibrium (or non-equilibrium, especially in supersaturated solutions). For example, recently we showed[1] that glycine mesospecies (nanodroplets) need to be significantly larger than their most typical average size in order to effectively participate in glycine nucleation, and that mechanical agitation can be used to enhance coalescence of smaller nanodroplets into larger ones.

1 A. Jawor-Baczynska, J. Sefcik, B. D. Moore, *Cryst. Growth Des.*, 2013, **13**, 470–478.

Professor Ben-Amotz remarked: It would be useful to more precisely define what we mean by "mesoscopic order" so as to distinguish the formation of such structures from micelle formation (and low-order aggregation processes), as well as and from the sort of fractal-like structures that would be expected to appear in non-aggregating random mixtures. For example, one might consider restricting the use of the term mesoscopic order to situations in which a system contains sub-micron structures that are significantly larger than 1 nm, with a well-resolved characteristic length scale (as opposed to a broad fractal-like distribution of aggregate sizes spanning the entire nm to micron regime). The latter characteristic length scale could perhaps further be associated with the existence of a well-resolved low-Q peak in X-ray or neutron scattering spectra. The latter definition might thus provide an experimental means of distinguishing unique mesoscopic structures from fractal-like random contact structures, which would presumably not give rise to a well defined low-Q peak, but would rather produce a broad and relatively featureless low-Q background.

Professor Anisimov replied: I agree with your comment.

Dr Sefcik addressed Dr Alexander: In your paper you mentioned Rayleigh scattering measurements on urea aqueous solutions, revealing presence of particles at concentrations on the order of 10^8 cm^{-3}. Have you been able to estimate their hydrodynamic diameters? Also, have there been any previous DLS measurements on urea solutions? Regarding filtration of hot urea solutions, have you tried different filters (in terms of filter materials and pore sizes) and if yes, were there any effects on your observations? Your observations of 2-photon fluorescence decay in concentrated glycine solutions is intriguing and while it may indeed be due to trace impurities as mentioned, it would be worth following up.

Dr Alexander replied: The analysis of our scattering measurements on nearly saturated solutions of urea is still in progress. We have found a very recent study that we did not cite in our paper: this article reports on static and dynamic light scattering for dilute aqueous solutions of urea ($S \sim 0.04$).[1] The conclusions strongly support our own findings, in particular the number densities of the native particles that they observe. The study also concludes that these particles are not nanobubbles. We have tried different types of filter. Smaller-pore filters (20 nm) appear to decrease the number of particles. Different filter materials also appear to have an effect, *e.g.*, poly(tetrafluoroethene) filters appear to reduce the number of particles, but the particles are never removed entirely. All of these results are in accordance with your own measurements on amino acids.[2,3]

1 M. Sedlak, D. Rak, *J. Phys. Chem. B*, 2013, **117**, 2495, DOI: 10.1021/jp4002093.
2 A. Jawor-Baczynska, J. Sefcik and B. D. Moore, *Cryst. Growth Des.*, 2013, **13**, 470–478.
3 A. Jawor-Baczynska, B. D. Moore, H. S. Lee, A. V. McCormick and J. Sefcik, *Faraday Discuss.*, 2013, **167**, DOI: 10.1039/C3FD00066D.

Dr Vila Verde addressed Dr Sefcik: Proving your system is free from impurities, and thus that the mesoscale structures that you see are not related to the presence of impurities, is difficult. It would be instructive to take the opposite approach and put small amounts of impurities in your systems, and investigate what happens to the size and stability of the mesostructures.

Dr Sefcik responded: This is indeed worth investigating, although one would need to choose a specific impurity (or impurities) to focus on. However, we also assessed potential effects of impurities on formation of mesoscale species indirectly by observations of mesospecies in solutions prepared after purification by nanofiltration and/or recrystallization.

Professor Wales asked: What sort of impurities did you consider?

Dr Sefcik replied: We assessed potential effects of impurities on formation of mesoscale species indirectly by observations of mesospecies in solutions prepared after purification by nanofiltration and/or recrystallization. As reported in our paper here,[1] filtration of DL-alanine solutions by either hydrophobic (PTFE) or hydrophilic (alumina) nanofilters removes virtually all observable mesospecies and yet they reappear after a certain period of time with the same mean size and number concentration as before. If any impurities were associated with mesospecies in original solutions, they would be expected to be at least partially removed together with original mesospecies and therefore their total amount should decrease. Significantly, filtration does not lead to a quantitative change in mesospecies populations once equilibrium is re-established (see Fig. 4 and 5 in our paper). Furthermore, as we reported previously,[2] purification of glycine by repeated recrystallization was performed to check for effects of impurities on mesospecies formation in glycine solutions, and again the same size and number concentration of mesospecies were found irrespective of the degree of purification. While we were not able to detect any specific impurities by the HPLC analysis, it can be reasonably expected that concentrations of any impurities present in glycine would significantly decrease after each recrystallization step, and therefore we believe that it is unlikely that impurities play a significant role in formation of mesospecies in these solutions.

1 A. Jawor-Baczynska, B. D. Moore, H. S. Lee, A. V. McCormick and J. Sefcik, *Faraday Discuss.*, 2013, **167**, DOI: 10.1039/C3FD00066D.
2 A. Jawor-Baczynska, J. Sefcik and B. D. Moore, *Cryst. Growth Des.*, 2013, **13**, 470–478.

Dr Perera commented: I would like to draw the attention of everyone on the remarkable work of Yoshikata Koga[1] from UBC, Vancouver, Canada, who is a calorimetrist. He measures the vapour pressure and enthalpy of solutions, and computes second derivatives of enthalpy with respect to the concentration of solutes. He reports an anomalous behaviour of such derivative at very low concentration of solute, under the form of a peak. When all anomalies for a variety of solutes are plotted in the (T, x) plane, where T is the temperature and x the solute mole fraction, then the different lines seem to converges to a single temperature for $x = 0$, independent of the nature of the solute, and pointing to an intrinsic property of water at $T = 70$ °C approximately. This temperature seems very close to the sound anomaly in water, but the relation to this anomaly and solute universal clustering scheme is unclear at present.

1 Y. Koga, *Solution Thermodynamics and its Application to Aqueous solutions*, Elsevier, Amsterdam, 2007.

Professor Molinero addressed Dr Sweatman: Is there an agreement on what is the physical origin of the stabilization of mesoscopic aggregates and whether these aggregates are stable only for specific intermolecular interactions?

Dr Sweatman communicated in reply: I am not aware of broad agreement at this early stage, but our work with the SALR model, which extends the work of others on the SALR model to lower density regions of the phase diagram where micelle-like behaviour is observed, demonstrates that competition between short-range attraction and long-range repulsion does generate thermodynamically stable mesostructures.

We have submitted this work for peer review elsewhere. We expect the same basic physics underlies many aggregation phenomena in liquid mixtures, and suggest that it can also provide an explanation for reversible non-classical, or two-step, crystal nucleation processes (reversible in the sense that SALR clusters can be thermodynamically stable, so that by adjusting solvent conditions one can go from macroscopic crystal to a micelle-like phase, in principle). The physical origin of these competing short and long-range effective interactions could be quite varied, depending on the specific chemistry of each system.

Regarding some of the details of our work, we devised a thermodynamic model to describe the cluster fluid phase of the SALR system. We discovered later that our thermodynamic model is similar to theories of micellization, which are usually used to model surfactant aggregation. However, unlike typical micelle theories, our approach has no empirical or fitted parameters – it is entirely self-contained. The model parameterizes the cluster fluid phase in terms of 4 parameters: the overall bulk density, the cluster density, the liquid "body" density of a cluster, and the "background" vapour density.

Each cluster is taken to be spherical and identical. Entropy and energy density functions are designed that when minimised yield the equilibrium behaviour. The model predicts, as you might expect, micelle-like behaviour.

We map the phase diagram at low density for some combinations of SALR fluid parameters. We find the cluster fluid phase is stable over a range of SALR parameters, and locate the "critical cluster concentration", analogous to the critical micelle concentration, below which a uniform vapour phase with pre-clusters is stable. We also locate a cluster vapour to cluster liquid first-order transition, which is driven by a depletion effect (vapour is depleted between clusters as they approach each other). The cluster size and body-density hardly vary with bulk density, but they are sensitive to the relative strength of attraction *vs.* repulsion; their size gradually diverges and their body-density increases as attractions become dominant. We suggest that at higher bulk densities the cluster fluid will transform to some form of modulated aggregate phase, perhaps a cluster solid. We also suggest that when the cluster body-density becomes sufficiently high (when attractions dominate) that clusters will be solid. Finally, we provide a simple approximate formula for the cluster size, which depends on the pair potential and radial distribution (rdf) function of a bulk system with the same density as the cluster body-density. Because of this explicit link between cluster size and rdf we argue that the cluster size will change, and possibly diverge, when clusters solidify; *i.e.* we provide a mechanism for a non-classical crystal nucleation process from within liquid-like clusters.

That is, we suggest macroscopic crystals can grow out of microscopic liquid-like clusters. We compare the model's predictions at one state point (deep within the cluster fluid phase) with Monte Carlo simulations, which demonstrate the cluster fluid is thermodynamically stable.

Professor Evans continued: There has been much discussion about the type of model used in the nice poster by Dr Sweatman *et al.* describing the phase behaviour of a simple (one-component) fluid shown to exhibit an interesting "cluster phase" at low densities. I am familiar with the model and its origins. The idea of using a "mermaid" pair potential, *i.e.* one with an attractive head and a repulsive tail is certainly not new in colloid science. This type of potential is also termed SALR. Steric effects in colloids plus depletant give rise to a short ranged, strongly attractive effective colloid–colloid interaction. And if the colloids are charged screening effects will give rise to a slowly decaying repulsive tail whose decay length depends on the Debye length of the solvent. It is not difficult to see that such competing effective interactions will have two length scales associated with them: the range of the attraction and that of the repulsion. The interesting cases have the attraction shorter ranged than the repulsion. Dr Sweatman cites the paper by Archer and Wilding.[1] This paper provides a fine introduction to the relevant literature and how such models came about. Papers by Imperio and Reatto[2] were important in illustrating the range of (meso)phases than can occur for such models. Archer and Wilding discuss a vapour to spherical cluster phase and, importantly, the sensitivity of phase behaviour to the height of the repulsive tail. Dr Sweatman appears to have taken substantial steps forward by identifying further transitions. Of course the key issue is whether this type of model is relevant to the systems investigated in several papers that were presented here. Are there ingredients in the physical chemistry of these particular aqueous solutions that would lead one to argue for a mapping to the type of model considered by the colloid community? This is not at all clear to me.

1 A. J. Archer and N. B. Wilding, *Phys. Rev. E*, 2007, **76**, 031501.
2 For example, A. Imperio and L. Reatto, *J. Chem. Phys.*, 2006, **124**, 164712.

Dr Perera commented: Following the remark of Professor Evans on the clustering in SALR interactions (Short range Attraction and Long range Repulsion) in one component systems, I would like to point out that we tried to extract such interaction as a bridge term from integral equation techniques (using the $g(r)$ from simulations) from aqueous-binary mixtures. Unfortunately, our findings were inconclusive, to our disappointment, since the molecular structure induces oscillations that prevent us from observing the long range repulsion in an unambiguous fashion. However, since the SALR repulsion is often very small compared to the short range attraction, we do not exclude that a LR mechanism is hidden behind the clustering observed in aqueous mixtures.

Professor Evans replied: It is important to recognize that the LR tail can have a very small amplitude but still influence significantly the phase behaviour and in particular the preponderance for clustering. That is why the studies of simple models are important in identifying the underlying physics. I am not surprised that it is difficult to pick up such subtleties in simulation or integral equation analyses of 'realistic' aqueous mixtures. It is an optimistic man or woman who attempts to invert pair correlation functions in such systems in order to identify a slow repulsive decay.

Dr Meech concluded the discussion of Dr Alexander's paper and communicated: In Fig. 4 the case is made for a distinction between the hyper-Rayleigh scattering and the second harmonic emission from the urea solution. Since an arbitrary scale factor is involved and the shapes of the model and the data are not radically different some additional mechanism for the distinction might help. The measurement recalls an earlier study second harmonic generation from an isotopic solution of structures which locally lack an inversion centre – in that case the purple membrane.[1] In that case the SHG was distinguished from HRS through the polarisation dependence and the spatial distribution of the emission (relative to the fundamental pump direction). Similar measurements could provide useful support for the mechanism proposed here.

1 P. Allcock, D. L. Andrews, S. R. Meech and A. Wigman, Doubly Forbidden Second Harmonic Generation From Isotropic Suspensions of The Purple Membrane, *Phys. Rev. A*, 1996, **53**, 2788–2791.

Mr Ward communicated in reply: Professors Meech and Walker make important points in this discussion. The experimental signals are quite challenging to collect, but even simply obtaining the angular dependence of the scattering could distinguish between incoherent hyper-Rayleigh scattering (HRS) and coherent second-harmonic generation (SHG) contributions. In our paper we were only able to infer the latter mechanism from the observation that large signals appear to come from floating particles.

Miss Rhys concluded the discussion of the paper by Dr Sefcik by communicating: The methods that have been used in this paper have either precluded or provided a more vague measure of the hydrodynamic diameter of any small molecular clusters. For the dynamic light scattering measurements, the paper states that the peak at 0.005 ms, found in Fig. 1, corresponds to objects up to 1 nm, and so this would mean that you are likely unable to accurately distinguish between monomers and the formation of very small clusters. As you quote, the technique is more sensitive towards larger sizes. Glycine is an amino acid known to favour the formation of dimers, which is mentioned in the references you quote (ref. 5 and 39), and so should be contributing to this higher small cluster peak seen in the dynamic light results. In any work your group has completed, or in any other studies you are aware of, is small-scale clustering such as that seen in glycine also known/thought to be formed in D/L-alanine? In reference to Fig. 1, it is interesting that the comparison between D/L-alanine and glycine solutions, though having nearly the same *S* value, is between one solution that is under the solubility limit for the amino acid (glycine), and one that is over the solubility limit (D/L-alanine). With that in mind and the fact that Fig. 2 suggests a higher concentration of mesospecies for D/L-alanine, has a critical concentration for formation of mesospecies been established for glycine and what would be the likely cause for any similarity/difference to the value determined for D/L-alanine (~10 mg ml^{-1})?

Dr Sefcik communicated in reply: Dynamic light scattering is able to detect hydrated monomers and/or small molecular clusters in DL-alanine solutions. This can be seen from a (relatively small) autocorrelation function decay at delay times below 0.01 ms in Fig. 1a. This decay can be seen much better after filtration of mesoscale clusters, which is similar to observations from glycine solutions.[1] It is an interesting question whether it would be possible to quantitatively distinguish contributions of dimers (or larger molecular clusters) from those of hydrated monomers, by analysis of autocorrelation function decays in filtered DL-alanine solutions. We have not attempted that but it might be worth trying.

In Fig. 1 we show DLS autocorrelation functions for undersaturated DL-alanine and glycine solutions at a similar value of supersaturation (S = 0.75). The concentration of DL-alanine was 125 mg ml^{-1} (not 225 as erroneously stated in the originally submitted manuscript) and the concentration of glycine was 180 mg ml^{-1} as indicated. The critical concentration for formation of mesospecies in DL-alanine solutions was found to be around 17 g ml^{-1} (corresponding to S = 0.1), while for glycine it is in a similar region of concentrations.[2]

1 A. Jawor-Baczynska, J. Sefcik and B. D. Moore, *Cryst. Growth Des.*, 2013, **13**, 470–478.
2 Manuscript in preparation.

Faraday Discussions

RSC Publishing

PAPER

Corresponding states for mesostructure and dynamics of supercooled water

David T. Limmer and David Chandler*

Received 6th May 2013, Accepted 30th May 2013
DOI: 10.1039/c3fd00076a

Water famously expands upon freezing, foreshadowed by a negative coefficient of expansion of the liquid at temperatures close to its freezing temperature. These behaviors, and many others, reflect the energetic preference for local tetrahedral arrangements of water molecules and entropic effects that oppose it. Here, we provide theoretical analysis of mesoscopic implications of this competition, both equilibrium and non-equilibrium, including mediation by interfaces. With general scaling arguments bolstered by simulation results, and with reduced units that elucidate corresponding states, we derive a phase diagram for bulk and confined water and water-like materials. For water itself, the corresponding states cover the temperature range of 150 K to 300 K and the pressure range of 1 bar to 2 kbar. In this regime, there are two reversible condensed phases – ice and liquid. Out of equilibrium, there is irreversible polyamorphism, *i.e.*, more than one glass phase, reflecting dynamical arrest of coarsening ice. Temperature–time plots are derived to characterize time scales of the different phases and explain contrasting dynamical behaviors of different water-like systems.

1 Introduction

Supercooled liquids exist in a metastable equilibrium made possible by a separation of timescales between local liquid equilibration and global crystallization.[1] Supercooled water is no different in this regard. However, the magnitude of the separation of timescales in supercooled water is of particular relevance due to speculation regarding the behavior of the thermodynamic properties of liquid water at very low temperatures.[2] In this work, we describe a theory for corresponding states that relates the low temperature, low pressure phase diagram with the time–temperature–transformation diagram for supercooled water and water-like systems. Our derivations use scaling theories with assumptions tested against molecular simulation. The relationships elucidate the connections between behaviors found for different molecular simulation models of water and for different water-like substances.

Department of Chemistry, University of California, Berkeley, California, 94702, USA. E-mail: chandler@berkeley.edu

Fig. 1 shows the portion of the phase diagram for supercooled water relevant to this paper. Temperature, T, ranges from ambient conditions to deep into the supercooled regime, and pressure, p, ranges from atmospheric conditions through the range of stability for ordinary hexagonal ice. Experimentally, this region corresponds to $150\ \mathrm{K} < T < 300\ \mathrm{K}$ and $0\ \mathrm{kbar} < p < 2\ \mathrm{kbar}$.[14] The locations of specific features relative to experiment vary from one molecular model to another.[15] This variability reflects a delicate competition between entropy and energy that is intrinsic to any reasonable model of water or water-like system.

One manifestation of this competition is the existence of the temperature of maximum density. We use T_{max} to denote the value of this temperature at ambient (*i.e.*, low pressure) conditions. For experimental water, $T_{\mathrm{max}} \approx 277$ K. The energy–entropy balance manifested in the density maximum is shifted to lower temperatures as elevated pressures favor denser packing. A measure of this shift is provided by the slope of the melting line or in terms of a reference pressure $p_{\mathrm{o}} = -\Delta H/10\ \Delta V$, where ΔH and ΔV are, respectively, the enthalpy and volume changes upon melting. For experimental water, $p_{\mathrm{o}} \approx 3.7$ kbar. We use T_{max} and p_{o} to compare the properties of different water models as well as to enable comparison with experiment.[3–7,7,14,16] Thus, the phase diagram in Fig. 1 employs the reduced variables

$$\tilde{p} = p/p_{\mathrm{o}} \text{ and } \tilde{T} = T/T_{\mathrm{max}}. \tag{1}$$

In this way, Fig. 1 relates results from various models and experiments for the onset temperature, T_{o}, and the homogeneous nucleation temperature, T_{s}. The former, T_{o}, marks the crossover to correlated (*i.e.*, hierarchical) dynamics,[17] and it plays a central role in formulas for the temperature dependence of liquid relaxation rates, as discussed below. The latter, T_{s}, marks the crossover to liquid instability,[18–20] and it plays a central role in the temperature dependence of ice

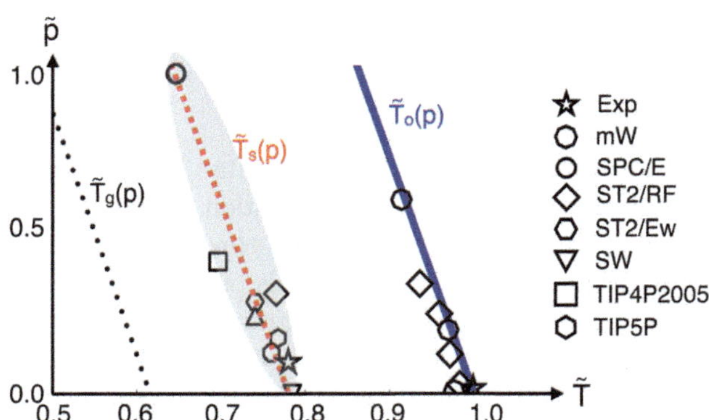

Fig. 1 The $\tilde{p}$–$\tilde{T}$ phase diagram for supercooled water. These symbols refer to the pressure, p, and the temperature T, in units of the reference pressure, p_{o}, and the reference temperature, T_{max}, for each specific material,[3–7] see text. The lines refer to theoretically expected trends as functions of pressure for the dynamical onset temperature, T_{o}, for the liquid limit of stability temperature, T_{s}, and for a glass transition temperature, T_{g}. Symbols indicate locations of these temperatures in the reduced units as obtained from experiment and from different molecular simulation models, where the key employs standard acronyms for each system.[8–13] For the glass transition, the line marks the stage where the liquid reorganization time exceeds $10^{14}\ \tau_{\mathrm{o}}$, where the reference time, τ_{o}, is that reorganization time at the onset to correlated dynamics ($\tau_{\mathrm{o}} \approx 1$ ps for liquid water).

coarsening, also as discussed below. These temperatures are material properties. Fig. 1 also shows a reduced glass transition temperature, $\tilde{T}_g = T_g/T_{max}$. This temperature is defined as that where the reversible structural relaxation time of liquid water equals 100 s.

Glass phases of water, where aging occurs on time scales of 100 s or longer, are not generally accessible by straightforward supercooling because bulk liquid water spontaneously freezes into crystal ice at temperatures below $T_s = T_s(p)$. Freezing in this regime occurs on milli-second or shorter time scales.[21] An amorphous solid can be reached with a cooling trajectory that is initially fast enough to arrest crystallization, and finally cold enough to produce very slow aging. Alternatively, an amorphous solid can be reached by cooling while perturbing water with surfaces that inhibit crystallization. A glass transition temperature of water is therefore not a material property because its value depends upon the protocol by which the material is driven out of equilibrium. Surface mediated approaches to amorphous solids can yield T_g's that are higher than those produced by rapid temperature quenches. The T_g graphed in Fig. 1 is necessarily an upper bound to those glass transition temperatures.

Dynamics in the vicinity of $T_s(p)$ exhibits a two-step coarsening of the crystal phase.[20,22,23] First, disperse nano-scale domains of local crystal order form throughout the melt; second, on a much longer time scale, the nano-scale domains meld into much larger ordered domains. These steps are arrested when forming glass.[24] Configurations appearing at the initial stages of this coarsening are often observed in computer simulations of water. These configurations are transient states that are almost as often confused with the presence of two distinct supercooled liquid phases.† There is one exceptional case, a water-like model where transient states have not yet been observed. Specifically, Giovambattista and co-workers[25] suggest that the SPC/E model does not exhibit this phenomenon. In that particular case, we shall see, it seems that the transient states have been overlooked simply because the relevant part of the phase diagram near $T_s(p)$ has been overlooked.

Two distinct reversible liquids in coexistence would imply the existence of a second critical point of the sort suggested by Stanley and his co-workers.[26] Not surprisingly, all reports of a low-temperature critical point in water or water-like systems locate a point on or near $T_s(p)$. In fact, all the simulation points clustered around that line in Fig. 1 have been incorrectly identified as low-temperature critical points.[8–13] Similarly, in experimental work, Mishima locates a putative critical point[27] close to the experimental limit of liquid stability.[28] We have analyzed and disproved this notion of a liquid–liquid transition for several different computer simulation models of water and water-like systems.[20]

We mention the disproved notion only to emphasize that the equilibrium phase diagram by itself gives an incomplete picture of the behavior of supercooled water. Supercooled water, being a metastable state, behaves reversibly for only finite observation times. The specific length of that time depends on the separation of timescales between local equilibration of liquid configurations and global crystallization. When the gap between these timescales is no longer large

† The list of representative papers is long. We have provided a summary elsewhere.[20]

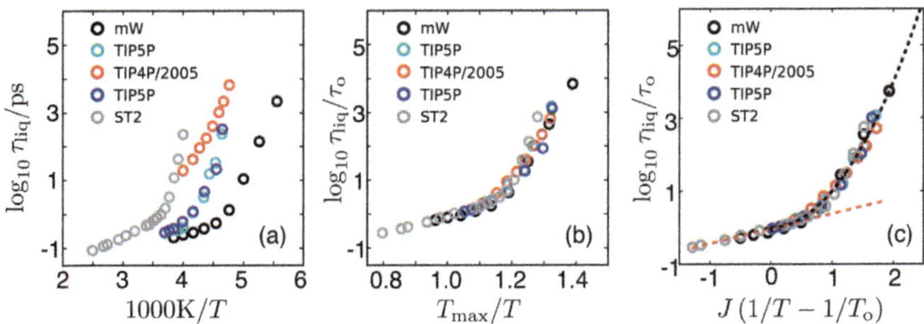

Fig. 2 Low temperature variation of structural relaxation time, τ_{liq}, for different models of water. (a) Logarithm of that time as a function of absolute temperature T. (b) The same data, now in units of the reference time τ_o, and as a function of reduced temperature $\tilde{T} = T/T_{max}$. The reference time, τ_o, is the structural relaxation time at the onset temperature, T_o. (c) The same data, now collapsed to the parabolic form, eqn (2), where the dashed black line is the prediction of that equation, and the dashed red line is the Arrhenius form that holds for temperatures above T_o.

enough, as occurs for either $T < T_s$ or $T \approx T_s$, time-independent thermodynamic properties are no longer well defined.

2 Universal temperature dependence of liquid relaxation times for supercooled water

In order to construct a scaling theory for the liquid relaxation time, τ_{liq}, we follow our previous work[19] in adopting a perspective of dynamic facilitation theory.[17] An important aspect of this perspective is that it supplies a universal form for the relaxation time as a function of temperature. This form, known as the "parabolic law", is

$$\log_{10}(\tau_{liq}/\tau_o) = J^2(1/T - 1/T_o)^2 \text{ for } T < T_o, \tag{2}$$

where J is an energy scale of hierarchical dynamics, T_o is the temperature below which that dynamics sets in, and τ_o is the liquid relaxation time at the onset temperature T_o.‡ This form has been used to collapse large and seemingly disparate collections of experimental and simulation data.[29] Fig. 2 illustrates the nature of this collapse for the structural relaxation times of several models of water.[30–32]

For the data shown in Fig. 2, the relaxation times have been calculated from the self-correlation function,

$$F_s(k,t) = \left\langle e^{i\mathbf{k}\cdot[\mathbf{r}_1(t) - \mathbf{r}_1(0)]} \right\rangle, \tag{3}$$

for wave vectors of magnitude $k \approx 2\pi/2.8$ Å. Here, $\mathbf{r}_1(t)$ denotes the position of a tagged molecule at time t, and the angle brackets stand for the equilibrium average over initial conditions. The time at which this $F_s(k,t)$ decays to 1/e is defined as τ_{liq}. In all of the models studied, the temperature dependence of this time crosses over from a weak Arrhenius temperature dependence to a super-Arrhenius temperature dependence. The location where the crossover occurs, the onset temperature T_o,

‡ In general, τ_o may itself multiply an Arrhenius temperature dependent factor, but we neglect this quantitative detail here because it is a small effect compared to the super-Arrhenius behavior at low temperature.

varies from model to model as does the reference timescale τ_o. This variability in part reflects quantitative differences between phase diagrams for each of the models. Indeed, Fig. 2b shows that the temperature dependence of the relaxation times can be collapsed by referencing the data to the temperature of maximum density. It is a remarkable result given the wide variation of T_{max}, ranging from 250 K to 320 K for the different models.[5]

In addition, Fig. 2c shows that the relaxation time data also collapses when referenced to the onset temperature, T_o, and that the collapsed data obeys the parabolic law for all $T < T_o$. This finding establishes that

$$T_o \approx T_{max}. \tag{4}$$

Further, collapsing data to the parabolic law reveals that J/T_o varies between models of water by no more that 5% and on average by only 1%. This typical value is $J/T_o \approx 7.4$. For the models considered here, τ_o varies between 0.3 ps and 8.0 ps, and this variability largely reflects the density differences between models at low pressure.

The collapse of the relaxation times for the difference models as a function of T_{max} implies a universality in the behavior of the glass transition for these models and, by proxy, for experiment. As we have done previously,[19] we can define a locus of laboratory glass transitions as the locations in the phase diagram where the liquid relaxation time is equal to 10^{14} τ_o, which for many of the models implies $\tau_{liq}(T_g) \approx 100$ s, so that with eqn (2) we have,

$$T_g/T_o \approx \left(\sqrt{14}\, T_o/J + 1\right)^{-1}. \tag{5}$$

Taking $J/T_o \approx 7.4$ and $T_o \approx T_{max}$, we therefore conclude that for water and water-like models, $T_g \approx 0.62T_{max}$. This value yields the glass-transition line graphed in Fig. 1, where the slope of the line is the same as that for $T_o(p)$. Experimentally, the density maximum for water occurs at $T_{max} \approx 277$, therefore our predicted glass transition is $T_g \approx 172$ K. This temperature agrees with our previous work inferring the glass transition temperature from relaxation data of confined water.[19] It also provides an upper bound to values for T_g obtained with other experimental protocols.[33,34]

3 Molecular theory for J, τ_o and T_o

The energy, time and temperature parameters in eqn (2) can be computed from microscopic theory following the procedures of Keys *et al.*[35] The procedures are based upon mapping the dynamics of atomic degrees of freedom to dynamics of a kinetically constrained East model.[36] The parabolic law, eqn (2), is a consequence of that mapping.

To illustrate the procedure for water, we have carried out molecular dynamics simulations of equilibrated water models to determine the net number of enduring displacements of length a appearing in N-molecule trajectories that run for observation times t_{obs}. This number of displacements is

$$C_a = \sum_{i=1}^{N} \sum_{j=0}^{t_{obs}/\Delta t} \Theta\left(\left|\bar{r}_i(j\Delta t + \Delta t) - \bar{r}_i(j\Delta t)\right| - a\right) \tag{6}$$

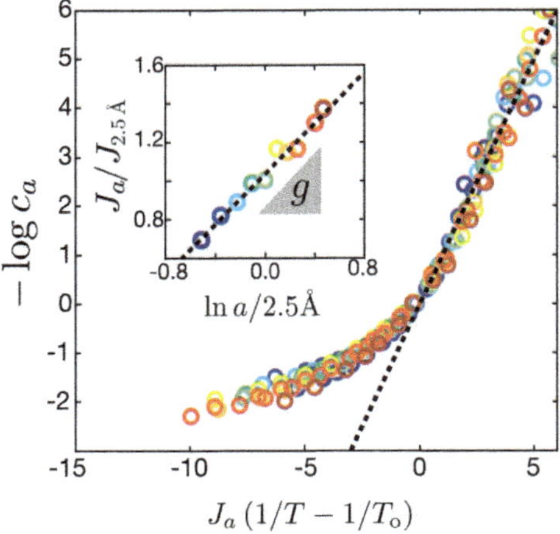

Fig. 3 Excitation concentration for the mW model at ambient pressure for enduring displacement lengthscales, a, between 1.5 and 3.5 Å. The dashed line has unit slope illustrating the Boltzmann scaling, eqn (8). The inset shows the logarithmic scaling of J_a with a. The dashed line is a fit to the data for eqn (9) with $g = 0.625$.

where $\Theta(x)$ is 1 for $x > 0$ and zero otherwise, Δt is the mean instanton time for enduring displacements of length a, and $\bar{r}_i(t)$ is the position of particle i averaged over the time interval $t - \delta t/2$ to $t + \delta t/2$. The averaging over δt coarse-grains out irrelevant vibrational motions. The instanton time, Δt, is taken to be large enough that non-enduring transitions are also removed from consideration. The two times, $\Delta t > \delta t$, are determined as prescribed by Keys *et al.*[35]

The mean mobility (or excitation concentration) is the net number of enduring transitions per molecule per unit time, *i.e.*,

$$c_a = \langle C_a \rangle / (N t_{\text{obs}}/\Delta t). \tag{7}$$

Its dependence upon temperature and displacement length is illustrated in Fig. 3. According to facilitation theory, c_a should have a Boltzmann temperature dependence, with an energy scale that grows logarithmically with displacement length. That is,

$$c_a \propto \exp[-J_a(1/T - 1/T_o)] \text{ for } T < T_o, \tag{8}$$

and

$$J_{a'} = J_a - g\, J_\sigma \ln(a'/a), \tag{9}$$

where σ is a reference molecular length and g is a system-dependent constant.§ The data graphed in Fig. 3 shows that for the model considered, the mW model of water, the theoretical expectations are obeyed. We have adopted the reference length $\sigma = 2.5$ Å, which is close to the diameter of the molecule in the mW model, and find $g = 0.625$, $T_o = 244$ K $\approx T_{\max} = 250$ K, and $J_\sigma/T_o = 23$.

§ Keys *et al.*[35] use the symbol γ for what we call g. We use γ to refer to the surface tension.

According to facilitation theory,[35] eqn (8) and (9) imply

$$\ln(\tau_{\mathrm{liq}}/\tau_{\mathrm{o}}) = (J_\sigma^2 g/d_{\mathrm{f}})(1/T - 1/T_{\mathrm{o}})^2 \text{ for } T < T_{\mathrm{o}}, \tag{10}$$

where d_{f} is the fractal dimension of dynamic heterogeneity, which for $d = 3$ is about 2.6. Eqn (10) therefore yields

$$J = J_\sigma \sqrt{g/2.3\, d_{\mathrm{f}}} \tag{11}$$

where the factor of 2.3 in the square-root accounts for the conversion between base e and base 10 logarithms.¶ Applying eqn (11) with the computed parameters yields $J/T_{\mathrm{o}} = 7.4$, in good agreement with the empirical value for this ratio reported in the previous section. The empirical value was obtained from fitting data for various water models. Here, we have succeeded at deriving this value from a molecular calculation.

4 Theory for crystallization time

To estimate the timescale for crystallization, τ_{xtl}, we start with the usual expression,

$$\tau_{\mathrm{xtl}} = \nu^{-1} \mathrm{e}^{\Delta F/T} \tag{12}$$

where ΔF is the free energy cost for growing a nascent crystal and ν^{-1} is the timescale for adding material to the burgeoning phase. Typical forms for ΔF can be motivated by classical nucleation theory, which has been shown previously to yield accurate results for nucleation rate of models of water at moderate supercooling.[37] In general, this free energy can be written as

$$\Delta F/T = \Phi(\gamma/\Delta h)(T/T_{\mathrm{m}} - 1)^{-2}, \tag{13}$$

where γ is surface tension for liquid–crystal coexistence, and $\Phi(\gamma/\Delta h)$ is a function of the ratio of γ to the difference of enthalpy density between those phases. The ratio is approximately temperature independent.[1,19] The temperature-dependent factor, $(T/T_{\mathrm{m}} - 1)^{-2}$, comes from expanding the chemical potential difference to lowest non-trivial order in $T - T_{\mathrm{m}}$.

The timescale for adding material to a growing cluster, ν^{-1}, is expected to be approximately athermal for $T > T_{\mathrm{o}}$, but to become significantly temperature dependent below that onset temperature. In particular, we expect $\nu^{-1} \propto D$, where D is the molecular self-diffusion constant, and below the onset temperature, supercooled liquids generically obey a fractional Stokes–Einstein relationship,[38]

$$D \propto \tau_{\mathrm{liq}}^{-z}. \tag{14}$$

For $T > T_{\mathrm{o}}$, $z = 1$.[39] On the other hand, for $T < T_{\mathrm{o}}$, $z \approx 3/4$.[38] This value for the exponent is predicted by the East model.[40] Adopting a fractional Stokes–Einstein relation with eqn (2) implies a super-Arrhenius form for ν^{-1},

¶ Keys *et al.*[35] employ natural logarithms in their use of the parabolic law, and thus the factor of 2.3 does not appear in their equations.

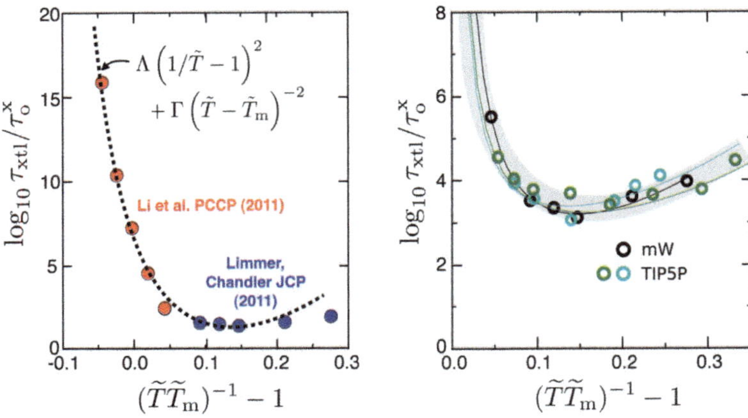

Fig. 4 Collapse of the crystallization times for models of supercooled water. (**left**) Crystallization times calculated for the mW model spanning the nucleation and growth regimes. Red markers are taken from Li *et al.*,[37] blue markers are taken from Limmer and Chandler.[18] The dashed black line is given by eqn (16) with $\Lambda = 94$, $\Gamma = 1.3$ and $\tau_o^x = 0.1$ ps. (**right**) Crystallization times for different models collapse. Black markers are our data for the mW model. Cyan and green markers are for two distinct sets of data for the TIP5P model taken from Yamada *et al.*[31] The three lines are eqn (16) with $\Lambda = 94$ and with Γ and τ_o^x adjusted for best fits to each of the three different data sets. To within 10%, for each of the three cases, $\Gamma \approx 0.5$ and $\tau_o^x \approx 8$ ps.

$$\nu^{-1} \propto \exp[2.3zJ^2(1/T - 1/T_o)^2]. \tag{15}$$

By combining eqn (12)–(15), and $T_o \approx T_{max}$, we obtain

$$\ln(\tau_{xtl}/\tau_o^x) = \Lambda(1/\tilde{T} - 1)^2 + \Gamma(\tilde{T} - \tilde{T}_m)^{-2}, \tag{16}$$

where $\Lambda = 2.3z(J/T_o)^2$, $\Gamma = \Phi(\gamma/\Delta h)\tilde{T}_m^2$ and τ_o^x is the proportionality constant in eqn (15).

Eqn (16) can be used to fit crystallization rates in terms of the constants Λ and Γ, and the values of these constants can be estimated *a priori*. From the universal East model value for z and the typical value of J/T_o for water-like systems (*i.e.*, $z \approx 0.75$ and $J/T_o \approx 7.4$), we have $\Lambda \approx 94$. Further, by approximating critical nuclei as spherical and mono-disperse,

$$\Gamma \approx \frac{4\pi}{3}\left(\frac{2\gamma}{\Delta h}\right)^2 \frac{\gamma}{T_m}. \tag{17}$$

For the mW model we have previously determined[19] all the quantities on the right-hand side of eqn (17), yielding for that model $\Gamma \approx 1.3$. Eqn (16) is plotted along side the numerical data in Fig. 4 with this parameterization. The agreement is good over a range of 10 orders of magnitude, spanning nanoseconds to seconds. The worst agreement is at the lowest temperature, where the rate is the most sensitive to the preparation of the initial state, as the liquid is no longer metastable at this condition. The next section of this paper expands upon this point.

5 Time–temperature–transformation diagrams

At conditions of liquid metastability, where a free energy barrier separates liquid and crystal basins, nucleation is the rate-determining step to form the equilibrium phase. Two data sets taken from the literature have used different rare-event sampling techniques to compute these times for a range of temperatures for the mW model. Limmer and Chandler[18] computed the nucleation rate constant following a standard Bennett–Chandler procedure.[41] Li *et al.*[37] calculated the nucleation rate using forward-flux sampling[42] with an order parameter based on a crystalline cluster sizes. To the extent that the kinetics is nucleation limited, both calculations are expected to have the same temperature dependence. However, because each used different order parameters and basin definitions, the prefactors can be different. In order to compare both data sets, we determine the ratio of prefactors by equating the rate at $T = 220$ K, which was calculated in both studies. These data sets are shown in left panel of Fig. 4.

For lower temperatures, we take data sets for crystallization times computed from first-passage calculations.[43] In the calculations of Limmer and Chandler,[18] the first-passage time is taken from simulations with the mW model. Similar calculations by Yamada *et al.*[31] based on first passage times have been calculated for the TIP5P model. In the latter case, the potential energy and structure factor were used as an order parameters for distinguishing crystallization. While T_m for the polymorph that TIP5P freezes into is not known, these data are taken sufficiently far away from any singular response that the fit to eqn 16 is insensitive to its precise value. Therefore we use the $\tilde{T}_m$ for ice 1h. Both of these calculations are shown in Fig. 5.

The crystallization times shown in Fig. 4 and 5 illustrate the non-monotonic temperature dependence predicted from eqn (16). In the higher-temperature regime, nucleation rates increase because the barrier to nucleation decreases in size. In the lower-temperature regime, the process of crystallization is slowed by the onset of glassy dynamics. At conditions where the amorphous phase is unstable, τ_{xtl}

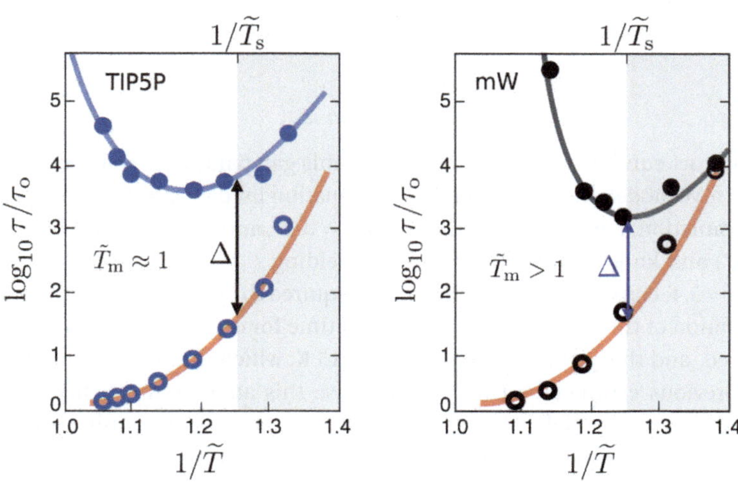

Fig. 5 Time scales of supercooled TIP5P water (left) and mW water (right) at 1 bar. Open circles are computer simulation results for structural relaxation times, τ_{liq}. Filled circles are computer simulation results for crystallization times, τ_{xtl}. One standard-deviation error estimates for the mW simulations (our results) are the size of the symbols. Error estimates for the TIP5P simulations[31] are unknown. The red line is the parabolic law, eqn (2), and the blue and black lines are fits to eqn (16) with $\Lambda = 94$, as predicted by theory. The grey region is where the liquid is unstable. At the boundary of liquid stability, the separation between times τ_{liq} and τ_{xtl} defines the gap parameter, Δ.

Table 1 Summary of properties for water and water models, with standard acronyms identifying different models.[3–7,14,16,19] All properties are measured at ambient pressure, and ice properties refer to ice 1h. See text for the meanings of symbols and the methods by which the properties are determined

Model	T_{max}/K	p_o/kbar	$\tilde{T}_m$	$\tilde{T}_o$	J/T_o	τ_o/ps	τ_o^x/ps	Γ	Δ
Experiment	277	3.7	0.99	0.98	7.4	1.0	0.3	1.2	3.4
mW	250	10.0	1.09	0.98	7.0	0.6	0.1	1.3	1.6
SW	1350	16.6	1.20	—	—	—	—	—	—
SPC/E	241	2.7	0.89	1.03	7.7	0.4	—	—	—
ST2	320	3.4	0.94	0.95	7.6	3.0	—	—	2.4
TIP4P	253	3.7	0.92	—	—	—	—	—	—
TIP4P/2005	277	3.4	0.90	0.99	7.5	9.0	—	—	—
TIP5P	285	19.4	0.96	0.98	7.6	0.2	8.0	0.5	2.1

becomes limited by mass diffusion, which from eqn (14) is proportional to τ_{liq}. In this region of the phase diagram, the liquid state is no longer physically realizable.

In plotting τ_{xtl} in Fig. 4, we use a different reduced temperature scale than previous plots of τ_{liq}. The different scale is chosen to emphasize the crossover region, where nucleation and growth compete. This particular scale also allows for crystallization times to be collapsed for different models because, to first order in Λ/Γ, this scale locates the minimum in τ_{xtl}. The location is the solution to a quadratic polynomial found by equating the nucleation and growth terms.|| This scaling holds only for T far below T_m. Away from the singular response at $T = T_m$, this form is conserved from model to model as it reflects the crossover to universal structural relaxation times away from the nucleation dominated regime.

In Fig. 5 we show both τ_{liq} and τ_{xtl} to illustrate how the separation of timescales evolves as a function of temperature for different models of water. By considering two cases, where $T_o \approx T_m$ and where $T_o < T_m$, we see large variation in the time-scale gap between liquid relaxation fastest crystallization. To quantify this variation between models, we define

$$\Delta = \log_{10} \left.\frac{\tau_{xtl}}{\tau_{liq}}\right|_{T=T_s}, \tag{18}$$

and subtract eqn (2) from 16 to predict how this gap parameter changes with $\tilde{T}_m$. For the mW model, the parameters in this equation predict $\Delta \approx 1.6$, in agreement with simulation. For experimental water, $\tilde{T}_m = 0.99$, and Γ can be computed using eqn (17) and known values for γ and Δh, yielding $\Gamma \approx 1.2$.** As such, eqn (18) gives $\Delta = 3.4$, consistent with cooling rates required to bypass crystal nucleation.[21] The location of the minimum crystallization time for experiment can be similarly predicted, and this yields $\tilde{T} = 0.77$ or $T \approx 215$ K, which is close to, though lower than, previous estimates.[22] One may also use this analysis to predict the time scales on which models of water will exhibit complex coarsening dynamics resulting in artificial polyamorphism.[20]

Table 1 summarizes material properties noted in this and preceding sections. Blanks (—) in the table refer to properties that have not yet been determined. Viewing the variability between models for the values for T_{max} and p_o elucidates

|| The solution for the minima is $\tilde{T} = (\tilde{T}_m + \omega)/2 + [(\omega + \tilde{T}_m)^2 - 4\tilde{T}_m]^{1/2}/2$, where $\omega = 1 - (\Lambda/\Gamma)^{1/2}$.

** Eisenberg and Kauzmann[14] gives $\Delta h \approx 3.0 \times 10^5$ kJ m^{-3}, and as we have done previously[19] we take γ to be an average of the values presented in Granasy *et al.*[44] or $\gamma = 32$ mJ m^{-2}.

how apparent different behaviors of different models can simply reflect different corresponding states.

6 Mesostructured supercooled water

The lengthscale over which the arguments presented in the previous sections are applicable to supercooled water reflects the lengthscale over which orientational order is correlated. We have previously studied these correlations using the phenomenological Hamiltonian of the form,

$$\mathcal{H}[q(\boldsymbol{r})] = \int_{\boldsymbol{r}} \left(f[q(\boldsymbol{r})] + \frac{m}{2} |\nabla q(\boldsymbol{r})|^2 \right), \tag{19}$$

where $q(\boldsymbol{r})$ is an order parameter that measures the amount of local orientational order at a point $\boldsymbol{r}$, $f(q)$ is a free energy density,

$$f(q) = a(T - T_s)q^2/2 - wq^3 + uq^4, \tag{20}$$

and a, w, u and m are positive constants determined by Δh, T_m and γ.[19] This Hamiltonian is isomorphic with that of van der Waals for liquid–vapor coexistence.[45] Consequently, mean profiles for $q(\boldsymbol{r})$ subject to external boundary conditions yield smooth order-parameter profiles like those at a liquid–vapor interface. Instantaneously, this field can be represented in a discrete basis and sampled with an interacting lattice gas.[46] Such a coarse-grained representation is amenable to large-scale computations, beyond what are tractable with atomistic models.

One case of water interacting with mesoscopic inhomogeneities that we have studied previously is water confined to hydrophilic nanopores.[19] For nanopores

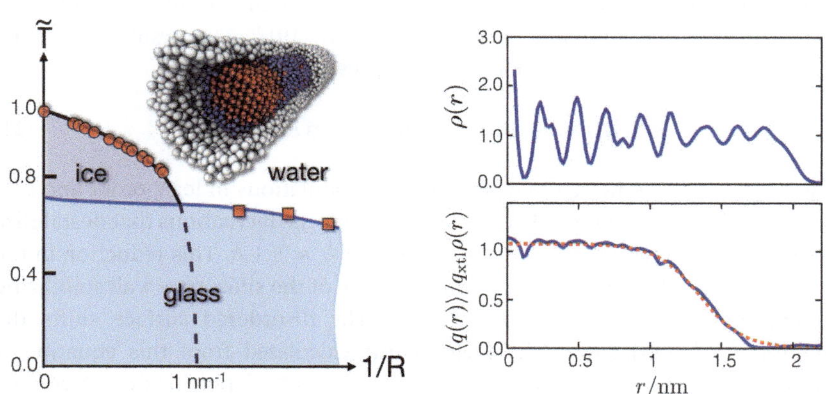

Fig. 6 Liquid–solid transitions of water confined to hydrophilic nanopores. (**left**) The phase diagram for water confined to hydrophilic nanopores divides into regions of liquid, glass and crystal-like states. The black line locates $T_m(R)$ while the blue locates $T_g(R)$. Red markers are experimental data for the melting line (circles)[48] and glass transition line (squares).[47] The top part of this panel illustrates a typical configuration of ice-like water, shown in red, in contact with the hydrophilic nanopore, shown in grey, mediated with a premelting layer, shown in blue. The configuration is taken from molecular dynamics calculations of the mW model.[19] (**right**) The mean molecular density (top), and mean local orientational order parameter (bottom), for mW water confined to a $R = 20$ Å pore for $T < T_m$, where q_{xtl} is the mean value of the order parameter in the center of the pore. The red dashed line in the bottom right panel is the theoretical prediction from the square-gradient theory in eqn (19).

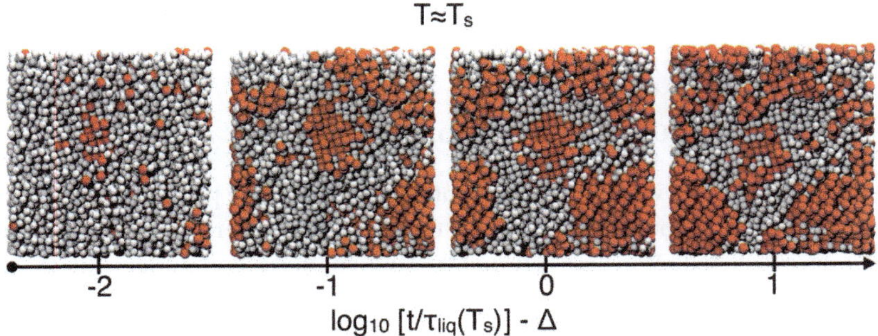

Fig. 7 Example of a coarsening trajectory from molecular dynamics simulations of the mW model liquid quenched to a temperature $T \approx T_s$. The pressure is fixed at $p = 0$ throughout the trajectory. At these conditions the initial liquid configuration is unstable, and the system evolves slowly to the crystal. The time over which this coarsening occurs is much longer than that of liquid relaxation time, as emphasized by the logarithmic time scale. To the extent that $\tau_o^x \approx \tau_o$, the universal form of the logarithmic scale can be used to predict coarsening time scales for other models and experiment. Red spheres locate the positions of the molecular centers that are locally crystal-like and grey spheres locate the positions of the molecular centers that are locally liquid-like. The pictures are from cuts through the simulation box at the times indicated by the tick marks on the time line. The simulation employs periodic boundary conditions.

with radii greater than, $R > 1$ nm, the properties of the water enclosed in the pore are sufficiently bulk-like that these scaling relations hold up to a perturbation due to the surface. Indeed using the expression in eqn (5) we have shown that the locations of glass transitions in $\tilde{p}$–R plane can be predicted.[19] These results are summarized in Fig. 6 which shows a $\tilde{p} = 0$ cut through the $\tilde{p}$–R plane. The location of the glass transition, T_g for finite pores has been measured.[47] These points are included in Fig. 6 and fall on our predicted glass transition line.

We have also previously computed the melting temperature in confinement from the partition function prescribed by eqn (19).[19] The resulting melting temperature as a function of pore size and pressure is given by

$$T_m(p,R) = T_m(p)[1 - \ell_m/R - \ell_s^2/8\pi(R - \ell_s)R] \tag{21}$$

where $\ell_m = 2\gamma/\Delta h$ reflects the typical spatial modulations in local order and $\ell_s = \ell_m/(1 - T_s/T_m)$ is the renormalized length that reflects fluctuations that destabilize order. For experimental water, $\ell_m \approx 2.1$ Å, and $\ell_s \approx 9.1$ Å. This reduction in the melting temperature, eqn (21), is a consequence of the silica pore wall stabilizing an adjacent disordered surface of water. The disordered surface shifts the conditions of coexistence. The melting line calculated from this equation is plotted in Fig. 6 and compared with the locations of previously determined freezing temperatures for water in silica nanopores.[48] As with the glass transition line, there is good agreement with experimental data. In our prior work,[19] we have also used this understanding of the phase diagram to explain the existence of a dynamic crossover and recent observations of hysteresis in density measurements for water confined to MCM-41 silica nanopores.[49]

We mention that explanation here because it relates to another instance of water evolving into mesoscopic structures, specifically the recent experimental observations by Murata and Tanaka.[50] Complex structure emerges from a mixture of water and gylcerol as it is quenched to low temperatures. The patterns observed depend on the depth of the quench and the relative concentrations of the two components. These patterns are reminiscent of the early stages of coarsening that we have found from theory for pure water near T_s. A specific example of such structural evolution is illustrated in Fig. 7, where the bulk free energy barrier to crystallization disappears.

Bear in mind that a local order parameter,[51] like that used in Fig. 7 to distinguish crystal and liquid domains, is not by itself a suitable reaction coordinate for dynamics in this situation. Nucleation occurs throughout the system and domain growth becomes rate limiting. This growth occurs through movement of interfaces, and it is reflected in the gap in timescales between density and long ranged order evolution, as quantified by Δ. Combining the quantitative understanding of timescales developed in this work with the understanding of how ice surfaces are modulated according to the phenomenological Hamiltonian in eqn (19) may admit a simple explanation for the observations of Murata and Tanaka.[50]

Acknowledgements

Work on this project was supported by the Helios Solar Energy Research Center, which is supported by the Director, Office of Science, Office of Basic Energy Sciences of the U.S. Department of Energy under Contract No. DE-AC02-05CH11231.

References

1 P. G. Debenedetti, *Metastable liquids: concepts and principles*, Princeton University Press, 1996.
2 C. A. Angell, *Annual Review of Physical Chemistry*, 1983, **34**, 593–630.
3 F. Stillinger and T. Weber, *Physical Review B*, 1985, **31**, 5262.
4 F. Stillinger and A. Rahman, *Journal of Chemical Physics*, 1974, **60**, 1545.
5 C. Vega and J. L. Abascal, *Journal of Chemical Physics*, 2005, **123**, 144504.
6 V. Molinero and E. B. Moore, *Journal of Physical Chemistry B*, 2009, **113**, 4008.
7 J. Abascal and C. Vega, *Journal of Chemical Physics*, 2005, **123**, 234505.
8 J. L. Abascal and C. Vega, *Journal of Chemical Physics*, 2010, **133**, 234502.
9 P. H. Poole, R. K. Bowles, I. Saika-Voivod and F. Sciortino, *Journal of Chemical Physics*, 2013, **138**, 034505.
10 Y. Liu, A. Z. Panagiotopoulos and P. G. Debenedetti, *Journal of Chemical Physics*, 2009, **131**, 104508.
11 I. Brovchenko, A. Geiger and A. Oleinikova, *Journal of Chemical Physics*, 2003, **118**, 9473.
12 V. Vasisht, S. Saw and S. Sastry, *Nature Physics*, 2011, **7**, 549–553.
13 L. Xu and V. Molinero, *Journal of Physical Chemistry B*, 2011, **115**, 14210–14216.
14 D. S. Eisenberg and W. Kauzmann, *The structure and properties of water*, Clarendon Press, London, 2005.
15 E. Sanz, C. Vega, J. L. F. Abascal and L. G. MacDowell, *Physical Review Letters*, 2004, **92**, 255701.
16 J. Broughton and X. Li, *Physical Review B*, 1987, **35**, 9120.
17 D. Chandler and J. P. Garrahan, *Annual Review of Physical Chemistry*, 2010, **61**, 191.
18 D. T. Limmer and D. Chandler, *Journal of Chemical Physics*, 2011, **135**, 134503.
19 D. T. Limmer and D. Chandler, *Journal of Chemical Physics*, 2012, **137**, 044509.
20 D. T. Limmer and D. Chandler, *Journal of Chemical Physics*, 2013, **138**, 214504.
21 T. Koop, B. Luo, A. Tsias and T. Peter, *Nature*, 2000, **406**, 611–614.
22 E. B. Moore and V. Molinero, *Nature*, 2011, **479**, 506–508.
23 E. B. Moore and V. Molinero, *Journal of Chemical Physics*, 2010, **132**, 244504.
24 D. T. Limmer and D. Chandler, Theory of amorphous ices, 2013, arXiv:1306.4728.
25 N. Giovambattista, T. Loerting, B. R. Lukanov and F. W. Starr, *Scientific Reports*, 2012, **2**, 390.
26 P. Poole, F. Sciortino, U. Essmann and H. Stanley, *Nature*, 1992, **360**, 324–328.
27 O. Mishima, *Journal of Chemical Physics*, 2010, **133**, 144503.
28 R. Speedy and C. Angell, *Journal of Chemical Physics*, 1976, **65**, 851.

29 Y. S. Elmatad, D. Chandler and J. P. Garrahan, *Journal of Physical Chemistry B*, 2010, **114**, 17113–17119.
30 K. T. Wikfeldt, C. Huang, A. Nilsson and L. G. Pettersson, *Journal of Chemical Physics*, 2011, **134**, 214506.
31 M. Yamada, S. Mossa, H. E. Stanley and F. Sciortino, *Physical Review Letters*, 2002, **88**, 195701.
32 P. H. Poole, S. R. Becker, F. Sciortino and F. W. Starr, *Journal of Physical Chemistry B*, 2011, **115**, 14176–14183.
33 S. Capaccioli and K. L. Ngai, *Journal of Chemical Physics*, 2011, **135**, 104504.
34 C. A. Angell, *Chemical Reviews*, 2002, **102**, 2627–2650.
35 A. S. Keys, L. O. Hedges, J. P. Garrahan, S. C. Glotzer and D. Chandler, *Physical Review X*, 2011, **1**, 021013.
36 J. Jäckle and S. Eisinger, *Zeitschrift für Physik B Condensed Matter*, 1991, **84**, 115–124.
37 T. Li, D. Donadio, G. Russo and G. Galli, *Physical Chemistry Chemical Physics*, 2011, **13**, 19807–19813.
38 M. Ediger, *Annual Review of Physical Chemistry*, 2000, **51**, 99–128.
39 J.-P. Hansen and I. R. McDonald, *Theory of simple liquids*, Academic press, 2006.
40 Y. Jung, J. P. Garrahan and D. Chandler, *Physical Review E*, 2004, **69**, 061205.
41 D. Frenkel and B. Smit, *Understanding molecular simulation: from algorithms to applications*, Academic press, 2001.
42 R. J. Allen, C. Valeriani and P. R. ten Wolde, *Journal of Physics: Condensed Matter*, 2009, **21**, 463102.
43 N. G. Van Kampen, *Stochastic processes in physics and chemistry*, North holland, 1992, vol. 1.
44 L. Gránásy, T. Pusztai and P. F. James, *Journal of Chemical Physics*, 2002, **117**, 6157.
45 J. S. Rowlinson and B. Widom, *Molecular theory of capillarity*, Courier Dover Publications, 2002, vol. 8.
46 N. Goldenfeld, *Lectures on phase transitions and the renormalization group*, Addison-Wesley, Advanced Book Program, Reading, 1992.
47 M. Oguni, Y. Kanke, A. Nagoe and S. Namba, *Journal of Physical Chemistry B*, 2011, **115**, 14023–14029.
48 G. H. Findenegg, S. Jähnert, D. Akcakayiran and A. Schreiber, *ChemPhysChem*, 2008, **9**, 2651–2659.
49 C. E. Bertrand, Y. Zhang and S.-H. Chen, *Physical Chemistry Chemical Physics*, 2013, **15**, 721–745.
50 K.-I. Murata and H. Tanaka, *Nature Materials*, 2012, **11**, 436–443.
51 A. Reinhardt, J. P. K. Doye, E. N. Noya and C. Vega, *J. Chem. Phys.*, 2012, **137**, 194504.

Faraday Discussions RSC Publishing

PAPER

Mesoscopic structural organization in triphilic room temperature ionic liquids

Olga Russina,[a] Fabrizio Lo Celso,[b] Marco Di Michiel,[c] Stefano Passerini,[d] Giovanni Battista Appetecchi,[e] Franca Castiglione,[f] Andrea Mele,*[f] Ruggero Caminiti[a] and Alessandro Triolo*[g]

Received 22nd April 2013, Accepted 29th July 2013
DOI: 10.1039/c3fd00056g

Room temperature ionic liquids are one of the most exciting classes of materials in the last decade. The interest for these low melting, ionic compounds stems from both their technological impact and the stimulating plethora of structural and dynamic peculiarities in the mesoscopic space–time scales. It is nowadays well-established that they are characterised by an enhanced degree of mesoscopic order originating from their inherent amphiphilicity. In this contribution we highlight the existence of a further degree of mesoscopic complexity when dealing with RTILs bearing a medium length fluorous tail: such triphilic materials (they simultaneously contain polar, hydrophobic and fluorophilic moieties that mutually segregate from each other) turn out to be highly structurally compartmentalised at the mesoscopic level, thus paving the way to new smart applications for this new class of RTILs.

Introduction

Room temperature ionic liquids (RTILs) represent an exciting class of materials, which are composed solely of ionic species and have low melting points (<100 °C).[1–3] One of their most thought-provoking features is the ability to self-organize into highly structured mesoscopic arrangements, due to their inherently amphiphilic nature, *via* charged moieties being covalently bound to apolar alkyl tails in RTILs.[4–32] In other words, the strong, long range coulombic forces tend to spatially distribute the charged moieties in a typical onion-like morphology.

[a]*Department of Chemistry, University of Rome "Sapienza", Rome, Italy*
[b]*Dipartimento di Chimica "Stanislao Cannizzaro", viale delle Scienze, ed. 17, 90128 Palermo, Italy*
[c]*European Synchrotron Radiation Facility, 6 Rue Jules Horowitz, 38000 Grenoble, France*
[d]*Institute of Physical Chemistry and MEET, University of Muenster, 48149 Muenster, Germany*
[e]*ENEA, Agency for New Technologies, Energy and Sustainable Economic Development, UTRINN-IFC, Rome, Italy*
[f]*Department of Chemistry, Materials and Chemical Engineering "G. Natta", Politecnico di Milano, Milano, Italy. E-mail: andrea.mele@polimi.it*
[g]*Laboratorio Liquidi Ionici, Istituto Struttura della Materia, CNR, Rome, Italy. E-mail: triolo@ism.cnr.it*

Conversely, apolar chains tend to spatially segregate from the charged moieties into mesoscopic domains.

Fluorous compounds[33] also represent a hot field of research in several areas including bio-medicine,[34] synthesis and separation.[35] Indications of self-assembly of fluorous tails in selective solvents or in the crystalline state have been provided.[36] Several fluorous anion-based RTILs have been developed[37,38] and they are proposed for smart applications including electrochemistry, tribology and catalysis.

In this study we investigate RTILs bearing fluorous tails aiming to probe their self-assembly as a counterpart of the alkyl chain clustering that has been observed in more conventional RTILs. This discovery represents a major result: the structure of triphilic RTILs turns out to be highly compartmentalized at the nm-scale, thus providing the possibility of finely dispersing polar, hydrophilic and fluorophilic compounds, which in the bulk state would be incompatible, into nano-pools that are only few Å apart from each other. There can be major applicative implications for such proximity in the fields of catalysis, synthesis, separation *etc.*

Since the first computational studies, it appeared that the clustering phenomenon in RTILs might be strongly affected by temperature and that, at sufficiently high temperatures, the domains might melt leading to a more homogeneous distribution.[9] Experimental evidence of tail segregation was provided *e.g.* by X-ray/neutron diffraction experiments, where the domains are fingerprinted by low Q amorphous peaks[11–20,39–43] So far however no experimental indication could be extracted of the domains melting, that would correspond to the disappearing of the low Q peak: even experiments at temperatures as high as 150 °C indicated the persistence of low Q features in samples such as C10mimBr.[12] In the present study we will also show indication of the melting of the fluorous domains at high enough temperature.

We will explore a series of model RTILs with a common anion, namely (nonafluorobutanesulfonyl) (trifluoromethanesulfonyl) imide [IM_{14}], and four cations

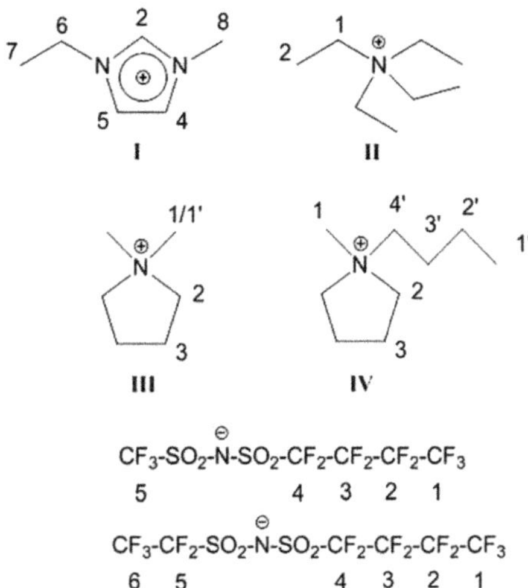

Fig. 1 Molecular formulae and atom numbering for cations **I–IV** and the [IM_{14}] and [IM_{24}] anions.

bearing an alkyl chain sufficiently short to prevent segregation (see Fig. 1): 1-ethyl-3-methyimidazolium $[IM_{14}]$ (EMIM–IM_{14}, or C_2mim–IM_{14}) (**I**), tetraethylammonium [IM14] (Et_4N-IM_{14}) (**II**), *N*,*N*-dimethylpyrrolidinium $[IM_{14}]$ (PYR_{11}–IM_{14}) (**III**) and *N*-butyl-*N*-methylpyrrolidinium $[IM_{14}]$ (PYR_{14}–IM_{14}) (**IV**). Furthermore we will also report results on a sample where the cation is *N*-butyl-*N*-methylpyrrolidinium and the anion is (nonafluorobutanesulfonyl) (pentafluoroethanesulfonyl) imide $[IM_{24}]$ (PYR_{14}–IM_{24}) (**V**), where the anion bears two perfluorinated chains, namely an ethyl and a butyl group. The rationale for the choice of these RTILs is the following: i) in order to make the experimental results independent of the cation geometry (with no spurious contribution such as hydrogen bonds, tight ion-pairs and/or steric contributions), **I–IV** show four different features: they are planar (**I**), non-cyclic symmetrical (**II**), cyclic symmetrical (**III**) and cyclic, non-symmetrical quaternary (**IV**) ammonium ions; ii) the anions are perfluorinated and the negative charge is delocalized onto the $(SO_2)_2N$ moiety, in turn flanked by two perfluoro alkyl groups; iii) two different kinds of perfluorinated anions are considered in order to probe the role played by the increasing level of fluorine content, namely $[IM_{14}]$ and $[IM_{24}]$; and finally iv) the chosen RTILs are *not* protic. These structural features are crucial for the interpretation of the results. As a matter of fact, protic RTILs with anions showing the negative charge localised onto functional groups placed on one terminal of the perfluoroalky chain – *e.g.* perfluorocarboxylates – are expected to show distinct R_3N–H···OOC–R_f ion pairing and/or hydrogen bond assembly motives, which are strong enough to drive the self-association of the R_f tails into domains, as elegantly demonstrated by Drummond and Greaves (DG)[17] (*vide infra*). The presence of such strong attractive interactions (that typically leads to crystalline ILs at ambient temperature) can thus bias the formation of the fluorous domains. However, in our present choice of salts, it has been recently pointed out that the interaction energy (*ca.* 15 kJ mol^{-1}) for ion pairs is at a lower end for ILs.[44] The limited tendency of RTILs **I–IV**

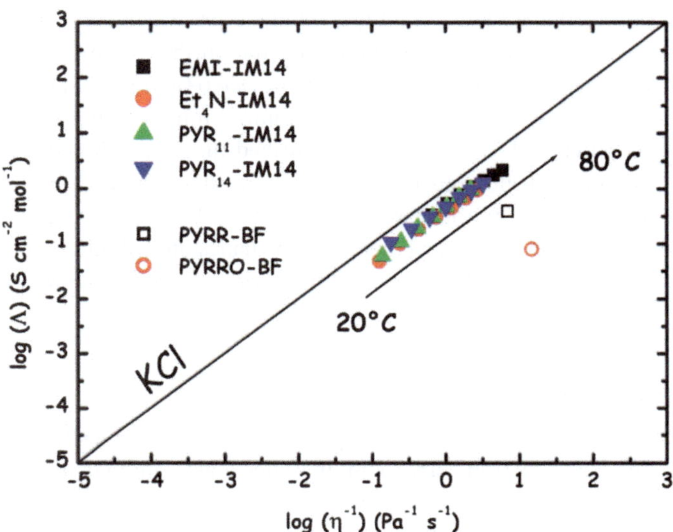

Fig. 2 Walden plot comparing ILs **I–IV** with selected fluorinated ILs investigated by Drummond and Greaves.[17]

to organize in ion pairs is confirmed by their Walden plot: in Fig. 2 we show log Λ (the equivalent ionic conductivity) *vs.* log f (fluidity or inverse viscosity) for salts **I–IV** and for two of DG's fluorinated salts.[17] It clearly appears that while the latter behave as "poor ILs", thus witnessing the formation of ion pairs, **I–IV** behave as "good ILs", with a high conductivity at a given viscosity.

Cations **I–IV**, when coupled with other symmetric perfluoroalkylsulfonyl imide anions tend to crystallise.[45] However, in this present case, where a strongly asymmetric perfluorinated anion is used, only a glass transition was identified (T_g(**I**) = 195 K, T_g(**II**) = 214 K, T_g(**III**) = 213 K, T_g(**IV**) = 201 K).[45] Similar behavior has been found on the basis of conductivity measurements, confirming that these $[IM_{14}]$-based salts remain liquid.[45]

In recent reports some of us highlighted the 'possible formation of mesoscopic fluorous domains', due to the segregation of fluorous tails into nano-scale compartments, on the basis of nuclear Overhauser enhancements (NOEs).[46] This hypothesis was later supported by the experimental work of Yoshida and Saito who studied RTILs based on the symmetric $(C_nF_{2n+1}SO_2)_2N$ anions, with different C_nF_{2n+1} chain lengths ($0 \leq n \leq 4$), proposing the 'possible aggregation of perfluoroalkyl groups'.[47] In 2010, Smith *et al.*, using MD simulations, observed that fluorinated tails connected to imidazolium rings segregate into domains, similarly to what is found for alkyl tails.[48] Recently DG reported on a series of protic ILs with a fluorinated carboxylate anion whose self-assembled nanostructure was explored using SWAXS data.[17] Of course, the occurrence of fluorous moieties segregation into clusters is well known in the case of solid state.[49–52]

Materials and methods

Ionic liquid materials

Bis(perfluoroalkylsulfonyl)imide anion-based ionic liquids were synthesized through a procedure developed at ENEA and described in detail elsewhere.[53,45] The chemicals *N*-methylpyrrolidine (97 wt.%), 1-methylimidazole (99 wt.%), bromomethane (99 wt.%), bromoethane (99 wt.%), 1-bromobutane (99 wt.%), ethyl acetate (ACS grade, >99.5 wt.%) were purchased from Aldrich and purified (with the exception of ethyl acetate) through activated carbon (Aldrich, Darco-G60) and alumina (acidic, Aldrich Brockmann I). The tetraethylammonium bromide (99 wt.%), Et_4NBr, precursor was purchased from Aldrich and purified through activated carbon and alumina. The acidic (trifluoromethanesulfonyl)(nonafluorobutanesulfonyl) imide, HIM_{14} (59 wt.% solution in water), was provided by 3M and used as received. The lithium (nonafluorobutanesulfonyl)(pentafluoroethanesulfonyl)imide salt, $LiIM_{24}$, was synthesized on a lab-scale as reported in previous work.[54] Deionized H_2O was obtained with a Millipore ion-exchange resin deionizer.

Physicochemical characterization

The ionic conductivity was determined by a conductivity meter AMEL 160. The IL samples were housed (within a dry-room) in sealed glass conductivity cells (AMEL 192/K1) equipped with two porous platinum electrodes (cell constant equal to 1.00 ± 0.01 cm). The conductivity tests were run in the temperature range from −40 °C to 100 °C by using a climatic test chamber. In order to fully crystallize the materials, the cells were immersed in liquid nitrogen for a few seconds and, then, transferred into the climatic chamber at −40 °C.[55] Nevertheless, after a few minutes of storage at this temperature, all samples, with the exception of $PYR_{14}IM_{24}$, turned back into a liquid (from solid) even after repeating this procedure various times, indicating a melting point below −40 °C. After storage at −40 °C for at least eighteen hours the conductivity of the materials was measured by running a heating scan at 1 °C h^{-1}.

The rheological properties were carried out using a rheometer (HAAKE RheoStress 600) located in the dry-room. The tests were performed from 20 °C to 80 °C (1 °C min^{-1} heating rate) in the 100 s^{-1} to 2000 s^{-1} rotation speed range. Measurements were taken after 10 °C steps.

The density measurements were performed from 90 °C to 20 °C at 10 °C step using a density meter (Mettler Toledo DE40) in the dry-room. The samples were previously degassed under vacuum at 50 °C overnight to avoid bubble formation during the cooling scan tests.

SWAXS measurements

The Small-Wide Angle X-ray Scattering (SWAXS) experiments were conducted at the high energy beam line ID15b, European Synchrotron Radiation Facility (ESRF), Grenoble, France, using an instrumental setup which allows covering the momentum range Q between 0.1 and 20 $Å^{-1}$, with a wavelength $\lambda = 0.22$ Å (Energy = 55.6 keV). Measurements were collected on a sample kept inside a 1 mm diameter capillary into a LINKAM THMS600 cryostat that allows temperature control between 173 and 373° K, using a liquid nitrogen flow. The corresponding empty cell contribution was subtracted.

NMR measurements

The ^{1}H and ^{19}F NMR spectra were recorded on a Bruker Avance 500 spectrometer operating at 500 MHz proton frequency equipped with a QNP four nuclei switchable probe. The NMR samples were prepared in a dry-room to avoid any contamination, transferred in a 5 mm NMR tube equipped with a co-axial capillary containing DMSO-d6 for locking and immediately flame-sealed. Heteronuclear $\{^{1}H–^{19}F\}$HOESY experiments were acquired using the inverse-detected pulse sequence with 512 increments in the $t1$ dimension and sixteen scans for each experiment. Qualitative spectra were acquired with mixing times of 20, 30 and 40 ms. Homonuclear $\{^{19}F–^{19}F\}$NOESY experiments were acquired by using a standard gradient -selected NOESY pulse sequence with mixing times ranging from 50 to 150 ms. 512 increments in the $t1$ dimension with sixteen scans for each experiment were used. For all the experiments the temperature was set at 305 K and controlled with an air flow of 535 l h^{-1}.

Results and discussion

SWAXS data from tetraethylammonium $[IM_{14}]$ (**II**), at temperatures in the range 173–323 K, are shown in Fig. 3.

The data over the whole Q range (data not shown) indicate that, across the probed temperature interval, no crystallization (that would be fingerprinted by Bragg peaks) occurred. The experimental data below $Q = 4$ $Å^{-1}$ (that are relevant to the present discussion) are characterized by three amorphous halos. Peak 1 is

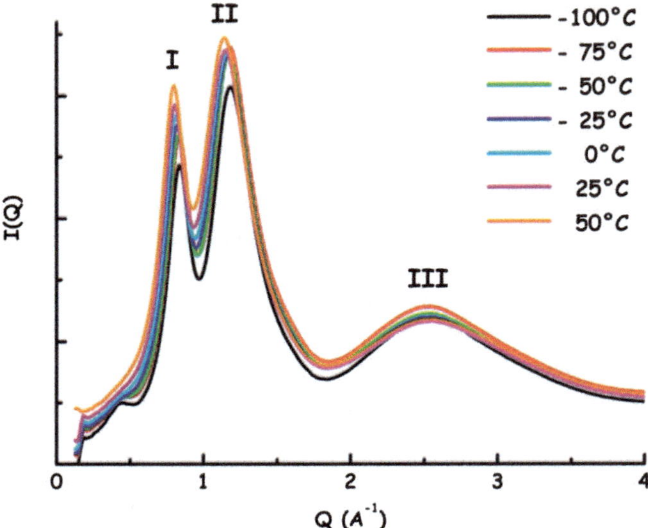

Fig. 3 SWAXS data from tetraethylammonium [IM_{14}] (**II**), at different temperatures (173–323 K).

centred at approx. 0.8 $Å^{-1}$ and it slightly depends on the temperature: its amplitude decreases and shifts to higher *Q* values upon decreasing *T*. Peak 2 is centred at *ca.* 1.15 $Å^{-1}$ and changes minimally with temperature, similarly to peak 3 that is centred at *ca.* 2.5 $Å^{-1}$. Peaks centred approximately at these positions are commonly found in RTILs bearing [IM_{XY}]-like anions, such as TFSI or BETI, [IM_{11}] or [IM_{22}], respectively, as they correspond to different, mostly intermolecular correlations in these classes of RTILs. Their shift towards higher *Q* values with decreasing temperature is consistent with the corresponding density increase that leads to a shortening of neighbour distances.

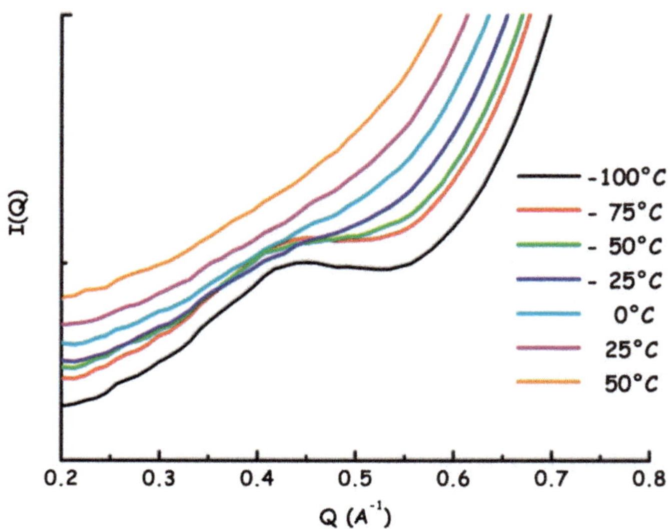

Fig. 4 Low *Q* portion of the SWAXS data from tetraethylammonium [IM_{14}] (**II**), at different temperatures (173–323 K).

We focus now our attention on the low Q portion of the SWAXS spectra (Fig. 4), in the range between 0.2 and 0.7 Å^{-1}.

There is now a large amount of literature converging on the finding that RTILs bearing short enough alkyl tails (such as the ethyl one) are characterized by essentially featureless SWAXS patterns in this momentum transfer range.[43,56–59] This means essentially that the RTIL structure at the microscopic level can be envisaged as a typical onion-like alternation of oppositely charged layers around a central one that quite rapidly smears out into an isotropic spatial organization, similar to conventional inorganic molten salts.

However an inspection of Fig. 4 indicates that while at room temperature (and above) the patterns are featureless, as can be expected, on the other hand, at low enough temperature (in the present case at −25 °C) a distinct diffraction feature emerges at *ca.* 0.43 Å^{-1} (corresponding to a Bragg characteristic size $D = 2\pi/Q =$ 14.6 Å). The feature becomes more and more detectable at lower temperatures. This behaviour is also observed in the SWAXS patterns from samples **I** and **III** at 173 K (Fig. 5). It is clear then that samples **I–III**, at the lowest probed temperature (which is below their glass transition, when diffusive dynamics are frozen), show a low Q diffraction feature, whose position is only slightly affected by the change of cation nature ($Q \sim 0.43$ Å^{-1}). This observation strongly suggests the development, at low temperatures, of structural heterogeneities which are related to the segregation of the fluorous chains.

As a further confirmation for this interesting behaviour we also present SWAXS data from sample **V**, [PYR_{14}][IM_{24}], that is characterized by a longer side alkyl chain in the cation and by the presence of a perfluoroethyl side chain in the anion rather than a perfluoromethyl (Fig. 6). The latter change leads to an enhanced tendency for the fluorinated tails to segregate.

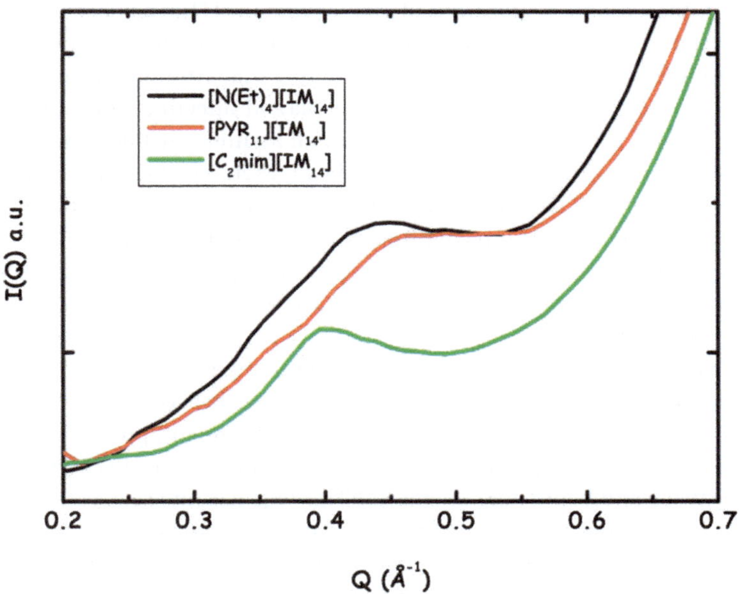

Fig. 5 Low Q portion of the SWAXS patterns from samples **I–III** at −100 °C.

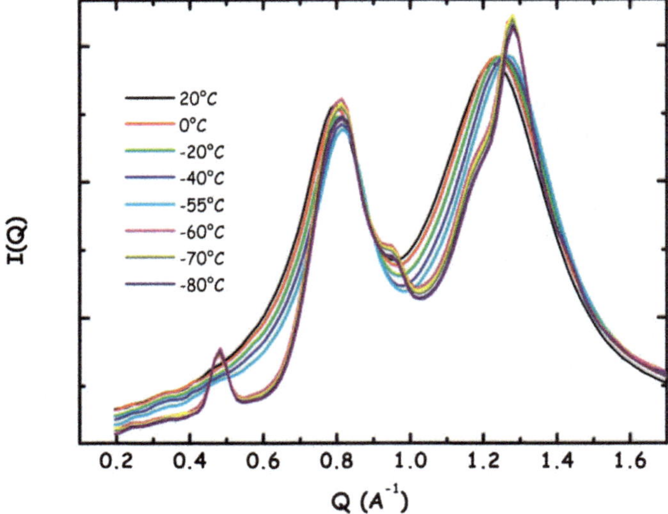

Fig. 6 SWAXS data from 1-butyl,3-methylpyrrolidinium [IM_{24}] (**V**), at different temperatures (193–293 K).

As a matter of fact at high temperature (*e.g.* at RT) the sample is characterized by lack of diffraction features below *ca.* 0.6 Å^{-1}. However, as the temperature decreases, hints of the existence of fluorous tailed clusters can be detected at −55 °C and the latter become a strong amorphous peak centred at 0.48 Å^{-1} at −60 °C. While the sample remains in the liquid amorphous state across the whole temperature range (as no indications of crystalline Bragg peaks occur), we also notice that the shape of the higher *Q* peaks is affected by the structural re-organization consequent to the fluorous tails clustering.

So far, mesoscopic segregation of side chains in RTILs has been observed only for the case of alkyl chains and for protic ILs with fluorinated tails, with the above mentioned caveat concerning the role that strongly hydrogen bond mediated cation–anion correlations might play in affecting tail segregation. Here we are reporting on aprotic RTILs with an asymmetric fluorinated anion, which show no tendency to crystallize and can, then, be studied in the amorphous state over a wide temperature range. These RTILs *do not* show evidence of tail segregation – on the nm length scale – at ambient conditions. Only below *ca.* −25 °C low *Q* diffraction features appear, fingerprinting the self-assembly phenomenon. However this mesoscopic structural organization vanishes at higher temperatures.

Indeed, while low *Q* diffraction peaks have been reported by several groups to exist in RTILs (when a polar *vs.* apolar difference exists between covalently bound moieties[12,40,18]), this is the first report accounting for the disappearance of such diffraction features at higher temperatures, thus supporting the CG-MD proposal made by Wang and Voth.[9] Alkyl chains in RTILs organize into segregated clusters that persist even at $T \approx 150$ °C.[12] In the case of fluorinated tails, the diffraction patterns indicate that mesoscopic structuring occurs at very low temperatures (<−50 °C) but, as soon as the temperature increases above *ca.* −25 °C, this order is lost in favour of a more homogeneous tail distribution.

Likely the differences with respect to the cases reported by DG[17] (where quite intense low Q peaks are observed in the liquid state even at temperatures as high as 50 °C) are due to the substantially different coulombic interactions occurring between cations and either the [IM_{14}] (present study) or DG's perfluoroalkylcarboxyl anions. In the latter case the carboxyl negative head strongly interacts with the cation head, while the [IM_{14}] anion, due to its low symmetry, flexibility, bulky geometry, sterically hindered pairing geometry and extensive charge delocalization can less efficiently interact with the cations. This results in remarkable differences in the phase diagram (*e.g.*, no crystalline phases for the case of [IM_{14}] salts *vs.* melting points above RT for several of DG's ILs) and in the stability of the segregated fluorous tail domains. Such a conclusion prompted us to further investigate at the atomistic level the structural correlations that might lead to such behaviour.

As previously reported, useful information for the assessment of the local structure of pure ionic liquids can be obtained by 2D NMR correlation spectroscopy based on the nuclear Overhauser enhancement (NOE).[60,61] The local organization of the ions at a smaller internuclear length scale compared to the one probed by the SAXS technique can be obtained by the analysis of the intermolecular heteronuclear ^{1}H–^{19}F NOE, homonuclear ^{1}H–^{1}H NOE and ^{19}F–^{19}F NOE patterns.

It is worthy of mention that the RTILs in Fig. 1 are characterized by a perfluorinated anion and non-fluorinated cations as positive counterparts. Thus, the experiments mentioned above give insights, in the same order, into the cation–anion, cation–cation and anion–anion correlations in the bulk liquid, in the examined temperature range and on the NMR spectroscopy time-scale.[46,50,62,63] In the case of RTILs **I–III** however, the hydrocarbon chains attached to the positively charged N atoms are not long enough to allow for the extraction of significant intermolecular NOEs, as shown in the case of *N*-butyl substituted imidazolium,[64,65] pyrrolidinium[46,50,63] and the brand new class of pyrazolium based ILs.[66] Therefore, the homonuclear ^{1}H–^{1}H NOE experiments will not be reported and discussed here for **I–III**.

The heteronuclear ^{1}H–^{19}F NOE correlation experiment (HOESY) performed on compound **I** (Fig. 7) shows strong NOE contacts between the perfluorobutyl chain of the anion and all the protons of the cation, the only exception is observed for the CF_2(3) atoms towards the imidazole aromatic protons H(2), H(5) and H(4). This pattern of selectivity can be ascribed to a preferential, non-random local organization of the anion with respect to the cation.

Similar results were obtained for compound **II**: the HOESY spectrum reported in Fig. 8 shows strong NOEs between the methyl groups of the cation and all the fluorine atoms of the anion, while no NOEs cross-peaks are observed between the CF_2(3) of the anion and the cation CH_2(1), thus revealing a certain degree of selectivity between the cation short ethyl chain and anion perfluorobutyl chain.

In this case, however, the observed selectivity can, in principle, be caused by the overall tetrahedral symmetry of the cation. Indeed, the N–CH_2 groups are not exposed to the anions due to the steric hindrance of the methyl groups of the ethyl chains. The perfluorinated chain of the anion mainly interacts with the methyl groups of the cation with a minor possibility to reach the N–CH_2 groups for both the low affinity of the perfluoroalkyl chain with the ethyl groups and for steric reasons. A more defined situation is found in compound **III**. The HOESY on **III**

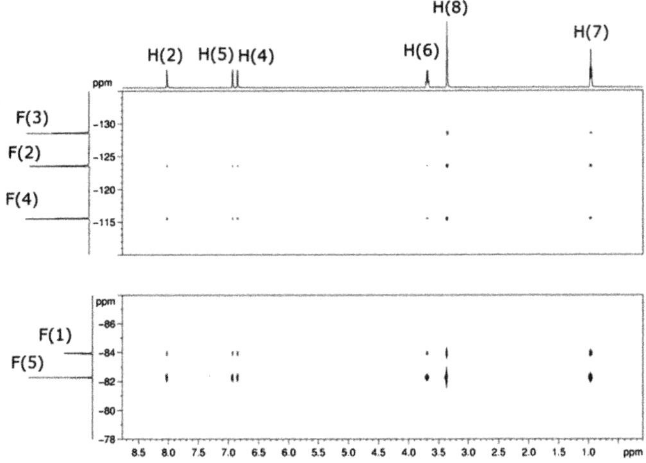

Fig. 7 Contour plot of the {^{1}H–^{19}F} HOESY experiment on ionic liquid **I**.

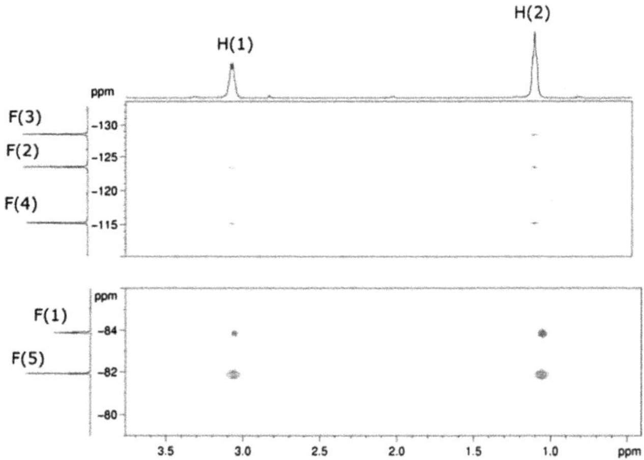

Fig. 8 Contour plot of the {^{1}H–^{19}F} HOESY experiment on ionic liquid **II**.

(Fig. 9) shows strong NOE correlation of the H nuclei of the *N*-methyl groups of the cation (H1/H1′) with all the anion's F atoms, thus meaning that the part of the pyrrolidinium cation close to the positively charged N atom is near to the whole anion. Conversely, the CH_2 groups in the α- and β-position (H2 and H3, respectively) do give rise to small but detectable NOEs only with the isolated CF_3 group of IM_{14} (CF_3 in position 5): no appreciable NOE between the perfluorobutyl chain of the anion and the CH_2–CH_2 frame of the cation is observed.

Accordingly, the experiment indicates that the perfluorobutyl chain does not interact with the less polar part of the pyrrolidinium molecular frame, namely the dimethylene C(2)–C(3) part of the five-membered ring. We note that the experiment carried out on *N*-butyl-*N*-methylpyrrolidinium IM_{14} (compound **IV**) showed a similar pattern of selectivity.[46] In this latter case, the anion's perfluorobutyl chain interacts with the cation's H atoms in the proximity of the positively

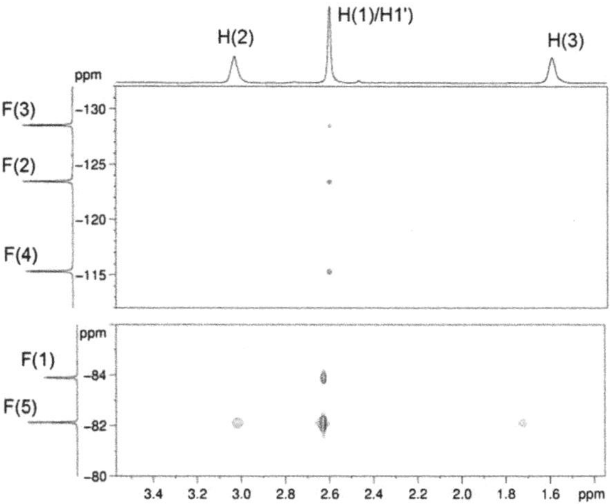

Fig. 9 Contour plot of the $\{^1H-^{19}F\}$ HOESY experiment on ionic liquid **III**.

charged quaternary N atom, but it does not interact with the hydrogen atoms on the *N*-butyl side-chain. The comparison of the HOESY data of compound **III** (this work) and **IV**[46] prompts for a general trend: the local order of anion–cation is such that the anion's fluorous chain undergoes the maximum repulsion with the less polar part of the counterion, namely the CH_2–CH_2 frame of **III** and the *N*-butyl chain in **IV**, respectively.

Under this point of view, the NOE selectivity indicates a pre-organization in terms of hydrocarbon/perfluorocarbon mutual segregation. Such a structural motif expands toward the formation of the nano-sized domains once the thermal motion is cooled down, as in the SAXS experimental conditions. This interpretation is consistent with the observed high selectivity of $CF_3(1)$, located at the end of the perfluorobutyl chain of IM_{14}, and lower selectivity shown by the isolated $CF_3(5)$ of the same anion with respect to the PYR_{14} cation. The $^1H-^{19}F$ HOESY NMR approach is powerful in proving the mutual segregation of hydrocarbon and fluorocarbon moieties provided the hydrocarbon chains (or the saturated ring) and the perfluorocarbon chains are sufficiently large, as simultaneously found in compound **III**.

On the other hand, **I**, **II** and **III** share the same perfluorinated anion and, as such, are supposed to show similar anion–anion organization. Indeed, $^{19}F-^{19}F$ NOESY experiments (Fig. 10) carried out on **I**, **II** and **III** gave the same picture of the perfluorocarbon chain aggregation features. Discrimination between intra- and intermolecular NOEs was done following Osteryoung *et al.*[67] In synthesis, the discrimination between intra- and intermolecular NOE can be accomplished by considering the threshold distance of 5 Å for a vanishing NOE between dipolar coupled nuclei. Thus, the observation of NOE contacts of nuclei belonging to the same moiety (*e.g.* two protons on the same cation or two fluorine atoms on the same anion) but separated by more than 5 Å as intramolecular distance can be confidently assigned to intermolecular NOE. Generally, the internuclear and intramolecular distances are taken either from single crystal X-ray structures (if available) or by calculation (DFT, molecular mechanics or dynamics).

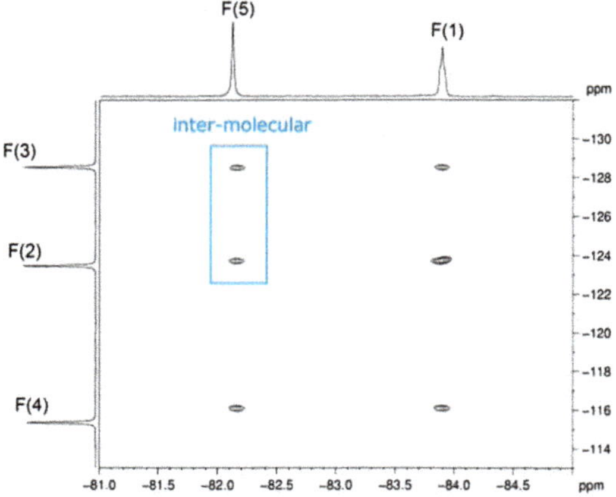

Fig. 10 Expansion of the contour plot of the ^{19}F–^{19}F NOESY spectrum on ionic liquid **III**. The box contains intermolecular cross-peaks related to the local anion–anion correlation. The $CF_3(1)$–$CF_3(5)$ intermolecular cross-peak is not shown in this expansion.

An example of a DFT protocol for the calculation of H–H distances and the use of calculated data for the interpretation of NOE data can be found in ref. 66. The interpretation of ^{19}F–^{19}F NOESY experiments shown in Fig. 10 was indeed supported by DFT calculated F–F distances for the IM_{14} anion. The average distance between groups of equivalent nuclei (*e.g.* $CF_2\cdots CF_3$) was achieved as described in ref. 64. The conformational search pointed out the existence of two low energy conformations, shown in Fig. 11.

Considering the data in Table 1, the ^{19}F–^{19}F NOESY experiments provide quite a clear picture of the anion–anion correlation. The cross-peak correlating $CF_3(1)$ to $CF_3(5)$ is a clear example of intermolecular contact: indeed the two CF_3 groups on the same ion are beyond the length threshold for the observable NOE in both of the conformational states. Thus, the presence of such a cross peak indicates a

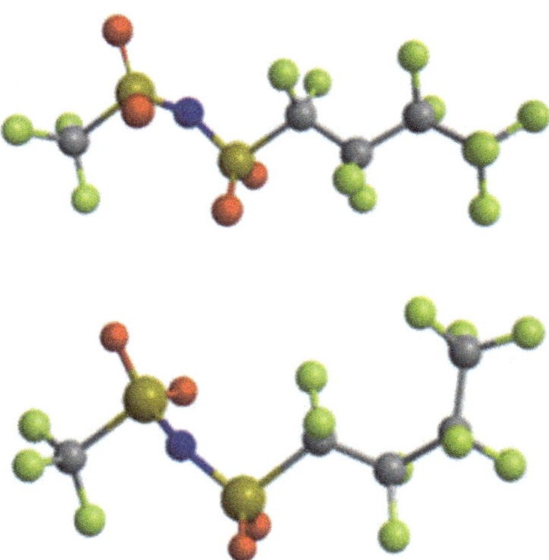

Fig. 11 DFT calculated conformations for $[IM_{14}]$. Top: all-*trans* (tt); bottom: *trans–gauche* (tg). The average distances (Å) are reported, for both conformations, in Table 1.

Table 1 Calculated average[a] distances ⟨*d*⟩ (Å) of equivalent F atoms groups in [IM_{14}]. Both *trans–trans* (tt) and *trans–gauche* (tg) conformations are considered. Distances in boldface are beyond the threshold for observable intramolecular NOEs

	$CF_3(1)–CF_2(3)$	$CF_3(1)–CF_2(4)$	$CF_3(1)–CF_3(5)$	$CF_2(2)–CF_3(5)$	$CF_2(3)–CF_3(5)$	$CF_2(4)–CF_3(5)$
tt	2.6	4.0	**6.3**	**6.8**	**5.5**	4.7
tg	3.0	2.7	**6.3**	**7.1**	**5.5**	4.8

[a] The average distance ⟨*d*⟩ for dipolar interactions is calculated according to the following equation: $\langle d \rangle = \left(\sum_i r_{ij}^{-3}/3 \right)^{-1/3}$.

spatial proximity of at least two different anions. In a similar fashion, the cross-peaks connecting $CF_3(5)$ to $CF_2(2)$ and $CF_2(3)$ are also intermolecular and strongly suggest the spatial proximity of adjacent perfluorinated chains. In conclusion, even at room temperature and on the NOE distance scale, the anions undergo spatial arrangements consistent with the formation of the perfluorocarbon domains observed, on a larger spatial scale, by the SAXS experiments.

Conclusion

It is nowadays well established that room temperature ionic liquids are far from being structurally homogeneous, but rather are characterised by an enhanced degree of mesoscopic heterogeneity. This stems from the role played by the self-excluding moieties building up the RTILs ions, where charged polar heads are covalently bound to apolar, lipophilic tails, thus delivering characteristic amphiphilic behaviour.

Here we have discussed new triphilic RTILs that bear a medium length fluorous tail, thus introducing a further element of structural incompatibility between the coexisting moieties. We reported complementary X-ray and NOE-NMR data that clearly portrays the phenomenon of the fluorous tails' self-assembly into mesoscopic size domains in this class of RTILs. This behaviour, when considered in combination with the already proven alkyl tail segregation into a three-dimensional matrix of charged heads in conventional RTILs, prompts the definition of a highly compartmentalized structural scenario at the mesoscopic scale. The simultaneous coexistence in these triphilic materials of a charged polar matrix, lipophilic alkyl tail clusters and fluorophilic domains that are just a few nm apart paves the way to new smart applications involving chemical reactivity, transport properties, catalysis and separation.

Acknowledgements

We acknowledge the European Synchrotron Radiation Facility for provision of synchrotron radiation facilities. We would like to thank Professor L. Conte for the kind supply of Li[IM_{24}]. A. T. acknowledges financial support from: FIRB "Futuro in Ricerca" (RBFR086BOQ) and PRIN2009 (2009WHPHRH). S. P., G. A., F. C. and A. M. wish to thank the EU for funding (FP7 – GreenLion Project GC.NMP.2011-1-Grant 285268). A. T. wishes to dedicate this contribution to the memory of Prof. Roberto Triolo.

References

1 T. Welton, *Chem. Rev.*, 1999, **99**, 2071–2083.
2 K. R. Seddon, *J. Chem. Technol. Biotechnol.*, 1997, **68**, 351–356.
3 N. V. Plechkova and K. R. Seddon, *Chem. Soc. Rev.*, 2008, **37**, 123–150.
4 C. Chiappe, *Monatsh. Chem.*, 2007, **138**, 1035–1043.
5 H. Weingärtner, *Angew. Chem., Int. Ed.*, 2008, **47**, 654–70.
6 E. W. Castner and J. F. Wishart, *J. Chem. Phys.*, 2010, **132**, 120901.
7 S. M. Urahata and M. C. C. Ribeiro, *J. Chem. Phys.*, 2004, **120**, 1855–1863.
8 Y. T. Wang and G. A. Voth, *J. Am. Chem. Soc.*, 2005, **127**, 12192.
9 Y. Wang and G. A. Voth, *J. Phys. Chem. B*, 2006, **110**, 18601.
10 J. N. Canongia Lopes and A. A. H. Padua, *J. Phys. Chem. B*, 2006, **110**, 3330.
11 A. Triolo, O. Russina, H.-J. Bleif and E. Di Cola, *J. Phys. Chem. B*, 2007, **111**, 4641–4.
12 O. Russina, A. Triolo, L. Gontrani and R. Caminiti, *J. Phys. Chem. Lett.*, 2012, **3**, 27.
13 C. Hardacre, J. D. Holbrey, C. L. Mullan, T. G. a. Youngs and D. T. Bowron, *J. Chem. Phys.*, 2010, **133**, 074510.
14 H. K. Kashyap, C. S. Santos, H. V. R. Annapureddy, N. S. Murthy, C. J. Margulis and E. W. Castner, *Faraday Discuss.*, 2012, **154**, 133–143.
15 R. Atkin and G. G. Warr, *J. Phys. Chem. B*, 2008, **112**, 4164–6.
16 T. L. Greaves, D. F. D. F. Kennedy, A. Weerawardena, N. M. K. Tse, N. Kirby and C. J. Drummond, *J. Phys. Chem. B*, 2011, **115**, 2055–66.
17 Y. Shen, D. F. Kennedy, T. L. Greaves, A. Weerawardena, R. J. Mulder, N. Kirby, G. Song and C. J. Drummond, *Phys. Chem. Chem. Phys.*, 2012, **14**, 7981–92.
18 A. Triolo, O. Russina, R. Caminiti, H. Shirota, H. Y. Lee, C. S. Santos, N. S. Murthy and E. W. Castner, *Chem. Commun.*, 2012, **48**, 4959–61.
19 H. K. Kashyap, J. J. Hettige, H. V. R. Annapureddy and C. J. Margulis, *Chem. Commun.*, 2012, **48**, 5103–5.
20 H. V. R. Annapureddy, H. K. Kashyap, P. M. De Biase and C. J. Margulis, *J. Phys. Chem. B*, 2010, **114**, 16838–16846.
21 H.-O. Hamaguchi and R. Ozawa, in *Advances in Chemical Physics*, 2005, vol. 131, pp. 85–104.
22 K. Iwata, H. Okajima, S. Saha and H.-O. Hamaguchi, *Acc. Chem. Res.*, 2007, **40**, 1174–1181.
23 A. Samanta, *J. Phys. Chem. Lett.*, 2010, **1**, 1557–1562.
24 K. Fruchey, C. M. Lawler and M. D. Fayer, *J. Phys. Chem. B*, 2012, **116**, 3054–64.
25 B. G. Nicolau, A. Sturlaugson, K. Fruchey, M. C. C. Ribeiro and M. D. Fayer, *J. Phys. Chem. B*, 2010, **114**, 8350–6.
26 R. Hayes, S. Imberti, G. G. Warr and R. Atkin, *Phys. Chem. Chem. Phys.*, 2011, **13**, 3237–47.
27 B. L. Bhargava, Y. Yasaka and M. L. Klein, *Chem. Commun.*, 2011, **47**, 6228–41.
28 D. a. Turton, J. Hunger, A. Stoppa, G. Hefter, A. Thoman, M. Walther, R. Buchner and K. Wynne, *J. Am. Chem. Soc.*, 2009, **131**, 11140–6.
29 S. Patra and A. Samanta, *J. Phys. Chem. B*, 2012, **116**, 12275.
30 A. a. H. Pádua, M. F. Costa Gomes and J. N. a. Canongia Lopes, *Acc. Chem. Res.*, 2007, **40**, 1087–96.
31 Y. Wang, W. E. I. Jiang, T. Yan and G. A. Voth, *Acc. Chem. Res.*, 2007, **40**, 1193–1199.
32 T. L. Greaves and C. J. Drummond, *Chem. Soc. Rev.*, 2012, **42**, 1096–1120.
33 J. Gladysz and D. P. Curran, *Tetrahedron*, 2002, **58**, 3823–3825.
34 J. G. Riess, *Tetrahedron*, 2002, **58**, 4113–4131.
35 I. T. Horvath and J. Rabai, *Science*, 1994, **266**, 72.
36 P. Lo Nostro, *Curr. Opin. Colloid Interface Sci.*, 2003, **8**, 223–226.
37 H. Xue, R. Verma and J. M. Shreeve, *J. Fluorine Chem.*, 2006, **127**, 159–176.
38 K. Matsumoto and R. Hagiwara, *J. Fluorine Chem.*, 2007, **128**, 317–331.
39 T. L. Greaves, D. F. D. F. Kennedy, N. Kirby and C. J. Drummond, *Phys. Chem. Chem. Phys.*, 2011, **13**, 13501–9.
40 O. Russina, A. Triolo, B. Fazio, G. B. Appetecchi, M. Carewska, S. Passerini, N. S. Murthy, G. A. Baker and E. W. Castner, *Faraday Discuss.*, 2012, **154**, 97–109.
41 D. Xiao, J. R. Rajian, A. Cady, S. Li, R. A. Bartsch and E. L. Quitevis, *J. Phys. Chem. B*, 2007, **111**, 4669–4677.
42 O. Russina, L. Gontrani, B. Fazio, D. Lombardo, A. Triolo and R. Caminiti, *Chem. Phys. Lett.*, 2010, **493**, 259–262.
43 C. S. Santos, N. S. Murthy, G. a. Baker and E. W. Castner, *J. Chem. Phys.*, 2011, **134**, 121101.
44 P. Johansson, L. E. Fast, A. Matic, G. B. Appetecchi and S. Passerini, *J. Power Sources*, 2010, **195**, 2074–2076.

45 G. B. Appetecchi, M. Montanino, M. Carewska, M. Moreno, F. Alessandrini and S. Passerini, *Electrochim. Acta*, 2011, **56**, 1300–1307.
46 F. Castiglione, M. Moreno, G. Raos, A. Famulari, A. Mele, G. B. Appetecchi and S. Passerini, *J. Phys. Chem. B*, 2009, **113**, 10750–9.
47 Y. Yoshida and G. Saito, *Phys. Chem. Chem. Phys.*, 2011, **13**, 20302.
48 G. D. Smith, O. Borodin, J. J. Magda, R. H. Boyd, Y. Wang, J. E. Bara, S. Miller, D. L. Gin and R. D. Noble, *Phys. Chem. Chem. Phys.*, 2010, **12**, 7064–76.
49 A. Babai and A.-V. Mudring, *Inorg. Chem.*, 2006, **45**, 3249–55.
50 K. Reichenbächer, H. I. Süss and J. Hulliger, *Chem. Soc. Rev.*, 2005, **34**, 22–30.
51 O. Jeannin and M. Fourmigué, *Chem.-Eur. J.*, 2006, **12**, 2994–3005.
52 S. R. Halper and S. M. Cohen, *Inorg. Chem.*, 2005, **44**, 4139–41.
53 G. B. Appetecchi, C. Tizzani, S. Scaccia, F. Alessandrini and S. Passerini, *J. Electrochem. Soc.*, 2006, **153**, A1685.
54 L. Conte, G. P. Gambaretto, G. Caporiccio, F. Alessandrini and S. Passerini, *J. Fluorine Chem.*, 2004, **125**, 243–252.
55 W. A. Henderson and S. Passerini, *Chem. Mater.*, 2004, **16**, 2881.
56 B. Aoun, A. Goldbach, M. a. González, S. Kohara, D. L. Price and M.-L. Saboungi, *J. Chem. Phys.*, 2011, **134**, 104509.
57 E. Bodo, L. Gontrani, R. Caminiti, N. V. Plechkova, K. R. Seddon and A. Triolo, *J. Phys. Chem. B*, 2010, **114**, 16398–407.
58 O. Russina, A. Triolo, L. Gontrani, R. Caminiti, D. Xiao, L. G. Hines Jr, R. A. Bartsch, E. L. Quitevis, N. V. Plechkova, K. R. Seddon and L. G. Hines Jr, *J. Phys.: Condens. Matter*, 2009, **21**, 424121.
59 W. Zheng, A. Mohammed, L. G. Hines, D. Xiao, O. J. Martinez, R. a. Bartsch, S. L. Simon, O. Russina, A. Triolo and E. L. Quitevis, *J. Phys. Chem. B*, 2011, 6572–6584.
60 S. Puttick, A. L. Davis, K. Butler, L. Lambert, J. El harfi, D. J. Irvine, A. K. Whittaker, K. J. Thurecht and P. Licence, *Chem. Sci.*, 2011, **2**, 1810.
61 A. Mele, *Chimica Oggi/Chemistry Today*, 2012, **28**, 48–55.
62 F. Castiglione, G. Raos, G. B. Appetecchi, M. Montanino, S. Passerini, M. Moreno, A. Famulari and A. Mele, *Phys. Chem. Chem. Phys.*, 2010, **12**, 1784–92.
63 F. Castiglione, E. Ragg, A. Mele, G. B. Appetecchi, M. Montanino and S. Passerini, *J. Phys. Chem. Lett.*, 2011, **2**, 153–157.
64 A. Mele, G. Romanò, M. Giannone, E. Ragg, G. Fronza, G. Raos and V. Marcon, *Angew. Chem., Int. Ed.*, 2006, **45**, 1123–6.
65 A. Mele, C. D. Tran and S. H. De Paoli Lacerda, *Angew. Chem., Int. Ed.*, 2003, **42**, 4364–6.
66 C. Chiappe, A. Sanzone, D. Mendola, F. Castiglione, A. Famulari, G. Raos and A. Mele, *J. Phys. Chem. B*, 2013, **117**, 668–76.
67 R. A. Mantz, P. C. Trulove, R. T. Carlin and R. A. Osteryoung, *Inorg. Chem.*, 1995, **34**, 3846–3847.

Faraday Discussions RSC Publishing

PAPER

Individual gold nanorods report on dynamical heterogeneity in supercooled glycerol

Haifeng Yuan,†[a] Saumyakanti Khatua,†[a] Peter Zijlstra[b] and Michel Orrit*[a]

Received 14th May 2013, Accepted 18th June 2013
DOI: 10.1039/c3fd00091e

Building upon previous reports of dynamical heterogeneity in supercooled glycerol probed with single dye molecules, we have investigated the rotational diffusion in glycerol of much larger objects, single gold nanorods. Because of their size and larger hydrodynamic volume, nanorods probe very different length and time scales than dyes. They are best suited for studies around the cross-over temperature of supercooled glycerol. Our results show that heterogeneity appears at temperatures lower than 225 K, even on the length scale of 30 nm, the averaged size of the nanorods. Moreover, the persistence of heterogeneity on the measurement time scale proves that the exchange times are much longer than the rotation time of the rods.

1 Introduction

Supercooled liquids have drawn much attention because of the fundamental importance of the glass transition, and because of the ubiquitous technological applications of glasses and amorphous materials. It is now widely accepted that, at temperatures close to their glass transition, supercooled liquids, while remaining amorphous, become heterogeneous.[1–4] Heterogeneity is believed to be a key element for a deeper mechanistic understanding of the glass transition of supercooled liquids. Although first observed in various bulk experiments,[5–12] heterogeneity in supercooled liquids remains difficult to observe and characterize at the ensemble level. In particular, the spatial and temporal scales associated to heterogeneity in supercooled liquids are still under debate. In simple glass-forming molecular liquids such as glycerol and *o*-terphenyl, different spatial and temporal scales have been reported in different bulk experiments. Nuclear magnetic resonance (NMR) and dielectric relaxation experiments indicate heterogeneity on a length scale of several nanometers and a time scale

[a]*MoNOS, Leiden Institute of Physics, Leiden University, Leiden, The Netherlands. E-mail: orrit@physics.leidenuniv.nl*

[b]*Molecular Biosensors for Medical Diagnostics, Department of Applied Physics, Eindhoven University of Technology, The Netherlands*

† These authors have contributed equally.

comparable to the alpha-relaxation time.[11–16] On the contrary, light scattering and fluorescence microscopy experiments indicate heterogeneity on length scales up to several hundred nanometers and orders of magnitude slower relaxation.[17–20] Without a clear agreement on the spatial and temporal scale(s) of heterogeneity, deeper mechanistic questions will remain open. Therefore, we have undertaken this study of rotational diffusion of nanoparticles to explore larger length and time scales in the hope of directly accessing the local properties of glass-forming liquids at molecular and nanometer scales.

Suppressing ensemble averaging, single-molecule studies on fluorophores doped in supercooled liquids started to provide detailed information at a molecular level.[21–23] Tracking rotational motions of individual molecules embedded in supercooled *o*-terphenyl, Deschenes and Vanden Bout found a broad distribution of molecular tumbling rates and revealed the heterogeneity experienced by single molecules.[24] In experiments on glycerol, Zondervan *et al.* found that the memory time of heterogeneity could exceed by many orders of magnitude the relaxation time of individual glycerol molecules.[25] These surprisingly long times may have been a consequence of the thermal history of that particular experiment, but in a more recent work, Mackowiak *et al.* have confirmed the presence of temporal heterogeneity with a memory time two orders of magnitude longer than the tumbling time of glycerol molecules.[26] These observations suggest the existence of slowly relaxing regions, or even of solid-like networks within supercooled liquids close to the glass transition. Moreover, these studies reveal that these structures persist well above the structural glass transition temperature of glycerol (190 K). A study of the tumbling rates of the molecules suggests the disparition of heterogeneity at a temperature of about 230 K for glycerol,[25] tentatively close to the cross-over temperature (T_c) predicted theoretically by mode-coupling theory.[1,27,28] A direct experimental demonstration of the merging of heterogeneous environments into a single liquid at T_c, however, could not be achieved in these first experiments because of the very fast rotational diffusion of the probe dye molecules at that high temperature, compounded by the inherent blinking and bleaching of single molecules. Hence, gold nanorods, which offer much larger volume and perfect photo-stability, are attractive probes of heterogeneities at higher temperatures.

The original properties of gold nanorods make them very attractive probes of heterogeneity on large length and time scales. First of all, they are easy to detect one by one thanks to the strong interaction of their longitudinal plasmon resonance with light. They have a much larger volume than single molecules and this volume can be easily tuned by chemical synthesis.[29,30] They do not blink nor bleach and thus can be detected continuously for practically unlimited times. They are easily detected optically, either by scattering,[31,32] by absorption,[33] or by luminescence (one-photon- and two-photon-excited).[34,35] Although its quantum yield is very low,[36] one-photon-excited luminescence of single gold nanorods is easily detected because of their large absorption cross-section. Moreover, the anisotropic shape of gold nanorods makes their optical response sensitive to orientation and is thus a direct probe of their rotational diffusion.

In the present work, we propose the use of individual gold nanorods as local viscosity reporters for studying the dynamics of glass-forming liquids. This demonstration experiment allows us to follow the rotational diffusion times of individual gold nanorods from 238 K to 222.5 K. The rotational diffusion time of a nanorod is deduced from autocorrelation functions of the linear dichroism provided by one-photon luminescence time-traces. We first measure the gold nanorods' rotational times at high temperature, where glycerol still appears homogeneous. While cooling, we follow the rotational times of the same gold nanorods at different temperatures and compare their rotational times with calculations based on the bulk viscosity of glycerol.

2 Experimental

Glycerol of spectrophotometric grade was purchased from Sigma-Aldrich and used without further purification. Gold nanorods were prepared with the seed-mediated growth method as described in the literature.[37] The gold nanorods used in this study had averaged dimensions of 14 nm × 35 nm, and are depicted in the SEM micrograph of Fig. 1a. Their longitudinal plasmon resonates around 650 nm in water, as shown in Fig. 1b. The gold nanorods were then dispersed into glycerol by first washing them, *i.e.*, by removing the excess amount of cetyl-trimethylammonium bromide (CTAB) through centrifugation and redispersion into MilliQ water, followed by subsequent dilution with glycerol to a volume ratio of 1 : 400 (water : glycerol). The gold nanorod suspension was then spin-coated onto an ozone-cleaned glass coverslide at 6000 rpm for 60 s, yielding a film of 2–3 μm thickness. We then mounted the sample into the cryostat of our home-built beam-scanning confocal microscope. The position of the sample is controlled by a three-dimensional set of piezo stages with feedback loop. Before cooling down, the sample was dried in the cryostat by repeatedly pumping and flushing with helium gas. Furthermore, we kept it under a dry and inert helium atmosphere throughout the entire experiment. The sample was directly cooled down to 238 K with a cooling rate of 5 K per hour. We followed the same gold nanorods at 238 K, 235 K, 232 K, 229 K, 226 K, and 222.5 K. During each measurement, the temperature inside the cryostat was regulated within ±0.1 K. Previous experiments[20,38,39] have shown that glycerol subject to such a thermal history remains amorphous, with no sign of nano-crystallites.

Single-particle measurements were performed on a home-built confocal laser-scanning microscope with the objective lens inside a flow cryostat. The detailed

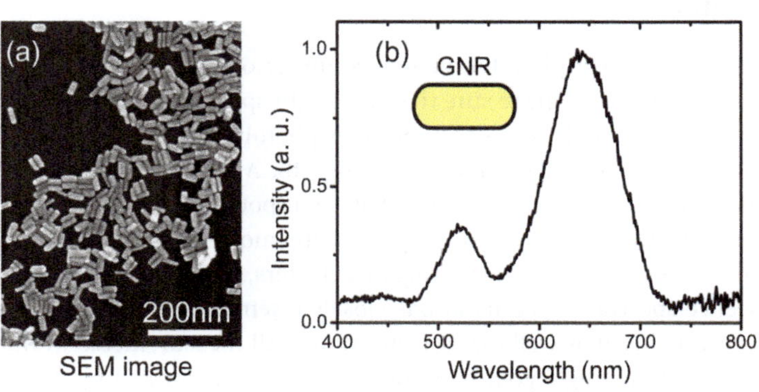

Fig. 1 (a) A SEM image of the gold nanorods (14 ± 3 nm × 35 ± 5 nm). (b)The UV-Vis spectrum of gold nanorods in water. It shows two plasmon resonances. The one around 650 nm is the longitudinal surface plasmon resonance. The other one around 520 nm is the transverse surface plasmon resonance.

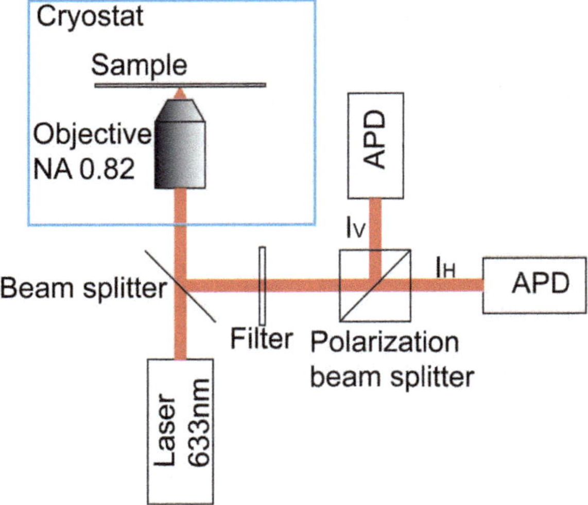

Fig. 2 Simplified scheme of the optical setup. Circularly polarized laser excitation at 633 nm is sent into the cryostat and then focused into the sample film by a fused-silica objective with a NA of 0.82. Photoluminescence from gold nanorods is collected by the same objective. After the 633 nm notch filter, the photo-luminescence is split into two parts according to polarization. Two Perkin-Elmer APDs are employed to measure the intensities in each channel.

description of the experimental setup can be found elsewhere.[40,41] A simplified scheme of the optical setup is shown in Fig. 2. A 633 nm He–Ne laser was used as the excitation source. The circularly polarized excitation beam was focused into the glycerol film through a fused-silica objective with a numerical aperture of 0.82. The photoluminescence signal was collected by the same objective and was separated from the excitation laser using a 633 nm notch filter. For polarization-sensitive measurements, the photoluminescence signal from the individual gold nanorods was split by a polarizing beam-splitter and then detected by two single-photon avalanche photodiodes (APDs) for the two orthogonal polarizations. Photoluminescence spectra were recorded by a liquid-nitrogen-cooled CCD spectrometer (Horiba Scientific).

3 Results

Fig. 3a shows a typical photoluminescence image of individual gold nanorods dispersed in glycerol. To make sure that the bright spots on the image stem from individual gold nanorods, we measured their photoluminescence spectra. A few typical spectra are shown normalized in Fig. 3b. A Lorentzian lineshape with narrow linewidth (34 ± 2 nm) confirms that each spot indeed arises from a single gold nanorod. Although scattering spectra are more commonly used in the literature to identify single nanorods against clusters, we have recently shown that luminescence spectra of gold nanorods closely resemble their scattering spectra and thus can be used as well for identification.[36] All measurements in this work were performed on single gold nanorods which were identified by their photoluminescence spectra. We note that there is a broad distribution of intensities among the bright spots even though they are all single gold nanorods. This intensity distribution results from two factors. (i) The nanorods produced by the

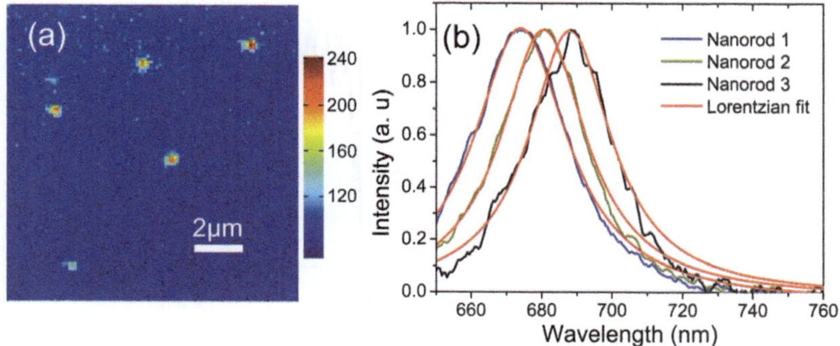

Fig. 3 (a) One-photon-excited luminescence image of individual gold nanorods dispersed in glycerol at 232 K. The excitation wavelength was 633 nm and the power density was 10 kW cm^{-2}. (b) Three examples of photoluminescence spectra of the bright spots in (a). The narrow Lorentzian line-shape of their spectra identifies the emitters as individual gold nanorods.

chemical synthesis may have different volumes. This is supported by the SEM image of Fig. 1a. (ii) The surface plasmon resonances of the individual nanorods, related to their aspect ratio, are also distributed, leading to a distribution of absorption cross sections at the excitation wavelength. The latter effect is particularly important as the excitation laser wavelength (633 nm) is close to the SPR of the bulk nanorod sample.

We note that individual nanorods in Fig. 3a are not stationary. This is quite evident from their photoluminescence time traces measured along two orthogonal polarizations. A typical photoluminescence time trace is shown in Fig. 4a. The anti-correlated intensity fluctuations along the orthogonal polarizations directions clearly indicate in-plane rotation of individual nanorods. Fluctuation of the total intensity is related to the out-of-plane motion of the nanorods. To quantify the rotational diffusion of the nanorods, we calculate the linear dichroism (D) from the photoluminescence time-trace using the following equation:

$$D = \frac{I_{\mathrm{H}} - I_{\mathrm{V}}}{I_{\mathrm{H}} + I_{\mathrm{V}}}, \tag{1}$$

where I_{H} and I_{V} are the intensities of horizontal and vertical polarizations, respectively. Fig. 4b shows the linear dichroism time trace calculated from the polarized photoluminescence time trace shown in Fig. 4a. We note that, even though linear dichroism only characterizes the in-plane projection of the long axis of the nanorods, the rotational diffusion times can be deduced from this two-dimensional projection linear dichroism, assuming that the nanorods' diffusion is isotropic.[24]

As in single-molecule experiments, we deduce rotational diffusion times of individual nanorods from the autocorrelation of the linear dichroism time-traces:[25]

$$C_{\mathrm{D}}' = \frac{\langle (D(t'+t)+1)(D(t')+1)\rangle}{\langle D(t')+1\rangle^2} - 1 \approx \frac{1}{2}\exp\left(-\frac{t}{\tau_{\mathrm{rot}}}\right). \tag{2}$$

The autocorrelation is then fitted with a single-exponential decay, where τ_{rot} is the rotational correlation time. Fig. 4c (black line) shows the autocorrelation of

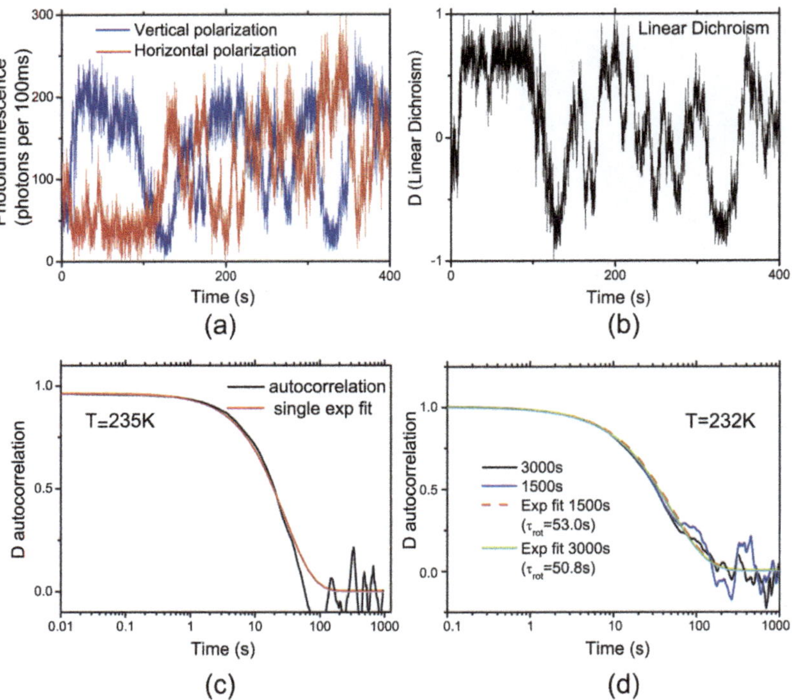

Fig. 4 (a) Typical photo-luminescence intensity time trace of a single gold nanorod rotating in supercooled glycerol. The excitation intensity of circularly polarized 633 nm excitation light was 200 W cm^{-2}. The integration time was 100 ms. Blue and red traces represent the intensities of the two orthogonal polarizations. (b) The linear dichroism trace calculated from (a) according to eqn (1). (c) Autocorrelation of the linear dichroism trace shown in (b). A single-exponential fit (red) yields a rotational correlation time of 24 s at 235 K. (d) Autocorrelation of linear dichroism traces of different lengths recorded for the same gold nanorod at 232 K. The blue curve is from a 1500 s timetrace yielding a rotational correlation time of 53 s (the red dashed line). The black curve is from a 3000 s timetrace yielding a rotational correlation time of 50.8 s (the green line).

the linear dichroism time trace shown in Fig. 4b (note the logarithmic time axis). The single-exponential fit of eqn (2) yields a rotational diffusion time of 24 s at 235 K. It is known that the finite length of the linear dichroism time-trace can affect the fitted diffusion time, and in particular that a too short measurement time leads one to incorrectly estimate the diffusion time.[26,42] Therefore, to check the reliability of our procedure, we compared the recovered rotational diffusion times for two linear dichroism time traces acquired on the same nanorod for different durations. Fig. 4d shows examples of linear dichroism autocorrelation functions acquired on the same rod for durations of 1500 (blue line) and 3000 (black line) seconds. The fitted rotational diffusion times are 53 s (for 1500 s, $30\tau_{rot}$) and 51 s (for 3000 s, $60\tau_{rot}$) respectively. From such measurements on several nanorods we estimate that the uncertainty in the recovered rotational diffusion time is at most ±15%, provided that the time traces are longer than 30 times τ_{rot}.[26] All further measurements of linear dichroism traces were done on time traces at least 50 times longer than the measured rotational diffusion time. Such measurements can require hours and were possible only thanks to the exceptional photostability of gold nanorods. Individual nanorods could be followed easily over such long times because translational diffusion through the focal spot was very slow, almost three orders of magnitude slower than rotational diffusion.

The rotational diffusion time of a nanorod can be directly related to the local viscosity it experiences *via* the Debye–Stokes–Einstein relation:

$$\tau_{\rm rot} = \frac{V_{\rm h}\eta}{k_{\rm B}T} \tag{3}$$

with V_h the hydrodynamic volume of nanorods, η the local viscosity, k_B Boltzmann's constant and T the temperature. The hydrodynamic volume of gold nanorods can be calculated using the equation described below:[34]

$$V_{\rm h} = \frac{4}{3}\pi(R_{\rm h})^3 = \frac{4}{3}\pi\left(\frac{L}{2\ln\left(\frac{L}{d}\right)+q}\right)^3 \tag{4}$$

with L the length of the nanorod, d the diameter, and $q = 0.312 + 0.565\ (d/L) - 0.1\ (d/L)^2$.

Looking at the distribution of observed rotational diffusion times of single nanorods at 238 K in Fig. 5a, we observe a large dispersion, by a factor of about 3. Much of this dispersion of course arises from the large size distribution of the nanorods, obvious in the SEM image. This dispersion is a fundamental difference from single-molecule experiments. Individual probe molecules are all chemically identical, and therefore have exactly the same hydrodynamic volume. Variations in tumbling times for different molecules directly reflect variations in local viscosities. To compare rotational diffusion of different nanorods, however, we first have to get rid of the trivial variation in their hydrodynamic volumes.

Before estimating the volume dispersion, we have to consider two other possible sources of variation: (i) some nanorods may approach interfaces between glycerol and glass or air, and (ii) laser excitation may induce nanorod-dependent local heating. As the excitation light is focused into the liquid film with depth of about 1 μm, the nanorods in our confocal measurements are far from both interfaces. Gold nanorods can indeed cause local heating at moderate laser intensities due to their large absorption cross-section.[43–45] To exclude heating effects, we followed the rotational diffusion of several nanorods at different laser excitation powers at a given cryostat temperature (data not shown). From this measurement, we estimate the temperature rise due to local heating to be less than 0.02 K at the excitation intensity of 200 W cm^{-2} at 633 nm, in good agreement with the calculated value 0.025 K based on the absorption cross-section.[46,47] Such small temperature rises are less than the uncertainty and stability in cryostat temperature. Furthermore, as the laser is applied throughout all measurements, variations of such small effects from rod to rod may safely be neglected. We thus mainly attribute the distribution of rotational times at 238 K to the volume distribution of gold nanorods. The average rotational time of 19 nanorods at 238 K is 8.5 ± 2.3 s. This value is close to that calculated for a 13 nm × 33 nm nanorod (8.4 s) at the same temperature, which is in a good agreement with the average dimensions of 14 ± 3 nm × 35 ± 5 nm from our SEM measurement, as shown in the inset of Fig. 5a.

This volume dispersion can be estimated in two independent ways. The SEM micrograph gives a dispersion of ±50%, (1.4 ± 0.6) × 10^4 nm^3, shown in

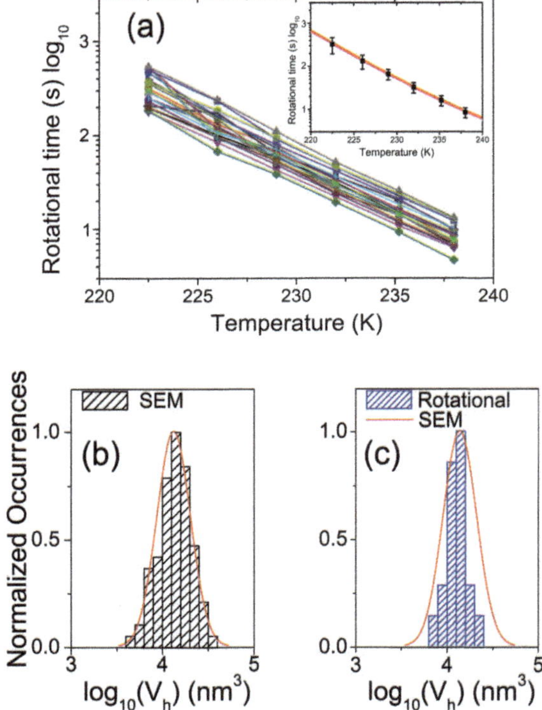

Fig. 5 (a) Rotational correlation times of 19 individual gold nanorods against temperature. The inset shows the calculated values for 13 nm × 33 nm nanorods (red line) and the average values from 19 nanorods (black squares). (b) The histogram of nanorods' hydrodynamic volumes calculated using eqn (4) from SEM measurements on 80 gold nanorods. The red curve is the Gaussian fit. The average hydrodynamic volume is $(1.4 \pm 0.6) \times 10^4$ nm^3. (c) The histogram of nanorods' hydrodynamic volumes deduced from rotational diffusion measurements on 19 nanorods at 238 K. It yield an average hydrodynamic volume of $(1.1 \pm 0.4) \times 10^4$ nm^3. The red curve is the same as the one in (b).

Fig. 5b. A second estimate is obtained by assuming that dynamical heterogeneity of glycerol is negligible at the highest measured temperature, 238 K. This assumption allows us to relate the time distribution to the volume distribution. The volume distribution estimated from the rotational diffusion, as shown in Fig. 5c, gives a half-maximum of ±40%. The distribution is narrower than that measured from SEM images (the red curve in Fig. 5c). This difference in volume distribution is probably due to a bias towards particles with a significant absorption cross-section at the excitation wavelength used for photoluminescence imaging. The estimated hydrodynamic volume of $(1.1 \pm 0.4) \times 10^4$ nm^3 is in good agreement with the SEM estimate. Therefore, in the following, we suppose that the distribution of tumbling times at the highest temperature (238 K) is exclusively due to volume dispersion. Previous studies have indicated that heterogeneity is probably absent from single-molecule measurements at 238 K.[8,20,21,25] This assumption is further supported by observations at 235 K and 232 K, to be discussed below, which show no big change in the width of the tumbling time distribution. With this assumption, we will deduce the hydrodynamic volume of each rod from its tumbling time at the higher temperatures.

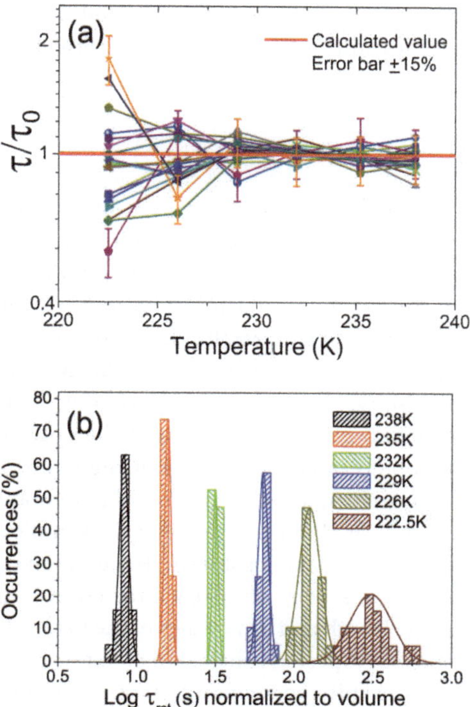

Fig. 6 (a) The rotational times of the same 19 nanorods are normalized to their hydrodynamic volumes and to the bulk viscosity of glycerol. The hydrodynamic volume of each nanorod is fitted using eqn (3) and (4) from its data points at 238 K, 235 K, and 232 K, where we do not see apparent broadening. After such normalization, we would expect a line ($\tau/\tau_0 = 1$, red line) within the measurement uncertainty (error bar) if there was no heterogeneity. (b) Histograms of nanorods' rotational times, normalized to their volumes, at different temperatures. At 238 K, 235 K, 232 K and 229 K, the distributions are narrow, broadened by experimental error only. Below 229 K, the distributions clearly broaden.

Rotational times normalized to the bulk viscosity of glycerol and to each nanorods' volume are plotted against temperature in Fig. 6a. The volume of each nanorod is fitted from the rotational times at 238 K, 235 K, 232 K and 229 K, neglecting any dispersion of the local viscosity in this temperature range. We use this procedure to reduce and also to characterize the uncertainty in our measurements. The whole variation in data points for temperatures above 229 K reflects the uncertainty of our measurements (±15%, see size of error bars in Fig. 6). Below 226 K, the tumbling times of several nanorods start to depart from the average by more than the measurement uncertainty. At 222.5 K, several tumbling times are more than 50% off the expected line based on the bulk viscosity. This leads to a broadening in the histogram of rotational times at this temperature, as shown in Fig. 6b. These observations unambiguously indicate deviations of the local viscosity of glycerol from the average, *i.e.*, they are unmistakable signatures of dynamical heterogeneity. These deviations start to appear already at 226 K. It would be very interesting to see whether the distribution further broadens at still lower temperatures. However, it was difficult to follow such large nanorods at lower temperatures in our experiments, as this would have required days of measurements for each point. The 10 K gap in temperature between single-molecule experiments and this work could be filled in future experiments with still smaller nanorods, or with enhanced fluorescence experiments.[48]

4 Discussion

The main result of this work is the evidence it gives for heterogeneity in a domain of surprisingly high temperatures. As heterogeneity is sometimes associated to the formation of nanocrystallites,[49] let us briefly rule out this phenomenon in the present experiments. Firstly, the lowest temperature upon first cooling in this study was 222.5 K, more than 30 K above the glass transition of glycerol. Previous experiments of our group have shown that crystallization of glycerol critically depends on thermal history, and only takes place after long enough aging at 195 K.[39] Secondly, if crystal growth occurred and the critical nucleus size was exceeded, the sample would rapidly crystallize at the temperature of 230 K, or upon warming up, and the nanorods would soon be immobilized. Instead, the nanorods remained mobile at all times, even after the sample was cooled further and was aged at 205 K for 10 h (data not shown). Hence, even if nanocrystals form transiently in our experiments, they never reach the critical nucleus size.

Let us now discuss the length and time scales of the heterogeneity as felt by the gold nanorods. The most unexpected conclusion of our work is that, below 230 K, dynamical heterogeneity is still apparent and pronounced on the length scale of nanorods (about 30 nm). This length presumably extends even further, up to the length scale a nanorod explores by translational diffusion during the measurement. This is typically 10 times the rod's dimensions, as typical traces were recorded for about 100 reorientations. These distances are more than 30 times larger than earlier estimates of the length scales for dynamical heterogeneity in supercooled liquids, some nanometers.[9,16,50] Our results thus show for the first time that dynamical heterogeneity not only persists at the scale of a large dye molecule (one nanometer), but is still not averaged out over distances of several tens of nm. Presumably, the length scales of heterogeneity may extend even further, as there was no fundamental reason for the choice of this particular size of nanorods. Future experiments with larger rods will be undertaken to answer this question.

We now turn to the time scale of the heterogeneity. Even after an integration time of hours, heterogeneity was not time-averaged. This proves that the exchange times for fluctuations with correlation lengths comparable to the rod's size have to be comparable to, or longer than, 100 times the rod's tumbling time. When translated into the tumbling time of individual glycerol molecules, this time corresponds to more than 10^7 tumbling steps of a glycerol molecule. It thus appears no longer relevant to discuss relaxation of heterogeneity at such large scales in terms of molecular tumbling times. Such long times would rather correspond to correlated rearrangements of a large number of molecules, in so-called cooperatively rearranging regions (CRR's). These observations remind us of earlier findings by Patkowski *et al.*,[17] who found large-scale, long-lived density fluctuations in glass-forming liquids. From small-angle X-ray scattering, they found a wide distribution of length scales, which led them to describe these structures as fractal. Such structures present heterogeneity at all length scales, and presumably also show wide distributions of relaxation times.

Our second key finding is the disappearance of heterogeneity when temperature is raised above 230 K. This value matches an earlier estimate by Zondervan *et al.*[25] based on observations of single dye molecules at temperatures between 205 and 212.5 K. We may tentatively identify this transition temperature with the critical temperature of mode coupling theory,[27,28] at which the relaxation time of the slowest fluctuation mode diverges. This temperature is also expected to lie around 230 K for glycerol.[1] We can thus speculate on a possible scenario for the emergence of heterogeneity upon cooling down. When the first slow mode "freezes in" and leads to a long-lived density fluctuation, different regions of the glass-forming liquid cannot equilibrate any more, even after arbitrarily long waiting times. The way is then free for heterogeneities to develop independently from each other as temperature is lowered further, eventually leading to the dynamical mosaic of the low-temperature glass. The coincidence of this transition to heterogeneous behavior as observed on different length scales, from those of molecules to those of nanoparticles, is a strong indication that heterogeneity appears at various length scales simultaneously, *i.e.*, at the same temperature or in a narrow range of temperatures.

Our results add to a large body of literature that conveys a view of glass-forming liquids as very complex and flexible systems. Depending on the thermal history of the sample, large structures up to microns can form,[20,51,52] which can be imaged by fluorescence of dye impurities. It is natural to assume that similar structures can form on sub-micron scales too. It would be interesting to image them by superresolution microscopy,[53–55] although the high doping densities required by these techniques might perturb the intrinsic structure of the pure glass-forming liquid. Using single molecules and single nanoparticles, as done here, has the advantage that the concentration of probes is so low that probe–probe interactions are negligible. Therefore, by comparing different individual probes with each other, one effectively removes the effect of probe–host interactions and one can concentrate on differences that must arise from the host itself.

5 Conclusions

We have demonstrated the first application of individual gold nanorods as local viscosity reporters in a supercooled liquid (glycerol) in a new range of temperatures (from 238 K to 222.5 K). The average rotational times, however, agree very well with calculations based on the bulk viscosity of glycerol in that temperature range. After removing the trivial effect of the volume distribution of gold nanorods, we found that their rotational dynamics present clear differences from nanorod to nanorod in their temperature dependencies. In this first study on 19 nanorods, cooling from 238 K to 222.5 K led to a significant distribution of local viscosities at temperatures lower than about 225 K. This result is surprising, because our nanorods are much larger (length 35 nm, diameter 14 nm) than the dye molecules used in previous studies. Moreover, because rotational diffusion times scale with volume, heterogeneity still survives averaging over very long times, several million times longer than molecular rotation times. The existence of such long memory times indicates that rearrangements must involve correlated motions of many molecules, in so-called cooperatively rearranging regions. Our results qualitatively match earlier observations of heterogeneity done on single molecules on shorter length scales and at lower temperatures, up to 212.5 K. For a quantitative comparison, it would be interesting to bridge the two temperature domains and to follow the development of dynamical heterogeneity over a temperature range including the homogeneous liquid at high temperature and the glassy mosaic at low temperature. However, it is difficult to follow rotational diffusion of these large gold nanorods at lower temperatures because their tumbling becomes extremely slow for the serial measurements done here. Parallel measurements of many rods simultaneously[26,56,57] would be a big advantage in this respect. To fill the temperature and dimension gap between this work and previous single-molecule experiments, using smaller gold nanorods would be an obvious solution. Alternatively, one could look at the rotational and translational diffusion of single dye molecules near plasmonic nanostructures, as was recently demonstrated with gold bowtie antennas[58] and gold nanorods.[48] Whichever way the study of glass formers will develop, dynamical heterogeneity is the central feature that has to be characterized and understood. Its unsuspected extent in space and time demonstrates once again the capacity of the nanoworld to surprise us by its intricacy and subtlety.

Acknowledgements

H. Y., S. K., and M. O. acknowledge financial support from the European Research Council (Advanced Grant SiMoSoMa). P. Z. acknowledges financial support from the Netherlands Organization for Scientific Research (Veni Fellowship).

References

1 H. Sillescu, *J. Non-Cryst. Solids*, 1999, **243**, 81–108.
2 M. D. Ediger, *Annu. Rev. Phys. Chem.*, 2000, **51**, 99–128.
3 P. G. Debenedetti and F. H. Stillinger, *Nature*, 2001, **410**, 259–267.
4 M. D. Ediger and P. Harrowell, *J. Chem. Phys.*, 2012, **137**, 080901.
5 F. Fujara, W. Petry, R. M. Diehl, W. Schnauss and H. Sillescu, *Europhys. Lett.*, 1991, **14**, 563–568.
6 J. Wuttke, J. Hernandez, G. Li, G. Coddens, H. Z. Cummins, F. Fujara, W. Petry and H. Sillescu, *Phys. Rev. Lett.*, 1994, **72**, 3052–3055.
7 K. Schroter and E. Donth, *J. Chem. Phys.*, 2000, **113**, 9101–9108.
8 R. Zondervan, T. Xia, H. van der Meer, C. Storm, F. Kulzer, W. van Saarloos and M. Orrit, *Proc. Natl. Acad. Sci. U. S. A.*, 2008, **105**, 4993–4998.
9 V. M. Syutkin, V. L. Vyazovkin, V. V. Korolev and S. Y. Grebenkin, *Phys. Rev. Lett.*, 2012, **109**, 137801.
10 M. Saito, S. Kitao, Y. Kobayashi, M. Kurokuzu, Y. Yoda and M. Seto, *Phys. Rev. Lett.*, 2012, **109**, 115705.
11 U. Tracht, M. Wilhelm, A. Heuer, H. Feng, K. Schmidt-Rohr and H. Spiess, *Phys. Rev. Lett.*, 1998, **81**, 2727–2730.
12 F. Qi, T. El Goresy, R. Bohmer, A. Doss, G. Diezemann, G. Hinze, H. Sillescu, T. Blochowicz, C. Gainaru, E. Rossler and H. Zimmermann, *J. Chem. Phys.*, 2003, **118**, 7431–7438.
13 R. Böhmer, G. Hinze, G. Diezemann, B. Geil and H. Sillescu, *Europhys. Lett.*, 1996, **36**, 55.
14 G. Diezemann, R. Bhmer, G. Hinze and H. Sillescu, *J. Non-Cryst. Solids*, 1998, **235–237**, 121–127.
15 U. Schneider, P. Lunkenheimer, R. Brand and A. Loidl, *J. Non-Cryst. Solids*, 1998, **235–237**, 173–179.
16 S. A. Reinsberg, X. H. Qiu, M. Wilhelm, H. W. Spiess and M. D. Ediger, *J. Chem. Phys.*, 2001, **114**, 7299–7302.
17 A. Patkowski, T. Thurn-Albrecht, E. Banachowicz, W. Steffen, P. Bösecke, T. Narayanan and E. W. Fischer, *Phys. Rev. E: Stat. Phys., Plasmas, Fluids, Relat. Interdiscip. Top.*, 2000, **61**, 6909–6913.
18 A. Patkowski, E. W. Fischer, W. Steffen, H. Gläser, M. Baumann, T. Ruths and G. Meier, *Phys. Rev. E: Stat. Phys., Plasmas, Fluids, Relat. Interdiscip. Top.*, 2001, **63**, 061503.

19 A. S. Bakai and E. W. Fischer, *J. Chem. Phys.*, 2004, **120**, 5235–5252.
20 T. Xia, L. T. Xiao and M. Orrit, *J. Phys. Chem. B*, 2009, **113**, 15724–15729.
21 L. J. Kaufman, *Annu. Rev. Phys. Chem.*, 2013, **64**, 177–200.
22 F. Kulzer, T. Xia and M. Orrit, *Angew. Chem., Int. Ed.*, 2010, **49**, 854–866.
23 M. Orrit, *Angew. Chem., Int. Ed.*, 2013, **52**, 163–166.
24 L. Deschenes and D. Vanden Bout, *J. Phys. Chem. B*, 2002, **106**, 11438–11445.
25 R. Zondervan, F. Kulzer, G. C. G. Berkhout and M. Orrit, *Proc. Natl. Acad. Sci. U. S. A.*, 2007, **104**, 12628–12633.
26 S. A. Mackowiak, T. K. Herman and L. J. Kaufman, *J. Chem. Phys.*, 2009, **131**, 244513.
27 S. P. Das, *Rev. Mod. Phys.*, 2004, **76**, 785–851.
28 D. R. Reichman and P. Charbonneau, *J. Stat. Mech.: Theory Exp.*, 2005, **2005**, P05013.
29 P. Zijlstra and M. Orrit, *Rep. Prog. Phys.*, 2011, **74**, 106401.
30 H. Chen, L. Shao, Q. Li and J. Wang, *Chem. Soc. Rev.*, 2013, **42**, 2679–2724.
31 J. Rodríguez-Fernández, J. PérezJuste, L. M. LizMarzán and P. R. Lang, *J. Phys. Chem. C*, 2007, **111**, 5020–5025.
32 N. K. Reddy, J. Pérez-Juste, I. Pastoriza-Santos, P. R. Lang, J. K. G. Dhont, L. M. Liz-Marzán and J. Vermant, *ACS Nano*, 2011, **5**, 4935–4944.
33 W.-S. Chang, J. W. Ha, L. S. Slaughter and S. Link, *Proc. Natl. Acad. Sci. U. S. A.*, 2010, **107**, 2781–2786.
34 A. Tcherniak, S. Dominguez-Medina, W.-S. Chang, P. Swanglap, L. S. Slaughter, C. F. Landes and S. Link, *J. Phys. Chem. C*, 2011, **115**, 15938–15949.
35 H. Wang, T. B. Huff, D. A. Zweifel, W. He, P. S. Low, A. Wei and J.-X. Cheng, *Proc. Natl. Acad. Sci. U. S. A.*, 2005, **102**, 15752–15756.
36 M. Yorulmaz, S. Khatua, P. Zijlstra, A. Gaiduk and M. Orrit, *Nano Lett.*, 2012, **12**, 4385–4391.
37 B. Nikoobakht and M. El-Sayed, *Chem. Mater.*, 2003, **15**, 1957–1962.
38 M. E. Mobius, T. Xia, W. van Saarloos, M. Orrit and M. van Hecke, *J. Phys. Chem. B*, 2010, **114**, 7439–7444.
39 H.-F. Yuan, T. Xia, M. Plazanet, B. Demé and M. Orrit, *J. Chem. Phys.*, 2012, **136**, 041102.
40 R. Zondervan, F. Kulzer, H. van der Meer, J. A. J. M. Disselhorst and M. Orrit, *Biophys. J.*, 2006, **90**, 2958–69.
41 H. Yuan, T. Xia, B. Schuler and M. Orrit, *Phys. Chem. Chem. Phys.*, 2011, **13**, 1762–1769.
42 C.-Y. Lu and D. A. Vanden Bout, *J. Chem. Phys.*, 2006, **125**, 124701.
43 H. Petrova, J. Perez Juste, I. Pastoriza-Santos, G. V. Hartland, L. M. Liz-Marzan and P. Mulvaney, *Phys. Chem. Chem. Phys.*, 2006, **8**, 814–821.
44 H. Ma, P. M. Bendix and L. B. Oddershede, *Nano Lett.*, 2012, **12**, 3954–3960.
45 Z. J. Coppens, W. Li, D. G. Walker and J. G. Valentine, *Nano Lett.*, 2013, **13**, 1023–1028.
46 P. V. Ruijgrok, N. R. Verhart, P. Zijlstra, A. L. Tchebotareva and M. Orrit, *Phys. Rev. Lett.*, 2011, **107**, 037401.
47 D. Rings, D. Chakraborty and K. Kroy, *New J. Phys.*, 2012, **14**, 053012.
48 H. Yuan, S. Khatua, P. Zijlstra, M. Yorulmaz and M. Orrit, *Angew. Chem., Int. Ed.*, 2013, **52**, 1217–1221.
49 J. D. Stevenson and P. G. Wolynes, *J. Phys. Chem. A*, 2011, **115**, 3713–3719.
50 X. Qiu and M. D. Ediger, *J. Phys. Chem. B*, 2003, **107**, 459–464.
51 H. Tanaka, R. Kurita and H. Mataki, *Phys. Rev. Lett.*, 2004, **92**, 025701.
52 R. Kurita and H. Tanaka, *Science*, 2004, **306**, 845–848.
53 M. P. Backlund, M. D. Lew, A. S. Backer, S. J. Sahl, G. Grover, A. Agrawal, R. Piestun and W. E. Moerner, *Proc. Natl. Acad. Sci. U. S. A.*, 2012, **109**, 19087–19092.
54 J. Foelling, M. Bossi, H. Bock, R. Medda, C. A. Wurm, B. Hein, S. Jakobs, C. Eggeling and S. W. Hell, *Nat. Methods*, 2008, **5**, 943–945.
55 M. Heilemann, P. Dedecker, J. Hofkens and M. Sauer, *Laser Photonics Rev.*, 2009, **3**, 180–202.
56 S. A. Mackowiak, L. M. Leone and L. J. Kaufman, *Phys. Chem. Chem. Phys.*, 2011, **13**, 1786–1799.
57 L. M. Leone and L. J. Kaufman, *J. Chem. Phys.*, 2013, **138**, 12A524.
58 A. A. Kinkhabwala, Z. Yu, S. Fan and W. Moerner, *Chem. Phys.*, 2012, **406**, 3–8.

Faraday Discussions RSC Publishing

PAPER

Water's non-tetrahedral side

Richard H. Henchman*[abc] and Stuart J. Cockram[b]

Received 8th May 2013, Accepted 17th June 2013
DOI: 10.1039/c3fd00080j

The case for liquid water having non-tetrahedral as well as tetrahedral coordination is put forward. Given the dependence of structure on the hydrogen bond definition, a recent conceptual breakthrough has been the topological hydrogen bond definition which overcomes the shortcomings of traditional cut-off-based hydrogen bond definitions. It identifies the hydrogen bonds in water's first coordination shell using assumed transition states as boundaries instead of fixed cut-offs. Here, the topological definition is applied to liquid water to characterise the distances, angles and energies of the hydrogen bonds for the different types of coordinations found. These coordinations include bent, trigonal, tetrahedral, trigonal bipyramidal, and octahedral structures, as well as bifurcated hydrogens, bifurcated oxygens and cyclic dimers, and larger polygons. All species are shown to have properties consistent with their classification, justifying their assignments, and supporting the structure of water as a continuous, single phase mixture. However, a detailed analysis to assess the existence of the assumed transition states reveals the remarkable finding that hydrogen bond switching *via* a bifurcated hydrogen under certain circumstances is a barrierless process. The likelihood of a switch depends on both the acceptor numbers and on the proximity of a donor to its acceptor. Specifically, a donor in an acceptor's outermost subshell switches uphill to an acceptor of the same or higher coordination to the starting acceptor, downhill to an acceptor of lower coordination by two or more, or sits bifurcated between two acceptors if the new acceptor has a coordination lower by only one. Which it is depends intimately on the donor molecule's oscillations and on other hydrogen bond switches that control the nearby acceptors' coordinations. Finally, a search is conducted for long-range structure in water in terms of asymmetry in the distribution of the donor–acceptor bias but none is found.

1 Introduction

There is still considerable disagreement about the nature of short-range structure in water, let alone that of longer-range structure. The prevailing, textbook view is that water's first coordination shell is tetrahedral, with molecules engaged in four

[a]*Manchester Institute of Biotechnology, The University of Manchester, 131 Princess Street, Manchester, M1 7DN, United Kingdom. E-mail: henchman@manchester.ac.uk; Fax: +44 161 306 5201; Tel: +44 161 306 5194*
[b]*School of Chemistry, The University of Manchester, Oxford Road, Manchester, M13 9PL, United Kingdom*
[c]*Heidelberg Institute for Theoretical Studies, Schloss-Wolfsbrunnenweg 35, 69118, Heidelberg, Germany*

hydrogen bonds (HBs), two of them donors and two of them acceptors. This is chiefly supported by the radial distribution functions from X-ray and neutron diffraction which show that oxygen atoms are surrounded on average by two hydrogens and four to five oxygens.[1–3] This structure is also commonly rationalised by the two lone pairs on oxygen accepting one donor each. To account for the greater disorder in liquid water than in ice, those who propose tetrahedral structure add the descriptors of "distorted", "random" or "on average".[4–9] However, in doing so, they implicitly admit the existence of some degree of non-tetrahedral water: "distorted" is imprecise and the four nearest neighbours, not necessarily equating to the first hydration shell, are always distorted tetrahedral to some extent; "random" omits that some tetrahedra are likely to be quite distorted; "on average" neglects the possibility of opposing fluctuations that may be undetectable by the method of analysis, and nor does such cancellation always occur, as in the case of water's oxygen–oxygen (OO) coordination number exceeding four, and in the more detailed 3D oxygen–hydrogen (OH) and OO distributions first presented by Kusalik and Svishchev.[10,11] They reveal a broad band of oxygens and hydrogens around the central oxygen rather than two distinct peaks as would be expected for tetrahedral structure.

Proponents of tetrahedral structure maintain that because water is a single continuous phase rather than a mixture, then only one coordination is possible and only tetrahedral coordination fits the data.[7–9,12] However, this neglects the possibility of a continuous single phase consisting of multiple coordinations mixed at the molecular scale. One only has to look at the case of two miscible liquids: they are considered to constitute a single continuous phase and yet neutron diffraction has demonstrated that even these do not mix completely at the molecular scale.[13] Indeed, evidence for patches of non-tetrahedral structure in water has been found. Many theories dating back to Röntgen[14] and earlier find that mixture models better explain the maxima or minima in many of water's properties.[15–17] Molecular-scale inhomogeneities in water have long been detected in numerous simulations going back to the work of Stanley and Teixeira,[18] and more recently using X-ray scattering.[19]

Proponents of tetrahedral structure argue that non-tetrahedral fluctuations at short length and time scales are not significant because water molecules are rapidly exchanging their HBs.[7–9,12] Any non-tetrahedral structure would be short-lived, such that molecules are tetrahedral on average. Moreover, such fluctuations are an artifact of the myriad of cut-offs commonly used to identify HBs.[20] These arguments may be challenged on numerous grounds. Firstly, a state's stability depends on the ratio of dynamic quantities rather than directly on the dynamic quantities themselves. Secondly, the distribution of all possible coordinations gives the more complete picture of structure than the average. Thirdly, HB cut-offs may well bring about artificial short-lived non-tetrahedral structure but this does not imply that all non-tetrahedral structure is a consequence of this problem. Fourthly, non-tetrahedral water has been shown to be the very means to facilitate water's rapid dynamics;[21–31] if only tetrahedral water were present, then multiple molecules would have to interconvert simultaneously in order to preserve tetrahedrality,[16,32,33] an unlikely event. Fifthly, neutron diffraction shows that the structure of supercritical water is better described as trigonal rather than tetrahedral before breaking down entirely at higher pressure,[34,35] so it is unlikely that trigonal structure entirely disappears at more ambient conditions. Even at ambient conditions, a range of structures are consistent with neutron-diffraction data.[36,37] How solutes perturb water's structure complicates the debate even more. Concerning the entropy loss of hydrophobic hydration, if small, hydrophobic solutes make water more tetrahedral, as much evidence suggests,[38–43] then conversely pure water must be more non-tetrahedral. Furthermore, the hydration of molecules with an uneven number of donors and acceptors requires non-tetrahedral water to balance up the number of donors and acceptors, as is seen in X-ray crystal structures of hydrated solutes[44,45] and molecular dynamics simulations.[46] Finally, for ion solvation the much debated terms of "structure-making" and "structure-breaking" rest on even weaker foundations while there is uncertainty about water itself.[47,48]

Despite the conflicting evidence, the tetrahedral view of water still holds sway. One explanation for this is the lack of a consensus on a non-tetrahedral component. While many candidates have been put forward, including various clathrates and crystalline forms, which have largely been dismissed,[15,17] the recurring contender has been the broken HB,[18,32,49] also referred to as a dangling or free hydrogen. However, its cause has not been helped by a wide variety of suggested values and justifications,[15,50] usually involving a cut-off in some property. The most recent well-known example, as derived from X-ray absorption, emission and scattering studies, estimated that more than half the HBs are broken.[19,51] This was immediately opposed because it was inconsistent with numerous data such as the radial distribution functions described earlier,[1–3] water's single-phase structure inferred from the temperature dependence of Raman spectra[12] and the transient life-times of broken HBs from femtosecond 2D IR spectroscopy.[20,52] Recent electronic structure calculations indicated that X-ray absorption spectra are only sensitive to the strongest one or two HBs in the first hydration shell which invariably change rapidly due to molecular oscillation.[53] Thus the spectra do not conflict with tetrahedral structure, but neither do they exclusively confirm it. In any case, Nilsson, Pettersson and co-workers soon after their original study[51] had proposed that the non-tetrahedral component was closer-packed with strongly distorted rather than broken HBs.[54] A less commonly considered non-tetrahedral alternative is a water molecule with more than four HBs.[21–25,27–31] The lower popularity of this structure probably reflects the common expectation that water has a maximum of four HBs, while broken HBs should be more common if HBs are frequently breaking. Giguère proposed that the a bifurcated hydrogen (BH), or in other words, a triple donor, was the main non-tetrahedral state in water, backed up by the temperature dependence of Raman and neutron-diffraction spectra.[21,22] Sciortino *et al.*, however, studying the OO coordination, proposed that the extra fifth neighbour of a water molecule catalysed HB switching.[23–25] They discussed how this species could manifest itself as a triple donor or triple acceptor but their data, ignoring hydrogens, did not make this explicit. Significant evidence that the BH was a transition state was found by a number of groups.[11,20,28,29,52] Later work showed that the triple acceptor was much more prevalent than the BH.[30,31,55] Nonetheless, the idea that water can have more than four HBs is still not widely considered.

To resolve the riddle of water's structure in the first hydration shell, the following questions must be answered: where are the HBs, what are broken HBs, can molecules have more than four HBs, where is the boundary between the first and second hydration shells, and what are the transition states of HB switching?

We next explain how the use of cut-offs to define HBs is responsible for much of this uncertainty and demonstrate that a new approach to determine HBs based on the topology of the potential energy surface is able to answer these questions consistently.

1.1 Traditional hydrogen bond definitions and water structure

The standard way to identify the HBs in the first hydration shell uses cut-offs in one or more quantities that relate to the strength of the donor–acceptor interaction. The most direct quantities are distributions in some coordinate(s) from either X-ray diffraction,[1] neutron diffraction[3] or computer simulation[51,56–60] or the energy distribution from computer simulation.[18,61–63] Beyond these cut-offs, HBs are usually regarded as being broken. Cut-offs have their origins in the analysis of structures in which the HBs are stable, regular, and consequently sharply resolved, such as in gas-phase clusters, minimised geometries or crystal structures. For liquids, on the other hand, in which the incessant dynamics and disorder means that HBs are continuously distributed, it is not obvious where to place the cut-off. The resulting HB distribution is highly sensitive to the choice of cut-off, leading to unavoidable ambiguity in water's structure as has already been described. The most robust types of cut-off have a reasonable justification, such as being at the first minimum in the radial distribution function[1,3,35,58–60] or at a point invariant to temperature.[61] However, many other definitions are more elaborate and arbitrary. There are a range of other techniques to determine first hydration shell structure indirectly from other data such as IR, Raman[7,12] or X-ray spectra[51,54] but these make more assumptions and we will not consider them further.

Regardless of how cut-offs are chosen, their main weakness is that they cannot accurately separate the overlap between the first and second-shell distributions that invariably occurs in a liquid. If the cut-off is too small, then oscillating molecules in the first hydration shell sometimes exceed the cut-off, leading to broken HBs and single acceptors. Conversely, if the cut-off is too large, then more distant, second-shell molecules may stray temporarily within the cut-off, leading to BHs and triple acceptors. These two effects artificially broaden the HB distribution, masking any real variation that may be present in the first hydration shell. For these reasons, it has often been said that instantaneous HB determination is impossible and that the structure depends on the time scale considered.[17,20,52,53,60] The cut-off problem is exacerbated when it comes to water dynamics because additional measures are usually required to determine if a HB switch has taken place, such as tighter HB criteria, averaging over a set time, or considering the time for the HB to break for the last time.[28,29,59,64–66] In any case, most HB definitions that are commonly used err on the conservative side and yield about 10% broken HBs at ambient conditions, with few molecules having more than four HB. This value is in accord with the greater disorder expected in the liquid than in ice and was rationalised by Pauling in terms of the ratio of the enthalpy of sublimation to that of ice.[67] However, given that water's OO coordination number is larger than that of ice,[1–3] very few broken HBs might also be expected. In fact, femtosecond 2D IR spectroscopy suggests that HB switches are very rapid, such that there should be a negligible number of broken HBs or BHs in water.[20,52] Moreover, in a remarkably under-reported inconsistency,[46] 10% of broken HBs is surprisingly comparable to the 10–12% broken HBs at the air–water interface (equivalent to 20–25% of molecules having a broken HB) measured by vibrational sum-frequency generation,[68] a system which should surely have many more broken HBs than the bulk.

All this points towards conceptual flaws in the nature of a broken HB as defined by a cut-off. Desiraju has expressed the view that the cut-off criterion "has caused more havoc to the field of hydrogen bonding than any other factor".[69] HBs, being predominantly electrostatic interactions, do not suddenly drop to zero at a cut-off, as "broken" or even the word "bond" implies. Rather, they extend out continuously and diminishingly to infinity. It is understandable that most donor–acceptor interactions lying outside a reasonably chosen cut-off should not be considered as HBs because they are weaker and most such donors already have a HB to a closer acceptor. However, it seems overly restrictive that a donor–acceptor interaction lying just outside the cut-off for a donor without any acceptor is said to have a broken HB. Likewise, it would appear not restrictive enough if a donor–acceptor interaction lies just inside the cut-off when the donor already has a closer acceptor, bringing about a BH or over-coordinated oxygen. Surely what has a greater effect on whether a donor–acceptor interaction might qualify as a HB is not some fixed cut-off but rather a consideration of the interactions with neighbouring molecules, an issue we take up later.

A single test can resolve both the reality of the broken HB as well as the extent to which the HB distribution is artificially broadened by the fixed cut-off, discussed earlier. A simulation of water at ambient conditions may be quenched to remove the confounding thermal vibration but largely preserve the local HBs and structure.[70] The results are that the spread of the HB distribution is reduced but not removed, and the broken HBs almost entirely vanish.[20,31] Both results happen for the same reason because broken HBs are the main culprit for the additional broadening. This implies that a broken HB as defined by a cut-off is merely a distortion and not a distinctly different entity, failing the requirement set forth by Eisenberg and Kauzmann for a broken HB.[15] Even more significantly, that the distribution does not collapse to a purely tetrahedral coordination reveals that non-tetrahedral structure is present in liquid water. This discrepancy between the quenched and ambient structures convincingly exposes the flaws of traditional HB definitions in understanding water structure. As we now explain, the best way to remove the dependence on the fixed cut-off is to remove it.

1.2 Topological hydrogen bond definition and water structure

The recently proposed topological HB definition is able to address many of the limitations of traditional HB definitions described above and provides a more robust determination of the structure of water's first coordination shell.[31] The definition is simple and intuitive: an electropositive hydrogen donates to the most attractive non-covalently bonded acceptor, unless there is a more attractive donor–acceptor interaction between the molecules involved, in which case the donor forms a broken HB. No cut-offs are specified in the definition, which distinguishes it from its two fore-runners that assume a HOO angle of less than 30° [12] or a maximum of two acceptors[71] and are based on the nearest-acceptor-to-a-donor rather than the more general most-attractive-force. The topological definition resembles nearest-neighbour approaches that are commonly used in analysing variable structures in the condensed phase or in chemical reactions. It is applicable to any set of force field parameters and any combination of donors and acceptors, although it may be advisable to remove the repulsive Lennard-Jones term from hydrogens that have them in order to avoid short-range repulsion. It could be used in an *ab initio* simulation but would require a method to define the charge density of an atom such as Atoms in Molecules[72] or only involve distances instead of forces.[73] It is generally consistent with the IUPAC HB definition[74] and addresses its open-ended requirement of bond formation by assuming that a donor forms a stable interaction with its nearest acceptor.

The feature of the topological definition that explains its name is that it places the cut-off of a HB at the point where the donors are equidistant from two acceptors and that this BH structure is presumed to lie close to the transition state surface separating the two different HB states.[20,23–25,28,29,52] Transition state surfaces are well-defined topological features of the potential energy surface. When a donor hydrogen is pulled equally by two acceptors, these opposing forces sum to zero in the direction of the switch, bringing about an estimate of a point lying on the full $(6N - 1)$-dimensional transition state surface, where N is the number of water molecules. This approximation is useful because it would otherwise be intractable to locate the full, exact transition state surface together with its saddle point for a single HB switch let alone for all of them. A molecular dynamics trajectory with its discrete time-step is already approximate, and it is only necessary to know when the surface has been crossed. Other longer-range interactions contribute but these are weaker and their perturbations, being positive or negative, would cancel to a large degree. Determining HBs based on transition state surfaces is particularly advantageous in studying HB kinetics.[31] Artificial recrossings are reduced without including additional procedures to ensure whether a switch has occurred, as are necessary when HB cut-offs are used.[28,29,59,64–66] Traditional HB definitions fare worse because they effectively assume that transition state surfaces, marked by a cut-off, are the same for every single HB switch. Because this assumption is clearly false in a liquid, this brings into being the enigmatic, cut-off-dependent broken HBs and over-coordinated water molecules. Not taking this variability into account has arguably been one of the main obstacles preventing a clear view of the structure of water. For conciseness, consistency with usage in the field, and because ensembles are more relevant to this analysis than specific stationary points, we will refer to transition state surfaces as transition states, in the same way that stable states are often used to refer to the ensemble rather than to only the local minimum.

By the topological definition for water, a donor hydrogen forms a HB to an oxygen acceptor depending on the strengths of three other types of interaction: the donor with the donor molecule's other hydrogen donating to this acceptor, either of the two hydrogens of the donor's acceptor donating back to the donor molecule, and all other acceptors. In turn, there are assumed to be three matching types of transition state which depend on whether the new acceptor is the switching water's initial acceptor, donor or neither. These are the bifurcated oxygen (BO), cyclic dimer (CD), and BH already mentioned.[31] The BO has the two donors of the same water pointing to a single acceptor while the CD has a donor from each water both trying to donate to the other molecule. The BO and CD have been observed before in gas-phase dimers[75–77] and larger clusters[78] and found to be transition states in those cases. They have also shown up in proposed mechanisms of water[15,65] and ice dynamics,[79,80] in simulations of water in carbon nanotubes,[81] and in electronic structure calculations in which they contribute to the enhanced pre-edge of the X-ray absorption spectrum.[53] Most works have not considered them because the strained structures are outside the cut-offs of traditional HB definitions or because studies of HB networks usually assume only one HB between water molecule pairs. While BHs and symmetric BOs and CDs exist instantaneously, BOs and CDs also have asymmetric states with non-zero lifetimes comparable to half a water's libration.[20,50,82] The weaker donor–acceptor interaction in these structures can be defined as broken.[31] Even though the interaction strength of the broken HB is not zero, if the broken HB tries to strengthen, then the other HB has to weaken and the total energy rises. This inability for a donor to point towards its preferred acceptor, owing to a local deficiency of acceptors, meets the criterion laid out by Eisenberg and Kauzmann that a broken HB must be distinctly different to that of a HB.[15] The fraction of broken HBs under ambient conditions is found to be only ~1%, which is consistent with the transient breakage time measured by experiment.[20,52] At the air–water interface ~20% of water molecules are predicted to have a broken HB,[46] again consistent with the 20–25% detectable from experiment.[68] Despite these successes, there is one complication. The logic to define broken HBs for dimers implies that the strained HBs in larger polygons such as trimers and tetramers found in water[83–85] could also be regarded as broken. Beyond water, multiple, strong HBs can readily exist between various pairs of molecules such as DNA bases or acetic acid dimers. This demonstrates that there is a whole spectrum of reasons to decide if a HB is broken. The case of the polygons we consider later.

Concerning the acceptor HB distribution, the topological definition predicts that about 70% of water molecules at ambient conditions have the tetrahedral coordination of two donors and two acceptors,[31] otherwise referred to as a double acceptor (DA). The main non-tetrahedral species are the single (SA) and triple acceptors (TA), which, because there are usually two donors, may be identified as having trigonal and trigonal bipyramidal coordinations. Trace amounts of non-acceptors (NA) and quadruple acceptors (QA) are also detected, with bent and octahedral coordinations. All such species are not new, having been seen in numerous computer simulations[18,23–26,30,51,56–63,65] and X-ray crystal structures of hydrated solutes.[86,87] but the difference here is that these SA and TA structures are substantial and almost equal to each other. The small difference between their populations is almost entirely due to the small fraction of broken HBs. Such species in many ways are consistent with the "tetrahedral on average" description of water because many of the trigonal and trigonal bipyramidal properties are opposing, making their combined effect harder to distinguish from tetrahedral structure by most forms of analysis, a prime example being the ordering of water by small hydrophobic molecules.[38–40,42,43,46] However, unlike the purely tetrahedral model, the combined effect of all coordinations can better explain other properties. These include the well-known broad band of oxygen or hydrogen density around the central oxygen in 3D probability distributions[10,11] because of the planar contribution,[31] and why water's OO coordination number lies between four and five.[1–3] This latter effect arises because of the five-fold trigonal bipyramidal coordination combined with the greater ability of non-HB oxygens, the so-called interstitials, to approach within the coordination cut-off at the exposed upper and lower faces of the trigonal coordination.[11,31]

Contributing to the strength of the assignment of HBs, the topological HB distributions are highly resistant to quenching, unlike the distributions described earlier using traditional HB definitions.[31] Trigonal and trigonal bipyramidal coordinations have respectively a more positive or more negative energy than the tetrahedral coordination, correlating with the number of HBs, while the HBs themselves are correspondingly stronger or weaker because more HBs induce more strain. The combined energy of the trigonal and trigonal bipyramidal coordinations is slightly more positive than that of the tetrahedral coordination but significantly less than the broken HB states described earlier.[31] Their lifetimes are only about a third that of the tetrahedral coordination but much longer than those of the BOs and CDs. The non-tetrahedral coordinations provide a substantial number of defects that permit many more types of single-step HB switches[27,30,31,55] than those that involve a donor jumping from an over to an under-coordinated acceptor.[28,29] Finally, it should be noted that water's oxygen is not limited to accepting two donors because it has two lone pairs, as is commonly supposed. Based on Bernal and Fowler's early model for water[4] as well as observations of X-ray crystal structures,[88] electronic structure calculations[89,90] and comparisons of three and four-point water models,[91] Finney has emphasised that the lone pairs are largely delocalized,[92] thus permitting the trigonal geometry. The same arguments also allow plenty of space on the oxygen to accept more than two HBs, particularly in water's unusually low OO coordination environment, thus permitting the trigonal bipyramidal geometry. Additional, well-recognised evidence for a HB is that it is a pre-cursor state to full proton transfer.[74] Indeed, the single and triple acceptors are respective pre-cursors of the Eigen forms of the hydronium and hydroxide ions which are known to be three[93] and five-coordinated.[94]

One of the main criticisms which has been levelled at the topological definition is that it is too generous in assigning HBs. In particular, it is possible for two isolated water molecules to form a HB at any distance or angle. This stretches the common view of HBs as being short-ranged and directional, as made clear by the many cut-off definitions described earlier. The IUPAC HB definition is less proscriptive and concerning stability only requires that there be "evidence of bond formation".[69,74] This is not merely to be compatible with the term "bond", but it helps make the HB detectable and brings about structure in the first hydration shell. Such evidence qualitatively relates to HB strength such as distance, directionality, polarity, infrared stretch modulation or protein deshielding. The one exception is the suggestion that the Gibbs energy of HB formation be greater than the thermal energy. This is a disguised cut-off in that the equilibrium constant for HB formation should exceed base e, although it is softened by the absence of the specification of standard states of the donor and acceptor prior to HB formation. Arunan has advocated that the potential barrier should exceed the zero-point energy[95] while other sources with a focus on weak HBs advise having no lower limit on HB strength.[44,87,96] In any case, the IUPAC HB definition does not stipulate complete adherence to every criteria but rather that the more criteria that are met, the more reliable the HB characterisation. A second, less commonly made criticism against the topological definition, and rather more a source of discomfort to its proposers, is that it has a disguised cut-off in allowing no more than one acceptor per donor. In other words, the possibility of a BH donating to two acceptors[21,22] is not permitted to be a stable state, even though such structures are known to be present in X-ray crystal structures.[86,87] It would be valuable if such a limitation could be removed.

In light of these criticisms, in this work we examine the nature of the non-tetrahedral components of water in a molecular dynamics simulation. The objectives are first of all to assess the validity of the topological HB classification and the resulting structure of water. Stability is assessed by the donor–acceptor distance r_{OH}, the HOO angle θ_{HOO} and the pairwise energy between two molecules U_{ij}. We examine three influences on HB stability: the size of the closed HB polygons to which a HB belongs, the acceptor type, and the relative closeness of multiple donors to an acceptor. Secondly, the three assumed transition states are examined in more detail to determine if they really are transition states.

2 Methods

2.1 Molecular dynamics simulations

A cubic periodic box containing 1000 TIP4P/2005 water molecules[97] and having volume approximately $3 \times 3 \times 3$ nm^3 is simulated using the sander module of the AMBER 12 software package.[98] 500 steps of steepest-descent minimisation are followed by a molecular dynamics simulation in the NVT ensemble at 298 K using a Langevin thermostat for 100 ps and in the NPT ensemble using a Berendsen barostat at 1 bar for a further 2 ns. Data collection proceeds over a further 1 ns under the same NPT conditions with SHAKE applied to all bonds, a 2 fs timestep, periodic boundary conditions, an 8 Å cutoff and particle mesh Ewald with default AMBER parameters. Coordinates are saved every 1 ps for analysis.

2.2 Hydrogen bond analysis

HBs are analysed using the topological definition.[31] The oxygen's charge of TIP4P/2005 is taken as the position of the acceptor as opposed to the mass and van der Waals centre as had been done in the original implementation based on distance. This is to be consistent with the general definition based on donor–acceptor force.[31,44,99,100] As noted earlier,[46] this makes little difference to the HB distribution, yielding marginally more broken HBs. For each HB coordination, we analyse the probability distributions in distance r_{OH} between the donor hydrogen and acceptor oxygen, the angle θ_{HOO} subtended by the HB r_{OH} vector and the donor's HO bond vector, and the interaction energy U_{ij} between the two molecules, calculated by summing over all electrostatic and Lennard-Jones energies of each atom pair. Three different comparisons are made:

1. We consider every donor and calculate the distributions of its nearest and next-nearest acceptors measured by the r_{OH} distance as a function of the number of acceptors of the nearest acceptor. We also examine these distributions according to whether the donor is a BO or CD, and when the next-nearest acceptor does not have a HB with the donor molecule.

2. We consider the reverse situation by taking each acceptor and finding the series of nearest donors up to a maximum of the number of acceptors plus one. Again, the nearest non-donor that does not already have a HB with the acceptor molecule is also analysed separately.

3. We consider how the HBs are strained by the constraints of the polygons to which they belong in the HB network. Polygons are defined using the algorithm of

Chihaia *et al.*[101] Different to previous work,[83–85,101] dimers, comprising BOs and CDs, are also considered.

The probability distribution for each HB species is normalised to one. The exceptions are the distributions of neighbours which exclude a HB to the donor's water to make it easier to assess the effect of this exclusion.

2.3 Order parameters

In addition to the widely used tetrahedral order parameter,[102] denoted here as q_4, two new analogous order parameters are considered: the trigonal and trigonal bipyramidal order parameters, denoted by q_3 and q_5, respectively. All three are calculated using the equations

$$q_3 = 1 - \frac{4}{7}\sum_{j=1}^{2}\sum_{k=j+1}^{3}\left(\cos\psi_{jk} + 1/2\right)^2 \tag{1}$$

where the ideal angle is 120°,

$$q_4 = 1 - \frac{3}{8}\sum_{j=1}^{3}\sum_{k=j+1}^{4}\left(\cos\psi_{jk} + 1/3\right)^2 \tag{2}$$

where the ideal angle is 109.5°, and

$$q_5 = 1 - \frac{6}{35}\sum_{i=1}^{2}\sum_{j=i+1}^{3}\left(\cos\psi_{ij} + 1/2\right)^2 - \frac{3}{10}\sum_{i=1}^{2}\sum_{k=1}^{3}\left(\cos\psi_{ik}\right)^2 - \frac{3}{40}\left(\cos\psi_{kk} + 1\right)^2 \tag{3}$$

where the ideal angles are 120° for the three equatorial angles, 90° for the six axial–equatorial angles, and 180° for the one axial–axial angle. Indices i and j indicate equatorial positions and k indicates axial positions, and the angles ψ are calculated over all oxygen–oxygen–oxygen angles centred on the acceptor of consideration and involving the nearest three oxygens for q_3, four oxygens for q_4, and five oxygens for q_5. The reference angles could be chosen more accurately for the expected water geometry, such as having a tetrahedral value for the HOH angle instead of 120°, but a more symmetrical set was chosen for simplicity. The equations for q_3 and q_5 are derived using a similar method to that of Chau and Hardwick[103] but normalised to one for perfect order and zero for a random distribution.[102] The three equatorial oxygens in the case of q_5 are those whose angles between their OO vector and the water molecule's HOH plane are closest to zero. The remaining two are axial. Each order parameter is applied to SAs, DAs and TAs. They are normalised for each species so they can be more easily compared.

2.4 Triple and quadruple acceptor geometry

Because the acceptor positions in TAs and QAs are non-equivalent, the probability distributions of the OH distances for the axial and equatorial donors to these two acceptors are calculated, together with those of the angles between their OH vectors and the water molecule's HOH plane, termed θ_{HOp}. The equatorial donor to a TA is determined as the one whose angle between its OH vector and the molecule's HOH plane is the closest to zero. Similarly for QAs, the two equatorial donors are those with these angles closest to zero. In each case the remaining two donors are axial.

2.5 HB switching profiles

The HB switching profiles *via* the three structures, BH, BO and CD, are plotted using the potential of mean force (PMF), defined as

$$\mathrm{PMF} = -RT \ln P(\zeta) \tag{4}$$

where R is the universal gas constant, T is temperature, ζ is the reaction coordinate and $P(\zeta)$ is its probability. Probabilities are normalised to one for all species combined in each graph. Two choices of ζ are considered. They are designed such that $\zeta = 0$ at the putative transition state. One is in terms of distance $\Delta r = |r_{\mathrm{OH1}} - r_{\mathrm{OH2}}|$ and the other in angle $\Delta\theta = |\theta_{\mathrm{HOO1}} - \theta_{\mathrm{HOO2}}|$. The subscript "1" designates the donor and its nearest acceptor for all transition states. "2" differs for each structure. For the BH it designates the donor and its next-nearest acceptor, for the BO it designates the donor molecule's other hydrogen and both donors' acceptor, and for the CD it designates the nearest donor of the donor molecule's acceptor and the donor molecule's oxygen. For the BO and CD structures, two specific cases of PMF are considered: one when the donor is broken and one when it is intact. For the BH structure, a whole series of PMFs are considered to examine the effects of three factors: the number of acceptors of the nearest and next-nearest acceptors, and the proximity of the donor to its acceptor. Proximity of a donor to an acceptor is assessed by assigning each donor to a subshell ranked according to r_{OH}. The subshell ranges from 1 up to the number of acceptors for the closest to furthest donors, respectively.

2.6 Donor–acceptor bias

To detect the presence of any long-range structuring in terms of an imbalance of donors and acceptors, the cumulative donor–acceptor bias[46,104] $\Delta N(r_{\mathrm{OO}})$ is integrated from the origin to a distance r_{OO} from a water molecule:

$$\Delta N(r_{\mathrm{OO}}) = \int_0^{r_{\mathrm{OO}}} \sum_i (d_i(r'_{\mathrm{OO}}) - a_i(r'_{\mathrm{OO}}))dr'_{\mathrm{OO}} \tag{5}$$

where $d_i(r'_{\mathrm{OO}})$ and $a_i(r'_{\mathrm{OO}})$ are the numbers of donors and acceptors of water species i at distance r'_{OO} from each water.

3 Results

3.1 Hydrogen bond analysis

The HB distributions produced by the topological definition are 0.2% NA, 15.0% SA, 71.5% DA, 13.0% TA, 0.3% QA, 0.0% non-donor, 0.3% BO, 1.5% CD, and 98.2% double donor at 298 K and 1 bar as reported before.[31] Similar populations are produced using other force fields[31] and from *ab initio* simulation, although there is a dependence on the size of the dispersion term.[73] Given that BOs are responsible for about half as many HB switches as CDs,[31] the much smaller percentage of BOs implies that the reaction coordinate for switching *via* a BO must be shorter than that *via* a CD. This is consistent with a BO switch requiring the rotation of only one molecule through the tetrahedral angle whereas the CD switch requires the rotation of two molecules through the tetrahedral angle.

Fig. 1 plots the probability distributions in r_{OH}, θ_{HOO} and U_{ij} of a donor's nearest acceptor and next-nearest acceptor for each acceptor species and for the broken donor of the BO and CD species. NAs are not included because by definition they are almost never a donor's preferred acceptor. They are only ever so when the NA's donor HB back to the molecule is stronger, leading to a broken HB to the NA. The distributions for QAs and the broken HB species are slightly noisier because of the much lower populations of these species. The main result is that the data support the assumption that a donor donates strongly to its nearest acceptor. The values of distance, angle or energy are low for the nearest acceptor, and there is a large difference going to the next-nearest acceptor. This demonstrates that a donor's HB to the nearest acceptor is usually much stronger than its HB to the next-nearest acceptor. This distinction is sharpest for the species with fewer acceptors, which are able to accept HBs more optimally because there are fewer constraints and less crowding brought about by other donors. It is also sharpest for the distance distributions of SAs, DAs and TAs. The stability of the donors to QAs is marginal. It is inevitable that there is some overlap between the nearest and next-nearest peaks because the donors exchange and spend some time at intermediate distances. The distributions for donors having broken HBs in BOs and CDs resemble those of the next-nearest acceptors, signifying much weaker donor–acceptor interactions as expected. The double-hump structure

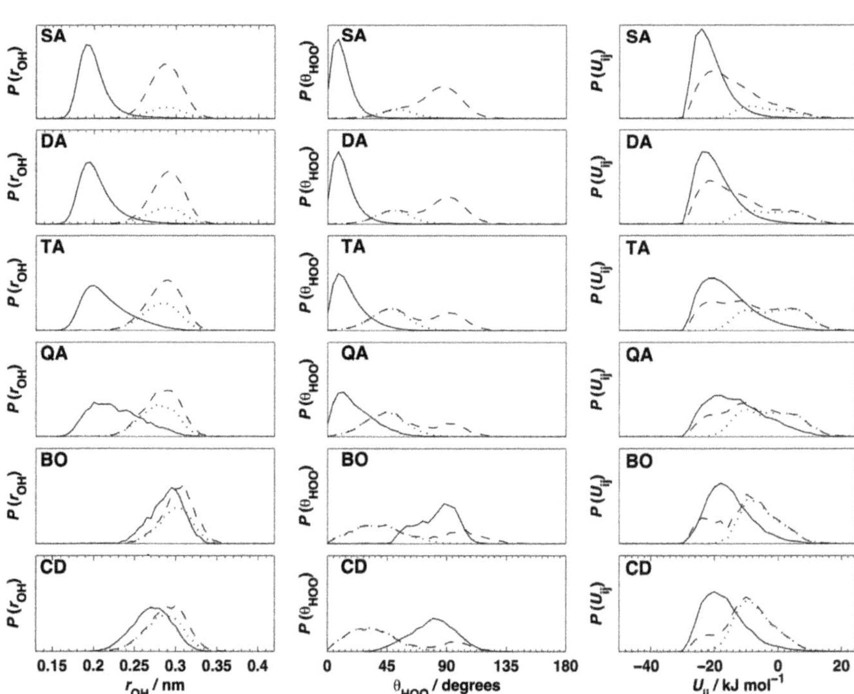

Fig. 1 Probability distributions $P(r_{OH})$ of the OH distance r_{OH} from a donor to its nearest (solid) acceptor, $P(\theta_{HOO})$ of the HOO angle θ_{HOO} subtended by the donor–acceptor vector and the donor's OH bond vector, and $P(U_{ij})$ of the pairwise energy U_{ij} of the donor–acceptor pair of water molecules. The acceptor is either a SA, DA, TA, QA, BO or CD (see text for abbreviations). Also plotted are the values for the donor's next-nearest acceptor (dashed) and the next-nearest acceptor without a HB with the donor molecule (dotted).

evident in the angle and energy distributions of the next-nearest acceptors can be understood as arising from two types of water molecules: those already in the first hydration shell with another HB to the donor and those without. This is made clear by the plots which exclude other first-shell water molecules in the next-nearest distribution. They have similar distances but lower angles peaked around 50° and energies much closer to zero, consistent with them not being in the first hydration shell.

Next we turn to Fig. 2, which plots the probability distributions in r_{OH}, θ_{HOO} and U_{ij} for the series of nearest donors for each type of acceptor. The main result here is that it supports the assignment of different HB coordinations in the first hydration shell. It would be far-fetched to conclude that the nearest two donors are always within the first hydration shell and that no others are, as would be expected for tetrahedral coordination. Generally, the largest transition from one donor to the next occurs when going to the donor number that exceeds the species' number of acceptors. This is especially so for the angles, indicating that there is much directionality for the HBs, even up to the fourth-nearest HB of QAs. As seen in Fig. 1 and 2, the HBs for species with more acceptors have longer distances, larger angles and higher energies. The nearest donor outside the coordination shell is either near the tetrahedral angle, suggesting that it already donates to the acceptor, or it lies in a broad distribution centred at 50°, which are weakly interacting water molecules as noted before. The equivalent distributions using traditional HB definitions with a cut-off in r_{OH}[58] or energy[18,61–63] would

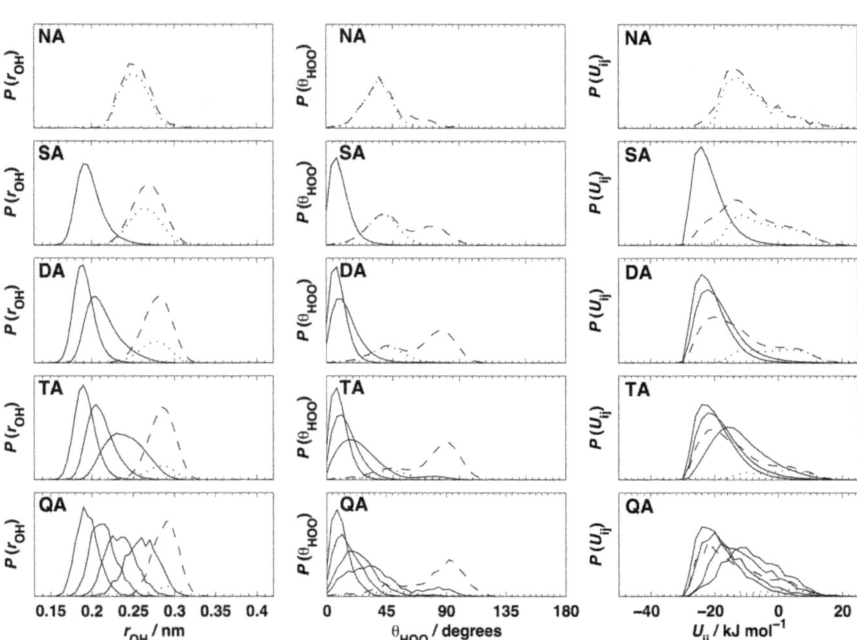

Fig. 2 Probability distributions $P(r_{OH})$, $P(\theta_{HOO})$ and $P(U_{ij})$ from an acceptor to its nearest, expected number of donors (solid). The distributions are ranked according to r_{OH}. The acceptor is either a NA, SA, DA, TA, or QA. Also plotted are the values for the next-nearest donor beyond the expected acceptor number for that species (dashed) and the next-nearest donor that has no HB to the acceptor molecule (dotted).

invariably have a sharp discontinuity at the cut-off. Such sharp transitions are softened in these one-dimensional projections for definitions with two or more criteria such as those based on a distance and angle.[49,53,56,57,60] Nevertheless, a sharp discontinuity would still occur in the full, higher-dimensional probability distribution. These features would be absent using the nearest-acceptor and two-nearest-donors definition[71] but the TA and QA species would be entirely absent because only a maximum of two acceptors are allowed per water molecule.

The third factor examined to assess HB strength and in particular, broken HBs, is the effect of the polygon ring size to which a HB belongs. The same distance, angle and energy distributions are plotted in Fig. 3. We do not consider the many permutations of HB directions in the ring,[17] as had been done for the BO and CD structures before in Fig. 1. Moreover, we suspend here the assignment of broken HBs brought about by the dimers. It can be seen that trimers induce a moderate amount of distortion, especially for the weakest HB. The distributions for the tetramers and pentamers are very similar and show a gradual increase in strain from the strongest to the weakest HB, suggesting that rings of these sizes induce negligible strain. These results support a possible case for considering the weakest HB in a trimer as broken. The number of trimers calculated per water molecule is 0.017, comparable to the 0.018 for dimers. Thus there would be twice as many broken HBs if trimer distortion were also regarded as leading to a broken HB. This makes clear that the distinction between broken and intact HBs is less distinct, even by the topological definition. The different levels of distortion are discrete for the smallest polygons but become more continuously distributed for larger polygons.

3.2 Order parameters

As another measure of coordination, the trigonal, tetrahedral and trigonal bipyramidal order parameters are plotted in Fig. 4(a)–(c) for SAs, DAs, and TAs. Each

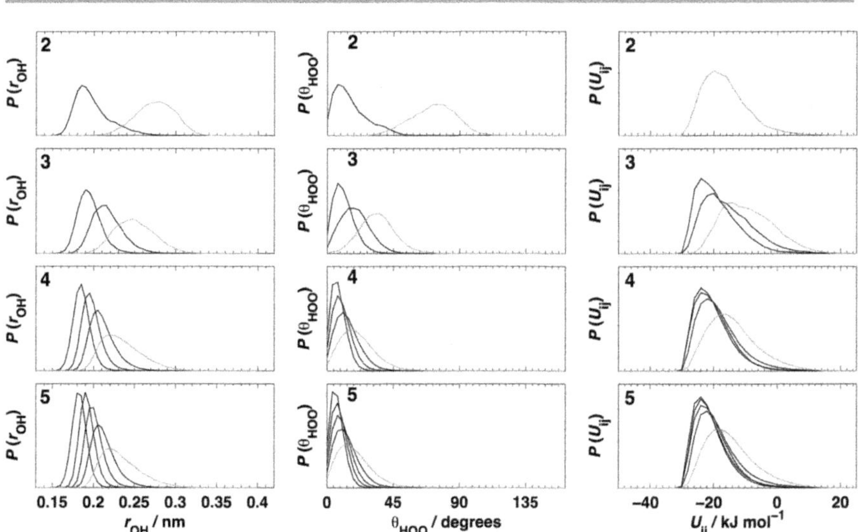

Fig. 3 Probability distributions $P(r_{OH})$, $P(\theta_{HOO})$ and $P(U_{ij})$ for all HB interactions in a polygon (solid). The polygons have sides from two to five. All distributions are ranked by r_{OH}, and the weakest donor–acceptor distribution is shaded grey.

parameter is largest for its respective species, confirming that their geometries mostly match that of the appropriate order parameter. In particular, the SAs and TAs are seen to be responsible for the shoulder in the q_4 plots that are typically rationalised as being less tetrahedral. The only species which are not well distinguished are the SAs and DAs by the q_3 order parameter. This indicates that the nearest three waters for tetrahedral coordination are often very close to trigonal.

3.3 Triple and quadruple acceptor geometry

To examine further the donor distributions around the acceptors with more than one donor type, namely TAs and QAs, Fig. 5(a)–(d) illustrate that the axial and equatorial OH distances and donor angles to the plane θ_{HOp} are very similar. Axial distances are marginally shorter by ~0.01 nm than equatorial ones, TA's equatorial donor is more in the plane than the QA ones, and the angles of QA's axial donors are marginally more axial than the TA ones. All these trends are consistent with the expected geometries.

3.4 HB switching profiles

Here we assess whether the BH, BO and CD structures are transition states. Fig. 6 illustrates the donor–acceptor PMFs for each of the three cases. The transition states would lie where the value of the reaction coordinate equals zero. The plots are not continued beyond this point because they are symmetric about zero. All the PMF curves display a maximum at the origin. This would appear to validate the assumption made in the topological definition and elsewhere[20,23–26,28,29,52] that these geometries are transition states. Another observation is that there is a high barrier near the transition state for the BO and CD PMFs when the donor HB is still intact. Once the donor HB has broken, the barrier for switching becomes much lower and both these structures are more delocalised, facilitating the switch.

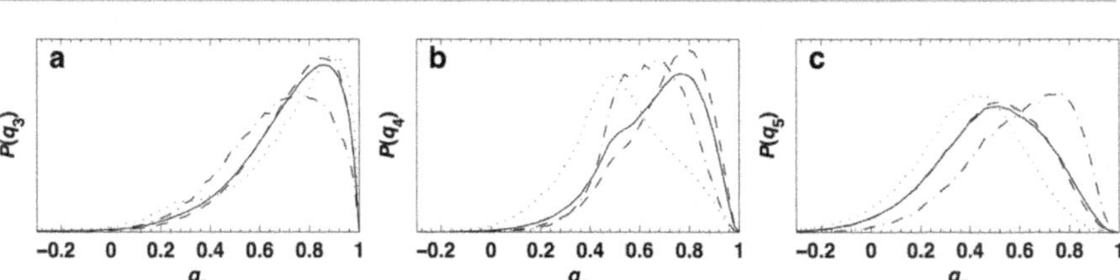

Fig. 4 (a) Trigonal q_3, (b) tetrahedral q_4 and (c) trigonal bipyramidal q_5 order parameters applied to SAs (dotted), DAs (dashed), TAs (dot-dashed) and all water (solid).

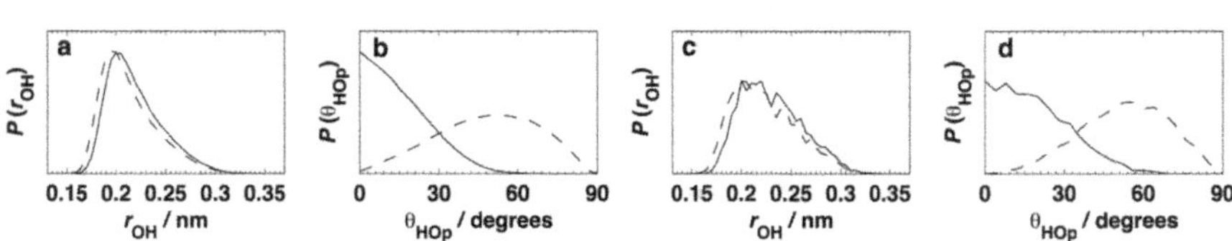

Fig. 5 Probability distributions of TAs (a) distances r_{OH} of the planar (solid) and axial (dashed) hydrogens donating to its oxygen, and (b) the angles θ_{HOp} of the respective OH vectors to the water's plane. (c) and (d) are the equivalent plots for QAs.

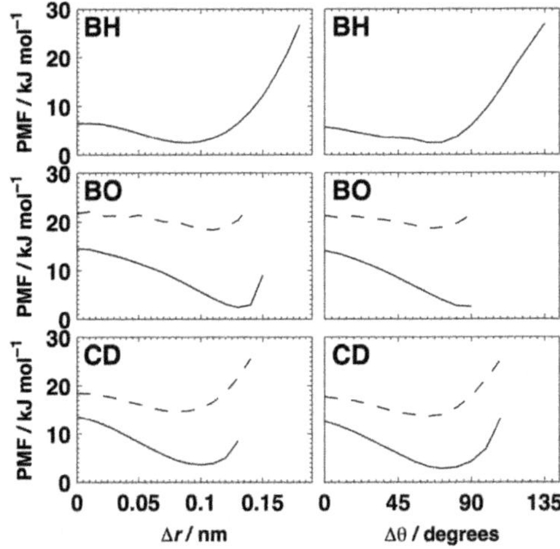

Fig. 6 PMF profiles for the three transition states BH, BO and CD as a function of the distance Δr and angle $\Delta\theta$ reaction coordinates. The BO and CD plots are for donors with a HB (solid) and those with a broken HB (dashed).

Now we examine the more detailed PMFs in terms of the number of acceptors of a donor's nearest and next-nearest acceptor as well as the donor's subshell to its nearest acceptor. These PMFs are displayed in Fig. 7 as a function of Δr or $\Delta\theta$. The plots are not extended to negative values of the reaction coordinate because the acceptor types would change. What these PMFs quantify is how easy it is for a donor to move from a given subshell of one acceptor type to another acceptor type. A much more intricate and intriguing picture emerges than what was seen in Fig. 6. In the outermost subshell of DAs and TAs, it can be seen that the barrier for switching to a lower-acceptor species is essentially removed! To consider all the possible cases, we define the acceptor differential Δa as the number of acceptors of the next-nearest acceptor minus that of the nearest acceptor. The more negative Δa, the more

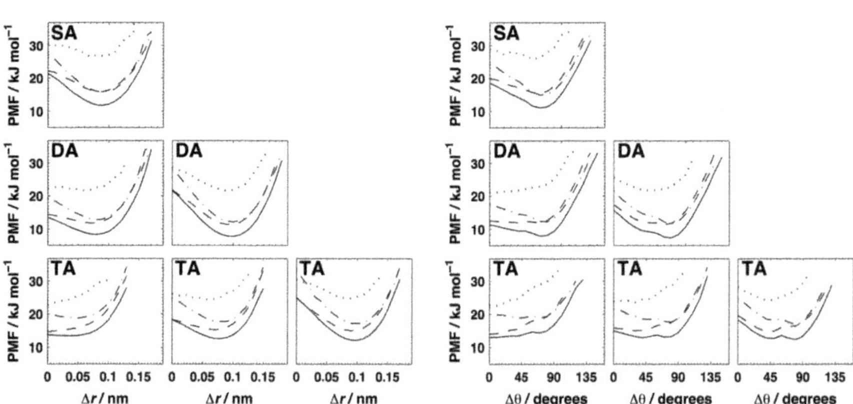

Fig. 7 BH PMF profiles as a function of the distance Δr and angle $\Delta\theta$ reaction coordinates for the donor to a SA, the outer and inner donors to a DA (left to right) and the outer, middle and inner donors to a TA (left to right). The next-nearest acceptor is either a NA (dotted), SA (dashed), DA (solid) or TA (dot-dashed).

favourable becomes the PMF for a switch. For switches with $\Delta a = -1$, such as a DA to a SA, or a TA to a DA, the reactant and product states are identical by symmetry. Such a situation is the hallmark of a stable BH, as proposed by Giguère.[21,22] The percentage of donors existing as a BH can be estimated to be 10% by taking the 15% and 13% of SAs and TAs, dividing by two to convert to a percentage of donors, and noting that their neighbour would be a DA about 70% of the time. This number may seem large but it includes asymmetric BH structures as well as the more symmetric ones. For switches with $\Delta a < -2$, such as when a donor switches from a TA to a SA, there is no barrier and a continual decrease in PMF that leads to a more stable product state. For switches with $\Delta a \geq 0$, then the switch is unfavourable. The most common such switch is two DAs going to a SA and TA. It nonetheless proceeds because of the high concentration of DA and the donor's penetration into the inner subshells of the TA.

These results may be surprising in light of the predominance of work suggesting that BHs are transition states.[20,23–25,28,29,52] However, they become more intuitive upon further reflection. The fewer donors an acceptor has, the easier it is for another donor to donate to it and *vice versa*. What is noteworthy is that this trend only applies to donors in an acceptor's outermost subshell. For other types of switch, the barriers are substantially higher. Evidently a donor that is interacting more strongly with its acceptor is less likely to leave. Conversely, a donor is more likely to leave if its acceptor is more strongly interacting with other molecules. This emphasises the point that if the different subshells had not been considered individually, then these effects would have been missed and barriers would have appeared for all acceptor types. In any case, rapid whole-molecule vibration continually changes the ranking of the nearest donors to an acceptor, giving each donor in turn a chance to switch. Interestingly, the lifetimes of SAs and TAs, determined earlier to be about $\sim$0.3 ps, match the timescale of whole-molecule vibration.[31] We do not need to consider the ranking of a donor to its next-nearest acceptor because it should almost always be further out than any of this acceptor's donors. The final piece in the mechanism is to recognise that because a donor's preferred position depends upon Δa and on its ranking to its nearest acceptor, when either of its two nearest neighbours change acceptor number or when its ranking to its acceptor changes, then so does the donor's preferred position. For example, consider a donor pointing to a DA with another DA nearby. As long as this donor is the outermost donor to its DA, if this nearby DA becomes an SA, then the donor becomes bifurcated between the two. If the nearby SA drops to a NA, then the bifurcated donor switches entirely to the NA, making it a full SA. This demonstrates how switches are fully at the mercy of their neighbours' switches in a complex recursive fashion.

The implications of these findings are profound. Firstly, it reveals that the nature of non-tetrahedral water is quite different to what is usually envisaged. Its identity is intimately coupled to its nearest neighbours, sometimes with a donor directed to an acceptor and sometimes bifurcated. In particular, it clarifies that BHs may be caused by either SAs or TAs but explains why not all SAs or TAs lead to BHs. It is also understandable why these non-tetrahedral structures with their donor–acceptor imbalances are largely resistant to quenching.[31] Secondly, a question mark is raised about the usefulness of the topological HB definition whose main assumption is that the donor donates to its nearest acceptor. Now there is a dependence as well on the next-nearest acceptor. Consequently, a more sophisticated procedure is required to allow for the possibility of a BH. The definition would need the qualification that as well as donating to its nearest acceptor, a donor may also donate to its second-nearest acceptor if it is the outermost donor to its closest acceptor and if the second-nearest acceptor has fewer acceptors than the nearest acceptor! This wordy definition reflects the greater complexity of HB structure in water. The presence of the BHs also makes the classification of the species by donor and acceptor number alone more complex. One might have a partial DA, for example, when one of the donors to it is bifurcated. Thirdly, HB switches *via* a BH for an acceptor's outermost donor are either uphill, downhill, or largely flat. There is effectively no barrier for these types of switch. It is the combination of all of these processes as seen in Fig. 6 that creates the illusion of a barrier, particularly the contribution from the more stable, inner subshells that do have a substantial barrier. Inner subshell donors can only switch if the outer donor, being further away, is also switching. Fourthly, BHs lead on average to more than four HBs per molecule in its first coordination shell and contribute to water's OO coordination number being 4–5. Fifthly, BHs even have the effect of making water structure look more tetrahedral. SAs are supplemented by a BH and one of the TA's HBs is weaker and bifurcated. Even so, the non-tetrahedral structures not engaged in BHs persist. Sixthly, an alternative interpretation of BHs is that their HBs are distorted or broken, not because of some arbitrary criteria, but because the donor is delocalised between two equally receptive acceptors. Thus the structure presented here shares features with all the main structures proposed for water.

3.5 Donor–acceptor bias

The final question to be examined, and incidentally the original objective of this work, is whether there is long-range structure in water. The measure for this considered here is the cumulative donor–acceptor bias $\Delta N(r_{OO})$ which is plotted in Fig. 8. While there is an imbalance at short-range, with initially fewer donors than acceptors at a distance corresponding to the minimum in the first hydration shell, this trend reverses further away and oscillates with decreasing amplitude, reaching zero comfortably within half the box-length. Thus there is no long-range structuring of water by this measure. Similar results for the distribution of individual HB species[46] also turn up similar null long-range results (data not shown). This is not surprising, given that water on average has the same number of donors and acceptors with itself. Nevertheless, it still needed to be confirmed. There is certainly structuring at short-range in water by this measure, with tetrahedral water displaying a slight preference for itself, as is well-known,[18,49] and high and low acceptor species such as TAs and SAs tending to attract each other more than average to alleviate imbalances in donors and acceptors.[31] However, these biases decay rapidly with distance beyond a few hydration shells. Thus it can be concluded that the presence of a solute or interface must be required to bring about long-range structuring in water,[46] at least in the type of structuring considered here. Such long-range structuring occurs because the short-range perturbation induced by the solute or interface in the donor–acceptor balance of water's HBs is best accommodated by spreading the compensating donor–acceptor balance over as many water molecules as possible. This capability of water presumably contributes to its ability to solvate molecules. The distribution

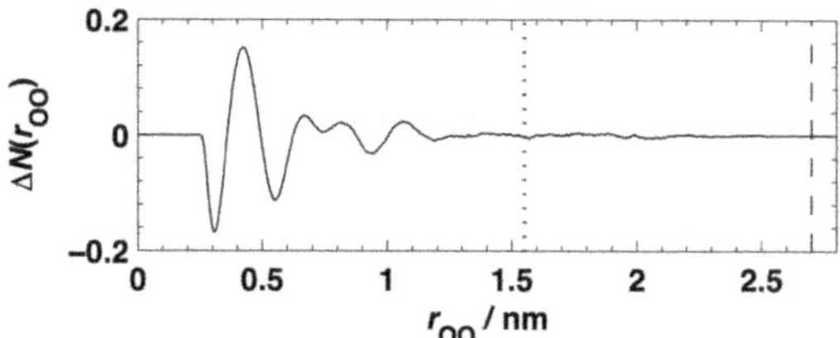

Fig. 8 Cumulative donor–acceptor bias $\Delta N(r_{OO})$ *versus* r_{OO} averaged over all water molecules. The vertical lines are half the box edge (dotted) and half the box long diagonal (dashed).

of HB species with different numbers of donors and acceptors as proposed here would be essential in accommodating the solute's HB preferences and present interesting possibilities at many levels of structure.

4 Conclusions

The case for non-tetrahedral water has been put forward, based primarily on the greater reliability of the topological HB definition which resolves HBs based on the transition states between them rather than using cut-off-based HB definitions with their unavoidable ambiguities. This is bolstered here by new evidence from molecular dynamics simulations, which confirms the identities of the non-tetrahedral components. This evidence includes distance, angle and energy distributions, new order parameters for the trigonal and trigonal bipyramidal coordinations, and an analysis of the HB distortion induced by closed HB polygons. However, in seeking to verify the positions of these transition states, it was found that HB switching *via* the BH is a barrierless process for an acceptor's outermost donor when its next-nearest acceptor has lower coordination than its nearest acceptor. This is in contrast to most studies which imply that the BH is always a transition state. While no long-range structuring was detected in water in terms of a donor–acceptor imbalance, the non-tetrahedral species presented here are expected to play a major role in long-range structuring around dissolved solutes, as has already been seen.[46] Such structuring efficiently dissipates the perturbation induced by the solute over the rest of the water. The ramifications of this new picture of water may be equally far-reaching.

Acknowledgements

We are indebted to E Arunan, G Desiraju, Prabal Maiti, Noam Agmon, Arieh Ben-Naim, Raphael Levine, Mark Tuckerman, John Finney, Imre Bako, David Chandler, Frank Stillinger, Frauke Gräter, Ulf Hensen, Anupam Chatterjee, Hlengisizwe Ndlovu and Sheeba Jem Irudayam for helpful discussions about hydrogen bonds and water structure, not to mention instructive comments by the anonymous reviewer. This work was carried out with support from the Alexander von Humboldt Foundation, the German Academic Exchange Service, the Klaus Tschira Foundation, the Lady Davis Fellowship Trust, the Indian Institute of Science, the EPSRC-funded Manchester Chemical Biology Network (EP/I037253/1), and BBSRC grant BB/K001558/1.

References

1 A. H. Narten and H. A. Levy, *Science*, 1969, **165**, 447–454.
2 F. H. Stillinger, *Science*, 1980, **209**, 451–457.
3 A. K. Soper and M. G. Phillips, *Chem. Phys.*, 1986, **107**, 47–60.
4 J. D. Bernal and R. H. Fowler, *J. Chem. Phys.*, 1933, **1**, 515–548.
5 J. A. Pople, *Proc. R. Soc. London, Ser. A*, 1951, **205**, 163–178.
6 M. G. Sceats and S. A. Rice, *J. Chem. Phys.*, 1980, **72**, 6183–6191.
7 Y. Marechal, *The hydrogen bond and the water molecule*, Elsevier, Oxford, 2007.
8 A. K. Soper, *Pure Appl. Chem.*, 2010, **82**, 1855–1867.
9 G. N. I. Clark, C. D. Cappa, J. D. Smith, R. J. Saykally and T. Head-Gordon, *Mol. Phys.*, 2010, **108**, 1415–1433.
10 I. M. Svishchev and P. G. Kusalik, *Chem. Phys. Lett.*, 1993, **215**, 596–600.
11 P. G. Kusalik and I. M. Svishchev, *Science*, 1994, **265**, 1219–1221.
12 J. D. Smith, C. D. Cappa, K. R. Wilson, R. C. Cohen, P. L. Geissler and R. J. Saykally, *Proc. Natl. Acad. Sci. U. S. A.*, 2005, **102**, 14171–14174.
13 S. Dixit, J. Crain, W. C. K. Poon, J. L. Finney and A. K. Soper, *Nature*, 2002, **416**, 829–832.
14 W. K. Röntgen, *Ann. Phys.*, 1892, **281**, 91–97.
15 D. Eisenberg and W. Kauzmann, *The structure and properties of water*, Oxford University Press, Oxford, 1969.
16 R. Ludwig, *Angew. Chem., Int. Ed.*, 2001, **40**, 1808–1827.
17 G. Malenkov, *J. Phys.: Condens. Matter*, 2009, **21**, 283101.
18 H. E. Stanley and J. Teixeira, *J. Chem. Phys.*, 1980, **73**, 3404–3422.
19 C. Huang, K. T. Wikfeldt, T. Tokushima, D. Nordlund, Y. Harada, U. Bergmann, M. Niebuhr, T. M. Weiss, Y. Horikawa, M. Leetmaa, M. P. Ljungberg, O. Takahashi, A. Lenz, L. Ojamae, A. P. Lyubartsev, S. Shin, L. G. M. Pettersson and A. Nilsson, *Proc. Natl. Acad. Sci. U. S. A.*, 2009, **106**, 15214–15218.
20 J. D. Eaves, J. J. Loparo, C. J. Fecko, S. T. Roberts, A. Tokmakoff and P. L. Geissler, *Proc. Natl. Acad. Sci. U. S. A.*, 2005, **102**, 13019–13022.
21 P. A. Giguère, *J. Raman Spectrosc.*, 1984, **15**, 354–359.
22 P. A. Giguère, *J. Chem. Phys.*, 1987, **87**, 4835–4839.
23 F. Sciortino, A. Geiger and H. E. Stanley, *Phys. Rev. Lett.*, 1990, **65**, 3452–3455.
24 F. Sciortino, A. Geiger and H. E. Stanley, *Nature*, 1991, **354**, 218–221.
25 F. Sciortino, A. Geiger and H. E. Stanley, *J. Chem. Phys.*, 1992, **96**, 3857–3865.
26 A. Geiger, M. Kleene, D. Paschek and A. Rehtanz, *J. Mol. Liq.*, 2003, **106**, 131–146.
27 N. Grishina and V. Buch, *Chem. Phys. Lett.*, 2003, **379**, 418–426.
28 D. Laage and J. T. Hynes, *Science*, 2006, **311**, 832–835.
29 D. Laage and J. T. Hynes, *J. Phys. Chem. B*, 2008, **112**, 14230–14242.
30 O. Markovitch and N. Agmon, *Mol. Phys.*, 2008, **106**, 485–495.
31 R. H. Henchman and S. J. Irudayam, *J. Phys. Chem. B*, 2010, **114**, 16792–16810.
32 G. Nemethy and H. A. Scheraga, *J. Chem. Phys.*, 1962, **36**, 3382–3400.
33 I. Ohmine, *J. Phys. Chem.*, 1995, **99**, 6767–6776.
34 A. Botti, F. Bruni, M. A. Ricci and A. K. Soper, *J. Chem. Phys.*, 1998, **109**, 3180–3184.
35 M. Bernabei, A. Botti, F. Bruni, M. A. Ricci and A. K. Soper, *Phys. Rev. E*, 2008, **78**, 021505.
36 M. Leetmaa, K. T. Wikfeldt, M. P. Ljungberg, M. Odelius, J. Swenson, A. Nilsson and L. G. M. Pettersson, *J. Chem. Phys.*, 2008, **129**, 084502.
37 K. T. Wikfeldt, M. Leetmaa, M. P. Ljungberg, A. Nilsson and L. G. M. Pettersson, *J. Phys. Chem. B*, 2009, **113**, 6246–6255.
38 H. S. Frank and M. W. Evans, *J. Chem. Phys.*, 1945, **13**, 507–532.
39 J. C. Owicki and H. A. Scheraga, *J. Am. Chem. Soc.*, 1977, **99**, 7403–7412.
40 S. Swaminathan, S. W. Harrison and D. L. Beveridge, *J. Am. Chem. Soc.*, 1978, **100**, 5705–5712.
41 S. J. Irudayam and R. H. Henchman, *J. Phys.: Condens. Matter*, 2010, **22**, 284108.
42 J. G. Davis, K. P. Gierszal, P. Wang and D. Ben-Amotz, *Nature*, 2012, **491**, 582–585.
43 N. Galamba, *J. Phys. Chem. B*, 2013, **117**, 589–601.
44 G. R. Desiraju, *J. Chem. Soc., Chem. Commun.*, 1991, 426–428.
45 L. Infantes, L. Fabian and W. D. S. Motherwell, *CrystEngComm*, 2007, **9**, 65–71.
46 S. J. Irudayam and R. H. Henchman, *J. Chem. Phys.*, 2012, **137**, 034508.
47 P. Ball, *Chem. Rev.*, 2008, **108**, 74–108.
48 Y. Marcus, *Chem. Rev.*, 2009, **109**, 1346–1370.
49 H. S. Frank and W. Y. Wen, *Discuss. Faraday Soc.*, 1957, **24**, 133–140.
50 M. Falk and T. A. Ford, *Can. J. Chem.*, 1966, **44**, 1699–1707.

51 P. Wernet, D. Nordlund, U. Bergmann, M. Cavalleri, M. Odelius, H. Ogasawara, L. A. Naslund, T. K. Hirsch, L. Ojamae, P. Glatzel, L. G. M. Pettersson and A. Nilsson, *Science*, 2004, **304**, 995–999.
52 S. T. Roberts, K. Ramasesha and A. Tokmakoff, *Acc. Chem. Res.*, 2009, **42**, 1239–1249.
53 T. D. Kuhne and R. Z. Khaliulin, *Nat. Commun.*, 2013, **4**, 1450–1456.
54 A. Nilsson and L. G. M. Pettersson, *Chem. Phys.*, 2011, **389**, 1–34.
55 N. Agmon, *Acc. Chem. Res.*, 2012, **45**, 63–73.
56 A. C. Belch, S. A. Rice and M. G. Sceats, *Chem. Phys. Lett.*, 1981, **77**, 455–459.
57 M. Mezei and D. L. Beveridge, *J. Chem. Phys.*, 1981, **74**, 622–632.
58 V. Buch, *J. Chem. Phys.*, 1992, **96**, 3814–3823.
59 A. Luzar and D. Chandler, *Phys. Rev. Lett.*, 1996, **76**, 928–931.
60 R. Kumar, J. R. Schmidt and J. L. Skinner, *J. Chem. Phys.*, 2007, **126**, 204107.
61 A. Rahman and F. H. Stillinger, *J. Chem. Phys.*, 1971, **55**, 3336–3359.
62 P. J. Rossky and M. Karplus, *J. Am. Chem. Soc.*, 1979, **101**, 1913–1937.
63 W. L. Jorgensen, *Chem. Phys. Lett.*, 1980, **70**, 326–329.
64 D. C. Rapaport, *Mol. Phys.*, 1983, **50**, 1151–1162.
65 G. Malenkov, D. Tytik and E. Zheligovskaya, *J. Mol. Liq.*, 1999, **82**, 27–38.
66 H. F. Xu, H. A. Stern and B. J. Berne, *J. Phys. Chem. B*, 2002, **106**, 2054–2060.
67 L. Pauling, *The nature of the chemical bond*, Cornell, Ithaca, 3rd ed., 1960.
68 Q. Du, E. Freysz and Y. R. Shen, *Science*, 1994, **264**, 826–828.
69 G. Desiraju, *Angew. Chem., Int. Ed.*, 2011, **50**, 52–59.
70 F. H. Stillinger and T. A. Weber, *Phys. Rev. A*, 1982, **25**, 978–989.
71 A. D. Hammerich and V. Buch, *J. Chem. Phys.*, 2008, **128**, 111101.
72 R. F. W. Bader, *Chem. Rev.*, 1991, **91**, 893–928.
73 Z. H. Ma, Y. L. Zhang and M. E. Tuckerman, *J. Chem. Phys.*, 2012, **137**, 044506.
74 E. Arunan, G. R. Desiraju, R. A. Klein, J. Sadlej, S. Scheiner, I. Alkorta, D. C. Clary, R. H. Crabtree, J. J. Dannenberg, P. Hobza, H. G. Kjaergaard, A. C. Legon, B. Mennucci and D. J. Nesbitt, *Pure Appl. Chem.*, 2011, **83**, 1637–1641.
75 K. Morokuma and L. Pedersen, *J. Chem. Phys.*, 1968, **48**, 3275–3282.
76 P. A. Kollman and L. C. Allen, *J. Chem. Phys.*, 1969, **51**, 3286–3293.
77 T. R. Dyke, K. M. Mack and J. S. Muenter, *J. Chem. Phys.*, 1977, **66**, 498–510.
78 F. N. Keutsch and R. J. Saykally, *Proc. Natl. Acad. Sci. U. S. A.*, 2001, **98**, 10533–10540.
79 N. Bjerrum, *Science*, 1952, **115**, 385–390.
80 R. Podeszwa and V. Buch, *Phys. Rev. Lett.*, 1999, **83**, 4570–4573.
81 B. Mukherjee, P. K. Maiti, C. Dasgupta and A. K. Sood, *J. Phys. Chem. B*, 2009, **113**, 10322–10330.
82 J. J. Loparo, C. J. Fecko, J. D. Eaves, S. T. Roberts and A. Tokmakoff, *Phys. Rev. B*, 2004, **70**, 180201.
83 A. Rahman and F. H. Stillinger, *J. Am. Chem. Soc.*, 1973, **95**, 7943–7948.
84 A. C. Belch and S. A. Rice, *J. Chem. Phys.*, 1987, **86**, 5676–5682.
85 P. E. Mason and J. W. Brady, *J. Phys. Chem. B*, 2007, **111**, 5669–5679.
86 M. Falk and O. Knop, in *Water: A Comprehensive Treatise*, ed. F. Franks, Plenum Press, 1973, pp. 55–113.
87 G. A. Jeffrey and W. Saenger, *Hydrogen bonding in biological structures*, Springer, Berlin, 1991.
88 I. Olovsson and P. Jonsson, in *The hydrogen bond. Recent developments in theory and experiments*, ed. P. Schuster, G. Zandel, and C. Sandorfy, North Holland, 1976, pp. 393–456.
89 G. F. Diercksen, *Theor. Chim. Acta*, 1971, **21**, 335–367.
90 K. Hermansson, *The electron distribution in the bound water molecule*, PhD thesis, University of Uppsala, Sweden, 1984.
91 J. L. Finney, J. E. Quinn, and J. O. Baum, in *Water Science Reviews*, ed. F. Franks, Cambridge University Press, 1985, pp. 93–170.
92 J. L. Finney, *Philos. Trans. R. Soc. London, Ser. B*, 2004, **359**, 1145–1163.
93 N. Agmon, *Chem. Phys. Lett.*, 1995, **244**, 456–462.
94 M. E. Tuckerman, D. Marx and M. Parrinello, *Nature*, 2002, **417**, 925–929.
95 M. Goswami and E. Arunan, *Phys. Chem. Chem. Phys.*, 2009, **11**, 8974–8983.
96 G. Desiraju and T. Steiner, *The weak hydrogen bond*, Oxford University Press, Oxford, 1999.
97 J. L. F. Abascal and C. Vega, *J. Chem. Phys.*, 2005, **123**, 234505.
98 D. A. Case, T. A. Darden, T. E. Cheatham III, C. I. Simmerling, J. Wang, R. E. Duke, R. Luo, K. M. Merz, D. A. Pearlman, M. Crowley, R. C. Walker, W. Zhang, B. Wang, S. Hayik, A. Roitberg, G. Seabra, K. F. Wong, F. Paesani, X. Wu, S. Brozell, V. Tsui, H. Gohlke, L. Yang, C. Tan, J. Mongan, V. Hornak, G. Cui, P. Beroza, H. Mathews,

D. C. Schafmeister, W. S. Ross, and P. A. Kollman, *AMBER 12*, University of California, San Francisco, 2012.
99 S. J. Irudayam, R. D. Plumb and R. H. Henchman, *Faraday Discuss.*, 2010, **145**, 467–485.
100 S. J. Irudayam and R. H. Henchman, *Mol. Phys.*, 2011, **109**, 37–48.
101 V. Chihaia, S. Adams and W. F. Kuhs, *Chem. Phys.*, 2005, **317**, 208–225.
102 J. R. Errington and P. G. Debenedetti, *Nature*, 2001, **409**, 318–321.
103 P. L. Chau and A. J. Hardwick, *Mol. Phys.*, 1998, **93**, 511–518.
104 R. Vacha, O. Marsalek, A. P. Willard, D. J. Bonthuis, R. R. Netz and P. Jungwirth, *J. Phys. Chem. Lett.*, 2012, **3**, 107–111.

Faraday Discussions RSC Publishing

PAPER

The study of correlations between hydrogen bonding characteristics in liquid, sub- and supercritical methanol. Molecular dynamics simulations and Raman spectroscopy analysis†

Abdenacer Idrissi,*[a] Roman D. Oparin,[b] Sergey P. Krishtal,[b] Sergey V. Krupin,[b] Evgeny A. Vorobiev,[b] Andrey I. Frolov,[b] Leo Dubois[a] and Mikhail G. Kiselev*[b]

Received 3rd July 2013, Accepted 24th July 2013
DOI: 10.1039/c3fd00103b

Molecular dynamics (MD) studies of hydrogen bonding (H-bonding) in liquid, sub- and supercritical methanol have been performed in a wide range of thermodynamic parameters of state, using various potential models and two H-bond criteria. It was shown that there is the universal correlation between the average number of H-bonds per molecule (n_{HB}) and the mole fraction of H-bonded molecules (X_{HB}) for the studied thermodynamic parameters of state. The same feature was observed for the correlations between fractions of molecules forming one (f_1), two (f_2), three (f_3) H-bonds and $X_{HB.}$ These correlations served to fit experimental Raman spectra of methanol recorded under sub- and supercritical conditions. The advantage of the approach used here is that f_1, f_2, f_3 values have a clear physical meaning and are dependent on the values of state parameters.

Introduction

Structure and properties of aliphatic alcohols in a supercritical state are a subject of continuous scientific interest from both fundamental and applied points of view. The latter implies extensive use of simple supercritical alcohols in the chemical industry, for example, as solvents and cosolvents in chemical synthesis (biodiesel production, metal nanoparticle synthesis), supercritical fluid extraction, and so on.[1–5]

One of the most challenging research prospects in this area is related to studying hydrogen-bonded (H-bonded) interactions. It is ubiquitously acknowledged nowadays that H-bonding still persists in sub- and supercritical alcohols,

[a]Laboratoire de Spectrochimie Infrarouge et Raman (UMR CNRS A8516), Université des Sciences et Technologies de Lille, 59655 Villeneuve d'Ascq Cedex, France

[b]Institute of Solution Chemistry of the RAS, Akademicheskaya st.1, 153045, Ivanovo, Russia

† Electronic supplementary information (ESI) available. See DOI: 10.1039/c3fd00103b

the number of H-bonds decreasing with an increase of temperature or with a decrease of density. Furthermore, many experimental[6–8] and theoretical[9–11] works revealed some universal signatures of H-bonding in aliphatic alcohols and water in a wide range of state parameters. NMR studies of chemical shift in supercritical water, methanol and ethanol by Hoffman and Conradi demonstrated linear behavior of the extent of H-bonding (η) as a function of the chemical shift of the hydroxyl group.[6] IR studies of liquid and supercritical ethanol, 2-propanol, and 1-butanol by Barlow *et al.*[7] showed that the correlations between the coefficient of integrated absorbance (A_b), the position of OH vibrational mode ($\bar{\nu}_{OH}$) and the mole fraction of H-bonded molecules (X_{HB}) can be approximated by a function with a bend at $X_{HB}^{max} \sim 0.6$–0.7. It was concluded that only one type of H-bonded species (presumably dimers) exists in these alcohols at $X_{HB} < X_{HB}^{max}$, while larger H-bonded oligomers start to form at $X_{HB} > X_{HB}^{max}$. Barlow *et al.* proposed that this "seems to be a general rule for all alkanols" and made a premise that such a threshold can be found in other H-bonded systems. Extensive molecular dynamics simulations (MD) studies of sub- and supercritical methanol by Krishtal *et al.*[9–11] confirmed some of the findings of Barlow *et al.* In particular, it was shown that the correlation between the peak position in the H-bond energy distribution and X_{HB} for methanol can be approximated by a function with a sharp bend at $X_{HB} = 0.7$–0.8. This was found to be associated with the destruction of H-bonded trimers and higher oligomers. Furthermore, MD studies by Krishtal *et al.* revealed the universal character of the correlation between the average number of H-bonds per methanol molecule (n_{HB}) and X_{HB}. Such universality has also been observed in the case of correlations between the mole fractions of molecules forming one (f_1) and two (f_2) H-bonds and X_{HB}. Importantly, many similarities in the behavior of H-bonding in sub- and supercritical water and methanol have also been highlighted.[11] For example, it was shown that n_{HB} temperature dependences of both fluids behave in a very similar way and that the correlations between n_{HB}, f_1, f_2 and X_{HB} are very similar at $X_{HB} < 0.8$ for water and methanol. Recently, combined room temperature X-ray Raman scattering (XRS)/classical MD–DFT studies of linear (methanol to butanol) and branched (isopropanol, isobutanol, 2-butanol) alcohols by Pylkkänen *et al.*[8] showed that H-bonding produces a clear spectral signature that is nearly identical in the oxygen K-edge spectrum of these alcohols. While the cited experimental and theoretical works brought significant evidence for similarities in H-bonding of liquid and supercritical aliphatic alcohols, systematic studies are yet to be carried out in order to draw a holistic picture of the observed phenomena. In particular, the universal character of the correlations between n_{HB}, f_1, f_2 and X_{HB} in methanol was studied only with a rigid non-polarisable methanol model, so its dependence on a chosen interaction potential was not studied until now. It remained also unverified as to whether these correlations would work in the same way for other alcohols.

The first aim of the present work was to investigate by means of computer simulations the dependence of the correlations between n_{HB}, f_1, f_2, f_3 (fraction of molecules having three H-bonds) and X_{HB} on methanol interaction potential models, including polarisable ones, as well as to clarify whether these universal correlations can be affected by the choice of H-bond criterion. To acknowledge this, MD simulations of methanol were performed along different isochors, isobars and isotherms, using various potential models and two H-bond criteria. The second aim was to show how the discussed correlations can be used to find the various H-bonded methanol molecules spectral contributions to the OH vibration mode of the Raman spectra. Finally, MD simulations of liquid, sub- and supercritical ethanol and 1-propanol with polarisable and non-polarisable models were performed along selected phase lines to verify if it is possible to extend the correlations between n_{HB},f_1,f_2,f_3 and X_{HB} observed in methanol to the case of other alcohols.

The paper is organized as follows. In the next section, we describe the computational details of our computer simulations. It is followed by the discussion of the correlations between n_{HB}, f_1, f_2, f_3 and X_{HB} for the case of different methanol phase lines/potential models/H-bond criteria. Further, we present the results of the Raman scattering experiments on sub- and supercritical methanol, where we show how the values of f_1,f_2,f_3 can help to analyze the Raman spectra in terms of H-bonding spectral contributions. The last section, augmented with the results of MD simulations of liquid and supercritical ethanol and 1-propanol, expresses our views regarding the existence of similar correlations in other aliphatic alcohols.

Computational details

We performed MD simulations for methanol, ethanol and 1-propanol under different thermodynamic conditions in liquid, sub- and supercritical states (see Table 1). To reveal how the choice of the potential model affects the simulation results, we used different potential models: H1,[12] Charmm22,[13] as well as a polarisable model by Anisimov *et al.*[14] (see the details on the potential models in the ESI†).

Simulations of methanol, ethanol and 1-propanol along the isobar 110 bar with the Charmm22 model[13] and polarisable model of Anisimov *et al.*[14] were performed in the NAMD simulation package.[15] For these simulations we used the Brunger–Brooks–Karplus (BBK) integrator[16] with a time step of 1 fs. We applied the Langevin thermostat for all non-hydrogen atoms with damping coefficient of 5 ps^{-1} in the case of the Charmm22 model and for all atoms with a damping coefficient of 1 ps^{-1} in the case of the polarisable model (a separate Langevin thermostat with a reference temperature of 1 K was applied for the Drude

Table 1 Thermodynamic state parameters for the studied alcohols as well as the potential models used in the simulations

	Compound	Potential model	Pressure range (bar)	Temperature range (K)
Isochor 650 kg m^{-3}	Methanol	H1	—	400–600
Isobar 150 bar	Methanol	H1	—	400–600
Isotherm 516 K	Methanol	H1	100–700	—
Isobar 110 bar (performed with NAMD)	Methanol, ethanol, 1-propanol	Charmm22 and polarisable model of Anisimov *et al.*	—	300–650
Isobar 110 bar (performed with CHARMM)	methanol, ethanol	Charmm22	—	300–620 (for methanol), 300–590 (for ethanol)

particles subsystem). We applied the Langevin piston barostat with an oscillation timescale of 200 ps and a damping timescale of 100 ps.[17] All bonds were constrained by the SHAKE method[18] with a tolerance of 10^{-8}. The electrostatic interactions were evaluated with the particle mesh Ewald method.[19] A neighbor list with 12.0 Å and 13.5 Å cutoff distances updated every ten and twenty steps was used for the short-range interactions in the simulations with the Charmm22 model and polarisable model, respectively. The "shifted-force" method[20] with the real space cut-off of 10.0 Å (in the case of the Charmm22 model) and 12.0 Å (in the case of the polarisable model) were used for the short-range screened Coulomb potential in real space. The switching function was applied to the Lennard-Jones interaction potential in the range of 8.0–10.0 Å (in the case of the Charmm22 model) and 10.5–12.0 Å (in the case of the polarisable model). The cubic simulation cell contained 128 molecules (we also performed several simulations with 1000 molecules, all of them reproduced the results of simulations with 128 molecules). Periodic boundary conditions were applied. The simulation time of the production run was 400 ps, coordinate frames were saved each 1 ps for further analysis.

Simulations of methanol with the H1 model (isotherm 516 K, isobar 150 bar, isochor 650 kg m^{-3}) were performed using MODYS software.[21] For these simulations we used the Verlet method to integrate the equations of motion, with a time step of 2 fs. All bonds were constrained by the SHAKE method with a tolerance of 10^{-8}. The cubic simulation cell contained 256–512 molecules. Periodic boundary conditions were applied. The NVT ensemble was used for all thermodynamic states. The size of the cell in the case of simulations under isobaric conditions was chosen in such a way that the density of methanol for the given number of molecules corresponded to experimental data taken from ref. 22. We applied the Berendsen thermostat. The minimum image convention was obeyed to calculate interactions in the system. The simulation time of production runs was 200–500 ps each. The time interval for saving coordinate frames was 10 fs. Further details on simulations with the H1 model can be found elsewhere.[9–11]

Simulations of methanol and ethanol on the 110 bar isobar with the Charmm22 force field[13] were performed in the CHARMM simulation package.[13] For these simulations we used Anderson–Hoover equations for constant pressure and temperature simulations with an integration timestep of 1 fs[23] (see ref. 24 for the implementation details). The characteristic response times for the thermostat and barostat were 0.1 and 2.0 ps, correspondingly. The electrostatic interactions were evaluated with the smooth particle mesh Ewald method,[25] with a real space cut-off of 10.0 Å and spline order of 6. The "shifted-force" and "shifted-potential" methods[26] with a real space cut-off of 10.0 Å were used for the short-range screened Coulomb potential in real space and Lennard-Jones interaction potential, correspondingly. The cubic simulation cell contained 500 molecules. Periodic boundary conditions were applied. The simulation time of the production run was 3 ns, coordinate frames were saved each 1 ps for further analysis.

H-bond criteria

In order to investigate the effect of various H-bond definitions on the results of H-bonding analysis, we utilized two different criteria to gather H-bond statistics from computer simulations. The first one is a hybrid geometry–energy criterion, according to which two aliphatic alcohol molecules are considered H-bonded if intermolecular distance between O and H ($r_{O\cdots H}$) is less than 2.6 Å and the interaction energy of the two molecules (E_{HB}) is less than $-3k_BT$, where k_B is the Boltzmann constant. The accuracy of this criterion was confirmed by combined classical MD–*ab initio* studies[9–11]of sub- and supercritical methanol, in which good agreement between calculated and experimental n_{HB} values was also shown. The second one is a purely geometric criterion with the following threshold values: $r_{O\cdots O} \leq 3.5$ Å and $\angle OHO \geq 150°$.[27] While both criteria give the same qualitative results associated with the dependence of the degree of H-bonding on the thermodynamic conditions, they produce somewhat different absolute values of n_{HB} and related properties. Hybrid criterion was applied in simulations of methanol with H1 and Charmm22 models, and the geometric one in simulations of methanol, ethanol, and 1-propanol with the Charmm22 model and the polarisable one.

Results and discussion

a. Correlations between n_{HB}, f_1, f_2, f_3 and X_{HB} for methanol

Analyzing the results of our simulations carried out at various thermodynamic conditions (see Table 1) and for various potential models of methanol, we found that the family of n_{HB} *versus* X_{HB} curves obtained along different phase lines fall into the same curve (Fig. 1), confirming the universal character of this correlation previously observed in MD simulations with the H1 methanol model.[9–11] Moreover, the same phenomenon also manifests itself in the correlations between fractions of molecules forming one, two and three H-bonds and X_{HB} (Fig. 2). We would like to stress that our data thoroughly reveal the known features of the methanol structure in a wide range of state parameters.[6 28 29] At low X_{HB} values ($X_{HB} < 0.2$, this corresponds to high temperature and low density) there are almost no molecules forming more than one H-bond in the system, meaning only

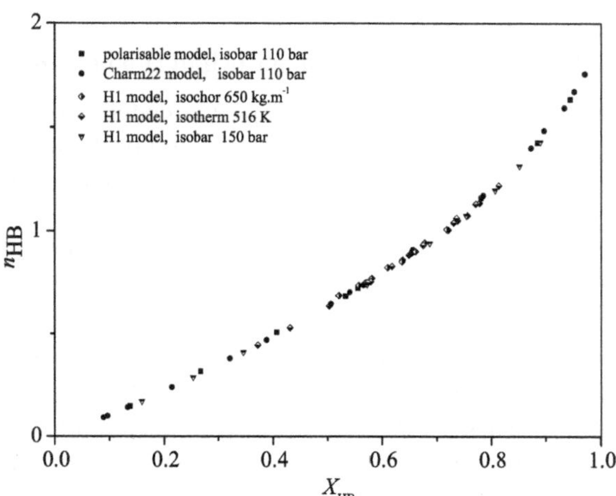

Fig. 1 The correlation between the average number of H-bonds per molecule (n_{HB}) and the mole fraction of H-bonded molecules (X_{HB}) in methanol at various thermodynamic conditions and for various potential models.

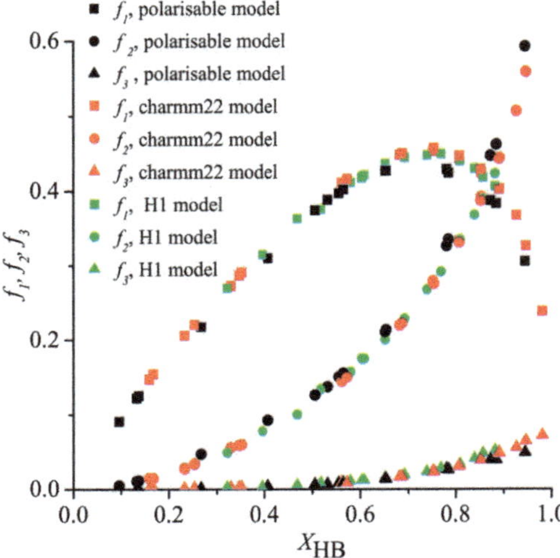

Fig. 2 The correlations between fractions of molecules forming one (f_1), two (f_2), three (f_3) H-bonds and the mole fraction of H-bonded molecules (X_{HB}) in methanol at various thermodynamic conditions and for various potential models.

monomers and dimers are present (see Fig. 2). At moderate X_{HB} values (this corresponds to low temperature and high density) we observe that there is a large number of molecules forming two H-bonds, which means that the H-bonded molecules arrange in chains. Moreover, we note that there is a small percentage of molecules forming three H-bonds which monotonically increases with the increase of X_{HB} and becomes as large as 5–8% at $X_{HB} > 0.9$. Obviously, the non-negligible values of f_3 show that there are branched structures in the system. From the correlations shown in Fig. 1 and 2 it follows that once X_{HB} is estimated at given state parameters, one can calculate n_{HB}, f_1, f_2, f_3, using the universal expressions relating them with X_{HB}. One could consider X_{HB} as a collective variable, which uniquely determines the discussed characteristics of H-bonding in methanol.

b. Analysis of the distance threshold in our hybrid H-bond criterion

The correlations between n_{HB}, f_1, f_2, f_3 and X_{HB} are related to the changes of the average values of the distance, of the orientation and of the interaction energy between two methanol molecules due to changes in thermodynamic conditions. These parameters are used in the criteria of H-bond formation between two methanol molecules and their threshold values have a decisive role in describing the H-bonding and thus in determining the values of n_{HB}, f_1, f_2, f_3 and X_{HB}. The purpose of this part of the paper is to investigate in detail the dependence of the average distance between two atoms involved in H-bonding on the thermodynamic conditions. To access this information one needs to know the probability density of finding the nearest neighbor at a given distance from a reference atom. This information is of fundamental importance as it allows, using the H-bond criteria, to find the first nearest neighbor molecules that are H-bonded to reference ones. Therefore, we calculated the nearest neighbor radial distribution functions (nnrdf). This approach has already been shown to give insight into the local structure of hard sphere systems,[30–32] of supercooled liquids,[33,34] of supercritical CO_2[35,36] and of binary mixtures.[37,38] In this approach, the neighbors of a central atom are sorted by distance into the first neighbors, second neighbors, *etc.* The nearest neighbor radial distribution functions, $p_{a\cdots b}(r, n)$, can be defined then for each set of nearest neighbor atoms b (n is the order of the nearest neighbor) with respect to reference atoms a. The sum of $p_{a\cdots b}(r, n)$ is equal to the corresponding total radial distribution function, $g_{a\cdots b}(r)$. The parameter chosen for the analysis of these functions is the average distance $<r_{a\cdots b}(n)>$ between a reference atom a and an atom b being the nth neighbor atom. The radial distribution functions of reference oxygen atoms of methanol molecules and the first ($p_{O\cdots H}(r, n = 1)$) and the second ($p_{O\cdots H}(r, n = 2)$) nearest hydrogen atoms at various temperatures T are plotted in Fig. 3a and Fig. 4, respectively. These functions were calculated along the isobar 110 bar and in the temperature range between 300 K and 620 K, using the Charmm22 potential model of methanol. The average distance ($<r_{O\cdots H}(n = 1)>$) between reference oxygen atoms and their first nearest hydrogen atoms as a function of temperature calculated using these distributions is given in the inset of Fig. 3a. The first peak in ($p_{O\cdots H}(r, n = 1)$) is interpreted as a signature of H-bond formation. Increasing the temperature in the range between 300 K and 530 K reduces the intensity of this peak. Concomitantly, the average distance increases, but remains lower than 2.6 Å, which is the distance threshold of the hybrid H-bond criterion used in our simulations. For higher temperatures between 530 K and 620 K, the average distance $<r_{O\cdots H}(n = 1)>$ is higher than 2.6 Å. In order to estimate the fraction of the first nearest hydrogen atoms which can be involved in H-bonding with reference oxygen atoms, we calculated the area under $p_{O\cdots H}(r < 2.6, n = 1)$ for distances up to 2.6 Å. This fraction will be referred to as X_{HB}^{nnrdf}. The values of X_{HB}^{nnrdf} are shown in Fig. 3b. Inspection of this figure shows clearly that below 530 K, 95% to 67% of methanol molecules having hydrogen atoms classified as the first nearest neighbors to reference oxygen atoms have on average H-bond lengths within the distance threshold of 2.6 Å. This fraction decreases from 60% to 35% for the subsequent temperatures. These results show that as far as the $r_{O\cdots H}$ threshold value in our hybrid H-bond criterion is considered, the percentage of methanol molecules having hydrogen atoms defined as the first nearest neighbors and thus being candidates to form H-bonds with reference molecules is dependent on thermodynamic state parameters. The dependence of X_{HB}^{nnrdf} and X_{HB} on temperature shows similar behavior (see Fig. 3b). These data underline the importance of the $r_{O\cdots H}$ threshold value in the H-bond criteria for analyzing the H-bonding.

A methanol molecule can form two H-bonds as an acceptor which makes the formation of branched H-bonded structures possible. To analyze formation of branched H-bonded structures, we calculated the second nearest neighbor distribution functions $p_{O\cdots H}(r, n = 2)$ (see Fig. 4). Below $T = 530$ K, the first peak in $p_{O\cdots H}(r, n = 2)$ is located at distances below the threshold value of the hybrid H-bond criterion. This indicates that the second nearest hydrogen atoms within the threshold value can be involved in H-bonding with the reference oxygen atoms. This shows that some number of methanol molecules can form two H-bonds as acceptors and thus branched H-bonded structures should be present in the system. At higher temperatures, this peak is almost smeared out indicating that methanol molecules do not form the second H-bond as acceptors. These results show that branched H-bonded structures are mainly destroyed above 530 K.

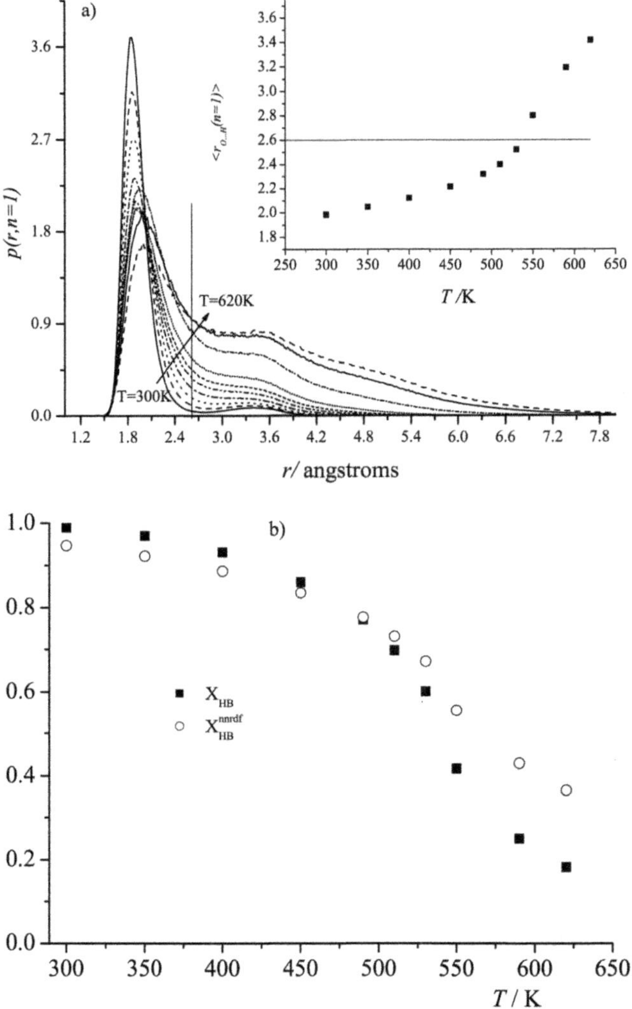

Fig. 3 (a) The first nearest neighbor radial distribution functions, $p_{O\cdots H}(r, n = 1)$, for temperatures $T =$ 300 K, 350 K, 400 K, 450 K, 490 K, 510 K, 530 K, 550 K, 590 K, 620 K along the 110 bar isobar. The vertical full line illustrates the distance threshold value of the hybrid H-bond criterion used in this study. The average distance ($<r_{O\cdots H}(n = 1)>$) as a function of temperature is given in the inset. (b) Fraction of the first nearest neighbors associated with the O···H distance criterion which can form H-bond to reference oxygen atoms (X_{HB}^{nnrdf}) and the mole fraction of H-bonded molecules (X_{HB}) as functions of temperature.

c. Analysis of the OH band of the Raman spectra

In this part of our paper we want to show how our findings on the correlations between f_1, f_2, f_3 and X_{HB} discussed in section a) can help to analyze the Raman spectra of methanol. It should be mentioned that usually in the analysis of Raman or IR spectra of alcohols, the first estimates of spectral contribution parameters (such as peak positions and activities) assigned to H-bonded species are obtained from quantum calculations.[39] During the fitting procedure of the OH band, the peak positions are being improved iteratively to get good reproducibility of the experimental spectral band, while the activity parameters are kept constant.

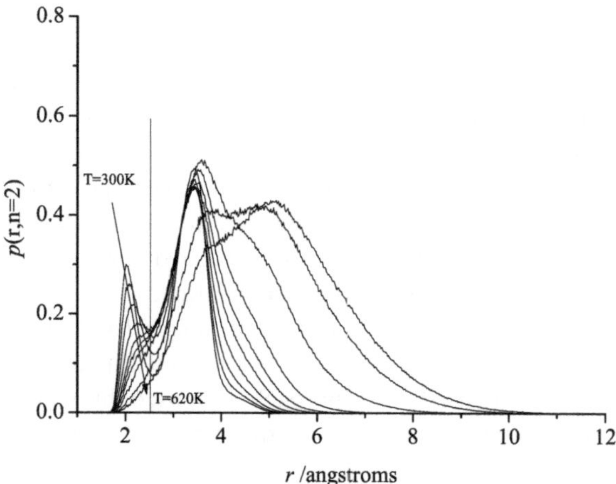

Fig. 4 The second nearest neighbor radial distribution functions ($p_{O\cdots H}(r,\ n = 2)$) for different temperatures (see legend of Fig. 3a).

The initial guess data *e.g.* peak positions, can be also obtained from additional experiments as shown for example in ref. 40. In our case we use f_1, f_2, f_3 values obtained from the correlations discussed in section a) instead of activities as the weighting parameters of the H-bonded molecules spectral contributions (approximated by Gaussian profiles) to the Raman OH band.

Details of the experiment. In order to analyze the H-bonding in methanol, we measured the Raman spectra for a wide range of thermodynamic parameters similar to those used in our simulations. The spectra were recorded using a LabRam Jobin Yvon spectrometer. The setup is based on a probe called "power head".[41] The spectroscopic cell, specially designed for high pressures and high temperatures and equipped with sapphire optical windows was used. Heating the cell by means of four cartridge heaters is implemented and managed by a controller, which allows the desired temperature to be maintained with an accuracy of 0.5 °C. The pressure in the system was created by a manual press and controlled by piezo-sensors with an accuracy of 1 bar. Finally, the spectra were registered in a range of wave numbers 2500–3900 cm^{-1} with a resolution of 2 cm^{-1}.

H-bonding in methanol as studied by the combination of Raman spectroscopy and computer simulations. The presence of H-bonded clusters of different sizes and topologies in the fluids significantly complicates the analysis of the vibration spectra. Indeed, the spectral contributions that are associated with the presence of various H-bonded clusters are situated fairly close to each other such that the difference in their positions is considerably smaller compared to their dispersion. In IR spectroscopy, the contribution associated with free OH groups is not clearly resolved.[42] This makes the deconvolution of the OH band in IR spectra into contributions associated with the free OH groups and those associated with bonded OH groups very difficult to perform. Therefore, in this work we used Raman spectroscopy. On the one hand, Raman spectroscopy is less sensitive to the existence of H-bonds. On the other hand, the spectral band assigned to the free OH groups is resolved enough even under ambient conditions. In addition, utilization of Raman spectroscopy eliminates the need to use an extremely small optical path length, which is difficult to control using IR spectroscopy. Raman spectra of methanol were recorded along two isobars 83 bar and 110 bar, and in the temperature range between 313 K and 553 K. These spectra are shown in Fig. 5a and 5b. The spectral band at around 3658 cm^{-1} associated with the free OH molecules is quite pronounced for supercritical conditions, demonstrating the existence of monomers in the system. The intensity of this band smoothly decreases with the decrease of temperature and becomes almost negligible at around 400 K for both studied isobars. The other spectral band, which is associated with H-bonded methanol molecules, is blue-shifted with increase of temperature.

1) Estimation of the mole fraction of free OH molecules. The first task in the analysis of the Raman spectra was to estimate the mole fraction of free methanol molecules. For this, we used the approach proposed by Gorbaty *et al.*[7] We recorded the Raman spectra along two isotherms T = 553 K and T = 573 K at various pressures (see Fig. 6a and 6b). One can see from Fig. 7 that the area of the CH_3 spectral contribution is proportional to the molar density. Assuming that there is a linear correlation between these two parameters, we normalized all the

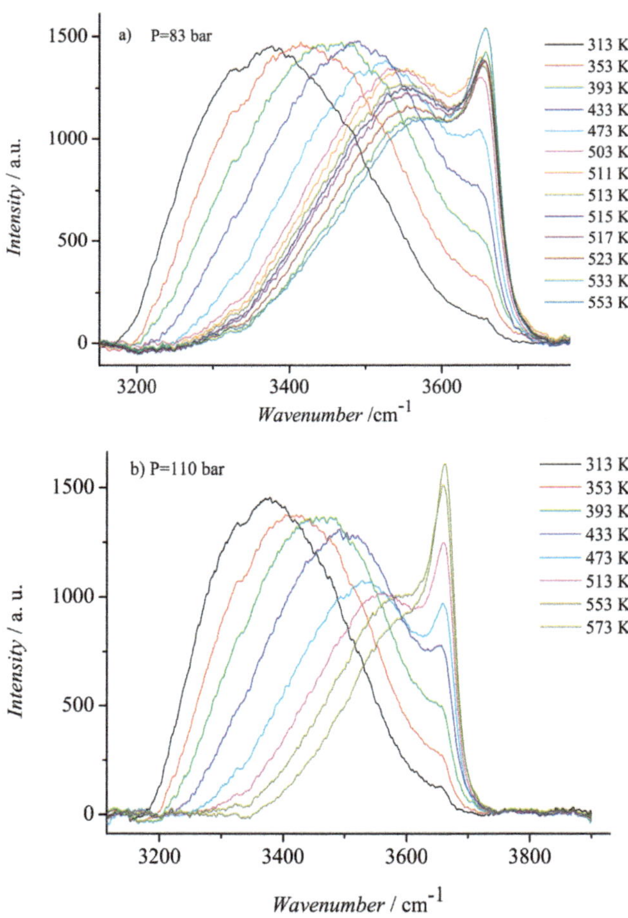

Fig. 5 OH vibration mode of Raman spectra of methanol along two isobars a) 83 bar and b) 100 bar in the temperature range between 313 K and 553 K.

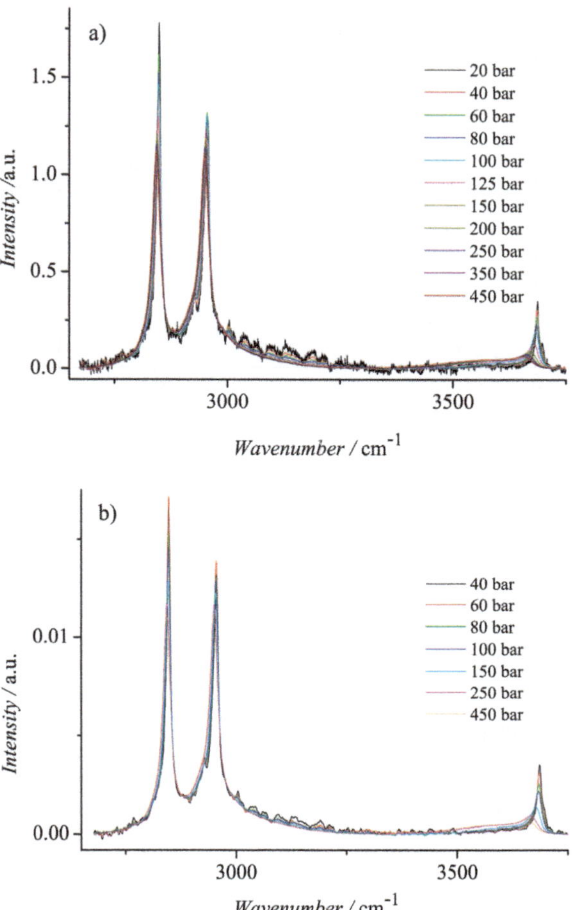

Fig. 6 Raman spectra of the OH vibration mode of methanol along the isotherms a) $T = 573$ K and b) $T = 553$ K and for various values of the pressure.

experimental spectra by the area of the CH_3 spectral contribution. The integral intensities associated with the contribution assigned to the free OH groups were calculated. Their behavior with respect to the pressure is presented in Fig. 8. Furthermore, the extrapolation of these values using an exponential decay to $P =$ 0 bar makes it possible to estimate for the two isotherms the "virtual" integral intensity (A_0) associated with free methanol molecules. The comparison between the two extrapolated values allowed us to estimate the error bars of A_0 to be $\pm$1.73%. Using this A_0 value we can estimate for each thermodynamic point the fraction of free molecules (X_{free}) by dividing the integral intensity of the profile associated with the free molecules (A_{free}) by A_0. Further, we can estimate X_{HB} as 1-X_{free}. This allows to calculate the f_1, f_2 and f_3 values from the correlations determined in our MD simulations.

2) Deconvolution of the spectral profile associated with H-bonded methanol molecules into spectral contributions associated with molecules forming one, two, and three H-bonds. We carried out a fitting procedure where the positions and the dispersions of the various spectral contributions associated with methanol

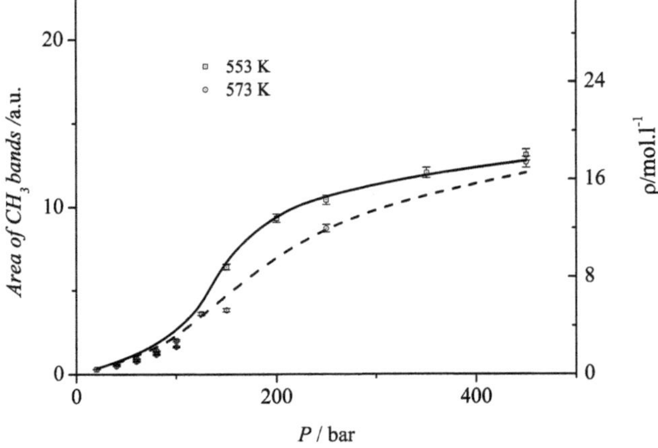

Fig. 7 Area of the CH_3 vibration mode intensity along the two isotherms 553 K and 573 K at various pressure values. The full and dashed lines represent the variation of the experimental densities as obtained from the NIST database.[43]

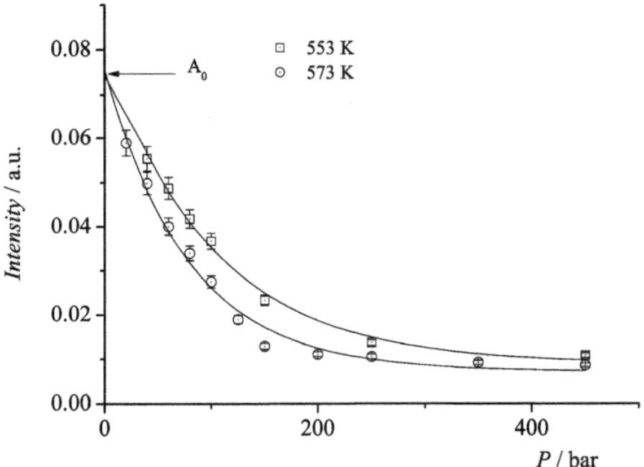

Fig. 8 Integral intensity values of the free OH groups spectral contributions as a function of pressure along the two isotherms 553 K and 573 K. The arrow in the figure indicates the "virtual" integral intensity of the free OH groups.

molecules forming one, two and three H-bonds were the only variable parameters. An example of fitting one can see in the ESI (Figure S1†). The peak positions (ν_{max}) of these spectral profiles are shown in Fig. 9 for both isobars. This figure shows that all the peak positions are blue-shifted with increasing the temperature. This fact validates the procedure used in this paper and particularly the weighting of the spectral contributions of molecules forming one, two, and three H-bonds. The advantage of the approach used here is that f_1, f_2, f_3 values have a clear physical meaning and are dependent on the values of state parameters.

d. Correlations between f_1, f_2, f_3 and X_{HB} in ethanol and 1-propanol

In this part of the paper, we report MD studies of the correlations between n_{HB}, f_1, f_2, f_3 and X_{HB} for ethanol and 1-propanol and compare the obtained results

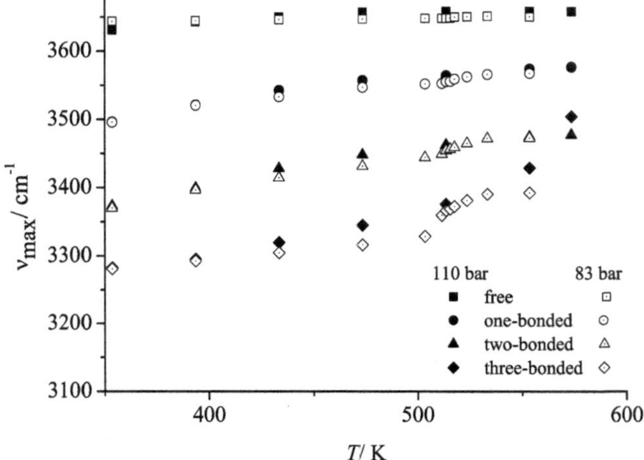

Fig. 9 The peaks positions as a function of temperature for free, one-bonded, two-bonded and three-bonded methanol molecules spectral contributions along the isobars 110 bar and 83 bar and in the temperature range between 350 K and 553 K.

with those for methanol shown in Fig. 1 and 2. The outcome is presented on Fig. 10 and 11.

As seen from Fig. 10, computer simulations confirm that the correlation between n_{HB} and X_{HB} seems to be a general feature of methanol, ethanol and 1-propanol. This work clearly shows that these alcohols exhibit much stronger similarities in H-bonding in a very broad range of thermodynamic conditions. This premise is in full agreement with conclusions drawn in IR studies by Barlow *et al.*[7] Fig. 11 shows the correlations between f_1, f_2, f_3 and X_{HB} for the considered alcohols. It can be seen that there is a similar trend in the behavior of f_1, f_2, f_3 at $X_{HB} < 0.6$ for all considered alcohols/potential models. At $X_{HB} > 0.6$, however, the clear distinctions can be observed, especially for the polarisable models of

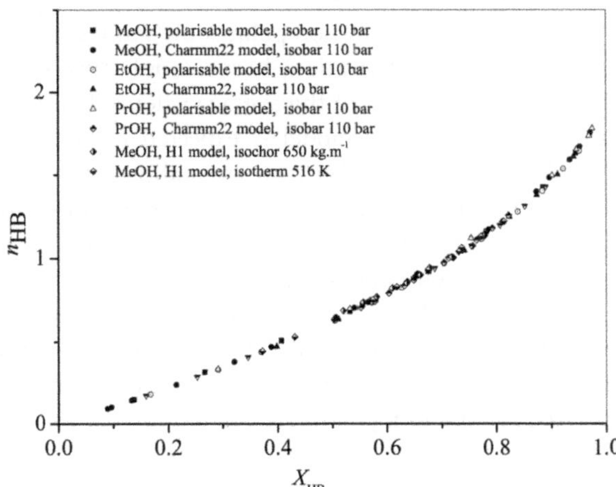

Fig. 10 The correlation between the average number of H-bonds per molecule (n_{HB}) and the mole fraction of H-bonded molecules (X_{HB}) in methanol, ethanol and 1-propanol at various thermodynamic conditions and for various potential models.

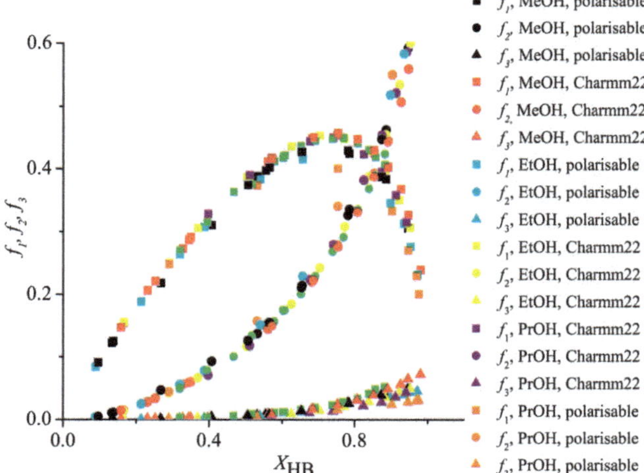

Fig. 11 The correlations between fractions of molecules forming one (f_1), two (f_2), three (f_3) H-bonds and the mole fraction of H-bonded molecules (X_{HB}) in methanol, ethanol and 1-propanol at various thermodynamic conditions and for various potential models.

ethanol and 1-propanol. Thus, it can be seen that at given X_{HB} values, the polarisable models of ethanol and propanol yield a remarkably lower number of one- and of three-bonded, but more two-bonded molecules than those obtained in the corresponding non-polarisable models. It should be also mentioned that differences for f_1 and f_2 values between polarisable and non-polarisable models of ethanol and 1-propanol become again much less pronounced at $X_{HB} > 0.9$, whereas those differences for f_3 values demonstrate opposite behavior. In the case of methanol, polarisable and non-polarisable models produce very similar results.

The differences in f_1, f_2, f_3 values between polarisable and non-polarisable models of alcohols at $X_{HB} > 0.6$ can be likely understood in terms of different polarisabilities of considered alcohols. In a dense fluid one expects significant polarisation of alcohol molecules. In non-polarisable models there is a definite lack in describing molecular polarization, because only "geometrical polarization" allowed by the model: change of the average valence bond length, valence angle, *etc.* In turn, the polarisable model is expected to give more details in describing molecular polarization, because in addition to the "geometrical polarization" it allows the effective electron density modeled by Drude particles to be polarized by surrounding molecules. Therefore, one can expect remarkable differences in f_1, f_2, f_3 between polarisable and non-polarisable models at high $X_{HB} > 0.6$. Notwithstanding the above differences in the fractions of bonded molecules, the mutual dependence of f_1, f_2, f_3 results in an identical (within a statistical error bar) n_{HB} *versus* X_{HB} curve for all considered alcohols and potential models, both polarisable and non-polarisable ones.

Conclusions

We have analyzed the correlations between the average number of H-bonds per molecule (n_{HB}), the fractions of molecules forming one (f_1), two (f_2), three (f_3)

H-bonds and the mole fraction of H-bonded molecules (X_{HB}). The main conclusions are as follows:

1) We showed for various potential models as well as for two H-bond criteria that the correlation between the average number of H-bonds per molecule (n_{HB}) and the mole fraction of H-bonded molecules (X_{HB}) is identical for all considered alcohols and thermodynamic states. Furthermore, in the case of methanol, we found that the correlations between f_1, f_2, f_3 and X_{HB} are universal as well, taking into account minor differences in the results from polarisable and non-polarisable models. This work clearly shows that small aliphatic alcohols exhibit much stronger similarities in H-bonding in a very broad range of thermodynamic conditions than was reported before.

2) We successfully decomposed the OH band in the Raman spectra of methanol at different thermodynamic parameters into contributions coming from molecules forming one, two, three H-bonds using as input parameters the values of f_1, f_2, f_3 from the discussed correlations. The advantage of the approach used here is that f_1, f_2, f_3 values have a clear physical meaning and are dependent on the values of state parameters. The observed correlations are also of particular importance because once the X_{HB} value for a specific phase state is obtained in a theoretical or experimental study, one can easily estimate n_{HB}, f_1, f_2, f_3 based on the discussed correlations.

Acknowledgements

This project was supported by the Marie Curie program IRSES (International Research Staff Exchange Scheme, GAN°247500). AIF is thankful to the support of the Russian Foundation for Basic Research (grant № 12-03-31354). The authors are thankful for the support of the Russian Foundation for Basic Research (grant № 11-03-00122-a). The Institut du Développement et des Ressources en Informatique Scientifique (IDRIS), the Centre de Ressources Informatiques (CRI) de l'Université de Lille, and the Centre de Ressource Informatique de Haute-Normandie (CRIHAN) are thankfully acknowledged for the CPU time allocation.

References

1 S. Rudyk, S. Hussain and P. Spirov, *J. Supercrit. Fluids*, 2013, **78**, 63–69.
2 P. Valle, A. Velez, P. Hegel, G. Mabe and E. A. Brignole, *J. Supercrit. Fluids*, 2010, **54**, 61–70.
3 C. Slostowski, S. Marre, O. Babot, T. Toupance and C. Aymonier, *Langmuir*, 2012, **28**, 16656–16663.
4 J. Rincón, R. Camarillo, L. Rodríguez and V. Ancillo, *Ind. Eng. Chem. Res.*, 2010, **49**, 2410–2418.
5 Ö. Güçlü-Üstündağ and F. Temelli, *J. Supercrit. Fluids*, 2005, **36**, 1–15.
6 M. M. Hoffmann and M. S. Conradi, *J. Phys. Chem. B*, 1998, **102**, 263–271.
7 S. J. Barlow, G. V. Bondarenko, Y. E. Gorbaty, T. Yamaguchi and M. Poliakoff, *J. Phys. Chem. A*, 2002, **106**, 10452–10460.
8 T. Pylkkänen, J. Lehtola, M. Hakala, A. Sakko, G. Monaco and K. Hämäläinen, *J. Phys. Chem. B*, 2010, **114**, 13076.
9 S. Krishtal, M. Kiselev, A. Kolker and A. Idrissi, *Theor. Chem. Acc.*, 2007, **117**, 297–304.
10 S. Krishtal and M. Kiselev, *Russ. J. Phys. Chem.*, 2003, **77**, 1817.
11 S. Krishtal, PhD Thesis, Institute of Solution Chemistry of Russian Academy of Sciences, 2004.
12 M. Haughney, M. Ferrario and I. R. McDonald, *J. Phys. Chem.*, 1987, **91**, 4934–4940.
13 B. R. Brooks, R. E. Bruccoleri, B. D. Olafson, D. J. States, S. Swaminathan and M. Karplus, *J. Comput. Chem.*, 1983, **4**, 187–217.

14 V. M. Anisimov, I. V. Vorobyov, B. Roux and A. D. Mackerell, *J. Chem. Theory Comput.*, 2007, **3**, 1927–1946.
15 J. C. Phillips, R. Braun, W. Wang, J. Gumbart, E. Tajkhorshid, E. Villa, C. Chipot, R. D. Skeel, L. Kalé and K. Schulten, *J. Comput. Chem.*, 2005, **26**, 1781–1802.
16 A. Brünger, C. L. Brooks Iii and M. Karplus, *Chem. Phys. Lett.*, 1984, **105**, 495–500.
17 S. E. Feller, Y. Zhang, R. W. Pastor and B. R. Brooks, *J. Chem. Phys.*, 1995, **103**, 4613–4621.
18 J.-P. Ryckaert, G. Ciccotti and H. J. C. Berendsen, *J. Comput. Phys.*, 1977, **23**, 327–341.
19 T. Darden, D. York and L. Pedersen, *J. Chem. Phys.*, 1993, **98**, 10089–10092.
20 http://www.ks.uiuc.edu/Research/namd/.
21 I. Vaisman, M. Kiselev, Y. Pukhovsky, D. Ivlev and S. Krishtal, Program package for Molecular Dynamics, Ivanovo, 1988–2013.
22 V. N. Zubarev and A. V. Bagdonas, *Teploenergetika (Moscow)*, 1969, **16**, 88–91.
23 G. J. Martyna, D. J. Tobias and M. L. Klein, *J. Chem. Phys.*, 1994, **101**, 4177–4189.
24 G. Lamoureux and B. Roux, *J. Chem. Phys.*, 2003, **119**, 3025–3039.
25 U. Essmann, L. Perera, M. L. Berkowitz, T. Darden, H. Lee and L. G. Pedersen, *J. Chem. Phys.*, 1995, **103**, 8577–8593.
26 P. J. Steinbach and B. R. Brooks, *J. Comput. Chem.*, 1994, **15**, 667.
27 F. Bruni, M. A. Ricci and A. K. Soper, *Phys. Rev. B: Condens. Matter*, 1996, **54**, 11876–11879.
28 T. Yamaguchi, C. J. Benmore and A. K. Soper, *J. Chem. Phys.*, 2000, **112**, 8976–8976.
29 P. Gómez-Álvarez, L. Romaní and D. González-Salgado, *J. Chem. Phys.*, 2013, **138**, 044509–044509.
30 S. Mazur, *J. Chem. Phys.*, 1992, **97**, 9276–9282.
31 B. Bhattacharjee, *Phys. Rev. E: Stat. Phys., Plasmas, Fluids, Relat. Interdiscip. Top.*, 2003, **67**, 041208.
32 N. L. Lavrik and V. P. Voloshin, *J. Chem. Phys.*, 2001, **114**, 9489–9491.
33 T. Keyes, *J. Chem. Phys.*, 1999, **110**, 1097–1105.
34 A. M. Saitta, T. Strassle, G. Rousse, G. Hamel, S. Klotz, R. J. Nelmes and J. S. Loveday, *J. Chem. Phys.*, 2004, **121**, 8430–8434.
35 A. Idrissi, P. Damay and M. Kiselev, *Chem. Phys.*, 2007, **332**, 139–143.
36 A. Idrissi, I. Vyalov, P. Damay, A. Frolov, R. Oparin and M. Kiselev, *J. Phys. Chem. B*, 2009, **113**, 15820–15830.
37 A. Idrissi, M. Gerard, P. Damay, M. Kiselev, Y. Puhovsky, E. Cinar, P. Lagant and G. Vergoten, *J. Phys. Chem. B*, 2010, **114**, 4731–4738.
38 P. Sur and B. Bhattacharjee, *J. Chem. Sci.*, 2009, **121**, 929–934.
39 J. M. Andanson, J. C. Soetens, T. Tassaing and M. Besnard, *J. Chem. Phys.*, 2005, **122**, 174512–174512.
40 R. Oparin, T. Tassaing, Y. Danten and M. Besnard, *J. Chem. Phys.*, 2004, **120**, 10691.
41 B. Grignard, B. Gilbert, C. Malherbe, C. Jérôme and C. Detrembleur, *ChemPhysChem*, 2012, **13**, 2666–2670.
42 A. A. Dyshin, R. D. Oparin and M. G. Kiselev, *Russ. J. Phys. Chem. B*, 2012, **6**, 868–872.
43 http://webbook.nist.gov/chemistry/.

Faraday Discussions RSC Publishing

PAPER

A mesoscopic model for the rheology of soft amorphous solids, with application to microchannel flows

Alexandre Nicolas and Jean-Louis Barrat§*

Received 26th April 2013, Accepted 18th June 2013
DOI: 10.1039/c3fd00067b

We have studied a mesoscopic model for the flow of amorphous solids. The model is based on key features identified at the microscopic level, namely periods of elastic deformation interspersed with localised rearrangements of particles that induce long-range elastic deformations. These long-range deformations are derived following a continuum mechanics approach, in the presence of solid boundaries, and are included in full in the model. Indeed, they mediate spatial cooperativity in the flow, whereby a localised rearrangement may lead a distant region to yield. In particular, we have simulated a channel flow and found manifestations of spatial cooperativity that are consistent with published experimental observations for concentrated emulsions in microchannels. Two categories of effects are distinguished. On the one hand, the coupling of regions subject to different shear rates, for instance, leads to finite shear rate fluctuations in the seemingly unsheared "plug" in the centre of the channel. On the other hand, there is convincing experimental evidence of a specific rheology near rough walls. We discuss the diverse possible physical origins for this effect, and we suggest that it may be associated with the bumps of particles into surface asperities as they slide along the wall.

1 Introduction

The flow of simple fluids can be described microscopically as a succession of local, independent processes: collisions in the kinetic theory or hopping events in the classical Eyring description. As the temperature is lowered, or as the density increases, these processes tend to become more collective, with a dynamical length scale that increases as the glass transition is approached.[1,2] Eventually, the liquid falls out of equilibrium and acquires a non-zero shear modulus on any finite time scale, as well as a yield stress that must be overcome in order to initiate the flow. Similar changes take place in athermal materials when the jamming point is crossed following an increase in density. It is now quite well established[3]

Laboratoire Interdisciplinaire de Physique, Université Grenoble-Alpes, CNRS UMR 5588, BP 87, 38402 Saint-Martin d'Hres, France that the flow mechanisms of such amorphous solids are different in essence from those of liquids, as they involve elastic interactions (shear waves) that are transmitted through solids, but not through fluids. This results in non-local effects in the flow of soft jammed/glassy materials, in contrast with the case of a simple fluid.

In fact, the flow of these materials bears notable similarities with the dynamics of earthquakes,[4] in that it features a solid-like behavior at rest and local yielding above a given applied stress. Yielding is characterized by the emergence of local 'shear transformations' involving a few particles,[5] associated with a local fluidisation of the material. These structural rearrangements, hereafter named plastic events and also often referred to as shear transformations or shear transformation zones in the literature, induce long-range deformations. The microscopic details vary to some extent with the particular nature of the material. In the case of foams, they are identified as T1 events, in which the local change of the first neighbors is mediated by an unstable stage with four bubbles sharing one vertex. In colloidal pastes and in atomic systems, they involve relative displacements of limited magnitude within a small group of atoms, which lead to a new equilibrium configuration that is related to the original one by a shear deformation. In all cases, the stress that was originally supported by the particles partaking in the plastic event is transmitted to the surrounding medium, which behaves as an elastic continuum. The robustness of the above scenario for an extremely wide range of materials is striking. Ample evidence of the local plastic events and their long-range effects is indeed provided both by experiments using diverse materials and simulations.[6–8]

In the last two decades the modelling of flow in amorphous systems has evolved along two distinct, but related, lines. First, several models have been proposed that incorporate the flow scenario in an average description. Among these, the shear transformation zone[9] and the Soft Glassy Rheology (SGR)[10,11] models are the most sophisticated examples. Other simplified models falling into the same category are the fluidity model[12] or the very simple λ-model,[13] which describe the average evolution of a population of flow defects under an imposed strain rate in a mean-field-like manner. The effect of the elastic interactions between these defects is not directly accounted for, but is included in the models indirectly *via* the introduction of parameters such as an effective temperature associated with the mechanical noise. These approaches have been remarkably successful in describing at least some aspects of steady state flow curves, *e.g.*, the existence of a yield stress and the low shear rate behaviour, as well as transient or oscillatory responses in various systems, from metallic glasses to foams or colloidal pastes. However, due to their intrinsic mean-field nature, fluctuations and spatial correlations in the flow are discarded. Also, in their most simplified version they are unable to account for heterogeneities and strain localisation. To capture the latter phenomenon, extensions of the models have introduced the coupling of the mean-field description and the diffusive behaviour of the effective temperature, which again can be understood as a consequence of the non-local interactions between elementary flow events.[14–16]

An alternative line of modelling consists in implementing numerically the scenario of plastic events interacting through an elastic continuum in the form of a discrete lattice model. Such an approach was pioneered by Chen, Bak and Obukhov, in a model initially proposed for the description of earthquakes,[17] and by Argon and Bulatov.[18–20] A number of similar mesoscopic models based on the same physical scenario, but with different implementations, have been proposed and studied in the literature.[21–23] The models are able to produce flow curves sharing similarities with those observed experimentally, although significant differences are revealed by closer inspection; they can account for the strain localisation and its dependence on the local dynamical rules,[23,24] and allow one to explore the influence of parameters such as ageing or temperature. They also reproduce the dynamical heterogeneities observed in the flow, and their variation with the strain rate.[25] However, these comparisons have generally remained qualitative, since the models are in general rather schematic, ignoring in particular tensorial aspects or convection.

In this contribution, we present a detailed study of a mesoscopic model that incorporates these elements in a manner that allows a comparison with experimental data obtained in simple geometries. In particular, we will focus on the channel flow geometry and show that the model captures experimental observations, including fluctuations in the local shear rates arising even in seemingly quiescent regions. Such fluctuations are the hallmark of non-locality and spatial cooperativity in the flow, which can give rise to spectacular long-range fluidisation phenomena.

Section 2 introduces the continuum mechanics-based description of a plastic event and presents our mesoscopic model. Details of its numerical implementation are also provided. In Section 3, we fit the parameters of the model to experimental data for concentrated emulsions taken from the literature, and we present the general features observed in our numerical simulations of a channel flow. The last two sections focus on the manifestations of spatial cooperativity in this particular geometry: Section 4 tackles cooperativity in the bulk, whereas some aspects of the specific rheology near a wall are addressed in Section 5. A shorter account of some of these results has been described in ref. 26.

2 Continuum mechanics-based description of plastic events and presentation of the mesoscopic model

Under homogeneous driving conditions, simple fluids flow homogeneously. Amorphous solids, on the other hand, exhibit localised plastic events when they are forced to flow,[5,6,27–29] associated with local shear transformations. In this section, we use an approach rooted in continuum mechanics to describe the effect of a plastic event on the surrounding (elastic) medium, along with its time evolution. Then, we show how these results are integrated into a mesoscopic model. The presentation of the model is brought to completion by the choice of the relevant probabilities for the onset and end of a plastic event. This section is a detailed extension of previous reports.[22,26]

2.1 Description of a plastic event

Consider a rectangular system described by Cartesian coordinates (x,y), where $x \in [0,L_x]$ and $y \in [0,L_y]$ are the streamwise and crosswise coordinates, respectively. Should the system be unbound, the following results will be applicable provided that $L_x \to \infty$ and $L_y \to \infty$. For the time being, periodic boundary conditions are assumed.

On account of the solidity of the material (which is preserved at low shear rates), the response of the system to a perturbation can be modelled by Hooke's law, whereby the local elastic stress σ^{el} is related to the local (deviatoric) strain ε *via* $\sigma^{el} = \mathbf{C}\varepsilon$, where $\mathbf{C}$ is the stiffness matrix. Before a perturbation (here, a plastic event) sets in, the mechanical equilibrium requires that:

$$\nabla \cdot (\mathbf{C}\varepsilon^{(0)}) - \nabla p^{(0)} = 0, \tag{1}$$

where p is the pressure, and the (0) superscripts denote the initial state. In the following, the material will be considered incompressible, which implies that the displacement field u obeys $\nabla \cdot u = 0$, and isotropic, so that the elastic stress can be written, in condensed notation, as

$$\sigma^{el} = \mu \begin{pmatrix} \varepsilon_{xx} - \varepsilon_{yy} \\ \varepsilon_{yy} - \varepsilon_{xx} \\ 2\varepsilon_{xy} \end{pmatrix} = 2\mu \begin{pmatrix} \varepsilon_{xx} \\ -\varepsilon_{xx} \\ \varepsilon_{xy} \end{pmatrix} \equiv 2\mu\varepsilon, \tag{2}$$

where μ is the shear modulus.

Clearly, Hooke's law will only hold within a certain limit. Indeed, when the configuration is too strained locally, say, in a region $\mathscr{S}^{(0)}$, the particles rearrange so that the system is brought to a new local minimum: this is a plastic event. While this rearrangement occurs, the memory of the reference elastic configuration is lost, and, consequently, the local elastic stress vanishes. The region undergoing the rearrangement is therefore liquid-like and its stress will be mainly of dissipative origin. Following this line of thinking and neglecting inertia, the force equilibrium during the plastic rearrangement reads

$$\begin{cases} \nabla \cdot \sigma^{diss} - \nabla p = 0 & \text{in region } \mathscr{S}, \\ 2\mu \nabla \cdot \varepsilon - \nabla p = 0 & \text{outside region } \mathscr{S}. \end{cases} \tag{3}$$

Notice that the boundaries of the plastic region shall be deformed during the event and $\mathscr{S}$ refers to the deformed region. In eqn (3), the dissipative stress σ^{diss} was supposed to be mainly concentrated in the rearranging region. For simplicity, we further assumed that the dissipation is linear with respect to the strain rate, *viz.* $\sigma^{diss} = 2\eta_{eff}\dot{\varepsilon}$. This linearity is naturally to be understood as a simplification, and not as a claim of the existence of some universality regarding the dissipative mechanism (see ref. 30 for a non-linear law in the case of a foam). In addition to eqn (3), the force balance requires the continuity of the stress all along the boundary of region $\mathscr{S}$. If $\mathscr{S}$ is small enough so that the (plastic) deformation rate in this region can be considered homogeneous, *viz.*, $\dot{\varepsilon}(r) \equiv \dot{\varepsilon}^{pl}$ for $r \in \mathscr{S}$, the continuity of the stress all along the boundary $\partial\mathscr{S}$ of the plastic inclusion leads to:

$$\dot{\varepsilon}^{pl} = \frac{1}{\tau}\varepsilon_{\partial\mathscr{S}}, \tag{4}$$

where $\varepsilon_{\partial\mathscr{S}}$ refers to the (elastic) strain on the outer boundary $\partial\mathscr{S}$. The time scale $\tau \equiv \frac{\eta_{eff}}{\mu}$ for the viscous dissipation of the elastic energy has been made apparent.

The leading-order response of the system to the plastic event immediately follows from eqn (4): it simply comes down to a (plastic) strain rate $\dot{\varepsilon}^{pl}(t) = \frac{1}{\tau}\varepsilon^{(0)}_{\partial\mathscr{S}}$ affecting only region $\mathscr{S}$. In an unconstrained environment, the inclusion would therefore undergo a deformation $\dot{\varepsilon}^{(1)}dt$ in a time interval dt.

However, since the inclusion is embedded in a solid, the latter reacts to this plastic strain: supplementary elastic stress and pressure fields, $\dot{\sigma}^{(1)}dt = 2\mu\dot{\varepsilon}^{(1)}dt$ and $\dot{p}^{(1)}dt$ respectively, are thereby induced in the medium.† The derivation of the fields $\dot{\sigma}^{(1)}$ and $\dot{p}^{(1)}$ is presented in the next subsection. For the time being, let us remark that, thanks to the linearity of the equations, one can express the induced stress on the boundary $\partial\mathcal{S}$ as

$$\dot{\sigma}^{(1)}_{\partial\mathcal{S}} = 2\mu\mathcal{G}_0\dot{\varepsilon}^{pl}, \tag{5}$$

where $\mathcal{G}_0$ is a yet unknown tensor. Now, since the response of the solid is a reaction to an imposed shear strain $\dot{\varepsilon}^{(1)}dt$, it will oppose it, at least in the direct vicinity of the inclusion. Therefore, one expects the eigenvalues of $\mathcal{G}_0$ to be negative. Inserting eqn (2) and (4) into eqn (5) yields, after simplification:

$$\dot{\varepsilon}^{(1)}_{\partial\mathcal{S}}(t) = \frac{\mathcal{G}_0}{\tau}\varepsilon_{\partial\mathcal{S}}(t) \tag{6}$$

Eqn (6) expresses the fact that, up to a (potentially time-dependent) shape prefactor $\mathcal{G}_0$, the force driving the rearrangement is the elastic stress imposed on $\mathcal{S}$ by the rest of the system, and that, in opposing this force, dissipation sets a finite timescale τ to this plastic transformation.‡ Cloitre *et al.*[32] suggested that the duration of a rearrangement in soft colloidal pastes coincides with the shortest structural relaxation time τ_β, which also results from a "competition between elastic restoring forces and interparticle friction", and experimentally confirmed the proposed scaling $\frac{\eta_{eff}}{\mu}$ for the latter time (where η_{eff} is determined by the dissipation within lubrication films). This scaling was also used to collapse flow curves onto a single master curve, which bolsters its relevance for the rheology of these materials.

2.2 Calculation of the elastic deformation induced by a single plastic event (2D, tensorial)

Ref. 33 proposed a method to derive the fields $\dot{\varepsilon}^{(1)}dt$ and $\dot{p}^{(1)}dt$ induced by the plastic strain $\dot{\varepsilon}^{pl}dt$, in a simplified context. First, one considers the limit of an infinitely small plastic inclusion $\mathcal{S}$, $\varepsilon^{pl}(r) \rightarrow \varepsilon^{pl}\,a^2\,\delta(r - r_0)$, where r_0 is the centre of region $\mathcal{S}$, and a, the typical linear size of $\mathcal{S}$. (The dots indicating derivation w.r.t. time are omitted in this section). Secondly, by virtue of the linearity of the equations, the inclusion applies a stress $\sigma^{inc} = 2\mu\alpha\varepsilon^{pl}$ on its surrounding, where α is a scalar (instead of a tensor) because of symmetry arguments. At the expense of a renormalisation of the timescale τ appearing in the definition of $\dot{\varepsilon}^{pl}$, eqn (4), *viz.* $\tau \equiv \alpha^{-1}\frac{\eta_{eff}}{\mu}$, we can consider that $\alpha = 1$. The mechanical equilibrium in the solid then reads:

$$2\mu\nabla\cdot[\varepsilon^{(1)}] - \nabla p^{(1)} = 2\mu\nabla\cdot[\varepsilon^{pl}\,a^2\,\delta(r - r_0)] \tag{7}$$

† This deformation will, in turn, affect the plastic deformation rate $\dot{\varepsilon}^{pl}$, but these higher order effects are neglected here.

‡ The finite duration of a plastic rearrangement, which is neglected in the Soft Glassy Rheology model,[10] the Kinetic Elastoplastic model,[16] as well as in the mesoscopic models of ref. 21,23, may play a crucial role in the compressed exponential relaxation of different soft materials. For details, see ref. 31.

To pursue, eqn (7) is solved with the help of the Oseen–Burgers tensor O, expressed in Fourier coordinates $\underline{q} \equiv (p_m, q_n)$, where $p_m \equiv \frac{2\pi m}{L_x}$ and $q_n \equiv \frac{2\pi n}{L_y}$, with $m,n \in \mathbb{Z}$:

$$\hat{O}\left(\underline{q}\right) = \frac{1}{\mu \underline{q}^2}\left(1 - \frac{1}{\underline{q}^2}\,\underline{q} \otimes \underline{q}\right). \tag{8}$$

The Oseen–Burgers tensor is the elementary solution in terms of the displacement u of the equations $\{2\mu\nabla\cdot\varepsilon - \nabla p = \delta(r),\ \nabla\cdot u = 0\}$, where the linearised deformation tensor obeys $\varepsilon = \frac{\nabla u + (\nabla u)^T}{2}$, with the boundary conditions specified above. Therefore,

$$u^{(1)}\left(\underline{q}\right) = 2\mu\hat{O}\left(\underline{q}\right)\cdot\left(-\mathrm{i}\,\underline{q}\cdot\hat{\varepsilon}^{\mathrm{pl}}\right). \tag{9}$$

Recalling Hooke's law, $\hat{\sigma}^{(1)}(\underline{q}) = 2\mu\left[\mathrm{i}\frac{\underline{q}\otimes u^{(1)} + (\underline{q}\otimes u^{(1)})^T}{2} - \hat{\varepsilon}^{\mathrm{pl}}(\underline{q})\right]$, one finally arrives at:

$$\begin{pmatrix}\hat{\varepsilon}^{(1)}_{xx}\\ \hat{\varepsilon}^{(1)}_{xy}\end{pmatrix}\left(\underline{q}\right) = \hat{\mathscr{G}}^{\infty}\left(\underline{q}\right)\cdot\begin{pmatrix}\hat{\varepsilon}^{\mathrm{pl}}_{xx}\\ \hat{\varepsilon}^{\mathrm{pl}}_{xy}\end{pmatrix}\left(\underline{q}\right) \tag{10}$$

where the elastic propagator $\hat{\mathscr{G}}^{\infty}$ obeys:

$$\hat{\mathscr{G}}^{\infty}\left(\underline{q}\right) \equiv \frac{1}{\underline{q}^4}\begin{bmatrix} -\left(p_m{}^2 - q_n{}^2\right)^2 & -2p_m q_n\left(p_m{}^2 - q_n{}^2\right) \\ -2p_m q_n\left(p_m{}^2 - q_n{}^2\right) & -4p_m{}^2 q_n{}^2 \end{bmatrix}. \tag{11}$$

Eqn (10) and (11) express the elastic deformation field induced by a pointlike plastic event in a system with periodic boundary conditions. The corresponding stress field is straightforwardly obtained by multiplication with the shear modulus 2μ.

In real space, the propagator expressed in eqn (11) has a four-fold angular symmetry and decays as r^{-d}, where $d = 21$ is the spatial dimension. These properties are consistent with observations from atomistic simulations[6,28,34] as well as experiments.[35]

Note that the present treatment does not describe the dilational effects[36] possibly taking place during plastic events. These effects may naturally add quantitative corrections to the picture drawn here, but, along with the associated flow concentration coupling[37] and the free volume diffusion mechanisms (see ref. 14 and references therein), they are probably not of paramount importance in the high density–low temperature situations considered here,[38] where such effects are not always present.[39,40]

2.3 Implementation of parallel confining walls

In order to study a genuine channel geometry, the boundary conditions need to be adapted to take into account two infinite parallel walls, directed along e_x, bounding the flow, while keeping the periodicity in the e_x direction. The effect of the walls is modelled by imposing no-slip boundary conditions at their locations, in line with what is commonly done in fluid mechanics.

To implement the no-slip boundary conditions, we extended the treatment in ref. 33: the system is duplicated in the direction perpendicular to the walls, so that the region $y \in [0,L_y]$ describes the real system, while the region $y \in [-L_y,0]$ is fictitious. For each plastic event (in the real system), a symmetric 'image plastic event' is created in the fictitious region (Fig. 2). The y-component of the velocity field is thereby cancelled at the walls. To remove the x-component of the velocity, adequate forces directed along e_x are added along the walls. These (fictitious) forces add a corrective term $\hat{\varepsilon}^{\rm corr}$ to the deformation field $\hat{\varepsilon}^{\infty}$ obtained for the periodic boundary:

$$\hat{\varepsilon}(\underline{q}) = \hat{\varepsilon}^{\infty}(\underline{q}) + \hat{\varepsilon}^{\rm corr}(\underline{q}).$$

The calculation of $\hat{\varepsilon}^{\rm corr}(\underline{q})$ is presented in Appendix A and yields the following result:

$$\begin{pmatrix} \hat{\varepsilon}^{\rm corr}_{xx}(p_m,q_n) \\ \hat{\varepsilon}^{\rm corr}_{xy}(p_m,q_n) \end{pmatrix} = \begin{pmatrix} \dfrac{-2p_m {q_n}^2}{\underline{q}^4}\left[{\rm i}\sum_y \zeta_\delta(X)\ \mathscr{F}_x\varepsilon^{\rm pl}_{xy}(p_m,y) + 2\sum_y \xi_\delta(X)\mathscr{F}_x\varepsilon^{\rm pl}_{xx}(p_m,y)\right] \\ \dfrac{q_n({p_m}^2 - {q_n}^2)}{\underline{q}^4}\left[{\rm i}\sum_y \zeta_\delta(X)\ \mathscr{F}_x\varepsilon^{\rm pl}_{xy}(p_m,y) + 2\sum_y \xi_\delta(X)\mathscr{F}_x\varepsilon^{\rm pl}_{xx}(p_m,y)\right] \end{pmatrix}, \quad (12)$$

where Σ_y denotes an integral over all the streamlines y = cst and $\mathscr{F}_x$ indicates a Fourier transformation along the x-direction. X is used as a shorthand for $\left(\dfrac{\pi y_{\rm ev}}{L_y}, \dfrac{p_m L_y}{\pi}\right)$ and the analytical expressions of the functions $\zeta_\delta(X)$ and $\xi_\delta(X)$ can be found in Appendix A. In Fig. 4, the displacement field induced by a single plastic event is depicted to illustrate the result of the calculation.

Note that the corrective term couples the different Fourier modes so that the translation invariance of the propagator $\mathscr{G}$ is broken (in the y-direction). In particular, for a given plastic strain, the local strain response now depends upon the distance to the wall. The dependence on the distance for a plastic event $\{\varepsilon^{\rm pl}_{xx} = 0, \varepsilon^{\rm pl}_{xy} \neq 0\}$ is presented in Fig. 3 for a system that is coarse-grained into blocks of unit size (see next section). In particular, one can see that the local strain relaxation induced by a given plastic strain is around 35% higher in the direct vicinity of a wall than in the bulk case, owing to the vicinity of a solid boundary.

2.4 Dynamics of the model and space discretisation

At every point in space, the dynamics is governed by the following equation, including both the external driving $\dot{\Sigma}^{\rm ext}$ and the (local and non-local) stress redistribution due to plastic events:

$$\partial_t \sigma(r) = \dot{\Sigma}^{\rm ext}(r) + \int \mathscr{G}(r,r')\cdot 2\mu\dot{\varepsilon}^{\rm pl}(r'){\rm d}^2r', \quad (13)$$

where $\dot{\varepsilon}^{\rm pl}(r) = \dfrac{\sigma(r)}{2\mu\tau}$ if r is in a plastic region, $\dot{\varepsilon}^{\rm pl}(r) = 0$ otherwise, and the propagator $\mathscr{G}$ takes into account both the bulk (periodic) contribution and the corrections due to the presence of walls. The plastic activity is determined by checking at every time step and at every point in space, the elements that undergo a plastic event. The criterion for triggering plastic events will be discussed in the next section. The time derivative in eqn (13) is handled numerically with a Eulerian procedure, with time step $dt \leq 0.01\tau$.

The convolution part of eqn (13) is most easily solved in the Fourier space, where the convolution turns into a product involving the propagator derived previously (see eqn (11) and (13), for the two contributions to $\mathscr{G}$). To prepare the use of a Fast Fourier Transform routine, the system is spatially coarse-grained into a rectangular lattice of square-shaped blocks of unit size. Physically, the size of the blocks should correspond to the spatial extent of a plastic event.

Technically, the slow decay of $\hat{\sigma}_{xx}(q)$ with q generates some irregularities in the computation of the associated back-Fourier transform. Accordingly, for the sake of precision, we used a finer mesh for the computation of the Fourier transformations, *i.e.*, we divided each block into four subblocks, so that each plastic event now spans four subblocks. Thanks to this technical trick, a smooth stress field is obtained, as shown in Fig. 1.

Besides, the mechanical equilibrium requires the average of the shear stress over any streamline (or any line with a given direction) to be homogeneous. However, the assumption of pointwise plastic events combined with the discretisation of space is not entirely consistent, insofar as it results in moderate violations of the aforementioned homogeneity, when plastic events are far off the direction of macroscopic shear. In order to restore homogeneity, an *ad hoc* shear stress is added globally to every streamline. We have checked that this procedure has very little effect on the results presented below.

2.5 Coarse-grained convection

Although the presence of a lattice precludes a rigorous implementation of convection, a coarse-grained version can be introduced as follows: the average velocity of each streamline in the flow direction is rigorously calculated at each time step, *viz.*

$$\begin{aligned}\langle u_x\rangle_x(y_0) &\equiv \frac{1}{L_x}\int_{-L_x/2}^{L_x/2} u_x(x,y_0)\mathrm{d}x\\ &= \sum_{y_{\mathrm{ev}}}\left[\mathrm{sign}(y_0-y_{\mathrm{ev}})\cdot\left(1-\frac{|y_0-y_{\mathrm{ev}}|}{L_y}\right)+1-\frac{y_{\mathrm{ev}}}{L_y}-\frac{y_0}{L_y}\right]\mathscr{F}_x\varepsilon_{xy}^{\mathrm{pl}}(m=0,y_{\mathrm{ev}}).\end{aligned}$$

Details of the algebra are provided in Appendix B. The line displacement can thus be updated at each time step. Whenever the displacement on a line grows larger than a multiple of the unit block size, this line incrementally shifts the adequate number of units, as a whole. In so doing, the regularity of the lattice is preserved.

A technical detail might be worth mentioning: a naïve implementation of the above method results in a violation of the Galilean invariance, insofar as lines with lower velocity will be shifted less often than others (artificial pinning) and therefore will tend to conserve their neighbours (in the velocity gradient direction) for a longer time, whereas the motion with respect to neighbouring lines should be exclusively controlled by the local shear rate. It turns out that, in a simple shear situation, the system is quite sensitive to such a bias, which may lead

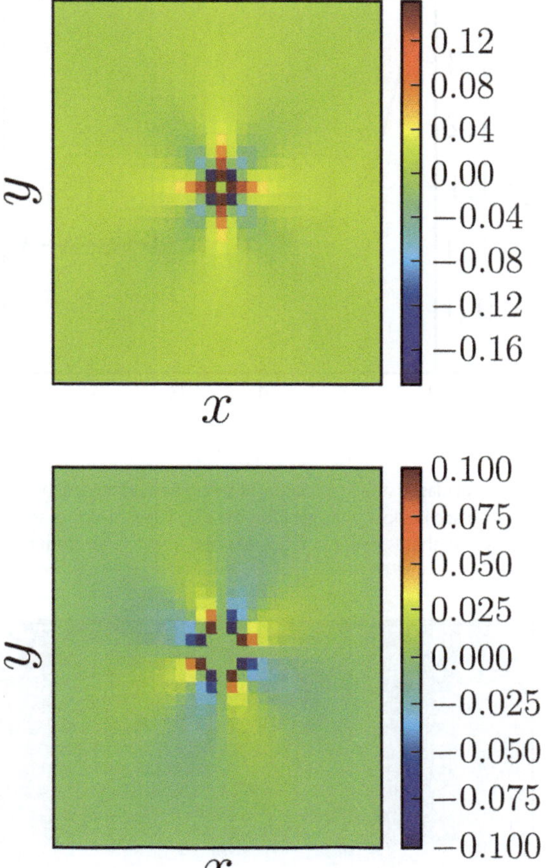

Fig. 1 Deformation field $\varepsilon^{(1)}$ induced by a single plastic event $\varepsilon_{xy}^{\rm pl}$. Top: $\varepsilon_{xy}^{(1)}$ component; bottom: $\varepsilon_{xx}^{(1)}$ component. The values are normalised by the absolute value of the locally induced deformation $\varepsilon_{xy}^{(1)}$. Because of the comparatively large (in magnitude) peak value at the origin, the central block has been artificially coloured.

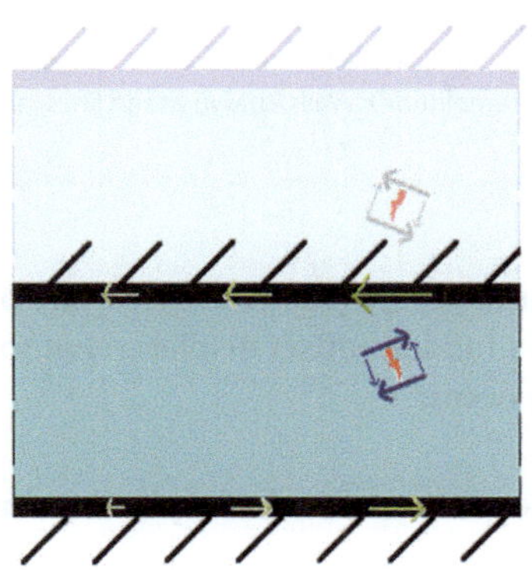

Fig. 2 Sketch of the duplicated system.

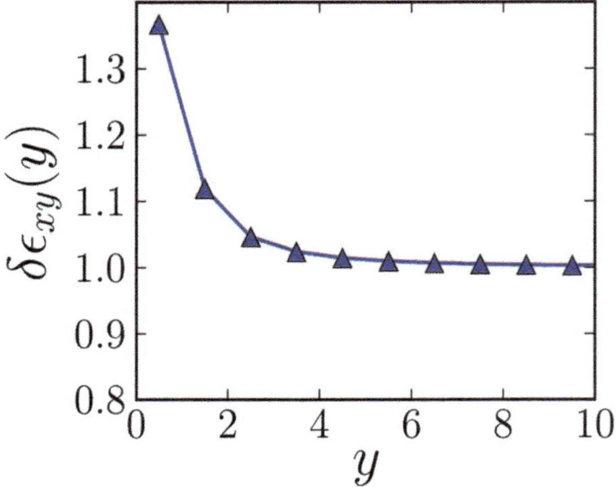

Fig. 3 Decrease of the local elastic strain $|\varepsilon_{xy}^{(1)}|$ induced by a given plastic strain $\varepsilon_{xy}^{\mathrm{pl}}$ as a function of the distance y to the wall (expressed in block units, which is the only relevant length scale). Values have been normalised to the 'bulk' value, that is, the quantity measured infinitely far from the wall.

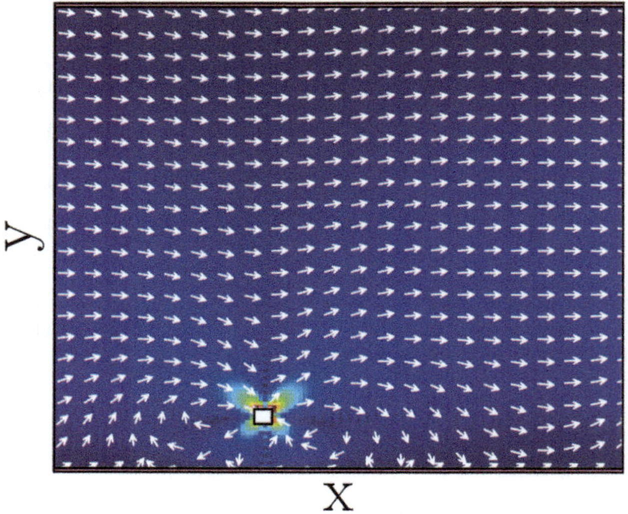

Fig. 4 Displacement field induced by a single plastic event (located in the white square). The white arrows show the direction of the field, while the colour code represents the displacement amplitude (brighter colours indicate higher amplitude). Walls, drawn as red lines, are present at the top and at the bottom of the system.

'pinned' lines to concentrate more plastic activity. The practical solution to this issue consists in adding a random offset displacement to all streamlines, so that no artificial pinning can occur.

2.6 Probabilities for the onset and end of a plastic event

So far, we have quantitatively described the effect of a plastic event and detailed its derivation from rather well established principles. In order to complete the model, criteria must now be fixed with regard to the onset and termination of a plastic event. Since the mesoscopic model is oblivious to the microscopic arrangement of the particles and their stability, the criteria will obviously be somewhat arbitrary. In the present model, they will involve two rates, $l(\sigma)$ and $e(\sigma)$, which govern, respectively, the transition from the elastic to the plastic regime and the recovery of elastic behavior after initiation of the plastic event,

$$\text{elastic regime} \underset{e(\sigma)}{\overset{l(\sigma)}{\rightleftharpoons}} \text{plastic event.}$$

The use of rates introduces a simple element of stochasticity in the model, and indirectly accounts for the variability of local environments.

Let us consider a mesoscopic region susceptible to undergo a plastic rearrangement. In the elastic regime, its configuration minimises the potential energy, under the stress/strain constraints imposed at the boundaries by the rest of the material. The minimum is stable as long as $E - E_{\text{constraint}} \le E_a$, where we assume the existence of an average energy barrier E_a. In an Eyring-like type of approach, the constraint is expressed as: $E_{\text{contraint}} \propto \sigma$, where σ is the local stress applied by the outer region, and we take an activation volume equal to unity. Consequently, the rate of plastic activation depends exponentially on the local stress. In the following discussion, we will use the expression

$$l(\sigma) = \Theta(\sigma - \sigma_{\mu y}) \exp\left(\frac{\sigma - \sigma_y}{x_{\text{loc}}}\right)\tau^{-1}.$$

Three parameters have been introduced in this expression: σ_y is the yield stress associated to the average energy barrier; x_{loc} is a material-dependent activation temperature. Unlike the effective temperatures used in the Soft Glassy Rheology model or the Eyring-like stress fluctuation approach,[41] x_{loc} only accounts for local microscopic effects and does not include the local stress fluctuations due to stress redistribution, *i.e.*, the mechanical noise: the latter should emerge naturally as a consequence of long range interactions between events, within the framework described above. Note that the limit $x_{\text{loc}} \to 0$ of the activation rate coincides with the usual von Mises yielding criterion, which states that the material yields if and only if $\sigma \ge \sigma_y$. However, under shear stress, the effective lowering of the energy barriers results in the necessity to preserve the possibility of activated events, even in materials usually referred to as athermal at rest. For instance, the occurrence of rearrangements in granular matter long after shear cessation supports this claim,[42] although the physical reason for such rearrangements is far from clear. Another possible justification for introducing x_{loc} may be that it effectively accounts for some dynamical disorder in the local yield stress. Finally, we found that introducing this fluctuating, apparently activated character in the yield criterion is the only way to obtain flow curves in reasonable agreement with experimental data, as shown below.

The parameter $\sigma_{\mu y}$ in the Heaviside function Θ is a critical stress, intended to be small ($\sigma_{\mu y} \ll \sigma_y$) and below which no rearrangement can occur. Clearly, this is an *ad hoc* approach to conserve a finite macroscopic yield stress in the limit of a vanishing shear rate $\dot{\gamma}_{\text{app}} \to 0$, when no ageing process is explicitly taken into account. Note that Amon *et al.*,[43] in a paper investigating the behaviour of granular matter on a tilted plane, recently called for a model displaying two critical stresses, with a microfailure stress in addition to the macroscopic one, although with a distinct definition.

A plastic event lasts until the local configuration gets trapped in a new potential well. This trapping is expected to occur when the local energy reaches low enough values, or, equivalently (recall that $\sigma^{\rm diss} \propto 2\mu\varepsilon_{\partial\mathcal{P}}$, see eqn (4)), when the dissipative stress is dominated by the local elastic stress (which was neglected in eqn (4)). Consequently, we define an associated threshold for the recovery of the elastic stability, whose value is set to $\sigma_{\mu y}$ in order to limit the number of parameters. The introduction of a new intensive parameter $x_{\rm res}$ allows us to write the rate at which elastic behaviour is recovered as: $e(\sigma) = \exp\left(\frac{\sigma_{\mu y}-\sigma}{x_{\rm res}}\right)\tau^{-1}$.

The definition of the rates e and l completes the description of the model. At each time step, the probability of failure of an elastic element is $l(\sigma){\rm d}t$, while the probability that a plastic element becomes elastic again is $e(\sigma){\rm d}t$.

3 Fitting of model parameters and general observations in a channel flow

In order to test the validity of the mesoscopic model presented in the previous section, we started fitting the model parameters by comparing the flow curve obtained in simulations of a simple shear setup to the experimental results for a concentrated emulsion.

3.1 Fitting of model parameters

We use units of time and stress such that $\tau = 1$ and $\sigma_y = 1$, and we set $\frac{\mu}{\sigma_y} = 1$ (note that the value of $\frac{\mu}{\sigma_y} = 1$ only contributes to rescaling the shear rate if the convection is omitted). The model then involves three parameters, $\sigma_{\mu y}$, $x_{\rm loc}$ and $x_{\rm res}$.

In the following, our numerical simulations are compared to the experimental data for concentrated oil-in-water emulsions collected by two different groups, Goyon *et al.*[44] and Jop *et al.*[45] The experimental systems are concentrated emulsions of 6–7 μm droplets of silicon oil in a water–glycerol mixture at an oil volume fraction $\phi \sim 0.75$ significantly larger than the jamming volume fraction. Both groups report a Herschel–Bulkley dependence of the shear stress on the shear rate, that is, $\sigma = \sigma_{\rm d}\left[1 + (\tau_{\rm HB}\dot{\gamma})^n_{\rm app}\right]$, with an exponent $n \simeq 0.5$ in both cases.

This Herschel–Bulkley law allows us to fit the remaining model parameters. To do so, we simulated a simple shear flow by setting the driving force to $\dot{\sum}_{\rm app} = \mu\dot{\gamma}_{\rm app}$ in eqn (13), with a stressless state as the initial condition. By varying the parameters, we find that the combination $\{\sigma_{\mu y} = 0.17,\ x_l = 0.249,\ x_e = 1.66\}$ provides a quite satisfactory fitting of the flow curve, as shown in Fig. 5. Note that the model units of time and stress have been appropriately rescaled in the figure, to allow for comparison with the experimental values. Of course, one may argue that the fitting to the flow curve only loosely constrains the parameters, implying that other combinations of parameters could yield the same result. Nevertheless, we would like to mention that, when starting with a moderately different set of parameters and after fine-tuning it to better match the data, we have recovered parameters similar to those selected.

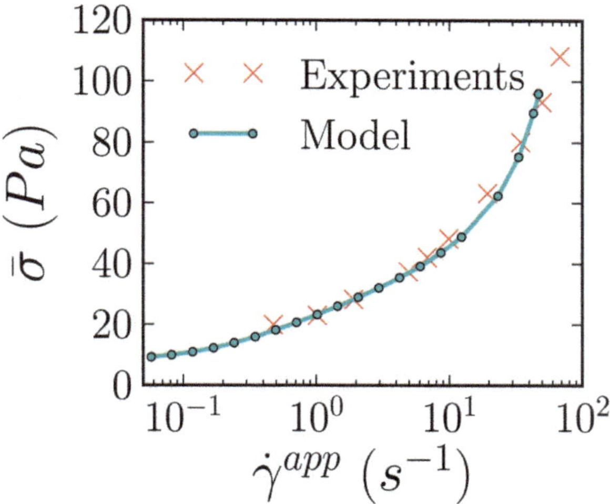

Fig. 5 Experimental (crosses) and simulated (dots) flow curves. The experimental data were obtained by Goyon *et al.* for an emulsion of $\sim$ 6.5 μm silicon oil droplets in a water–glycerin mixture at volume fraction $\phi = 75\%$. The solid line is a guide to the eye.

3.2 Channel flow: general observations

Having set the model, we now turn to the specific case of a channel flow. Indeed, many intriguing experimental results have been reported concerning the flow of soft jammed/glassy materials with that geometry,[39,44–48] which is also relevant for practical applications, in particular in the area of microfluidics.

First of all, it is important to realise that, unlike the simple shear case, the flow is pressure driven in a channel flow, instead of being strain driven. Recalling that the driving $\sum^{\text{ext}}$ in eqn (13) corresponds to the response of a purely elastic solid, it immediately follows that: $\dot{\sum}^{\text{ext}} = 0$, $\sigma_{xy}(x,y,t = 0) = \nabla p \cdot (y - L_y/2)$. Note the streamline-averaged stress conserves a linear profile throughout the simulation, because plastic events induce a homogeneous streamline-averaged stress, owing to the mechanical equilibrium.

We first discuss some general features of the flow of soft jammed solids in that geometry. Conspicuous is, in the first place, the presence of a "plug" in the centre of the channel, *i.e.*, a solid-like region in which the material is convected, but not sheared. The plug can clearly be seen in Fig. 6, which demonstrates a nice agreement between the numerical and the experimental (time averaged) velocity profiles across the channel. Note that showing the velocity differences with respect to the maximal velocity across the channel obviates the experimental issue of the determination of the wall slip.

However, averaging over time masks the temporal fluctuations of the flow. If one heeds the variations of the maximal streamline velocity of the simulated flow with respect to time, flow intermittency becomes evident.§ This phenomenon is more acute for narrow channels (see Fig. 8), in agreement with results from numerical simulations regarding the effect of the confinement (see ref. 40 and

§ However, these fluctuations would presumably vanish in our model if the channel were of infinite length.

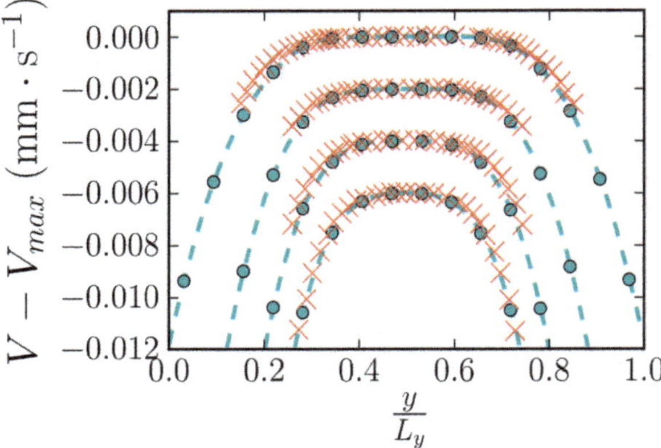

Fig. 6 Experimental (crosses) and simulated (dots) velocity profiles, for stresses at the wall $\sigma_{\mathrm{w}} = 141$ Pa, 188 Pa, 235 Pa, 282 Pa, corresponding to $\sigma_{\mathrm{w}} = 0.36, 0.48, 0.60, 0.72$ in model units, from top to bottom. The experimental data are courtesy of Jop *et al.*[45] The model time and stress units have been rescaled to match the experimental data.

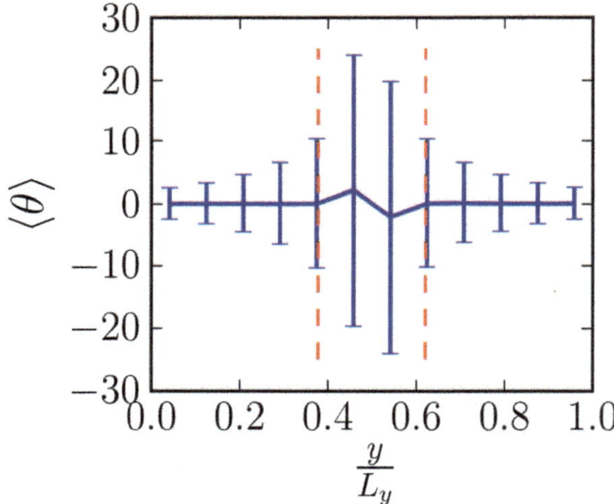

Fig. 7 Principal direction $\theta \in [-45°,45°]$ of plastic events as a function of the position in the channel. Channel width: 12. $\sigma_{\mathrm{w}} = 0.6$ in model units. The vertical dashed lines delimit the "plug", *i.e.*, the region where $|\langle\sigma_{xy}\rangle| \leq \sigma_{\mathrm{d}}$. The bars give the standard deviation, $\pm\langle\langle\theta^2\rangle - \langle\theta\rangle^2\rangle$.

references therein). Note that flow intermittency, or "stick-slip" behaviour, has often been reported experimentally, but it has been interpreted in various ways depending on the particular system under study: the creation and failure of force chains is put forward in the case of granular matter,[49,50] while variations in the local concentration of colloids and erosion by the solvent have been reported for concentrated colloidal suspensions.[48]

The spatial distribution of plastic events is also of interest. Indeed, although the plug remains virtually still on average, sparse plastic events are clearly seen to occur in that region, especially for narrow channels, and, consequently, below the

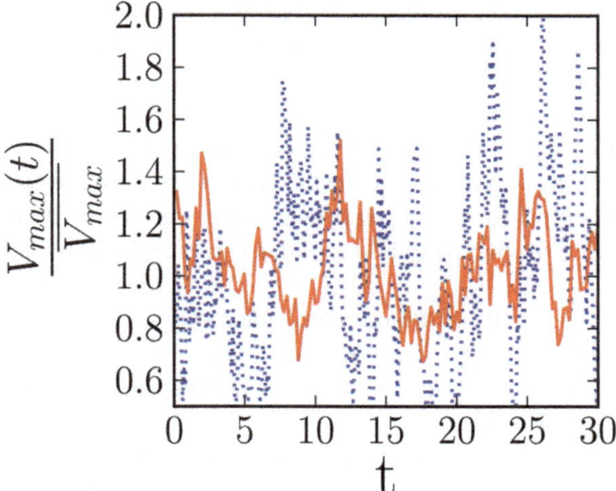

Fig. 8 Time variations of the maximal velocity in the channel, rescaled by its average value over time, for two channel widths: (solid red line) 24 blocks, (dotted blue line) 6 blocks.

bulk yield stress. Therefore, these plastic events essentially originate in cooperative effects, *via* the redistribution of stress generated by distant plastic events. Being of cooperative nature, the principal direction of their stress at the yielding point (the 'angle of yield' of the plastic event) is broadly dispersed, since it is not strongly biased by a fixed applied shear (see Fig. 7).

4 Cooperativity in the bulk flow: a manifestation of the coupling between heterogeneous regions

4.1 Origin and description of the non-locality in the flow

Spatial cooperativity is a hallmark of the flow of amorphous solids: because of the solidity of these materials, shear waves can propagate in the bulk. Accordingly, a plastic event induces a long-range deformation of the material and can thus set off other plastic events, possibly triggering an avalanche. However, the channel geometry is particular in that the driving is intrinsically inhomogeneous; therefore, the cooperativity couples regions (streamlines) subject to different stresses.

When considering a given region, one may then expect its behaviour to differ from what it would have in a homogeneous flow. This is a serious issue, since it undermines the paradigm that there exists a constitutive equation relating the local shear rate to the local shear stress, as explained by Goyon and colleagues.[39,44] (Note, however, that doubts regarding the existence of a single flow curve for concentrated emulsions had also been mentioned earlier, following experiments in a different geometry.[51])

To rationalise the deviations that they had observed experimentally, Goyon *et al.*[39] made use of a diffusion equation operating on the local fluidity, that is to say, the inverse viscosity $f(r) \equiv \frac{\dot{\gamma}(r)}{\sigma_{xy}(r)}$:

$$\xi^2 \Delta f - (f - f_{\text{bulk}}[\sigma(y)]) = 0, \tag{14}$$

where $f_{\mathrm{bulk}}(\sigma)$ denotes the fluidity measured in a homogeneous flow at applied stress σ. The length scale ξ is a cooperative length, that scales with the particle diameter.[39,52] This diffusion equation is based on the idea that plastically active regions will fluidise their neighbours, and inversely. In ref. 16, Bocquet *et al.* showed that this equation can formally be derived from a Hebraud–Lequeux fluidity model,[12] provided heterogeneities are taken into account. However, the limitations imposed by the analytical treatment required to cut off the propagator beyond the first neighbours, and to consider the limit of the vanishing shear rate.

Nevertheless, eqn (14) was found to provide a very satisfactory description of experimental and numerical data in several cases,[39,40,44–46,52] provided that the parameters, that is, the cooperativity length ξ and the value f_{wall} of the fluidity at the wall, are carefully fitted.

Assuming that this equation offers a valid first-order approximation of the flow, we have used it to assess the amplitude of the expected deviations from bulk behaviour.

To do so, we quantified the extent of the coupling by estimating the relative deviation $\delta f(y) \equiv f(y) - f_{\mathrm{bulk}}$ of the fluidity. This defines a dimensionless number, the Babel number $\mathrm{Ba} \equiv \frac{\delta f}{f}$. In Appendix C, we show that, under the assumption of a Herschel–Bulkley constitutive equation, Ba is of the order of $\left(\xi \frac{||\nabla\sigma||}{\sigma - \sigma_{\mathrm{d}}}\right)^2$, that is, $\left(\xi \frac{||\nabla p||}{\sigma - \sigma_{\mathrm{d}}}\right)^2$ for a channel flow.

Noteworthy is the (quadratic) dependence of the Babel number on the stress gradient, *i.e.*, the pressure gradient in a Poiseuille flow. Indeed, it is generally several orders of magnitude larger in microchannels than in their larger counterparts, which explains why striking manifestations of cooperativity have been observed only in the former. The Babel number is also negligible in the wide-gap Taylor–Couette geometry. For instance, a rough estimation yields $\mathrm{Ba} \sim 10^{-9}$ at most in the wide-gap setup used by Ovarlez *et al.*[53] where no deviations from macroscopic rheology were reported.

The denominator of Ba, $(\sigma - \sigma_{\mathrm{d}})^2$, also deserves a comment: at high applied stresses, when the material is more fluid-like, relative deviations become less significant. We should however say that, to measure relative deviations, the absolute fluidity deviations are divided by the fluidity, which increases as σ increases.

4.2 Non-local effects in the velocity profiles

Following the above considerations, we expect deviations from macroscopic rheology to increase with confinement, at a fixed wall stress.

Indeed, Goyon's experiments on emulsions confined in microchannels with smooth walls tend to indicate that the deviations of the velocity with respect to the bulk predictions follow such a trend. However, overall, these deviations were found to be rather small. The mentioned effect of the confinement is also confirmed by Chaudhuri *et al.* with atomistic simulations of a Poiseuille flow with biperiodic boundary conditions with atomistic simulations.[40]

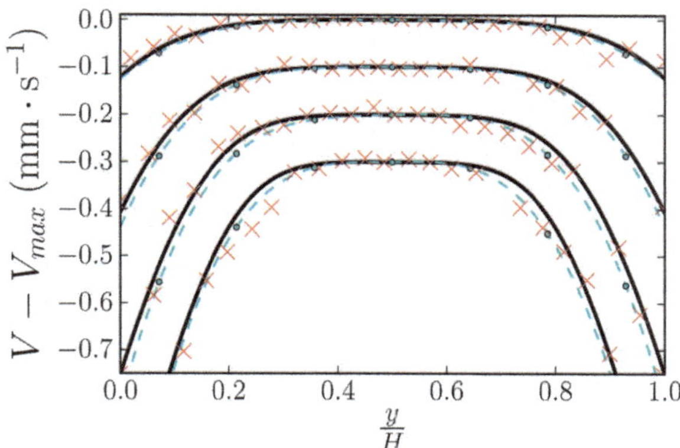

Fig. 9 Velocity profiles across the channel, for $\sigma_w = 45, 60, 75, 91$ Pa, *i.e.*, $\sigma_w = 0.75, 1.0, 1.25, 1.52$ in model units, from top to bottom: (dashed line) simulation results, (solid line) predictions based on the numerical bulk flow curve. The crosses are the experimental data obtained by Goyon *et al.*[44] Note that the curves have been shifted with respect to each other for clarity.

Fig. 9 shows a comparison between the actual velocity profile obtained with simulations of the mesoscopic model and the predictions from the (bulk) flow curve. As in the experiments, small deviations can be observed. For the extent of these deviations to roughly match that in the experiments, the channel width must be of the order of 7–10 block units. From this we deduce a first estimate for the linear size N_ϕ of a mesoscopic block in terms of particle diameters: $N_\phi \approx 2$, which is comparable to the experimental values found in the literature.[35]

Let us now investigate how compatible our simulation results are with the fluidity diffusion equation, eqn (14). To solve eqn (14), two boundary conditions are required: for symmetry reasons, we impose $f(y = 0) = f(y = L_y)$, and we set the fluidity at a point close to the wall to the value measured in the simulations. In addition, the shear-rate dependence of the cooperativity length ξ must be specified. Two possibilities are considered in Fig. 10: either, following Goyon *et al.*,[39] ξ is supposed independent of the shear rate, *i.e.*, $\xi = \xi_0$, or a power-law dependence is assumed, $\xi(\dot{\gamma}) = \xi_0(\dot{\gamma}\tau)^{-\frac{1}{4}}$, where $\dot{\gamma}$ is the product of the local shear stress and the fluidity, as derived in ref. 16 in the limit $\dot{\gamma} \to 0$, and in reasonable agreement with the data of ref. 45. In both cases, ξ_0 is adjusted by a least square minimization. Both cases give a reasonable fit, but neither matches our data accurately over the whole range of applied pressures. We ascribe this defect, among other details, to the approximation of long-range interactions by a diffusive term, and to the neglect of fluidity fluctuations.

In Fig. 11, we have assessed the predictive capability of the theoretically derived Babel number for our channel flow simulations by plotting the $\frac{\delta f}{f}$ obtained in our simulations as a function of $\mathrm{Ba} = \left(\xi \frac{||\nabla\sigma||}{\sigma - \sigma_d}\right)^2$. It shows a global trend towards larger relative deviations from macroscopic rheology for larger Ba, but the correlations are poor. Nevertheless, one may expect Ba to still be a valid predictor in practice, when widely different situations are considered.

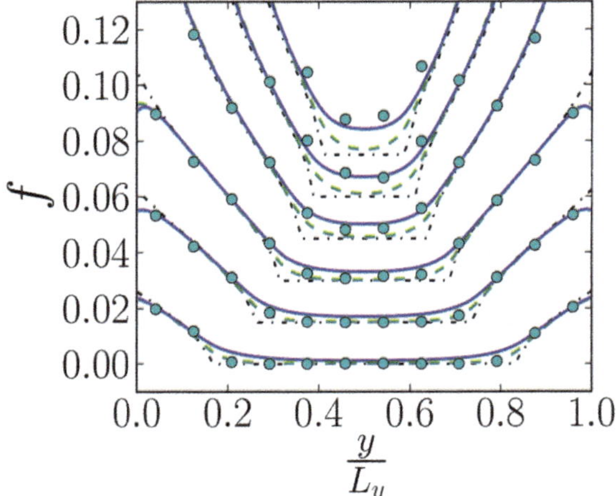

Fig. 10 Fluidity profiles for $N_y = 12$, for $\sigma_w = 0.20, 0.28, 0.36, 0.48, 0.60, 0.72$ in model units. Filled circles: numerical results, dashed green line: solution of eqn (14) with $\xi(\dot{\gamma}) = 0.03702$, solid blue line: solution of eqn (14) with $\xi(\dot{\gamma}) = 0.01146\dot{\gamma}^{-0.25}$. The thin dash-dotted lines represent the bulk fluidity f_{bulk}. Note that the curves are shifted with respect to each other for clarity.

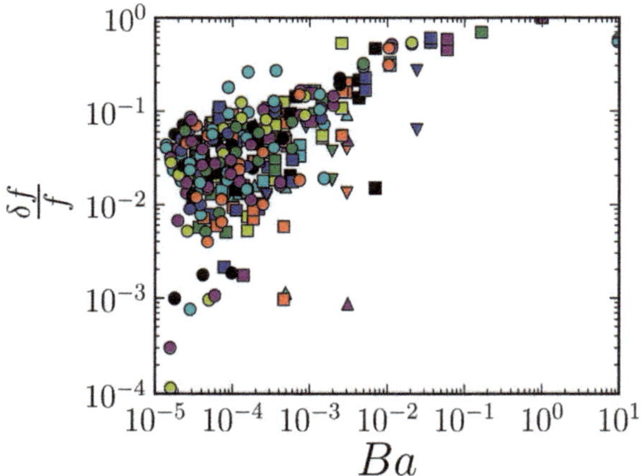

Fig. 11 Relative deviations $\frac{\delta f}{f}$ of the local fluidity f from the bulk fluidity $f_{\text{bulk}}(\sigma)$ measured in the simulations, where σ is the local shear stress, as a function of the estimated Babel number $\text{Ba} = \left(\xi_0 \frac{\nabla\sigma}{\sigma - \sigma_d}\right)^2$. We have set ξ_0 to 0.037 (see Fig. 10). Data only include regions where $\sigma > \sigma_d$, but cover various applied pressures and channel widths: (▼) 6 blocks, (▲) 10 blocks, (■) 16 blocks, (●) 24 blocks.

4.3 Shear rate fluctuations in the plug

Quite recently, Jop *et al.*[45] have showed experimentally that the seemingly quiescent plug in the centre of the microchannel actually sustains finite shear rate fluctuations. This observation is obviously consistent with the occurrence of sparse plastic events in the plug in our simulations.

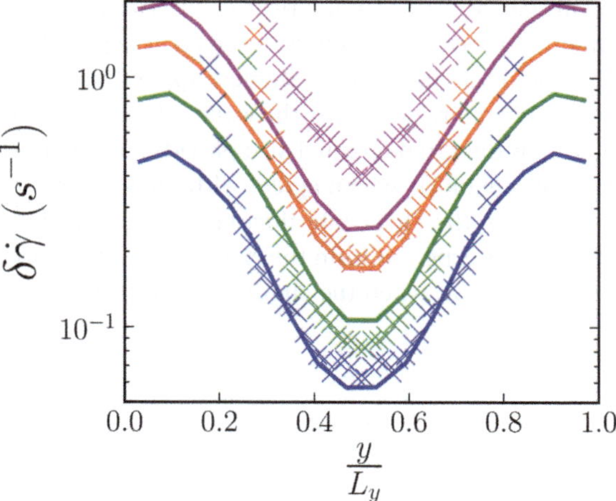

Fig. 12 Shear rate fluctuations $\delta\dot{\gamma}(y)$ (averaged along the x-direction), for $\sigma_w = 141$ Pa, 188 Pa, 235 Pa, 282 Pa (identical to Fig. 6), from bottom to top. Experimental data collected by Jop *et al.* ($\times$),[45] numerical results for $N_y = 16$ (solid lines).

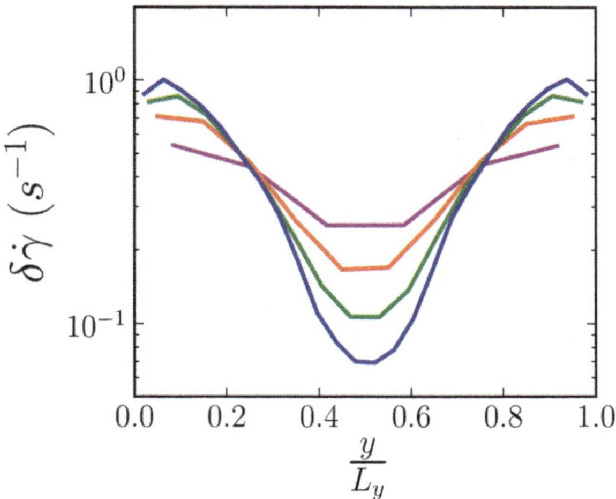

Fig. 13 Shear rate fluctuation profiles for a given stress at the wall, $\sigma_w = 0.48$ in model units, for different channel widths: (fuchsia) 6, (red) 10, (green) 16, and (blue) 24 blocks, in descending order of minimal values.

To go further than this qualitative agreement, we have directly compared the local shear rate fluctuations $\delta\dot{\gamma}(x,y) = \sqrt{\langle\dot{\gamma}(x,y)^2\rangle - \langle\dot{\gamma}(x,y)\rangle^2}$ to the experimental data,¶ with the parameters used to fit the associated velocity profiles (see Fig. 9). Here, $\dot{\gamma}\ (x,y)$ is the local shear rate at point (x, y); it is given by $\dot{\gamma}\ (x,y) = 2(\dot{\varepsilon}^{\mathrm{pl}}_{xy}(x,y) + \dot{\varepsilon}^{(1)}_{xy}(x,y))$ in the model and is therefore obtained directly, that is, without deriving the velocity with respect to space. Fig. 12 presents the experimental shear

¶ Note that we have discarded the two curves corresponding to the lowest applied pressures, which seem to plateau in the centre, because we were not entirely sure of the accuracy of these measurements.

rate fluctuation profiles and their numerical counterparts for $N_y = 16$ blocks crosswise. A semi-quantitative agreement is observed in regions far from the walls – apart from the large discrepancy at the highest applied pressure. The discrepancies in the highly-sheared regions near the walls will be considered below. It is interesting to note that the fitted channel size provides another estimate for the size N_ϕ of an elastoplastic block, which agrees with the first one, $N_\phi \approx 2$. Fig. 13 shows the dependence of the shear rate fluctuations on the channel size for a given stress at the wall. As expected from the expression of the Babel number, the fluctuations in the plug decay when the channel width is increased.

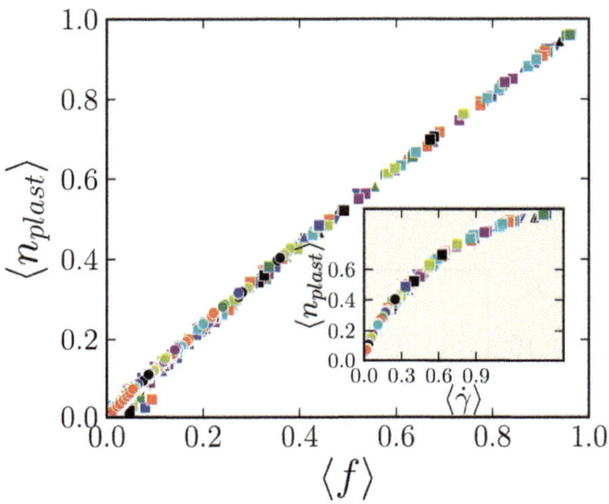

Fig. 14 Time-averaged fraction of plastic blocks $\langle n_{plast} \rangle$ on a given streamline as a function of the mean fluidity $\langle f \rangle$ on that line, for diverse applied pressures and various channel widths: (▼) 6 blocks, (▲) 10 blocks, (■) 16 blocks, (●) 24 blocks. Inset: $\langle n_{plast} \rangle$ *vs.* the mean shear rate $\langle \dot{\gamma} \rangle$ on the line. (Same symbols).

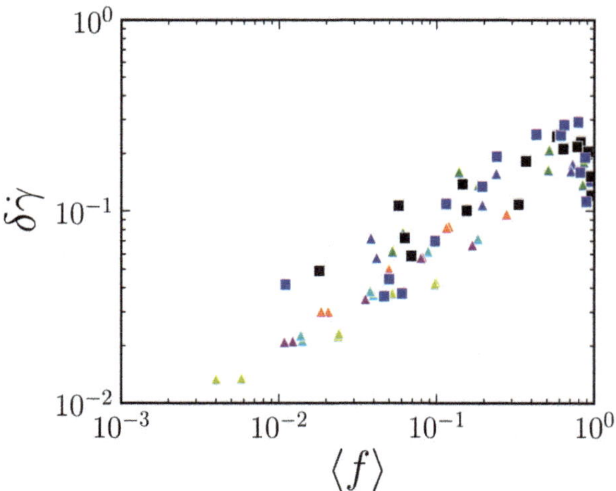

Fig. 15 Shear rate fluctuations $\delta\dot{\gamma}$ on a given streamline as a function of the mean fluidity $\langle f \rangle$ on that line, for diverse applied pressures and various channel widths: (▲) 8 blocks, (■) 16 blocks, (●) 24 blocks.

Let us note that the data collected by Jop and co-workers suggest a proportionality between the shear rate fluctuations and the local fluidity, implying that both are indicators of the intensity of the plastic activity. Fig. 14 shows that the line-averaged plastic activity does indeed depend linearly on the local fluidity in our channel flow simulations, despite some discrepancies at low values of the fluidity, that is, probably in the plug. However, the relation between the shear rate fluctuations and the mean fluidity is much less clear (see Fig. 15).

5 A specific rheology near the walls?

In the previous section, we have dealt with the flow cooperativity associated with the coupling of heterogeneous streamlines, leaving aside another potentially significant difference with the bulk homogeneous flow: the presence of walls bounding the flow, which is known to affect the flow of diverse complex fluids like wormlike micellar solutions,[54,55] laponite,[56] dense colloidal suspensions,[47] *etc.* Indeed, Goyon *et al.* provided experimental evidence of the occurrence of ample changes when rough walls are substituted for smooth walls.[39] Then, much larger deviations from bulk rheology are observed, especially at high applied pressures, and these deviations are maximal close to the walls, contrary to the predictions based on the Babel number.

5.1 Weak deviations due to the no-slip boundary condition

Remember that walls are described by a no-slip boundary condition in our model. This condition results in a significantly larger dissipation during plastic events in their vicinity. Is this sufficient to capture the very large deviations observed experimentally?

Fig. 16 shows the local flow curve for the simulations. To decouple to a certain extent the problem of wall rheology from the inhomogeneous driving, the latter being associated with large values of Ba, a relatively large channel is considered here. For each value of the wall stress, the points with the highest local shear rates in Fig. 16 are closest to the walls. We do observe some slight deviations,‖ but they are clearly much weaker than in Goyon's observations (see Fig. 6 of ref. 44 for instance). In this respect, they much better describe the situation for smooth walls, which, at first, might seem surprising given the no-slip boundary condition. Yet, in reality, the large slip observed at the smooth walls is not expected to give rise to significant changes: it only adds a simple global translation to the complex velocity field obtained with the no-slip boundary condition.

5.2 Physical effect of rough walls

As the deviations observed for rough walls are not captured by a simple no-slip boundary condition, we discuss here some physical mechanisms that may be responsible for the observed behaviour.

‖ Nevertheless, replacing the no-slip boundary condition with a periodic boundary condition will play a role if the Babel number is large enough. See ref. 40 for the effect of confinement on the observed yield stress in a biperiodic Poiseuille flow.

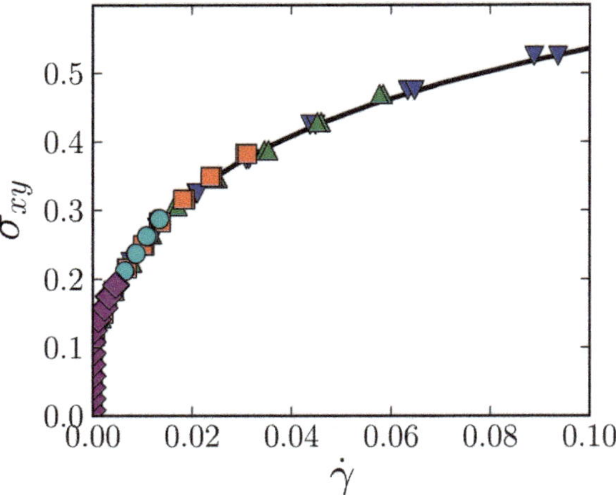

Fig. 16 Local shear stress as a function of the local shear rate, for various applied pressures for a channel width of 24 blocks. The corresponding stresses at the walls are: (purple rhombs) $\sigma_W = 0.2$; (cyan dots) $\sigma_W = 0.2$; (red squares) $\sigma_W \simeq 0.4$; (green upper triangles) $\sigma_W \simeq 0.5$; (blue lower triangles) $\sigma_W = 0.6$.

First, the static structure near the walls is known to differ from that in the bulk. For smooth, or not too rough, boundaries, stratification in layers is often reported over a distance of a few particle diameters,[57,58] though not systematically: Goyon *et al.*[44] actually observed no such layering in their experiments. Besides, the vicinity of a solid boundary hinders the mobility of Brownian particles.[59] But these structural changes for the material at rest imply a decrease of the fluidity at the wall, as opposed to the enhancement that is experimentally observed by Goyon[39] and Geraud[46] at high enough stresses, *i.e.*, where the largest deviations occur. Alternatively, the specific behaviour at the wall is often rationalised by the existence of a depleted 'lubrication layer' close to the wall, as is often found in sheared dispersions.[60–66] This phenomenon is more acute for deformable particles[62] undergoing high shear rates and/or high shear gradients; it generates an apparent wall slip. However, at the very high concentrations investigated here, owing to the large osmotic pressure, such a lubricating layer would have a thickness of the order of 100 nm or less[44,65,66] (if the lubricating layer is composed of pure solvent). Effectively, Goyon directly measured the concentration profile across the channel and was not able to detect any significant variation. This finding was corroborated by the absence of radial droplet migration for a similar material in a Taylor–Couette cell, even at high shear rates, as reported by Ovarlez *et al.*[53] Adding the fact that systems of soft particles have a much weaker viscosity dependence on concentration than their hard particle counterparts, the effects of concentration variations could be ruled out as regarded in Goyon's experiments. Nevertheless, we attempted to simulate a less viscous layer close to the wall by decreasing the yield stress of the associated mesoscopic blocks, but this only had little effect on the rest of the system. Therefore, one is led to seek another explanation.

An aspect that has been overlooked so far is the reported observation of wall slip in Goyon's, Geraud's and Jop and Mansard's works,[39,44,46,58] both with smooth and rough walls. In order to extract information that is relevant for the bulk flow, the authors measured the local velocities and shear rates in the channel by microscopic observation, so that the occurrence of slippage should *a priori* not affect their results. Indeed, in the presence of smooth surfaces, where wall slip accounts for around 30% of the maximal velocity at the typical pressures applied by Goyon *et al.*,[39] slip only results in a global translation of the system, that leaves the local flow curve strictly unaltered. For rough surfaces, let us first remark that the presence of wall slip is more surprising, since roughened surfaces** are often used to strongly suppress, or eliminate, the slip for the very same type of materials, which is monitored by rheological measurements, and then used as benchmarks for a system without slip.[61,64,67–69] However, in several cases, the measurement of local velocities in the flow, either by microvelocimetry with fluorescent tracers[44,46,70] or through direct visualisation with confocal microscopy,[45,58] demonstrated that concentrated emulsions may slip along rough surfaces in microchannels. A seemingly quadratic,[46,70] or linear,[58] dependence of the slip velocity on the shear stress at the wall was reported in these cases. As a side note, let us add that slip along a rough wall is not restricted to the microchannel geometry: for instance, Divoux *et al.* showed with ultrasonic speckle velocimetry that another yield stress fluid, namely carbopol, experiences a phase of total slip in a Taylor–Couette rheometer whose cylinders had been coated with sand paper.[71]

Now, when particles slide along a rough wall, they are expected to bump into, and be deformed by, the surface asperities. In the case of asperities that are large as compared to the "particle" size (~60 microns *vs.* from a few to 20 microns), this phenomenon is best exemplified by the spatio-temporal diagram acquired with ultrasonic velocimetry for a carbopol microgel, Fig. 6 of ref. 72, where one can see a large deformation of the material that originates at the rotor and propagates almost instantaneously into the bulk; this signal was interpreted by the authors as the signature of a "bump" into a surface protuberance. Albeit less visible, this effect should also appear with walls characterised by a smaller roughness, whereby rough walls in the presence of slip act as sources of mechanical noise and cause deviations from bulk rheology in their vicinity. This tentative scenario has the potential to explain why deviations may, or may not, be observed in the vicinity of rough surfaces: for instance, Goyon *et al.*[44] and Ovarlez *et al.*,[53] as well as Seth *et al.*,[73] have reported that the local flow curves obtained in wide-gap Taylor–Couette or plate–plate geometries with rough walls could be mapped onto the macroscopic flow curves; yet, they also indicated that, in those cases, no evidence of wall slip could be found. Very recently, Mansard endeavoured to investigate the impact of wall roughness by combining experiments and molecular dynamics simulations.[58] Non-monotonic variations of the wall fluidity as a function of the roughness were reported in the experiments, but the data did not allow for the extraction of the parameters responsible for the deviations from macroscopic rheology. Nevertheless, he noted that "the particles must jump over the patterns [on the walls]. This effect induces the rearrangements and increases the wall fluidity".

** Diverse methods are available for roughening a surface, such as sandblasting, covering it with sandpaper, or coating it with particles.

Naturally, this prompts the following question: what determines the occurrence of slip along rough walls? This question lies far beyond the scope of the present study. Let us simply note that in ref. 39, 44–46 the size of surface asperities was a couple of microns at most, that is, significantly less than the typical "particle" size, which plausibly favours slip, as well as the high shear rates experienced at the microchannel walls. Nevertheless, recent theories of slip along smooth walls involved, in addition, parameters such as the deformability of the droplets,[63,64] and the particle–wall interactions,[74] not to mention the presumably significant impact of Brownian motion in cases where it is relevant.[75,76] As far as we know, the somewhat daunting challenge to extend these theories to the case of rough walls still awaits a successful accomplisher.

In the above discussion, we have carefully eluded the question of the surface's chemistry and its interactions with the particles. However, Seth *et al.* showed that they can play a significant role; in particular, for the yield stress fluid they studied, smooth attractive surfaces were observed to induce deviations from macroscopic rheology relatively far into the bulk, whereas smooth repulsive surfaces induced none at all.

Finally, we would like to mention another possible impact of the confinement of the material between walls. The channel may be so narrow that the layers where the specific wall rheology dominates start overlapping. This situation, which is described as strong confinement, is expected to occur when the channel width becomes of the order of, or smaller than, the cooperativity length ξ. For the data of ref. 39, 44, 45 discussed above, this mechanism is therefore not relevant.

5.3 Fictitious plastic events along the wall as mechanical noise sources

As we have already noted, non-local effects leading to deviations from the macroscopic flow curve are often rationalised in terms of the fluidity diffusion equation, eqn (14) (see, *e.g.*, ref. 39, 40, 44–46, 52, 58, 73). In this approach, the fluidity at the wall is needed as an input parameter, whose precise value turns out to be determinant. Most likely, the suggested mechanical noise at the walls would be hidden in that value. (Note that, in Goyon's report[70] the fluidity at the rough walls, where larger deviations are observed, is indeed larger than that for smooth walls and larger than the bulk fluidity corresponding to the same shear stress.)

Our mesoscopic model is also oblivious to the microscopic details of the flow near a boundary and therefore cannot describe the effect of wall slip along a rough wall without further input. Nevertheless, since bumps act as sources of mechanical noise in the system, one can attempt to account for their occurrence by adding fictitious plastic events along the walls. Note that this *ad hoc* treatment is similar to imposing a wall fluidity larger than the bulk fluidity as a boundary condition when solving the fluidity diffusion equation, eqn (14).

More precisely, we have modified the implementation of the model slightly, so that a wall is now described as a line of plastically inert blocks: the bottom wall will, for instance, occupy the portion of space $0 \leq y \leq 1$, and the no-slip boundary condition is imposed at its centre, *i.e.*, $y = 0.5$. On this line, a fraction of blocks

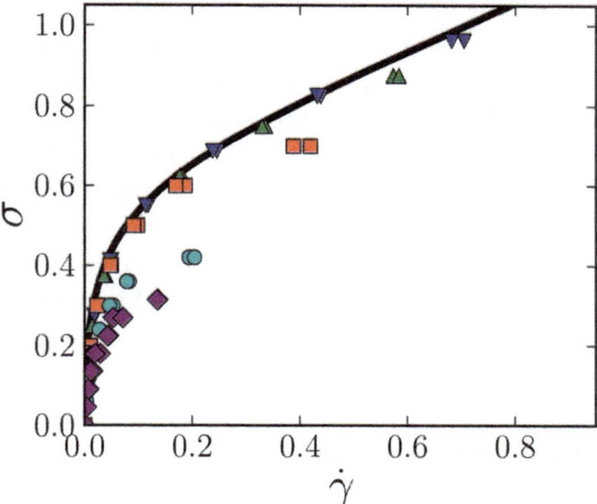

Fig. 17 Local shear rate $\sigma(y)$ *vs* local shear rate $\dot{\gamma}(y)$ (averaged on streamlines y = cst) in the microchannel, when fictitious mechanical noise sources of intensity $\dot{\varepsilon}_{xy}^{\text{fict pl}} = \pm 4.5$ are added on a fraction (1/3) of blocks on the wall lines. σ_{W} = (◆) 0.36, (●) 0.48, (■) 0.8, (▲) 1.0, (▼) 1.1 in model units. Solid line: macroscopic flow curve.

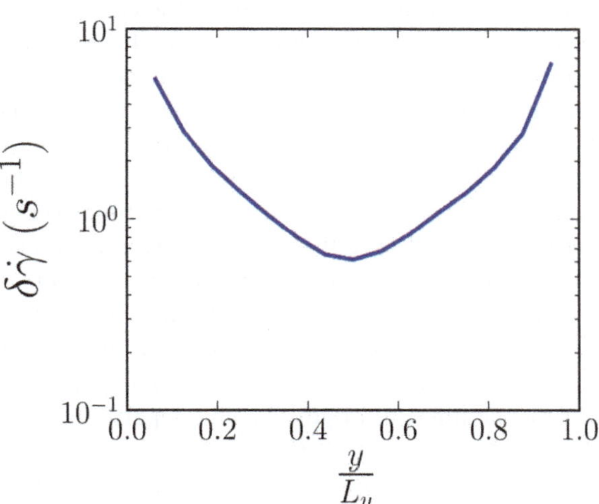

Fig. 18 Shear rate fluctuation profiles in the presence of fictitious plastic events along the walls. A third of the blocks on the wall lines have been randomly selected to release a constant plastic stress $\dot{\sigma}_{xy}^{\text{pl fict}}$ = 4.5 per unit time.

was selected†† at random to act as sources of mechanical noise, that is, to mimic, *e.g.*, bumps of particles into surface asperities. To do so, they shall release a constant plastic strain $\dot{\varepsilon}_{\text{fict}}^{\text{pl}}$ per unit time, along the direction of the macroscopic shear (for simplicity). We emphasise that the mechanical equilibrium is not violated by the addition of these fictitious plastic events.

†† Note that shuffling these blocks, *i.e.*, selecting new random blocks as noise sources, at low enough frequency, hardly affects the results presented below.

Fig. 17 shows the local flow curves obtained with this protocol. The observed deviations are qualitatively similar to those reported by Goyon (see Fig. 7 of ref. 44). However, we must note that a rather intense mechanical noise is required to get such deviations ($\dot{\varepsilon}^{\mathrm{pl}}_{\mathrm{fict}} \approx 5$). (As the value of $\dot{\varepsilon}^{\mathrm{pl}}_{\mathrm{fict}}$ is arbitrary, we do not seek quantitative agreement with the experimental data here.) In addition, these fictitious plastic events also alter the shear rate fluctuation profile, as shown in Fig. 18. Besides a global increase of the fluctuations, the profile no longer flattens in the vicinity of the walls, which renders it more consistent with the experimental results of Jop *et al.* (collected in a channel with rough walls).

6 Conclusions and outlook

In conclusion, we have derived analytical formulae from continuum mechanics for the effect and time evolution of a plastic event occurring in a two dimensional medium bounded by walls. We have integrated these formulae in a lattice model for the flow of amorphous solids, in which elastoplastic blocks receive stress from their surroundings and have a certain probability to become plastic; the chosen form of probabilities for the onset and end of a plastic event allowed us to match experimental flow curves for concentrated emulsions. Then, we turned to the simulation of flow in microchannels, where the most prominent feature is the existence of a seemingly unsheared "plug". Remarkable manifestations of spatial cooperativity in the flow had been unveiled experimentally, and we have proposed to distinguish those pertaining to cooperativity in the bulk from those pointing to the specific rheology near a solid boundary. For the former category, deviations of time-averaged quantities are generally weak, but could nevertheless be observed with our model. More strikingly, shear rate fluctuations were observed in the plug, consistent with the experiments. With regard to the specific wall rheology, it turned out that imposing a no-slip boundary condition at the walls in our model was not sufficient to capture the experimentally observed phenomena. We have discussed several possible physical origins for the departure from the observed macroscopic behaviour, above all, in the vicinity of rough surfaces; we have insisted in particular on a tentative scenario in which mechanical noise is created at the wall by, *e.g.*, bumps of particles into surface asperities as they slide along the wall. Finally, an *ad hoc* implementation of this mechanical noise was attempted.

Concerning our mesoscopic model, several improvements can be considered. First and foremost, regions undergoing plastic events are fluidised, and the presence of fluid-like regions is expected to damp shear waves and reduce the cooperativity. This point is not taken into account in the model. Also, the distinction between an activation temperature, of non-cooperative origin, and a more general effective temperature will be worth investigating further, both for thermal and 'athermal' soft solids under shear. In an unrelated way, it has been made apparent that, in spite of the vast amount of literature on the question of slip for soft solids and the recent progress made in that respect,[63] the issue of slip along rough surfaces and its consequences on the local fluidity remain quite challenging.

Appendices

A Derivation of the correction terms to the propagator for a system bounded by walls

The system covers the domain $(x,y) \in [0,L_x] \times [-L_y,L_y]$ and is periodically replicated throughout space. The region $y \in [0,L_y]$, bounded by walls at $y = 0$ and $y = L_y$ represents the real system, whereas the other half is a fictitious region introduced for the calculations.

For any plastic event $\varepsilon^{\mathrm{pl}} = (\varepsilon_{xx}^{\mathrm{pl}}, \varepsilon_{xy}^{\mathrm{pl}})^T$ occurring at position (x, y) in the real half, a 'symmetric' plastic event $\varepsilon^{\mathrm{pl}'} = (\varepsilon_{xx}^{\mathrm{pl}}, -\varepsilon_{xy}^{\mathrm{pl}})^T$ is created at location $(x,-y)$ in the fictitious region. For symmetry reasons, the y-component of the velocity field is thereby cancelled on lines $y = 0$ and $y = L_y$ (bear in mind that the $2L_y$-wide system is periodically replicated).

Let us now introduce forces $f_x^{(y=0)}$ and $f_x^{(y=L_y)}$ along the x-direction at the bottom $(y = 0)$ and top $(y = L_y)$ walls, respectively, to cancel the x-components. The Fourier transform of the force field reads:

$$f_x(m,n) = f_x^{(y=0)}(m) + (-1)^n f_x^{(y=L_y)}(m)$$

Note that we have simplified the notations by using the shorthand $g(m,n)$ for $\hat{g}(p_m,q_n)$, for any function g, where $p_m \equiv \frac{2\pi m}{L_x}$ and $q_n \equiv \frac{2\pi n}{L_y}$ are the Fourier wavenumbers.

With these forces, the Fourier-transformed displacement field turns into:

$$u^{(1)}(m,n) = \mathcal{G}^{\infty}(m,n) \cdot (\hat{\varepsilon}^{\mathrm{pl}}(m,n) + \hat{\varepsilon}^{\mathrm{pl}'}(m,n)) + O(m,n) \cdot f_x(m,n) \tag{15}$$

$$\equiv u^{*\infty}(m,n) + u^{\mathrm{corr}}(m,n), \tag{16}$$

where $\hat{u}^{\mathrm{corr}}$ is the contribution from the wall forces and $\hat{O}$ is the Oseen–Burgers tensor introduced in eqn (8). The star in $\hat{u}^{*\infty}$ only indicates that this symbol represents the velocity field induced by both the real plastic event and its 'symmetric' counterpart.

Remarking that the condition of zero velocity at the bottom and top walls reads, in terms of the Fourier components,

$$\forall m, \sum_n u^{(1)}(m,n) = 0$$

$$\text{and } \forall m, \sum_n (-1)^n u^{(1)}(m,n) = 0,$$

respectively, we obtain two equations on the f_x after insertion from eqn (15). Adding and subtracting these equations yield, for any m:

$$\sum_{n \in \mathrm{O}} u_x^{*\infty}(m,n) + O(m,n) \cdot \left(\left(\hat{f}_x^{(y=0)} - \hat{f}_x^{(y=L_y)} \right)(m) \right) = 0$$

$$\sum_{n\in \mathrm{E}} u_x^{*\infty}(m,n) + O(m,n)\cdot\left(\left(\hat{f}_x^{(y=0)} + \hat{f}_x^{(y=L_y)}\right)(m)\right) = 0$$

where $\mathrm{O} \equiv 2\mathbb{Z}+1$ is the set of odd integers, and $\mathrm{E} \equiv 2\mathbb{Z}$ is the set of even integers.

The solution of this linear system of equations is:

$$f(m\neq 0, n\in\delta) = \frac{-\mu}{e_\delta(m)}\sum_{n'\in\delta} u_x^{*\infty}(m,n'), \tag{17}$$

where the symbol δ stands for either E (even n's) or O (odd n's). The expressions for $m = 0$ are written separately:

$$f(0, n\in 2\mathbb{Z}) = 0$$

$$f(0, n\in \mathrm{O}) = \frac{-4\mu}{{L_y}^2}\sum_{n'\in \mathrm{O}} u_x^{*\infty}(m,n').$$

In eqn (17), we have introduced auxiliary functions $e_\mathrm{E}(m)$ and $e_\mathrm{O}(m)$, which satisfy:‡‡

$$e(m) \equiv \sum_{n\in Z}\frac{{q_n}^2}{({p_m}^2+{q_n}^2)^2} = \frac{{L_y}^2}{2\pi}\left[\frac{-\pi}{\sinh^2(2\pi L_y m/L_x)} + \frac{L_x}{2mL_y}\frac{1}{\tanh(2\pi L_y m/L_x)}\right]$$

$$e_\mathrm{E}(m) \equiv \sum_{n\in \mathrm{E}}\frac{{q_n}^2}{({p_m}^2+{q_n}^2)^2} = \frac{1}{4}\, e\left(\frac{m}{2}\right)$$

$$e_\mathrm{O}(m) \equiv \sum_{n\in \mathrm{O}}\frac{{q_n}^2}{({p_m}^2+{q_n}^2)^2} = e(m) - \frac{1}{4}e\left(\frac{m}{2}\right)$$

Now, the infinite summation in eqn (17) needs to be calculated. For a single plastic event located at $(x_\mathrm{ev}, y_\mathrm{ev})$, that is, $\hat{\varepsilon}^\mathrm{pl}(m,n) = e^{-\mathrm{i}p_m x_\mathrm{ev}}e^{-\mathrm{i}q_n y_\mathrm{ev}}(\varepsilon_{xx}^\mathrm{pl}, \varepsilon_{xy}^\mathrm{pl})^T$, the use of the expression for $\hat{u}_x^{*\infty}$ leads to:

$$\sum_{n'\in\delta}\hat{u}_x{}^{*\infty}(m,n') = 4e^{-\mathrm{i}p_m x_\mathrm{ev}}\left[\varepsilon_{xy}^\mathrm{pl}\left({p_m}^2\frac{{L_y}^3}{\pi^3}j_\delta(X) - \frac{L_y}{\pi}k_\delta(X)\right) - 2\mathrm{i}\varepsilon_{xx}^\mathrm{pl}p_m\frac{{L_y}^2}{\pi^2}s_\delta(X)\right], \tag{18}$$

where the δ-subscript stands for either E or O, and $X \equiv (x,\alpha) \equiv \left(\frac{\pi y_\mathrm{ev}}{L_y}, \frac{p_m L_y}{\pi}\right)$.

Inserting eqn (18) into eqn (17), summing the plastic activity of all y lines, *i.e.*,§§ $y = 0.5,\ldots,L_y - 0.5$ ($L_y \in \mathbb{N}^*$) in the discretised version, and Fourier transforming the results along the x-direction *via* the operator $\mathcal{F}_x$, defined by $\mathcal{F}_x\sigma = \frac{1}{L_x}\int\sigma(x)\mathrm{e}^{-\mathrm{i}p_m x}\mathrm{d}x$, one finally arrives at:

‡‡ The analytical calculations leading to the second part of the equality involve the decomposition into simple elements and the use of well established summation results.[77]

§§ The +0.5 term comes from the fact that the y-coordinate of a block (of unit size) is evaluated at its centre.

$$\underline{\hat{u}}^{corr}(m, n \in \delta) = \begin{pmatrix} \dfrac{-4{q_n}^2}{4\mu q^4} \cdot \left[\sum_y \overbrace{\left(\dfrac{{p_m}^2 {L_y}^2}{e_\delta(m)\pi^3} j_\delta(X) - \dfrac{1}{\pi} k_\delta(X) \right)}^{\equiv \zeta_\delta(X)} \mathscr{F}_x \sigma_{xy}^{\mathrm{pl}}(m, y) - 2\mathrm{i} \sum_y \overbrace{\left(\dfrac{p_m L_y}{e_\delta(m)\pi^2} s_\delta(X) \right) \mathscr{F}_x \sigma_{xx}^{\mathrm{pl}}(m, y)}^{\equiv \xi_\delta(X)} \right] \\ \dfrac{4 p_m q_n}{4\mu q^4} \left[\sum_y \left(\dfrac{{p_m}^2 {L_y}^2}{e_\delta(m)\pi^3} j_\delta(X) - \dfrac{1}{\pi} k_\delta(X) \right) \mathscr{F}_x \sigma_{xy}^{\mathrm{pl}}(m, y) - 2\mathrm{i} \sum_y \dfrac{p_m L_y}{e_\delta(m)\pi^2} s_\delta(X) \mathscr{F}_x \sigma_{xx}^{\mathrm{pl}}(m, y) \right] \end{pmatrix},$$

where new summations appear and can be expressed analytically *via* a decomposition into simple elements and the use of known summation formulae:[77]

$$j(x,\alpha)\equiv\sum_{k=-\infty}^{+\infty}\frac{k\sin(kx)}{(k^2+\alpha^2)^2}=\frac{\pi}{2\alpha^2}\frac{\sinh(\alpha(\pi-x))}{\sinh(\alpha\pi)}-\frac{1}{2\alpha^2}\mathscr{H}(x,\alpha)$$

$$j_{\mathrm{E}}(x,\alpha)=\frac{1}{8}j\left(2x,\frac{\alpha}{2}\right)$$

$$\mathscr{H}(x\neq 0,\alpha)\equiv\sum_{k=-\infty}^{+\infty}\frac{k\sin(kx)}{(k-\mathrm{i}\alpha)^2}=\frac{h(x,\alpha)+h(x,-\alpha)}{2}$$

$$h(x\neq 0,\alpha)\equiv-\mathrm{i}\sum_{k=-\infty}^{+\infty}\frac{k\exp(\mathrm{i}kx)}{(k-\mathrm{i}\alpha)^2}=\frac{\pi\exp(-x\alpha)}{1-\cosh(2\pi\alpha)}\Big[x\alpha\left(\mathrm{e}^{2\pi\alpha}-1\right)+2\pi\alpha-\left(\mathrm{e}^{2\pi\alpha}-1\right)\Big]$$

$$k(x,\alpha)\equiv\sum_{k=-\infty}^{+\infty}\frac{k^3\sin(kx)}{(k^2+\alpha^2)^2}=\frac{\pi}{2}\frac{\sinh(\alpha(\pi-x))}{\sinh(\alpha\pi)}+\frac{\mathscr{H}(x,\alpha)}{2}$$

$$k_{\mathrm{E}}(x,\alpha)=\frac{1}{2}k\left(2x,\frac{\alpha}{2}\right)$$

$$s(x,\alpha)\equiv\sum_{k=-\infty}^{+\infty}\frac{k^2\exp(\mathrm{i}kx)}{(k^2+\alpha^2)^2}=\frac{\pi}{2}\frac{\cosh(\alpha(\pi-x))}{\alpha\sinh(\alpha\pi)}+\frac{\pi}{4}u(x,\alpha)$$

$$s_{\mathrm{E}}(x,\alpha)=\frac{1}{4}s\left(2x,\frac{\alpha}{2}\right)$$

$$u(x,\alpha)\equiv\frac{2x\cosh\left(\alpha(x-2\pi)\right)+(2\pi-x)\cdot 2\cosh(\alpha x)}{\left(1-\cosh(2\pi\alpha)\right)}$$

The function j_{O} is obtained by writing $j(x,\alpha)=j_{\mathrm{O}}(x,\alpha)+j_{\mathrm{E}}(x,\alpha)$; the same applies for the other functions with subscripts O.

The coincidence of the infinite summations and their analytical expressions has been verified numerically for particular values of the parameters.

As a technical remark, we would like to mention that the preceding formulae are difficult to evaluate numerically for $|\alpha|\gg 1$, on account of the large arguments of the hyperbolic functions. Nevertheless, the following approximations provide very satisfactory results in the limit of large $\alpha(\alpha>0)$:

$$\frac{\sinh[\alpha(\pi - x)]}{\sinh(\alpha\pi)} \approx \exp(-x\alpha) - \exp(\alpha(x - 2\pi))$$

$$\frac{\cosh[\alpha(\pi - x)]}{\sinh(\alpha\pi)} \approx \exp(-x\alpha) + \exp(\alpha(x - 2\pi))$$

$h(x,\alpha) \approx -2\pi \exp(-x\alpha)[x\alpha - 1]$

$u(x,\alpha) \approx -2[x \exp[\alpha(x - 4\pi)] + x \exp(-\alpha x) + (2\pi - x) \cdot \exp[\alpha(x - 2\pi)]]$.

Our final result is:

$$\begin{pmatrix} \sigma_{xx}^{\text{corr}}(m,n) \\ \sigma_{xy}^{\text{corr}}(m,n) \end{pmatrix} = \begin{pmatrix} \dfrac{-2p_m q_n^2}{q^4}\left[\mathrm{i}\sum_y \zeta_\delta(X)\mathscr{F}_x\sigma_{xy}^{\text{pl}}(m,y) + 2\sum_y \xi_\delta(X)\mathscr{F}_x\sigma_{xx}^{\text{pl}}(m,y)\right] \\ \dfrac{q_n(p_m^2 - q_n^2)}{q^4}\left[\mathrm{i}\sum_y \zeta_\delta(X)\mathscr{F}_x\sigma_{xy}^{\text{pl}}(m,y) + 2\sum_y \xi_\delta(X)\mathscr{F}_x\sigma_{xx}^{\text{pl}}(m,y)\right] \end{pmatrix}, \tag{19}$$

where we should note that $\zeta(0, n\in \mathrm{O}) = \dfrac{-2}{L_y^2}$.

As a computational detail, note that the y-coordinates are here integers shifted by half unity, *i.e.*, of the form $p + \frac{1}{2}$, $p\in\mathbb{N}$, whereas computational routines for Fast Fourier Transforms take as input an array with integer indices. It is therefore easier to suppose that the walls are at positions $y = -\frac{1}{2}$ and $y = L_y - \frac{1}{2}$. This translation is readily achieved by simply multiplying the Fourier components of the correction term, as given above, by prefactors $\exp\left(\frac{\mathrm{i}q_n}{2}\right)$.

Assuming a complexity $O(N \ln N)$ for the Fast Fourier Transform of an array of N cells, the number of operations performed at each time step of our algorithm is of order $O(L_x L_y^2 \ln L_x)$ for large integers L_y and L_x, as it is evident from eqn (19).

B Calculation of the line-averaged velocity

The mean velocity on a line $y = y_0$ reads:

$$\langle u_x\rangle_x(y_0) \equiv \frac{1}{L_x}\int_{-L_x/2}^{L_x/2} u_x(x, y_0)\mathrm{d}x$$

$$= \sum_{n=-\infty}^{+\infty} \hat{u}_x(m = 0, n)\mathrm{e}^{\mathrm{i}q_n y_0}$$

$$= \sum_{\substack{n=-\infty \\ n\neq 0}}^{+\infty} \hat{u}_x^{*\infty}\infty(0,n)\mathrm{e}^{\mathrm{i}q_n y_0} + \hat{u}_x^{*\infty}(0,0) - (1 - 2|y_0|/L_y)\sum_I \hat{u}_x^{*\infty}(0,\cdot) + \sum_P \overbrace{\hat{u}_x^{\text{corr}}(0,\cdot)}^{0}\mathrm{e}^{\mathrm{i}q_n y_0}$$

$$= \sum_{y_{\text{ev}}} \frac{a}{2\mu}\left[\text{sign}(y_0 - y_{\text{ev}}) \cdot \left(1 - \frac{|y_0 - y_{\text{ev}}|}{L_y}\right) + 1 - \frac{y_{\text{ev}}}{L_y} - \frac{y_0}{L_y}\right]\mathscr{F}_x\sigma_{xy}^{\text{pl}}(m = 0, y_{\text{ev}}),$$

where the last summation is performed over all streamlines $y_e v$, and $\hat{u}_x^{*\infty}$ is the bulk contribution in the duplicated system.

C Estimation of the deviations due to bulk cooperativity

Assume the fluidity diffusion equation is a valid approximation,

$\xi^2\Delta f - (f - f_{\text{bulk}}) = 0$ where $f = \frac{\dot{\gamma}}{\sigma}$ is the local fluidity, and ξ is a cooperativity length that may vary with the shear rate.

Let $\delta f = f - f_{\text{bulk}}$ be the deviation from the expected fluidity profile owing to cooperative effects between regions subject to different driving forces.

One now assumes $\delta f \ll f_{\text{bulk}}$ and $\Delta\delta f \ll \Delta f_{\text{bulk}}$.

To leading order, the fluidity diffusion equation reads

$\xi^2\Delta f_{\text{bulk}} = \delta f.$

The amplitude of the deviations due to cooperativity is given by the Babel number $\text{Ba} \equiv \frac{\delta f}{f} \approx \xi^2\frac{\Delta f_{\text{bulk}}}{f_{\text{bulk}}}$

If the flow curve follows a Herschel–Bulkley law: $\sigma(\dot{\gamma}) = \sigma_{\text{d}} + A\dot{\gamma}^n$, $f''_{\text{bulk}} = \frac{\sigma'^2}{A^{\frac{1}{n}}}$

$$\times\frac{\sigma^{n-1}(\sigma - \sigma_{\text{d}})^{\frac{1}{n}-1}}{n}\left[\left(\frac{1}{n}-1\right)\frac{\sigma^{-n}}{\sigma - \sigma_{\text{d}}}\left((1-n)+n\frac{\sigma_{\text{d}}}{\sigma}\right)^2 - n\sigma^{-n-1}\left(1-n+(1+n)\frac{\sigma_{\text{d}}}{\sigma}\right)\right]$$

Here, the primes denote derivatives with respect to the space coordinate. Then,

$$\frac{f''_{\text{bulk}}}{f_{\text{bulk}}} = \frac{\sigma'^2}{n(\sigma - \sigma_{\text{d}})}\left[\frac{\left(\frac{1}{n}-1\right)}{\sigma - \sigma_{\text{d}}}\left((1-n)+n\frac{\sigma_{\text{d}}}{\sigma}\right)^2 - \frac{n}{\sigma}\left(1-n+(1+n)\frac{\sigma_{\text{d}}}{\sigma}\right)\right]$$

To leading order, one finally arrives at $\frac{\delta f}{f} \sim \xi^2\frac{\sigma'^2}{(\sigma - \sigma_{\text{d}})^2}$.

Acknowledgements

We thank T. Divoux, K. Martens, P. Chaudhuri, S. Manneville, and M. Fardin for interesting discussions. JLB is supported by Institut Universitaire de France and by grant ERC-2011-ADG20110209.

References

1 L. Berthier, G. Biroli, J.-P. Bouchaud, L. Cipelletti, D. El Masri, D. L'Hôte, F. Ladieu and M. Pierno, *Science*, 2005, **310**, 1797–800.
2 C. Heussinger, P. Chaudhuri and J.-L. Barrat, *Soft Matter*, 2010, **6**, 3050–3058.
3 D. Rodney, A. Tanguy and D. Vandembroucq, *Modell. Simul. Mater. Sci. Eng.*, 2011, **19**, 083001.
4 K. A. Dahmen, Y. Ben-Zion and J. T. Uhl, *Nat. Phys.*, 2011, **7**, 554–557.
5 A. Argon and H. Kuo, *Mater. Sci. Eng.*, 1979, **39**, 101–109.
6 A. Lemaître and C. Caroli, *Phys. Rev. E: Stat., Nonlinear, Soft Matter Phys.*, 2007, **76**, 036104.
7 M. Tsamados, A. Tanguy, F. Léonforte and J.-L. Barrat, *Eur. Phys. J. E*, 2008, **26**, 283–93.
8 M. Falk and J. Langer, *Phys. Rev. E: Stat. Phys., Plasmas, Fluids, Relat. Interdiscip. Top.*, 1998, **57**, 7192.
9 M. L. Falk and J. S. Langer, *Annual Review of Condensed Matter Physics, vol 2*, 2011, vol. 2, pp. 353–373.
10 P. Sollich, F. Lequeux, P. Hébraud and M. Cates, *Phys. Rev. Lett.*, 1997, **78**, 2020–2023.
11 P. Sollich, *Phys. Rev. E: Stat. Phys., Plasmas, Fluids, Relat. Interdiscip. Top.*, 1998, **58**, 738.
12 P. Hébraud and F. Lequeux, *Phys. Rev. Lett.*, 1998, **81**, 2934–2937.
13 P. Coussot, Q. Nguyen, H. Huynh and D. Bonn, *Phys. Rev. Lett.*, 2002, **88**, 175501.

14 M. Manning, J. Langer and J. Carlson, *Phys. Rev. E: Stat., Nonlinear, Soft Matter Phys.*, 2007, **76**, 056106.
15 S. M. Fielding, M. E. Cates and P. Sollich, *Soft Matter*, 2009, **5**, 2378.
16 L. Bocquet, A. Colin and A. Ajdari, *Phys. Rev. Lett.*, 2009, **103**, 036001.
17 K. Chen, P. Bak and S. Obukhov, *Phys. Rev. A: At., Mol., Opt. Phys.*, 1991, **43**, 625–630.
18 V. V. Bulatov and A. S. Argon, *Modell. Simul. Mater. Sci. Eng.*, 1994, **2**, 167–184.
19 V. V. Bulatov and A. S. Argon, *Modell. Simul. Mater. Sci. Eng.*, 1994, **2**, 185–202.
20 V. V. Bulatov and A. S. Argon, *Modell. Simul. Mater. Sci. Eng.*, 1994, **2**, 203–222.
21 J.-C. Baret, D. Vandembroucq and S. Roux, *Phys. Rev. Lett.*, 2002, **89**, 195506.
22 G. Picard, A. Ajdari, F. Lequeux and L. Bocquet, *Phys. Rev. E: Stat., Nonlinear, Soft Matter Phys.*, 2005, **71**, 010501.
23 E. R. Homer and C. A. Schuh, *Acta Mater.*, 2009, **57**, 2823–2833.
24 K. Martens, L. Bocquet and J.-L. Barrat, *Soft Matter*, 2012, **8**, 4197.
25 K. Martens, L. Bocquet and J.-L. Barrat, *Phys. Rev. Lett.*, 2011, **106**, 156001.
26 A. Nicolas and J.-L. Barrat, *Phys. Rev. Lett.*, 2013, **110**, 138304.
27 H. Princen, *J. Colloid Interface Sci.*, 1985, **105**, 150–171.
28 C. Maloney and A. Lemaître, *Phys. Rev. E: Stat., Nonlinear, Soft Matter Phys.*, 2006, **74**, 016118.
29 A. Amon, V. Nguyen, A. Bruand, J. Crassous and E. Clément, *Phys. Rev. Lett.*, 2012, **108**, 135502.
30 M. Le Merrer, S. Cohen-Addad and R. Höhler, *Phys. Rev. Lett.*, 2012, **108**, 188301.
31 J. Bouchaud and E. Pitard, *Eur. Phys. J. E: Soft Matter Biol. Phys.*, 2001, **6**, 231–236.
32 M. Cloitre, R. Borrega, F. Monti and L. Leibler, *Phys. Rev. Lett.*, 2003, **90**, 068303.
33 G. Picard, A. Ajdari, F. Lequeux and L. Bocquet, *Eur. Phys. J. E*, 2004, **15**, 371–81.
34 F. Leonforte, R. Boissière, A. Tanguy, J. Wittmer and J.-L. Barrat, *Phys. Rev. B: Condens. Matter Mater. Phys.*, 2005, **72**, 224206.
35 P. Schall, D. A. Weitz and F. Spaepen, *Science*, 2007, **318**, 1895–9.
36 J. Bokeloh, S. V. Divinski, G. Reglitz and G. Wilde, *Phys. Rev. Lett.*, 2011, **107**, 235503.
37 R. Besseling, L. Isa, P. Ballesta, G. Petekidis, M. Cates and W. Poon, *Phys. Rev. Lett.*, 2010, **105**, 268301.
38 M. Talamali, V. Petäjä, D. Vandembroucq and S. Roux, *C. R. Mec.*, 2012, **340**, 275.
39 J. Goyon, A. Colin, G. Ovarlez, A. Ajdari and L. Bocquet, *Nature*, 2008, **454**, 84–7.
40 P. Chaudhuri, V. Mansard, A. Colin and L. Bocquet, *Phys. Rev. Lett.*, 2012, **109**, 036001.
41 O. Pouliquen, Y. Forterre and S. Le Dizes, *Advances in Complex Systems*, 2001, **04**, 441–450.
42 R. R. Hartley and R. P. Behringer, *Nature*, 2003, **421**, 928–31.
43 A. Amon, R. Bertoni and J. Crassous, *Phys. Rev. E*, 2013, **87**, 012204.
44 J. Goyon, A. Colin and L. Bocquet, *Soft Matter*, 2010, **6**, 2668.
45 P. Jop, V. Mansard, P. Chaudhuri, L. Bocquet and A. Colin, *Phys. Rev. Lett.*, 2012, **108**, 148301.
46 B. Geraud, L. Bocquet and C. Barentin, *Eur. Phys. J. E*, 2013, **36**, 9845.
47 L. Isa, R. Besseling and W. C. K. Poon, *Phys. Rev. Lett.*, 2007, **98**, 198305.
48 L. Isa, R. Besseling, A. Morozov and W. Poon, *Phys. Rev. Lett.*, 2009, **102**, 058302.
49 O. Pouliquen and R. Gutfraind, *Phys. Rev. E: Stat. Phys., Plasmas, Fluids, Relat. Interdiscip. Top.*, 1996, **53**, 552.
50 R. Gutfraind and O. Pouliquen, *Mech. Mater.*, 1996, **24**, 273–285.
51 J.-B. Salmon, L. Bécu, S. Manneville and A. Colin, *Eur. Phys. J. E*, 2003, **10**, 209–221.
52 K. Kamrin and G. Koval, *Phys. Rev. Lett.*, 2012, **108**, 178301.
53 G. Ovarlez, S. Rodts, A. Ragouilliaux, P. Coussot, J. Goyon and A. Colin, *Phys. Rev. E: Stat., Nonlinear, Soft Matter Phys.*, 2008, **78**, 036307.
54 C. Masselon and A. Colin, *Phys. Rev. E: Stat., Nonlinear, Soft Matter Phys.*, 2010, **81**, 021502.
55 L. Bécu, S. Manneville and A. Colin, *Phys. Rev. Lett.*, 2004, **93**, 018301.
56 T. Gibaud, C. Barentin and S. Manneville, *Phys. Rev. Lett.*, 2008, **101**, 258302.
57 P. Ballesta, R. Besseling, L. Isa, G. Petekidis and W. Poon, *Phys. Rev. Lett.*, 2008, **101**, 258301.
58 V. Mansard, PhD thesis, Université de Bordeaux I, 2012.
59 E. Pagac, R. Tilton and D. Prieve, *Chem. Eng. Commun.*, 1996, **148**, 105–122.
60 A. Yoshimura and R. K. Prud'homme, *J. Rheol.*, 1988, **32**, 53.
61 H. A. Barnes, *J. Non-Newtonian Fluid Mech.*, 1995, **56**, 221–251.
62 J. Franco, C. Gallegos and H. Barnes, *J. Food Eng.*, 1998, **36**, 89–102.
63 S. Meeker, R. Bonnecaze and M. Cloitre, *Phys. Rev. Lett.*, 2004, **92**, 198302.
64 S. P. Meeker, R. T. Bonnecaze and M. Cloitre, *J. Rheol.*, 2004, **48**, 1295.
65 J.-B. Salmon, A. Colin and S. Manneville, *Phys. Rev. Lett.*, 2003, **90**, 228303.
66 L. Bécu, P. Grondin, A. Colin and S. Manneville, *Colloids Surf., A*, 2005, **263**, 146–152.

67 T. Mason, J. Bibette and D. Weitz, *J. Colloid Interface Sci.*, 1996, **179**, 439–448.
68 M. Sanchez, C. Valencia, J. Franco and C. Gallegos, *J. Colloid Interface Sci.*, 2001, **241**, 226–232.
69 G. H. Meeten, *J. Non-Newtonian Fluid Mech.*, 2004, **124**, 51–60.
70 J. Goyon, PhD thesis, Université de Bordeaux I, 2008.
71 T. Divoux, C. Barentin and S. Manneville, *Soft Matter*, 2011, **7**, 9335–9349.
72 T. Divoux, C. Barentin and S. Manneville, *Soft Matter*, 2011, **7**, 8409–8418.
73 J. R. Seth, C. Locatelli-Champagne, F. Monti, R. T. Bonnecaze and M. Cloitre, *Soft Matter*, 2012, **8**, 140.
74 J. R. Seth, M. Cloitre and R. T. Bonnecaze, *J. Rheol.*, 2008, **52**, 1241.
75 R. Besseling, L. Isa, E. R. Weeks and W. C. Poon, *Adv. Colloid Interface Sci.*, 2009, **146**, 1–17.
76 P. Ballesta, G. Petekidis, L. Isa, W. C. K. Poon and R. Besseling, *J. Rheol.*, 2012, **56**, 1005.
77 I. Gradshteyn and I. Ryzhik, *Tables of Integrals, Series and Products*, (edited by A. Jeffrey), 1994.

Faraday Discussions RSC Publishing

DISCUSSIONS

General discussion

DOI: 10.1039/C3FD90040A

Professor Molinero opened the discussion of the paper by Professor Chandler: What is the scope of applicability of the theory of corresponding states? Most if not all models of water display a temperature of maximum density, but not all favor an open tetrahedral crystal as the ground state at room pressure. TIP5P, for example favors the high density ice II over the low-density ice I at low but positive pressures and crystallizes into a non natural high density polymorph in the high pressure simulations of Yamada *et al.*[1] referred to in your paper. Does the theory apply to any model that presents a density maximum and is a poor glass former, irrespective of the crystal they form?

1 M. Yamada, S. Mossa, H. Stanley and F. Sciortino, Interplay between time-temperature transformation and the liquid–liquid phase transition in water, *Phys. Rev. Lett.*, 2002, **88**, 195701.

Professor Chandler replied: I believe it is important to the behavior of water and water-like systems that the shadow of a low-density crystal is found in the low-pressure liquid. To the extent that such a shadow is present in a model of water, there will be a density maximum and I believe our analysis should be valid. Whether the temperature of maximum density is above or below the melting temperature seems less important, and whether the low pressure of the low density crystal is positive or negative seems less important too.

Professor Molinero continued: Fig. 4 shows remarkable agreement in the curves of reduced crystallization times for the TIP5P and mW water models, although I understand that for TIP5P the melting temperature T_m of hexagonal ice at 1 bar was used in lieu of the T_m of the high density polymorph that spontaneously formed in the high pressure simulations. Is the good agreement a result of a fortuitous coincidence of the melting point of the high density ice polymorph at high pressure and the one of ice I at 1 bar, or the scaling is somehow insensitive to the value of T_m?

Professor Chandler responded: Yes, for the conditions considered (*i.e.*, away from coexistence), there is not much sensitivity to the precise value of the melting temperature, though I expect similar values for the melting temperatures of different low-pressure crystals of TIP5P. In the case of the mW model, where we have computed ice–liquid surface tension, heat of fusion, and so forth, there are no fitting of parameters in our analysis. For TIP5P we know less, so some fitting is required. The specific values of parameters used to compare numerical data with your equations are provided, as you note, in Table 1.

Professor Anisimov asked: This approach is very straightforward. We know that any corresponding state "law" cannot be exact. Can you identify a parameter that could make this "law" not exactly universal, depending on the model?

Professor Chandler answered: I appreciate that you agree that the approach Limmer and I have followed is straight-forward. I agree with you that a corresponding states relationship, like the parabolic law, eqn 2 of our paper, cannot be exact, but in the sense of dynamical crossovers, not in the sense of critical phenomena. For one thing, eqn 2 holds only at temperatures below the onset to correlated hierarchical dynamics (*i.e.*, $T < T_o$), and we have no theory that predicts the crossover to the normal liquid regime at higher temperatures. We also have no theory to predict why $T_o \approx T_{max}$. That relationship is an empirical finding about water. On the other hand, we understand how the parabolic law emerges as a universal result from hierarchical dynamics characteristic of glass formers, we have theory for computing the parameters of the parabolic law, and that theory has significant predictive power as Limmer and I have illustrated.

Professor Caupin noted: On p. 3, you write that "Two distinct reversible liquids in coexistence would imply the existence of a low temperature critical point of the sort suggested by Stanley and his coworkers." I would like to point out that it is not a necessary condition that a liquid–liquid transition gives a critical point at low temperature, near the homogenous crystallization temperature. There are other thermodynamically consistent possibilities. In simple water-like models,[1,2] depending on the parameter values, the critical point associated to the liquid–liquid transition in these models can be moved to zero temperature (which corresponds to the singularity free interpretation), or on the contrary to high temperature, even exceeding the liquid–vapour spinodal line. In the latter case, the liquid–vapour spinodal would be replaced at low temperature by the liquid–liquid spinodal, that reaches positive pressure; this case would be qualitatively equivalent to the stability limit conjecture.

1 P. H. Poole *et al.*, *Phys. Rev. Lett.*, 1994, **73**, 1632–1635.
2 K. Stokely *et al.*, *Proc. Natl. Acad. Sci. U. S. A.*, 2010, **107**, 1301–1306.

Professor Chandler responded: Thank you for allowing me to clarify. While symmetry dictates that liquid–liquid co-existence is accompanied by a critical point, you are correct that it need not be at a low temperature. Indeed, it is possible to construct models that yield a second critical point above the melting temperature. Stanley and his co-workers have detailed phase behavior of models that exhibit all sorts of behaviors. But the context of the remark you quote has little if anything to do with those models. Specifically, our remark refers either to water or to some reasonable approximation of water. The models you cite are not so in that they do not describe a liquid with local tetrahedral structure that freezes into an ice-like crystal.

Real water and water-like models do exhibit polyamorphism, but it occurs far from equilibrium. This irreversible behavior refers to amorphous solids of ice.

Those solids have a time symmetry that is distinct from that of an ergodic liquid. As such, the line separating the high-density amorphous solid from the low-density amorphous solid ends at a triple point, not at a critical point.[1] The reversible liquid–liquid transitions that appear in the models you cite seem to be nothing like what is found in water.

1 D. T. Limmer and D. Chandler, 2013, arXiv:1306.4728.

Professor Caupin continued: You use the value 32 mN m^{-1} for the liquid–solid interfacial tension γ of water, citing the paper by Granasy *et al.* (Ref. 44). However, this paper gives a review of proposed values for γ, which is a quantity difficult to measure. 32 mN m^{-1} is one of these values, but there is a large scatter, see their Table IV. Classical nucleation theory, and your parameter Γ, are very sensitive to the choice of γ. Also it is not clear if the value of γ at equilibrium can be used far in the metastable region.

Dr Limmer replied: Prof. Caupin is entirely correct, at present there is a large experimental uncertainty associated with the liquid–solid surface tension. The value of γ = 32 mN m^{-1} we use is actually the mean of all of the measurements presented by Granasy *et al.* (see Ref. 44 of our Faraday contribution). This point was elaborated on in our previous work,[1] and will be clarified in our manuscript before the final submission. As to the latter point regarding the use of the equilibrium surface tension deep into the supercooled regime, we have explored this assumption explicitly in previous work[1] and found that while the surface tension can indeed exhibit a strong temperature dependence, this temperature dependence is the same as the temperature dependence of the enthalpy difference between liquid and crystal (following the so-called Turnbull relation). As ratios of these two quantities enter into the classical nucleation rate expression, there is a cancelation of errors associated with this assumption.

1 D. T. Limmer and D. Chandler, Phase diagram of supercooled water confined to hydrophilic nanopores, *J. Chem. Phys.*, 2012, **137**, 045509.1–11.

Professor Wang addressed Professor Chandler: In your paper, you mentioned that T_o is approximately T_{max}. When you derive the super-Arrhenius to weak Arrhenius transition, you used the fractional Stokes–Einstein relationship for supercooled liquid at $T < T_o$. Does this suggest your derivation requires T_{max} to be close to T_m to be valid?

Professor Chandler replied: The quick answer is "no", but I think I detect confusion that perhaps I can address.

The onset to correlated dynamics, which begins around the temperature $T = T_o$, refers to dynamics and not thermodynamics. In contrast, the melting temperature, T_m, and the temperature of maximum density, T_{max}, refer to thermodynamics. T_m is the temperature of a transition (the reversible thermodynamic transition between liquid water and ice). In contrast, neither T_{max} nor T_o are "transition" temperatures (*i.e.*, neither are associated with a singularity in a response function). Both can be regarded as "crossover" temperatures.

In water and water-like models, T_{max} and T_m are reasonably close in value. In some such models, $T_{max} < T_m$ and in others, $T_{max} > T_m$. In either case, for all water-like models by definition, there is preferentially local tetrahedral structure in the liquid, and there is global tetrahedral order when the liquid transitions to ice. Further, and this is one of the principal points of Limmer's and my paper, $T_o \approx T_{max}$.

This approximate but accurate connection between T_o and T_{max} is an empirical result that we have discovered by examining various water-like models and also by examining experimental water. It is a fact about water. Similarly, T_{max} being reasonably close in value to T_m is also a fact about water.

All glass-forming liquids exhibit a crossover from normal to correlated hierarchical dynamics at an onset temperature, T_o.[1,2] In some cases $T_o > T_m$, and in other cases $T_o < T_m$. Which is to say, the onset to correlated dynamics is not directly linked to the thermodynamics of melting and freezing.

1 Y. Elmatad, D. Chandler and J. P. Garrahan, *J. Phys. Chem. B*, 2009, **113**, 5563.
2 Y. Elmatad, D. Chandler and J. P. Garrahan, *J. Phys. Chem. B*, 2010, **114**, 171113.

Professor Wang continued: On page 7, you mentioned that the time scale for adding material to a growing cluster has a super-Arrhenius behavior at $T < T_o$. In bulk water, it is postulated there is a strong to fragile transition in the supercooled regime. The transition described in your paper at T_o is in the fragile regime. Is there any experimental evidence for this transition?

Professor Chandler responded: This question would seem to suggest that T_o is the temperature of a "fragile–strong" transition, but that suggestion would not be correct. First and foremost, as I noted in my response to your prior question, T_o locates a crossover region, not a transition. Experimental evidence of an onset temperature crossover in glass-formers is abundant.[1,2] Second, in most glass-forming liquids, the temperature dependence of transport in the normal liquid regime, $T > T_o$, is sub-Arrhenius. Water and some alcohols and models of liquid mixtures studied in simulations are exceptions in that transport in the normal liquid regime for those liquids is Arrhenius.

On the other hand, while it is not the onset crossover, there is a well-documented non-equilibrium fragile–strong transition. It is the glass transition. The temperature of that transition, T_g, depends upon the protocols by which the glass is prepared and by which its transition temperature is measured. While not a reversible thermodynamic transition, for a fixed set of preparation and measurement protocols, the glass transition is nevertheless a point of singular response. Limmer and I have recently presented an explicit demonstration of the nature of that singularity for the special case of supercooled water.[3]

For $T_o > T > T_g$, relaxation is reversible and hierarchical. As such, the parabolic law is obeyed (*i.e.*, transport is super Arrhenius) in that regime. For $T < T_g$, relaxation is no longer hierarchical, but rather dominated by motions on one principal length scale (and thus one principal energy scale). As such, relaxation is Arrhenius in that regime.

Thus, there are three general regimes for liquid and glass behaviors: (a) $T > T_o$, relaxation is normal, *i.e.*, neither highly correlated nor hierarchical. Here, relaxation times are possibly, but not necessarily, Arrhenius; (b) $T_o > T > T_g$, relaxation remains reversible but hierarchical. Here, relaxation times are super-Arrhenius, obeying the parabolic law, eqn 2 in our paper; (c) $T_g > T$, relaxation is irreversible and Arrhenius. The change from (a) to (b) is a reversible crossover; the change from (b) to (c) is a non-equilibrium transition – the irreversible glass transition.

Finally, you refer to a "postulated" strong-to-fragile in supercooled water. The phenomena discussed in that regard are multifaceted. In cases where the experimental phenomena appear to be reproducible, Limmer and I have been able to relate the phenomena to one of two processes: the water–ice transition and the water–glass transition.[4]

1 Y. Elmatad, D. Chandler and J. P. Garrahan, *J. Phys. Chem. B*, 2009, **113**, 5563.
2 Y. Elmatad, D. Chandler and J. P. Garrahan, *J. Phys. Chem. B*, 2010, **114**, 171113.
3 D. T. Limmer and D. Chandler, 2013, arXiv:1306.4728.
4 D. T. Limmer and D. Chandler, *J. Chem. Phys.*, 2012, **137**, 045509.

Dr Royall inquired: As Prof. Molinero has pointed out, in Fig. 5b of your paper, you suggest that the structural relaxation time will, at some point exceed the time to crystallise. As far as I can see, such behaviour would likely be seen in many systems. For example we have seen it in hard spheres.[1]

Do you argue that such behaviour occurs in water and water-like systems, and that the observation in hard spheres is atypical, or would you say it is quite a usual phenomenon? In fact is this (partly) what defines a marginal glassformer?

1 Fig. 7b of Taffs *et al.*, *Soft Matter*, 2013, **9**, 297.

Professor Chandler replied: Yes, the idea of using time–temperature plots for understanding whether a material is a marginal or good glass former is fairly standard and should apply rather generally. That said, when the time scales for crystallization and liquid relaxation become of the same order, the two classes dynamics are not easily disentangled, and the quantitative meaning of a time–temperature plot ceases to be reliable.

Dr Billard opened the discussion of the paper by Professor Triolo: Could you suggest any compound which properties (for example, solubility) could be influenced by the presence or absence of such polar, apolar and fluorinated domains, as an indirect proof of these structural organisation within the solvent?

Dr Triolo replied: Of course this question might open a wide discussion on the new applicative potentialities of highly nanostructured ILs. For us it is clear that the compartmentalised morphology that has been detected in ILs might represent a great opportunity to imagine and invent specific systems that might serve in as wide fields as separation, synthesis, catalysis *etc.*

The most straightforward idea that we plan to explore in this respect is to mix the proposed ILs with different, mutually incompatible compounds such as alkanes, perfluoroalkanes and inorganic salts, in order to probe if their location inside the IL domains succeeds in keeping them just a few Å apart, despite their reciprocal repulsion. Of course the potential mixtures that might have huge applicative impact are numerous and right now we are trying to explore the stability of a few of them.

Professor Wynne asked: In Fig. 6 of your paper,[1] one can see a sudden change in the structure factor at −60 °C from relatively broad bands to much narrower bands accompanied by the appearance of a relatively sharp peak at 0.5 Å^{-1}. This suggests that this is associated with a liquid–liquid phase transition. Would it be possible to confirm this as a LLPT? Can you exclude an isotropic to nematic (or similar) transition? Can you exclude the formation of metastable nano-crystallites, which have been proposed as an alternative explanation [2,3] for (what appears to be a) LLPT?

1 O. Russina, F. Lo Celso, M. Di Michiel, S. Passerini, G. B. Appetecchi, F. Castiglione, A. Mele, R. Caminiti and A. Triolo, *Faraday Discuss.*, 2013, **167**, DOI: 10.1039/C3FD00056G
2 A. Wypych, Y. Guinet and A. Hedoux, *Phys. Rev. B*, 2007, **76**, 144202.
3 A. Hedoux, Y. Guinet, P. Derollez, O. Hernandez, L. Paccou and M. Descamps, *J. Non-Cryst. Solids*, 2006, **352**, 4994–5000.

Dr Triolo replied: The results shown in Fig. 6 of the paper represent a further step in the fluorous tails segregation behaviour observed in Fig. 3 and 5 for samples with a slightly differently fluorinated anion. In this framework, we believe that the overall structural features of these materials should be pretty similar. As a matter of fact we find that upon cooling the sample in Fig. 6, before the development of the distinct low Q feature, a small bump, directly resembling the feature in Fig. 5, shows up (*e.g.* at −55 °C). Only at lower temperatures (or possibly, after longer aging time) the distinct peak at low Q develops. This leads us to envisage that the two structural features (low Q bump and peak) are correlated and represent one evolution of the other, upon appropriate temperature/aging conditions.

Of course these are preliminary data and we are in the progress of further exploring these features. However some additional comments can be made.

At odds with the diffraction data reported in the above mentioned reference,[1] the present data sets do not show any appreciable indication of Bragg peaks that might fingerprint the formation of crystallites. Over the whole Q range accessible to our measurements (*i.e.* from 0.2 to 20 Å^{-1}), no indication of Bragg peaks could be detected for all the temperatures probed in the present data set.

At the present stage we are not in a condition of determining whether these structural features are related to a LLPT. Our calorimetric characterizations do not indicate appreciable thermal events across the relevant temperature range. A possible explanation might be the occurrence of some collective re-organization, maybe related to conformational transitions in the fluorinated anion. Further studies are in progress to verify this hypothesis.

1 A. Hedoux, Y. Guinet, P. Derollez, O. Hernandez, L. Paccou and M. Descamps, *J. Non-Cryst. Solids*, 2006, **352**, 4994–5000.

Dr Perera inquired: The segregated domain formation reported in ILs bears many striking resemblance with the water–solute segregation reported previously in aqueous mixtures. Moreover, the IL systems show a nice pronounced cluster pre-peak in the scattered intensity $I(q)$, while this feature is generally very hard to see in aqueous mixtures, thus inducing some difficulties in the proper interpretation of clustering in these latter systems (Anisomov, Ben-Amotz, Soper, *etc*...). This is very intriguing, in view of the visual similarity in segregated clustering observed in the simulations of both systems. However, I observe that the $I(q)$ reported here for ILs has small low-q scattering, as opposed to those in aqueous mixtures, indicating small concentration fluctuations ($S(q = 0)$) for ILs, opposite to high $S(q = 0)$ found in aqueous mixtures. Since these IL systems are charged, I would like to hypothesize that charge ordering takes place (through the well known second Stillinger–Lovett sum rule in charged systems), which screens the concentration fluctuations. Do you think that such mechanism could help understand why these systems have almost no fluctuations, as opposed to aqueous mixtures where no direct charge ordering takes place?

Dr Triolo answered: This is a very stimulating question that will be worth further studies. As a matter of fact so far there is not an analytical model that has been proposed and compared with the large plethora of experimental data on the low Q features in ILs.

It is well established experimentally that at lower Q values than *ca.* 0.3 $Å^{-1}$ these systems do not show further diffraction features, thus confirming the observation reported in the question.

An accurate modelling in terms of atomistic contributions to the experimental $S(Q)$ has been conducted in the last few years by the group of Margulis.[1,2] Such studies highlighted that the diffraction pattern arising from paradigmatic ILs can be rationalised (in its low Q portion) as due to three major contributions. Features at *ca.* 1.5 $Å^{-1}$ are associated to first neighbour correlations (*e.g.* chain–chain), a peak or shoulder at *ca.* 0.8 $Å^{-1}$ fingerprints alternating positive–negative layers as can be expected in a molten salt and the peak centred at *ca.* 0.3–0.6 $Å^{-1}$ is associated to the polar–apolar moieties alternation. Of course this kind of diffraction pattern reflects the complex chemical structure of ILs that bear not only net charges in the cation and anion head but also side alkyl tails that tend to segregate from the charged moieties. Accordingly the structural picture would be one where charge ordering plays a major role in affecting the structure, but also steric effects are very effective in influencing in a non-trivial way the mesoscopic morphology.

I would also comment that, of course, when dealing with binary mixtures of ILs with other molecular liquids, such as water or alcohols, direct evidence of concentration fluctuations related to local demixing might be well evident as was observed in some cases, using small angle neutron scattering[3–5] or small angle X-ray scattering.[6] These systems are clearly characterised by excess low Q amplitude that reflects the existence of clustering that has a different origin than the polar–apolar dichotomy in neat ILs.

1 H. K. Kashyap, J. J. Hettige, H. V. R. Annapureddy, and C. J. Margulis, *Chem. Commun.*, 2012, **48**, 5103–5105.
2 J. J. Hettige, H. K. Kashyap, H. V. R. Annapureddy, and C. J. Margulis, *J. Phys. Chem. Lett.*, 2013, **4**, 105–110.
3 L. Almásy, M. Turmine, and A. Perera, *J. Phys. Chem. B*, 2008, **112**, 2382–2387.
4 T. Shimomura, K. Fujii, and T. Takamuku, *Phys. Chem. Chem. Phys.*, 2010, **12**, 12316–12324.
5 T. Takamuku, Y. Honda, K. Fujii, and S. Kittaka, *Anal. Sci.*, 2008, **24**, 1285–1290.
6 O. Russina, B. Fazio, G. Di Marco, R. Caminiti, A. Triolo, "Structural Organization in Neat Ionic Liquids and in Their Mixtures" in *The Structure of Ionic Liquids*, ed. R. Caminiti and L. Gontrani, Springer, 2014.

Dr Perkin asked: You have presented a Walden plot in the paper, with comparison to your ionic liquids to those of Drummond *et al.* described as 'good' and 'poor' ionic liquids in relation to their dissociation. Your results showing flourous aggregation involve 'good' ionic liquids, whereas Drummond's demonstration of separate alkyl and fluorous domains used 'poor' ionic liquids. Is there any *a priori* reason why poor ionic liquids should more easily give rise to triphilic segregation?

Dr Triolo responded: Our present understanding of the high level of structuring in these triphilic compounds is that long enough chains will tend to segregate into an affine environment, in such a way that alkyl tails will form oily clusters and fluorinated moieties will form fluorous ones. For some reasons, most probably the chain length is not long enough, our IM14-based ILs do not show distinct aggregation evidence at room conditions (as fingerprinted by the diffraction features at RT) and only at low enough temperature the fluorous tails aggregation seems to occur. On the other hand the systems proposed by Drummond's group show evidence of fluorous domains even at temperatures as high as 50 °C. This goes together with the facts that these salts are crystalline at RT and there is strong evidence of hydrogen bonding between the counter ions in these salts.

Our interpretation is then that, in Drummond's class of salts, the strong HB between ions bearing non compatible moieties is the driving force for the stable formation of the separate alkyl and fluorinated clusters. Of course this implies that similar classes of salts (strong correlation between ions bearing non-compatible moieties) should have comparable structural organization.

However this is not supposed to be the only criterion to determine whether these domains can form at RT conditions or not. Most probably, when long enough tails are involved, they will segregate even in the absence of such a driving force as the one that we mentioned above.

Professor Drummond commented: To my knowledge, the effect of segmentation regions in mesostructured liquids on the transport properties of species in the liquids has not been analysed. However I suggest that it is likely that liquid mesostructure would result in negative deviations from the 'good' ionic liquid region of the Walden plot.

What's your feeling/speculation on the structure of the mesoscale objects in this media?

Dr Triolo replied: The structure of ILs bearing long enough side chains is characterised by the occurrence of low *Q* diffraction peaks that fingerprint the existence of alternating polar–apolar domains. This is nowadays well established both in ref. 1–5 and computationally.[6–8] The alkyl tails, due to the different nature and extent of interaction with respect to the charged moieties will tend to segregate into more separated domains than the polar portions. Modifications of the chemical nature of the side tails can induce non negligible structural effects at a mesoscopic level. For example it is now well understood that side chains containing either terminal hydroxyl groups or ether moieties show much lower tendency to develop segregated clusters, as can be observed from X-ray scattering data that do not show (or show little) evidence of the low *Q* diffraction.[9–14] On the other hand the recent reports from Drummond's group[15,16] and our present contribution show that the presence of fluorinated tails can induce a further level of structure complexity as mutual repulsion between charged, alkyl and fluorous moieties will eventually lead to the formation of different compartments where affine moieties tend to cluster. Molecular Dynamics studies show that these clusters are very ill-shaped and so far no clear indication could be experimentally extracted both on the shape and the life time of these domains. We can speculate that alkyl and fluorous tails will tend to conformationally organise in a fashion that should resemble their state in the neat (alkane or perfluoro alkane) liquids, but no conclusive information exists so far on this issue (see for example the discussion at a recent Faraday Discussion meeting[17]). At this stage we can infer that for example upon addition of compounds, these will distribute preferentially inside the affine domains thus altering their average size. This preferential solubility will have major consequences in terms of bringing into relatively close contact (just a few Å apart) compounds (*e.g.* ions, alkanes, perfluoroalkanes) that would otherwise tend to avoid each other, thus opening the way to several applications.

1 A. Triolo, O. Russina, H.-J. Bleif, and E. Di Cola, *J. Phys. Chem. B*, 2007, **111**, 4641–4644.
2 C. S. Santos, N. S. Murthy, G. A. Baker, and E. W. Castner, *J. Chem. Phys.*, 2011, **134**, 121101.
3 B. Aoun, A. Goldbach, M. a González, S. Kohara, D. L. Price, and M.-L. Saboungi, *J. Chem. Phys.*, 2011, **134**, 104509.
4 R. Hayes, S. Imberti, G. G. Warr, and R. Atkin, *Phys. Chem. Chem. Phys.*, 2011, **13**, 3237–3247.
5 T. L. Greaves and C. J. Drummond, *Chem. Soc. Rev.*, 2012, **42**, 1096.
6 H. V. R. Annapureddy, H. K. Kashyap, P. M. De Biase, and C. J. Margulis, *J. Phys. Chem. B*, 2010, **114**, 16838–16846.
7 H. K. Kashyap, J. J. Hettige, H. V. R. Annapureddy, and C. J. Margulis, *Chem. Commun.*, 2012, **48**, 5103–5105.
8 K. Fujii, R. Kanzaki, T. Takamuku, Y. Kameda, S. Kohara, M. Kanakubo, M. Shibayama, S. Ishiguro, and Y. Umebayashi, *J. Chem. Phys.*, 2011, **135**, 244502.
9 O. Russina, A. Triolo, L. Gontrani, and R. Caminiti, *J. Phys. Chem. Lett.*, 2012, **3**, 27.
10 O. Russina and A. Triolo, *Faraday Discuss.*, 2012, **154**, 97–109.
11 A. Triolo, O. Russina, R. Caminiti, H. Shirota, H. Y. Lee, C. S. Santos, N. S. Murthy, and E. W. Castner, *Chem. Commun.*, 2012, **48**, 4959–4961.
12 H. Y. Lee, H. Shirota, and E. W. Castner, *J. Phys. Chem. Lett.*, 2013, **4**, 1477.
13 H. K. Kashyap, C. S. Santos, R. P. Daly, J. J. Hettige, N. S. Murthy, H. Shirota, E. W. Castner, and C. J. Margulis, *J. Phys. Chem. B*, 2013, **117**, 1130–1135.
14 K. Shimizu, C. E. S. Bernardes, A. Triolo, and J. N. C. Lopes, *Phys. Chem. Chem. Phys.*, 2013, **15**, 16256–16262.
15 Y. Shen, D. F. Kennedy, T. L. Greaves, A. Weerawardena, R. J. Mulder, N. Kirby, G. Song, and C. J. Drummond, *Phys. Chem. Chem. Phys.*, 2012, **14**, 7981–7992.
16 T. L. Greaves, D. F. Kennedy, Y. Shen, A. Hawley, G. Song, and C. J. Drummond, *Phys. Chem. Chem. Phys.*, 2013, **15**, 7592–7598.
17 M. F. C. Gomes, J. N. C. Lopes, A. A. H. Padua, and A. Triolo, *Faraday Discuss.*, 2012, **154**, 81–96.

Professor Ben-Amotz opened the discussion of the paper by Professor Orrit: Can you clarify the significance of the apparent convergence temperature of the molecular and nano-rod reorientation times? It seems that the way you normalized the nano-rod results forced them to cross at around 235 K. So, it is not clear to me how the apparent crossing temperatures of the molecular rotation times and the assumed crossing temperature of the nano-rod results are related to each other.

Professor Orrit replied: Let me first explain in more detail how the normalization was done. We start from the hypothesis that heterogeneity is absent or negligible at the highest temperature (238 K). When we normalize all times (corresponding to hydrodynamic volumes) to those at 238 K, we find that the three histograms at 235 K, 232 K, and 229 K are narrow, and those at 226 K and 222.5 K are broader. Our next approximation then is to assume that heterogeneity is negligible at 238, 235, 232 and 229 K. We now normalize by the average of the hydrodynamic volumes found at these four temperatures (this trick allows us to minimize the effect on noise on these normalization factors). The result of this second procedure is shown in Fig. 6, with the associated histograms. You can see that indeed, only the two histograms at the lowest temperatures are significantly broader than the other ones.

Indeed, we have used a set of data (either at 238 K in the first procedure, or the average of the four higher temperatures in the second procedure), which of course forces all the points towards the average. However, the three other sets, which are still independent of this forcing, are consistent with the starting hypothesis. In other words, the four narrow histograms at the highest temperatures may be seen as tests that no large deviation from the average occurs for those four data sets. In contrast to these, large deviations appear at the two lower temperatures.

It is true that, as you note, we used the hypothesis that heterogeneity is absent above 229 K in this normalization procedure. However, this assumption can only reduce the apparent heterogeneity, not enhance it. For a completely homogeneous system, the normalized histograms should all present the same width. Therefore, our conclusion that heterogeneity (which is very clear at temperatures around 210 K from single molecules) has still survived at 222.5 and 235 K is robust against the initial hypothesis.

Professor Chandler noted: Your interesting measurements might be interpretable in terms of facilitation theory. Here, I refer to the analysis of decoupling phenomena found through application of a model of a diffusing probe[1] coupled to the East model.[2] The East model exhibits hierarchical dynamics in a form that is strikingly similar to that observed in molecular simulation,[3] in equilibrium transport[4] and in glass transitions.[5] The diffusing probe model would need generalization to describe rotations of long particles like those you have studied.

1 Y. Jung, J. P. Garrahan and D. Chandler, *Phys. Rev. E*, 2004, **69**, 061205.1–7.
2 J. Jckle and S. Eisinger, *Z. Phys. B*, 1991, **84**, 115124.
3 A. S. Keys, L. O. Hedges, J. P. Garrahan, S. C. Glotzer, and D. Chandler, *Phys. Rev. X*, 2011, **1**, 021013.
4 Y. S. Elmatad, D. Chandler, and J. P. Garrahan, *J. Phys. Chem. B*, 2009, **113**, 5563–5567.
5 A. S. Keys, J. P. Garrahan, and D. Chandler, *Proc. Natl. Acad. Sci. U. S. A.*, 2013, **110**, 4482–4487.

Professor Orrit responded: The facilitation theory indeed seems relevant to describe the rotational diffusion of the probes we use. I can just mention that the gold nanorods we use are not very elongated (aspect ratio 2.5) so that the diffusion of a sphere may not be too bad an approximation. It is possible to coat the rods with additional shells, of silica for example, to make them more spherical.

Dr Royall asked: What is the glass transition of glycerol and thus how close to the transition are your measurements?

Professor Orrit responded: The glass transition of glycerol is around 190 K (literature values range from 187 to 192 K. The viscosity of glycerol at 197 K is about 10^9 Pa s. Therefore, our experiments with molecules are 20 K above the glass transition, the experiments with nanorods are 40 K above the glass transition. Note that the melting temperature of glycerol is 291 K, so that all experiments are done on supercooled, metastable glycerol.

Dr Royall continued: In your very interesting experiments, you have rods of 30 nm in length. You argue that dynamic heterogeniety has a spatial extent of 30 nm at least, at temperatures around 20–30 degrees above the glass transition.

However, you say that the rods have remained fixed for hours, which is surely much longer than the structural relaxation at the glass transition, which I presume is ~100 s.

30 nm is very much larger than other measurements, notably the method introduced by Berthier *et al.*[1] where the dynamic length is an order of magnitude smaller – at least – at the glass transition. Is it in fact possible that the rods are larger that the lengthscale of dynamic heterogeniety, and thus almost always, they are sitting in dynamically inactive regions? This would lead to the rotational dynamics that are much slower than the structural relaxation time that you find.

This could be tested with rods of differing lengths, which if their rotational diffusion did not scale as expected, would indicate that the lengthscale of dynamic heterogeniety is less than the rod length.

1 L. Berthier *et al.*, Science, 2005, **310**, 1797–1800.

Professor Orrit answered: I comment on the different parts of your question. In some of my answers, I distinguish between a first part based on observations, and a second, more speculative part.

A. The rotational correlation time of glycerol molecules at 230 K (mid-range of the nanorods measurements) is about 0.2 ms. The volume of a nanorod is about 3×10^3 nm^3, 60 000 times larger than of a glycerol molecule, leading to a rotational time of about 10 s. The rods would need some 10 000 tumbling events to leave the focal spot of the microscope (about 500 nm in diameter), which would thus take several days. Yet, although the rods remain at the same spatial location within 500 nm, they perform a large number of reorientations, which allows us to record enough statistics to evaluate the tumbling time (about at least 50 reorientations per trace).

B. The big surprise of these measurements is that heterogeneity does not appear to be averaged, neither on the length scale of a nanorod (30 nm in length, a little less in hydrodynamic radius), nor even on the enormous time scale of the measurement of their rotational diffusion, 100 s, almost a million times longer than the rotational correlation time of glycerol molecules. The heterogeneity therefore extends much beyond the size of a dye molecule (typically 10 times larger in volume than a glycerol molecule). Heterogeneities are spatially extended and long-lived.

Speculation: The four-point correlation function of Berthier *et al.* is a higher moment of a distribution. Isn't it possible that this moment, which would faithfully reflect the extent of heterogeneity for a normal distribution, severely underestimate it if the distribution presents slowly decreasing tails? In my mind, heterogeneities resemble the fractal, transient aggregates postulated by Fischer (Fischer clusters), so that ensemble averaging removes most of their solid-like or liquid-like character. Local single-molecule or single-nanoparticle measurements remove any ensemble averaging and can reveal the true inhomogeneous distribution, tails and all.

C. If the rods were larger than the length scale of heterogeneity, each rod would see the same average effective medium. Moreover, by moving through this medium over minutes to hours, their diffusion behavior would appear even more uniform upon time-averaging. This appears to be true at 238 K, and down to 229 K maybe, but something definitely different happens at 222.5 K and 226 K. The large spread of (mean) times at these two low temperatures indicates that different rods sample different local viscosities at least during the measurement time. This behavior is quite similar to that observed for dye probes at much lower temperatures. On the longer time scale required to change and stabilize the cryostat temperature, however, the average time of the rods appears to change (see Fig. 6a).

Speculation: The change in behavior between 226 K and 229 K resembles what I would expect for the crossover temperature predicted by mode-coupling theory (MCT). As correctly pointed out by David Chandler during the discussion, MCT is a mean-field theory and cannot describe the glass-forming liquid once ergodicity is lost. However, MCT's prediction in the normal liquid, above the crossover temperature, is deduced from thermodynamic arguments and should still be correct. It leads to averaging of spatial and temporal heterogeneities. This resembles what we observe for temperatures higher than 230 K.

D. We already have an indication in this direction from the comparison of nanorods (volume about 5 000 nm^3) and dye molecules (volume about 0.5 nm^3). Although the temperature ranges are very different (because the experimental time scales are more or less fixed by fluorescence count rates), we see that the amount of heterogeneity on a logarithmic scale does not vary appreciably between 212 K and 222 K. It would be very interesting to fill the gap between these two temperatures with other objects or other methods, for example plasmon-enhanced fluorescence, which probes smaller volumes with translational diffusion. We plan such measurements in the future.

Speculation: I do not think that spectacular changes in heterogeneity will be found between 212 K and 222 K. Again, the picture in my mind is that of a disordered system with multiple length and time scales (see above). However, supercooled glass formers have surprised us in the past and might still do so.

Professor Angell asked: This is a question about the spreading out of distribution of time scales at low T for glycerol. Most of the fragile liquid anomalies, have been interpreted in these terms, but glycerol is one of the much less pronounced cases. Are you limited to glycerol, or could you study more fragile liquids like sorbitol that might have a much broader distribution?

Professor Orrit replied: We only studied glycerol so far with single molecules and single nanorods. Similar studies have been published by D. A. Vanden Bout on single molecules in *ortho*-terphenyl.[1] These authors also find spreads of rotation times, but the temperature dependence seems larger than the one we found in glycerol (see Fig.7). The memory time (or environmental exchange time) found in this paper, however, was much shorter than the one we found in glycerol. We attribute this difference to the ageing of the sample, which was extremely long in our case (weeks). I agree that sorbitol and other glass formers would be very interesting to explore, but we are interested in glycerol for practical reasons.

L. A. Deschenes, D. A. V. Bout, Heterogeneous dynamics and domains in supercooled o-terphenyl: A single molecule study, *J. Phys. Chem. B*, 2002, **106**, 11438–11445.

Dr Soper said: The results that you describe are truly fascinating, but also quite baffling. Can you give us any physical picture of what these heterogeneities are like? Glycerol is complicated by having multiple hydrogen bonding sites which presumably gives rise to the high viscosity under ambient conditions. What might be happening under the supercooled conditions that makes them become heterogeneous?

Professor Orrit responded: Indeed, we also were very surprised by our first experiments on single molecules (ref. 25 of our paper), which we interpreted as a coexistence of liquid-like and solid-like regions in the supercooled liquid close to the glass transition. The new results with nanorods are even more surprising because they are now done at a much higher temperature and on a much larger length scale and longer timescale relative to the alpha relaxation time (although the absolute time scale, seconds, is the same in both experiments for practical reasons).

My naïve interpretation of these results is that density or packing fluctuations occur over a wide range of length (and therefore time) scales. This picture follows closely that of Fischer clusters, with length scales spread from microns to nanometers, therefore giving rise to scattering of light as well as X-rays (see ref. 17 and 18 of our paper). Clearly, many more experiments with well controlled thermal histories are needed to understand the nature of heterogeneity better.

Professor Chandler asked Professor Orrit: I believe that some of the previous questions on this paper overlook the fact that glass-forming materials are fluctuation dominated. It is not suffcient to think only about mean structural relaxation times or mean dynamic heterogeneity lengths. The materials examined by Professor Orrit possess relaxation processes over a broad range of time scales, and dynamical heterogeneity over a broad range of length scales.[1] His experiments are therefore probes of the wings of distributions of dynamical activity. As such, his measurements can provide pictures of the hierarchical dynamics that underlies the dynamics of glass formers.

1 See, for example, J. P. Garrahan and D. Chandler, *Phys. Rev. Lett.*, 2002, **89**, 035704.

Professor Orrit replied: Although my view of that material is very naïve and intuition-driven, I agree with this comment on the nature of heterogeneity and the broad range of scales, both temporal and spatial.

Professor Chandler continued: In light of comments we have just heard about fragility and strength of glass-forming materials, I believe it is important to recall that all glass formers exhibit a single seemingly universal form of hierarchical dynamics. The East model[1] provides its simplest caricature. A result of this hierarchical dynamics is the parabolic law, which states that mean structural relaxation time, τ, as a function of temperature, T, (or as functions of osmotic pressure over temperature, in cases of colloidal systems) is given by $\log(\tau/\tau_o) = J^2(1/T - 1/T_o)^2$ for $T < T_o$, where T_o is the temperature (or temperature over osmotic pressure) below which the glass former exhibits super-cooled behavior. Ref. 2 shows that literally thousands of pieces of data for a wide range of systems extending over the entire range of glass-forming behaviors collapse to this universal form. Glass-forming materials differ only in as much as J's differ and T_o's differ. That is the entire story of so-called "fragility".

Further, and this point is crucial, there is a theoretical underpinning to the parabolic form, so that J and T_o are not simply fitting parameters, but rather quantities that are computable from molecular theory and simulation.[3] The same is not true for fitting parameters associated with model coupling theory, or Adam–Gibbs theory, or random-first-order-theory, none of which have underlying microscopic models. And indeed, when tested, the formulas from those theories are found to have little or no predictive power.[4] In contrast, illustrations of the predictive power of the parabolic law and its theoretical underpinning are found in my Faraday Discussion paper with Dr Limmer.[5]

1 J. Jckle and S. Eisinger, *Z. Phys. B*, 1991, **84**, 115124.
2 Y. S. Elmatad, D. Chandler and J. P. Garrahan, *J. Phys. Chem. B*, 2009, **113**, 5563–5567.
3 A. S. Keys, L. O. Hedges, J. P. Garrahan, S. C. Glotzer and D. Chandler, *Phys. Rev. X*, 2011, **1**, 021013.
4 Y. S. Elmatad, J. P. Garrahan and D. Chandler, *J. Phys. Chem. B*, 2010, **114**, 17113–17119.
5 D. T. Limmer and D. Chandler, *Faraday Discuss.*, 2013, **167**, DOI: 10.1039/C3FD00076A

Professor Orrit replied: Thank you for the reference and comparison to different theoretical descriptions of glass formers in the literature.

Professor Angell expanded: Professor Chandler's comment that "all glass formers exhibit a single seemingly universal form of hierarchical dynamics", and his demonstration that viscosity data can be scaled onto a master plot between some upper temperature T_o and a lower temperature (related to an energy scale in their case), is not a new recognition. The first such master plot was given by Roessler and Sokolov in 1996 and is shown below.[1] These authors might well have written at that time, as Professor Chandler has in his comment, "Glass-forming materials differ only in as much as the T_g's (J's) differ and T_c's (T_o's) differ. That is the entire story of so-called 'fragility'."

This is a fairly dismissive statement, but Roessler and Sokolov's demonstration (in Fig. 1 below) does not seem to have diminished interest in the origin of the large differences that exist between the upper and lower temperatures involved in

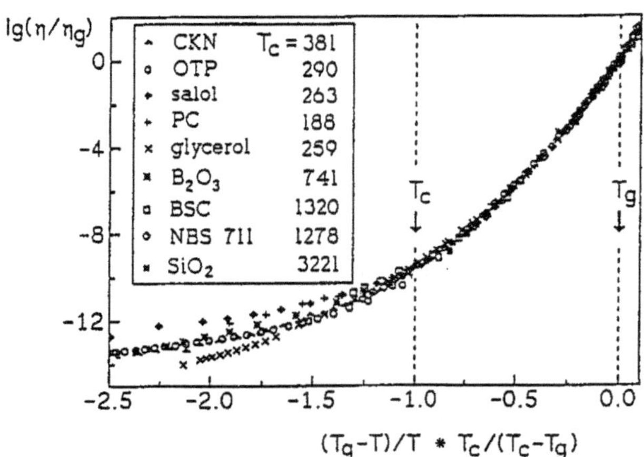

Fig. 1 Viscosity data master plot from Roessler and Sokolov[1]

the scaling. It is these differences (scaled by T_g itself) that need to be explained. In another part of the discussion, Professor Chandler states that he has no theory for the T_o values that different liquids exhibit, which would imply that the liquid fragility question is still an open one.

In Roessler and Sokolov's figure, the upper scaling temperature that collapsed the viscosity data so nicely was found to lie close to the critical temperature T_c of mode coupling theory, hence the use of T_c in their scaling formula. They suggested it could be related to an upper percolation temperature for configurationally excited states. I have given a summary of the current state of thinking about liquid fragility, including the athermal manifestations seen in the study of colloidal systems and hard particle simulations, in a current symposium proceedings (of unusual form) on this subject.[2]

1 E. Rossler, and A.P. Sokolov, *Chem. Geol.*, 1996, **128**, 143–153.
2 "The fragility of glassforming liquids: thermal *vs* athermal systems, and kinetic *vs* thermodynamic origins", C. A. Angell, in *Fragility of glass forming liquids*, ed. A. L. Greer, K. F. Kelton, S. Sastry, Hindustan Book Agency, New Delhi, India, in press.

Professor Mount communicated: Your work raises an interesting question as to what extent the nanorods are simply reporters of the glycerol dynamics, or whether their presence also spatially influences the distribution and/or form of the heterogeneity *e.g.* through their shape and size. It seems that one could probe in principle whether this influence was significant by carrying out measurements with both molecular and nanorod reporters present and determining the effect of the parameters of each on the other. Is this the case and have you carried out or are you planning to carry out such measurements?

Professor Orrit communicated in reply: Of course, any probe may always influence the system it is supposed to report on. Gold nanorods are no exception to this rule. However, as briefly discussed in the paper, comparing the responses of the system to different individual probes makes it in principle possible to eliminate the probe–system interaction, and to concentrate on differences arising from the system itself. In our case, the differences in rotational diffusion times we observe cannot be attributed to differences in hydrodynamic volumes of the individual rods. Therefore, in the simple Debye–Stokes–Einstein model, those differences have to stem from viscosity differences of the glycerol itself. Now, you can argue that other differences from rod to rod (*e.g.* the surface coating or the shape) could lead to spreads that cannot be represented by spreads of hydrodynamic volume). In that case, those changes would have to be temperature-dependent, since we observe a variation of the spread with temperature. We find it much more plausible and likely that the glycerol itself is the source of the spread.

I agree with you that we need much more measurements to confirm this first result, not only with more rods of given size and shape, but with more sizes, shapes, surface functionalizations, *etc.*, as has been done by Dr Kaufman for molecules (see ref. 56 of our paper). Unfortunately, for practical reasons, we haven't been able to do more measurements yet since the submission of our paper. Let me add that rotational diffusion experiments on rods and molecules involve very different times (because of the large difference in volume). However, rotational diffusion of rods could be measured simultaneously with translational diffusion of molecules thanks to the fluorescence enhancement effect that we have described in ref. 48 of our paper.

Dr Vila Verde opened the discussion of the paper by Dr Henchman: Is the BH state never a transition state according to your model, or did I misunderstand? If it is not a transition state, how does one reconcile this fact with that that hydrogen-bond exchange is an activated process?

Dr Henchman replied: Based on the general view, our expectation prior to this research was that under ambient conditions linear hydrogen bonds were the stable arrangements and bifurcated hydrogen bonds were transition states; consequently water dynamics would be an activated process. Surprisingly, the data we obtained in the paper suggest otherwise: some degree of hydrogen bifurcation is the norm in liquid water. Thus no additional activation is required to induce bifurcation. The resolution of the seeming conflict between this finding and the view that switching is activated lies in how one defines the starting point of switching. If switching is defined as a donor going from the outermost donor of one acceptor to another acceptor of lower coordination, then the data here show that there is a negligible barrier. On the other hand, there is a barrier if the donor starts as an inner donor to an acceptor. This barrier is comparable to the average vibrational energy of the molecules involved and reinforces how hydrogen-bond switching is intimately coupled to whole-molecule vibration. The other scenario for which there is a barrier is when the donor switches to an acceptor of the same or higher coordination. In this case, the donor must penetrate into the inner coordination of the new acceptor in order to form a stable hydrogen bond.

Dr Vila Verde continued: For the future, it would be useful if you could use your hydrogen bond definition to investigate the mechanism of hydrogen bond exchange events. Recent simulation work, using standard hydrogen bond definitions based on interatomic distances and angles, strongly indicates that hydrogen bond exchange proceeds through large-amplitude angular jumps[1] but other models for hydrogen bond exchange also exist, such as the flickering clusters mechanism, the diffusion of Bjerrum orientational defects[2] or Debye's small step diffusion model.[3] It would be interesting to check if the jump mechanism still holds using your hydrogen bond definition or if a different physical picture arises.

1 D. Laage and J. T. A. Hynes, Molecular jump mechanism of water reorientation, *Science*, 2006, **311**, 832–835.
2 D. Eisenberg, W. Kauzmann, The Structure and Properties of Water; Oxford Clarendon Press, London, 1969.
3 P. Debye, Polar Molecules; The Chemical Catalog Company, New York, 1929.

Dr Henchman replied: Indeed, in this work we have only examined equilibrium distributions and it would certainly be interesting to extend it to kinetics and switching mechanisms. We have made some progress already applying the topological definition to hydrogen-bond switching.[1] This revealed the distribution of possible switches that can take place as well as two mechanisms of hydrogen-bond switching *via* the bifurcated oxygen and cyclic dimer transition states. I expect that Laage and Hynes's analysis of hydrogen bond switching would be slightly different with the topological hydrogen bond definition but I doubt that their overall findings would change. The number of hydrogen bonds is likely to be slightly larger. In particular, the change in hydrogen bond number for each acceptor at the switch would be the expected 0.5 rather than the smaller value of 0.3 that they get. The value of 0.5 is because the first acceptor loses one hydrogen bond at the exact same time as when the second acceptor gains one, and we divide by two to get hydrogen bonds per molecule. Their hydrogen bond definition is probably not detecting all distorted hydrogen bonds during the switch. Regarding the other mechanisms that you mention, Eisenberg and Kauzmann largely dismissed the flickering cluster model based on the unreasonably large energy required to break multiple hydrogen bonds when the clusters dissolve. The Bjerrum defect model is useful for ice because these defects can be readily identified in the tetrahedral ice lattice. However, in water essentially all molecules are defects and there are more types of defects than Bjerrum defects *e.g.* single and triple acceptors that are also important in dynamics. One can equate cyclic dimer structures with D defects and adjacent, non-hydrogen-bonded water molecules as L defects. Debye's diffusion model neglects hydrogen bonds and their associated correlations.

1 R. H. Henchman and S. J. Irudayam, *J. Phys. Chem. B*, 2010, **114**, 16792–16810.

Mr Menzl asked: The topological hydrogen bond definition described in your paper produces convincing results for condensed bulk systems. How does the detected hydrogen bond structure at liquid–vapor interfaces compare to cutoff-based definitions and experimental data?

Dr Henchman responded: We have examined this issue in another publication.[1] While there is some arbitrariness about how one defines surface water, the topological definition predicts that 20% of surface water molecules are single donors which agrees well with 25% detected by vibrational sum-frequency generation.[2] However, the distance-angle cut-off method (OO cut-off of 0.35 nm and HOO cut-off of 30 degrees) gives a much larger value of 47% single donors. This is not so surprising when one considers that the distance-angle definition already yields about 20% single donors in bulk water, corresponding to the commonly assumed value of 10% broken hydrogen bonds. We surmise that this sizeable discrepancy at the air–water interface has been overlooked in a number of publications because the number of broken hydrogen bonds (24% by the distance–angle definition) has been erroneously equated with the number of single donors, leading to seeming agreement with experiment. We take this as further proof that the distance–angle definition overestimates the number of broken hydrogen bonds in bulk water.

1 S. J. Irudayam and R. H. Henchman, *J. Chem. Phys.*, 2012, **137**, 034508.
2 Q. Du, E. Freysz, and Y. R. Shen, Science, 1994, **264**, 826–828.

Professor Wang asked: Since the topological definition for hydrogen bonds does not have a distance cutoff, the definition may be problematic when investigating water clustering in the gas phase. In this case a hydrogen bond may be assigned between water molecules are separated far apart.

Dr Henchman replied: This is one of the key points differentiating the topological definition from other definitions. Firstly, let me point out there is no conflict with expectations for small water clusters when the donor–acceptor distances and associated angles are reasonably close to equilibrium values. For example, for a dimer the topological definition yields only one hydrogen bond which lies between the nearest donor and acceptor. The other three donor–acceptor pairs are assumed to be weaker; they cannot strengthen without disrupting the strongest donor–acceptor interaction and so are defined as broken hydrogen bonds. For a trimer the definition predicts three hydrogen bonds and three broken hydrogen bonds, and so on for larger clusters. The greater conflict with expectations arises when the nearest donor–acceptor distances or associated angles deviate substantially from the equilibrium values. As you say, the topological definition assigns a hydrogen bond between the nearest donor and acceptor of two isolated water molecules no matter how far apart they are. Similarly, a hydrogen bond is assigned no matter what their relative orientations. I agree that such interactions are certainly highly strained. However, if they are only large distortions away from the equilibrium geometry, then there is nothing fundamentally different about them to justify the use of a different label such as broken. Moreover, important interactions may be missed as our paper makes clear. We examine these issues extensively in our paper.

Dr Vila Verde commented: It would be useful to extend your work to study the hydrogen bond exchange mechanism near hydrophobic solutes and near ions, and try to address some outstanding issues on the topic. Laage, Hynes and co-workers have done extensive work on the topic but using a different approach, and have developed a model whereby the slowdown of OH rotation (a) in the first hydration shell of hydrophobic solutes is explained in purely entropic terms,[1] and (b) in the first hydration shell of anions has the same entropic contribution plus an enthalpic contribution arising from the different strength of the OH···Ion hydrogen bond.[2] However, their model does not explain (a) the temperature dependence of the activation energy of water reorientation near hydrophobic solutes observed in experiment, or (b) the slowdown of OH rotation near cations.

1 D. Laage, G. Stirnemann and J. T. Hynes, Why Water Reorientation Slows without Iceberg Formation around Hydrophobic Solutes, *J. Phys. Chem. B*, 2009, **113**, 2428–2435.
2 D. Laage J. T. Hynes, Reorientional dynamics of water molecules in anionic hydration shells, *Proc. Natl. Acad. Sci. U. S. A.*, 2007, **104**, 11167–11172.

Dr Henchman added: We agree that it would be useful to extend this analysis to solutions. We have already looked at the hydrogen-bond structure around anions, cations, and non-polar atoms.[1] However, a further analysis of cations would ideally require a topological way to define cation–oxygen interactions without the use of cut-off parameters. This is an issue we are currently considering. Somewhat related to dynamics, we have developed methods to calculate the entropy of water molecules around non-polar atoms[2] and ions.[3] These results show distinctly different mechanisms of entropy reduction of water for the different solutes. For anions, water molecules are more confined vibrationally. For non-polar atoms, water molecules have fewer hydrogen-bond arrangements, reflecting an excluded volume effect by the solute. For cations, water molecules have even fewer hydrogen-bond arrangements. This only partially relates to dynamics and I cannot comment reliably on how they would relate to the problematic cases to which you refer.

1 S. J. Irudayam and R. H. Henchman, *J. Chem. Phys.*, 2012, **137**, 034508.
2 S. J. Irudayam and R. H. Henchman, *J. Phys.: Condens. Matter*, 2010, **22**, 284108.
3 S. J. Irudayam and R. H. Henchman, *Mol. Phys.*, 2011, **109**, 37–48.

Mr Wexler communicated: Long range ordering is a primary focus of your work and from the perspective of the cumulative donor–acceptor bias it is concluded that no long range order is apparent from this metric. However, the influence of solutes and interfaces on forcing long range order as a means of distributing the local donor-acceptor imbalances is posited. Considering the dipole nature of both water molecules and hydrogen bonds within the complex topology encountered in your model is it reasonable to expect that these dipoles would radiate during configurational translations (*e.g.* order parameter changes) driven by network fluctuations? Furthermore, could this radiated electromagnetic field force the harmonic selection of certain modes over others and thus give rise to long range order as the extended local network would be forced to likewise oscillate and further reinforce the induced perturbation?

Dr Henchman communicated in reply: Long-range ordering, or the absence of it in pure water by the measure we used, turned out to be a minor part of this paper. It was in another work[1] where we had found evidence of long-range ordering in water around atomic solutes or at the air–water interface. This ordering is in terms of water's donor–acceptor difference to compensate for the perturbation induced by the solute. To answer the first part of your question, yes, I would expect that the long-range ordering in water would persist regardless of solute translation. These effects have the same inverse square dependence as the ion's electric field, although a discrete-solvent effect also appears to be present. I do not think it best to refer to this ordering as a radiated electromagnetic field and I do not follow the second part of your question to be able to answer it.

1 S. J. Irudayam and R. H. Henchman, *J. Chem. Phys.*, 2012, **137**, 034508.

Professor Soper opened the discussion of the paper by Professor Idrissi: With regard to the assignment of frequencies for the different structural species, can you explain parameters you used? How do you know those are 1, 2 and 3 H bonded? Have you done simulations? Do they predict frequencies *etc*?

Professor Idrissi replied: The spectral assignment of clusters of methanol can be found in the literature[1–3] (there are many other works). These results were used to rank the spectral contribution of f_1, f_2 and f_3 hydrogen bonded methanol molecules.

1 P. Lalanne, J. M. Andason, J.-C. Soetens, T. Tassaing, Y. Danten, M. Besnard, *J. Phys. Chem. A*, 2004, **108**, 3902.
2 R. A. Provencal, J. B. Paul, K. Roth, C. Chapo, R. N. Casaes, R. J. Saykally, G. S. Tschumper, and H. F. Schaefer, *J. Chem. Phys.*, 1999, **110**, 4258.
3 I. Doroshenko, V. Pogrorelov, V. Sablinska, V. Balevicius, *J. Mol. Liq.*, 2010, **157**, 142–145.

Professor Soper continued: As regards calculating the fraction of 1, 2, 3, *etc.* hydrogen bonded molecules from the OH vibrational spectrum, I believe you have assumed Gaussian contributions with different excitation frequencies for each species. However I also believe that Jim Skinner and others have shown that the contribution of different structural species to the Raman spectra can be quite broad and non-Gaussian. Hence how reliable is this reconstruction?

Professor Ben-Amotz added: Your analysis of the OH band of methanol as a superposition of Gaussian sub-bands is not consistent with theoretical analysis of the OH stretch Raman band of water by Yang and Skinner,[1] as well as earlier work by Torri.[2] These analyses imply that the low frequency shoulder of the water OH stretch band is strongly influenced by intermolecular resonance coupling (or equivalently coherent energy transfer) between strongly coupled OH groups on different water molecules, and cannot be represented by a Gaussian sub-band. Moreover, even in the absence of resonance coupling (in systems such as dilute HOD in D_2O), computer simulations suggest that sub-populations corresponding to water molecules with different numbers of H-bonds are expected to produce non-Gaussian OH stretch sub-bands.[3] Although the above results all pertain to water rather than methanol, they are consistent with your assumption that increasing the number of H-bonds to a given water molecule is expected to produce a decrease in the corresponding mean OH frequency. However, the above results also suggest that it is unlikely that a Gaussian sub-band decomposition can be relied upon to quantify the corresponding populations, or the associated mean OH frequencies and widths.

1 M. Yang and J. L. Skinner, *Phys. Chem. Chem. Phys.*, 2010, **12**, 982–991.
2 H. Torii, *J. Mol. Liq.*, 2007, **136**, 274–280.
3 J. Tomlinson-Phillips, J. Davis, D. Ben-Amotz, D. Spangberg, L. Pejov and K. Hermansson, *J. Phys. Chem. A*, 2011, **115**, 6177–6183.

Professor Idrissi responded: The Gaussian form that we used in our fitting procedure has no physical meaning. The purpose was to get information on the behavior of the position of the spectral contribution of one , two and three hydrogen bonded methanol molecules. The frequency values of the spectral contribution of these hydrogen bonded entities is consistent with many previous spectral and theoretical calculations.[1–3] Furthermore, the OH spectral decomposition in terms of one, two, and three hydrogen bonded molecules was not used to quantify the corresponding population. However, we used the percentage f_1, f_2 and f_3 of hydrogen bonded methanol molecules determined from the molecular simulations as input parameters in the fitting procedure of the OH spectral region. Indeed, we considered a simple model to fit the experimental spectra involving one, two, and three hydrogen bonded methanol molecules. We assumed that within each contribution, we have the same parameters (frequency position, and fluctuation) and that these contributions are weighted by a factors defined by the percentage f_1, f_2 and f_3 determined from our simulations. This constitutes a difference with previous model used to fit the OH spectral region, where the weighting parameters are defined by the activities of the different clusters determined from quantum calculations. At the end we want to clarify that the analysis of OH spectral region given in the cited references, indicating that all vibrational spectra showing a signature of coherent vibrational energy transfer is based on solid arguments. However, it is not straightforward from the experimental point of view, to analyse the spectra using this interpretation.

1 P. Lalanne, J. M. Andason, J.-C. Soetens, T. Tassaing, Y. Danten, M. Besnard, *J. Phys. Chem. A*, 2004, **108**, 3902.
2 R. A. Provencal, J. B. Paul, K. Roth, C. Chapo, R. N. Casaes, R. J. Saykally, G. S. Tschumper and H. F. Schaefer, *J. Chem. Phys.*, 1999, **110**, 4258.
3 I. Doroshenko, V. Pogrorelov, V. Sablinska, V. Balevicius, *J. Mol. Liq.*, 2010, **157**, 142–145.

Dr Henchman asked: Following on from my paper, Giguere proposed that bifurcated hydrogens are even more likely to be present in methanol solutions.[1] Have you considered this structural possibility in interpreting your spectra?

1 P. A. Giguere and M. Pigeon-Gosselin, *J. Sol. Chem.*, 1988, **17**, 1007–1014.

Professor Idrissi replied: No, we didn't consider the spectral contribution of bifurcated hydrogen bonds in analyzing the methanol spectra in supercritical conditions.

Professor Chandler opened the discussion of the paper by Professor Barrat: Does the model you use, with elasticity deformations, contradict the finding[1] that excitations in glass-forming materials are localized without significant static correlations between excitations? I would have thought that elastic deformations, when significant, would correlate regions over long distances.

1 A. S. Keys, L. O. Hedges, J. P. Garrahan, S. C. Glotzer, and D. Chandler, Excitations are localized and relaxation is hierarchical in glass-forming liquids, *Phys. Rev. X*, 2011, **1**, 021013.

Professor Barrat replied: The authors of ref. 1 you mentioned observed that, in equilibrium conditions, (moderately or deeply) supercooled liquids exhibit localised excitations, *i.e.*, large and persistent correlated displacements of a "handful of particles". They further noted that these excitations "facilitate dynamics of neighbouring excitations", that several seemingly uncorrelated excitation nuclei coexist in the system, and need to connect so that the system can relax. For the sheared materials that we consider, we and others[2] also have observed, in microscopic simulations, localised excitations, and have shown that these localised plastic events induce a deformation field in the surrounding regions that can be described by elasticity theory, and therefore is long ranged. This deformation field can, in turn, induce plastic events far away from the initial one. Accordingly, "plastic facilitation" not only affects the direct neighbours, but can also operate on distant regions, and this is the physics that is incorporated in the mesoscopic model discussed here. The plastic facilitation leads to complex spatio-temporal patterns that can be characterized using four point correlations,[3] and in some cases evolve into a permanent localisation of the plastic activity in the form of shear bands. The sensitivity of shear band formation to the long range character of the stress redistribution has been discussed in ref. 3. The question raised here is whether static correlations can be observed, *i.e.* if the spatial distribution of the plastically active regions at a given time differs from that of an ideal gas. This is clearly the case, as illustrated by the possible existence of shear bands in which all plastic activity "condensates". Even in systems where the deformation remains homogeneous, such static correlations exist, in experiments such as in microscopic or mesoscopic simulations.[4] That being said, we would like to emphasize that the difference between the shear conditions investigated in our work and the equilibrium ones studied in ref. 1 is not peripheral. Indeed, shear favours a particular type of localised "excitations", namely shear transformations (which we refer to as plastic events), instead of the broad class of string-like excitations found in Ref. 1. These shear transformations induce a well-known characteristic deformation field on average, which has the same symmetries as those predicted by continuum elasticity. It seems unlikely that the same deformation field would be observed around the excitations described in ref. 1. Indeed, it is also likely that predictions based on elasticity theory in a uniform medium underestimate the spatial decay of this field, on account of the neglect of both spatial heterogeneities (in shear moduli, for instance) and the damping of shear waves due to dissipative mechanisms, thereby biasing the competition between correlations and factors of disorder (thermal fluctuations, structural spatial heterogeneities *etc.*) in favour of the former. Application of our model (in its present state) therefore requires that the material under consideration be deep within its solid phase, so that these omissions are not critical. The good agreement between our predictions and experimental results on rather subtle observables tends to indicate that this was indeed the case for the systems we have considered. However the applicability of our description to a viscous system such as a supercooled liquid would be very questionable. The issue of the crossover between the low temperature, elastic solid and the supercooled liquid, whether at rest or in the presence of an externally applied stress, remains an open one.

1 A. S. Keys, L. O. Hedges, J. P. Garrahan, S. C. Glotzer, and D. Chandler, Excitations are localized and relaxation is hierarchical in glass-forming liquids, *Phys. Rev. X*, 2011, **1**, 021013.
2 (a) M. L. Falk and J. S. Langer, *Phys. Rev. E*, 1998, **57**, 7192; (b) C. Maloney and A. Lemaitre, *Phys. Rev. E*, 2006, **74**, 016118; (c) F. Leonforte, A.Tanguy and J.-L. Barrat, *Eur. Phys. J. E: Soft Matter Biol. Phys.*, 2006, **20**, 355.
3 (a) K. Martens, L. Bocquet and J.-L. Barrat, *Phys. Rev. Lett.*, 2011, **106**, 156102; (b) K. Martens, L. Bocquet and J.-L. Barrat, *Soft Matter*, 2012, **8**, 197.
4 (a) V. Chikkadi, G. Wegdam, D. Bonn, B. Nienhuis and P. Schall, *Phys. Rev. Lett.*, 2011, **107**, 198303; (b) A. Nicolas, J. Rottler, and J.-L. Barrat, in preparation.

Professor Wynne opened the discussion of the Concluding Remarks by Professor Angell: I would like to ask you approximately the same question as Prof. Hajime Tanaka at the start of the meeting: why are we not seeing many more examples of liquid–liquid phase transitions? If there are in fact many more examples of LLPTs but they are all in the supercooled-liquid regime, why is this? Elongated anisotropic molecules show many examples of isotropic to nematic (or similar) liquid crystalline transitions. Again, why do we not see similar transitions in liquids consisting of more symmetric molecules?

Professor Angell answered: A partial answer is contained in Fig. 7 of my concluding commentary on this overall problem where we reproduce the findings of the Fayer group showing that the scattering associated with the development of orientational correlations in rodlike molecular liquids anticipating the isotropic nematic transition passes over, as a function of decreasing molecular aspect ratio, into the scattering found common to glassforming liquids as they pass from the high temperature Arrhenius domain into the pre-glass transition super-Arrhenius domain. We make reference there to the observations (by Popova and Sturetsov) of the rapid increase in the Landau–Placzek ratio, as the concentration of locally favored structures that Hajime Tanaka (and later Patrick Royall) described to us, builds up. The more fragile the liquid the more rapid the build-up. The glass-forming ability is perhaps the reflection of an inability at lower aspect ratios, to relieve the frustration by first order transition. On the other hand, the glass-forming metallic liquids may provide the examples you are looking for.

Professor Anisimov communicated: You have addressed an important feature of thermodynamics anomalies near phase transitions, emphasizing the fact that anomalies of the isobaric heat-capacity are the indicators of strong entropy fluctuations. While this is definitely correct, there is a more subtle, but nevertheless important, issue regarding the origin and the strength of such anomalies. In particular, the entropy fluctuations in fluids may diverge strongly at a critical point, being associated with long-range near-critical fluctuations of the order parameter, or may diverge weakly if the entropy is weakly coupled with the order parameter. The strong divergence of C_p is demonstrated by fluids near the vapor-liquid critical point. The weak divergence of C_p is observed at the Curie point of ferromagnetic materials and in binary fluids at the liquid–liquid critical point. Furthermore, there are other examples of C_p divergent behaviour. One is the "square-root" divergence of C_p in the ordered phase, upon approaching the tri-critical point. This divergence exists even in the mean-field (Landau) approximation and is observed near phase transitions in ferroelectrics and in some liquid crystals. Another kind originates from pre-crystallization fluctuations and is observed near the so-called "weak" first order transitions, such as the nematic-isotropic transition in liquid crystals. Careful analysis of the strength and shape of the observed heat-capacity anomalies in supercooled aqueous systems can elucidate the nature of these phenomena.

Professor Angell concluded the discussion: I think your comments here are very helpful. Indeed, in my concluding commentary I give several examples of the sort of weaker coupling effects you have described. Even when the coupling is weak one is able to observe a variety of effects that are provocative and helpful. Of course the heat capacity measurements lack the sensitivity that can be obtained from observation of refractive index fluctuations of which some interesting examples have been presented at this meeting.

Faraday Discussions RSCPublishing

PAPER

Fluctuations, clusters, and phase transitions in liquids, solutions, and glasses: from metastable water to phase change memory materials

C. Austen Angell and Zuofeng Zhao

Received 11th November 2013, Accepted 11th November 2013
DOI: 10.1039/c3fd00111c

We revisit the relations between clustering, fluctuations and thermodynamics for a range of clustering and disordering phenomena in liquids, seeking commonalities and links to phenomena reported at this meeting.

1. Introduction: clustering, thermodynamics, and scattering

In my concluding remarks I have taken, as my main connecting thread, the phenomenon of clustering of molecules in the liquid state. But since the subject material of the meeting has been broadened to include specifically the matter of crystallization and liquid structure I will also branch into that when it seems appropriate. The two are often interwoven.

As Anisimov has reminded us at this meeting,[1] the presence of large scale clustering usually has consequences in the density or enthalpy and on their intrinsic fluctuations. Fluctuations in these extensive properties are manifested in the familiar response functions, heat capacity, expansivity and compressibility, and the precise relations were written down in the famous Landau–Lifschitz text on Statistical Physics.[2] These relations are very well known and are reproduced below. They tell us that clustering of molecules, ions *etc.*, in liquids is likely to be associated with anomalies in the response functions.

The isothermal compressibility reflects directly the mean square volume fluctuation, $\langle(\Delta V)\rangle^2$, through the relation,

$$\kappa_T = \langle(\Delta V)\rangle^2 / V k_B T \tag{1}$$

while the heat capacities, C_v and C_p, are related to the corresponding kinetic energy fluctuations

$$C_v = kT^2 / \langle(\Delta T)\rangle^2 \text{ (at constant volume)} \tag{2}$$

Dept. of Chem.-Biochem., Arizona State University, Tempe AZ, 85287-1604, USA and entropy fluctuations,

$$C_p = kT^2/\langle(\Delta S)\rangle^2 \text{ (at constant volume)} \tag{3}$$

Since the isothermal compressibility is also related to the long wavelength limit of the isobaric structure factor, this scattering quantity is also directly related to the mean square density fluctuation.[3] The relation is the well-known

$$S(0) = \langle(\Delta N)^2\rangle/N = \rho k_B T \kappa_T \tag{4}$$

where N is the number of particles, $\langle(\Delta N)^2\rangle$ is the mean square fluctuation in N, ρ is the number density, and κ_T is the isothermal compressibility defined above, and by the thermodynamic relation $\kappa_T = (\partial^2 G/\partial p^2)_T/V$.

To these familiar relationships we need to add the relations connecting fluctuations in concentration to thermodynamic functions of *multicomponent* systems. These fluctuations manifest themselves spectacularly in many binary and multicomponent systems where the components mix with positive deviations from ideality, and tend to split into two phases at lower temperatures. As is well known, the concentration–concentration fluctuation lengthscales and times both diverge, in many cases, at the *consolute* temperature, which is a critical point in every sense of the word. To my surprise we have not seen much, if any, mention of the corresponding relation between these fluctuations and the solution thermodynamics, at this meeting though they surely belong at the forefront of our thinking in discussion of clusters in liquid solutions. The thermodynamic equivalent of the compressibility for concentration fluctuations is the activity coefficient, and of course it is this quantity to which the scattering related quantity, S_{cc}, is related.

The relations laid out by Bhatia and Thornton[4] in a famous paper dealing with metal alloy thermodynamics, are:

$$S_{cc}(0) = \mathrm{N}\langle(\Delta c)^2\rangle,$$

where $\langle(\Delta c)^2\rangle$ is the mean square fluctuation in the concentration, and is further related to the inverse of the second derivative of the Gibbs free energy with respect to concentration by

$$S_{cc}(0) = Nk_B T/[\partial^2 G/\partial c^2)_{T,P,N}]$$

which means that the concentration fluctuations, hence the related scattering intensities are giving information on the thermodynamic activity coefficients of the components.

They have been particularly useful in relating excess neutron scattering data to thermodynamic observations for cases of critical demixing, *e.g.*, in metal–ammonia (deuterated) solutions[5] and metal–molten salt solutions.[6]

Away from the binary solution critical point (consolute temperature) the fluctuations do not occur on all length scales up to the correlation length, but still occur, leading to anomalies in the various measurable quantities, activity coefficients, heat capacities and compressibilities (hence also sound velocities, *etc.*). In this concluding talk of Faraday Discussion 167, I want to range over these signs of anomalous structuring in the liquids in a number of different contexts, relating them, where I can, to the subjects that have been addressed by our contributors here, and inviting their study where they have not.

2. Structural questions, and the heat capacity as a diagnostic

Of course the question of structure is always a tricky one in liquids. When the same phenomena are seen in solids it is usually possible to decide what, in the way of structural entities, might be responsible. For this reason I will start this contribution by reference to a couple of cases where phenomena similar to those that are seen in liquids, can occur. This meeting, like so many others, has given a lot of attention to the subject of water so I will use water as an example to make my point.

Water is famous for its extraordinary, highly anomalous, heat capacity in the supercooled state, $T \ll 0$ °C. Only liquid silicon and liquid sulfur have anything like it. We show the data below, but first we want to look at a case for which there is much understanding. An excellent example of a system that (i) has a major anomaly in the heat capacity, and also (ii) has a well-defined glass transition at lower temperature, where a state of disorder is frozen in, and (iii) that is well-understood structurally, is the simple ordered alloy Co–Fe.[7,8] This is one of the simplest of a very broad variety of cooperative transitions falling in the Ising model universality class. The lambda form of the heat capacity is very well known and approximately derivable by Ising model theoretical treatments made famous by Onsager and others. Popularly known as order–disorder transitions, these convert ordered or phase-separated systems at low temperature into infinite clusters at the critical point and then to smaller clusters on the other side of the critical point. At the lambda temperature the order parameter characterizing the state of order of the system, falls to zero.

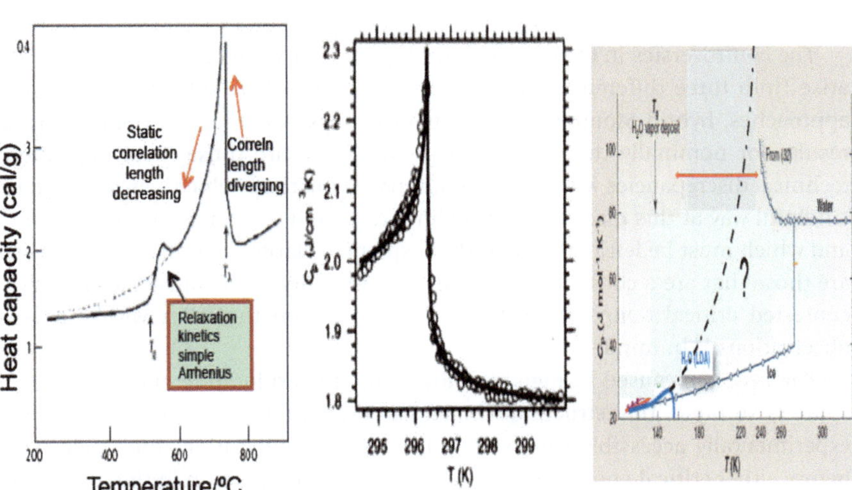

Fig. 1 Heat capacity forms for cooperative disordering systems generating infinte clusters at a critical point. From left, (i) solid Co–Fe ordered alloy (two interpenetrating simple cubic lattices, becoming random BCC above critical temperature T_l) (ii) separate isoctane and perfluoroheptane phases becoming a single phase solution with infinite clusters at the critical temperature (consolute temperature at 296.4 K) and (iii) rationalization of the observed heat capacity behavior of water, on either side of "No-man's land" (indicated by the horizontal red bar) where only very short time scale measurements can be made, see text.

The third of the systems in Fig. 1 is the case of water's heat capacity as published in 2008 (in Science[9]), and elaborated upon more recently in ref. 10, as a way of explaining the weird combination of vanishingly small change in heat capacity at the glass transition at 136 K, and the exponentially increasing heat capacity as the liquid approaches its homogeneous nucleation temperature at 235 K during cooling.[11]

Our point is, of course, that the strangeness of supercooled water can be understood if it is recognized as the high temperature arm of a critical order–disorder process, just like the well-studied liquid–gas phenomenon at 374 °C but involving the angle disordering of the open hydrogen bond network during heating, or its re-establishment during cooling. Note that this phenomenon lies entirely in the volumetric consequences of the ordering. No unusual heat capacity changes occur, and no excess scattering (due to structural fluctuations) would be seen, if the volume is held constant.[11] To continue the discussion, we put aside for the moment the current controversy over whether or not there can be two coexisting liquid phases on time scales short with respect to crystallization, supposing for the sake of argument, that there are (in which case, with the right combination of pressure and composition manipulations,[12] it will prove possible to see them by freeze fracture electron microscopy – and of course their absence would bolster the case for the converse).

Continuing, it would of course be strange indeed if the critical point for this special liquid (and this special disordering process) should lie exactly at ambient pressure. The form of heat capacity would be modified somewhat to a rounded form if the T_c lies at small positive pressure, or alternatively, it would have first order transition character (like that of liquid silicon) if the critical point should lie at negative pressure, as various lines of evidence suggest could be the case.[9,10] Measurements of heat capacity during rapid scans on thin glassy water films[13] would support the latter[10] but, for the purposes of this meeting, this question is best left aside.

The controversies in the case of water,[14] particularly the most recent ones,[15–19] arise from three different sources. There are those where different simulation approaches, hybrid Monte Carlo *vs.* metadynamics, for instance, yield different results for nominally the same state points, there are those involving purely technical discrepancies with the same simulation method (which were detailed in a helpful way at this discussion [by Palmer *et al.* and the pertinent discussion[20]] and which must be left to the algorithm experts to clarify), and then finally there are those that are a consequence of the high mobility of the molecules near the contested critical temperature. We will comment on the latter with the new observations,[12] in mind.

The problem caused by high mobility is that the exploration of configuration space is so rapid, that structural changes in the liquid become inseparable, on experimentally accessible time scales, from crystal nucleation time scales. *i.e.* before either critical point or first order liquid–liquid transition (if they exist) can be observed, the system has crystallized. Such crystallization is greatly repressed if the cooling is carried out at constant volume, when it would occur closer to −90 °C (since the pressure increases during constant volume cooling) or if it is carried out at temperatures near the glass transition itself, where the outcome of a cooling process can easily be preserved for inspection. This is the case we wish to consider here.

Let us consider the relation between the nucleation and crystallization temperatures for a well studied glassforming system, $Ca(NO_3)_2$–KNO_3 near the limit of its glassforming range.[21] Fig. 2 shows a relation between internal relaxation time (liquid equilibration time) and crystallization time (the latter being the time needed to crystallize 45% of the liquid and generate a peak in the crystallization exotherm – see ref. 22 for details). The experiments can distinguish the time scale for nucleation from the time scale for the nucleation + growth. The relationship of these to the internal relaxation time (τ_{in}) is very similar to that shown for water (using an inverse temperature axis) by Limmer and Chandler in Fig. 5 of their paper in this Discussion.[23] However, because in the CKN case the data are taken near T_g where everything is slowed down, there is no problem in bypassing the crystallization to obtain the glassy state, just by minor manipulation of the cooling rate (there is no critical point to observe in this case). Structural changes within the liquid state can then be observed at leisure during annealing, the nucleation being greatly delayed by the longer length scale of nucleus-forming fluctuations relative to liquid structure-changing fluctuations.

The same result is obtained by adding salt or other molecular solvents to water, breaking down the network and lowering the temperature, and removing the possibility of large cluster build up, until temperatures near the glass temperature can be reached. One of the early observations of residual clustering in cold aqueous solutions was in the light scattering study of the LiCl–H_2O system by Hsich *et al.*,[24] recently revisited by the Discussion Chairman Klaas Wynne and his group,[25] see also, ref. 26 where an interesting new anomaly arises amongst the hydrate compositions.

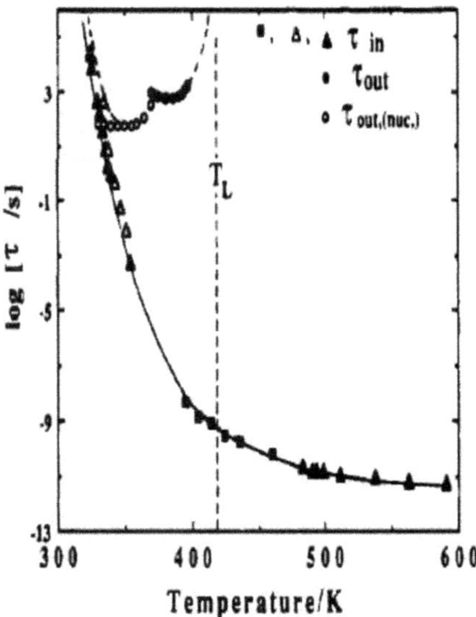

Fig. 2 Relation between internal relaxation time and nucleation time (or crystallization time, which is time for nucleation + growth) obtained near the edge of the glassforming range of the well known model glassformer system, calcium potassium nitrate (4 : 6), by means of one and two-step differential scanning calorimetry techniques (ref. 14). Compare with Fig. 5 in ref. 23. Reproduced from ref. 14 with permission of the American Chemical Society.

The liquid–liquid separation that occurs in the LiCl–H_2O system is of an unusual type which has now been explored rather thoroughly in the laboratory by Mishima,[27] and in modelling by Le and Molinero[28] (and Bullock and Molinero at this meeting[29]). We note that the second component can serve as a sort of microscope for the tendencies of the solvent to cluster. Indeed the much less pronounced tendencies of the classical glassformer, SiO_2, to generate network clusters during cooling can be enhanced by addition of second components, and a well-established but puzzling phenomenon of essentially pure SiO_2 micro-droplets forming in silica-rich alkali silicate glasses has been re-interpreted in these terms.[30]

Rather than pursue this avenue, it is better to turn attention to cases of the more conventional multicomponent systems where there are clustering phenomena that result in critical solution temperatures of both upper and lower types, and then also recognize the existence of "hidden" or "disappearing" types of the same phenomenon which leave behind only a non-critical memory of what has now moved out of reach (rather like the end of the "10 green bottles" song…"There was nothing but a sme-ell, a-hangin' on the wall"). Anisimov has drawn attention, in one of his discussion comments, to the case of 3-methyl pyridine (3-picoline) in water at 70 °C, and since the water +2-picoline system has a closed loop like those considered next, the thermodynamic spike at disappearance, anticipated below, could be explored systematically.

3. Clustering and its consequences near binary solution critical points

Fig. 3 shows the interesting and well-known system, water + nicotine, which exhibits both upper and lower critical temperatures. We have already seen, in Fig. 1, the lambda-like form of the heat capacity through the familiar upper

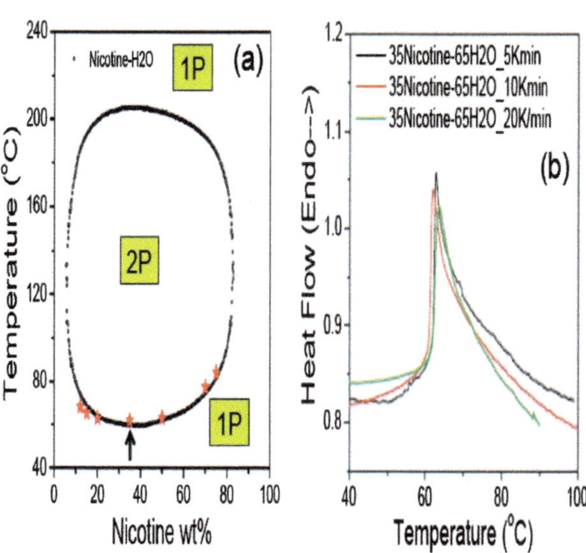

Fig. 3 (LHS) Phase diagram for the water + nicotine system, showing lower consolute point at 35 wt % nicotine (arrow), and (RHS) DSC scans showing the different (mirror image) form of the cooperative excitation profile that distinguishes this case from that seen in Fig. 1 for the upper critical point. Red crosses on LH figure are from DSC endotherms in Fig. 5.

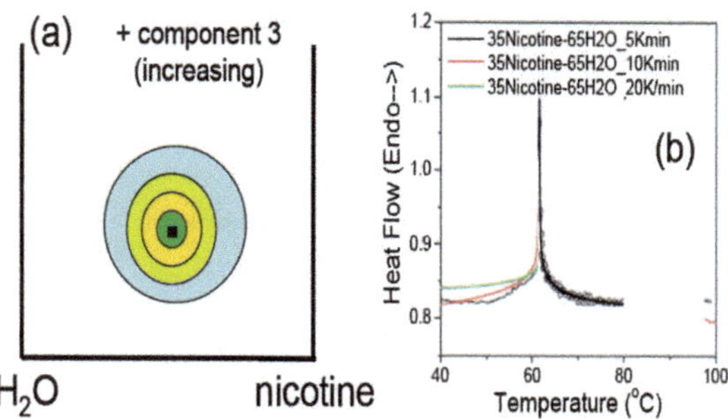

Fig. 4 (a). Phase diagram for a system with a vanishing two-phase loop as a 3rd component is blended in. (b) The heat capacity behavior as temperature is scanned though the composition at which the immiscibility zone disappears.†

critical temperature (UCT) case. Now we show how, as we scan up in temperature from the one phase into the two-phase domain of the (closed loop) water + nicotine system, through the LCT, we obtain just the opposite form of heat capacity with of course some time-dependent effects due to the slow kinetics of two-phase composition equilibration. Precise heat capacity data for this LCT case, were reported by Thoen *et al.*,[31] for the system triethylamine–D_2O, and are reviewed by Anisimov and Thoen (see Fig. 1.4.10) in Chapter 14 of ref. 32.

The question naturally arises as to what happens when a third component is added to reduce the incompatibility of the two phases at intermediate temperatures such that the two phase region shrinks first to a point and then vanishes altogether.

This scenario is envisaged in Fig. 4 where we can see how an unusual form of heat capacity would appear for the composition where the two phase domain contracts to a point. For a scan at this composition, the form to be seen is illustrated in Fig. 4(b), constructed from the single phase branches of Fig. 3 and its equivalent for an upper critical temperature (obtained from part 2 of Fig. 1 (which we confirmed by scans through the consolute point of the water–propylene carbonate system – $T_{\text{consolute}} = 61.1$ °C – see below).

Then, for compositions a little further removed from the two phase domain, only a weak heat capacity anomaly, and weak excess neutron scattering, would be observed. Critical fluctuations die off rather rapidly as the distance from a critical point increases, so the chances of encountering such anomalies, without guidance from systems in which the two phase domain is observable, would be small, but may suffice to explain some unexplained observations, suggesting heterogeneities, in multicomponent systems.

† A practical example of this sort of system was suggested by M. Anisimov in the General Discussion of Professor Debenedetti's paper. It is the 3-methyl pyridine (3MP) + water system, with a hidden critical point at about 70 °C that can be "brought out" by small additions of salt (*e.g.* 0.4 % NaCl). Confusingly, it is also reported by Jung and Jhon[33] that the β-picoline + water system has a large immiscibility loop. β-Picoline and 3-methylpyridine are the same compound. We observed liquid immiscibility in this binary system ourselves, though had made no special effort to purify the starting materials. The complexity of this system and the need for special attention to equilibration is brought out in the detailed study of the three component system water–3MP–NaBr.[34]

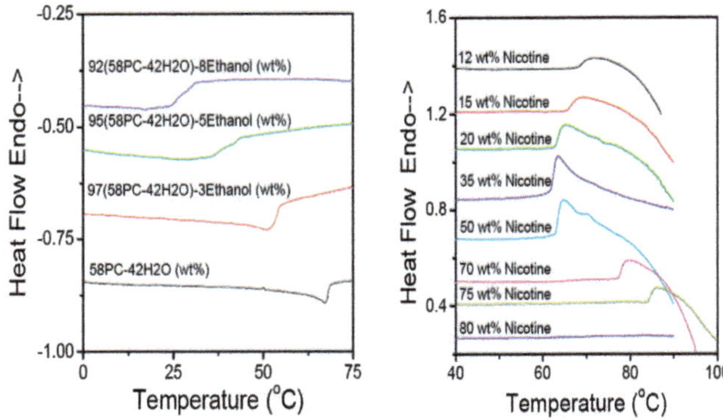

Fig. 5 (a) and (b) effects of composition changes on the heat capacity behavior in systems with binary solution critical points with associated clustering phenomena.

When, instead of diminishing the entire two-phase domain to a point, one merely pushes the probed composition towards the edge of that domain by adding a third component, a different heat capacity effect is seen. Fig. 5(a) takes the case of the system water + propylene carbonate (PC), which has a consolute point at 61.1 °C and 19.3 mol % (57.5 wt %) PC,[35] and shows the effect on the DSC *downscans* (from the single phase region into the two-phase region) of adding ethanol as third component. The change in studied composition, away from criticality towards the homogeneous solution domain, systematically eliminates cooperativity from the phenomenon. The conversion of the lambda like-transition into a glass-like transition is probably of significance to the understanding of both types of transition. Of course, in the glass transition, the decrease in heat capacity at low temperatures results from the kinetic elimination of a degree of freedom. (However, see the interesting and anomalous case of LiCl–H_2O at hydrate compositions[26]). In the case of Fig. 5, it is an ergodic effect that is being observed, where, for structural reasons, the high temperature state has additional degrees of freedom, due to the composition fluctuations that are permitted.

The same effect may be seen in upscans in the two-component water + nicotine system for which the phase diagram was shown in Fig. 3. Fig. 5(b) shows DSC *upscans* as composition is changed in each direction from that of the lower consolute point. The most dilute solution (12% nicotine) is still inside the two-phase domain and the composition that is tangent to the 2-phase area, (known as the *plait* point) is unfortunately inaccessible, being well above the boiling point of water. The case referred to in the introduction to this section would provide a better case to characterize the thermodynamic signature of plait point clustering.

Heat capacity measurements are only viable detectors of inhomogeneity (clustering) effects in liquids if the inhomogeneities involve a substantial enthalpy of assembly. For many phenomena, much more sensitive methods of detection of heterogeneities are available. These are the light scattering methods, based on scattering from refractive index fluctuations. Refractive index fluctuations themselves are intimately related to the density fluctuations that determine the response function compressibility as given in eqn (1). The inhomogeneities in focus in the next sections have been revealed by light scattering techniques. We note also the relevance of neutron scattering techniques when the contrast in neutron scattering lengths of the solution atomic species is favorable,[6] and Raman scattering techniques using multivariate analysis, as described by Ben-Amotz in his paper presented in the Discussion.[36]

4. Micelles and mesophases, vesicles

Related to the above-discussed area, are the fields of micelles and mesophases, with their critical micelle concentrations, and critical divergences underlying weak first order phase transitions – for which there exists a huge literature. This is not to be discussed here except for the boundary between the domain of nematic liquid crystals and glassforming liquids, as it relates to molecular aspect ratios, which has a special interest for the present authors.[37] This boundary has been studied by the same optical Kerr effect (OKE) spectroscopy used by Torre, in this Discussion,[38] to analyze the buildup of inhomogeneities in water as it supercools towards the controversial singularity at 227 K. OKE is a time domain measurement covering the time scale range from ps (below the librational resonance time in the IR) down to ns. It is the *time domain* equivalent of the *frequency domain* depolarized light scattering phenomenon.

Fayer and coworkers[39,40] have used OKE spectroscopy to study the power law governing orientational fluctuation times in systems undergoing the (weakly first order) isotropic–nematic transition, (for which NMR studies yield the famous Landau–de Gennes singularity slightly below the liquid crystal boundary – *which, like the water 2nd critical point, can only be identified and quantified by theoretical analysis, because of pre-emption by the first order I–N phase transition*). These authors showed that the exponent in the power law describing the reorientation time scale, increases towards unity as the aspect ratio of the molecule under study decreases towards unity. For aspect ratios below 2.5, no liquid crystal phases were found, but the power law describing the OKE scattering continued to describe the single phase liquid

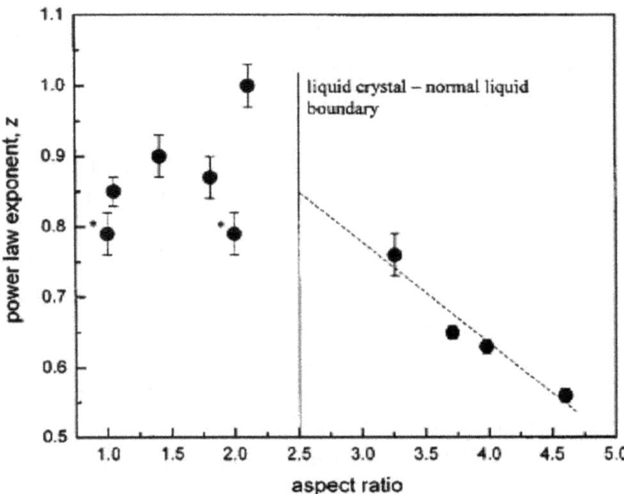

Fig. 6 Exponents of the relaxation time power law deduced from OKE scattering. Compare the cases in which the system approaches an isotropic–nematic first order phase transition temperature, with those for molecules of lower aspect ratio that are merely glassforming.

behavior, with exponents ranging from 0.8 to unity. What sort of clustering does that imply for the smaller aspect ratio liquids? The Fayer group findings are reproduced in Fig. 6. The glassforming ability of the liquids to the left of the vertical line in Fig. 6 is perhaps the reflection of an inability, at lower aspect ratios, to relieve the frustration described by Tanaka, (in the opening lecture[41]) by first order liquid–liquid transition. Indeed, for van der Waals ellipsoids at aspect ratio 1.45, it seems that there is no crystal phase more stable than the liquid, down to 0 K – which is the definition of the "ideal glassformer".[37]

The relation of the OKE studies to the frequency domain depolarized light scattering (DLS), has been examined in detail Brodin and Rössler[42] using some of the same liquids studied by Fayer and co-workers. From the comparison, it seems that OKE provides a wider dynamic range than DLS, but probably is not examining anything more cluster-related than that provided by the excess wing absorption of dielectric spectroscopy (about which little is understood).

An alternative light scattering technique, Raleigh–Brillouin scattering, that may be more helpful with respect to giving insight into cluster build-up, is discussed in a final section of this paper, after giving some attention to a fascinating type of inhomogeneity, apparently of very large spatial dimensions (but of minimal thermodynamic significance) that has been introduced to this meeting by Sefcik and coworkers.

5. Feeble clusters

An interesting and provocative type of clustering has been described at the Discussion meeting by Sefcik *et al.*[43] The phenomena that Sefcik and co-workers describe might have importance reaching well beyond their particular study of simple binary aminoacids (glyine or alanine) in water. A key feature of their clusters is that they could be easily disrupted by mechanical forces such as stirring and filtering, from which rupture they re-establish themselves on very long time scales. This was also the finding of Fischer, Patkowski and colleagues at the Max Planck Institute for polymer research who conducted a number of investigations into what have since become known, in the field of supercooled liquids, as "Fischer clusters".

Fischer clusters are the slowest-forming, slowest-relaxing, longest correlation length, inhomogeneities ever discussed in the liquids field – by many orders of magnitude in each case. They are observed in both polymers[44] and glassforming liquids:[45] for a short review see ref. 46. While controversial, most workers concede they are real, and the principal authors[45,47] believe they are to be understood in hierarchical terms with "fluctuons", of nm dimensions, as the correlated entities.[47] When the hierarcharchical structure is fully developed, speckle patterns may be observed, *e.g.* in *o*-terphenyl at 293 K, see Fig. 3 of ref. 47. The Fischer clusters are, however, thermodynamically *inconsequential* since the alpha relaxation time (always very sensitive to thermodynamic state because of the inverse exponential dependence on configurational entropy[48,49]) proves *not* to depend on whether the clusters have developed or not.[44] It would be of great interest to determine whether or not this is true of the alanine and glycine mesostructures of ref. 43 as well, by determining, for instance, the dielectric relaxation time before and after a filtration process (or an ultrasonic agitation process) that temporarily removes (or disrupts) the mesostructures.

6. Clustering in single component liquids

The Fischer clusters discussed above, actually belong in this present section as they have been studied most carefully in single component systems. However, we preferred to give a separate section to the "feeble cluster" type of phenomenon.

Light scattering studies of the Rayleigh–Brillouin type have an important role to play in the study of inhomogeneous structure development in single phase liquids as they supercool towards the glass transition. It is known that, in the Rayleigh–Brillouin spectrum of glassforming liquids in their high temperature Arrhenius, the ratio of the central Raleigh line intensity to the sum of the two Brillouin line intensities (known as the Landau–Placzek (LP) ratio) has a well-defined theoretical value determined by the state variables density and entropy.[50] It is argued, for instance by Popova and Surovtsev,[51] that any departure from this theoretical behavior implies the need for an additional order parameter, and thus is a direct indication of inhomogeneity onset in the liquid.

An increase in the LP ratio was part of the evidence provided by Fischer and coworkers for the anomalous clustering found, on sufficiently long standing, in the case of the model glassformer orthoterphenyl (see Figure 13 in ref. 46 for the LP ratio in OTP with and without Fischer clusters). A detailed study of this blow-up of the LP ratio on cooling into the super-Arrhenius domain of glassforming liquids, has been provided in ref. 51 for a number of the liquids studied by OKE and DLS that were discussed in a previous section for systems near their isotropic–nematic liquid crystal boundary. Some data illustrating the dramatic

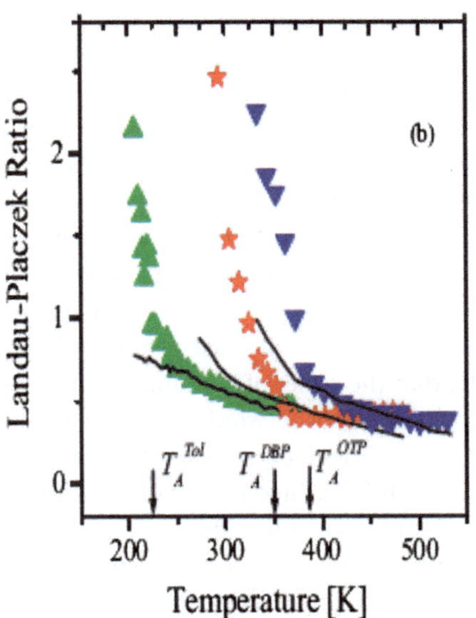

Fig. 7 Temperature dependences of the Landau–Placzek ratio (b) for dibutyl phthalate DBT (red stars), for toluene (Tol) (green up triangles), and for OTP (blue down triangles). Lines correspond to the theoretical estimations for simple (unclustered) liquids. The limits of high temperature Arrhenius behavior are marked T_A in each case.

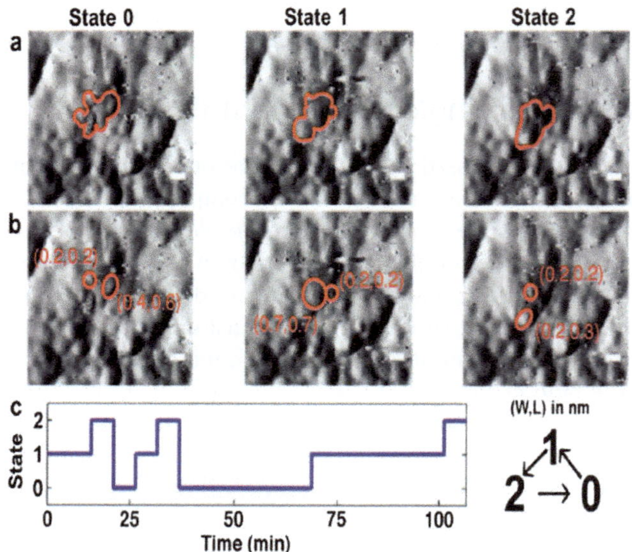

Fig. 8 Visualization using STM images, of two-state rearrangement events, by means of which the glass structure can reorganize (from ref. 53 by permission of Amer. Inst. Phys.).

departure of the LP ratio from the simple two-parameter domain, as the liquid enters the super-Arrhenius state, is shown in Fig. 7. The authors interpret this in terms of scattering from the inhomogeneities that arise, as "locally preferred structures" of the type discussed in Tanaka's opening lecture,[41] start to form. Clearly they form with rapidly increasing population as the system cools into the low temperature domain. It should be stressed that this is not *in any way* associated with ergodicity-breaking, which only occurs at much lower temperature as the *glass-transformation range* is entered. The glass transformation range is entered when, during cooling at a "standard" 20 K min^{-1}, relaxation times within ~2.3 decades of the 100s characteristic of the "standard" glass transition temperature, are reached. Clearly this transformation range (across which the system passes from 99% structurally equilibrated to 1% equilibrated) will depend on the cooling rate. The *temperature range* over which the ergodicity-breaking occurs, depends critically on the liquid "fragility" – and to a lesser extent on the non-exponentiality of the relaxation process[46] – which will not be discussed here. It is a frequent misconception, (or at least a terminology misuse) particularly among simulationists, that the onset of dynamic heterogeneity is somehow also the onset of "glassy" dynamics. The word "glassy" has always implied broken ergodicity, and therefore time-dependence of physical properties, whereas dynamic heterogeneity, and any related static cluster statistics, are strictly part of the ergodic behavior of supercooled liquids.

The growing-in of such locally preferred structures was nicely illustrated by Royall and co-workers in their paper presented at the meeting,[52] using simulations to identify the populations of various low-energy packing motifs. An important feature was the observation that the lifetime of clusters of energetically favored topology, can greatly exceed the alpha relaxation time. Furthermore there was no correlation between static and dynamic correlation lengths defined for the population of anticrystalline-structured clusters. The question of a connection between cluster energetics, correlation lengths (static or dynamic), and overall liquid slowdown (α relaxation time) evidently remains very open.

What cannot be seen in these simulations even with a bias to select the lowest energy trajectories, is the final state, in which a state of vanishing diffusivity ($D = 10^{-22}$ m^2 s^{-1}) is reached at the experimental glass transition temperature. The simulation time scales stop short of those explored in the experiments, by some *eight* orders of magnitude.

However, the type of packings found in experimental glasses, and the manner in which *configurational excitations* can lead to rearrangements of the particles within them (and so to the laboratory glass transition), is now being visualized directly, thanks to high resolution scanning tunnelling microscopy (STM) studies of glass surfaces that are currently being executed by Ashtekar *et al.*[53] of the Gruebele group. These workers take advantage of the atomic simplicity of metallic glasses to monitor the surface packings and their two-state reorganizational dynamics (for which movies are available). These workers have also studied cerium-based glasses[54] and amorphous silicon[55] at temperatures that they believe lie below the experimental glass transition temperatures for the bulk amorphous phases, though the estimate used for the glass temperature in the latter case is probably rather inaccurate (if we judge by the known diffusivities of *crystalline* silicon[56]).

7. Concluding case. A major change within a single phase, single component, liquid and a connection to technology

This contribution commenced with discussion of the case of water where, despite opinions to the contrary, cooling at one atm. pressure (or decrease in pressure from high pressure low temperature states), appears to be pushing the system into another thermodynamic phase of lower density. It can also be proposed that such changes are pushing the liquid across a supercritical extension of a coexistence line, depending on where one supposes that the second critical point lies. It seems appropriate, then, to conclude our contribution with an example of a liquid in

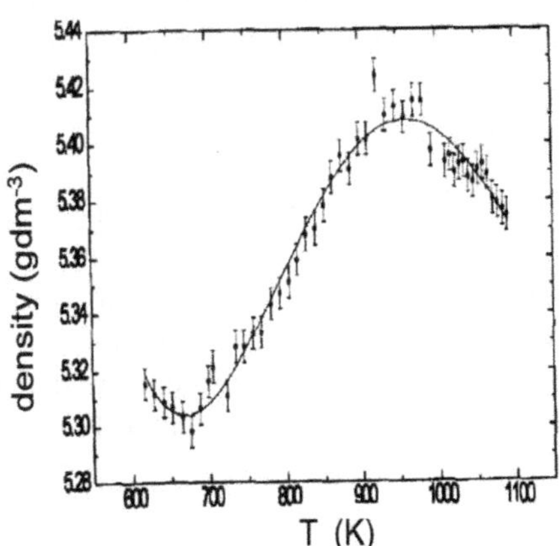

Fig. 9 Anomalous volume *vs.* temperature relations for liquid As_2Te_3. The density minimum occurs shortly above the melting point at 640 K.

which both a density maximum and a density minimum exist, and both lie within the stable liquid state, hence beyond the reach of controversy.[57] The case in question is an interesting liquid, a molten chalcogenide, As_2Te_3 which passes through a semiconductor-to-metal transition at the point of maximum negative expansivity in Fig. 9 – which is also a point of maximum heat capacity. Clearly this is a liquid in which there are major structural adjustments being promoted by change of temperature – but without any first order phase change. Is it a "Widom-line crossing"? Probably not. The high entropy phase in this case is, as in the case of water, a high density phase (Fig. 9). It is noted in ref. 57 that the integral over the heat capacity in the anomalous region is close to the enthalpy of a known phase change in the crystalline state, from monoclinic to rhombohedral, that occurs during heating. The high temperature crystalline phase is also of higher density, and is more metallic in character. Thus one has a good idea of the structural origin of the anomaly in the liquid. In this case, however, the increasing volume during cooling does not seem to be accompanied by any extraordinary viscosity effects as in the case of water. Rather it behaves as a typical fragile liquid, yielding a glassy state if the cooling is fast enough. Pure tellurium is also strikingly water-like, an interesting comparison having been provided by Kanno.[58] However, like As_2Te_3, Te is not as cooperative as water, at least not under ambient pressure conditions.

It is interesting that As_2Te_3 is one of three components comprising a technologically exciting ultrafast phase-change switching composition used in the fabrication of light- or heat-activated computer memory devices, pioneered by Ovshinsky and coworkers.[59] Interestingly enough, a second of the three components in the remarkable phase change switching compositions, is germanium, which has a water-like liquid–liquid phase change – a transition that has been caught in progress by hyperquenching.[60]

One gets the impression from Ovshinsky's writings that the fastest phase change switches occur in systems in which there can be two amorphous phases, and one crystalline phase trapped in the glassy matrix, all separated by only small energy barriers and capable of switching from one to the other by small voltage, or light, pulses. How strange it would be if some clustered aqueous phase, like Tanaka's glycerol–water solution near its T_g,[61] with triple free energy basins like those seen in the figures of Palmer's paper at this meeting,[20] and also transiently in Limmer and Chandler's,[23] should finally serve to model the behavior of one of computing technologies important phenomenologies!

Concluding remarks

The study of inhomogeneous aspects of liquid and polymer structures, and the associated thermodynamics (where there is a need for additional parameters beyond the usual P and T to properly specify the state of the system) has many intriguing aspects which this Discussion has helped to expose and clarify. Hopefully many of these complexities will become better understood in the coming decade as additional experimental tools and theoretical concepts are applied.

Acknowledgements

The authors acknowledge the support of the NSF Chemistry division under (collaborative research) grant CHE1213265. We are grateful to colleagues Pablo

Debenedetti and Gene Stanley for monthly discussions of cooperative structuring problems of the type discussed in this contribution.

References

1 D. Subramanian, C. T. Boughter, J. B. Klauda, B. Hammoudac and M. A. Anisimov, *Faraday Discuss.*, 2013, **167**, DOI: 10.1039/C3FD00070B.

2 L. Landau & E. M. Lifschitz. *Statistical Physics Sec.*, Pergamon and Addison-Wesley, Reading, MA, 1958, 111, p. 350.

3 P. A. Egelstaff. *An introduction to the Iiquid state*, Academic Press, New York, 114, 1967.

4 A. B. Bhatia and D. E. Thornton, Structural Aspects of the Electrical Resistivity of Binary Alloys, *Phys. Rev. B*, 1970, **2**, 3004–3012.

5 P. Chieux and P. Damay, On the long wavelength thermodynamic limit of a neutron diffraction experiment in the vicinity of a liquid–liquid critical point. Application to the concentration flucturations in the Li-ND3 system, *Chem. Phys. Lett.*, 1978, **58**, 619–621.

6 G. Chabrier, J. F. Jal, P. Chieux and J. Dupuy, A neutron scattering investigation of the structural order of Rb-RbBr solutions, *Phys. Lett. A*, 1982, **93**, 47–51.

7 S. Kaya and H. Sato, Superstructuring in the iron-cobalt system and their magnetic properties, *Proc. PhysicoMath. Soc. Japan*, 1943, **25**, 261.

8 S. Wei, I. Gallino, R. Busch and C. A. Angell, Glass Transition during ordering of $Fe_{50}Co_{50}$ and its relation to strong glassformers, *Nat. Phys.*, 2011, **7**, 178–182.

9 C. A. Angell, Insights into Phases of Liquid Water from Study of Its Unusual Glass-Forming Properties, *Science*, 2008, **319**, 582–587.

10 S. L. Meadley and C. A. Angell, Water and its closest relatives: insights from metastable state studies, *Nuovo Cimento (Enrico Fermi summer school on water (to be published*, 2014.

11 C. A. Angell, J. Shuppert and J. C. Tucker, Anomalous Properties of Supercooled Water: Heat Capacity. Expansivity. and PMR Chemical Shift from 0–38 °C., *J. Phys. Chem.*, 1973, 77, 3092.

12 Indeed, we will soon be publishing new observations of water-like heat capacity anomalies observed during cooling of aqueous solutions of non-disruptive ionic liquid solutes that push the anomalous heat capacity increase down to the temperature range of the glass transition - which makes the observations on crystallization kinetics, given below, of much relevance.

13 A. Sepúlveda, *et al.* Glass transition in ultrathin films of amorphous solid water, *J. Chem. Phys.*, 2012, **137**, 244506.

14 D. H. Rasmussen, A. P. Mackenzie, J. C. Tucker and C. A. Angell, Anomalous Heat Capacities of Supercooled Water, D20, *Science*, 1973, **181**, 342–344.

15 D. T. Limmer and D. Chandler, The Putative Liquid–Liquid Transition is a Liquid-Solid Transition in Atomistic Models of Water, *J. Chem. Phys.*, 2011, **135**, 134503.

16 Y. Liu, J. C. Palmer, A. Z. Panagiotopoulos and P. G. Debenedetti, Liquid–liquid transition in ST2 water, *J. Chem. Phys.*, 2012, **137**, 214505.

17 T. A. Kesselring, G. Franzese, S. V. Buldyrev and H. E. Stanley, *Sci. Rep.*, 2012, **2**, 474.

18 P. H. Poole, R. K. Bowles, I. Saika-Voivod and F. Sciortino, *J. Chem. Phys.*, 2013, **138**, 034505.

19 D. T. Limmer and D. Chandler, The putative liquid–liquid transition is a liquid-solid transition in atomistic models of water. II., *J. Chem. Phys.*, 2013, **138**, 214504.

20 J. C. Palmer, R. Car and P. G. Debenedetti, *Faraday Discuss.*, 2013, **167**, DOI: 10.1039/C3FD00074E.

21 H. Senapati, K. K. Kadiyala and C. A. Angell, Single and Two-step Calorimetry Studies of Homogeneous Nucleation and Growth Processes in Supercooled Ionic Glass-Forming Liquids: The Ca(NO3)2-KNO3 System, *J. Phys. Chem.*, 1991, **95**, 7050–7054.

22 D. R. MacFarlane, K. Kadiyala and C. A. Angell, Homogeneous Nucleation and Growth of Ice from Solutions: TTT Curves, the Nucleation Rate and the Stable Glass Criterion, *J. Chem. Phys.*, 1983, **79**, 3921–3927.

23 D. T. Limmer and D. Chandler, *Faraday Discuss.*, 2013, **167**, DOI: 10.1039/C3FD00076A.

24 S. Y. Hsich, R. W. Gammon, P. B. Macedo and C. J. Montrose, Light Scattering in Aqueous LiCl Solutions; Evidence for a Low Temperature Immiscibility, *J. Chem. Phys.*, 1972, **56**, 1663.

25 D. A. Turton, C. Corsaro, D. F. Martin, F. Mallamace and K. Wynne, The dynamic crossover in water does not require bulk water, *Phys. Chem. Chem. Phys.*, 2012, **14**, 8067–8073.

26 M. Kobayashi and H. Tanaka, Relationship between the Phase Diagram, the Glass-Forming Ability, and the Fragility of a Water/Salt Mixture, *J. Phys. Chem. B*, 2011, **115**, 14077–14090.

27 O. Mishima, Phase separation in dilute LiCl–H2O solution related to the polyamorphism of liquid water, *J. Chem. Phys.*, 2007, **126**, 244507.

28 L. Le and V. Molinero, Nanophase Segregation in Supercooled Aqueous Solutions and Their Glasses Driven by the Polyamorphism of Water, *J. Phys. Chem. A*, 2011, **115**, 5900–5907.

29 G. Bullock and V. Molinero, *Faraday Discuss.*, 2013, **167**, DOI: 10.1039/C3FD00085K.

30 C. A. Angell, P. H. Poole & M. Hemmati, in *12th East European Glass Conference*, ed. B. Samunova & Y Demetriew, pp. 100–109.

31 J. Thoen, E. Bloemen and W. Van Dael, *J. Chem. Phys.*, 1978, **68**, 735.

32 Heat Capacities: Liquids, Solutions and Vapours, ed. E. Wilhelm and T. M. Letcher, Royal Society of Chemistry, 2010.

33 Jung and Jhon, *Korean J. Chem. Eng.*, 1983, **1**, 59–63.

34 A. F. Kostko, M. A. Anisimov, and J. V. Sengers, Criticality in aqueous solutions of 3-methylpyridine and sodium bromide, *Phys. Rev. E*, 2004, 700261118.

35 N. F. Catherall and A. G. Williamson, Mutual Solubilities of Propylene Carbonate and Water, *J. Chem. Eng. Data*, 1971, **16**, 335–336.

36 D. S. Wilcox, B. M. Rankin and D. Ben-Amotz, *Faraday Discuss.*, 2013, **167**, DOI: 10.1039/C3FD00086A.

37 V. Kapko, Z.-F. Zhao, D. V. Matyushov and C. A. Angell, Ideal glassformers *vs.* "ideal glasses": Studies of crystal-free routes to the glass by "potential tuning" MD, and laboratory calorimetry, *J. Chem. Phys.*, 2013, **138**, 12A549.

38 A. Taschin, P. Bartolini, A. Marcelli, R. Righini and R. Torre, *Faraday Discuss.*, 2013, **167**, DOI: 10.1039/C3FD00060E.

39 S. D. Gottke, H. Cang, B. Bagchi and M. D. Fayer, Comparison of the ultrafast to slow time scale dynamics of three liquid crystals in the isotropic phase, *J. Chem. Phys.*, 2002, **116**, 6339.

40 H. Cang, J. Li, V. N. Novikov and M. D. Fayer, Dynamics in supercooled liquids and in the isotropic phase of liquid crystals: A comparison, *J. Chem. Phys.*, 2003, **118**, 9303–9311.

41 H. Tanaka, *Faraday Discuss.*, 2013, **167**, DOI: 10.1039/C3FD00110E.

42 A. Brodin and E. A. Rössler, Depolarized light scattering *versus* optical Kerr effect spectroscopy of supercooled liquids: Comparative analysis, *J. Chem. Phys.*, 2006, **125**, 114502.

43 A. Jawor-Baczynska, B. Moore, H. S. Lee, A. McCormick and J. Sefcik, *Faraday Discuss.*, 2013, **167**, DOI: 10.1039/C3FD00066D.

44 T. Kanaya, *et al.* Light-Scattering Studies of Short- and Long-Range Density and Anisotropy Fluctuations in a Bulk Polysiloxane, *Macromolecules*, 1995, **28**, 7831.

45 A. Patkowski, *et al.* Long-range density fluctuations in orthoterphenyl as studied by means of ultrasmall-angle X-ray scattering, *Phys. Rev. E*, 2000, **61**, 6909–6913.

46 C. A. Angell, K. L. Ngai, G. B. McKenna, P. F. McMillan and S. W. Martin, Relaxation in glassforming liquids and amorphous solids, *J. Appl. Phys.*, 2000, **88**, 3113–3157.

47 A. S. Bakai and E. H. Fischer, Nature of long-range correlations of density fluctuations in glass-forming liquids, *J. Chem. Phys.*, 2004, **120**, 5235–5252.

48 G. Adam and J. H. Gibbs, On the Temperature Dependence of Cooperative Relaxation Properties in Glass-Forming Liquids, *J. Chem. Phys.*, 1965, **43**, 139.

49 S. Mossa, *et al.* Dynamics and configurational entropy in the Lewis-Wahnstrom model for supercooled orthoterphenyl, *Phys. Rev. E*, 2002, **65**, 041205.

50 H. Z. Cummins and R. W. Gammon, Rayleigh and Brillouin scattering in liquids: the Landau-Plazek ratio, *J. Chem. Phys.*, 1966, **44**, 2785.

51 V. A. Popova and N. V. Surovtsev, Temperature dependence of the Landau-Placzek ratio in glass forming liquids, *J. Chem. Phys.*, 2011, **135**, 134510.

52 A. Malins, J. Eggers, H. Tanaka and C. P. Royall, *Faraday Discuss.*, 2013, **167**, DOI: 10.1039/C3FD00078H.

53 S. Ashtekar, J. Lyding and M. Gruebele, Temperature-Dependent Two-State Dynamics of Individual Cooperatively Rearranging Regions on a Glass Surface, *Phys. Rev. Lett.*, 2012, **109**, 166103.

54 S. Ashtekar, *et al.* An obligatory glass surface, *J. Chem. Phys.*, 2012, **137**, 141102.

55 S. Ashtekar, G. Scott, J. Lyding and M. Gruebele, Direct Imaging of Two-State Dynamics on the Amorphous Silicon Surface, *Phys. Rev. Lett.*, 2011, **106**, 235501.

56 C. A. Angell, C. T. Moynihan and M. Hemmati, 'Strong' and 'superstrong' liquids, and an approach to the perfect glass state *via* phase transition, *J. Non-Cryst. Solids*, 2000, **274**, 319–331.

57 Y. S. Tver'yanovich, V. M. Ushakov and A. Tverjanovich, Heat of structural transformation at the semiconductor-metal transition in As2Te3 liquid, *J. Non-Cryst. Solids*, 1996, **235–237**.
58 H. Kanno, H. Yokoyama and Y. Yoshimura, A new interpretation of anomalous properties of water based on Stillinger's postulate, *J. Phys. Chem.*, 2001, **B105**, 2019.
59 S. Ovshinsky. Chapter 121. *Insulating and Semiconducting Glasses*, Ed. P. Boolchand, World Scientific, Singapore, Chapter 121, pp. 729–779, 2001.
60 H. Bhat, *et al.* Vitrification of a monatomic metallic liquid, *Nature*, 2007, **448**(7155), 787–791.
61 K.-I. Murata and H. Tanaka, Liquid–liquid transition without macroscopic phase separation in a water–glycerol mixture, *Nat. Mater.*, 2012, **11**, 436–442.

Poster titles

Terahertz spectroscopy of hydrogen-bonded glass-forming liquids, **J. Sibik and J. A. Zeitler**, *University of Cambridge, UK*

The cluster fluid phase of fluids with competing short and long-ranged interactions, **M. Sweatman, R. Fartaria and L. Lue**, *University of Edinburgh, UK*

Different behavior of water in confined solutions of high and low solute concentrations, **K. Elamin, H. Jansson, S. Kittaka and J. Swenson**, *Chalmers University of Technology, Sweden*

Fluids confined between a solvophilic and a solvophobic substrate: interfaces, local structure and layering transitions, **R. Evans and M. C. Stewart**, *University of Bristol, UK*

Liquid–liquid phase transition in supercooled water suggested by microsecond simulations from a first-principle based force field, **Y. Li, J. Li and F. Wang**, *University of Arkansas, USA*

Experiments at negative pressure to probe nanoscale structures in water, **M. A. Gonzalez, J. F. Abascal, J. L. Aragones, C. Valeriani, G. Palleres, M. El Mekki Azouzi, S. Meadly, C. Shekhar, P. Tripathi, F. Caupin and J.-F. Lenain**, *Université Claude Bernard Lyon, France*

Micrõsctructure analysis of confined liquids in porous media, **N. Y. Tan, P. Bräuer, C. D'Agostino, L. F. Gladden and J. A. Zeitler**, *University of Cambridge, UK*

Dynamics of monohydroxy alcohol 2-propanol in the liquid and supercooled state, **A. Faraone, M. Nagao, K. Nakajima and T. Kikuchi**, *University of Maryland & NIST Center for Neutron Research, USA*

Charge-dense inorganic ions induce coorperative slowdown of water rotation: a molecular dynamics study, **A. C. A. Vila Verde and R. Lipowsky**, *Max Planck Institute of Colloids and Interfaces, Germany*

Dynamic properties of water–xylitol solutions, **K. Elamin, S. Cazzato, J. Sjöström, S. M. King and J. Swenson**, *Chalmers University of Technology, Sweden*

Chlorine–chlorine interactions and dipole alignments in liquid chloroform, **J. J. Shephard, J. S. O. Evans, S. Imberti, A. K. Soper and C. G. Salzmann**, *Durham University and University College London, UK*

A dipolar mean-field theory for electron transfer in solvent mixtures, **B. Zhuang and Z.-G. Wang**, *California Institute of Technology, USA*

Dynamics and structure of a typical room temperature ionic liquid across glass transition: a multinuclear NMR study, **T. Endo, D, C. Kaseman and S. Sen**, *University of California, Davis, USA*

Cavitation of water at negative pressures, **G. Menzl, P. Geiger and C. Dellago**, *University of Vienna, Austria*

Long-range hydrogen-bond structure in aqueous solutions, **S. J. Irudayam and R. H. Henchman**, *The University of Manchester, UK*

Correlation of ionic liquid mesostructure, cohesive energy density and their ability to promote amphiphile self-assembly, **T. L. Greaves and C. J. Drummond**, *CSIRO Materials Science and Engineering, Australia*

NMR relaxation study of ionic liquids in bulk and under confinement in porous media, **A. O. Seyedlar, S. Stapf and C. Mattea**, *Ilmenau University of Technology, Germany*

Critical micelle concentration (CMC) of amphiphiles in molecular solvents and protic ionic liquids (PILs), **E. C. Wijaya, T. L. Greaves and C. J. Drummond**, *CSIRO, Australia*

Structural relaxation and glass transition of low-density amorphous ice, **C. G. Salzmann, J. J. Shephard and J. S. O. Evans**, *University College London, UK*

Supression of hard sphere nucleation by increase in local fivefold symmetry, **J. Taffs and C. P. Royall**, *University of Bristol, UK*

Development of an electrical method for inducing mesoscale state transitions in polar liquids, **A. D. Wexler, S. Drusová, E. C. Fuchs and J. Woisetschläger**, *Wetsus - Center for Sustainable Water Technology, The Netherlands*

Structural examination of the interactions of glutamine: a model system for understanding hydrophilic association, **N. H. Rhys, A. K. Soper and L. Dougan**, *University of Leeds, UK*

Fluorescence lifetime imaging of liquid–liquid phase transition of triphenyl phosphite, **J. Mosses, C. D. Syme and K. Wynne**, *University of Glasgow, UK*

Second-harmonic scattering in aqueous urea solutions: evidence for solute clusters? **M. R. Ward, S. W. Botchway, A. D. Ward and A. J. Alexander**, *University of Edinburgh, UK*

Lipidic mesophases as tools for biological investigation, **T. G. Meikle, C. E. Conn, F. Separovic and C. J. Drummond**, *CSIRO/University of Melbourne, Australia*

Phase transitions and self-assembly in liquids confined to thin films, **A. M. Smith, K. R. J. Lovelock, N. N. Gosvami and S. Perkin**, *University of Oxford, UK*

Understanding the ionic liquid/gas surface structure: a combined approach, **K. R. J. Lovelock, A. J. S. McIntosh, S. Fearn and A. M. Smith**, *Imperial College London, UK*

The CO_2–water equilibrium at partial pressure below 101.325 kPa: eighty years of measurements, **B. G. Durham**, *University of Oxford, UK*

Coarsening and nonequilibrium relaxation in supercooled water, **D. Limmer and D. Chandler**, *University of California Berkeley, USA*

The Skinner Prize for the best poster was awarded to Mr Georg Menzl of the University of Vienna, Austria, for his poster on cavitation of water at negative pressures.

List of participants

Dr A. Alexander, *University of Edinburgh, United Kingdom*
Professor A. Angell, *Arizona State University, U. S. A.*
Professor M. Anisimov, *University of Maryland, College Park, U. S. A.*
Professor J.-L. Barrat, *Université Joseph Fourier, Grenoble, France*
Professor D. Ben-Amotz, *Purdue University, U. S. A.*
Dr I. Billard, *CNRS IPHC UMR7178, France*
Professor Dr F. Caupin, *Université Claude Bernard Lyon 1, France*
Professor D. Chandler, *University of California, Berkeley, U. S. A.*
Mr S. Cockram, *The University of Manchester, United Kingdom*
Mr S. Cox, *University College London, United Kingdom*
Professor C. Drummond, *CSIRO, Australia*
Mr A. Dunleavy, *University of Bristol, United Kingdom*
Mr B. Durham, *University of Oxford, United Kingdom*
Mr K. Elamin, *Chalmers University of Technology, Sweden*
Dr T. Endo, *University of California, Davis, U. S. A.*
Professor R. Evans, *University of Bristol, United Kingdom*
Dr A. Faraone, *University of Maryland & NIST Center for Neutron Research, U. S. A.*
Dr T. Greaves, *CSIRO Materials Science and Engineering, Australia*
Mr P. K. Gupta, *University of Basel, Switzerland*
Dr A. Haji-Akbari, *Princeton University, U. S. A.*
Dr R. Henchman, *The University of Manchester, United Kingdom*
Dr K. Holt, *University College London, United Kingdom*
Professor A. Idrissi, *LASIR, France*
Mr S. Jackson, *University of Sheffield, United Kingdom*
Miss D. Lesnicki, *École normale supérieure, France*
Mr D. Limmer, *UC Berkeley, U. S. A.*
Dr K. Lovelock, *Imperial College London, United Kingdom*
Miss K. Lunnon, *Royal Society of Chemistry, United Kingdom*
Professor F. Mallamace, *Messina University, Italy*
Professor S. Meech *UEA, United Kingdom*
Mr T. Meikle *University of Melbourne & CSIRO, Australia*
Mr G. Menzl *Faculty of Physics, University of Vienna, Austria*
Professor M. Meuwly *University of Basel, Switzerland*
Professor V. Molinero *The University of Utah, U. S. A.*
Mrs J. Mosses *University of Glasgow, United Kingdom*
Professor A. Mount *The University of Edinburgh, United Kingdom*
Dr. K. Muirhead *Royal Society of Chemistry, United Kingdom*
Dr H. Ndlovu *The University of Manchester, United Kingdom*
Mr A. Ordikhani Seyedlar *Department of Technical Physics II / Polymer Physics, Institute of Physics, Ilmenau University of Technology, Germany*
Professor M. Orrit *Leiden University, Netherlands*
Dr J. Palmer *Princeton University, U. S. A.*
Dr A. Perera *CNRS, France*

Dr S. Perkin *Oxford University, United Kingdom*
Miss R. Pinney *University of Bristol, United Kingdom*
Miss N. Rhys *University of Leeds, United Kingdom*
Dr C. Royall *University of Bristol, United Kingdom*
Dr C. Salzmann *University College London, United Kingdom*
Professor K. Seddon *QUILL Centre, Queen's University Belfast, United Kingdom*
Dr J. Sefcik *University of Strathclyde, United Kingdom*
Mr J. Shephard *Durham University and University College London, United Kingdom*
Mr J. Sibik *University of Cambridge, United Kingdom*
Mr M. Smith *Royal Society of Chemistry, United Kingdom*
Mr A. M. Smith *University of Oxford, United Kingdom*
Dr A. Soper *STFC Rutherford Appleton Laboratory, United Kingdom*
Dr M. Sweatman *University of Edinburgh, United Kingdom*
Ms J. Taffs *University of Bristol, United Kingdom*
Mr N. Tan *University of Cambridge, United Kingdom*
Professor Dr H. Tanaka *Institute of Industrial Science, University of Tokyo, Japan*
Dr R. Torre *European Lab. for Non-Linear Spectroscopy (LENS), University of Firenze, Italy*
Mr J. Towey *University of Leeds, United Kingdom*
Dr A. Triolo *ISM-CNR, Italy*
Dr A. Vila Verde *Max Planck Institute of Colloids and Interfaces, Germany*
Professor G. Voth *The University of Chicago, U. S. A.*
Professor D. Wales *University of Cambridge, United Kingdom*
Professor R. Walker *Montana State University, U. S. A.*
Professor Dr F. Wang *University of Arkansas, U. S. A.*
Mr M. Ward *University of Edinburgh, United Kingdom*
Mr A. Wexler *Wetsus - Center for Sustainable Water Technology, Netherlands*
Miss E. C. Wijaya *CSIRO, Australia*
Prof. K. Wynne *University of Glasgow, United Kingdom*
Dr A. Zeitler *University of Cambridge, United Kingdom*
Ms B. Zhuang *California Institute of Technology, U. S. A.*

Index of contributors*

*The page numbers in **bold** type indicate papers submitted for discussions.